15 Years'

UNSOLVED QUESTION PAPERS

2006 - 2020

IIT JEE

(JEE Main & Advanced)

15 Years'

UNSOLVED QUESTION PAPERS

2006-2020

IIT JEE

(JEE Main & Advanced)

PHYSICS
CHEMISTRY
MATHEMATICS

arihant

ARIHANT PRAKASHAN (Series), MEERUT

arihant

ARIHANT PRAKASHAN (Series), MEERUT

Administrative & Production Offices

Regd. Office

'Ramchhaya' 4577/15, Agarwal Road, Darya Ganj, New Delhi -110002
Tele: 011- 47630600, 43518550

Head Office

Kalindi, TP Nagar, Meerut (UP) - 250002
Tel: 0121-7156203, 7156204

Sales & Support Offices

Agra, Ahmedabad, Bengaluru, Bareilly, Chennai, Delhi, Guwahati, Hyderabad, Jaipur, Jhansi, Kolkata, Lucknow, Nagpur & Pune.

ISBN 978-93-25293-29-8

PRICE ₹305.00

PO No : TXT-XX-XXXXXXX-X-XX

Published by Arihant Publications (India) Ltd.

For further information about the books published by Arihant, log on to www.arihantbooks.com or e-mail at info@arihantbooks.com

Follow us on

Contents

JEE Main & Advanced, IIT JEE

Unsolved Questions Papers (2006-2020)

IIT JEE

UNSOLVED PAPER 2006

Physics

SECTION A

1. A solid sphere of radius R has moment of inertia I about its geometrical axis. If it is melted into a disc of radius r and thickness t. If its moment of inertia about the tangential axis (which is perpendicular to plane of the disc), is also equal to I, then the value of r is equal to

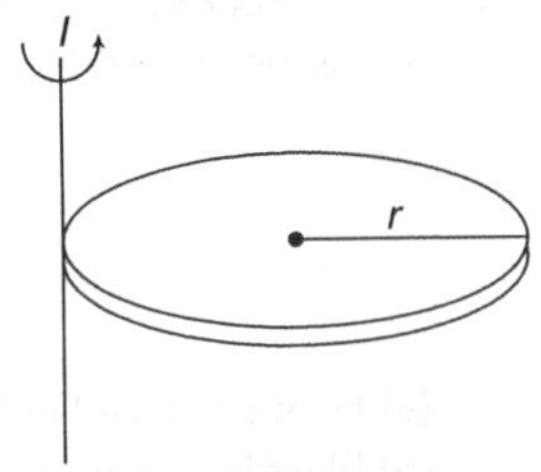

(a) $\frac{2}{\sqrt{15}}R$ (b) $\frac{2}{\sqrt{5}}R$

(c) $\frac{3}{\sqrt{15}}R$ (d) $\frac{\sqrt{3}}{\sqrt{15}}R$

2. System shown in figure is in equilibrium and at rest. The spring and string are massless, now the string is cut. The acceleration of mass $2m$ and m just after the string is cut will be

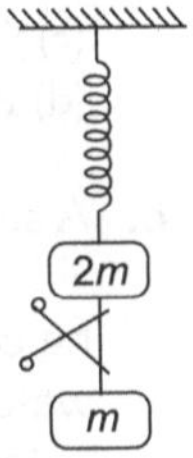

(a) $g/2$ upwards, g downwards

(b) g upwards, $g/2$ downwards

(c) g upwards, $2g$ downwards

(d) $2g$ upwards, g downwards

3. A massless rod BD is suspended by two identical massless strings AB and CD of equal lengths. A block of mass 'm' is suspended point P such that BP is equal to 'x', if the fundamental frequency of right wire, then the value of x is

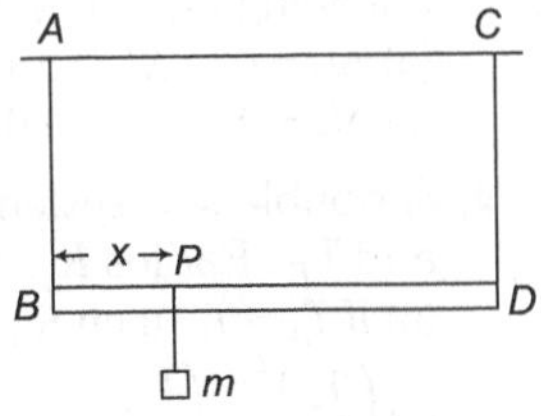

(a) $l/5$ (b) $l/4$

(c) $4l/5$ (d) $3l/4$

4. A point object is placed at distance of 20 cm from a thin planoconvex lens of focal length 15 cm. The plane surface of the lens is now silvered. The image created by the system is at

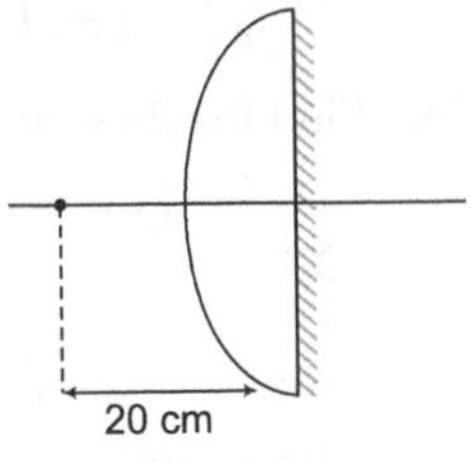

(a) 60 cm to the left of the system

(b) 60 cm to the right of the system

(c) 12 cm to the left of the system

(d) 12 cm to the right of the system

5. Half-life of a radio active substance 'A' is 4 days. The probability that a nucleus will decay in two half-lives is

(a) $\frac{1}{4}$ (b) $\frac{3}{4}$ (c) $\frac{1}{2}$ (d) 1

6. The graph between object distance u and image distance v for a lens is given below. The focal length of the lens is

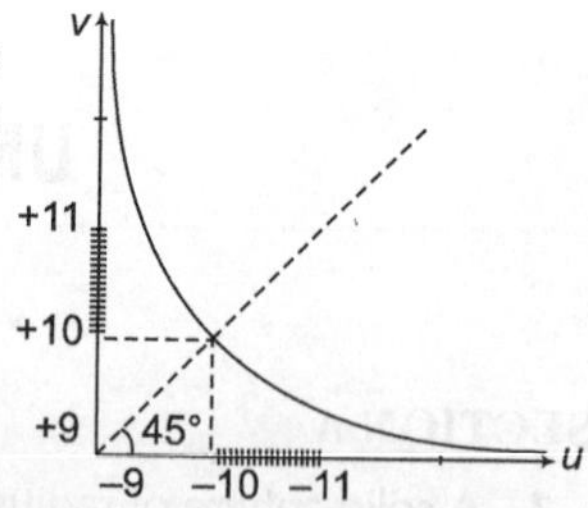

(a) 5 ± 0.1
(b) 5 ± 0.05
(c) 0.5 ± 0.1
(d) 0.5 ± 0.05

7. Two bars of radius 'r' and '$2r$' are kept in contact as shown. An electric current I is passed through the bars. Which one of the following is correct?

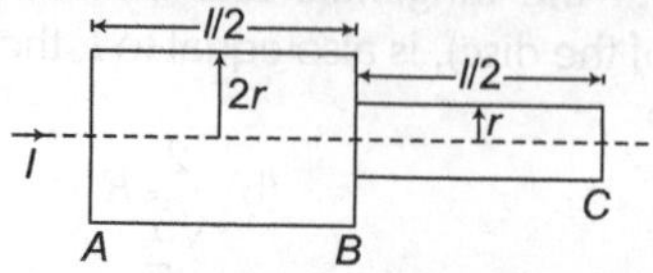

(a) Heat produced in bar BC is 4 times the heat produced in bar AB
(b) Electric field in both halves is equal
(c) Current density across AB is doubled that of across BC
(d) Potential difference across AB is 4 times that of across BC

8. A student performs an experiment for determination of $g\left(=\dfrac{4\pi^2 l}{T^2}\right)$, $l \approx 1$ m, and he commits an error of Δl. For T he takes the time of n oscillations with the stop watch of least count ΔT and he commits a human error of 0.1 s. For which of the following data, the measurement of g will be most accurate?

(a) $\Delta L = 0.5, \Delta T = 0.1, n = 20$
(b) $\Delta L = 0.5, \Delta T = 0.1, n = 50$
(c) $\Delta L = 0.5, \Delta T = 0.01, n = 20$
(d) $\Delta L = 0.1, \Delta T = 0.05, n = 50$

9. A double star system consists of two stars A and B which have time period T_A and T_B. Radius R_A and R_B and mass M_A and M_B. Choose the correct option.

(a) If $T_A > T_B$, then $R_A > R_B$
(b) If $T_A > T_B$, then $M_A > M_B$
(c) $\left(\dfrac{T_A}{T_B}\right)^2 = \left(\dfrac{R_A}{R_B}\right)^3$
(d) $T_A = T_B$

10. Find the time constant for the given RC circuits in correct order.

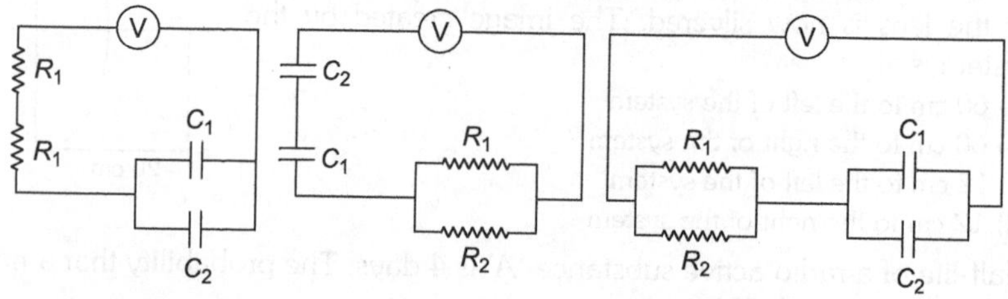

$R_1 = 1\Omega, R_2 = 2\Omega, C_1 = 4\mu F, C_2 = 2\mu F$

(a) 18, 4, 8/9
(b) 18, 8/9, 4
(c) 4, 18, 8/9
(d) 4, 8/9, 18

11. A biconvex lens of focal length f forms a circular image of radius r of sun in focal plane. Then, which option is correct?

(a) $\pi r^2 \propto f$

(b) $\pi r^2 \propto f^2$

(c) If lower half part is covered by black sheet, then area of the image is equal to $\pi r^2 / 2$

(d) If f is doubled, intensity will increase

12. The circular scale of a screw gauge has 50 divisions and pitch of 0.5 mm. Find the diameter of sphere. Main scale reading is 2.

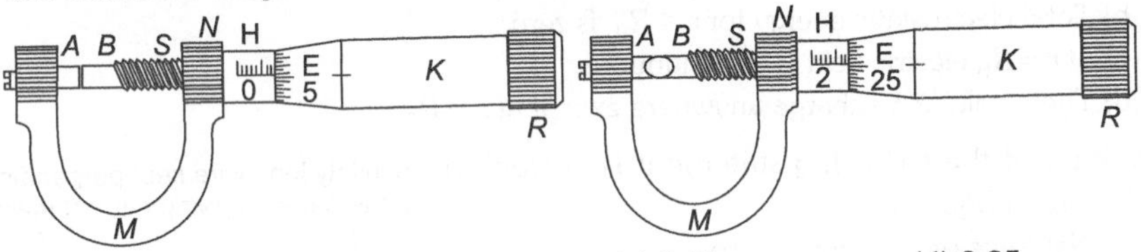

(a) 1.2 (b) 1.25 (c) 2.20 (d) 2.25

SECTION B

13. The graph between $1/\lambda$ and stopping potential (V) of three metals having work function ϕ_1, ϕ_2 and ϕ_3 in an experiment of photoelectric effect is plotted as shown in the figure. Which of the following statement(s) is/are correct? [Here λ is the wavelength of the incident ray].

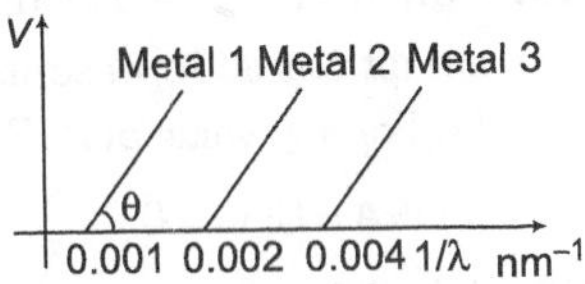

(a) Ratio of work functions $\phi_1 : \phi_2 : \phi_3 = 1:2:4$

(b) Ratio of work functions $\phi_1 : \phi_2 : \phi_3 = 4:2:1$

(c) $\tan\theta$ is directly proportional to hc/e, where h is Planck's constant and c is the speed of light

(d) The violet colour light can eject photoelectrons from metals 2 and 3

14. A field line is shown in the figure. This field cannot represent

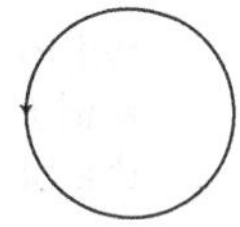

(a) magnetic field (b) electrostatic field

(c) induced electric field (d) gravitational field

15. A solid sphere is in pure rolling motion on an inclined surface having inclination θ.

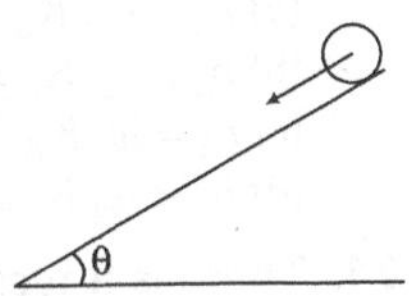

(a) Frictional force acting on sphere is $f = \mu\, mg \cos\theta$

(b) f is disspative force

(c) Friction will increase its angular velocity and decreases its linear velocity

(d) If θ decreases, friction will decrease

16. A black body of temperature T is inside chamber of temperature T_0. Now, the closed chamber is slightly opened to sun such that temperature of black body (T) and chamber (T_0) remains constant.

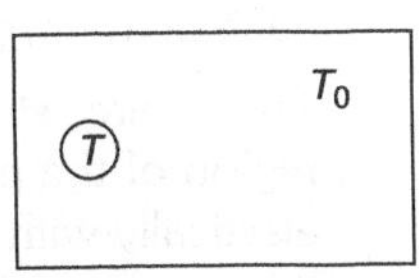

(a) Black body will absorb more radiation

(b) Black body will absorb less radiation

(c) Black body emit more energy

(d) Black body emit energy equal to energy absorbed by it

17. For spherical symmetrical charge distribution, variation of electric potential with distance from centre is given in diagram. Given that:

$$V = \frac{q}{4\pi\varepsilon_0 R_0} \quad \text{for } r \le R_0$$

and $$V = \frac{q}{4\pi\varepsilon_0 r} \quad \text{for } r \ge R_0$$

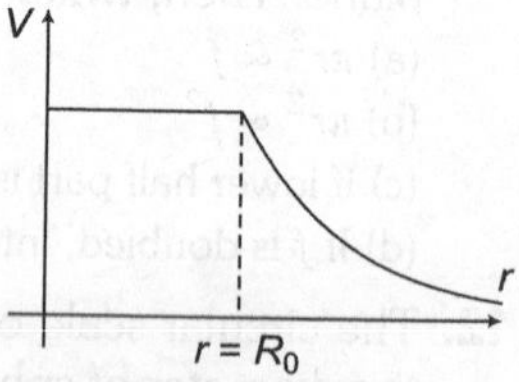

Then, which option(s) is (are) correct

(a) Total charge within $2R_0$ is q

(b) Total electrostatic energy for $r \le R_0$ is zero

(c) At $r = R_0$ electric field is discontinuous

(d) There will be no charge anywhere except at $r = R_0$

18. Which of the following statement is correct in the given figure?

(a) Net force on the loop is zero

(b) Net torque on the loop is zero

(c) Loop will rotate clockwise about axis OO' when seen from O

(d) Loop will rotate anti-clockwise about OO' when seen from O

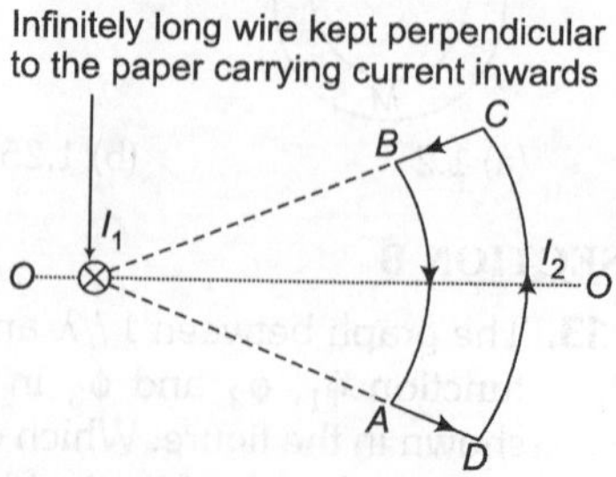

19. Function $x = A\sin^2 \omega t + B\cos^2 \omega t + C \sin \omega t \cos \omega t$ represents SHM.

(a) For any value of A, B and C (except $C = 0$)

(b) If $A = -B$, $C = 2B$, amplitude $= |B\sqrt{2}|$

(c) If $A = B$; $C = 0$

(d) If $A = B$; $C = 2B$, amplitude $= |B|$

20. A ball moves over a fixed track as shown in the figure. From A to B the ball rolls without slipping. If surface BC is frictionless and K_A, K_B and K_C are kinetic energies of the ball at A, B and C respectively, then

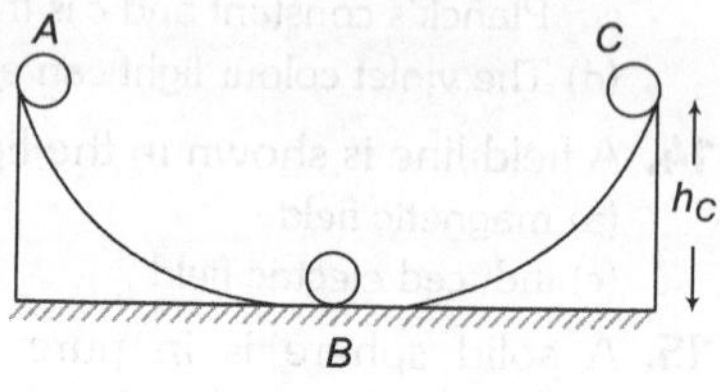

(a) $h_A > h_C$; $K_B > K_c$

(b) $h_A > h_c$; $K_c > K_A$

(c) $h_A = h_c$; $K_B = K_c$

(d) $h_A < h_c$; $K_B > K_c$

21. There is a rectangular plate of mass M kg of dimensions $(a \times b)$. The plate is held in horizontal position by striking n small balls each of mass m per unit area per unit time.

These are striking in the shaded half region of the plate. The balls are colliding elastically with velocity v. What is v?

It is given $n = 100$, $M = 3$ kg, $m = 0.01$ kg; $b = 2$ m; $a = 1$ m; $g = 10\text{ m/s}^2$.

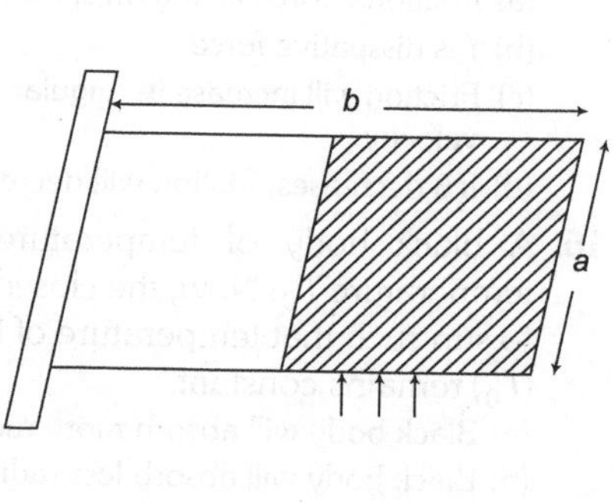

22. A circular disc with a groove along its diameter is placed horizontally. A block of mass 1 kg is placed as shown. The coefficient of friction between the block and all surfaces of groove in contact is $\mu = 2/5$. The disc has an acceleration of 25 m/s^2. Find the acceleration of the block with respect to disc.

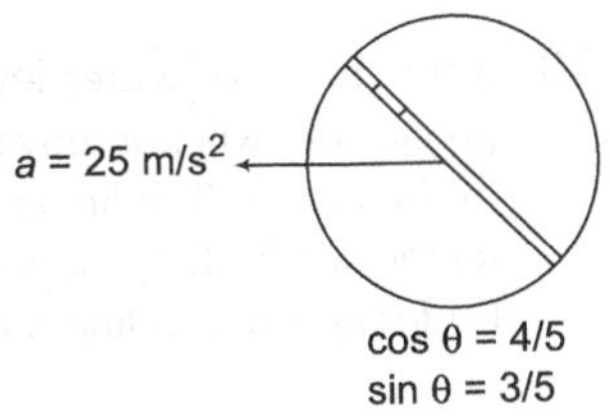

23. In an insulated vessel, 0.05 kg steam at 373 K and 0.45 kg of ice at 253 K are mixed. Find the final temperature of the mixture (in kelvin).

Given, $L_{\text{fusion}} = 80\text{ cal/g} = 336\text{ J/g}$, $L_{\text{vaporization}} = 540\text{ cal/g} = 2268\text{ J/g}$,

$S_{\text{ice}} = 2100\text{ J/kg}, K = 0.5\text{ cal/gK}$ and $S_{\text{water}} = 4200\text{ J/kg}, K = 1\text{ cal/gK}$.

24. If the wavelength of the nth line of Lyman series is equal to the de-Broglie wavelength of electron in initial orbit of a hydrogen like element ($Z = 11$). Find the value of n.

SECTION C

Passage 1

Two plane harmonic sound waves are expressed by the equations.

$y_1(x, t) = A\cos(0.5\pi x - 100\pi t)$

$y_2(x, t) = A\cos(0.46\pi x - 92\pi t)$ (All parameters are in MKS)

25. How many times does an observer hear maximum intensity in one second?

(a) 4 (b) 10 (c) 6 (d) 8

26. What is the speed of the sound?

(a) 200 m/s (b) 180 m/s (c) 192 m/s (d) 96 m/s

27. At $x = 0$, how many times the amplitude of $y_1 + y_2$ is zero in one second at $x = 0$?

(a) 192 (b) 48 (c) 100 (d) 96

Passage 2

A wooden cylinder of diameter $4r$, height H and density $\rho/3$ is kept on a hole of diameter $2r$ of a tank, filled with liquid of density ρ as shown in the figure.

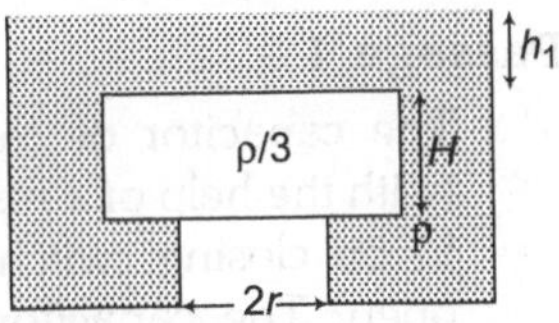

28. Now level of the liquid starts decreasing slowly. When the level of liquid is at a height h_1 above the cylinder, the block starts moving up. At what value of h_1, will the block rise?

(a) $\frac{4h}{9}$ (b) $\frac{5h}{9}$ (c) $\frac{5h}{3}$ (d) remains same

29. The block in the above question is maintained at the position by external means and the level of liquid is lowered. The height h_2 when this external force reduces to zero, is

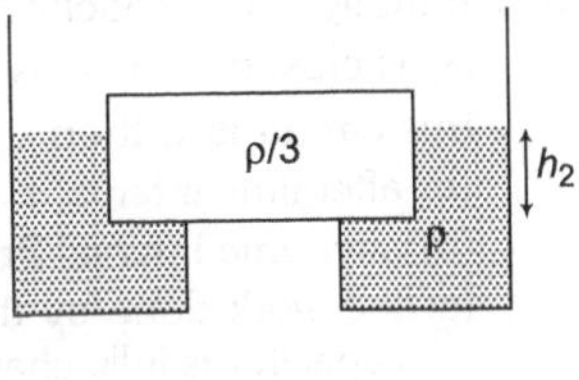

(a) $\frac{4h}{9}$ (b) $\frac{5h}{9}$

(c) remains same (d) $\frac{2h}{3}$

30. If height h_2 of water level is further decreased, then
(a) cylinder will not move up and remains at its original position
(b) for $h_2 = h/3$, cylinder again starts moving up
(c) for $h_2 = h/4$, cylinder again starts moving up
(d) for $h_2 = h/5$ cylinder again starts moving up

Passage 3

Modern trains are based on Maglev technology in which trains are magnetically leviated, which runs its EDS Maglev system. There are coils on both sides of wheels. Due to motion of train, current induces in the coil of track which levitate it. This is in accordance with Lenz's law. If trains lower down then due to Lenz's law a repulsive force increases due to which train gets uplifted and if it goes much high then there is a net downward force disc to gravity. The advantage of Maglev train is that there is no friction between the train and the track, thereby reducing power consumption and enabling the train to attain very high speeds.

Disadvantage of Maglev train is that as it slows down the electromagnetic forces decreases and it becomes difficult to keep it leviated and as it moves forward according to Lenz law, there is an electromagnetic drag force.

31. What is the advantage of this system?
(a) No friction hence no power consumption (b) No electric power is used
(c) Gravitation force is zero (d) Electrostatic force draws the train

32. What is the disadvantage of this system?
(a) Train experiences upward force according to Lenz's law
(b) Friction force create a drag on the train
(c) Retardation
(d) By Lenz's law train experience a drag

33. Which force causes the train to elevate up?
(a) Electrostatic force (b) Time varying electric field
(c) Magnetic force (d) Induced electric field

Passage 4

The capacitor of capacitance C can be charged (with the help of a resistance R) by a voltage source V, by closing switch S_1 while keeping switch S_2 open. The capacitor can be connected in series with an inductor L by closing switch S_2 and opening S_1.

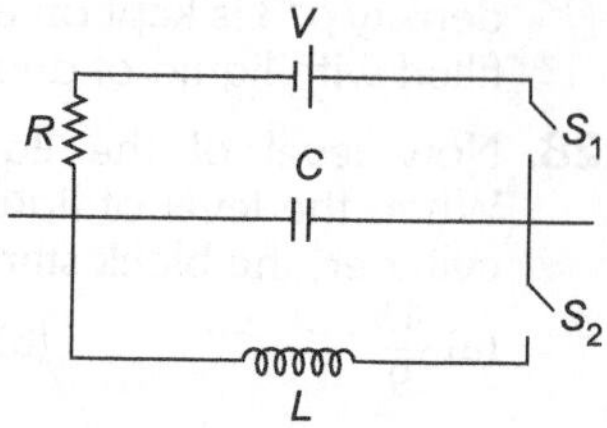

34. Initially, the capacitor was uncharged. Now, switch S_1 is closed and S_2 is kept open. If time constant of this circuit is τ, then
(a) after time interval τ, charge on the capacitor is $CV/2$
(b) after time interval 2τ, charge on the capacitor is $CV/(1-e^{-2})$
(c) the work done by the voltage source will be half on the heat dissipated when the capacitor is fully charged
(d) after time interval 2τ, charge on the capacitor is $CV(1-e^{-1})$

35. After the capacitor gets fully charged, S_1 is opened and S_2 is closed so that the inductor is connected in series with the capacitor. Then,

(a) at $t = 0$, energy stored in the circuit is purely in the form of magnetic energy

(b) at any time $t > 0$, current in the circuit is in the same direction

(c) at $t > 0$, there is no exchange of energy between the inductor and capacitor

(d) at any time $t > 0$, maximum instantaneous current in the circuit may $V\sqrt{\frac{C}{L}}$

36. If the total charge stored in the LC circuit is Q_0, then for $t \geq 0$

(a) the charge on the capacitor is $Q = Q_0 \cos\left(\frac{\pi}{2} + \frac{t}{\sqrt{LC}}\right)$

(b) the charge on the capacitor is $Q = Q_0 \cos\left(\frac{\pi}{2} - \frac{t}{\sqrt{LC}}\right)$

(c) the charge on the capacitor is $Q = -LC\frac{d^2Q}{dt^2}$

(d) the charge on the capacitor is $Q = -\frac{1}{\sqrt{LC}}\frac{d^2Q}{dt^2}$

37. Match the following for the given process

Column I	Column II
(a) Process $J \to K$	(p) $Q > 0$
(b) Process $K \to L$	(q) $W < 0$
(c) Process $L \to K$	(r) $Q > 0$
(d) Process $M \to J$	(s) $Q < 0$

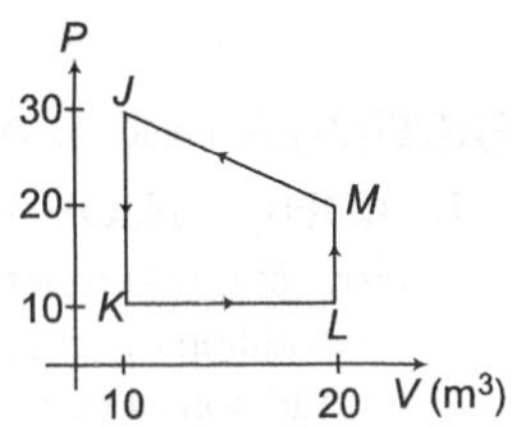

38. Match the following columns

Column I	Column II
(a) Nuclear fusion	(p) Converts some matter into energy
(b) Nuclear fission	(q) Generally possible for nuclei with low atomic number
(c) β-decay	(r) Generally possible for nuclei with higher atomic number
(d) Exothermic nuclear reaction	(s) Essentially proceeds by weak nuclear forces

39. Match the following columns

Column I	Column II
(a) Dielectric ring uniformly charged	(p) Time independent electrostatic field out of system
(b) Dielectric ring uniformly charged rotating with angular velocity ω	(q) Magnetic field
(c) Constant current in ring i_0	(r) Induced electric field
(d) $i = i_0 \cos \omega t$	(s) Magnetic moment

40. A simple telescope used to view distant objects has eyepiece and objective lens of focal lengths f_e and f_0, respectively. Then

Column I	Column II
(a) Intensity of light received by lens	(p) radius of aperture (R)
(b) Angular magnification	(q) dispersion of lens
(c) Length of telescope	(r) focal length f_0, f_e
(d) Sharpness of image	(s) spherical aberration

ANSWERS

1. (a) **2.** (a) **3.** (a) **4.** (c) **5.** (b) **6.** (b)
7. (a) **8.** (d) **9.** (d) **10.** (b) **11.** (b) **12.** (a)
13. (a, c) **14.** (a, d) **15.** (c, d) **16.** (*) **17.** (*) **18.** (a, c)
19. (b, d) **20.** (a) **21.** 10 m/s **22.** 10 m/s^2 **23.** 273 K **24.** $n = 24$
25. (a) **26.** (a) **27.** (d) **28.** (c) **29.** (a) **30.** (a)
31. (a) **32.** (d) **33.** (c) **34.** (b) **35.** (d) **36.** (c)
37. (a) s (b) p, r (c) r (d) q, s **38.** (a) p, q (b) p, r (c) p, s (d) p, q, r
39. (a) p (b) p, q, s (c) p, s (d) q, r, s **40.** (a) p (b) r (c) r (d) p, q, s

Chemistry

SECTION A (Only one Option is Correct)

1. $B(OH)_3 + NaOH \rightleftharpoons NaBO_2 + Na[B(OH)_4] + H_2O$.
How can this reaction is made to proceed in forward direction?
(a) Addition of *cis* 1, 2 diol (b) Addition of borax
(c) Addition of *trans* 1, 2 diol (d) Addition of Na_2HPO_4

2. A solution when diluted with H_2O and boiled, it gives a white precipitate. On addition of excess NH_4Cl / NH_4OH, the volume of precipitate decreases leaving behind a white gelatinous precipitate. Identify the precipitate which dissolves in NH_4OH/NH_4Cl.
(a) $Zn(OH)_2$ (b) $Al(OH)_3$ (c) $Mg(OH)_2$ (d) $Ca(OH)_2$

3. When benzene sulphonic acid and *p*-nitrophenol are treated with $NaHCO_3$, the gases released respectively are
(a) SO_2, NO_2 (b) SO_2, NO (c) SO_2, CO_2 (d) CO_2, CO_2

4. A monoatomic ideal gas undergoes a process in which the ratio of P to V at any instant is constant and equals to 1. What is the molar heat capacity of the gas?
(a) $\frac{4R}{2}$ (b) $\frac{3R}{2}$ (c) $\frac{5R}{2}$ (d) 0

5. (I) 1, 2-dihydroxy benzene (II) 1, 3-dihydroxy benzene
(III) 1, 4-dihydroxy benzene (IV) Hydroxy benzene
The increasing order of boiling points of above mentioned alcohols is
(a) I < II < III < IV (b) I < II < IV < III
(c) IV < I < II < III (d) IV < II < I < III

6. $CH_3—CH=CH_2 + NOCl \longrightarrow P$
Identify the adduct

(a) 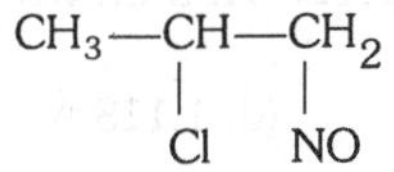

(b) 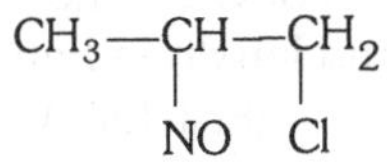

(c) 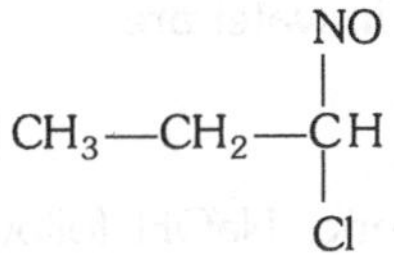

(d) 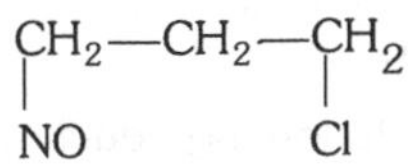

7. The IUPAC name of C_6H_5COCl is

(a) Benzoyl chloride
(b) Benzene chloro ketone
(c) Benzene carbonyl chloride
(d) Chloro phenyl ketone

8. $Ag^+ + NH_3 \rightleftharpoons [Ag(NH_3)]^+; k_1 = 3.5 \times 10^{-3}$
$[Ag(NH_3)]^+ + NH_3 \rightleftharpoons [Ag(NH_3)_2]^+; k_2 = 1.7 \times 10^{-3}$
Then the formation constant of $[Ag(NH_3)_2]^+$ is

(a) 6.08×10^{-6}
(b) 6.08×10^6
(c) 6.08×10^{-9}
(d) None of these

9. $CH_3NH_2 + CHCl_3 + KOH \rightarrow$ Nitrogen containing compound $+ KCl + H_2O$.
Nitrogen containing compound is

(a) $CH_3—C\equiv N$
(b) $CH_3—NH—CH_3$
(c) $CH_3—\overline{N}=\overline{C}$
(d) $CH_3\overset{+}{N}=\overline{C}$

10. $CuSO_4$ decolourises on addition of KCN, the product is

(a) $[Cu(CN)_4]^{2-}$
(b) Cu^{2+} get reduced to form $[Cu(CN)_4]^{3-}$
(c) $Cu(CN)_2$
(d) CuCN

11. The direct conversion of A to B is difficult, hence it is carried out by the following shown path.

$$\begin{array}{ccc} C & \longrightarrow & D \\ \downarrow & & \downarrow \\ A & & B \end{array}$$

Given $\Delta S_{(A\rightarrow C)} = 50$ eu; $\Delta S_{(C\rightarrow D)} = 30$ eu; $\Delta S_{(B\rightarrow D)} = 20$ eu
where eu is entropy unit, then $\Delta S_{(A\rightarrow B)}$ is

(a) + 100 eu
(b) + 60 eu
(c) – 100 eu
(d) – 60 eu

12. $N_2 + 3H_2 \rightleftharpoons 2NH_3$
Which is correct statement if N_2 is added at equilibrium condition?

(a) The equilibrium will shift to forward direction because according to IInd law of thermodynamics the entropy must increases in the direction of spontaneous reaction.
(b) The condition for equilibrium is $G_{N_2} + 3G_{H_2} = 2G_{NH_3}$ where G is Gibbs free energy per mole of the gaseous species measured at the partial pressure. The condition of equilibrium is unaffected by the use of catalyst, which increases the rate of both the forward and backward reactions to the same extent.
(c) The catalyst will increase the rate of forward reaction by α and that of backward reaction by β.
(d) Catalyst will not alter the rate of either of the reaction.

SECTION B (May have more than one option correct)

13. If the bond length of CO bond in carbon monoxide is 1.128 Å, then what is the value of CO bond length in $Fe(CO)_5$?

(a) 1.15 Å (b) 1.128 Å (c) 1.72 Å (d) 1.118 Å

14. The species present in solution when CO_2 is dissolved in water are

(a) CO_2, H_2CO_3, HCO_3^-, CO_3^{2-} (b) H_2CO_3, CO_3^{2-}

(c) CO_3^{2-}, HCO_3^- (d) CO_2, H_2CO_3

15. Which of the following reactants on reaction with conc. NaOH followed by acidification gives following lactone as the

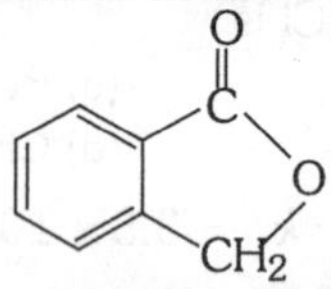

(a) 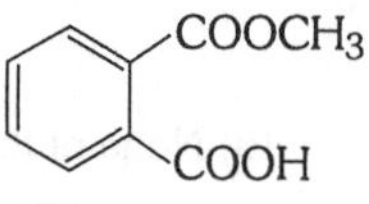

(b) 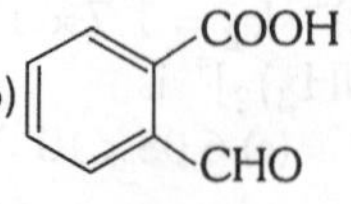

(c) 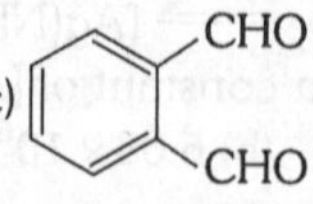

(d) 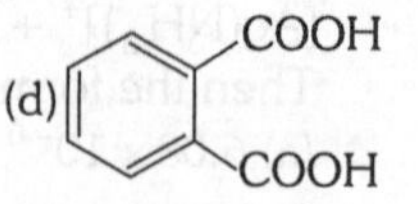

16.  $+ Cl—CH_2CH_2—CH_3 \xrightarrow{AlCl_3} P \xrightarrow[\text{(ii) } H_3O^+]{\text{(i) } O_2/D} Q$ + Phenol

The major products *P* and *Q* are

(a) 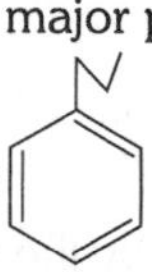 and CH_3CH_2CHO (b) 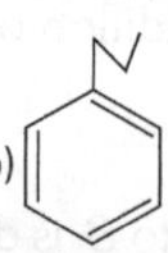 and CH_3COCH_3

(c) 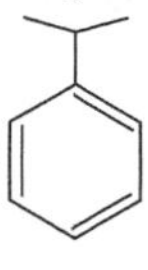 and CH_3COCH_3 (d) 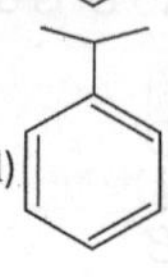 and CH_3CH_2CHO

17. The given graph represent the variations of compressibility factor $(Z) = \frac{PV}{nRT}$ *versus P*, for three real gases *A*, *B* and *C*. Identify the only incorrect statement

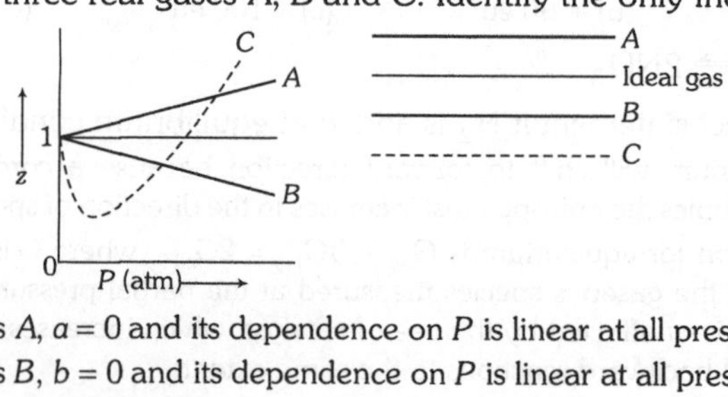

(a) For the gas *A*, $a = 0$ and its dependence on *P* is linear at all pressure.

(b) For the gas *B*, $b = 0$ and its dependence on *P* is linear at all pressure.

(c) For the gas *C*, which is typical real gas for which neither *a* nor $b = 0$. By knowing the minima and point of the intersection, with $Z = 1$, *a* and *b* can be calculated.

(d) At high pressure, the slope is positive for all real gases.

18. The smallest ketone and its next homologoue are reacted with NH_2OH to form oxime

(a) two different oximes are formed (b) three different oximes are formed

(c) two oximes are optically active (d) all oximes are optically active

19.

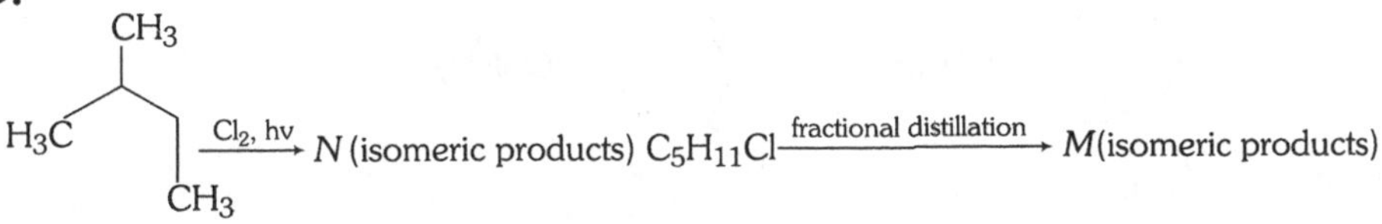

What are N and M?

(a) 6, 6 (b) 6, 4 (c) 4, 4 (d) 3, 3

20. $MgSO_4$ on reaction with NH_4OH and Na_2HPO_4 forms a white crystalline precipitate. What is its formula?

(a) $Mg(NH_4)PO_4$ (b) $Mg_3(PO_4)_2$ (c) $MgCl_2 \cdot MgSO_4$ (d) $MgSO_4$

SECTION C

Comprehension-I

$RCONH_2$ in converted into RNH_2 by means of Hofmann bromamide degradation.

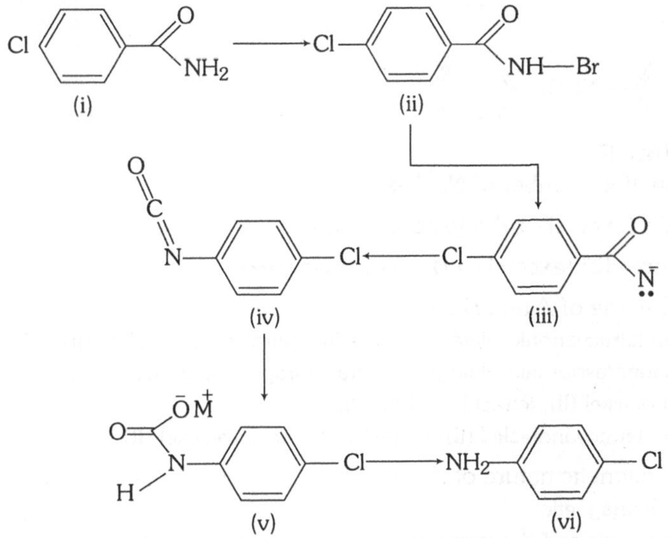

In this reaction, $RCONHBr$ is formed from which this reaction has derived its name. Electron donating group at phenyl activates the reaction. Hofmann degradation reaction is an intramolecular reaction.

21. How can the conversion of (i) to (ii) be brought about?

(a) KBr (b) $KBr + CH_3ONa$

(c) $KBr + KOH$ (d) $Br_2 + KOH$

22. Which is the rate determining step in Hofmann bromamide degradation?

(a) Formation of (i)
(b) Formation of (ii)
(c) Formation of (iii)
(d) Formation of (iv)

23. What are the constituent amines formed when the mixture of (i) and (ii) undergoes Hofmann bromamide degradation?

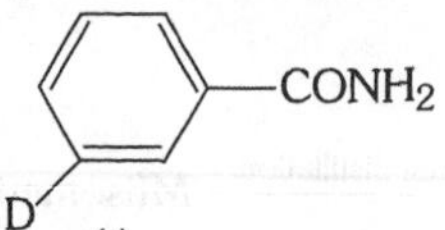
(i)

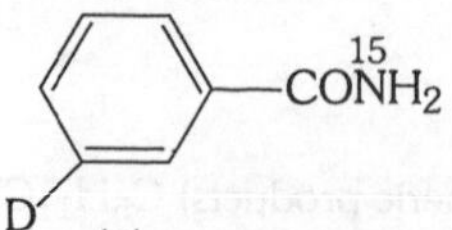
(ii)

(a) 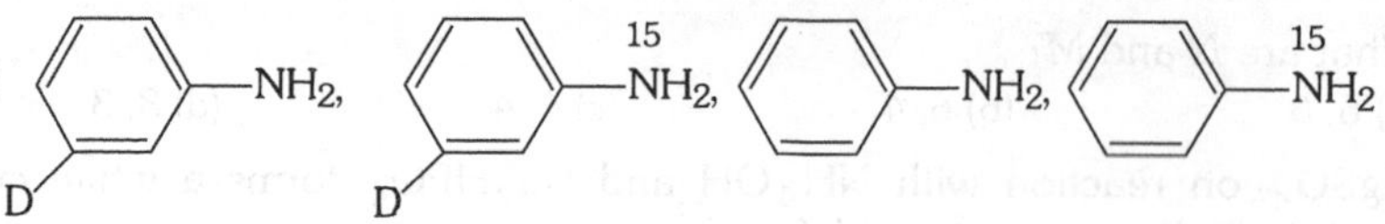

(b) 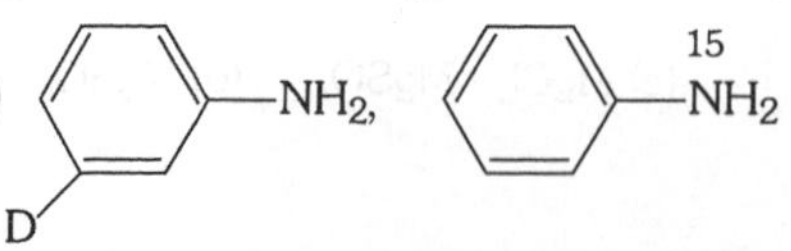

(c) 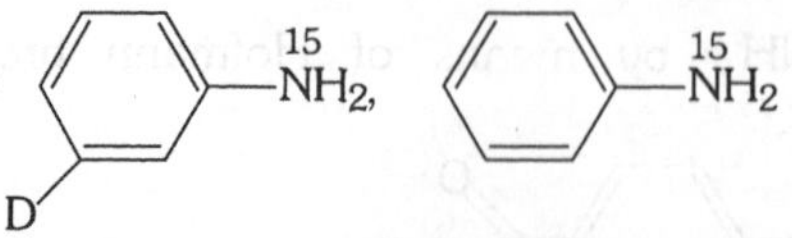

(d) 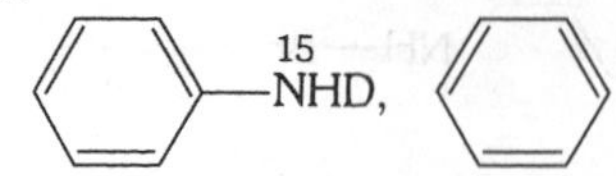

Comprehension-II

The coordination number of Ni^{2+} is 4.

$NiCl_2 + KCN \text{ (excess)} \rightarrow A$ (cyano complex)

$NiCl_2 + \text{conc. } HCl \text{ (excess)} \rightarrow B$ (chloro complex)

24. The IUPAC name of *A* and *B* are

(a) Potassium tetracyanonickelate (II), potassium tetrachloronickelate (II)
(b) Tetracyanopotassiumnickelate (II), tetrorachloropotassiumnickelate (II)
(c) Tetracyanonickel (II), tetrachloronickel (II)
(d) Potassium tetracyanonickel (II), potassium tetrachloronickel (II)

25. Predict the magnetic nature of *A* and *B* are

(a) both are diamagnetic
(b) *A* is diamagnetic and *B* is paramagnetic with one unpaired electron
(c) *A* is diamagnetic and *B* is paramagnetic with two unpaired electrons
(d) both are paramagnetic

26. The hybridization of *A* and *B* are

(a) dsp^2, sp^3
(b) sp^3, sp^3
(c) dsp^2, dsp^2
(d) sp^3d^2, d^2sp^3

Comprehension-III

Carbon-14 is used to determine the age of organic material. The procedure is based on the formation of ^{14}C by neutron capture in the upper atmosphere.

$$^{14}_{7}N + ^{1}_{0}n \longrightarrow ^{14}_{6}C + {}_{1}n^{1}$$

^{14}C is absorbed by living organisms during photosynthesis. The ^{14}C content is constant in living organism once the plant or animal dies, the uptake of carbon dioxide by it ceases and the level of ^{14}C in the dead being, falls to the decay which ^{14}C undergoes.

$$^{14}_{6}C \longrightarrow ^{14}_{7}N + \beta^-$$

The half life period of ^{14}C is 5770 years. The decay constant (λ) can be calculated by using the following formula $\lambda = \frac{0.693}{t_{1/2}}$.

The comparison of the β^- activity of the dead matter with that of the carbon still in circulation enables measurement of the period of the isolation of the material from the living cycle. The method however, ceases to be accurate over periods longer than 30,000 years. The proportion of ^{14}C to ^{12}C in living matter is $1:10^{12}$.

27. Which of the following option is correct?

(a) In living organisms, circulation of ^{14}C from atmosphere is high so the carbon content is constant in organism.

(b) Carbon dating can be used to find out the age of earth crust and rocks.

(c) Radioactive absorption due to cosmic radiation is equal to the rate of radioactive decay, hence the carbon content remains constant in living organisms.

(d) Carbon dating can not be used to determine concentration of ^{14}C in dead beings.

28. What should be the age of fossil for meaningful determination of its age?

(a) 6 years (b) 6000 years

(c) 60,000 years (d) It can be used to calculate any age

29. A nuclear explosion has taken place leading to increase in concentration of C^{14} in nearby areas. C^{14} concentration is C_1 in nearby areas and C_2 in areas far away. If the age of the fossil is determined to be T_1 and T_2 at the places respectively then

(a) The age of the fossil will increase at the place where explosion has taken and $T_1 - T_2 = \frac{1}{\lambda} \ln \frac{C_1}{C_2}$

(b) The age of the fossil will decrease at the place where explosion has taken place and $T_1 - T_2 = \frac{1}{\lambda} \ln \frac{C_1}{C_2}$

(c) The age of fossil will be determined to be same

(d) $\frac{T_1}{T_2} = \frac{C_1}{C_2}$

Comprehension-IV

Tollen's reagent is used for the detection of aldehyde when a solution of $AgNO_3$ is added to glucose with NH_4OH then gluconic acid is formed

$$Ag^+ + e^- \rightarrow Ag; \; E^\circ_{red} = 0.8 \text{ V}$$

$$C_6H_{12}O_6 + H_2O \rightarrow \text{Gluconic acid } (C_6H_{12}O_7) + 2H^+ + 2e^-; \; -E^\circ_{red} = -0.05 \text{ V}$$

$$Ag(NH_3)_2^+ + e^- \rightarrow Ag(s) + 2NH_3; \; E^\circ_{red} = 0.337 \text{ V}$$

[Use $2.303 \times \frac{RT}{F} = 0.0592$ and $\frac{F}{RT} = 38.92$ at 298 K]

30. $2Ag^+ + C_6H_{12}O_6 + H_2O \rightarrow 2Ag(s) + C_6H_{12}O_4 + 2H^+$. Find ln K of the this reaction.

(a) 66.13 (b) 58.38 (c) 28.30 (d) 46.29

31. When ammonia is added to the solution, pH is raised to 11. Which half-cell reaction is affected by pH and by how much?

(a) E_{oxd} will increase by a factor of 0.65 from E°_{oxd}

(b) E_{oxd} will decrease by a factor of 0.65 from E°_{oxd}

(c) E_{red} will incresae by a factor of 0.65 from E°_{red}

(d) E_{red} will decrease by a factor of 0.65 from E°_{red}

32. Ammonia is always added in this reaction.Which of the following must be incorrect?

(a) NH_3 combines with Ag^+ to form a complex

(b) $Ag(NH_3)_2^+$ is a stronger oxidising reagent than Ag^+.

(c) In absence of NH_3 silver salt of gluconic acid is formed.

(d) NH_3 has affected the standard reduction potential of glucose/gluconic acid electrode.

SECTION D

33. 75.2 g of C_6H_5OH (phenol) is dissolved in a solvent of $K_f = 14$. If the depression in freezing point is 7 K, then find the % of phenol that dimerises.

34. For the reaction, $2CO + O_2 \rightarrow 2CO_2$; $\Delta H = -560$ kJ. Two moles of CO and one mole of O_2 are taken in a container of volume 1 L. They completely form two moles of CO_2, then gases deviate appreciably from ideal behaviour. If the pressure in the vessel changes from 70 to 40 atm, find the magnitude (absolute value) of ΔU at 500 K. (1 L-atm = 0.1 kJ)

35. We have taken a saturated solution of AgBr, K_{sp} is 12×10^{-14}. If 10^{-7} mole of $AgNO_3$ are added to 1 litre of this solution, find conductivity (specific conductance) of this solution in terms of 10^{-7} Sm^{-1} units.

Given, $\lambda^\circ_{(Ag^+)} = 6 \times 10^{-3}$ $Sm^2 \, mol^{-1}$, $\lambda^\circ_{(Br^-)} = 8 \times 10^{-3}$ $Sm^2 \, mol^{-1}$,

$\lambda^\circ_{(NO_3^-)} = 7 \times 10^{-3}$ $Sm^2 \, mol^{-1}$,

36. The edge length of unit cell of a metal having molecular weight 75 g/mol is 5 Å which crystallizes in cubic lattice. If the density is 2g/cc then find the radius of metal atom, ($N_A = 6 \times 10^{23}$). Give the answer in pm.

SECTION E

37. Match the extraction processes listed in column I with metals listed in column II.

Column-I	Column-II
(A) Self reduction	(P) Lead
(B) Carbon reduction	(Q) Silver
(C) Complex formation and displacement by metal	(R) Copper
(D) Decomposition of iodide	(S) Boron

38. Match the following

Column-I	Column-II
(A) $Bi^{3+} \rightarrow (BiO)^{+}$	(P) Heat
(B) $[AlO_2]^{-} \rightarrow Al(OH)_3$	(Q) Hydrolysis
(C) $SiO_4^{4-} \rightarrow Si_2O_7^{6-}$	(R) Acidification
(D) $(B_4O_7^{2-}) \rightarrow [B(OH)_3]$	(S) Dilution by water

39. According to Bohr's theory E_n = Total energy, K_n = Kinetic energy, V_n = Potential energy, r_n = Radius of nth orbit
Match the following

Column I	Column II
(A) $V_n / K_n = ?$	(P) 0
(B) If radius of nth orbit $\propto E_n^x$, $x = ?$	(Q) -1
(C) Angular momentum in lowest orbital	(R) -2
(D) $\frac{1}{r^n} \propto Z^y$, $y = ?$	(S) 1

40. Match the following

Column I	Column II
(A) CH_3—CHBr—CD_3 on treatment with alc. KOH gives $CH_2 = CH—CD_3$ as a major product.	(P) E1 reaction
(B) Ph—CHBr—CH_3 reacts faster than Ph—CHBr—CD_3	(Q) E2 reaction
(C) Ph—CH_2—CH_2Br on treatment with $C_2H_5OD / C_2H_5O^-$ gives Ph—CD= CH_2 as the major product	(R) E1cb reaction
(D) $PhCH_2CH_2Br$ and $PhCH_2CH_2Br$ react with same rate	(S) First order reaction

ANSWERS

1. (a)	**2.** (a)	**3.** (d)	**4.** (a)	**5.** (c)	**6.** (a)
7. (c)	**8.** (a)	**9.** (d)	**10.** (d)	**11.** (b)	**12.** (b)
13. (a)	**14.** (a)	**15.** (c)	**16.** (c)	**17.** (b)	**18.** (b)
19. (b)	**20.** (a)	**21.** (d)	**22.** (d)	**23.** (b)	**24.** (a)
25. (c)	**26.** (a)	**27.** (c)	**28.** (b)	**29.** (a)	**30.** (b)
31. (c)	**32.** (d)	**33.** 35%	**34.** – 557 kJ	**35.** 55 sm^{-1}	**36.** 216.5 Pm

37. (A) P, R (B) P (C) Q (D) S **38.** (A) Q, S (B) S (C) R (D) Q, R **39.** (A) R (B) Q (C) P (D) S
40. (A) Q (B) Q (C) R, S (D) P, S

Mathematics

SECTION I

1. Axis of a parabola is $y = x$ and vertex and focus are at a distance $\sqrt{2}$ and $2\sqrt{2}$ respectively, from the origin. Then, equation of the parabola is

(a) $(x - y)^2 = 8(x + y - 2)$ (b) $(x + y)^2 = 2(x + y - 2)$

(c) $(x - y)^2 = 4(x + y - 2)$ (d) $(x + y)^2 = 2(x - y + 2)$

2. Inradius of a circle which is inscribed in an isosceles triangle one of whose angle is $2\pi/3$, is $\sqrt{3}$, then area of triangle is

(a) $4\sqrt{3}$ (b) $12 - 7\sqrt{3}$

(c) $12 + 7\sqrt{3}$ (d) None of these

3. The value of $\int \frac{(x^2 - 1)dx}{x^3\sqrt{2x^4 - 2x^2 + 1}}$ is

(a) $2\sqrt{2 - \frac{2}{x^2} + \frac{1}{x^4}} + C$ (b) $2\sqrt{2 + \frac{2}{x^2} + \frac{1}{x^4}} + C$

(c) $\frac{1}{2}\sqrt{2 - \frac{2}{x^2} + \frac{1}{x^4}} + C$ (d) None of these

4. The set of values of θ satisfying the inequation $2\sin^2\theta - 5\sin\theta + 2 > 0$, where $0 < \theta < 2\pi$, is

(a) $\left(0, \frac{\pi}{6}\right) \cup \left(\frac{5\pi}{6}, 2\pi\right)$ (b) $\left[0, \frac{\pi}{6}\right] \cup \left[\frac{5\pi}{6}, 2\pi\right]$

(c) $\left[0, \frac{\pi}{3}\right] \cup \left[\frac{2\pi}{3}, 2\pi\right]$ (d) None of these

5. If a, b and c are the sides of a ΔABC such that $x^2 - 2(a + b + c)x + 3\lambda(ab + bc + ca) = 0$ has real roots, then

(a) $\lambda < \frac{4}{3}$ (b) $\lambda > \frac{5}{3}$ (c) $\lambda \in \left(\frac{4}{3}, \frac{5}{3}\right)$ (d) $\lambda \in \left(\frac{1}{3}, \frac{5}{3}\right)$

6. Let $\mathbf{a} = \hat{\mathbf{i}} + 2\hat{\mathbf{j}} + \hat{\mathbf{k}}$, $\mathbf{b} = \hat{\mathbf{i}} - \hat{\mathbf{j}} + \hat{\mathbf{k}}$, $\mathbf{c} = \hat{\mathbf{i}} + \hat{\mathbf{j}} - \hat{\mathbf{k}}$. A vector coplanar to $\mathbf{a}$ and $\mathbf{b}$ has a projection along $\mathbf{c}$ of magnitude $\frac{1}{\sqrt{3}}$, then the vector is

(a) $4\hat{\mathbf{i}} - \hat{\mathbf{j}} + 4\hat{\mathbf{k}}$ (b) $4\hat{\mathbf{i}} + \hat{\mathbf{j}} - 4\hat{\mathbf{k}}$

(c) $2\hat{\mathbf{i}} + \hat{\mathbf{j}} + \hat{\mathbf{k}}$ (d) None of these

7. If $f''(x) = -f(x)$, where $f(x)$ is a continuous double differentiable function and $g(x) = f'(x)$.

If $F(x) = \left(f\left(\frac{x}{2}\right)\right)^2 + \left(g\left(\frac{x}{2}\right)\right)^2$ and $F(5) = 5$, then $F(10)$ is

(a) 0 (b) 5 (c) 10 (d) 25

8. If $w = \alpha + i\beta$, where $\beta \neq 0$ and $z \neq 1$ satisfies the condition that $\left(\frac{w - \overline{w}z}{1 - z}\right)$ is purely real, then the set of values of z is

(a) $|z| = 1$ and $z \neq 2$
(b) $|z| = 1$ and $z \neq 1$
(c) $z = \overline{z}$
(d) None of these

9. For $x > 0$, $\lim_{x \to 0} \left((\sin x)^{1/x} + \left(\frac{1}{x}\right)^{\sin x}\right)$ is

(a) 0 (b) -1 (c) 1 (d) 2

10. If r, s, t are prime numbers and p, q are the positive integers such that LCM of p, q is $r^2 s^4 t^2$, then the number of ordered pairs (p, q) is

(a) 252 (b) 254 (c) 225 (d) 224

11. Let $\theta \in \left(0, \frac{\pi}{4}\right)$ and $t_1 = (\tan\theta)^{\tan\theta}$, $t_2 = (\tan\theta)^{\cot\theta}$ and $t_4 = (\cot\theta)^{\tan\theta}$, then

(a) $t_1 > t_2 > t_3 > t_4$ (b) $t_4 > t_3 > t_1 > t_2$ (c) $t_3 > t_1 > t_2 > t_4$ (d) $t_2 > t_3 > t_1 > t_4$

12. A plane passes through $(1, -2, 1)$ and is perpendicular to two planes $2x - 2y + z = 0$ and $x - y + 2z = 4$, then the distance of the plane from the point $(1, 2, 2)$ is

(a) 0 (b) 1 (c) $\sqrt{2}$ (d) $2\sqrt{2}$

SECTION II

13. Equation of common tangent of $y = x^2$, $y = -x^2 + 4x - 4$ is

(a) $y = 4(x - 1)$ (b) $y = 0$ (c) $y = -4(x - 1)$ (d) $y = -30x - 50$

14. If $f(x)$ is cubic polynomial which has local maximum at $x = -1$. If $f(2) = 18$, $f(1) = -1$ and $f'(x)$ has local minimum at $x = 0$, then

(a) the distance between $(-1, 2)$ and $(a, f(a))$ where $x = a$ is the point of local minima is $2\sqrt{5}$
(b) $f(x)$ is increasing for $x \in [1, 2\sqrt{5}]$
(c) $f(x)$ has local minima at $x = 1$
(d) the value of $f(0) = 5$

15. If e_1 is the eccentricity of the ellipse $\frac{x^2}{16} + \frac{y^2}{25} = 1$ and e_2 is the eccentricity of the hyperbola passing through the focii of the ellipse and $e_1 e_2 = 1$, then equation of the hyperbola is

(a) $\frac{x^2}{9} - \frac{y^2}{16} = 1$ (b) $\frac{x^2}{16} - \frac{y^2}{9} = -1$ (c) $\frac{x^2}{9} - \frac{y^2}{25} = 1$ (d) None of these

16. Tangent is drawn at any point P of a curve which passes through $(1, 1)$ cutting X-axis and Y-axis at A and B, respectively. If $AP : BP = 3 : 1$, then

(a) differential equation of the curve is $3x\frac{dy}{dx} + y = 0$

(b) differential equation of the curve is $3x\frac{dy}{dx} - y = 0$

(c) curve is passing through $\left(\frac{1}{8}, 2\right)$

(d) normal at (1, 1) is $x + 3y = 4$

17. If $f(x) = \min\{1, x^2, x^3\}$, then

(a) $f(x)$ is continuous everywhere

(b) $f(x)$ is continuous and differentiable everywhere

(c) $f(x)$ is not differentiable at two points

(d) $f(x)$ is not differentiable at one point

18. Internal bisector of $\angle A$ of ΔABC meets side BC at D. A line drawn through D perpendicular to AD intersects the side AC at E and side AB at F. If a, b and c represent sides of ΔABC, then

(a) AE is HM of b and c

(b) $AD = \frac{2bc}{b+c} \cos \frac{A}{2}$

(c) $EF = \frac{4bc}{b+c} \sin \frac{A}{2}$

(d) the ΔAEF is isosceles

19. If $f(x) = \begin{cases} e^x, & 0 \le x \le 1 \\ 2 - e^{x-1}, & 1 < x \le 2 \\ x - e, & 2 < x \le 3 \end{cases}$ and $g(x) = \int_0^x f(t)\, dt$, $x \in [1, 3]$, then

(a) $g(x)$ has local maxima at $x = 1 + \log_e 2$ and local minima at $x = e$

(b) $f(x)$ has local maxima at $x = 1$, and local minima at $x = 2$

(c) $g(x)$ has no local minima

(d) $f(x)$ has no local maxima

20. Let $\mathbf{A}$ be vector parallel to line of intersection of planes P_1 and P_2 through origin. P_1 is parallel to the vectors $2\hat{\mathbf{j}} + 3\hat{\mathbf{k}}$ and $4\hat{\mathbf{j}} - 3\hat{\mathbf{k}}$ and P_2 is parallel to $\hat{\mathbf{j}} - \hat{\mathbf{k}}$ and $3\hat{\mathbf{i}} + 3\hat{\mathbf{j}}$, then the angle between vector $\mathbf{A}$ and $2\hat{\mathbf{i}} + \hat{\mathbf{j}} - 2\hat{\mathbf{k}}$ is

(a) $\frac{\pi}{2}$ (b) $\frac{\pi}{4}$ (c) $\frac{\pi}{6}$ (d) $\frac{3\pi}{4}$

SECTION III

Read the following passage and answer the questions.

$A = \begin{bmatrix} 1 & 0 & 0 \\ 2 & 1 & 0 \\ 3 & 2 & 1 \end{bmatrix}$, if U_1, U_2 and U_3 are column matrix satisfying.

$AU_1 = \begin{bmatrix} 1 \\ 0 \\ 0 \end{bmatrix}$, $AU_2 = \begin{bmatrix} 2 \\ 3 \\ 0 \end{bmatrix}$, $AU_3 = \begin{bmatrix} 2 \\ 3 \\ 1 \end{bmatrix}$ and U is 3×3 matrix whose column are U_1, U_2 and U_3.

21. The value of $|U|$ is

(a) 3 (b) -3 (c) $\frac{3}{2}$ (d) 2

22. The value of the elements of U^{-1} is

(a) -1 (b) 0 (c) 1 (d) 3

23. The value of $[3 \quad 2 \quad 0]\, U \begin{bmatrix} 3 \\ 2 \\ 0 \end{bmatrix}$ is

(a) [5] (b) $\begin{bmatrix} \frac{5}{2} \end{bmatrix}$

(c) [4] (d) $\begin{bmatrix} \frac{3}{2} \end{bmatrix}$

Read the following passage and answer the questions.

For every function $f(x)$ which is twice differentiable, these will be good approximation of $\int_a^b f(x)\, dx \cong \left(\frac{b-a}{2}\right)\{f(a) + f(b)\}$. Now, if we take $c = \frac{a+b}{2}$, then using above again, we get $\int_a^b f(x)\, dx = \int_a^c f(x)\, dx + \int_c^b f(x)\, dx \cong \frac{b-a}{4}\{f(a) + f(b) + 2f(c)\}$ and so on.

We get approximation for value of $\int_a^b f(x)\, dx$.

24. Good approximation of $\int_0^{\pi/2} \sin x\, dx$, is

(a) $\frac{\pi}{4}$ (b) $\frac{\pi}{4}(\sqrt{2} + 1)$

(c) $\frac{\pi}{8}(\sqrt{2} + 1)$ (d) $\frac{\pi}{8}$

25. If $f''(x) < 0, \forall\, x \in (a, b)$, $c(c, f(c))$ is point of maxima where $c \in (a, b)$, then $f'(c)$ is

(a) $\frac{f(b) - f(a)}{b - a}$ (b) $3\left(\frac{f(b) - f(a)}{b - a}\right)$

(c) $2\left(\frac{f(b) - f(a)}{b - a}\right)$ (d) 0

26. If $\lim_{t \to a} \dfrac{\int_a^t f(x)\, dx - \frac{(t-a)}{2}\{f(t) + f(a)\}}{(t-a)^3} = 0$, then degree of polynomial function $f(x)$ at most is

(a) 0 (b) 1

(c) 3 (d) 2

Read the following passage and answer the questions.

There are n urns each containing $(n + 1)$ balls such that the ith urn contains i white balls and $(n + 1 - i)$ red balls. Let u_i be the event of selecting ith urn, $i = 1, 2, 3, \ldots, n$ and W denotes the event of getting a white balls.

27. If $P(u_i) \propto i$, where $i = 1, 2, 3, \ldots, n$, then $\lim_{n \to \infty} P(W)$ is equal to

(a) 1 (b) $\frac{2}{3}$ (c) $\frac{1}{4}$ (d) $\frac{3}{4}$

28. If $P(u_i) = c$, where c is a constant, then $P(u_i / W)$ is equal to

(a) $\frac{2}{n+1}$ (b) $\frac{1}{n+1}$ (c) $\frac{n}{n+1}$ (d) $\frac{1}{2}$

29. If n is even and E denotes the event of choosing even numbered urn $\left(P(u_i) = \frac{1}{n}\right)$, then the value of $P(W / E)$ is

(a) $\frac{n+2}{2n+1}$ (b) $\frac{n+2}{2(n+1)}$

(c) $\frac{n}{n+1}$ (d) $\frac{1}{n+1}$

Read the following passage and answer the questions.

Let $ABCD$ be a square of side length 2 units. C_2 is the circle through vertices A, B, C, D and C_1 is the circle touching all the sides of square $ABCD$. L is the line through A.

30. If P is a point on C_1 and Q is a point on C_2, then $\frac{PA^2 + PB^2 + PC^2 + PD^2}{QA^2 + QB^2 + QC^2 + QD^2}$ is equal to

(a) 0.75 (b) 1.25

(c) 1 (d) 0.5

31. A circle touches the line L and the circle C_1 externally such that both the circles are on the same side of the line, then the locus of centre of the circle is

(a) ellipse (b) hyperbola

(c) parabola (d) parts of straight line

32. A line M through A is drawn parallel to BD. Points S moves such that its distances from the line BD are the vertex A are equal. If locus of S cuts M at T_2 and T_3 and AC at T_1, then area of $\Delta T_1T_2T_3$ is

(a) $\frac{1}{2}$ sq unit (b) $\frac{2}{3}$ sq unit

(c) 1 sq unit (d) 2 sq unit

SECTION IV

33. The value of $\frac{(5050) \int_0^1 (1 - x^{50})^{100}\, dx}{\int_0^1 (1 - x^{50})^{101}\, dx}$

34. If $f(x)$ is twice differentiable function such that $f(a) = 0$, $f(b) = 2$, $f(c) = 1$, $f(d) = 2$, $f(e) = 0$, where $a < b < c < d < e$, then the minimum number of zeros of $g(x) = \{f'(x)\}^2 + f''(x) \cdot f(x)$ in the interval $[a, e]$ is

35. If $x^2 - 10ax - 11b = 0$ have roots c and d and $x^2 - 10cx - 11d = 0$ have roots a and b, then, find $a + b + c + d$.

36. Let $A_n = \left(\frac{3}{4}\right) - \left(\frac{3}{4}\right)^2 + \left(\frac{3}{4}\right)^3 + \ldots + (-1)^{n-1}\left(\frac{3}{4}\right)^n$ and $B_n = 1 - A_n$. Find a least odd natural number n_0, so that $B_n > A_n, \forall\, n \ge n_0$.

SECTION V

37. Match the column.

Normals at P, Q, R are drawn to $y^2 = 4x$ which intersect at (3, 0). Then,

(i) area of ΔPQR	(a) 2
(ii) radius of circumcircle of ΔPQR	(b) $\frac{5}{2}$
(iii) centroid of ΔPQR	(c) $\left(\frac{5}{2}, 0\right)$
(iv) circumcentre of ΔPQR	(d) $\left(\frac{2}{3}, 0\right)$

38. Match the following.

(i) $\int_0^{\pi/2} (\sin x)^{\cos x} (\cos x \cot x - \log (\sin x)^{\sin x})\, dx$	(a) 1
(ii) Area bounded by $-4y^2 = x$ and $x - 1 = -5y^2$	(b) 0
(iii) The angle of intersection of curves $y = 3^{x-1} \log x$ and $y = x^x - 1$	(c) $3e^{y/2}$
(iv) If $\frac{dy}{dx} = \frac{2}{x + y}$ passing through (1, 0), then $(x + y + 2)$ is	(d) $\frac{4}{3}$

39. Match the following

(i) Two rays in I quadrant $x + y = \lvert a \rvert$ and $ax - y = 1$ intersect each other in the interval $a \in (a_0, \infty)$, then value of $\left(\frac{2a_0}{3}\right)$ is	(a) 2
(ii) Point (α, β, γ) lie on the plane $x + y + z = 2$ let $\mathbf{a} = \alpha\hat{\mathbf{i}} + \beta\hat{\mathbf{j}} + \gamma\hat{\mathbf{k}}$, $\hat{\mathbf{k}} \times (\hat{\mathbf{k}} \times \hat{\mathbf{a}}) = 0$, then γ is	(b) $\frac{2}{3}$
(iii) $\left\lvert \int_0^1 (1 - y^2)\, dy \right\rvert + \left\lvert \int_1^0 (y^2 - 1) dy \right\rvert$	(c) $\left\lvert \int_0^1 \sqrt{1 - x}\, dx \right\rvert + \left\lvert \int_{-1}^0 \sqrt{1 + x}\, dx \right\rvert$
(iv) If $\sin A \sin B \sin C + \cos A \cos B$, then the value of $\sin C$ equals	(d) 1

40. Match the following.

(i) $\sum_{i=1}^{\infty} \tan^{-1}\left(\frac{1}{2i^2}\right) = t$, then $\tan t$ is — (a) $\frac{2}{3}$

(ii) Sides a, b, c of a ΔABC are in AP and $\cos\theta_1 = \frac{a}{b+c}$, $\cos\theta_2 = \frac{b}{a+c}$ and $\cos\theta_3 = \frac{c}{a+b}$, then $\tan^2\left(\frac{\theta_1}{2}\right) + \tan^2\left(\frac{\theta_3}{2}\right)$ is — (b) 1

(iii) A line is perpendicular to $x + 2y + 2z = 0$ and passes through (0, 1, 0). The perpendicular distance of this line from the origin is — (c) $\frac{\sqrt{5}}{3}$

ANSWERS

1. (a)	**2.** (c)	**3.** (c)	**4.** (a)	**5.** (a)	**6.** (a)
7. (b)	**8.** (b)	**9.** (c)	**10.** (c)	**11.** (b)	**12.** (d)
13. (a, b)	**14.** (b, c)	**15.** (b)	**16.** (a, c)	**17.** (a, d)	**18.** (a, b, c, d)
19. (a, b)	**20.** (b, d)	**21.** (a)	**22.** (a)	**23.** (a)	**24.** (c)
25. (d)	**26.** (b)	**27.** (b)	**28.** (a)	**29.** (b)	**30.** (a)
31. (c)	**32.** (c)	**33.** 5051	**34.** 6	**35.** 1210	**36.** 7

37. (i) (a) (ii) (b) (iii) (d) (iv) (c)
38. (i) (a) (ii) (d) (iii) (b) (iv) (c)
39. (i) (b) (ii) (a) (iii) (c) (iv) (d)
40. (i) (b) (ii) (a) (iii) (c)

* All options

IIT JEE

UNSOLVED PAPER 2007

Physics

Paper-I

Section I : *This section contains 9 multiple choice questions numbered 1 to 9. Each question has 4 choices (a), (b), (c) and (d), out of which* **ONLY ONE** *is correct.*

Straight Objective Type

1. A long, hollow conducting cylinder is kept coaxially inside another long, hollow conducting cylinder of larger radius. Both the cylinders are initially electrically neutral.

(a) a potential difference appears between the two cylinders when a charge density is given to the inner cylinder

(b) a potential difference appears between the two cylinders when a charge density is given to the outer cylinder

(c) no potential difference appears between the two cylinders when a uniform line charge is kept along the axis of the cylinders

(d) no potential difference appears between the two cylinders when same charge density is given to both the cylinders

2. In an experiment to determine the focal length (f) of a concave mirror by the u-v method, a student places the object pin A on the principal axis at a distance x from the pole P. The student looks at the pin and its inverted image from a distance keeping his/her eye in line with PA. When the student shifts his/her eye towards left, the image appears to the right of the object pin. Then

(a) $x < f$ (b) $f < x < 2f$

(c) $x = 2f$ (d) $x > 2f$

3. A circuit is connected as shown in the figure with the switch S open. When the switch is closed, the total amount of charge that flows from Y and X is

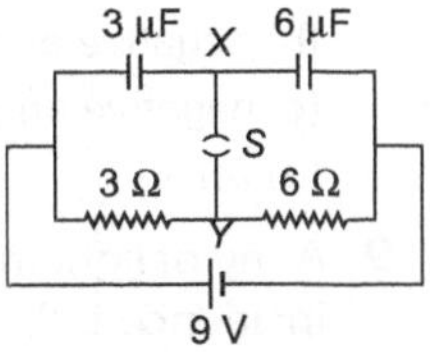

(a) zero (b) 54 μC

(c) 27 μC (d) 81 μC

4. A resistance of 2 Ω is connected across one gap of a meter-bridge (the length of the wire is 100 cm) and an unknown resistance, greater than 2 Ω, is connected across the other gap. When these resistances are interchanged, the balance point shifts by 20 cm. Neglecting any corrections, the unknown resistance is

(a) 3 Ω (b) 4 Ω

(c) 5 Ω (d) 6 Ω

5. In the options given below, let E denotes the rest mass energy of a nucleus and n a neutron. The correct option is

(a) $E({}^{236}_{92}U) > E({}^{137}_{53}I) + E({}^{97}_{36}Y) + 2E(n)$

(b) $E({}^{236}_{92}U) < E({}^{137}_{53}I) + E({}^{97}_{39}Y) + 2E(n)$

(c) $E({}^{236}_{92}U) < E({}^{140}_{56}Ba) + E({}^{94}_{36}Kr) + 2E(n)$

(d) $E({}^{236}_{92}U) < E({}^{140}_{56}Ba) + E({}^{94}_{36}Kr) + 2E(n)$

6. Two particles of mass m each are tied at the ends of a light string of length $2a$. The whole system is kept on a frictionless horizontal surface with the string held tight so that each mass is at a distance a from the centre P(as shown in the figure). Now, the mid-point of the string is pulled vertically upwards with a small but constant force F. As a result, the particles move towards each other on the surface.

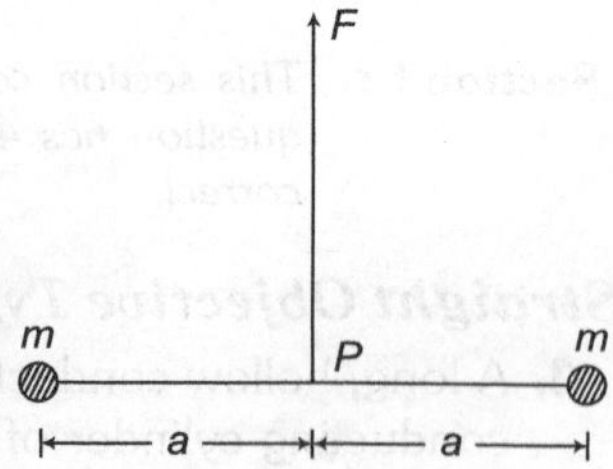

The magnitude of acceleration, when the separation between them becomes $2x$, is

(a) $\dfrac{F}{2m}\dfrac{a}{\sqrt{a^2 - x^2}}$

(b) $\dfrac{F}{2m}\dfrac{x}{\sqrt{a^2 - x^2}}$

(c) $\dfrac{F}{2m}\dfrac{x}{a}$

(d) $\dfrac{F}{2m}\dfrac{\sqrt{a^2 - x^2}}{x}$

7. The largest wavelength in the ultraviolet region of the hydrogen spectrum is 122 nm. The smallest wavelength in the infrared region of the hydrogen spectrum (to the nearest integer) is

(a) 802 nm
(b) 823 nm
(c) 1882 nm
(d) 1648 nm

8. Consider a neutral conducting sphere. A positive point charge is placed outside the sphere. The net charge on the sphere is then

(a) negative and distributed uniformly over the surface of the sphere
(b) negative and appears only at the point on the sphere closet to the point charge
(c) negative and distributed non-uniformly over the entire surface of the sphere
(d) zero

9. A ray of light travelling in water is incident on its surface open to air. The angle of incidence is θ, which is less than the critical angle. Then there will be

(a) only a reflected ray and no refracted ray
(b) only a refracted ray and no reflected ray
(c) a reflected ray and refracted ray and the angle between them would be less than $180° - 2\theta$
(d) a reflected ray and a refracted ray and the angle between them would be greater than $180° - 2\theta$

Section II : *This section contains 4 questions numbered 10 to 13. Each question contains Statement I (Assertion) and Statement II (Reason). Each question has 4 choices (a), (b), (c) and (d) out of which* ***ONLY ONE*** *is correct.*

Assertion and Reason Type

10. Statement I If the accelerating potential in an X-ray tube is increased, the wavelengths of the characteristic X-rays do not change.

Statement II When an electron beam strikes the target in an X-ray tube, part of the kinetic energy is converted into X-ray energy.

(a) Statement I is true, Statement II is true; Statement II is a correct explanation for Statement I

(b) Statement I is true, Statement II is true; Statement II is NOT a correct explanation for Statement I

(c) Statement I is true, Statement II is false

(d) Statement I is false, Statement II is true

11. Statement I In an elastic collision between two bodies, the relative speed of the bodies after collision is equal to the relative speed before the collision.

Statement II In an elastic collision, the linear momentum of the system is conserved.

(a) Statement I is true, Statement II is true; Statement II is a correct explanation for Statement I

(b) Statement I is true, Statement II is true; Statement II is NOT a correct explanation for Statement I

(c) Statement I is true, Statement II is false

(d) Statement I is false, Statement II is true

12. Statement I The formula connecting u, v and f for a spherical mirror is valid only for mirrors whose sizes are very small compared to their radii of curvature.

Statement II Laws of reflection are strictly valid for plane surfaces, but not for large spherical surfaces.

(a) Statement I is true, Statement II is true; Statement II is a correct explanation for Statement I

(b) Statement I is true, Statement II is true; Statement II is NOT a correct explanation for Statement I

(c) Statement I is true, Statement II is false

(d) Statement I is false, Statement II is true

13. Statement I A block of mass m starts moving on a rough horizontal surface with a velocity v. It stops due to friction between the block and the surface after moving through a certain distance. The surface is now tilted to an angle of 30° with the horizontal and the same block is made to go up on the surface with the same initial velocity v. The decrease in the mechanical energy in the second situation is smaller than that in the first situation.

Statement II The coefficient of friction between the block and the surface decreases with the increase in the angle of inclination.

(a) Statement I is true, Statement II is true; Statement II is a correct explanation for Statement I
(b) Statement I is true, Statement II is true; Statement II is NOT a correct explanation for Statement I
(c) Statement I is true, Statement II is false
(d) Statement I is false, Statement II is true

Section III : *This section contains 2 paragraphs* $\mathbf{P_{14-16}}$ *and* $\mathbf{P_{17-19}}$*. Based upon each paragraph, 3 multiple choice questions have to be answered. Each question has 4 choices (a), (b), (c) and (d), out of which* ***ONLY ONE*** *is correct.*

Linked Comprehension Type

$\mathbf{P_{14-16}}$ **: Paragraph for Questions Nos. 14 to 16** A fixed thermally conducting cylinder has a radius R and height L_0. The cylinder is open at its bottom and has a small hole at its top. A piston of mass M is held at a distance L from the top surface as shown in the figure. The atmospheric pressure is p_0.

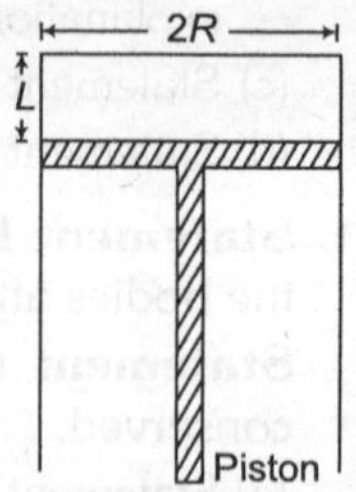

14. The piston is now pulled out slowly and held at a distance $2L$ from the top. The pressure in the cylinder between its top and the piston will then be

(a) p_0 (b) $\frac{p_0}{2}$ (c) $\frac{p_0}{2} + \frac{Mg}{\pi R^2}$ (d) $\frac{p_0}{2} - \frac{Mg}{\pi R^2}$

15. While the piston is at a distance $2L$ from the top, the hole at the top is sealed. The piston is then released, to a position where it can stay in equilibrium. In this condition, the distance of the piston from the top is

(a) $\left(\frac{2p_0\pi R^2}{\pi R^2 p_0 + Mg}\right)(2L)$ (b) $\left(\frac{p_0\pi R^2 - Mg}{\pi R^2 p_0}\right)(2L)$

(c) $\left(\frac{p_0\pi R^2 + Mg}{\pi R^2 p_0}\right)(2L)$ (d) $\left(\frac{p_0\pi R^2}{\pi R^2 p_0 - Mg}\right)(2L)$

16. The piston is taken completely out of the cylinder. The hole at the top is sealed. A water tank is brought below the cylinder and put in a position so that the water surface in the tank is at the same level as the top of the cylinder as shown in the figure. The density of the water is ρ. In equilibrium, the height H of the water column in the cylinder satisfies

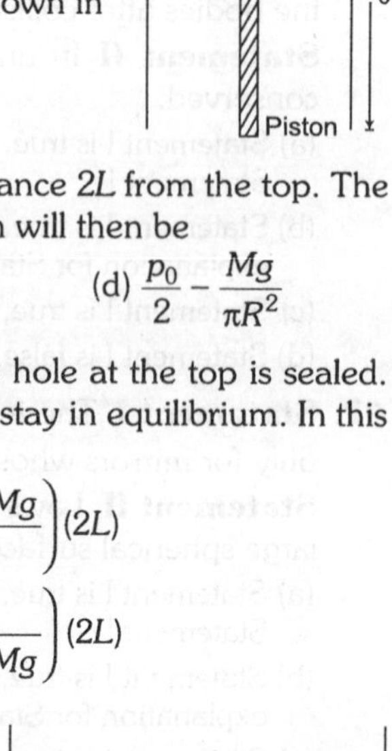

(a) $\rho g (L_0 - H)^2 + p_0 (L_0 - H) + L_0 p_0 = 0$
(b) $\rho g (L_0 - H)^2 - p_0 (L_0 - H) - L_0 p_0 = 0$
(c) $\rho g (L_0 - H)^2 + p_0 (L_0 - H) - L_0 p_0 = 0$
(d) $\rho g (L_0 - H)^2 - p_0 (L_0 - H) + L_0 p_0 = 0$

P_{17-19} **: Paragraph for Questions Nos. 17 to 19** *Two discs A and B are mounted coaxially on a vertical axle. The discs have moments of inertia I and 2I respectively about the common axis. Disc A is imparted an initial angular velocity 2ω using the entire potential energy of a spring compressed by a distance x_1. Disc B is imparted an angular velocity ω by a spring having the same spring constant and compressed by a distance x_2. Both the discs rotate in the clockwise direction.*

17. The ratio $\frac{x_1}{x_2}$ is

(a) 2 (b) $\frac{1}{2}$ (c) $\sqrt{2}$ (d) $\frac{1}{\sqrt{2}}$

18. When disc B is brought in contact with disc A, they acquire a common angular velocity in time t. The average frictional torque on one disc by the other during this period is

(a) $\frac{2I\omega}{3t}$ (b) $\frac{9I\omega}{2t}$ (c) $\frac{9I\omega}{4t}$ (d) $\frac{3I\omega}{2t}$

19. The loss of kinetic energy during the above process is

(a) $\frac{I\omega^2}{2}$ (b) $\frac{i\omega^2}{3}$ (c) $\frac{I\omega^2}{4}$ (d) $\frac{I\omega^2}{6}$

Section IV : *This section contains 3 questions. Each question contains Statements given in two columns which have to be matched. Statements (A, B, C, D) in Column I have to be matched with Statement (p, q, r, s) in Column II. The answers to these questions have to be appropriately bubbled as illustrated in the following example.*

If the correct matches are A–p, A–s, B–q, B–r, C–p, C–q and D–s, then the correctly bubbled 4 × 4 matrix should be as follows :

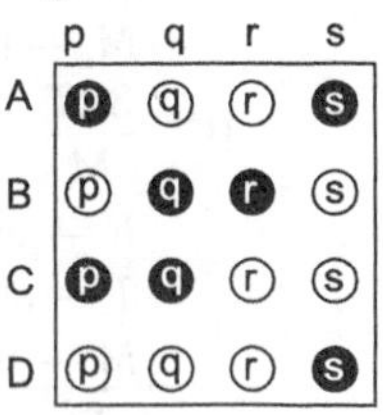

Matrix-Match Type

20. Some laws/processes are given in **Column I.** Match these with the physical phenomena given in **Column II** and indicate your answer by darkening appropriate bubbles in the 4 × 4 matrix given in the ORS.

	Column I		Column II
A.	Transition between two atomic energy levels	p.	Characteristic X-rays
B.	Electron emission from a material	q.	Photoelectric effect
C.	Mosley's law	r.	Hydrogen spectrum
D.	Change of photon energy into kinetic energy of electrons	s.	β-decay

21. Column I gives certain situations in which a straight metallic wire of resistance R is used and **Column II** gives some resulting effects. Match the statements in **Column I** with the statements in **Column II** and indicate your answer by darkening appropriate bubbles in the 4 × 4 matrix given in the ORS.

Column I	Column II
A. A charged capacitor is connected to the ends of the wire.	p. A constant current flows through the wire.
B. The wire is moved perpendicular to its length with a constant velocity in a uniform magnetic field perpendicular to the plane of motion.	q. Thermal energy is generated in the wire.
C. The wire is placed in a constant electric field that has a direction along the length of the wire.	r. A constant potential difference develops between the ends of the wire.
D. A battery of constant emf is connected to the ends of the wire.	s. Charges of constant magnitude appear at the ends of the wire.

22. Some physical quantities are given in **Column I** and some possible SI units in which these quantities may be expressed are given in **Column II**. Match the physical quantities in **Column I** with the units in **Column II** and indicate your answer by darkening appropriate bubbles in the 4 × 4 matrix given in the ORS.

Column I	Column II
A. GM_eM_s G — universal gravitational constant, M_e — mass of the earth, M_s — mass of the sun.	p. (volt) (coulomb) (metre)
B. $\frac{3RT}{M}$ R — universal gas constant, T — absolute temperature, M — molar mass	q. (kilogram)(metre)^3 (second)$^{-2}$
C. $\frac{F^2}{q^2B^2}$ F — force, q — charge, B — magnetic field.	r. (metre)^2 (second)$^{-2}$
D. $\frac{GM_e}{R_e}$ G — universal gravitational constant, M_e — mass of the earth, R_e — radius of the earth	s. (farad) (volt)^2 (kg)$^{-1}$

ANSWERS

1. (a)	**2.** (b)	**3.** (c)	**4.** (a)	**5.** (a)	**6.** (b)
7. (b)	**8.** (d)	**9.** (c)	**10.** (b)	**11.** (b)	**12.** (c)
13. (c)	**14.** (a)	**15.** (d)	**16.** (c)	**17.** (c)	**18.** (a)

19. (b) **20.** A–p, r; B–p; q, s; C–p; D–q **21.** A–q; B–r, s; C–s, D–p; q,r

22. A–p, q; B–r, s; C–r, s; D–r, s

Paper-II

Section I : *This section contains 9 multiple choice questions numbered 1 to 9. Each question has choices (a), (b), (c) and (d), out of which* ***ONLY ONE*** *is correct.*

Straight Objective Type

1. A student performs an experiment to determine the Young's modulus of a wire, exactly 2 m long, by Searle's method. In a particular reading, the student measures the extension in the length of the wire to be 0.8 mm with an uncertainty of ± 0.05 mm at a load of exactly 1.0 kg. The student also measures the diameter of the wire to be 0.4 mm with an uncertainty of ± 0.01 mm. Take $g = 9.8\ \text{m/s}^2$ (exact). The Young's modulus obtained from the reading is

(a) $(2.0 \pm 0.3)\, 10^{11}\ \text{N/m}^2$
(b) $(2.0 \pm 0.2) \times 10^{11}\ \text{N/m}^2$
(c) $(2.0 \pm 0.1) \times 10^{11}\ \text{N/m}^2$
(d) $(2.0 \pm 0.05) \times 10^{11}\ \text{N/m}^2$

2. A small object of uniform density rolls up a curved surface with an initial velocity v. It reaches up to a maximum height of $\frac{3v^2}{4g}$ with respect to the initial position. The object is

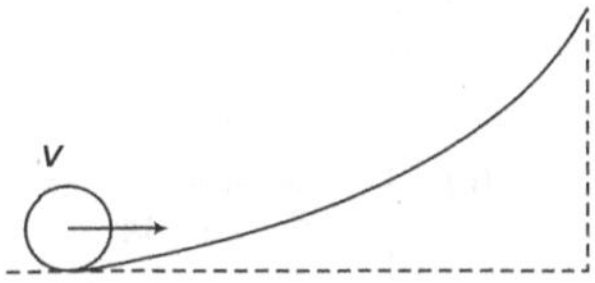

(a) ring
(b) solid sphere
(c) hollow sphere
(d) disc

3. In the experiment to determine the speed of sound using a resonance column :

(a) Prongs of the tuning fork are kept in a vertical plane
(b) Prongs of the tuning fork are kept in a horizontal plane
(c) In one of the two resonances observed, the length of the resonating air column is close to the wavelength of sound in air
(d) In one of the two resonances observed, the length of the resonating air column is close to half of the wavelength of sound in air

4. A spherical portion has been removed from a solid sphere having a charge distributed uniformly in its volume as shown in the figure. The electric field inside the emptied space is

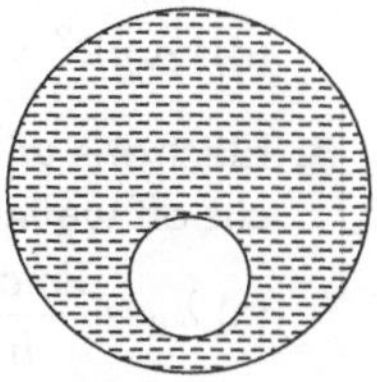

(a) zero everywhere
(b) non-zero and uniform
(c) non-uniform
(d) zero only at its centre

5. A particle moves in the X-Y plane under the influence of a force such that its linear momentum is $\mathbf{p}(t) = A[\hat{\mathbf{i}} \cos kt - \hat{\mathbf{j}} \sin kt]$, where A and k are constants. The angle between the force and the momentum is

(a) 0°
(b) 30°
(c) 45°
(d) 90°

6. A magnetic field $\mathbf{B} = B_0\hat{\mathbf{j}}$ exists in the region $a < x < 2a$ and $\mathbf{B} = B_0\ \hat{\mathbf{j}}$ in the region $2a < x < 3a$, where B_0 is a positive constant. A positive point charge moving with a velocity $\mathbf{v} = v_0\hat{\mathbf{i}}$, where v_0 is a positive constant, enters the magnetic field at $x = a$. The trajectory of the charge in this region can be like

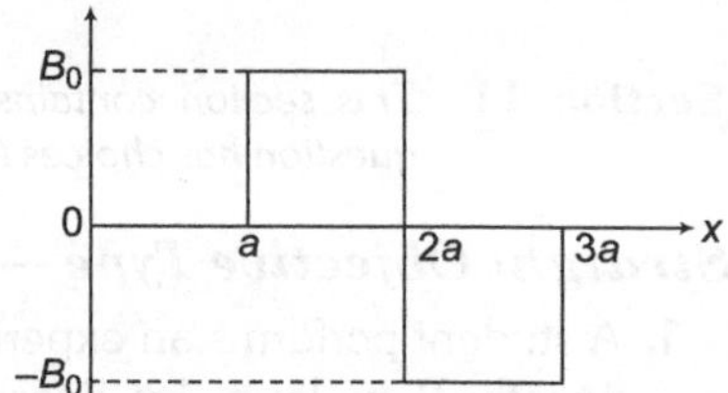

(a)

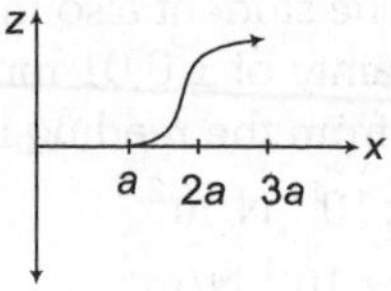

(b)

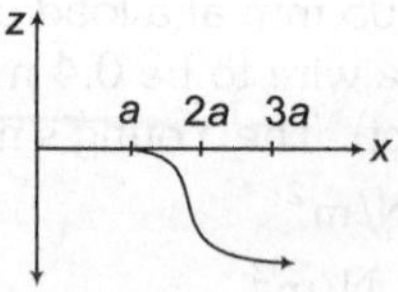

(c)

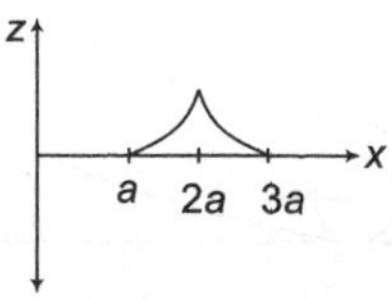

(d)

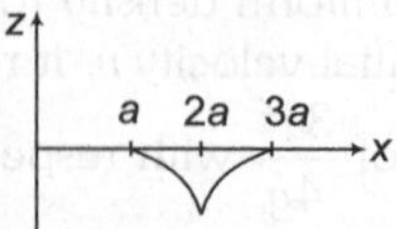

7. Water is filled up to a height h in a beaker of radius R as shown in the figure. The density of water is ρ, the surface tension of water is T and the atmospheric pressure is p_0. Consider a vertical section $ABCD$ of the water column through a diameter of the beaker. The force on water on one side of this section by water on the other side of this section has magnitude.

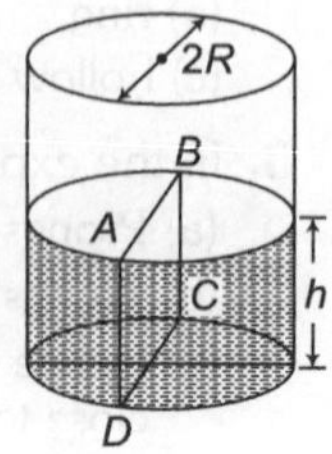

(a) $|2\,p_0Rh + \pi R^2\rho gh - 2RT|$

(b) $|2\,p_0Rh + R\rho gh^2 - 2RT|$

(c) $|\,p_0\pi R^2 + R\rho gh^2 - 2RT|$

(d) $|\,p_0\pi R^2 + R\rho gh^2 + 2RT|$

8. Electrons with de-Broglie wavelength λ fall on the target in an X-ray tube. The cut-off wavelength of the emitted X-rays is

(a) $\lambda_0 = \dfrac{2mc\lambda^2}{h}$ (b) $\lambda_0 = \dfrac{2h}{mc}$ (c) $\lambda_0 = \dfrac{2m^2c^2\lambda^3}{h^2}$ (d) $\lambda_0 = \lambda$

9. Positive and negative point charges of equal magnitude are kept at $\left(0,\ 0,\ \dfrac{a}{2}\right)$ and $\left(0,\ 0,\ \dfrac{-a}{2}\right)$, respectively. The work done by the electric field when another positive point charge is moved from $(-a,\ 0,\ 0)$ to $(0,\ a, 0)$ is

(a) positive

(b) negative

(c) zero

(d) depends on the path connecting the initial and final positions

Section II : *This section contains 4 questions numbered 10 to 13. Each question contains Statement I (Assertion) and Statement II (Reason). Each question has 4 choices (a), (b), (c) and (d) out of which* ***ONLY ONE*** *is correct.*

Assertion and Reason Type

10. Statement I A cloth covers a table. Some dishes are kept on it. The cloth can be pulled out without dislodging the dishes from the table.

Statement II For every action there is an equal and opposite reaction.

(a) Statement I is true, Statement II is true; Statement II is a correct explanation for Statement

(b) Statement I is true, Statement II is true; Statement II is NOT a correct explanation for Statement I

(c) Statement I is true, Statement II is false

(d) Statement I is false, Statement II is true

11. Statement I The total translational kinetic energy of all the molecules of a given mass of an ideal gas is 1.5 times the product of its pressure and its volume.

Statement II The molecules of a gas collide with each other and the velocities of the molecules change due to the collision.

(a) Statement I is true, Statement II is true; Statement II is a correct explanation for Statement I

(b) Statement I is true, Statement II is true; Statement II is NOT a correct explanation for Statement I

(c) Statement I is true, Statement II is false

(d) Statement I is false, Statement II is true

12. Statement I If there is no external torque on a body about its centre of mass, then the velocity of the centre of mass remains constant.

Statement II The linear momentum of an isolated system remains constant.

(a) Statement I is true, Statement II is true; Statement II is a correct explanation for Statement

(b) Statement I is true, Statement II is true; Statement II is NOT a correct explanation for Statement I

(c) Statement I is true, Statement II is false

(d) Statement I is false, Statement II is true

13. Statement I A vertical iron rod has a coil of wire wound over it at the bottom end. An alternating current flows in the coil. The rod goes through a conducting ring as shown in the figure. The ring can float at a certain height above the coil.

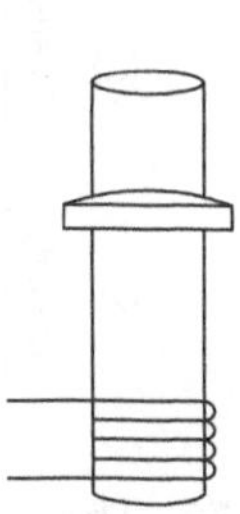

Statement II In the above situation, a current is induced in the ring which interacts with the horizontal component of the magnetic field to produce an average force in the the upward direction.

(a) Statement I is true, Statement II is true; Statement II is a correct explanation for Statement
(b) Statement I is true, Statement II is true; Statement II is NOT a correct explanation for Statement I
(c) Statement I is true, Statement II is false
(d) Statement I is false, Statement II is true

Section III : *This section contains 2 paragraphs* **P$_{14-16}$** *and* **P$_{17-19}$**. *Based upon each paragraph, 3 multiple choice questions have to be answered. Each question has 4 choices (a), (b), (c) and (d), out of which* ***ONLY ONE*** *is correct.*

Linked Comprehension Type

P$_{14-16}$: Paragraph for Questions Nos. 14 to 16

Two trains *A* and *B* are moving with speeds 20 m/s and 30 m/s, respectively in the same direction on the same straight track, with *B* ahead of *A*. The engines are at the front ends. The engine of train *A* blows a long whistle.

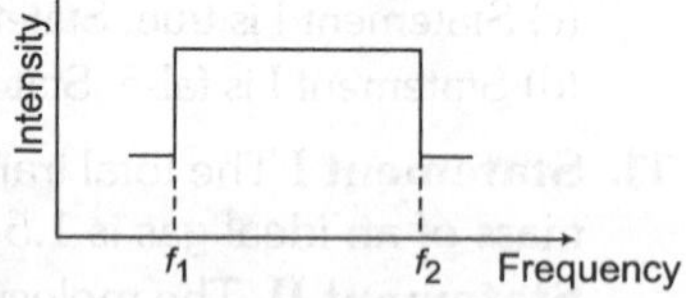

Assume that the sound of the whistle is composed of components varying in frequency from $f_1 = 800$ Hz to $f_2 = 1120$ Hz, as shown in the figure. The spread in the frequency (highest frequency – lowest frequency) is thus 320 Hz. The speed of sound in still air is 340 m/s.

14. The speed of sound of the whistle is
(a) 340 m/s for passengers in *A* and 310 m/s for passengers in *B*
(b) 360 m/s for passengers in *A* and 310 m/s for passengers in *B*
(c) 310 m/s for passengers in *A* and 360 m/s for passengers in *B*
(d) 340 m/s for passengers in both the trains

15. The distribution of the sound intensity of the whistle as observed by the passengers in train *A* is best represented by

(a) 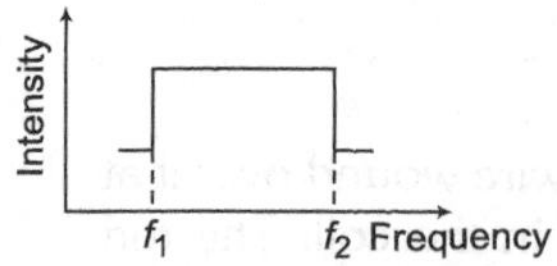

(b) 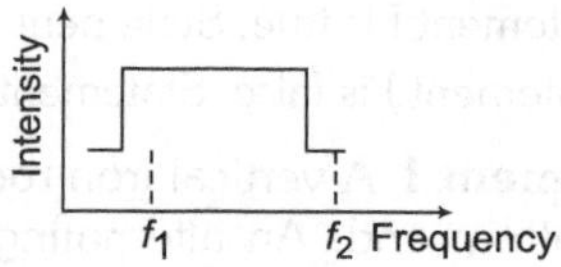

(c) 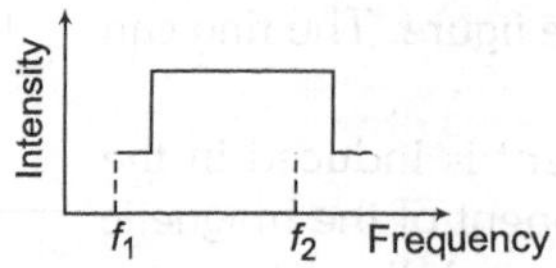

(d) 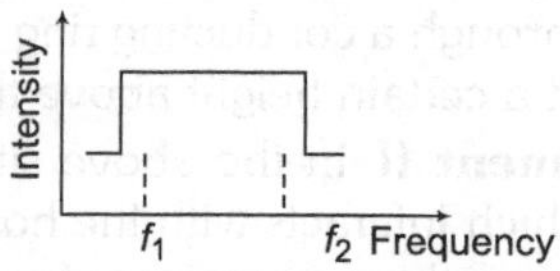

16. The spread of frequency as observed by the passengers in train *B* is
(a) 310 Hz
(b) 330 Hz
(c) 350 Hz
(d) 290 Hz

P_{17-19} : Paragraph for Questions Nos. 17 to 19 :

The figure shows a surface XY separating two transparent media, medium-1 and medium-2. The lines ab and cd represent wavefronts of a light wave travelling in medium-1 and incident on XY. The lines ef and gh represent wavefronts of the light wave in medium-2 after refraction.

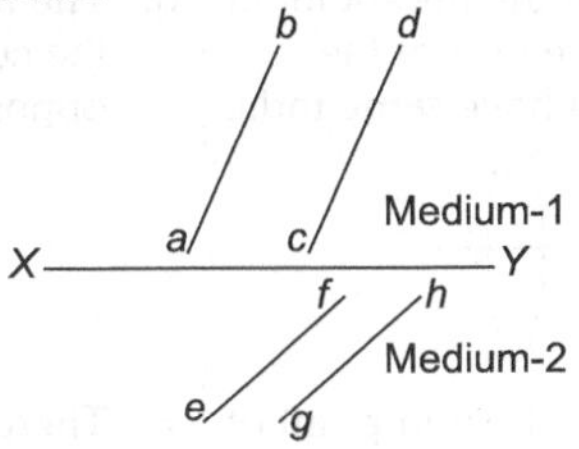

17. Light travels as a
(a) parallel beam in each medium
(b) convergent beam in each medium
(c) divergent beam in each medium
(d) divergent beam in one medium and convergent beam in the other medium

18. The phases of the light wave at c, d, e and f are ϕ_c, ϕ_d, ϕ_e and ϕ_f, respectively. It is given that $\phi_c \neq \phi_f$.
(a) ϕ_c cannot be equal to ϕ_d
(b) ϕ_d can be equal to ϕ_e
(c) $(\phi_d - \phi_f)$ is equal to $(\phi_c - \phi_e)$
(d) $(\phi_d - \phi_c)$ is not equal to $(\phi_f - \phi_e)$

19. Speed of light is
(a) the same in medium-1 and medium-2
(b) larger in medium-1 than in medium-2
(c) larger in medium-2 than in medium-1
(d) different at b and d

Section IV : *This section contains 3 questions. Each question contains statements given in two columns which have to be matched. Statements (A, B, C, D) in* **Column I** *have to be matched with statements (p, q, r, s) in* **Column II**. *The answer to these questions have to be appropriately bubbled as illustrated in the following example.*
If the correct matches are A–p, A–s, B–q, C–p, C–q and D–s, then the correctly bubbled 4 × 4 matrix should be as follows :

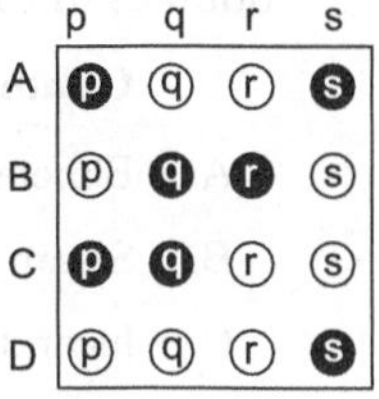

Matrix-Match Type

20. Two wires each carrying a steady current I are shown in four configurations in **Column I**. Some of the resulting effects are described in **Column II**. Match the statements in **Column I** with the statements in **Column II** and indicate your answer by darkening appropriate bubbles in the 4 × 4 matrix given in the ORS.

	Column I		Column II
A.	Point P is situated midway between the wires. 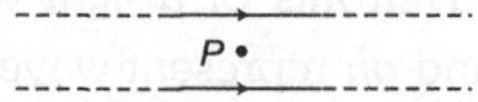	p.	The magnetic fields (B) at P due to the currents in the wires are in the same direction.
B.	Point P is situated at the mid-point of the line joining the centres of the circular wires, which have same radii. 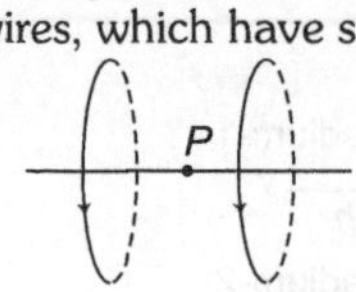	q.	The magnetic fields (B) at P due to the currents in the wires are in opposite directions.
C.	Point P is situated at the mid-point of the line joining the centres of the circular wires, which have same radii. 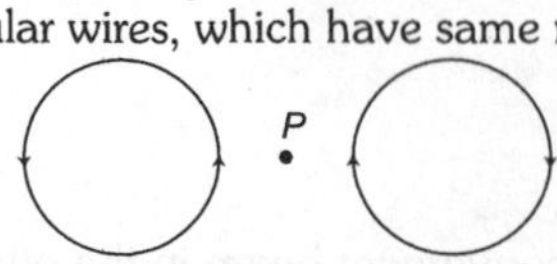	r.	There is no magnetic field at P.
D.	Point P is situated at the common centre of the wires. 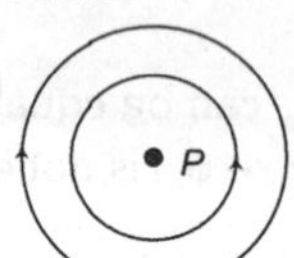	s.	The wires repel each other.

21. **Column I** gives some devices and **Column II** gives some processes on which the functioning of these devices depend. Match the devices in **Column I** with the processes in **Column II** and indicate your answer by darkening appropriate bubbles in the 4 × 4 matrix given in the ORS.

	Column I		Column II
A.	Bimetallic strip	p.	Radiation from a hot body
B.	Steam engine	q.	Energy conversion
C.	Incandescent lamp	r.	Melting
D.	Electric fuse	s.	Thermal expansion of solids

22. **Column I** describes some situations in which a small object moves. **Column II** describes some characteristics of these motions. Match the situations in **Column I** with the characteristics in **Column II** and indicate your answer by darkening appropriate bubbles in the 4 × 4 matrix given in the ORS.

	Column I		Column II
A.	The object moves on the x-axis under a conservative force in such a way that its "speed" and "position" satisfy $v = c_1\sqrt{c_2 - x^2}$, where c_1 and c_2 are positive constants.	p.	The object executes a simple harmonic motion.
B.	The object moves on the x-axis in such a way that its velocity and its displacement from the origin satisfy $v = -kx$, where k is a positive constant.	q.	The object does not change its direction.
C.	The object is attached to one end of a massless spring of a given spring constant. The other end of the spring is attached to the ceiling of an elevator. Initially, everything is at rest. The elevator starts going upwards with a constant acceleration a. The motion of the object is observed from the elevator during the period it maintains this acceleration.	r.	The kinetic energy of the object keeps on decreasing.
D.	The object is projected from the Earth's surface vertically upwards with a speed $2\sqrt{\frac{GM_e}{R_e}}$, where M_e is the mass of the Earth and R_e is the radius of the Earth. Neglect forces from objects other than the Earth.	s.	The object can change its direction only once.

ANSWERS

1. (b) **2.** (d) **3.** (a) **4.** (b) **5.** (d) **6.** (a)
7. (b) **8.** (a) **9.** (c) **10.** (b) **11.** (b) **12.** (d)
13. (a) **14.** (b) **15.** (a) **16.** (a) **17.** (a) **18.** (c)
19. (b) **20.** A–q, r; B–p, C–q; C–q, r; D–q, s
21. A–s, B–q; C–p, q; D–q **22.** A–p; B–q; C–p; D–q, r

IIT JEE
UNSOLVED PAPER 2007

Chemistry

Paper-I

Section I : *This section contains 9 multiple choice questions numbered 1 to 9. Each questions has 4 choices (a), (b), (c) and (d), out of which ONLY ONE is correct.*

Straight Objective Type

1. The species having bond order different from that in CO is

(a) NO^- (b) NO^+ (c) CN^- (d) N_2

2. Among the following, the paramagnetic compound is

(a) Na_2O_2 (b) O_3 (c) N_2O (d) KO_2

3. Extraction for zinc from zinc blende is achieved by

(a) electrolytic reduction

(b) roasting followed by reduction with carbon

(c) roasting followed by reduction with another metal

(d) roasting followed by self-reduction

4. In the following reaction,

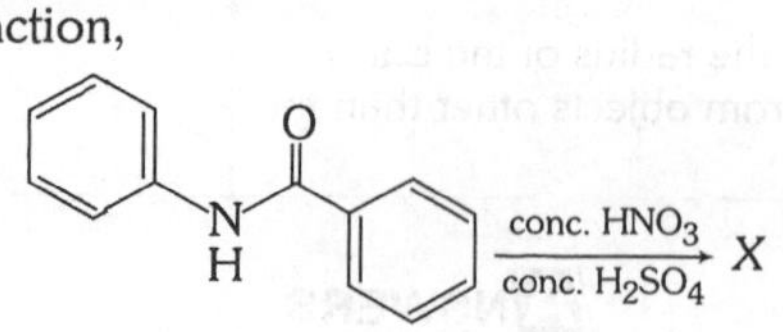

The structure of the major product X is

(a)
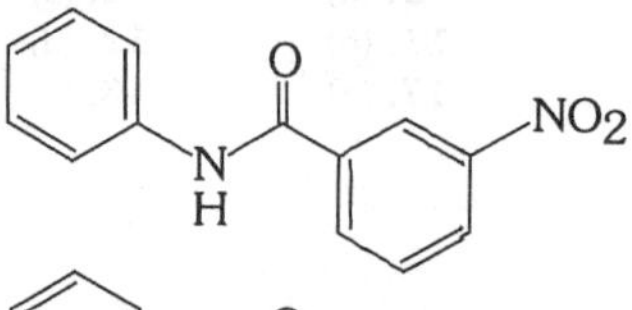

(b)
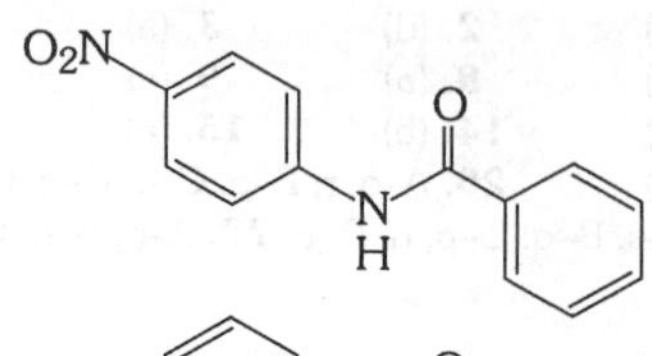

(c)
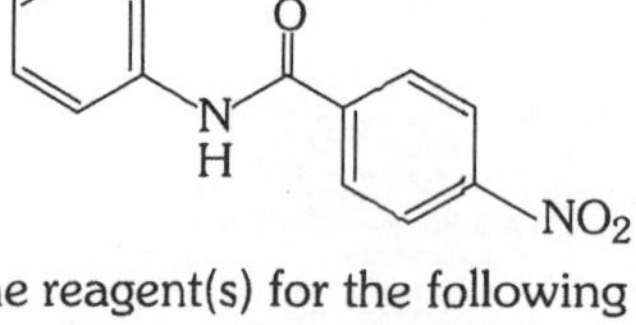

(d)
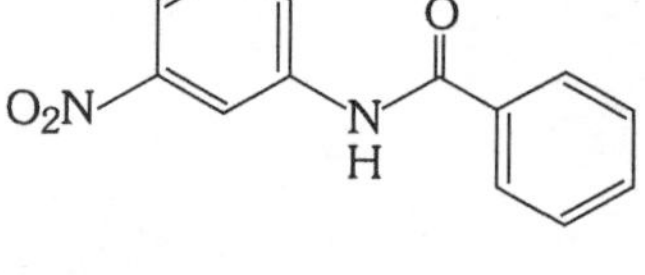

5. The reagent(s) for the following conversion,

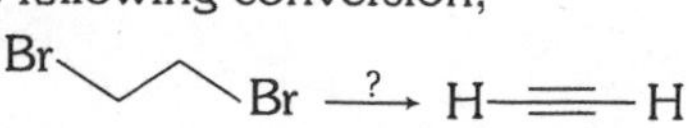

is/are

(a) alcoholic KOH (b) alcoholic KOH followed by $NaNH_2$

(c) aqueous KOH followed by $NaNH_2$ (d) Zn/CH_3OH

6. The number of structural isomers for C_6H_{14} is

(a) 3 (b) 4 (c) 5 (d) 6

7. The percentage of *p*-character in the orbitals forming P—P bonds in P_4 is

(a) 25 (b) 33 (c) 50 (d) 75

8. When 20 g of naphthoic acid ($C_{11}H_8O_2$) is dissolved in 50 g benzene ($K_f = 1.72$ K kg mol^{-1}), a freezing point depression of 2 K is observed. The van't Hoff factor (*i*) is

(a) 0.5 (b) 1 (c) 2 (d) 3

9. The value of $\log_{10} K$ for a reaction $A \rightleftharpoons B$ is (Given : $\Delta_r H^\circ_{298\,K} = -54.07$ kJ mol^{-1}, $\Delta_r S^\circ_{298\,K} = 10$ JK^{-1} mol^{-1} and $R = 8.314$ JK^{-1}mol^{-1}; $2.303 \times 8.314 \times 298 = 5705$)

(a) 5 (b) 10 (c) 95 (d) 100

Section II : *This section contains 4 questions numbered 10 to 13. Each question contains Statement-I (Assertion) and Statement-II (Reason). Each question has 4 choices (a), (b), (c) and (d) our of which **ONLY ONE** is correct.*

Assertion and Reason Type

10. Statement I Boron always forms covalent bond

Statement II The small size of B^{3+} favours formation of covalent bond.

(a) Statement I is true, Statement II is true, Statement II is a correct explanation for Statement I

(b) Statement I is true, Statement-II is true; Statement II is not a correct explanation for Statement II

(c) Statement I is true, Statement II is false

(d) Statement I is false, Statement II is true

11. Statement I In water, orthoboric acid behaves as a weak monobasic acid.

Statement II In water, orthoboric acid acts as proton donor.

(a) Statement I is true, Statement II is true, Statement II is a correct explanation for Statement I

(b) Statement I is true, Statement II is true; Statement II is not a correct explanation for Statement I

(c) Statement I is true, Statement II is false

(d) Statement I is false, Statement II is true

12. Statement I *p*-hydroxybenzoic acid has a lower boiling point than *o*-hydroxybenzoic acid.

Statement II *o*-hydroxybenzoic acid has intramolecular hydrogen bonding.

(a) Statement I is true, Statement II is true, Statement II is a correct explanation for Statement I

(b) Statement I is true, Statement II is true; Statement II is not a correct explanation for Statement I

(c) Statement I is true, Statement II is false

(d) Statement I is false, Statement II is true

13. Statement I Micelles are formed by surfactant molecules above the critical micellar concentration (CMC).

Statement II The conductivity of a solution having surfactant molecules deceases sharply at the CMC.

(a) Statement I is true, Statement II is true, Statement II is a correct explanation for Statement I

(b) Statement I is true, Statement II is true; Statement II is not a correct explanation for Statement I

(c) Statement I is true, Statement II is false

(d) Statement I is false, Statement II is true

Section III : *The section contains 2 paragraphs C_{14-16} and C_{17-19}. Based upon each paragraph, 3 multiple choice questions have to be answered. Each questions has 4 choices (a), (b), (c) and (d), out of which only one is correct.*

Linked Comprehension Type

C_{14-16} : *Paragraph for Questions Nos. 14 to 16*

The noble gases have closed-shell electronic configuration and are monoatomic gases under normal conditions. The low boiling points of the lighter noble gases are due to weak dispersion forces between the atoms and the abssence of other interatomic interactions.

The direct reaction of xenon with fluorine leads to a series of compounds with oxidation number +2, +4 and +6. XeF_4 reacts violently with water to give XeO_3. The compounds of xenon exhibit rich stereochemistry and their geometries can be deduced considering the total number of electron pairs in the valence shell.

14. Argon is used in are welding because of its

(a) low reactivity with metal

(b) ability to lower the melting point of metal

(c) flammability

(d) high calorific value

15. The structure of XeO_3 is

(a) linear (b) planar (c) pyramidal (d) T-shaped

16. XeF_4 and XeF_6 are expected to be

(a) oxidising (b) reducing (c) unreactive (d) strongly basic

C_{17-19} : *Paragraph for Questions Nos. 17 to 19*

Chemical reactions involve interaction of atoms and molecules. A large number of atoms/molecules (approximately 6.023×10^{23}) are present in a few grams of any chemical compound varying with their atomic/molecular masses. To handle such large number conveniently, the mole concept was introduced. This concept has implications in diverse areas such as analytical chemistry, biochemistry, electrochemistry and radiochemistry. The following example illustrates a typical case, involving chemical/electrochemical reaction, which requires a clear understanding of the mole concept.

A 4.0 molar aqueous solution of NaCl is prepared and 500 mL of this solution is electrolysed. This leads to the evolution of chlorine gas at one of electrodes
(atomic mass : Na = 23, Hg = 200; 1 faraday = 96500 coulombs).

17. The total number of moles of chlorine gas evolved is

(a) 0.5 (b) 1.0 (c) 2.0 (d) 3.0

18. If the cathode is a Hg electrode, the maximum weight (*g*) of amalgam formed from this solution is

(a) 200 (b) 225 (c) 400 (d) 446

19. The total charge (coulombs) required for complete electrolysis is

(a) 24125 (b) 48250 (c) 96500 (d) 193000

Section IV : *This section contains 3 questions. Each question contains statements given in two columns which have to be matched. Statement (A, B, C, D) in Column I have to be matched with statement (p, q, r, s) in Column II. The answers to these questions have to be appropriately bubbled as illustrated in the following example.*

If the correct matches are $A-p, A-s, B-q, B-r, C-p, C-q$ and $D-s$, then the correctly bubbled 4×4 matrix should be as follows :

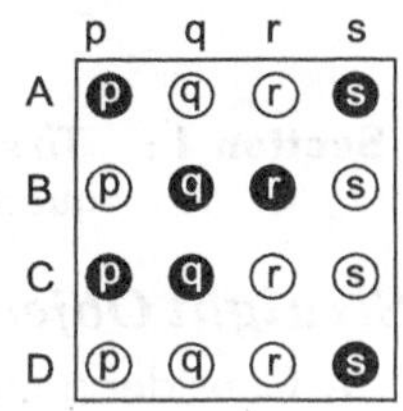

Matrix-Match Type

20. Match the complexes in Column I with their properties listed in Column II. Indicates your answer by darkening the appropriate bubbles of 4×4 matrix given in the ORS.

Column I	Column II
A. $[Co(NH_3)_4(H_2O)_2]\ Cl_2$	p. geometrical isomers
B. $[Pt(NH_3)_2Cl_2]$	q. paramagnetic
C. $[Co(H_2O)_5Cl]Cl$	r. diamagnetic
D. $Ni(H_2O)_6]Cl_2$	s. metal ion with $+2$ oxidation state

21. Match the gases under specified conditions listed in Column I with their properties/laws in Column II. Indicate your answer by darkening the appropriate bubbles of 4×4 matrix given in the ORS.

Column I	Column II
A. Hydrogen gas ($P = 200$ atm, $T = 273$ K)	p. compressibility factor $\neq 1$
B. Hydrogen gas ($P \sim 0, T = 273$ K)	q. attractive forces are dominant
C. CO_2 ($P = 1$ atm, $T = 273$ K)	r. $PV = nRT$
D. Real gas with very large molar volume	s. $P(V - nb) = nRT$

22. Match the chemical substance in Column I with type of polymers/type of bond in Column II. Indicate your answer by darkening the appropriate bubbles of 4×4 matrix given in the ORS.

Column I	Column II
A. Cellulose	p. natural polymer
B. Nylon-6, 6	q. synthetic polymer
C. Protein	r. amide linkage
D. Sucrose	s. glycoside linkage

ANSWERS

1. (a)	**2.** (d)	**3.** (b)	**4.** (b)	**5.** (b)	**6.** (c)
7. (d)	**8.** (a)	**9.** (b)	**10.** (a)	**11.** (c)	**12.** (d)
13. (b)	**14.** (a)	**15.** (c)	**16.** (a)	**17.** (b)	**18.** (d)
19. (d)	**20.** A–p, q, s; B–p, r, s; C–q, s; D q, s				

21. A–p, s; B–r; C–p, q; D–r

22. A–p, s; B–q, r; C–p, r; D–s

Paper-II

Section I : *This section contains 9 multiple choice questions numbered 1 to 9. Each questions has 4 choices (a), (b), (c) and (d), out of which only ONE is correct.*

Straight Objective Type

1. Consider a titration of potassium dichromate solution with acidified Mohr's salt solution using diphenylamine as indicator. The number of moles of Mohr's salt required per mole of dichromate is

(a) 3 (b) 4
(c) 5 (d) 6

2. Among the following metal carbonyls, the C—O bond order is lowest in

(a) $[MO(CO)_6]^+$ (b) $[Fe(CO)_5]$
(c) $[Cr(CO)_6]$ (d) $[V(CO)_6]^-$

3. A solution of a metal ion when treated with KI gives a red precipitate which dissolved in excess KI to give a colourless solution. Moreover, the solution of metal ion on treatment with a solution of cobalt (II) thiocyanate gives rise to a deep blue crystalline precipitqte. The metal ion is :

(a) Pb^{2+} (b) Hg^{2+}
(c) Cu^{2+} (d) Co^2

4. Cyclohexne on ozonoloysis followed by reaction with zinc dust and water gives compund *E*. Compound *E* on further treatment with aqueous KOH yields compound *F*. Compound *F* is :

(a) 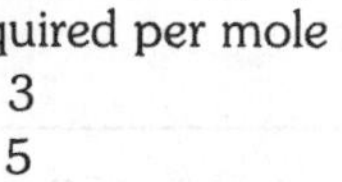

(b) 

(c) 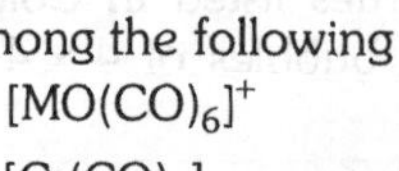

(b) 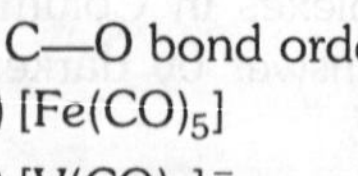

5. The number of stereoisomers obtained by bromiantion of trans-2-butene is

(a) 1 (b) 2
(c) 3 (d) 4

6. Among the following, the least stable resonance strcutgure is

(a) 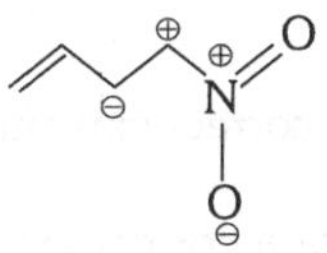 (b) 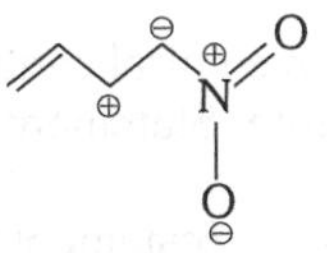 (c) 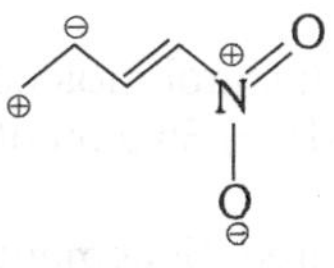 (d) 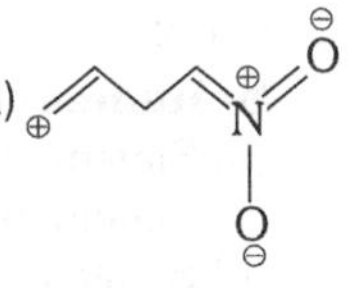

7. A positron is emitted from $^{23}_{11}Na$. The ratio of the atomic mass and atomic number of the resulting nuclide is

(a) 22/10 (b) 22/11 (c) 23/10 (d) 23/12

8. For the process $H_2O(l)$ (1 bar, 273 K) $\rightarrow H_2O(g)$ (1 bar, 373 K), the correct set of thermodynamic parameters is

(a) $\Delta G = 0, \Delta S = +ve$ (b) $\Delta G = 0, \Delta S = -ve$

(c) $\Delta G = +ve, \Delta S = 0$ (d) $\Delta G = -ve, \Delta S = +ve$

9. Consider a reaction $aG + bH \rightarrow$ products. When concentratin of both the reactants G and H is doubled, the rate increases by eight times. However, when concentration of G is doubled keeping the concentration of H fixed, the rate is doubled. The overall order of the reaction is :

(a) 0 (b) 1 (c) 2 (d) 3

Section II : *This section contains* 4 questions numbered 10 to 13. *Each question contains* Statement-I *(Assertion) and* Statement-II *(Reason). Each question has* 4 choices (a), (b), (c) and (d) out of which ONLY ONE is correct.

Assertion and Reason

10. Statement-I Alkali metals dissolve in liquid ammonia to give blue solutions.

Because

Statement II Alkali metals in liquid ammonia give solvated species of the type $(M(NH_3)_n]^+$ (M = alkali metals)

(a) Statement I is true, Statement II is true; Statement II is a correct explanation for Statement I

(b) Statement I is true, Statement II is true; Statement II is not a correct explanation for Statement I

(c) Statement I is true, Statement II is false

(d) Statement I is false, Statement II is true

11. Statement I Glucose gives a reddish-brown precipitate with Fehling's solution.

Statement II Rection of glucose with Fehling's solution gives CuO and gluconic acid.

(a) Statement I is true, Statement II is true; Statement II is a correct explanation for Statement I

(b) Statement I is true, Statement II is true; Statement II is not a correct explanation for Statement I

(c) Statement I is true, Statement II is false

(d) Statement I is false, Statement II is true

12. **Statement-I** Molecules that are not superimposable on their mirror images are chiral.

Statement II All chiral molecules have chiral centres.

(a) Statement I is true, Statement II is true; Statement II is a correct explanation for Statement I

(b) Statement I is true, Statement II is true; Statement II is not a correct explanation for Statement I

(c) Statement I is true, Statement II is false

(d) Statement I is false, Statement II is true

13. **Statement I** Band gap in germanium is small.

Statement II The energy spread of each germanium atomic energy level is infinitesimally small.

(a) Statement I is true, Statement II is true; Statement II is a correct explanation for Statement I

(b) Statement I is true, Statement II is true; Statement II is not a correct explanation for Statement I

(c) Statement-I is true, Statement II is false

(d) Statement-I is talse, Statement II is true

Section III : *This section contains 2 paragraphs C_{14-16} and C_{17-19}. Based upon each paragraph, 3 multiple choice questions have to answered. Each questions has 4 choices (a), (b), (c), (d) out of which only one is correct.*

Linked Comprehension

C_{14-16} : Paragraph for Question Nos. 14 to 16 : *Redox reaction play a pivotal role in chemistry and biology. The values of standard redox potential ($E°$) of two half-cell reactions decide which way the reaction is expected to proceed. A simple example is a Daniel cell in which zinc goes into solutio and copper gets deposited. Given below are a set of half-cell recations (acidic medium) along with their $E°$ (V with respect to normal hydrogen electrode) values. Using this data obtain the correct explanaitons to Questions 14–19.*

$I_2 + 2e^- \longrightarrow 2I^-$ $\quad E° = 0.54$

$Cl_2 + 2e^- \longrightarrow 2Cl^-$ $\quad E° = 1.36$

$Mn^{3+} + e^- \longrightarrow Mn^{2+}$ $\quad E° = 1.50$

$Fe^{3+} + e^- \longrightarrow Fe^{2+}$ $\quad E° = 0.77$

$O_2 + 4H^+ + 4e^- \longrightarrow 2H_2O$ $\quad E° = 1.23$

14. Among the following, identify the correct statement :

(a) Chloride ion is oxidised by O_2 (b) Fe^{2+} is oxidised by iodine

(c) Iodide ion is oxidised by chlorine (d) Mn^{2+} is oxidised by chlorine

15. While Fe^{3+} is stable, Mn^{3+} is not stable in acid solution because

(a) O_2 oxidises Mn^{2+} to Mn^{3+}

(b) O_2 oxidises both Mn^{2+} and Fe^{2+} to Fe^{3+}

(c) Fe^{3+} oxidises H_2O to O_2

(d) Mn^{3+} oxidises H_2O to O_2

16. Sodium fusion extract, obtained from aniline, on treatment with iron (II) sulphate and H_2SO_4 in presence of air gives a Prussian blue precipitate. The blue colour is due to the formation of

(a) $Fe_4[Fe(CN)_6]_3$ (b) $Fe_3[Fe(CN)_6]_2$

(c) $Fe_4[Fe(CN)_6]_2$ (d) $Fe_3[Fe(CN)_6]_3$

C_{17-19} : ***Paragraph for Questions Nos. 17 to 19*** *Reimer-Tiemann reaction introduces an aldehyde group, on to the aromatic ring of phenol, ortho to the hydroxyl group. This reaction involves electrophilic aromatic substitution. This is a general method for the synthesis of substituted salicylaldehydes as depicted below.*

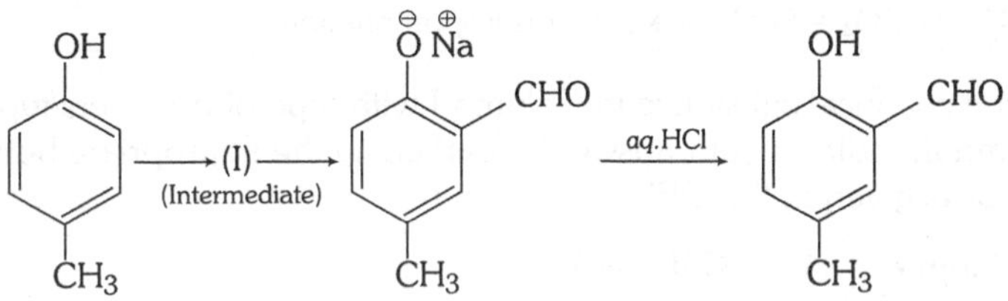

(I) (II) (III)

17. Which one of the following reagents is used the above reaction?

(a) *aq.* NaOH+ CH_3Cl (b) *aq.* NaOH+ CH_2Cl_2

(c) *aq.* NaOH+ $CHCl_3$ (d) *aq.* NaOH+ CCl_4

18. The electrophile in this reaction is

(a) :CHCL (b) $^{+}CHCl_2$

(c) : CCl_2 (d) • CCl_3

19. The structure of the intermediate I is :

(a) 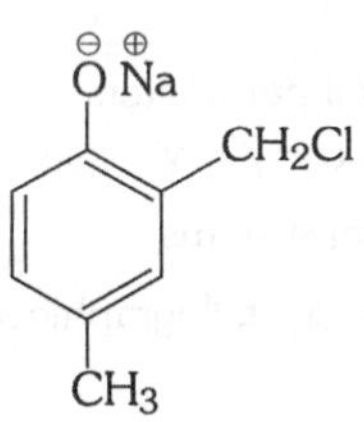

(b) 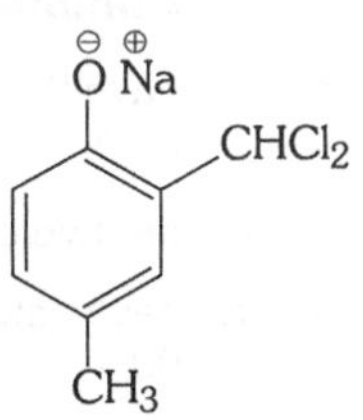

(c) 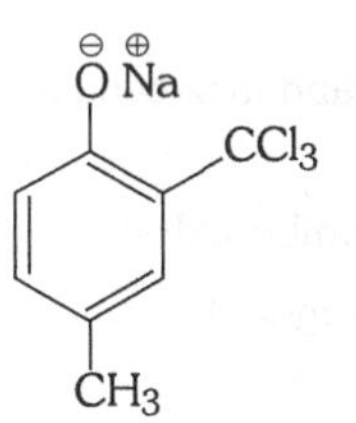

(d) 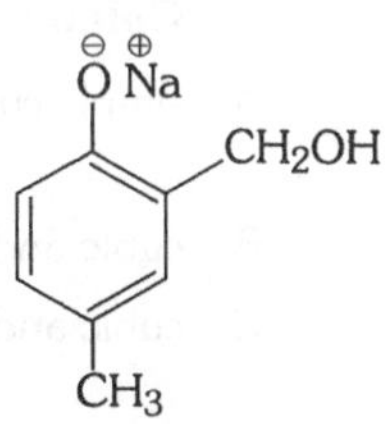

Section IV : *This section contains 3 questions. Each question contains statements given in two columns which have to be matched. Statement (A, B, C, D) in Column I have to be matched with statement (p, q, r, s) in Column II. The answers to these questions have to be appropriately bubbled as illustrated in the following example.*

If the correct matches are $A-p, A-s, B-q, B-r, C-p, C-q$ and D—s, then the correctly bubbled 4 × 4 matrix should be as follows :

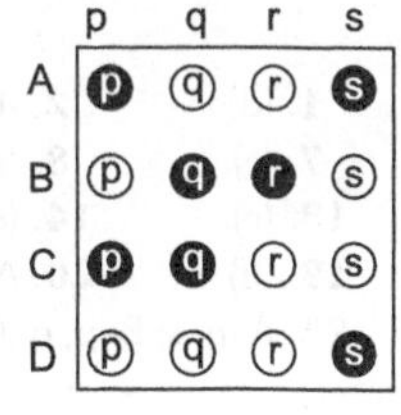

Matrix-Match Type

21. Match the complexes in Column I with their properties listed in Column II. Indicate your answer by darkening the appropriate bubbles of the 4 × 4 matrix given in the ORS.

Column I	Column II
A. $O_2^- \longrightarrow O_2 + O_2^{2-}$	p. redox reaction
B. $CrO_4^{2-} + H^+ \longrightarrow$	q. one of the products has trigonal planar structure
C. $MnO_4^- + NO_2^- + H^+ \longrightarrow$	r. dimeric bridge tetrahedral metal ion
D. $NO_3^- + H_2SO_4 + Fe^{2+} \longrightarrow$	s. disproportionation

21. Match the chemical substance in Column I with type of polymers/type of bonds in Column II. Indicate your answer by darkening the appropriate bubbles of the 4 × 4 matrix given in the ORS.

Column I	Column II
A. C_6H_5CHO	p. gives precipitate with 2, 4-dinitrophenylhydrazine
B. $CH_3C \equiv CH$	q. gives precipitate with $AgNO_3$
C. CN^-	r. is a nucleophile
D. I^-	s. is involved in cyanohydrin formation

22. Match the specified conditions in solids listed in Column I with their properties in Column II. Indicate your answer by darkening the appropriate bubbles of the 4 × 4 matrix given in the ORS.

Column I	Column II
A. simple cubic and face-centered cubic	p. have these cell parameters $a = b = c$ and $\alpha = \beta = \gamma$
B. cubic and rhombohedral	q. are two crystal systems
C. cubic and tetragonal	r. have only two crystallographic angles of 90°
D. hexagonal and monoclinic	s. belong to same crystal system

1. (d) **2.** (b) **3.** (b) **4.** (a) **5.** (a) **6.** (a)
7. (c) **8.** (a) **9.** (d) **10.** (b) **11.** (c) **12.** (c)
13. (c) **14.** (c) **15.** (d) **16.** (a) **17.** (c) **18.** (c)
19. (b) **20.** A–p, s; B–r; C–p, q; D–p,s **21.** A–p, q, s; B–q; C–r, s; D–r
22. A–p, s; B–p, q; C–q; D–p, r

IIT JEE

UNSOLVED PAPER 2007

Mathematics

Paper-I

Section I : *This section contains 9 multiple choice questions numbered 1 to 9. Each question has 4 choices (a), (b), (c) and (d), out of which* ***ONLY ONE*** *is correct.*

Straight Objective Type

1. A hyperbola, having the transverse axis of length $2\sin\theta$, is confocal with the ellipse $3x^2 + 4y^2 = 12$. Then, its equation is

(a) $x^2 \operatorname{cosec}^2\theta - y^2 \sec^2\theta = 1$ (b) $x^2 \sec^2\theta - y^2 \operatorname{cosec}^2\theta = 1$

(c) $x^2 \sin^2\theta - y^2 \cos^2\theta = 1$ (d) $x^2 \cos^2\theta - y^2 \sin^2\theta = 1$

2. The tangent to the curve $y = e^x$ drawn at the point (c, e^c) intersects the line joining the points $(c-1, e^{c-1})$ and $(c+1, e^{c+1})$

(a) on the left of $x = c$ (b) on the right of $x = c$

(c) at no point (d) at all points

3. A man walks a distance of 3 units from the origin towards the North-East (N 45° E) direction. From there, he walks a distance of 4 units towards the North-West (N 45° W) direction to reach a point P. Then, the position of P in the Argand plane is

(a) $3e^{i\pi/4} + 4i$ (b) $(3-4i)e^{i\pi/4}$ (c) $(4+3i)e^{i\pi/4}$ (d) $(3+4i)e^{i\pi/4}$

4. Let $f(x)$ be differentiable on the interval $(0, \infty)$ such that $f(1) = 1$, and

$$\lim_{t\to x} \frac{t^2 f(x) - x^2 f(t)}{t - x} = 1 \text{ for each } x > 0. \text{ Then, } f(x) \text{ is}$$

(a) $\frac{1}{3x} + \frac{2x^2}{3}$ (b) $-\frac{1}{3x} + \frac{4x^2}{3}$ (c) $-\frac{1}{x} + \frac{2}{x^2}$ (d) $\frac{1}{x}$

5. The number of solutions of the pair of equations $2\sin^2\theta - \cos 2\theta = 0$ and $2\cos^2\theta - 3\sin\theta = 0$ in the interval $[0, 2\pi]$ is

(a) zero (b) one (c) two (d) four

6. Let α, β be the roots of the equation $x^2 - px + r = 0$ and $\frac{\alpha}{2}, 2\beta$ be the roots of the equation $x^2 - qx + r = 0$. Then, the value of r is

(a) $\frac{2}{9}(p-q)(2q-p)$ (b) $\frac{2}{9}(q-p)(2p-q)$

(c) $\frac{2}{9}(q-2p)(2q-p)$ (d) $\frac{2}{9}(2p-q)(2q-p)$

7. The number of distinct real values of λ, for which the vectors $-\lambda^2\hat{\mathbf{i}}+\hat{\mathbf{j}}+\hat{\mathbf{k}}$, $\hat{\mathbf{i}}-\lambda^2\hat{\mathbf{j}}+\hat{\mathbf{k}}$ and $\hat{\mathbf{i}}+\hat{\mathbf{j}}-\lambda^2\hat{\mathbf{k}}$ are coplanar, is

(a) zero (b) one
(c) two (d) three

8. One Indian and four American men and their wives are to be seated randomly around a circular table. Then, the conditional probability that the Indian man is seated adjacent to his wife given that each American man is seated adjacent to his wife, is

(a) $\frac{1}{2}$ (b) $\frac{1}{3}$ (c) $\frac{2}{5}$ (d) $\frac{1}{5}$

9. $\lim_{x \to \frac{\pi}{4}} \dfrac{\int_2^{\sec^2 x} f(t)\,dt}{x^2 - \frac{\pi^2}{16}}$ equals

(a) $\frac{8}{\pi} f(2)$ (b) $\frac{2}{\pi} f(2)$ (c) $\frac{2}{\pi} f\left(\frac{1}{2}\right)$ (d) $4f(2)$

Section II : *This section contains 4 questions numbered 10 to 13. Each questions contains Statement-I (Assertion) and Statement II (Reason). Each question has 4 choices (a), (b), (c) and (d) out of which* ***ONLY ONE*** *is correct.*

Assertion and Reason Type

10. Let the vectors **PQ**, **QR**, **RS**, **ST**, **TU** and **UP** represent the sides of a regular hexagon.

Statement I $\mathbf{PQ} \times (\mathbf{RS} + \mathbf{ST}) \neq \mathbf{0}$

Statement II $\mathbf{PQ} \times \mathbf{RS} = \mathbf{0}$ and $\mathbf{PQ} \times \mathbf{SQ} \neq \mathbf{0}$

(a) Statement I is true, Statement II is true; Statement II is a correct explanation for Statement I
(b) Statement I is true, Statement II is true; Statement II is not a correct explanation for Statement I
(c) Statement I is true, Statement II is false
(d) Statement I is false, Statement II is true

11. Let $F(x)$ be an indefinite integral of $\sin^2 x$.

Statement I The function $F(x)$ satisfies $F(x+\pi) = F(x)$ for all real x.

Statement II $\sin^2(x+\pi) = \sin^2 x$ for all real x.

(a) Statement I is true, Statement II is true; Statement II is a correct explanation for Statement I
(b) Statement I is true, Statement II is true; Statement II is not a correct explanation for Statement I.
(c) Statement I is true, Statement II is false
(d) Statement I is false, Statement II is true

12. Let $H_1, H_2, \ldots, H_n$ be mutually exclusive events with $P(H_i) > 0$, $i = 1, 2, \ldots, n$. Let E be any other event with $0 < P(E) < 1$.

Statement I $P(H_i / E) > P(E / H_i) P(H_i)$ for $i = 1, 2, \ldots, n$.

Statement II $\sum_{i=1}^{n} P(H_i) = 1$

(a) Statement I is true, Statement II is true; Statement II is a correct explanation for Statement I

(b) Statement I is true, Statement II is true; Statement II is not a correct explanation for Statement II

(c) Statement I is true, Statement II is false

(d) Statement I is false, Statement II is true

13. Tangents are drawn from the point (17, 7) to the circle $x^2 + y^2 = 169$.

Statement I The tangents are mutually perpendicular.

Statement II The locus of the points from which mutually perpendicular tangents can be drawn to the given circle is $x^2 + y^2 = 338$

(a) Statement I is true, Statement II is true; Statement II is a correct explanation for Statement I

(b) Statement I is true, Statement II is true; Statement II is not a correct explanation for Statement I

(c) Statement I is true, Statement II is false

(d) Statement I is false, Statement II is true

Section III : *This section contains 2 paragraphs M_{14-16} and M_{17-19}. Based upon each paragraph. 3 multiple choice questions have to be answered. Each question has 4 choice (a), (b), (c) and (d), out of which **ONLY ONE** is correct.*

M_{14-16} : Paragraph for Question Nos. 14 to 16

Consider the circle $x^2 + y^2 = 9$ and the parabola $y^2 = 8x$. They intersect at P and Q in the first and the fourth quadrants, respectively. Tangents to the circle at P and Q intersect the X-axis at R and tangents to the parabola at P and Q intersect the X-axis at S.

14. The ratio of the areas of the ΔPQS and ΔPQR is

(a) $1 : \sqrt{2}$ (b) 1 : 2

(c) 1 : 4 (d) 1 : 8

15. The radius of the circumcircle of the ΔPRS is

(a) 5 (b) $3\sqrt{3}$

(c) $3\sqrt{2}$ (d) $2\sqrt{3}$

16. The radius of the incircle of the ΔPQR is

(a) 4 (b) 3 (c) $\frac{8}{3}$ (d) 2

M_{17-19} : Paragraph for Question Nos. 17 to 19 *Let V_r denotes the sum of the first r terms of an arithmetic progression (AP) whose first term is r and the common difference is $(2r-1)$. Let $T_r = V_{r+1} - V_r - 2$ and $Q_r = T_{r+1} - T_r$ for $r = 1, 2, \ldots$*

17. The sum $V_1 + V_2 + \ldots + V_n$ is

(a) $\frac{1}{12}n(n+1)(3n^2 - n + 1)$ (b) $\frac{1}{12}n(n+1)(3n^2 + n + 2)$

(c) $\frac{1}{2}n(2n^2 - n + 1)$ (d) $\frac{1}{3}(2n^3 - 2n + 3)$

18. T_r is always

(a) an odd number (b) an even number

(c) a prime number (d) a composite number

19. Which one of the following is a correct statement?

(a) $Q_1, Q_2, Q_3, \ldots$ are in AP with common difference 5

(b) $Q_1, Q_2, Q_3, \ldots$ are in AP with common difference 6

(c) $Q_1, Q_2, Q_3, \ldots$ are in AP with common difference 11

(d) $Q_1 = Q_2 = Q_3 = \ldots$

Section IV : *This section contains 3 questions. Each question contains statements given in two columns which have to be matched. Statements (A, B, C, D) in **Column I** have to be method with statements (p, q, r, s) in **Column II**.*

20. Consider the following linear equations

$$ax + by + cz = 0,\ bx + cy + az = 0,\ cx + ay + bz = 0$$

Match the conditions/expressions in **Column I** with statements in **Column II**.

Column I	Column II
A. $a+b+c \neq 0$ and $a^2 + b^2 + c^2 = ab + bc + ca$	p. the equations represent planes meeting only at a single point
B. $a+b+c = 0$ and $a^2 + b^2 + c^2 \neq ab + bc + ca$	q. the equations represent the line $x = y = z$
C. $a+b+c \neq 0$ and $a^2 + b^2 + c^2 \neq ab + bc + ca$	r. the equations represent identical planes
D. $a+b+c = 0$ and $a^2 + b^2 + c^2 = ab + bc + ca$	s. the equations represent the whole of the three dimensional space

21. Match the integrals in **Column I** with the values in **Column II**.

Column I	Column II
A. $\int_{-1}^{1} \frac{dx}{1+x^2}$	p. $\frac{1}{2}\log\left(\frac{2}{3}\right)$
B. $\int_{0}^{1} \frac{dx}{\sqrt{1-x^2}}$	q. $2\log\left(\frac{2}{3}\right)$

Column I	Column II
C. $\int_2^3 \frac{dx}{1-x^2}$	r. $\frac{\pi}{3}$
D. $\int_1^2 \frac{dx}{\sqrt{x^2-1}}$	s. $\frac{\pi}{2}$

22. In the following $[x]$ denotes the greatest integer less than or equal to x. Match the functions in **Column I** with the properties **Column II**.

Column I	Column II
A. $x\,|x|$	p. continuous in $(-1, 1)$
B. $\sqrt{|x|}$	q. differentiable in $(-1, 1)$
C. $x + [x]$	r. strictly increasing $(-1, 1)$
D. $|x-1| + |x+1|$	s. not differentiable at least at one point in $(-1, 1)$

ANSWERS

1. (a) **2.** (a) **3.** (d) **4.** (a) **5.** (c) **6.** (d)
7. (c) **8.** (c) **9.** (a) **10.** (c) **11.** (d) **12.** (d)
13. (a) **14.** (c) **15.** (b) **16.** (d) **17.** (b) **18.** (d)
19. (b) **20.** A–r; B–q; C–p; D–s **21.** A–s; B–s; C–p; D–r
22. A–q, p, r; B–p, s; C–r, s; D–p, q

Paper-II

Section I : *This section contains 9 multiple choice questions numbered 1 to 9. Each question has 4 choices (a), (b), (c) and (d), out of which* ***ONLY ONE*** *is correct.*

Straight Objective Type

1. Let $\mathbf{a}$, $\mathbf{b}$, $\mathbf{c}$ be unit vectors such that $\mathbf{a} + \mathbf{b} + \mathbf{c} = \mathbf{0}$. Which one of the following is correct?

(a) $\mathbf{a} \times \mathbf{b} = \mathbf{b} \times \mathbf{c} = \mathbf{c} \times \mathbf{a} = \mathbf{0}$

(b) $\mathbf{a} \times \mathbf{b} = \mathbf{b} \times \mathbf{c} = \mathbf{c} \times \mathbf{a} \neq \mathbf{0}$

(c) $\mathbf{a} \times \mathbf{b} = \mathbf{b} \times \mathbf{c} = \mathbf{a} \times \mathbf{c} = \mathbf{0}$

(d) $\mathbf{a} \times \mathbf{b}$, $\mathbf{b} \times \mathbf{c}$, $\mathbf{c} \times \mathbf{a}$ are mutually perpendicular

2. Let $f(x) = \dfrac{x}{(1 + x^n)^{1/n}}$ for $n \geq 2$ and $g(x) = \underbrace{(fofo\ldots of)}_{f \text{ occurs } n \text{ times}}(x)$. Then, $\int x^{n-2} g(x)\, dx$ equals

(a) $\dfrac{1}{n(n-1)}(1 + nx^n)^{1 - \frac{1}{n}} + k$ (b) $\dfrac{1}{n-1}(1 + nx^n)^{1 - \frac{1}{n}} + k$

(c) $\dfrac{1}{n(n+1)}(1 + nx^n)^{1 + \frac{1}{n}} + k$ (d) $\dfrac{1}{n+1}(1 + nx^n)^{1 + \frac{1}{n}} + k$

3. $\dfrac{d^2x}{dy^2}$ equals

(a) $\left(\dfrac{d^2y}{dx^2}\right)^{-1}$ (b) $-\left(\dfrac{d^2y}{dx^2}\right)^{-1}\left(\dfrac{dy}{dx}\right)^{-3}$

(c) $\left(\dfrac{d^2y}{dx^2}\right)\left(\dfrac{dy}{dx}\right)^{-2}$ (d) $-\left(\dfrac{d^2y}{dx^2}\right)\left(\dfrac{dy}{dx}\right)^{-3}$

4. The letters of the word COCHIN are permuted and all the permutations are arranged in an alphabetical order as in an English dictionary. The number of words that appear before the word COCHIN is

(a) 360 (b) 192 (c) 97 (d) 48

5. If $|z| = 1$ and $z \neq \pm 1$, then all the values of $\dfrac{z}{1 - z^2}$ lie on

(a) a line not passing through the origin (b) $|z| = \sqrt{2}$

(c) the X-axis (d) the Y-axis

6. Let $ABCD$ be a quadrilateral with area 18, with side AB parallel to the side CD and $AB = 2CD$. Let AD be perpendicular to AB and CD. If a circle is drawn inside the quadrilateral $ABCD$ touching all the sides, then its radius is

(a) 3 (b) 2 (c) $\dfrac{3}{2}$ (d) 1

7. Let $O(0, 0)$, $P(3, 4)$ and $Q(6, 0)$ be the vertices of the ΔOPQ. The point R inside the ΔOPQ is such that the ΔOPR, ΔPQR, ΔOQR are of equal area. The coordinates of R are

(a) $\left(\frac{4}{3}, 3\right)$ (b) $\left(3, \frac{2}{3}\right)$ (c) $\left(3, \frac{4}{3}\right)$ (d) $\left(\frac{4}{3}, \frac{2}{3}\right)$

8. The differential equation $\frac{dy}{dx} = \frac{\sqrt{1-y^2}}{y}$ determines a family of circles with

(a) variable radii and a fixed centre at (0, 1)
(b) variable radii and a fixed centre at (0, – 1)
(c) fixed radius 1 and variable centres along the X-axis
(d) fixed radius 1 and variable centres along the Y-axis

9. Let E^c denotes the complement of an event E. Let E, F, G be pairwise independent events with $P(G) > 0$ and $P(E \cap F \cap G) = 0$. Then, $P(E^c \cap F^c | G)$ equals

(a) $P(E^c) + P(F^c)$ (b) $P(E^c) - P(F^c)$
(c) $P(E^c) - P(F)$ (d) $P(E) - P(F^c)$

Section II : *This section contains 4 questions numbered 10 to 13. Each questions contains Statement I (Assertion) and Statement II (Reason). Each question has 4 choices (a), (b), (c) and (d) out of which* ***ONLY ONE*** *is correct.*

10. Let $f(x) = 2 + \cos x$ for all real x.

Statement I For each real t, there exists a point c in $[t, t + \pi]$ such that $f'(c) = 0$.
Statement II $f(t) = f(t + 2\pi)$ for each real t.

(a) Statement I is true, Statement II is true; Statement II is a correct explanation for Statement I
(b) Statement I is True, Statement II is true; Statement II is not a correct explanation for Statement I
(c) Statement I is true, Statement II is false
(d) Statement I is false, Statement II is true

11. Consider the planes $3x - 6y - 2z = 15$ and $2x + y - 2z = 5$.

Statement I The parametric equations of the line of intersection of the given planes are $x = 3 + 14t$, $y = 1 + 2t$ and $z = 15t$.

Statement II The vectors $14\hat{\mathbf{i}} + 2\hat{\mathbf{j}} + 15\hat{\mathbf{k}}$ is parallel to the line of intersection of the given planes.

(a) Statement I is true, Statement II is true; Statement II is a correct explanation for Statement I
(b) Statement I is true, Statement II is true; Statement II is not a correct explanation for Statement I
(c) Statement I is true, Statement II is false
(d) Statement I is false, Statement II is true

12. Lines $L_1 : y - x = 0$ and $L_2 : 2x + y = 0$ intersect the line $L_3 : y + 2 = 0$ at P and Q, respectively. The bisector of the acute angle between L_1 and L_2 intersects L_3 at R.

Statement I The ratio $PR : RQ$ equals $2\sqrt{2} : \sqrt{5}$.

Statement II In any triangle, bisector of an angle divides the triangle into two similar triangles.

(a) Statement I is true, Statement II is true; Statement II is a correct explanation for Statement I

(b) Statement I is true, Statement II is true; Statement II is not a correct explanation for Statement I

(c) Statement I is true, Statement II is false

(d) Statement I is false, Statement II is true

13. Statement I The curve $y = -\dfrac{x^2}{2} + x + 1$ is symmetric with respect to the line $x = 1$.

Statement II A parabola is symmetric about its axis.

(a) Statement I is true, Statement II is true; Statement II is a correct explanation for Statement I

(b) Statement I is true, Statement II is true; Statement II is not a correct explanation for Statement I

(c) Statement I is true, Statement II is false

(d) Statement I is false, Statement II is true

Section III : *This section contains 2 paragraphs M_{14-16} and M_{17-19}. Based upon each paragraph 3 multiple choice questions have to be answered. Each question has 4 choice (a), (b), (c) and (d), out of which* ***ONLY ONE*** *is correct.*

M_{14-16} : Paragraph for questions Nos. 14 to 16

If a continuous f defined on the real line R, assume positive and negative values in R, then the equation $f(x) = 0$ has a root in R. *For example*, if it is known that a continuous function f on R is positive at some point and its minimum values is negative, then the equation $f(x) = 0$ has a root in R.

Consider $f(x) = ke^x - x$ for all real x, where k is real constant.

14. The line $y = x$ meets $y = ke^x$ for $k \leq 0$ at

(a) no point

(b) one point

(c) two points

(d) more than two points

15. The positive value of k for which $ke^x - x = 0$ has only one root is

(a) $\frac{1}{e}$ (b) 1 (c) e (d) $\log_e 2$

16. For $k > 0$, the set of all values of k for which $ke^x - x = 0$ has two distinct roots, is

(a) $\left(0, \frac{1}{e}\right)$ (b) $\left(\frac{1}{e}, 1\right)$ (c) $\left(\frac{1}{e}, \infty\right)$ (d) (0, 1)

M_{17-19} : Paragraph for Question Nos. 17 to 19

Let A_1, G_1, H_1 denote the arithmetic, geometric and harmonic means respectively, of two distinct positive numbers. For $n \geq 2$, let A_{n-1} and H_{n-1} has arithmetic, geometric and harmonic means as A_n, G_n, H_n, respectively.

17. Which one of the following statements is correct?

(a) $G_1 > G_2 > G_3 > \dots$
(b) $G_1 < G_2 < G_3 < \dots$
(c) $G_1 = G_2 = G_3 = \dots$
(d) $G_1 < G_3 < G_5 < \dots$ and $G_2 > G_4 > G_6 > \dots$

18. Which of the following statements is correct?

(a) $A_1 > A_2 > \dots$
(b) $A_1 < A_2 < A_3 < \dots$
(c) $A_1 > A_3 > A_5 > \dots$ and $A_2 < A_4 < A_6 < \dots$
(d) $A_1 < A_3 < A_5 < \dots$ and $A_2 > A_4 > A_6 > \dots$

19. Which of the following statements is correct?

(a) $H_1 > H_2 > H_3 > \dots$
(b) $H_1 < H_2 < H_3 < \dots$
(c) $H_1 > H_3 > H_5 > \dots$ and $H_2 < H_4 < H_6 < \dots$
(d) $H_1 < H_3 < H_5 < \dots$ and $H_2 > H_4 > H_6 > \dots$

Section IV : *This section contains 3 questions. Each question contains statements given in two columns which have to be matched. Statements (A, B, C, D) in **Column I** have to be method with statements (p, q, r, s) in **Column II**.*

20. Let $f(x) = \dfrac{x^2 - 6x + 5}{x^2 - 5x + 6}$.

Match the conditions/expressions in **Column I** with statements in **Column II**.

Column I	Column II
A. If $-1 < x < 1$, then $f(x)$ satisfies	p. $0 < f(x) < 1$
B. If $1 < x < 2$, then $f(x)$ satisfies	q. $f(x) < 0$
C. If $3 < x < 5$, then $f(x)$ satisfies	r. $f(x) > 0$
D. If $x > 5$, then $f(x)$ satisfies	s. $f(x) < 1$

21. Let (x, y) be such that $\sin^{-1}(ax) + \cos^{-1}(y) + \cos^{-1}(bxy) = \frac{\pi}{2}$.

Match the statements in **Column I** with the values in **Column II**.

Column I	Column II
A. If $a = 1$ and $b = 0$, then (x, y)	p. lies on the circle $x^2 + y^2 = 1$
B. If $a = 1$ and $b = 1$, then (x, y)	q. lies on $(x^2 - 1)(y^2 - 1) = 0$
C. If $a = 1$ and $b = 2$, then (x, y)	r. lies on $y = x$
D. If $a = 2$ and $b = 2$, then (x, y)	s. lies on $(4x^2 - 1)(y^2 - 1) = 0$

22. Match the statements in **Column I** with the properties **Column II**.

Column I	Column II
A. Two intersecting circles	p. have a common tangent
B. Two mutually external circles	q. have a common normal
C. Two circles, one strictly inside the other	r. do not have a common tangent
D. Two branches of a hyperbola	s. do not have a common normal

ANSWERS

1. (b)	**2.** (a)	**3.** (d)	**4.** (c)	**5.** (d)	**6.** (b)
7. (c)	**8.** (c)	**9.** (c)	**10.** (b)	**11.** (d)	**12.** (b)
13. (a)	**14.** (b)	**15.** (a)	**16.** (a)	**17.** (c)	**18.** (a)
19. (b)	**20.** A–p; B–q; C–q; D–p		**21.** A–p; B–q; C–p; D–s		

22. A–q, p; B–p, q; C–q, r; D–q, r

IIT JEE
UNSOLVED PAPER 2008

Physics

Paper 1

Section I

Straight Objective Type

This section contains 6 multiple choice questions. Each question has 4 choices (a), (b), (c) and (d), out of which **ONLY ONE** is correct.

1. An ideal gas is expanding such that $pT^2 =$ constant. The coefficient of volume expansion of the gas is

(a) $\frac{1}{T}$ (b) $\frac{2}{T}$

(c) $\frac{3}{T}$ (d) $\frac{4}{T}$

2. Two beams of red and violet colours are made to pass separately through a prism (angle of the prism is 60°). In the position of minimum deviation, the angle of refraction will be

(a) 30° for both the colours
(b) greater for the violet colour
(c) greater for the red colour
(d) equal but not 30° for both the colours

3. A spherically symmetric gravitational system of particles has a mass density

$$\rho = \begin{cases} \rho_0 \text{ for } r \le R \\ 0 \text{ for } r > R \end{cases}$$

where ρ_0 is a constant. A test mass can undergo circular motion under the influence of the gravitational field of particles. Its speed v as a function of distance r $(0 < r < \infty)$ from the centre of the system is represented by

(a) 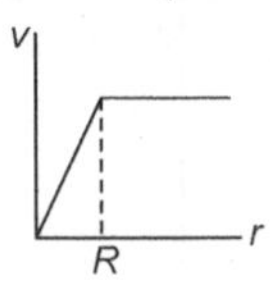

(b) 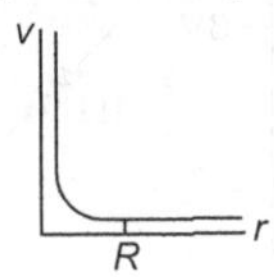

(c) 

(d) 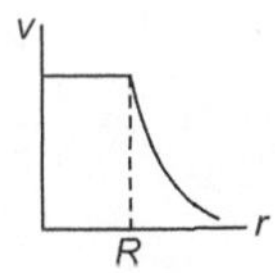

4. Which one of the following statement is **WRONG** in the context of X-rays generated from X-ray tube?

(a) Wavelength of characteristic X-rays decreases when the atomic number of the target increases

(b) Cut-off wavelength of the continuous X-rays depends on the atomic number of the target

(c) Intensity of the characteristics X-rays depends on the electrical power given to the X-ray tube

(d) Cut-off wavelength of the continuous X-rays depends on the energy of the electrons in the X-ray tube

5. Students I, II and III perform an experiment for measuring the acceleration due to gravity (g) using a simple pendulum. They use different lengths of the pendulum and/or record time for different number of oscillations. The observations are shown in the table.

Least count for length = 0.1 cm

Least count for time = 0.1 s

Student	**Length of the pendulum** (cm)	**Number of oscillations** (n)	**Total time for** (n) **oscillations** (in s)	**Time period** (s)
I	64.0	8	128.0	16.0
II	64.0	4	64.0	16.0
III	20.0	4	36.0	9.0

If E_I, E_{II} and E_{III} are the percentage errors in g, i.e., $\left(\frac{\Delta g}{g} \times 100\right)$, for students I, II and III, respectively.

(a) $E_I = 0$ (b) E_I is minimum (c) $E_I = E_{II}$ (d) E_{II} is maximum

6. Figure shows three resistor configurations R_1, R_2 and R_3 connected to 3V battery. If the power dissipated by the configuration R_1, R_2 and R_3 is P_1, P_2 and P_3, respectively, then

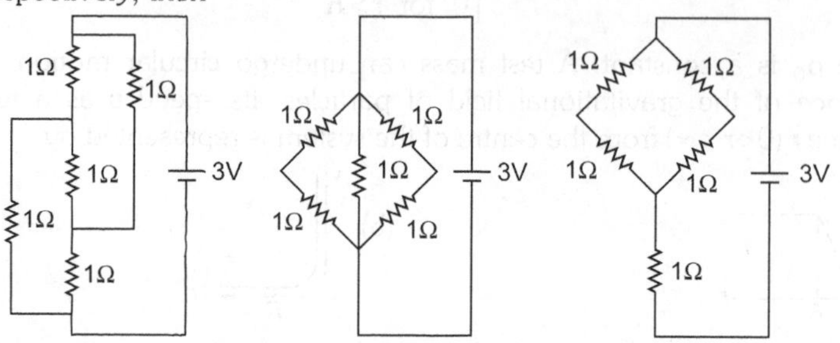

(a) $P_1 > P_2 > P_3$ (b) $P_1 > P_3 > P_2$

(c) $P_2 > P_1 > P_3$ (d) $P_3 > P_2 > P_1$

Section II

Multiple Correct Answers Type

This section contains 4 multiple correct answer(s) type questions. Each question has 4 choices (a), (b), (c) and (d) out of which **ONE OR MORE** is/are correct.

7. A particle of mass m and charge q, moving with velocity v enters Region II normal to the boundary as shown in the figure. Region II has a uniform magnetic field B perpendicular to the plane of the paper. The length of the Region II is l. Choose the correct choice(s).

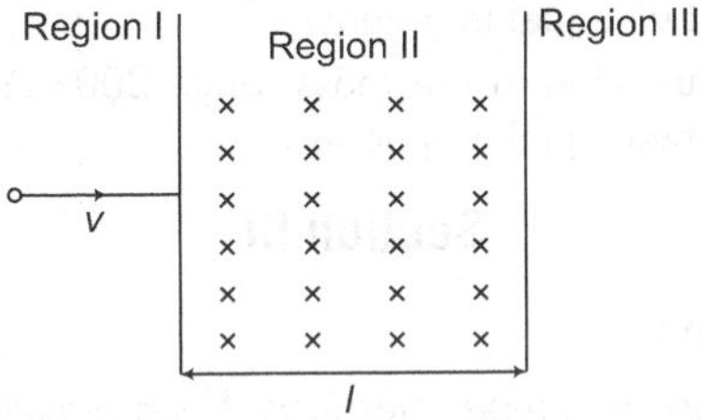

(a) The particle enters Region III only if its velocity $v > \frac{qlB}{m}$

(b) The particle enters Region III only if its velocity $v < \frac{qlB}{m}$

(c) Path length of the particle in Region II is maximum when velocity $v = \frac{qlB}{m}$

(d) Time spent in Region II is same for any velocity v as long as the particle returns to Region I

8. In a Young's double slit experiment, the separation between the two slits is d and the wavelength of the light is λ. The intensity of light falling on slit 1 is four times the intensity of light falling on slit 2. Choose the correct choice(s),

(a) If $d = \lambda$, the screen will contain only one maximum

(b) If $\lambda < d < 2\lambda$, at least one more maximum (besides the central maximum) will be observed on the screen

(c) If the intensity of light falling on slit 1 is reduced so that it becomes equal to that of slit 2, the intensities of the observed dark and bright fringes will increase

(d) If the intensity of light falling on slit 2 is increased so that it becomes equal to that of slit 1, the intensities of the observed dark and bright fringes will increase

9. Two balls, having linear momenta $\vec{\mathbf{p}}_1 = p\hat{\mathbf{i}}$ and $\vec{\mathbf{p}}_2 = -p\hat{\mathbf{i}}$, undergo a collision in free space. There is no external force acting on the balls. Let $\vec{\mathbf{p}}'_1$ and $\vec{\mathbf{p}}'_2$ be their final momenta. The following option(s) is/are **NOT ALLOWED** for any non-zero value of p, a_1, a_2, b_1, b_2, c_1 and c_2

(a) $\vec{\mathbf{p}}'_1 = a_1\hat{\mathbf{i}} + b_1\hat{\mathbf{j}} + c_1\hat{\mathbf{k}}$, $\vec{\mathbf{p}}'_2 = a_2\hat{\mathbf{i}} + b_2\hat{\mathbf{j}}$

(b) $\vec{\mathbf{p}}'_1 = c_1\hat{\mathbf{k}}$, $\vec{\mathbf{p}}'_2 = c_2\hat{\mathbf{k}}$

(c) $\vec{\mathbf{p}}'_1 = a_1\hat{\mathbf{i}} + b_1\hat{\mathbf{j}} + c_1\hat{\mathbf{k}}$

$\vec{\mathbf{p}}'_2 = a_2\hat{\mathbf{i}} + b_2\hat{\mathbf{j}} - c_1\hat{\mathbf{k}}$

(d) $\vec{\mathbf{p}}'_1 = a_1\hat{\mathbf{i}} + b_1\hat{\mathbf{j}}$

$\vec{\mathbf{p}}'_2 = a_2\hat{\mathbf{i}} + b_1\hat{\mathbf{j}}$

10. Assume that the nuclear binding energy per nucleon (B/A) *versus* mass number (A) is as shown in the figure. Use this plot to choose the correct choice (s) given below.

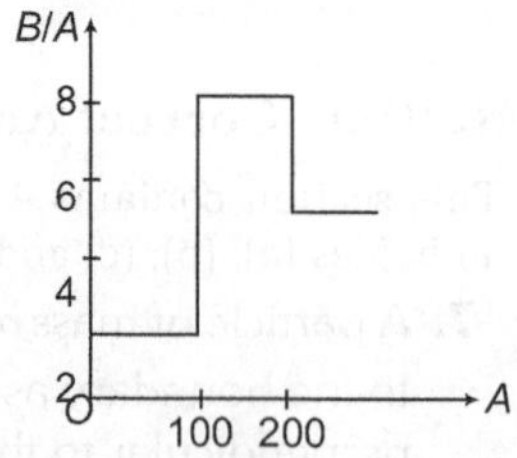

(a) Fusion of two nuclei with mass numbers lying in the range of $1 < A < 50$ will release energy

(b) Fusion of two nuclei with mass numbers lying in the range of $51 < A < 100$ will release energy

(c) Fission of a nucleus lying in the mass range of $100 < A < 200$ will release energy when broken into two equal fragments

(d) Fission of a nucleus lying in the mass range $200 < A < 260$ will release energy when broken into two equal fragments

Section III

Assertion-Reason Type

*This section contains 4 reasoning type questions. Each question has 4 choices (a), (b), (c) and (d), out of which **ONLY ONE** is correct.*

(a) Statement 1 is True, Statement 2 is True; Statement 2 is a correct explanation for Statement 1

(b) Statement 1 is True, Statement 2 is True; Statement 2 is not a correct explanation for Statement 1

(c) Statement 1 is True, Statement 2 is False

(d) Statement 1 is False, Statement 2 is True

11. **Statement 1** An astronaut in an orbiting space station above the earth experiences weightlessness.

and **Statement 2** An object moving around the earth under the influence of earth's gravitational force is in state of 'free fall'.

12. **Statement 1** Two cylinders, one hollow (metal) and the other solid (wood) with the same mass and identical dimensions are simultaneously allowed to roll without slipping down an inclined plane from the same height. The hollow cylinder will reach the bottom of the inclined plane first.

and **Statement 2** By the principle of conservation of energy, the total kinetic energies of both the cylinders are identical when they reach the bottom of the incline.

13. **Statement 1** The stream of water flowing at high speed from a garden hose pipe tends to spread like a fountain when held vertically up, but tends to narrow down when held vertically down.

and **Statement 2** In any steady flow of an incompressible fluid, the volume flow rate of the fluid remains constant.

14. Statement 1 In a Meter Bridge experiment, null point for an unknown resistance is measured. Now, the unknown resistance is put inside an enclosure maintained at a higher temperature. The null point can be obtained at the same point as before by

and **Statement 2** Resistance of a metal increase with increase in temperature.

Section IV

Linked Comprehension Type

This section contains 3 paragraph. Based upon each paragraph, 3 multiple choice questions have to be answered. Each question has 4 choices (a), (b), (c) and (d), out of which ***ONLY ONE*** is correct.

Paragraph for Question Nos. 15 to 17

In a mixture of H—H^+ *gas* (He^+ *is singly ionised* He *atom), H atoms and* He^+ *ions are excited to their respective first excited states. Subsequently, H atoms transfer their total excitation energy to* He^+ *ions (by collisions). Assume that the Bohr model of atom is exactly valid.*

15. The quantum number n of the state finally populated in He^+ ions is

(a) 2 (b) 3 (c) 4 (d) 5

16. The wavelength of light emitted in the visible region by He^+ ions after collision with H atoms is

(a) 6.5×10^{-7} m (b) 5.6×10^{-7} m

(c) 4.8×10^{-7} m (d) 4.0×10^{-7} m

17. The ratio of the kinetic energy of the $n = 2$ electron for the H atom to that of He^+ ion is

(a) $\frac{1}{4}$ (b) $\frac{1}{2}$ (c) 1 (d) 2

Paragraph for Question Nos. 18 to 20

A small block of mass M moves on a frictionless surface of an inclined plane, as shown in figure. The angle of the incline suddenly changes from 60° to 30° at point B. The block is initially at rest at A. Assume that collisions between the block and the incline are totally inelastic ($g = 10 \text{ m/s}^2$)

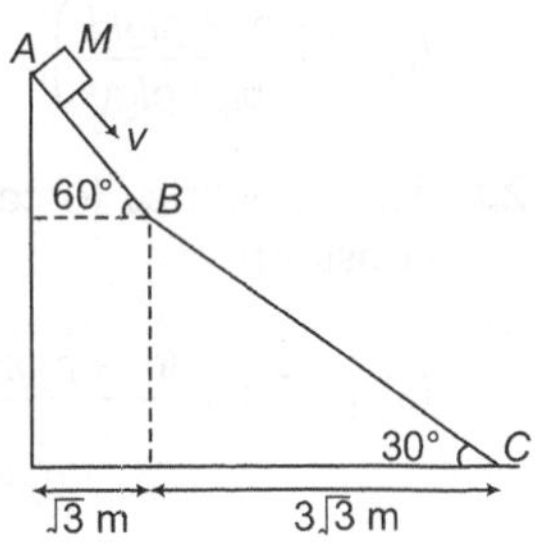

18. The speed of the block at point B immediately after it strikes the second incline is

(a) $\sqrt{60}$ m/s (b) $\sqrt{45}$ m/s

(c) $\sqrt{30}$ m/s (d) $\sqrt{15}$ m/s

19. The speed of the block at point C, immediately before it leaves the second incline is

(a) $\sqrt{120}$ m/s (b) $\sqrt{105}$ m/s (c) $\sqrt{90}$ m/s (d) $\sqrt{75}$ m/s

20. If collision between the block and the incline is completely elastic, then the vertical (upward) component of the velocity of the block at point B, immediately after it strikes the second incline is

(a) $\sqrt{30}$ m/s (b) $\sqrt{15}$ m/s (c) zero (d) $-\sqrt{15}$ m/s

Paragraph for Question Nos. 21 to 23

A small spherical monoatomic ideal gas bubble $\left(\gamma = \frac{5}{3}\right)$ *is trapped inside a liquid of density* ρ_l *(see figure). Assume that the bubble does not exchange any heat with the liquid. The bubble contains n moles of gas. The temperature of the gas when the bubble is at the bottom is* T_0*, the height of the liquid is H and the atmospheric pressure is* p_0 *(Neglect surface tension).*

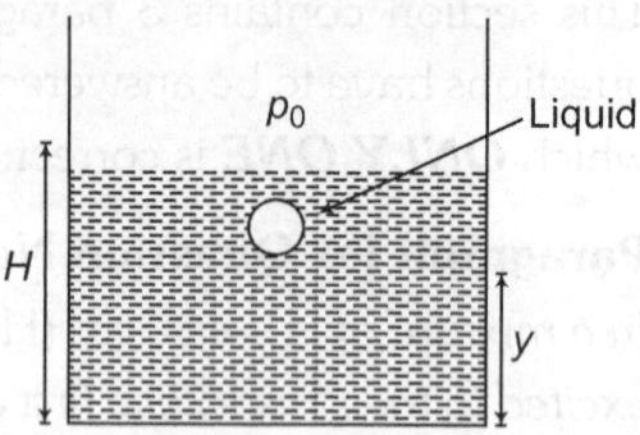

21. As the bubble moves upwards, besides the buoyancy force the following forces are acting on it.

(a) Only the force of gravity

(b) The force due to gravity and the force due to pressure of the liquid

(c) The force due to gravity, the force due to the pressure of the liquid and the force due to viscosity of the liquid

(d) The force due to gravity and the force due to viscosity of the liquid

22. When the gas bubble is at height y from the bottom, its temperature is

(a) $T_0\left(\dfrac{p_0+\rho l g H}{p_0+\rho l g y}\right)^{\frac{2}{5}}$ (b) $T_0\left(\dfrac{p_0+\rho l g(H-y)}{p_0+\rho l g H}\right)^{\frac{2}{5}}$

(c) $T_0\left(\dfrac{p_0+\rho l g H}{p_0+\rho l g y}\right)^{\frac{3}{5}}$ (d) $T_0\left(\dfrac{p_0+\rho l g (H-y)}{p_0+\rho l g H}\right)^{\frac{3}{5}}$

23. The buoyancy force acting on the gas bubble is (Assume R is the universal gas constant)

(a) $\rho_l n R g T_0 \dfrac{(p_0+\rho_l g H)^{\frac{2}{5}}}{(p_0+\rho_l g y)^{\frac{7}{5}}}$ (b) $\dfrac{\rho_l n R g T_0}{(p_0+\rho_l g H)^{\frac{2}{5}}\,[p_0+\rho_l g(H-y)]^{\frac{3}{5}}}$

(c) $\rho_l n R g T_0 \dfrac{(p_0+\rho_l g H)^{\frac{3}{5}}}{(p_0+\rho_l g y)^{\frac{8}{5}}}$ (d) $\dfrac{\rho_l n R g T_0}{(p_0+\rho_l g H)^{\frac{3}{5}}\,[p_0+\rho_l g (H-y)]^{\frac{2}{5}}}$

Paper 2

SECTION I

Straight Objective Type

This section contains 9 multiple choice questions. Each question has 4 choices (a), (b), (c), and (d), out of which ***ONLY ONE*** is correct.

1. A bob of mass M is suspended by a massless string of length L. The horizontal velocity v at position A is just sufficient to make it reach the point B. The angle θ at which the speed of the bob is half of that at A, satisfies

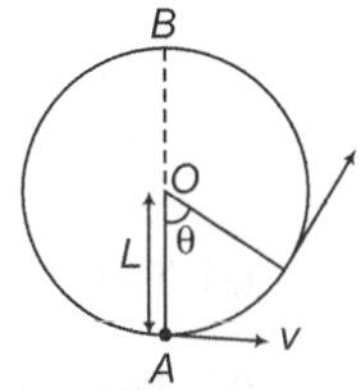

(a) $\theta = \frac{\pi}{4}$ (b) $\frac{\pi}{4} < \theta < \frac{\pi}{2}$

(c) $\frac{\pi}{2} < \theta < \frac{3\pi}{4}$ (d) $\frac{3\pi}{4} < \theta < \pi$

2. A block (B) is attached to two unstretched S_1 and S_2 with spring constants k and $4k$, respectively (see figure I). The other ends are attached to identical supports M_1 and M_2 not attached to the walls. The springs and supports have negligible mass. There is no friction anywhere. The block B is displaced towards wall 1 by a small distance x (figure II) and released. The block returns and moves a maximum distance y towards wall 2. Displacements x and y are measured with respect to the equilibrium position of the block B. The ratio $\frac{y}{x}$ is

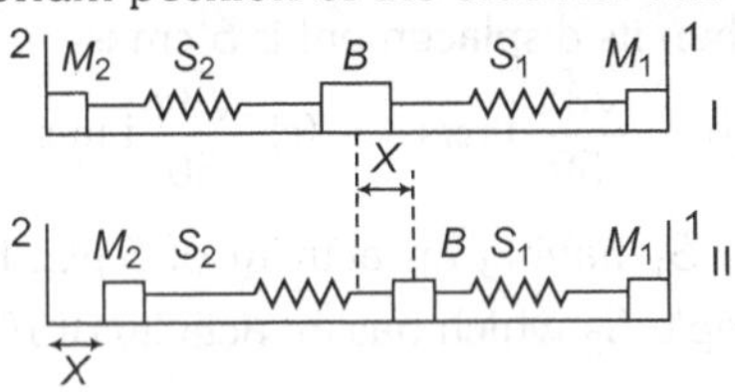

(a) 4 (b) 2 (c) $\frac{1}{2}$ (d) $\frac{1}{4}$

3. A glass tube of uniform internal radius (r) has a valve separating the two identical ends. Initially, the valve is in a tightly closed position. End 1 has a hemispherical soap bubble of radius r. End 2 has sub-hemispherical soap bubble as shown in figure. Just after opening the valve,

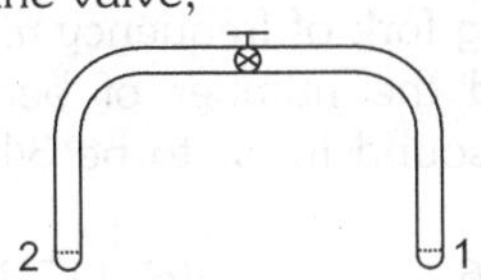

(a) air from end 1 flows towards end 2. No change in the volume of the soap bubbles

(b) air from end 1 flows towards end 2. Volume of the soap bubble at end 1 decreases

(c) no change occurs

(d) air from end 2 flows towards end 1. Volume of the soap bubble at end 1 increases

4. Consider a system of three charges $\frac{q}{3}, \frac{q}{3}$ and $-\frac{2q}{3}$ placed at points A, B and C, respectively, as shown in the figure. Take O to be the centre of the circle of radius R and angle $CAB = 60°$

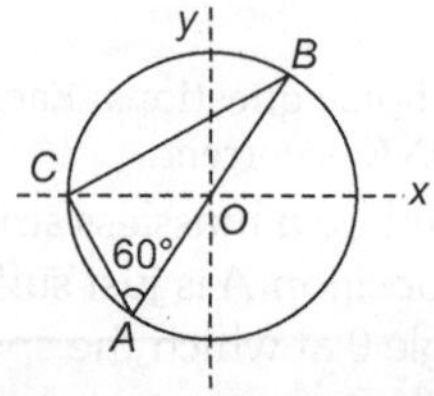

(a) The electric field at point O is $\frac{q}{8\pi\varepsilon_0 R^2}$ directed along the negative x-axis

(b) The potential energy of the system is zero

(c) The magnitude of the force between the charges at C and B is $\frac{q^2}{54\pi\varepsilon_0 R^2}$

(d) The potential at point O is $\frac{q}{12\pi\varepsilon_0 R}$

5. A transverse sinusoidal wave moves along a string in the positive x-direction at a speed of 10 cm/s. The wavelength of the wave is 0.5 m and its amplitude is 10 cm. At a particular time t, the snap-shot of the wave is shown in figure. The velocity of point P when its displacement is 5 cm is

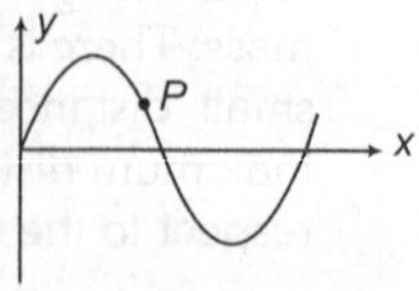

(a) $\frac{\sqrt{3}\pi}{50}\hat{j}$ m/s (b) $-\frac{\sqrt{3}\pi}{50}\hat{j}$ m/s (c) $\frac{\sqrt{3}\pi}{50}\hat{i}$ m/s (d) $-\frac{\sqrt{3}\pi}{50}\hat{i}$ m/s

6. A radioactive sample S_1 having an activity of 5 μCi has twice the number of nuclei as another sample S_2 which has an activity of 10 μCi. The half lives of S_1 and S_2 can be

(a) 20 yr and 5 yr, respectively (b) 20 yr and 10 yr, respectively
(c) 10 yr each (d) 5 yr each

7. A vibrating string of certain length l under a tension T resonates with a mode corresponding to the first overtone (third harmonic) of an air column of length 75 cm inside a tube closed at one end. The string also generates 4 beat/s when excited along with a tuning fork of frequency n. Now when the tension of the string is slightly increased the number of beats reduces to 2 per second. Assuming the velocity of sound in air to be 340 m/s, the frequency n of the tuning fork (in Hz) is

(a) 344 (b) 336 (c) 117.3 (d) 109.3

8. A light beam is travelling from Region I to Region IV (refer figure). The refractive index in Region I, II, III and IV are $n_0, \frac{n_0}{2}, \frac{n_0}{6}$ and $\frac{n_0}{8}$, respectively. The angle of incidence θ for which the beam just misses entering Region IV is

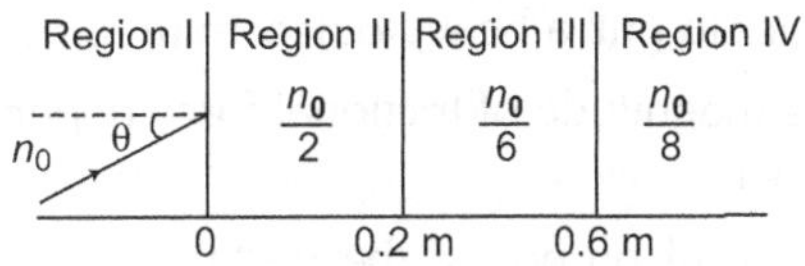

(a) $\sin^{-1}\left(\frac{3}{4}\right)$ (b) $\sin^{-1}\left(\frac{1}{8}\right)$ (c) $\sin^{-1}\left(\frac{1}{4}\right)$ (d) $\sin^{-1}\left(\frac{1}{3}\right)$

9. A parallel plate capacitor C with plates of unit area and separation d is filled with a liquid of dielectric constant $K = 2$. The level of liquid is $\frac{d}{3}$ initially. Suppose the liquid level decreases at a constant speed v, the time constant as a function of time t is

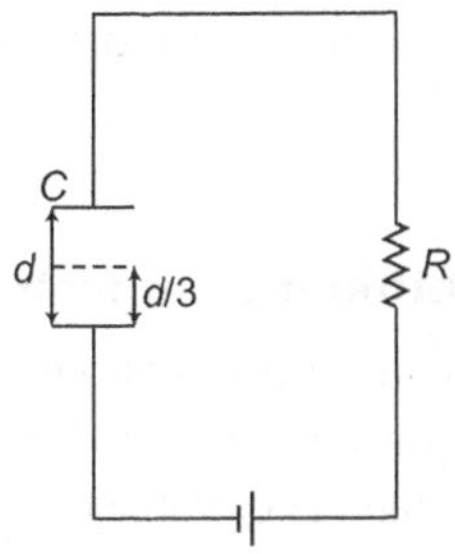

(a) $\frac{6\varepsilon_0 R}{5d + 3vt}$ (b) $\frac{(15d + 9vt)\varepsilon_0 R}{2d^2 - 3dvt - 9v^2t^2}$

(c) $\frac{6\varepsilon_0 R}{5d - 3vt}$ (d) $\frac{(15d - 9vt)\varepsilon_0 R}{2d^2 + 3dvt - 9v^2t^2}$

Section II

Assertion-Reason Type

*This section contains 4 reasoning type questions. Each question has 4 choices (a), (b), (c) and (d), out of which **ONLY ONE** is correct.*

(a) Statement 1 is True, Statement 2 is True; Statement 2 is a correct explanation for Statement 1.

(b) Statement 1 is True, Statement 2 is True; Statement 2 is NOT a correct explanation for Statement 1.

(c) Statement 1 is True, Statement 2 is False.

(d) Statement 1 is False, Statement 2 is True.

10. **Statement 1** The sensitivity of a moving coil galvanometer is increased by placing a suitable magnetic material as a core inside the coil.

and **Statement 2** Soft iron has a high magnetic permeability and cannot be easily magnetized or demagnetized.

11. **Statement 1** For an observer looking out through the window of a fast moving train, the nearby objects appear to move in the opposite direction to the train, while the distant objects appear to be stationary.

and **Statement 2** If the observer and the object are moving at velocities $\vec{\mathbf{v}}_1$ and $\vec{\mathbf{v}}_2$ respectively with reference to a laboratory frame, the velocity of the object with respect to the observer is $\vec{\mathbf{v}}_2 - \vec{\mathbf{v}}_1$.

12. **Statement 1** It is easier to pull a heavy object than to push it on a level ground.

and **Statement 2** The magnitude of frictional force depends on the nature of the two surfaces in contact.

13. **Statement 1** For practical purposes, the earth is used as a reference at zero potential in electrical circuits.

Statement 2 The electrical potential of a sphere of radius R with charge Q uniformly distributed on the surface is given by $\frac{Q}{4\pi\varepsilon_0 R}$.

Section III

Linked Comprehension Type

This section contains 2 paragraphs. Based upon each paragraph, 3 multiple choice questions have to be answered. Each question has 4 choices (a), (b), (c) and (d), out of which **ONLY ONE** is correct.

Paragraph for Question Nos. 14 to 16

A uniform thin cylindrical disk of mass M and radius R is attached to two identical massless springs of spring constant k which are fixed to the wall as shown in the figure. The springs are attached to the axle of the disk symmetrically on either side at a distance d from its centre. The axle is massless and both the springs and the axle are in a horizontal plane. The unstretched length of each spring is L. The disk is initially at its equilibrium position with its centre of mass (CM) at a distance L from the wall. The disk rolls without slipping with velocity $\mathbf{v}_0 = v_0\hat{\mathbf{i}}$. *The coefficient of friction is* μ.

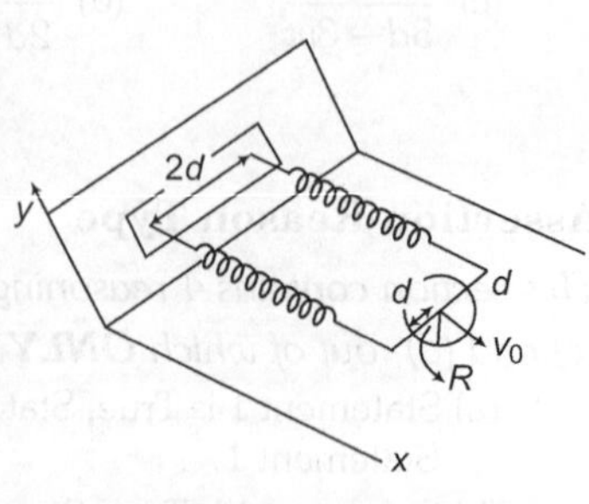

14. The net external force acting on the disk when its centre of mass is at displacement x with respect to its equilibrium position is

(a) $-kx$ (b) $-2kx$ (c) $-\frac{2kx}{3}$ (d) $-\frac{4kx}{3}$

15. The centre of mass of the disk undergoes simple harmonic motion with angular frequency ω equal to

(a) $\sqrt{\frac{k}{M}}$ (b) $\sqrt{\frac{2k}{M}}$ (c) $\sqrt{\frac{2k}{3M}}$ (d) $\sqrt{\frac{4k}{3M}}$

16. The maximum value of v_0 for which the disk will roll without slipping is

(a) $\mu g\sqrt{\frac{M}{k}}$ (b) $\mu g\sqrt{\frac{M}{2k}}$

(c) $\mu g\sqrt{\frac{3M}{k}}$ (d) $\mu g\sqrt{\frac{5M}{2k}}$

Paragraph for Question Nos. 17 to 19

The nuclear charge (Ze) is non-uniformly distributed within a nucleus of radius R. The charge density ρ(r) [charge per unit volume] is dependent only on the radial distance r from the centre of the nucleus as shown in figure. The electric field is only along the radial direction.

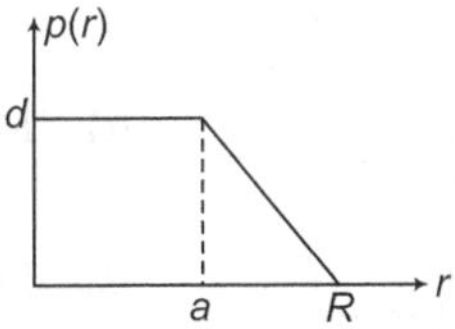

17. The electric field at $r = R$ is

(a) independent of a
(b) directly proportional to a
(c) directly proportional to a^2
(d) inversely proportional to a

18. For $a = 0$, the value of d (maximum value of ρ as shown in the figure) is

(a) $\frac{3Ze}{4\pi R^3}$
(b) $\frac{3Ze}{\pi R^3}$
(c) $\frac{4Ze}{3\pi R^3}$
(d) $\frac{Ze}{3\pi R^3}$

19. The electric field within the nucleus is generally observed to be linearly dependent on r. This implies

(a) $a = 0$
(b) $a = \frac{R}{2}$
(c) $a = R$
(d) $a = \frac{2R}{3}$

Section IV

Matrix Match Type

This section contains 3 questions. Each question contains statements given in two columns, which have to be matched. Statements in **Column I** are labelled as A, B, C and D whereas statements in **Column II** are labelled as p, q, r and s. The answers to these questions have to be appropriately bubbled as illustrated in the following example.

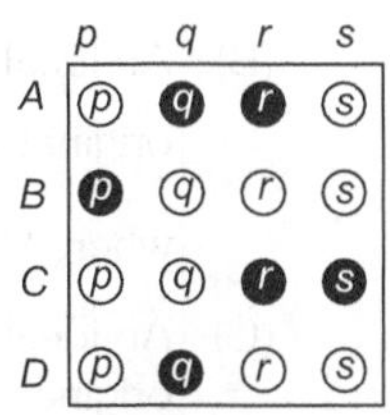

If the correct matches are A-q, A-r, B-p, B-s, C-r, C-s and D-q, then the correctly bubbled matrix will like the following :

22. An optical component and an object S placed along its optic axis are given in Column I. The distance between the object and the component can be varied. The properties of images are given in Column II. Match all the properties of images from Column II with the appropriate components given in Column I. Indicate your answer by darkening the appropriate bubble of the 4×4 matrix given in the ORS.

	Column I		Column II
(A)	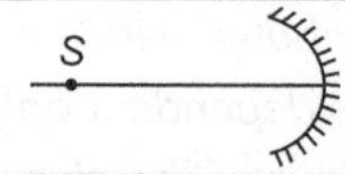	(p)	Real image
(B)	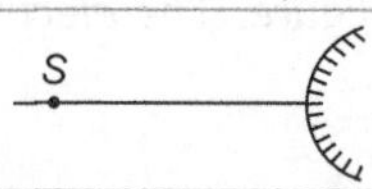	(q)	Virtual image
(C)	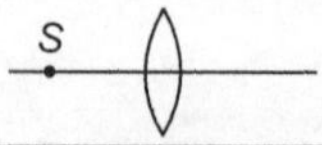	(r)	Magnified image
(D)	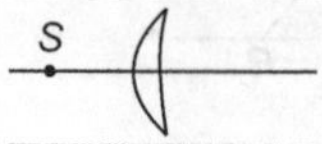	(s)	Image at infinity

21. Column I contains a list of processes involving expansion of an ideal gas. Match this with Column II describing the thermodynamic change during this process. Indicate your answer by darkening the appropriate bubbles of the 4×4 matrix given in the ORS.

	Column I		Column II
(A)	An insulated container has two chambers separated by a valve. Chamber I contains an ideal gas and the Chamber II has vacuum. The valve is opened. 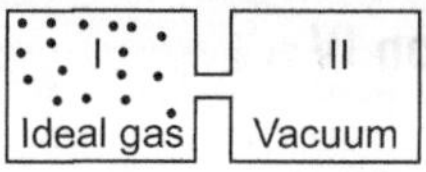	(p)	The temperature of the gas decreases
(B)	An ideal monoatomic gas expands to twice its original volume such that its pressure $p \propto \frac{1}{V^2}$, where V is its volume	(q)	The temperature of the gas increases or remains constant
(C)	An ideal monoatomic gas expands to twice its original volume such that is its pressure $p \propto \frac{1}{V^{4/3}}$ where V is its volume.	(r)	The gas loses heat
(D)	An ideal monoatomic gas expands such that its pressure p and volume V follows the behaviour shown in the graph 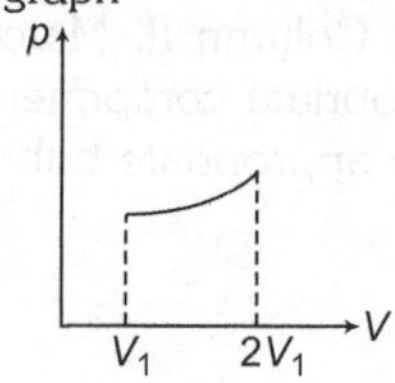	(s)	The gas gains heat

22. Column I gives a list of possible set of parameters measured in some experiments. The variations of the parameters in the form of graphs are shown in Column II. Match the set of parameters given in Column I with the graphs given in Column II. Indicate your answer by darkening the appropriate bubble of the 4×4 matrix given in the ORS.

	Column I		**Column II**
(A)	Potential energy of a simple pendulum (y-axis) as a function of displacement (x-axis)	(p)	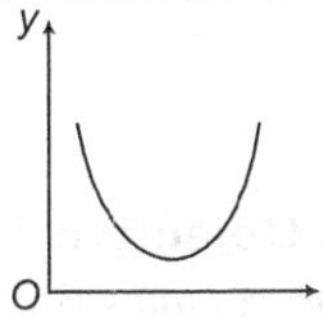
(B)	Displacement (y-axis) as a function of time (x-axis) for a one dimensional motion at zero or constant acceleration when the body is moving along the positive x direction.	(q)	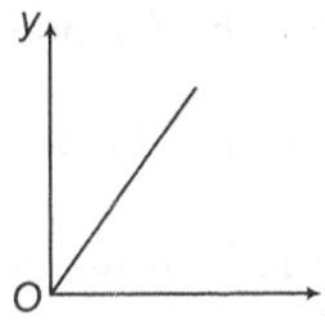
(C)	Range of a projectile (y-axis) as a function of its velocity (x-axis) when projected at a fixed angle.	(r)	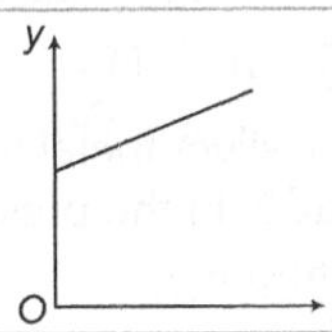
(D)	The square of the time period (y-axis) of a simple pendulum as a function of its length (x-axis)	(s)	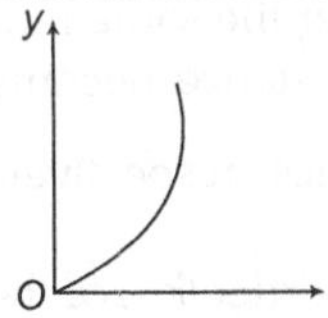

ANSWERS

Paper I

1. (c) 2. (a) 3. (c) 4. (b) 5. (b) 6. (c) 7. (a,c,d) 8. (a,b) 9. (a,d) 10. (b,d)
11. (a) 12. (d) 13. (a) 14. (d) 15. (c) 16. (c) 17. (a) 18. (b) 19. (b) 20. (c)
21. (d) 22. (b) 23. (b)

Paper II

1. (d) 2. (c) 3. (b) 4. (c) 5. (a) 6. (a) 7. (a) 8. (b) 9. (a) 10. (c)
11. (b) 12. (b) 13. (b) 14. (d) 15. (d) 16. (c) 17. (a) 18. (b) 19. (c)

20. (A)p, q, r, s (B) q (C) p, q, r, s (D) p, q, r, s
21. (A) q, s (B) q, p (C) p, s (D) q, s
21. (A) p or p, s (B) q, s or q, r, s (C) s (D) q

IIT JEE
UNSOLVED PAPER 2008

Chemistry

Paper 1

Section I

Straight Objective Type

This section contains 6 multiple choice questions numbered 1 to 6. Each question has 4 choices (a), (b), (c), and (d), out of which ***ONLY ONE*** *is correct.*

1. 2.5 mL of $\frac{2}{5}$ M weak monoacidic base ($K_b = 1 \times 10^{-12}$ at 25°C) is titrated with $\frac{2}{15}$ M HCl in water at 25°C. The concentration of H^+ at equivalence point is ($K_w = 1 \times 10^{-14}$ at 25°C)

(a) 3.7×10^{-13} M (b) 3.2×10^{-7} M (c) 3.2×10^{-2} M (d) 2.7×10^{-2} M

2. Native silver metal forms a water soluble complex with a dilute aqueous solution of NaCN in the presence of

(a) nitrogen (b) oxygen (c) carbon dioxide (d) argon

3. Under the same reaction conditions, initial concentration of 1.386 mol dm^{-3} of a substance becomes half in 40 s and 20 s through first-order and zero-order kinetics, respectively. Ratio $\left(\frac{k_1}{k_0}\right)$ of the rate constants for first-order (k_1) and zero-order (k_0) of the reaction is

(a) 0.5 $mol^{-1}dm^3$ (b) 1.0 mol dm^3 (c) 1.5 mol dm^3 (d) 2.0 $mol^{-1}dm^3$

4. The major product of the following reaction is

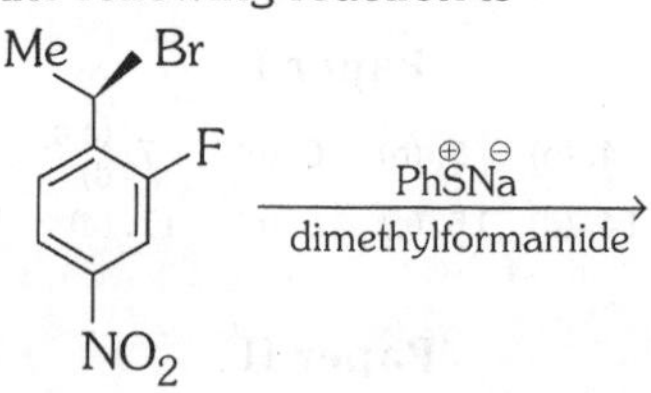

(a) 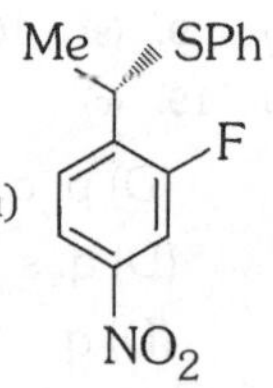

(b) 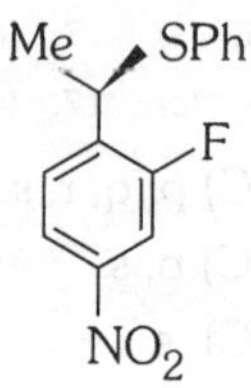

(c) 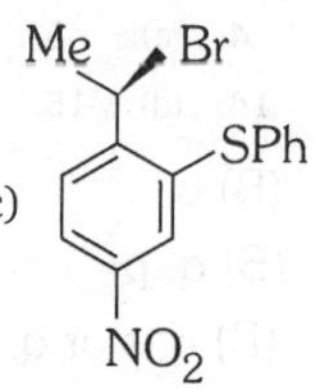

(d) 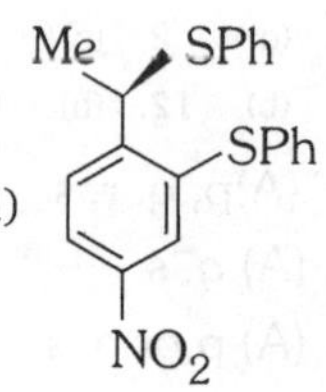

5. Aqueous solution of $Na_2S_2O_3$ on reaction with Cl_2 gives

(a) $Na_2S_4O_6$ (b) $NaHSO_4$ (c) NaCl (d) NaOH

6. Hyperconjugation involves overlap of the following orbitals

(a) $\sigma - \sigma$ (b) $\sigma - p$ (c) $p - p$ (d) $\pi - \pi$

Section II

Multiple Correct Answer Type

This section contains 4 multiple correct answer(s) type questions. Each question has 4 choices (a), (b), (c) and (d), out of which ***ONE OR MORE*** *is/are correct.*

7. A gas described by van der Waal's equation

(a) behaves similar to an ideal gas in the limit of large molar volums

(b) behaves similar to an ideal gas in the limit of large pressures

(c) is characterised by van der Waals' coefficients that are dependent on the identity of the gas but are independent of the temperature

(d) has the pressure that is lower than the pressure exerted by the same gas behaving ideally

8. A solution of colourless salt H on boiling with excess NaOH produces a non-flammable gas. The gas evolution ceases after sometime. Upon addition of Zn dust to the same solution, the gas evolution restarts. The colourless salt (s) H is (are)

(a) NH_4NO_3 (b) NH_4NO_2 (c) NH_4Cl (d) $(NH_4)_2SO_4$

9. The correct statement(s) about the compound given below is are

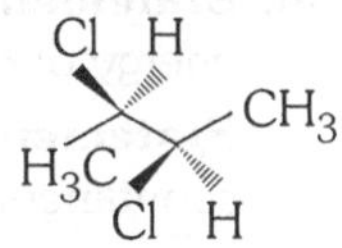

(a) the compound is optically active

(b) the compound possesses centre of symmetry

(c) the compound possesses plane of symmetry

(d) the compound possesses axis of symmetry

10. The correct statement(s) concerning the structures E, F and G is are

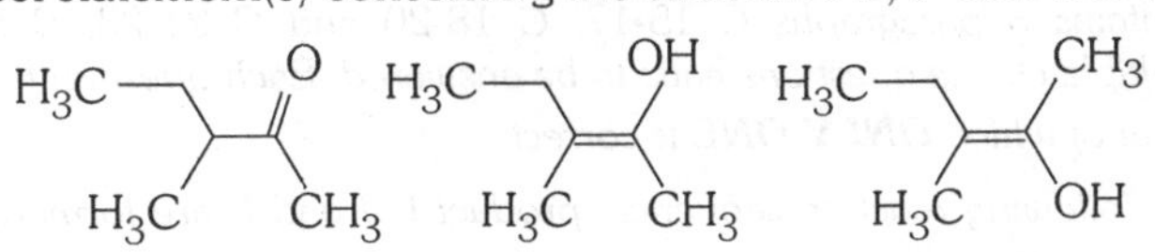

(E) (F) (G)

(a) E, F and G are resonance structures (b) E, F and E, G are tautomers

(c) F and G are geometrical isomers (d) F and G are diastereomers

Section III

Assertion-Reason Type

This section contains 4 reasoning type questions. Each question has 4 choices (a), (b), (c) and (d) of which ***ONLY ONE*** *is correct*

Directions *For the following questions, choose from the correct codes (a) (b) (c) and (d) defined as follows*

(a) Statements 1 is true, statement 2 is also true; statement 2 is the correct explanation of statement 1
(b) Statement 1 is true, statement 2 is also true; statement 2 is not the correct explanation of statement 1
(c) Statement 1 is true, statement 2 is false
(d) Statement 1 is false, statement 2 is true

11. Statement 1 : Pb^{4+} compounds are stronger oxidising agents than Sn^{4+} compounds.
Statement 2 : The higher oxidation states for the group 14 elements are more stable for the heavier members of the group due to 'inert pair effect'.

12. Statement 1 : The plot of atomic number (x-axis) *versus* number of neutrons (x-axis) for stable nuclei shows a curvature towards x-axis from the line of 45° slope as the atomic number is increased
Statement 2 : Proton-proton electrostatic repulsions begin to overcome attractive forces involving protons and neutrons in heavier nuclides.

13. Statement 1 : Bromobenzene, upon reaction with Br_2/Fe gives 1, 4-dibromobenzene as the major product.
Statement 2 : In bromobenzene, the inductive effect of the bromo group is more dominant than the mesomeric effect in directing the incoming electrophile.

14. Statement 1 : For every chemical reaction at equilibrium, standard Gibbs energy of reaction is zero.
Statement 2 : At constant temperature and pressure, chemical reactions are spontaneous in the direction of decreasing Gibbs energy.

Section IV

Linked Comprehension Type

This section contains 3 paragraphs C 15-17, C 18-20 and C 21-23. Based upon each paragraph, 3 multiple choice questions have to be answered. Each question has 4 choices (a), (b), (c) and (d) out of which ***ONLY ONE*** *is correct*

C_{15-17} *In the following reaction sequence, product I. J and L are formed. K represent a reagent*

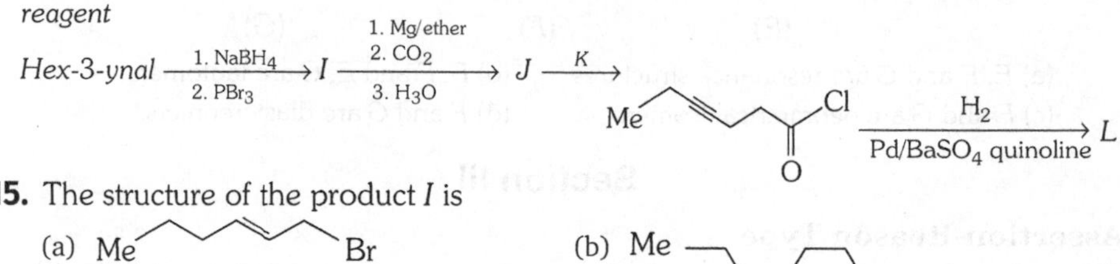

15. The structure of the product *I* is

(a) 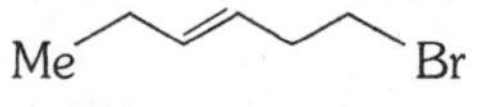

(b) 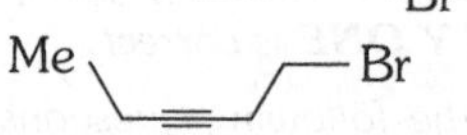

(c)

(d)

16. The structures of compounds *J* and *K*, respectively, are

(a) 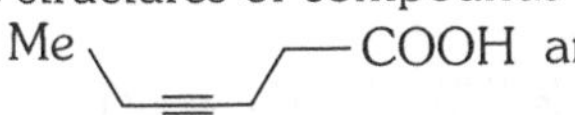 and SO_2Cl_2

(b) 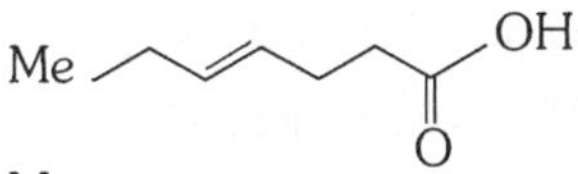 and SO_2Cl_2

(c) 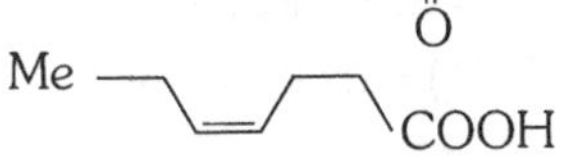 and $SOCl_2$

(d)  and CH_3SO_2Cl

17. The structure of product *L* is

(a) 

(b) 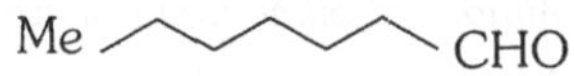

(c) 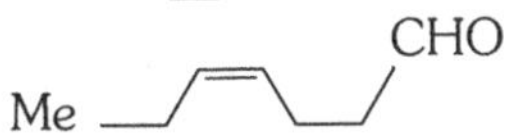

(d) 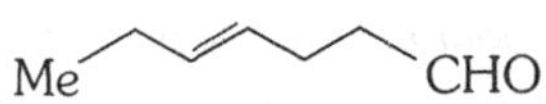

C₁₈₋₂₀ *There are some deposits of nitrates and phosphates in earth's crust. Nitrates are more soluble in water. Nitrates are difficult to reduce under the laboratory conditions but microbes do it easily. Ammonia forms large number of complexes with transition metal ions. Hybridization easily explains the ease of sigma donation capability of NH_3 and PH_3. Phosphine is a flammable gas and is prepared from white phosphorus.*

18. Among the following, the correct statement is

(a) Phosphates have no biological significance in humans

(b) Between nitrates and phosphates, phosphates are less abundant in earth's crust

(c) Between nitrates and phosphates, nitrates are less abundant in earth's crust

(d) Oxidation of nitrates is possible in soil

19. Among the following, the correct statement is

(a) Between NH_3 and PH_3, NH_3 is a better electron donor because the lone pair of electrons occupies spherical 's' orbital and is less directional

(b) Between NH_3 and PH_3, PH_3 is a better electron donor because the lone pair of electrons occupies sp^3 orbital and is more directional

(c) Between NH_3 and PH_3, NH_3 is a better electron donor because the lone pair of electrons occupies sp^3 orbital and is more directional

(d) Between NH_3 and PH_3, PH_3 is a better electron donor because the lone pair of electrons occupies spherical 's' orbital and is less directional

20. White phosphorus on reaction with NaOH gives PH_3 as one of the products. This is a

(a) dimerisation reaction

(b) disproportionation reaction

(c) condensation reaction

(d) precipitation reaction

C₂₁₋₂₃ *Properties such as boiling point, freezing point and vapour pressure of a pure solvent change when solute molecules are added to get homogeneous solution. These are called colligative properties. Applications of colligative properties are very useful in day-to-day life. One of its examples is the use of ethylene glycol and water mixture as anti-freezing liquid in the radiator automobiles.*

A solution M is prepared by mixing ethanol and water. The mole fraction of ethanol in the mixtrue is 0.9

Given : Freezing point depression constant of water (k^{water}) = 1.86 K kg mol^{-1}

Freezing point depression constant of ethanol ($k_f^{ethanol}$) = 2.0 K kg mol^{-1}

Boiling point elevation constant of water (k_b^{water}) = 0.52 K kg mol^{-1}

Boiling point elevation constant of ethanol ($k_b^{ethanol}$) = 1.2 K kg mol^{-1}

Standard freezing point of water = 273 K

Standard freezing point of ethanol = 155.7 K

Standard boiling point of water = 373 K

Standard boiling point of ethanol = 351.5 K

Vapour pressure of pure water = 32.8 mm Hg

Vapour pressure of pure ethanol = 40 mm Hg

Molecular weight of water = 18 g mol^{-1}

Molecular weight of ethanol = 46 g mol^{-1}

In answering the following questions, consider the solutions to be ideal dilute solutions and solutes to be non-volatile and non-dissociative

21. The freezing point of the solution *M* is

(a) 268.7 K (b) 268.5 K (c) 234.2 K (d) 150.9 K

22. The vapour pressure of the solution *M* is

(a) 39.3 mm Hg (b) 36.0 mm Hg (c) 29.5 mm Hg (d) 28.8 mm Hg

23. Water is added to the solution *M* such that the mole fraction of water in the solution becomes 0.9 The boiling point of this solutions is

(a) 380.4 K (b) 376.2 K (c) 375.5 K (d) 354.7 K

PAPER 2

Section I

Straight Objective Type

This section contains 9 multiple choice questions numbered 1 to 9. Each question has 4 choices (a), (b), (c) and (d), out of which ***ONLY ONE*** *is correct.*

1. Solubility product constant (K_{sp}) of salts of types MX, MX_2 and M_3X at temperature 'T' are 4.0×10^{-8}, 3.2×10^{-14} and 2.7×10^{-15}, respectively. Solubilities (mol dm^{-3}) of the salts at temperature 'T' are in the order

(a) $MX > MX_2 > M_3X$ (b) $M_3X > MX_2 > MX$

(c) $MX_2 > M_3X < MX$ (d) $MX > M_3X > MX_2$

2. Electrolysis of dilute aqueous NaCl solution was carried out by passing 10 mA current. The time required to liberate 0.01 mole of H_2 gas at the cathode is (1 Faraday = 96500 C mol^{-1})

(a) 9.65×10^4s (b) 19.3×10^4s (c) 28.95×10^4s (d) 38.6×10^4s

3. Among the following, the surfactant that will form micelles in aqueous solution at the lowest molar concentration at ambient conditions is

(a) $CH_3(CH_2)_{15}N^+(CH_3)_3Br^-$ (b) $CH_3(CH_2)_{11}OSO_3^-Na^+$
(c) $CH_3(CH_2)_6COO^-Na^+$ (d) $CH_3(CH_2)_{11}N^+(CH_3)_3Br^-$

4. Both $[Ni(CO)_4]$ and $[Ni(CN)_4]^{2-}$ are diamagnetic. The hybridisations of nickel in these complexes, respectively, are

(a) sp^3, sp^3 (b) sp^3, dp^2 (c) dsp^2, sp^3 (d) dsp^2, dsp^2

5. The IUPAC name of $[Ni(NH_3)_4]$ $[NiCl_4]$ is

(a) Tetrachloronickel (II) - tetraamminenickel (II)
(b) Tetraamminenickel (II) - tetrachloronickel (II)
(c) Tetraamminenickel (II) - tetrachloronickelate (II)
(d) Tetrachloronickel (II) - Tetraamminenickelate (0)

6. Among the following, the coloured compound is

(a) CuCl (b) $K_3[Cu(CN)_4]$
(c) CuF_2 (d) $[Cu(CH_3CN_4)]BF_4$

7. In the following reaction sequence, the correct structures of *E*, *F* and *G* are

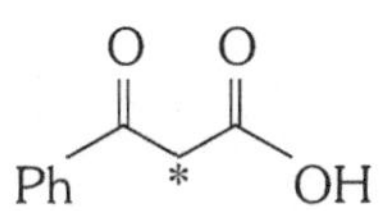 $\xrightarrow{\text{Heat}} [E] \xrightarrow[\text{NaOH}]{I_2} [F]+[G]$

(*implies ^{13}C labelled carbon)

(a) $E =$ 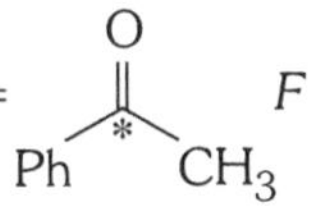 $F =$ 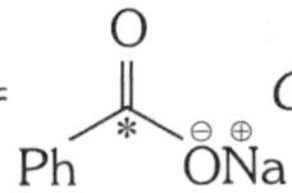 $G = CHI_3$

(b) $E =$ 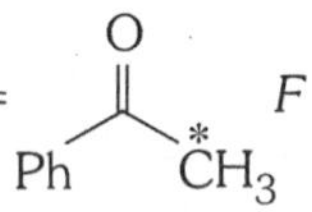 $F =$ 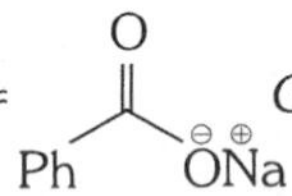 $G = CHI_3$

(c) $E =$ 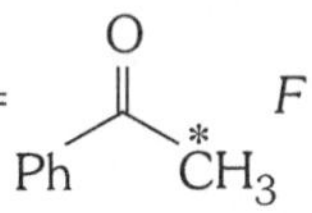 $F =$ 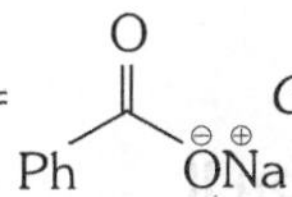 $G = \overset{*}{C}HI_3$

(d) $E =$ 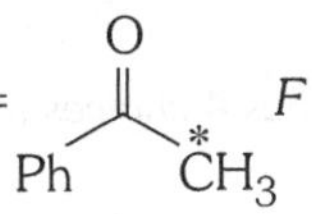 $F =$ 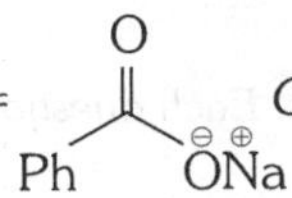 $G = \overset{*}{C}H_3I$

8. The correct stability order for the following species is

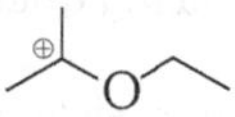
(I)

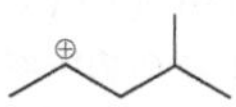
(II)

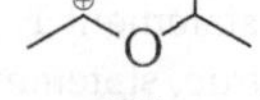
(III)

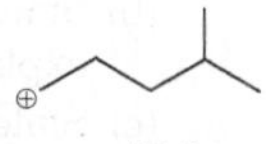
(IV)

(a) (II) > (IV) > (I) > (III)
(b) (I) > (II) > (III) > (IV)
(c) (II) > (I) > (IV) > (III)
(d) (I) > (III) > (II) > (IV)

9. Cellulose upon acetylation with excess acetic anhydride/H_2SO_4 (catalytic) gives cellulose triacetate whose structures is

(a) 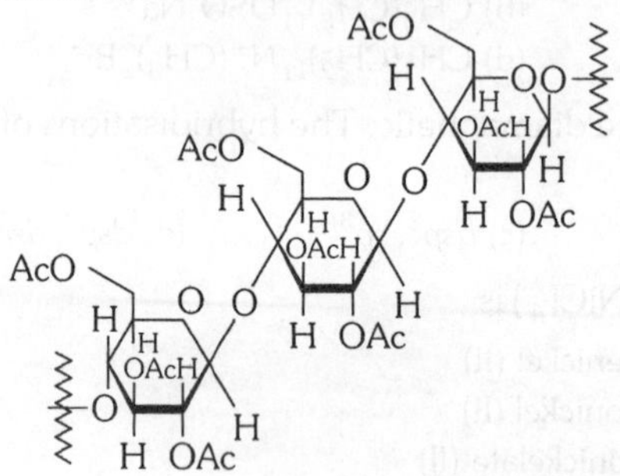

(b) 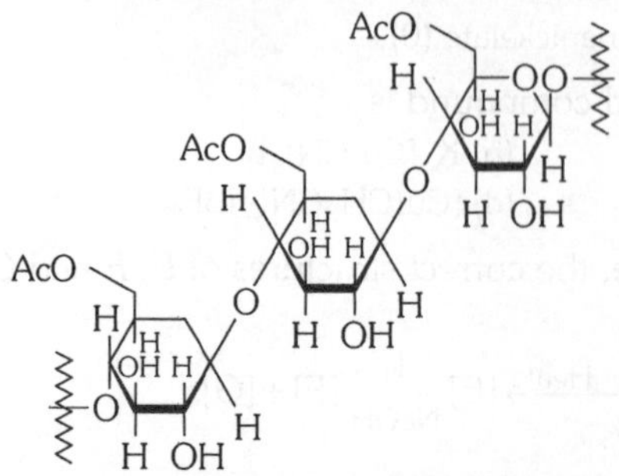

(c) 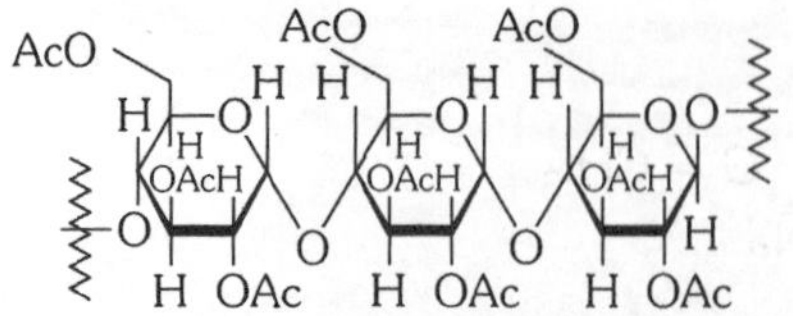

(d) 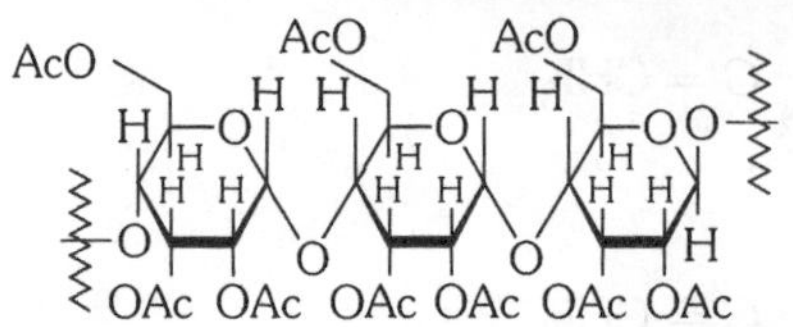

Section II

Assertion Reason Type

*This section contains 4 reasoning type questions. Each question has 4 choices (a), (b), (c) and (d) of which **ONLY ONE** is correct.*

Directions *For the following questions, choose from the correct codes (a), (b), (c) and (d) defined as follows :*

(a) Statements 1 is true, statement 2 is also true; statement 2 is the correct explanation of statement 1
(b) Statement 1 is true, statement 2 is also true; statement 2 is not the correct explanation of statement 1
(c) Statement 1 is true, statement 2 is false
(d) Statement 1 is false, statement 2 is true

10. **Statement 1 :** $[Fe(H_2O)_5NO]SO_4$ is paramagnetic

Statement 2 : The Fe in $[Fe(H_2O)_5NO]SO_4$ has three unpaired electrons.

11. **Statement 1 :** The geometrical isomer of the complex $[M(NH_3)_4Cl_2]$ are optically inactive

Statement 2 : Both geometrical isomers of the complex $[M(NH_3)_4Cl_2]$ possess axis of symmetry.

12. **Statement 1 :** There is a natural asymmetry between converting work to heat and converting heat to work.

Statement 2 : No process is possible in which the sole result in the absorption of heat from a reservoir and its complete conversion into work.

13. **Statement 1 :** Aniline on reaction with $NaNO_2$ / HCl at 0°C followed by coupling with β-naphthol gives a dark coloured precipitate.

Statement 2 : The colour of the compound formed in the reaction of aniline with $NaNO_2$ / HCl at 0°C followed by coupling with β-naphthol is due to the extended conjugation.

Section III

Linked Comprehension Type

This section contains 2 paragraphs C 14-16 and C 17-19. Based upon each paragraph, 3 multiple choice questions have to be answered. Each question has 4 choices (a), (b), (c) and (d) out of which ***ONLY ONE*** *is correct.*

C_{14-16} *In hexagonal system of crystals, a frequently encountered arrangement of atoms is described as a hexagonal prism. Here, the top and bottom of the cell are regular hexagons and three atoms are sandwiched in between them. A space-filling model of this structure, called hexagonal close-packed (HCP), is constituted of a sphere on a flat surface surrounded in the same plane by six identical spheres as closely as possible. Three spheres are then placed over the first layer so that they touch each other and represent the second layer. Each one of the three spheres touches three spheres of the bottom layer. Finally, the second layer is covered with a third layer that is identical to the bottom layer in relative position. Assume radius of every sphere to be 'r'.*

14. The number of atoms in this HCP unit cell is

(a) 4 (b) 6 (c) 12 (d) 17

15. The volume of this HCP unit cell is

(a) $24\sqrt{2}r^3$ (b) $16\sqrt{2}r^3$ (c) $12\sqrt{2}r^3$ (d) $\frac{64}{3\sqrt{3}}r^3$

16. The empty space in this HCP unit cell is

(a) 74% (b) 47.6% (c) 32% (d) 26%

C_{17-19} *A tertiary alcohol H upon acid catalysed dehydration gives a product I. Ozonolysis of I leads to compounds J and K. Compound J upon reaction with KOH gives benzyl alcohol and a compound L, where K on reaction with KOH gives only M.* $M =$

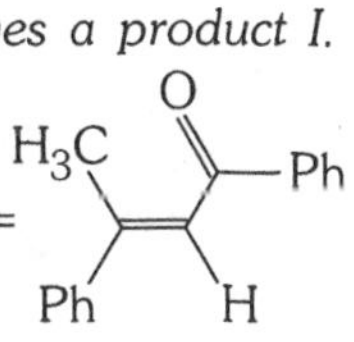

17. Compound H is formed by the reactions of

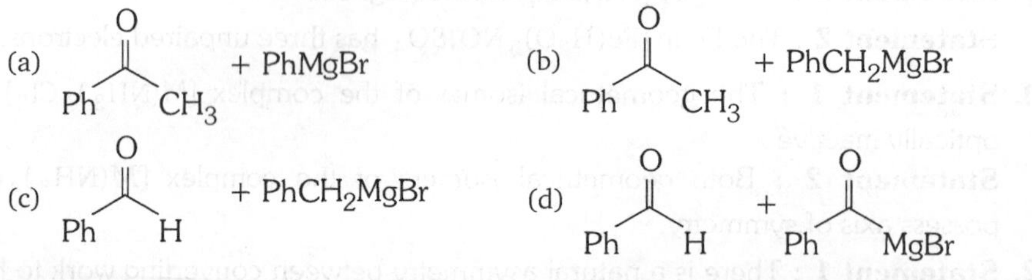

18. The structure of compound I is

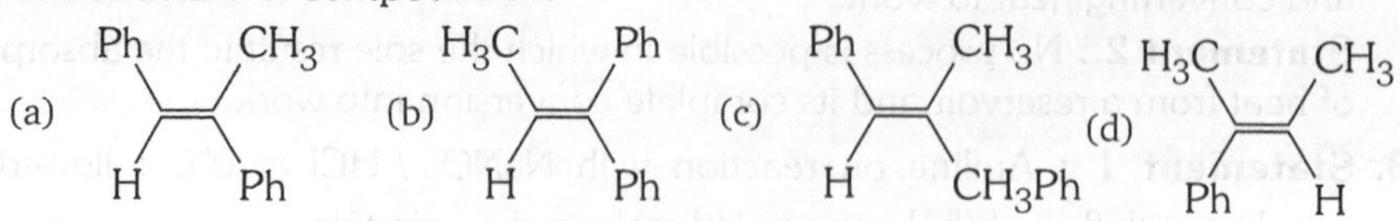

19. The structures of compounds J, K and L, respectively, are

(a) $PhCOCH_3$, $PhCH_2COCH_3$ and $PhCH_2COO^-K^+$

(b) $PhCHO$, $PhCH_2CHO$ and $PhCH_2COO^-K^+$

(c) $PhCOCH_3$, $PhCH_2CHO$ and $CH_3COO^-K^+$

(d) $PhCHO$, $PhCOCH_3$ and $PhCOO^-K^+$

Section IV

Matrix-match Type

*This section contains 3 questions. Each question contains statements given is two columns which have to be matched. Statements in **Column I** are labelled as A, B, C and D whereas statements in **Column II** are labelled as p, q, r and s. The answers to these questions have to be appropriately bubbled as illustrated in the following example*

If the correct matches are A-q, A-r, B-p, B-s, C-r, C-s and D-q, then the correctly bubbled matrix will look like the following

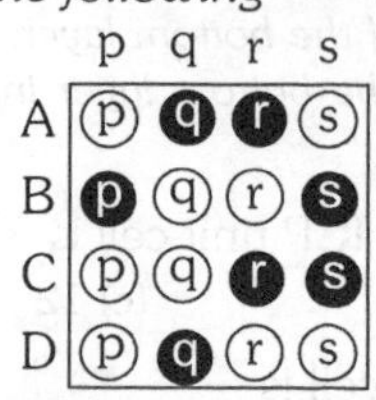

20. Match the conversions in **Column I** with the type(s) given in **Column II.** Indicate your answer by darkening the appropriate bubbles of the 4 × 4 matrix given in the ORS.

	Column I	Column II
A.	$PbS \longrightarrow PbO$	p. Roasting
B.	$CaCO_3 \longrightarrow CaO$	q. Calcination
C.	$ZnS \longrightarrow Zn$	r. Carbon reduction
D.	$Cu_2S \longrightarrow Cu$	s. Self reduction

21. Match the entries in **Column I** with the correctly related quantum number(s) in **Column II**. Indicate your answer by darkening the appropriate bubbles of the 4×4 matrix given in the ORS

	Column I		Column II
A.	Orbital angular momentum of the electron in a hydrogen-like atomic orbital.	p.	Principal quantum number
B.	A hydrogen-like one electron wave function obeying Pauli principle	q.	Azimuthal quantum number
C.	Shape, size and orientation of hydrogen-like atomic orbitals	r.	Magnetic quantum number
D.	Probability density of electron at the nucleus in hydrogen-like atom	s.	Electron spin quantum number

22. Match the compounds in **Column I** with their characteristic test(s)/reactions(s) given in **Column II.** Indicate your answer by darkening the appropriate bubbles of the 4×4 matrix given in the ORS.

Column I	Column II
A. $H_2N—\overset{\oplus}{N}H_3\,\overset{\ominus}{Cl}$	p. Sodium fusion extract of the compound gives Prussian blue colour with $FeSO_4$
B. 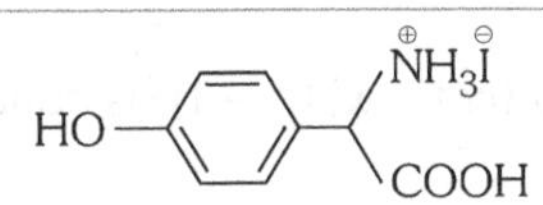	q. Gives positive $FeCl_3$ test
C. 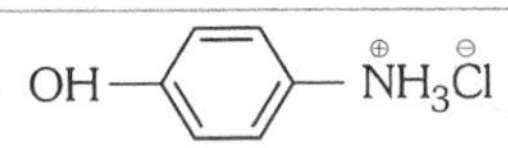	r. Gives white precipitate with $AgNO_3$
D. 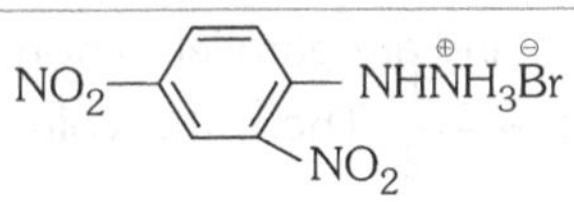	s. Reacts with aldehydes to form the corresponding hydrazone derivative

ANSWER

Paper-1

1. (d) **2.** (b) **3.** (a) **4.** (a) **5.** (b) **6.** (b)
7. (b, c, d) **8.** (a, b) **9.** (a, b) **10.** (b, c, d) **11.** (c) **12.** (a)
13. (c) **14.** (d) **15.** (d) **16.** (a) **17.** (b) **18.** (c)
19. (c) **20.** (b) **21.** (d) **22.** (b) **23.** (b)

Paper-2

1. (d) **2.** (b) **3.** (a) **4.** (b) **5.** (c) **6.** (c)
7. (c) **8.** (d) **9.** (a) **10.** (a) **11.** (a) **12.** (a)
13. (a) **14.** (b) **15.** (a) **16.** (d) **17.** (b) **18.** (a)
19. (d) **20.** A→ p; B→ q; C→ p, r; D→ p, r, s
21. A→ q; B→ p, q, r, s; C→ p, q, r; D→ p, q, r
22. A→ r, s; B→ p, q; C→ p, q, r; D→ p

IIT JEE
UNSOLVED PAPER 2008

Mathematics

Paper 1

Section I

Straight Objective Type

This section contains 6 multiple choice questions. Each question has 4 choices (a), (b), (c) and (d), out of which **ONLY ONE** is correct.

1. Consider the two curves

$C_1 : y^2 = 4x$

$C_2 : x^2 + y^2 - 6x + 1 = 0$, then

(a) C_1 and C_2 touch each other only at one point

(b) C_1 and C_2 touch each other exactly at two points

(c) C_1 and C_2 intersect (but do not touch) at exactly two points

(d) C_1 and C_2 neither intersect nor touch each other

2. If $0 < x < 1$, then $\sqrt{1+x^2}[\{x \cos(\cot^{-1} x) + \sin(\cot^{-1} x)\}^2 - 1]^{\frac{1}{2}}$ is equal to

(a) $\dfrac{x}{\sqrt{1+x^2}}$ (b) x

(c) $x\sqrt{1+x^2}$ (d) $\sqrt{1+x^2}$

3. The edges of a parallelopiped are of unit length and are parallel to non-coplanar unit vectors $\hat{\mathbf{a}}, \hat{\mathbf{b}}, \hat{\mathbf{c}}$ such that $\hat{\mathbf{a}} \cdot \hat{\mathbf{b}} = \hat{\mathbf{b}} \cdot \hat{\mathbf{c}} = \hat{\mathbf{c}} \cdot \hat{\mathbf{a}} = \frac{1}{2}$. Then, the volume of the parallelopiped is

(a) $\dfrac{1}{\sqrt{2}}$ cu units (b) $\dfrac{1}{2\sqrt{2}}$ cu units

(c) $\dfrac{\sqrt{3}}{2}$ cu units (d) $\dfrac{1}{\sqrt{3}}$ cu units

4. Let a and b be non-zero and real numbers. Then, the equation $(ax^2 + by^2 + c)(x^2 - 5xy + 6y^2) = 0$ represents

(a) Four straight lines, when $c = 0$ and a, b are of the same sign

(b) Two straight lines and a circle, when $a = b$ and c is of sign opposite to that of a

(c) Two straight lines and a hyperbola, when a and b are of the same sign and c is of sign opposite to that of a

(d) A circle and an ellipse, when a and b are of the same sign and c is of sign opposite to that of a

5. Let $g(x) = \dfrac{(x-1)^n}{\log \cos^m (x-1)}$; $0 < x < 2$, m and n are integers, $m \neq 0$, $n > 0$ and let p be the left hand derivative of $|x-1|$ at $x = 1$. If $\lim\limits_{x \to 1^+} g(x) = p$, then

(a) $n = 1, m = 1$ (b) $n = 1, m = -1$

(c) $n = 2, m = 2$ (d) $n > 2, m = n$

6. The total number of local maxima and local minima of the function $f(x) = \begin{cases} (2+x)^3; & -3 < x \le -1 \\ x^{\frac{2}{3}}; & -1 < x < 2 \end{cases}$ is

(a) 0 (b) 1 (c) 2 (d) 3

Section II

Multiple Correct Answers Type

This section contains 4 multiple correct answer(s) type questions. Each question has 4 choices (a), (b), (c) and (d), out of which **ONE OR MORE THAN ONE** is/are correct.

7. A straight line through the vertex P of a ΔPQR intersects the side QR at the point S and the circumcircle of the ΔPQR at the point T. If S is not the centre of the circumcircle, then

(a) $\dfrac{1}{PS} + \dfrac{1}{ST} < \dfrac{2}{\sqrt{QS \cdot SR}}$ (b) $\dfrac{1}{PS} + \dfrac{1}{ST} > \dfrac{2}{\sqrt{QS \cdot SR}}$

(c) $\dfrac{1}{PS} + \dfrac{1}{ST} < \dfrac{4}{QR}$ (d) $\dfrac{1}{PS} + \dfrac{1}{ST} > \dfrac{4}{QR}$

8. Let $P(x_1, y_1)$ and $Q(x_2, y_2)$, $y_1 < 0, y_2 < 0$, be the end points of the latusrectum of the ellipse $x^2 + 4y^2 = 4$. The equations of parabolas with latusrectum PQ are

(a) $x^2 + 2\sqrt{3}y = 3 + \sqrt{3}$ (b) $x^2 - 2\sqrt{3}y = 3 + \sqrt{3}$

(c) $x^2 + 2\sqrt{3}y = 3 - \sqrt{3}$ (d) $x^2 - 2\sqrt{3}y = 3 - \sqrt{3}$

9. Let $S_n = \sum\limits_{k=0}^{n} \dfrac{n}{n^2 + kn + k^2}$ and $T_n = \sum\limits_{k=0}^{n-1} \dfrac{n}{n^2 + kn + k^2}$, for $n = 1, 2, 3, \ldots$, then

(a) $S_n < \dfrac{\pi}{3\sqrt{3}}$ (b) $S_n > \dfrac{\pi}{3\sqrt{3}}$ (c) $T_n < \dfrac{\pi}{3\sqrt{3}}$ (d) $T_n > \dfrac{\pi}{3\sqrt{3}}$

10. Let $f(x)$ be a non-constant twice differentiable function defined on $(-\infty, \infty)$, such that $f(x) = f(1-x)$ and $f'\left(\dfrac{1}{4}\right) = 0$. Then,

(a) $f'(x)$ vanishes atleast twice on $[0, 1]$ (b) $f'\left(\dfrac{1}{2}\right) = 0$

(c) $\int_{-\frac{1}{2}}^{\frac{1}{2}} f\left(x + \dfrac{1}{2}\right) \sin x \, dx = 0$ (d) $\int_0^{\frac{1}{2}} f(t)\, e^{\sin \pi t} dt = \int_{\frac{1}{2}}^{1} f(1-t) e^{\sin \pi t} dt$

Section III

Assertion-Reason Type

This section contains 4 reasoning type questions. Each questions has 4 choices (a), (b), (c) and (d) , out of which **ONLY ONE** is correct.

(a) Statement 1 is true, Statement 2 is true; Statement 2 is a correct explanation for Statement 1
(b) Statement 1 is true, Statement 2 is true; Statement 2 is not a correct explanation for Statement 1
(c) Statement 1 is true, Statement 2 is false
(d) Statement 1 is false, Statement 2 is true

---

11. Let f and g be real valued functions defined on interval $(-1, 1)$ such that $g''(x)$ is continuous, $g(0) \neq 0$, $g'(0) = 0$, $g''(0) \neq 0$ and $f(x) = g(x) \sin x$.

Statement 1 $\lim_{x \to 0} [g(x) \cos x - g(0) \operatorname{cosec} x] = f''(0)$.

Statement 2 $f'(0) = g(0)$.

12. Consider three planes

$P_1 : x - y + z = 1$, $P_2 : x + y - z = -1$

and $P_3 : x - 3y + 3z = 2$

Let L_1, L_2 and L_3 be the lines of intersection of the planes P_2 and P_3, P_3 and P_1, P_1 and P_2, respectively.

Statement 1 Atleast two of the lines L_1, L_2 and L_3 are non-parallel.

Statement 2 The three planes do not have a common point.

13. Consider the system of equations

$x - 2y + 3z = -1$, $x - 3y + 4z = 1$ and $-x + y - 2z = k$

Statement 1 The system of equations has no solution for $k \neq 3$.

Statement 2 The determinant $\begin{vmatrix} 1 & 3 & -1 \\ -1 & -2 & k \\ 1 & 4 & 1 \end{vmatrix} \neq 0$, for $k \neq 3$.

14. Consider the system of equations

$$ax + by = 0,\ cx + dy = 0$$

Where $a, b, c, d \in \{0, 1\}$

Statement 1 The probability that the system of equations has a unique solution, is 3/8.

Statement 2 The probability that the system of equations has a solution, is 1.

Section IV

Linked Comprehension Type

This section contains 3 paragraphs. Based upon each paragraph, 3 multiple choice questions have to be answered. Each questions has 4 choices (a), (b), (c) and (d), out of which **ONLY ONE** is correct.

Paragraph for Questions Nos. 15 -17

A circle C of radius 1 is inscribed in an equilateral ΔPQR. The points of contact of C with the sides PQ, QR, RP are D, E, F respectively. The line PQ is given by the equation

$\sqrt{3}\,x + y - 6 = 0$ *and the point* D *is* $\left(\frac{3\sqrt{3}}{2}, \frac{3}{2}\right)$*. Further, it is given that the origin and the centre of* C *are on the same side of the line* PQ.

15. The equation of circle C is

(a) $(x - 2\sqrt{3})^2 + (y - 1)^2 = 1$ (b) $(x - 2\sqrt{3})^2 + \left(y + \frac{1}{2}\right)^2 = 1$

(c) $(x - \sqrt{3})^2 + (y + 1)^2 = 1$ (d) $(x - \sqrt{3})^2 + (y - 1)^2 = 1$

16. Points E and F are given by

(a) $\left(\frac{\sqrt{3}}{2}, \frac{3}{2}\right), (\sqrt{3}, 0)$ (b) $\left(\frac{\sqrt{3}}{2}, \frac{1}{2}\right), (\sqrt{3}, 0)$

(c) $\left(\frac{\sqrt{3}}{2}, \frac{3}{2}\right), \left(\frac{\sqrt{3}}{2}, \frac{1}{2}\right)$ (d) $\left(\frac{3}{2}, \frac{\sqrt{3}}{2}\right), \left(\frac{\sqrt{3}}{2}, \frac{1}{2}\right)$

17. Equations of the sides QR, RP are

(a) $y = \frac{2}{\sqrt{3}}x + 1, \; y = -\frac{2}{\sqrt{3}}x - 1$ (b) $y = \frac{1}{\sqrt{3}}x, \; y = 0$

(c) $y = \frac{\sqrt{3}}{2}x + 1, \; y = -\frac{\sqrt{3}}{2}x - 1$ (d) $y = \sqrt{3}x, \; y = 0$

Paragraph for Questions Nos.18-20

Consider the functions defined implicitly by the equation $y^3 - 3y + x = 0$ *on various intervals in the real line. If* $x \in (-\infty, -2) \cup (2, \infty)$, *the equation implicitly defined a unique real valued differentiable function* $y = f(x)$. *If* $x \in (-2, 2)$, *the equation implicitly defines a unique real valued differentiable function* $y = g(x)$, *satisfying* $g(0) = 0$.

18. If $f(-10\sqrt{2}) = 2\sqrt{2}$, then $f''(-10\sqrt{2})$ is equal to

(a) $\frac{4\sqrt{2}}{7^3 \cdot 3^2}$ (b) $-\frac{4\sqrt{2}}{7^3 \cdot 3^2}$ (c) $\frac{4\sqrt{2}}{7^3 \cdot 3}$ (d) $-\frac{4\sqrt{2}}{7^3 \cdot 3}$

19. The area of the region bounded by the curve $y = f(x)$, the X- axis and the lines $x = a$ and $x = b$, where $-\infty < a < b < -2$, is

(a) $\int_a^b \frac{x}{3[f(x)^2 - 1]}\,dx + b\,f(b) - a\,f(a)$ (b) $-\int_a^b \frac{x}{3[(f(x))^2 - 1]}\,dx + b\,f(b) - a\,f(a)$

(c) $\int_a^b \frac{x}{3[(f(x))^2 - 1]}\,dx - b\,f(b) + a\,f(a)$ (d) $-\int_a^b \frac{x}{3[f(x)^2 - 1]}\,dx - b\,f(b) - a\,f(a)$

20. $\int_{-1}^{1} g'(x)\,dx$ is equal to

(a) $2g(-1)$ (b) 0 (c) $-2g(1)$ (d) $2g(1)$

Paragraph for Question Nos. 21- 23

Let A, B, C *be three sets of complex numbers as defined below*

$A = \{z;\ \text{Im}\, z \geq 1\};\ B = \{z : |z - 2 - i| = 3\};\ C = \{z : \text{Re}\,((1 - i)z) = \sqrt{2}\}$.

21. The number of elements in the set $A \cap B \cap C$ is

(a) 0 (b) 1 (c) 2 (d) ∞

22. Let z be any point in $A \cap B \cap C$. The $|z+1-i|^2 + |z-5-i|^2$ lies between

(a) 25 and 29 (b) 30 and 34 (c) 35 and 39 (d) 40 and 44

23. Let z be any point in $A \cap B \cap C$ and let w be any point satisfying $|w-2-i| < 3$. Then, $|z| - |w| + 3$ lies between

(a) −6 and 3 (b) −3 and 6 (c) −6 and 6 (d) −3 and 9

Paper 2

Section I

Straight Objective Type

This section contains 9 multiple choice questions. Each question has 4 choices (a), (b), (c) and (d) , out of which **ONLY ONE** is correct

1. A particle P starts from the point $z_0 = 1 + 2i$, where $i = \sqrt{-1}$. It moves first horizontally away from origin by 5 units and then vertically away from origin by 3 units to reach a point z_1. From z_1 the particle moves $\sqrt{2}$ units in the direction of the vector $\hat{\mathbf{i}} + \hat{\mathbf{j}}$ and then it moves through an angle $\frac{\pi}{2}$ in anti-clockwise direction on a circle with centre at origin, to reach a point z_2. The point z_2 is given by

(a) $6 + 7i$ (b) $-7 + 6i$ (c) $7 + 6i$ (d) $-6 + 7i$

2. Let the function $g : (-\infty, \infty) \to \left(-\frac{\pi}{2}, \frac{\pi}{2}\right)$ be given by $g(u) = 2\tan^{-1}(e^u) - \frac{\pi}{2}$. Then, g is

(a) even and is strictly increasing in $(0, \infty)$

(b) odd and is strictly decreasing in $(-\infty, \infty)$

(c) odd and is strictly increasing in $(-\infty, \infty)$

(d) neither even nor odd, but is strictly increasing in $(-\infty, \infty)$

3. Consider a branch of the hyperbola $x^2 - 2y^2 - 2\sqrt{2}x - 4\sqrt{2}y - 6 = 0$ with vertex at the point A. Let B be one of the end points of its latusrectum. If C is the focus of the hyperbola nearest to the point A, then the area of the ΔABC is

(a) $1 - \sqrt{\frac{2}{3}}$ sq unit (b) $\sqrt{\frac{3}{2}} - 1$ sq unit (c) $1 + \sqrt{\frac{2}{3}}$ sq unit (d) $\sqrt{\frac{3}{2}} + 1$ sq unit

4. The area of the region between the curves $y = \sqrt{\frac{1+\sin x}{\cos x}}$ and $y = \sqrt{\frac{1-\sin x}{\cos x}}$ bounded by the lines $x = 0$ and $x = \frac{\pi}{4}$ is

(a) $\int_0^{\sqrt{2}-1} \frac{t}{(1+t^2)\sqrt{1-t^2}}\, dt$ (b) $\int_0^{\sqrt{2}-1} \frac{4t}{(1+t^2)\sqrt{1-t^2}}\, dt$

(c) $\int_0^{\sqrt{2}+1} \frac{4t}{(1+t^2)\sqrt{1-t^2}}\, dt$ (d) $\int_0^{\sqrt{2}+1} \frac{t}{(1+t^2)\sqrt{1-t^2}}\, dt$

5. Consider three points
$P = (-\sin(\beta-\alpha), -\cos\beta), Q = (\cos(\beta-\alpha), \sin\beta)$ and $R = (\cos(\beta-\alpha+\theta), \sin(\beta-\theta))$,
where $0 < \alpha, \beta, \theta < \frac{\pi}{4}$. Then,

(a) P lies on the line segment RQ (b) Q lies on the line segment PR
(c) R lies on the line segment QP (d) P, Q, R are non- collinear

6. An experiment has 10 equally likely outcomes. Let A and B be two non-empty events of the experiment. If A consists of 4 outcomes, the number of outcomes that B must have so that A and B are independent, is
(a) 2, 4 or 8 (b) 3, 6 or 9 (c) 4 or 8 (d) 5 or 10

7. Let two non-collinear unit vectors $\hat{\mathbf{a}}$ and $\hat{\mathbf{b}}$ form an acute angle.
A point P moves so that at any time t the position vector **OP** (where O is the origin) is given by $\hat{\mathbf{a}} \cos t + \hat{\mathbf{b}} \sin t$. When P is farthest from origin O, let M be the length of **OP** and $\hat{\mathbf{u}}$ be the unit vector along **OP**. Then,

(a) $\hat{\mathbf{u}} = \frac{\hat{\mathbf{a}}+\hat{\mathbf{b}}}{|\hat{\mathbf{a}}+\hat{\mathbf{b}}|}$ and $M = (1+\hat{\mathbf{a}}\cdot\hat{\mathbf{b}})^{1/2}$ (b) $\hat{\mathbf{u}} = \frac{\hat{\mathbf{a}}-\hat{\mathbf{b}}}{|\hat{\mathbf{a}}-\hat{\mathbf{b}}|}$ and $M = (1+\hat{\mathbf{a}}/\hat{\mathbf{b}})^{1/2}$

(c) $\hat{\mathbf{u}} = \frac{\hat{\mathbf{a}}+\hat{\mathbf{b}}}{|\hat{\mathbf{a}}+\hat{\mathbf{b}}|}$ and $M = (1+2\hat{\mathbf{a}}\cdot\hat{\mathbf{b}})^{1/2}$ (d) $\hat{\mathbf{u}} = \frac{\hat{\mathbf{a}}-\hat{\mathbf{b}}}{|\hat{\mathbf{a}}-\hat{\mathbf{b}}|}$ and $M = (1+2\hat{\mathbf{a}}\cdot\hat{\mathbf{b}})^{1/2}$

8. Let $I = \int \frac{e^x}{e^{4x}+e^{2x}+1}\,dx$, $J = \int \frac{e^{-x}}{e^{-4x}+e^{-2x}+1}\,dx$.

Then, for an arbitrary constant C, the value of $J - I$ equals

(a) $\frac{1}{2}\log\left|\frac{e^{4x}-e^{2x}+1}{e^{4x}+e^{2x}+1}\right|+C$ (b) $\frac{1}{2}\log\left|\frac{e^{2x}+e^{x}+1}{e^{2x}-e^{x}+1}\right|+C$

(c) $\frac{1}{2}\log\left|\frac{e^{2x}-e^{x}+1}{e^{2x}+e^{x}+1}\right|+C$ (d) $\frac{1}{2}\log\left|\frac{e^{4x}+e^{2x}+1}{e^{4x}-e^{2x}+1}\right|+C$

9. Let $g(x) = \log f(x)$, where $f(x)$ is a twice differentiable positive function on $(0, \infty)$ such that $f(x+1) = xf(x)$. Then, for $N = 1, 2, 3, \ldots g''\left(N+\frac{1}{2}\right) - g''\left(\frac{1}{2}\right)$ is equal to

(a) $-4\left\{1+\frac{1}{9}+\frac{1}{25}+\ldots+\frac{1}{(2N-1)^2}\right\}$ (b) $4\left\{1+\frac{1}{9}+\frac{1}{25}+\ldots+\frac{1}{(2N-1)^2}\right\}$

(c) $-4\left\{1+\frac{1}{9}+\frac{1}{25}+\ldots+\frac{1}{(2N+1)^2}\right\}$ (d) $4\left\{1+\frac{1}{9}+\frac{1}{25}+\ldots+\frac{1}{(2N+1)^2}\right\}$

Section III

Assertion-Reason Type

This section contains 4 reasoning type questions. Each question has 4 choices (a), (b), (c) and (d) out of which ONLY ONE one is correct.

(a) Statement 1 is true, Statement 2 is true, Statement 2 is a correct explanation for Statement 1

(b) Statement 1 is true , Statement 2 is true; Statement 2 is not a correct explanation for Statement 1

(c) Statement 1 is true , Statement 2 is false

(d) Statement 1 is false, Statement 2 is true

10. Suppose four distinct positive numbers a_1, a_2, a_3 and a_4 are in GP. Let $b_1 = a_1$, $b_2 = b_1 + a_2$, $b_3 = b_2 + a_3$ and $b_4 = b_3 + a_4$.

Statement 1 The numbers b_1, b_2, b_3 and b_4 are neither in AP nor in GP.

Statement 2 The numbers b_1, b_2, b_3 and b_4 are in HP.

11. Let a, b, c, p, q be real numbers.

Suppose, α, β are the roots of the equation $x^2 + 2px + q = 0$ and $\alpha, \frac{1}{\beta}$ are the roots of the equation $ax^2 + 2bx + c = 0$, where $\beta^2 \notin \{-1, 0, 1\}$.

Statement 1 $(p^2 - q)(b^2 - ac) \geq 0$.

Statement 2 $b \neq pa$ or $c \neq qa$.

12. Consider $L_1: 2x + 3y + p - 3 = 0$; $L_2: 2x + 3y + p + 3 = 0$, where p is a real number and $C: x^2 + y^2 + 6x - 10y + 30 = 0$.

Statement 1 If line L_1 is a chord of circle C, then line L_2 is not always a diameter of circle C.

Statement 2 If line L_1 is a diameter of circle C, then line L_2 is not a chord of circle C.

13. Let a solution $y = y(x)$ of the differential equation $x\sqrt{x^2 - 1}\, dy - y\sqrt{y^2 - 1}\, dx = 0$ satisfy $y(2) = \frac{2}{\sqrt{3}}$.

Statement 1 $y(x) = \sec\left(\sec^{-1} x - \frac{\pi}{6}\right)$

Statement 2 $y(x)$ is given by $\frac{1}{y} = \frac{2\sqrt{3}}{x} - \sqrt{1 - \frac{1}{x^2}}$

Section III

Linked Comprehension Type

This section contains 2 paragraphs. Based upon each paragraph, 3 multiple choice questions have to be answered. Each question has 4 choices (a), (b), (c) and (d), out of which ONLY ONE is correct.

Paragraph for Question Nos. 14-16

Consider the function $f: (-\infty, \infty) \to (-\infty, \infty)$ *defined by* $f(x) = \frac{x^2 - ax + 1}{x^2 + ax + 1}; 0 < a < 2$.

14. Which of the following is true ?

(a) $(2 + a)^2 f''(1) + (2 - a)^2 f''(-1) = 0$
(b) $(2 - a)^2 f''(1) - (2 - a)^2 f''(-1) = 0$
(c) $f'(1) f'(-1) = (2 - a)^2$
(d) $f'(1) f'(-1) = -(2 + a)^2$

15. Which of the following is true?

(a) $f(x)$ is decreasing on $(-1, 1)$ and has a local minimum at $x = 1$
(b) $f(x)$ is increasing on $(-1, 1)$ and has a local maximum at $x = 1$
(c) $f(x)$ is increasing on $(-1, 1)$ but has neither a local maximum nor a local minimum at $x = 1$
(d) $f(x)$ is decreasing on $(-1, 1)$ but has neither a local maximum nor a local minimum at $x = 1$

16. Let $g(x) = \int_0^{e^x} \frac{f'(t)}{1+t^2}\, dt$. Which of the following is true ?

(a) $g'(x)$ is positive on $(-\infty, 0)$ and negative on $(0, \infty)$
(b) $g'(x)$ is negative on $(-\infty, 0)$ and positive on $(0, \infty)$
(c) $g'(x)$ changes sign on both $(-\infty, 0)$ and $(0, \infty)$
(d) $g'(x)$ does not change sign on $(-\infty, \infty)$

Paragraph for Question Nos. 17-19

Consider the lines: $L_1: \frac{x+1}{3} = \frac{y+2}{1} = \frac{z+1}{2}$, $L_2: \frac{x-2}{1} = \frac{y+2}{2} = \frac{z-3}{3}$

17. The unit vector perpendicular to both L_1 and L_2 is

(a) $\frac{-\hat{\mathbf{i}}+7\hat{\mathbf{j}}+7\hat{\mathbf{k}}}{\sqrt{99}}$ (b) $\frac{-\hat{\mathbf{i}}-7\hat{\mathbf{j}}+5\hat{\mathbf{k}}}{5\sqrt{3}}$ (c) $\frac{-\hat{\mathbf{i}}+7\hat{\mathbf{j}}+5\hat{\mathbf{k}}}{5\sqrt{3}}$ (d) $\frac{-7\hat{\mathbf{i}}-7\hat{\mathbf{j}}-\hat{\mathbf{k}}}{\sqrt{99}}$

18. The shortest distance between L_1 and L_2 is

(a) 0 (b) $17/\sqrt{3}$ (c) $41/5\sqrt{3}$ (d) $17/5\sqrt{3}$

19. The distance of the point (1, 1, 1) from the plane passing through the point (−1, − 2, − 1) and whose normal is perpendicular to both the lines L_1 and L_2, is

(a) $2/\sqrt{75}$ (b) $7/\sqrt{75}$ (c) $13/\sqrt{75}$ (d) $23/\sqrt{75}$

Section IV

Matrix- Match Type

This section contains 3 questions. Each questions contains Statements given in two columns, which have to be matched. Statements in **Column I** are labelled as A, B, C and D, whereas Statements in **Column II** are labelled as p, q, r and s. The answers to these questions have to be appropriately bubbled as illustrated in the following example.

If the correct matches are A−q, A−r, B−p, B−s, C−r, C−s and D−q, then the correctly bubbled matrix will look like the following:

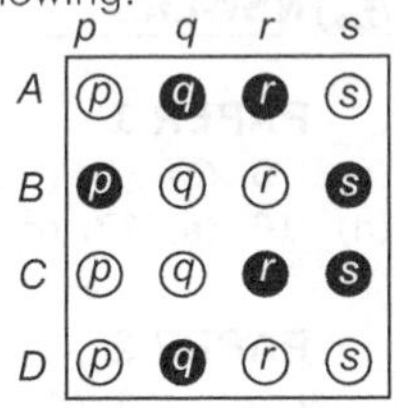

20. Consider the lines given by

$L_1: x + 3y - 5 = 0$, $L_2: 3x - ky - 1 = 0$, $L_3: 5x + 2y - 12 = 0$

Match the Statements/Expressions in Column I with the Statements/Expressions in Column II.

	Column I		Column II
(A)	L_1, L_2, L_3 are concurrent , if	(p)	$k = -9$
(B)	One of L_1, L_2, L_3 is parallel to atleast one of the other two, if	(q)	$k = -\frac{6}{5}$
(C)	L_1, L_2, L_3 form a triangle, if	(r)	$k = 5/6$
(D)	L_1, L_2, L_3 do not form a triangle, if	(s)	$k = 5$

21. Match the Statements/Expressions in Column I with the Statements/Expressions in Column II.

	Column I		Column II
(A)	The minimum value of $\frac{x^2+2x+4}{x+2}$ is	(p)	0
(B)	Let A and B be 3×3 matrices of real numbers, where A is symmetric, B is skew-symmetric and $(A+B)(A-B) = (A-B)(A+B)$. If $(AB)^t = (-1)^k AB$, where $(AB)^t$ is the transpose of the matrix AB, then possible values of k are	(q)	1
(C)	Let $a = \log_3 \log_3 2$. An integer k satisfying $1 < 2^{(-k+3^{-a})} < 2$, must be less than	(r)	2
(D)	If $\sin\theta = \cos\phi$, then the possible values of $\frac{1}{\pi}\left(\theta \pm \phi - \frac{\pi}{2}\right)$ are	(s)	3

22. Consider all possible permutations of the letters of the word ENDEANOEL. Match the Statements/Expressions in Column I with the Statements /Expressions in Column II.

	Column I		Column II
(A)	The number of permutations containing the word ENDEA, is	(p)	5!
(B)	The number of permutations in which the letter E occurs in the first and the last positions, is	(q)	$2 \times 5!$
(C)	The number of permutations in which none of the letters D, I, N occurs in the last five positions, is	(r)	$7 \times 5!$
(D)	The number of permutations in which the letters A, E, O occur only in odd positions, is	(s)	$21 \times 5!$

ANSWERS

PAPER 1

1. (b) **2.** (c) **3.** (a) **4.** (b) **5.** (c) **6.** (c) **7.** (b,d) **8.** (b,c) **9.** (a,d) **10.** (a,b,c,d)
11. (b) **12.** (d) **13.** (a) **14.** (b) **15.** (d) **16.** (a) **17.** (d) **18.** (b) **19.** (a) **20.** (d)
21. (b) **22.** (c) **23.** (d)

PAPER 2

1. (d) **2.** (c) **3.** (b) **4.** (b) **5.** (d) **6.** (d) **7.** (a) **8.** (c) **9.** (a) **10.** (c)
11. (b) **12.** (c) **13.** (c) **14.** (a) **15.** (a) **16.** (b) **17.** (b) **18.** (d) **19.** (c)

20. A $\to$ (s); B $\to$ (p,q); C $\to$ (r); D $\to$ (p,q,s)

21. A $\to$ (r); B $\to$ (q,s); C $\to$ (r); D $\to$ (p,r)

22. A $\to$ (p); B $\to$ (s); C $\to$ (q); D $\to$ (q)

IIT JEE
UNSOLVED PAPER 2009

Physics

Paper I

SECTION I

Single Correct Choice Type

This section contains 8 multiple choice questions. Each question has 4 choices (a), (b), (c) and (d) for its answer, out of which ONLY ONE is correct.

1. Look at the drawing given in the figure, which has been drawn with ink of uniform line-thickness.

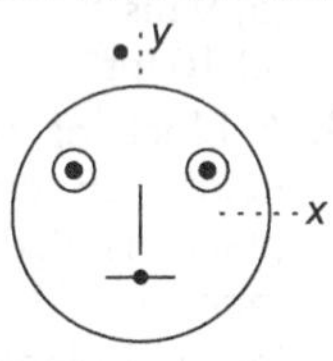

The mass of ink used to draw each of the two inner circles, and each of the two line segments is m. The mass of the ink used to draw the outer circle is $6\,m$. The coordinates of the centres of the different parts are, outer circle (0, 0), left inner circle (0, 0), right inner circle (a, a), vertical line (0, 0) and horizontal line $(0, -a)$. The y-coordinate of the centre of mass of the ink in this drawing is

(a) $\frac{a}{10}$ (b) $\frac{a}{8}$ (c) $\frac{a}{12}$ (d) $\frac{a}{3}$

2. The figure shows certain wire segments joined together to form a coplanar loop. The loop is placed in a perpendicular magnetic field in the direction going into the plane of the figure. The magnitude of the field increases with time. I_1 and I_2 are the currents in the segments ab and cd. Then,

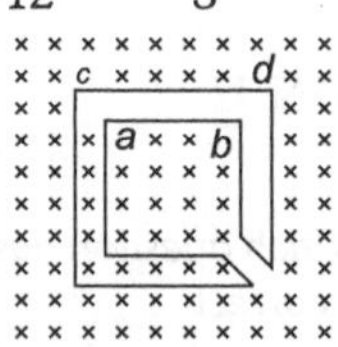

(a) $I_1 > I_2$ (b) $I_1 < I_2$

(c) I_1 is in the direction ba and I_2 is in the direction cd

(d) I_1 is in the direction ab and I_2 is in the direction dc

3. Two small particles of equal masses start moving in opposite directions from a point A in a horizontal circular orbit. Their tangential velocities are v and $2v$ respectively, as shown in the figure. Between collisions, the particles move with constant speeds. After making how many elastic collisions, other than that at A, these two particles will again reach the point A?

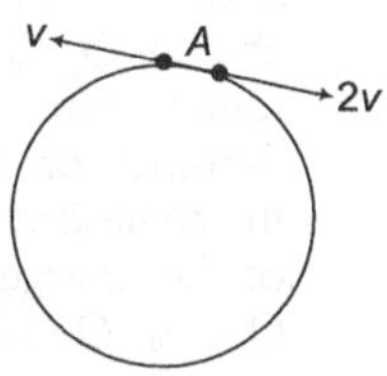

(a) 4 (b) 3 (c) 2 (d) 1

4. A disk of radius $\frac{a}{4}$ having a uniformly distributed charge 6 C is placed in the $x-y$ plane with its centre at $\left(\frac{-a}{2}, 0, 0\right)$. A rod of length a carrying a uniformly distributed charge 8 C is placed on the x-axis from $x = \frac{a}{4}$ to $x = \frac{5a}{4}$. Two point charges -7 C and 3 C are placed at $\left(\frac{a}{4}, \frac{-a}{4}, 0\right)$ and $\left(\frac{-3a}{4}, \frac{3a}{4}, 0\right)$, respectively. Consider a cubical surface formed by six surfaces $x = \pm\frac{a}{2}, y = \pm\frac{a}{2}, z = \pm\frac{a}{2}$.

The electric flux through this cubical surface is

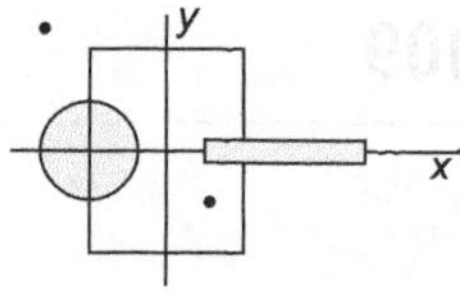

(a) $\frac{-2C}{\varepsilon_0}$ (b) $\frac{2C}{\varepsilon_0}$

(c) $\frac{10C}{\varepsilon_0}$ (d) $\frac{12C}{\varepsilon_0}$

5. Three concentric metallic spherical shells of radii R, $2R$, $3R$ are given charges Q_1, Q_2, Q_3, respectively. It is found that the surface charge densities on the outer surfaces of the shells are equal. Then, the ratio of the charges given to the shells, $Q_1 : Q_2 : Q_3$, is

(a) 1 : 2 : 3 (b) 1 : 3 : 5

(c) 1 : 4 : 9 (d) 1 : 8 : 18

6. The $x-t$ graph of a particle undergoing simple harmonic motion is shown below. The acceleration of the particle at $t = \frac{4}{3}$ s is

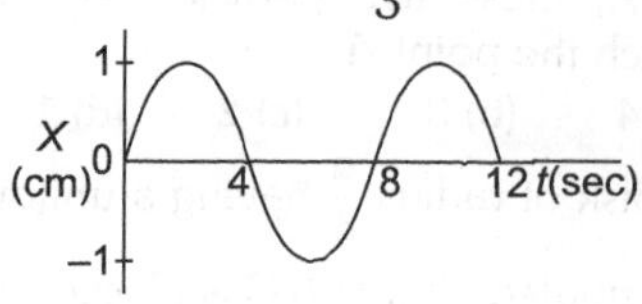

(a) $\frac{\sqrt{3}}{32}\pi^2$ cms^{-2} (b) $\frac{-\pi^2}{32}$ cms^{-2}

(c) $\frac{\pi^2}{32}$ cms^{-2} (d) $-\frac{\sqrt{3}}{32}\pi^2$ cms^{-2}

7. A ball is dropped from a height of 20 m above the surface of water in a lake. The refractive index of water is $\frac{4}{3}$. A fish inside the lake, in the line of fall of the ball, is looking at the ball. At an instant, when the ball is 12.8 m above the water surface, the fish sees the speed of ball as

(a) 9 ms^{-1} (b) 12 ms^{-1}

(c) 16 ms^{-1} (d) 21.33 ms^{-1}

8. A block of base 10 cm × 10 cm and height 15 cm is kept on an inclined plane. The coefficient of friction between them is $\sqrt{3}$. The inclination θ of this inclined plane from the horizontal plane is gradually increased from 0°. Then,

(a) at $\theta = 30°$, the block will start sliding down the plane

(b) the block will remain at rest on the plane up to certain θ and then it will topple

(c) at $\theta = 60°$, the block will start sliding down the plane and continue to do so at higher angles

(d) at $\theta = 60°$, the block will start sliding down the plane and on further increasing θ, it will topple at certain θ

SECTION II

Multiple Correct Choice Type

This section contains 4 multiple choice questions. Each question has 4 choices (a), (b), (c) and (d) for its answer, out of which ***ONE OR MORE*** *is/are correct.*

9. For the circuit shown in the figure

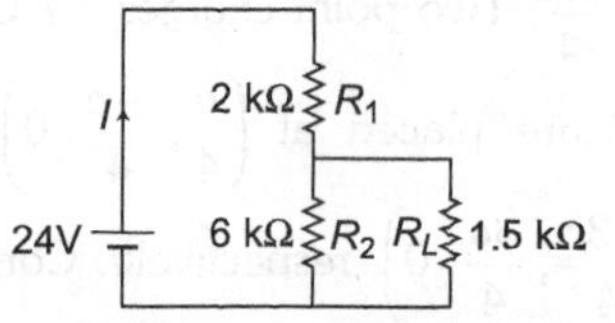

(a) the current I through the battery is 7.5 mA

(b) the potential difference across R_L is 18 V

(c) ratio of powers dissipated in R_1 and R_2 is 3

(d) if R_1 and R_2 are interchanged, magnitude of the power dissipated in R_L will decrease by a factor of 9.

10. C_V and C_p denote the molar specific heat capacities of a gas at constant volume and constant pressure, respectively. Then

(a) $C_p - C_V$ is larger for a diatomic ideal gas than for a monoatomic ideal gas

(b) $C_p + C_V$ is larger for a diatomic ideal gas than for a monoatomic ideal gas.

(c) $\frac{C_p}{C_V}$ is larger for a diatomic ideal gas than for a monoatomic ideal gas

(d) $C_p \cdot C_V$ is larger for a diatomic ideal gas than for a monoatomic ideal gas

11. A student performed the experiment of determination of focal length of a concave mirror by $u - v$ method using an optical bench of length 1.5 m. The focal length of the mirror used is 24 cm. The maximum error in the location of the image can be 0.2 cm. The 5 sets of (u, v) values recorded by the student (in cm) are : (42, 56), (48, 48), (60, 40), (66, 33), (78, 39). The data set(s) that cannot come from experiment and is (are) incorrectly recorded, is(are)

(a) (42, 56) (b) (48, 48)
(c) (66, 33) (d) (78, 39)

12. If the resultant of all the external forces acting on a system of particles is zero, then from an inertial frame, one can surely say that

(a) linear momentum of the system does not change in time

(b) kinetic energy of the system does not change in time

(c) angular momentum of the system does not change in time

(d) potential energy of the system does not change in time

SECTION III

Comprehension Type

*This section contains 2 groups of questions. Each group has 3 multiple choice questions based on a paragraph. Each question has 4 choices (a), (b), (c) and (d) for its answer, out of which **ONLY ONE** is correct.*

Paragraph for Question Nos. 13 to 15

When a particle is restricted to move along x-axis between $x = 0$ and $x = a$, where a is of nanometer dimension, its energy can take only certain specific values. The allowed energies of the particle moving in such a restricted region, correspond to the formation of standing waves with nodes at its ends $x = 0$ and $x = a$. The wavelength of this standing wave is related to the liner momentum p of the particle according to the de Broglie relation. The energy of the particle of mass m is related to its linear momentum as $E = \frac{p^2}{2m}$. Thus, the energy of the particle can be denoted by a quantum number n taking values 1, 2, 3, ... ($n = 1$, called the ground state) corresponding to the number of loops in the standing wave.

Use the model described above to answer the following three questions for a particle moving in the line $x = 0$ to $x = a$. [Take $h = 6.6 \times 10^{-34}$ Js and $e = 1.6 \times 10^{-19}$ C]

13. The allowed energy for the particle for a particular value of n is proportional to

(a) a^{-2} (b) $a^{-3/2}$
(c) a^{-1} (d) a^{2}

14. If the mass of the particle is $m = 1.0 \times 10^{-30}$ kg and $a = 6.6$ nm, the energy of the particle in its ground state is closest to

(a) 0.8 meV (b) 8 meV
(c) 80 meV (d) 800 meV

15. The speed of the particle that can take discrete values is proportional to

(a) $n^{-3/2}$ (b) n^{-1} (c) $n^{1/2}$ (d) n

Paragraph for Question Nos. 16 to 18

Scientists are working hard to develop nuclear fusion reactor. Nuclei of heavy hydrogen, ${}_1^2\text{H}$ known as deuteron and denoted by D can be thought of as a candidate for fusion reactor. The D-D reaction is ${}_1^2\text{H}+{}_1^2\text{H}\rightarrow{}_2^3\text{He}+n+$ energy. In the core of fusion reactor. A gas of heavy hydrogen is fully ionized into deuteron nuclei and electrons. This collection of ${}_1^4\text{H}$ nuclei and electrons is known as plasma. The nuclei move randomly in the reactor core and occasionally come close enough for nuclear fusion to take place. Usually, the temperatures in the reactor core are too high and no material wall can be used to confine the plasma. Special techniques are used which confine the plasma for a time t_0 before the particles fly away from the core. If n is the density (number/volume) of deutrons, the product nt_0 is called Lawson number. In one of the criteria, a reactor is termed successful if Lawson number is greater than 5×10^{14} scm^{-3}.

It may be helpful to use the following : Boltzmann constant $k = 8.6\times10^{-5}$ eV/K; $\frac{e^2}{4\pi\varepsilon_0} = 1.44\times10^{-9}$ eVm.

16. In the core of nuclear fusion reactor, the gas becomes plasma because of

(a) strong nuclear force acting between the deuterons

(b) Coulomb fore acting between the deuterons

(c) Coulomb force acting between deuteron-electron pairs

(d) the high temperature maintained inside the reactor core

17. Assume that two deuteron nuclei in the core of fusion reactor at temperature T are moving towards each other, each with kinetic energy $1.5\,kT$, when the separation between them is large enough to neglect Coulomb potential energy. Also neglect any interaction from other particles in the core. The minimum temperature T required for them to reach a separation of 4×10^{-15} m is in the range

(a) $1.0\times10^9\text{ K} < T < 2.0\times10^9\text{ K}$

(b) $2.0\times10^9\text{ K} < T < 3.0\times10^9\text{ K}$

(c) $3.0\times10^9\text{ K} < T < 4.0\times10^9\text{ K}$

(d) $4.0\times10^9\text{ K} < T < 5.0\times10^9\text{ K}$

18. Results of calculations for four different designs of a fusion reactor using D-D reaction are given below. Which of these is most promising based on Lawson criterion?

(a) Deuteron density $= 2.0\times10^{12}$ cm^{-3}, confinement time $= 5.0\times10^{-3}$ s

(b) Deuteron density$= 8.0\times10^{14}$ cm^{-3}, confinement time $= 9.0\times10^{-1}$ s

(c) Deuteron density $= 4.0\times10^{23}$ cm^{-3}, confinement time $= 1.0\times10^{-11}$ s

(d) Deuteron density $= 1.0\times10^{24}$ cm^{-3}, confinement time $= 4.0\times10^{-12}$ s

SECTION IV

Matrix-Match Type

*This section contains 2 questions. Each question contains Statements given in two Columns which have to be matched. The Statements in Column I are labelled (A), (B), (C) and (D), while the Statements in Column II are labelled p, q, r, s and t. Any given Statement in Column I can have correct matching with **ONE OR MORE** statement(s) in Column II. The appropriate bubbles corresponding to the answers to these questions have to be darkened as illustrated in the following example.*

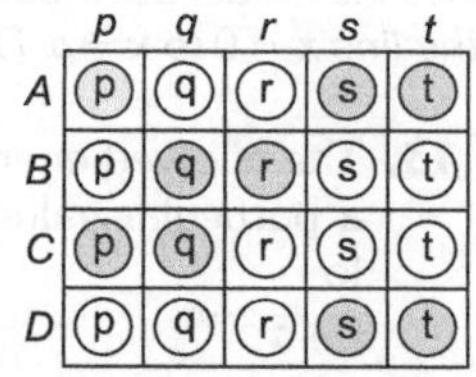

If the correct matches are A-p, s and t; B-q and r; C-p and q; and D-s and t; then the correct darkening of bubbles will look like the following.

19. Column II shows five systems in which two objects are labelled as X and Y. Also in each case a point P is shown. Column-I gives some statements about X and/or Y. Match these statements to the appropriate system(s) from Column-II.

Column I		Column II	
A. The force exerted by X on Y has a magnitude Mg.	(p)	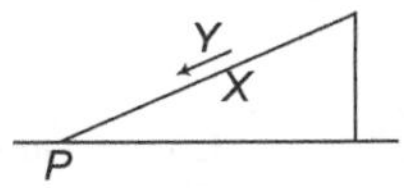	Block Y of mass M left on a fixed inclined plane X, slides on it with a constant velocity.
B. The gravitational potential energy of X is continuously increasing.	(q)	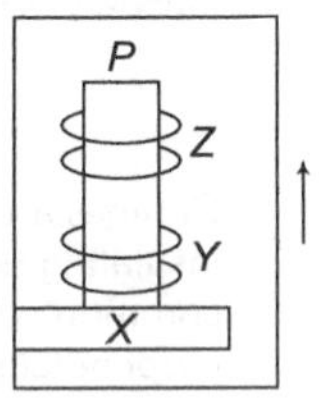	Two ring magnets Y and Z each of mass M, are kept in frictionless vertical plastic stand so that they repel each other. Y rests on the base X and Z hangs in air in equilibrium. P is the topmost point of the stand on the common axis of the two rings. The whole system is in a lift that is going up with a constant velocity.
C. Mechanical energy of the system $X + Y$ is continuously decreasing.	(r)	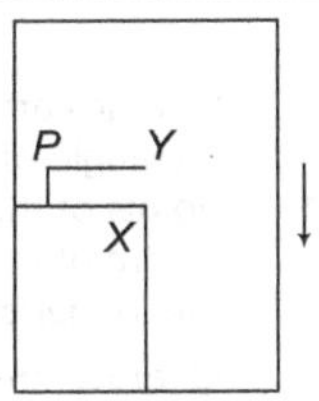	A pulley Y of mass m_0 is fixed to a table through a clamp X. A block of mass M hangs from a string that goes over the pulley and is fixed at point P of the table. The whole system is kept in a lift that is going down with a constant velocity.
D. The torque of the weight of Y about point P is zero.	(s)	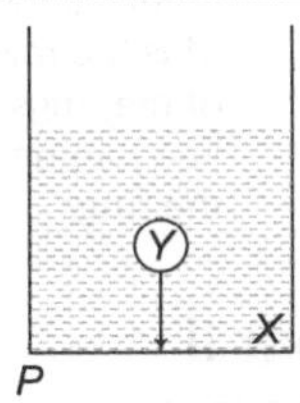	A sphere Y of mass M is put in a non-viscous liquid X kept in a container at rest. The sphere is released and it moves down in the liquid.
	(t)	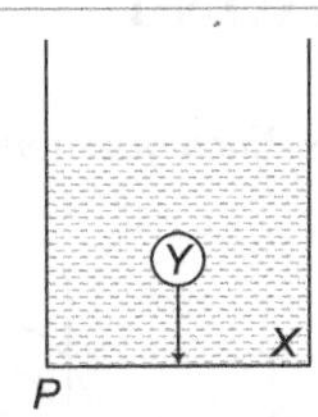	A sphere Y of mass M is falling with its terminal velocity in a viscous liquid X kept in a container.

20. Six point charges, each of the same magnitude q, are arranged in different manners as shown in Column-II. In each case, a point M and a line PQ passing through M are shown. Let E be the electric field and V be the electric potential at M (potential at infinity is zero) due to the given charge distribution when it is at rest. Now, the whole system is set into rotation with a constant angular velocity about the line PQ. Let B the magnetic field at M and μ be the magnetic moment of the system in this condition. Assume each rotating charge to be equivalent to a steady current.

Column I		Column II
A. $E = 0$	(p) 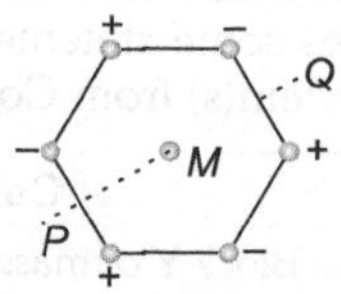	Charges are at the corners of a regular hexagon. M is at the centre of the hexagon. PQ is perpendicular to the plane of the hexagon.
B. $V \neq 0$	(q) 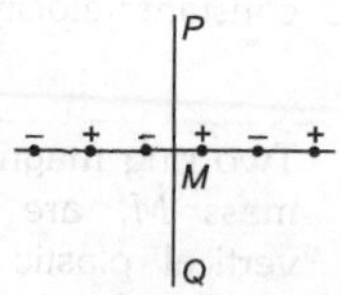	Charges are on a line perpendicular to PQ at equal intervals. M is the mid-point between the two innermost charges.
C. $B = 0$	(r) 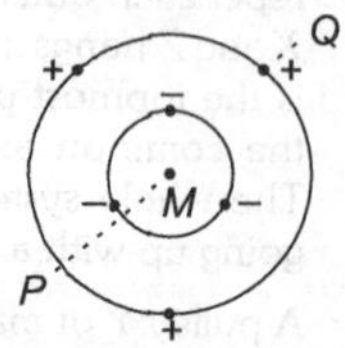	Charges are placed on two coplanar insulating rings at equal intervals. M is the common centre of the rings. PQ is perpendicular to the plane of the rings.
D. $\mu \neq 0$	(s) 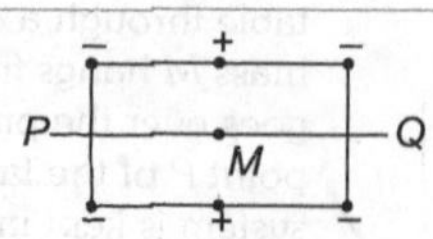	Charges are placed at the corners of a rectangle of sides a and $2a$ and at the mid points of the longer sides. M is at the centre of the rectangle. PQ is parallel to the longer sides.
	(t) 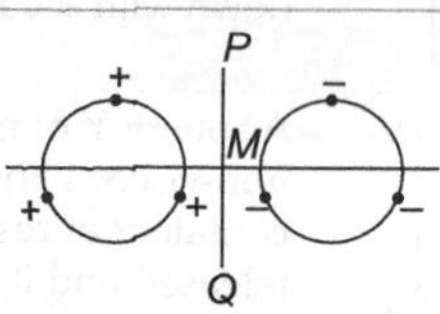	Charges are placed on two coplanar, identical insulting rings at equal intervals. M is the mid points between the centres of the rings. PQ is perpendicular to the line joining the centres and coplanar to the rings.

Paper II

SECTION I

Single Correct Choice Type

This section contains 4 multiple choice questions. Each question has 4 choices (a), (b), (c) and (d) for its answer, out which ***ONLY ONE*** *is correct.*

1. Photoelectric effect experiments are performed using three different metal plates p, q and r having work functions $\varphi_p = 2.0$ eV, $\varphi_q = 2.5$ eV and $\varphi_r = 3.0$ eV, respectively. A light beam containing wavelengths of 550 nm, 450 nm and 350 nm with equal intensities illuminates each of the plates. The correct $I - V$ graph for the experiment is

(a) 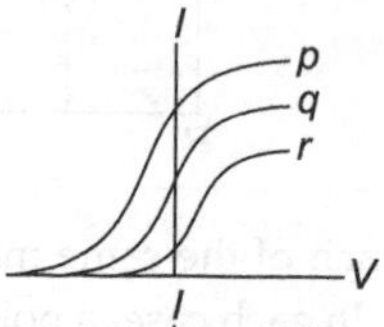

(b) 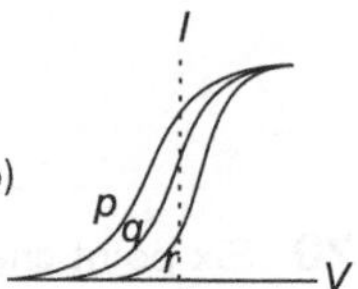

(c) 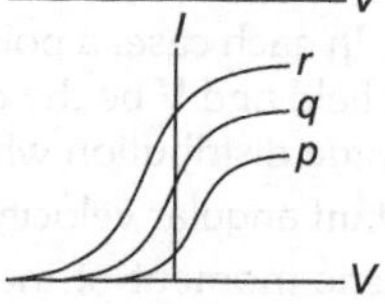

(d) 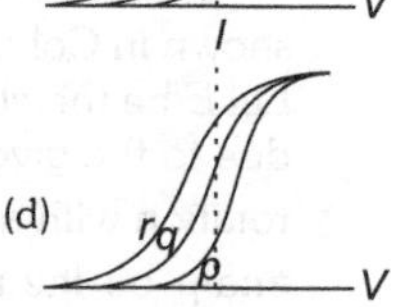

2. A uniform rod of length L and mass M is pivoted at the centre. Its two ends are attached to two springs of equal spring constants k. The springs are fixed to rigid supports as shown in the figure, and rod is free to oscillate in the horizontal plane. The rod is gently pushed through a small angle θ in one direction and released. The frequency of oscillation is

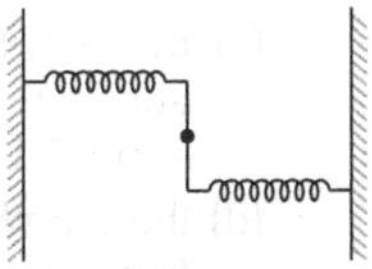

(a) $\frac{1}{2\pi}\sqrt{\frac{2k}{M}}$ (b) $\frac{1}{2\pi}\sqrt{\frac{k}{M}}$

(c) $\frac{1}{2\pi}\sqrt{\frac{6k}{M}}$ (d) $\frac{1}{2\pi}\sqrt{\frac{24k}{M}}$

3. A piece of wire is bent in the shape of a parabola $y = kx^2$ (y-axis vertical) with a bead of mass m on it. The bead can slide on the wire without friction. It stays at the lowest point of the parabola when the wire is at rest. The wire is now accelerated parallel to the x-axis with the constant acceleration a. The distance of the new equilibrium position of the bead, where the bead can stays at rest with respect to the wire, from the y-axis is

(a) $\frac{a}{gk}$ (b) $\frac{a}{2gk}$ (c) $\frac{2a}{gk}$ (d) $\frac{a}{4gk}$

4. The mass M shown in the figure oscillates in simple harmonic motion with amplitude A. The amplitude of the point P is

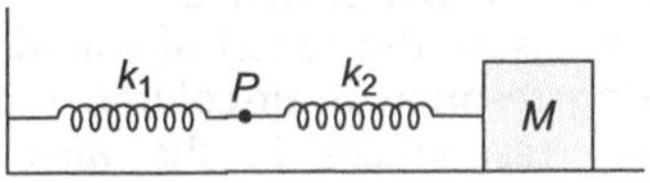

(a) $\frac{k_a A}{k_2}$ (b) $\frac{k_2 A}{k_1}$

(c) $\frac{k_1 A}{k_1 + k_2}$ (d) $\frac{k_2 A}{k_1 + k_2}$

SECTION II

Multiple Correct Choice Type

This section contains 5 multiple choice questions. Each question has 4 choices (a), (b), (c) and (d) for its answer, out of which ***ONE OR MORE*** *is/are correct.*

5. Under the influence of the coulomb field of charge $+Q$, a charge $-q$ is moving around it in an elliptical orbit. Find out the correct statement(s).

(a) The angular momentum of the charge $-q$ is constant

(b) The linear momentum of the charge $-q$ is constant

(c) The angular velocity of the charge $-q$ is constant

(d) The linear speed of the charge $-q$ is constant

6. The figure shows the $p-V$ plot of an ideal gas taken through a cycle $ABCDA$. The part ABC is a semi-circle and CDA is half of an ellipse. Then,

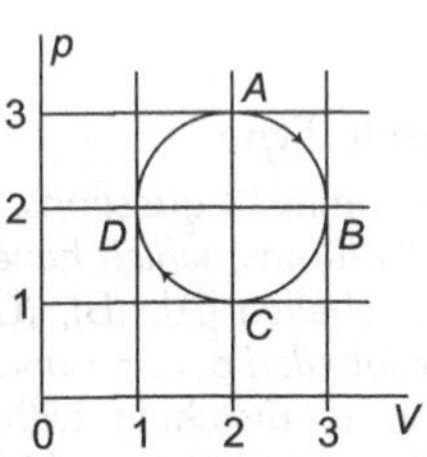

(a) the process during the path $A \to B$ is isothermal

(b) heat flows out of the gas during the path $B \to C \to D$

(c) work done during the path $A \to B \to C$ is zero

(d) positive work is done by the gas in the cycle $ABCDA$.

7. A sphere is rolling without slipping on a fixed horizontal plane surface. In the figure, A is the point of contact. B is the centre of the sphere and C is its topmost point. Then,

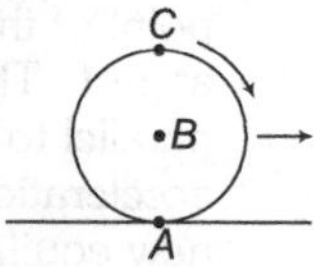

(a) $\vec{\mathbf{v}}_C - \vec{\mathbf{v}}_A = 2(\vec{\mathbf{v}}_B - \vec{\mathbf{v}}_C)$

(b) $\vec{\mathbf{v}}_C - \vec{\mathbf{v}}_B = \vec{\mathbf{v}}_B - \vec{\mathbf{v}}_A$

(c) $|\vec{\mathbf{v}}_C - \vec{\mathbf{v}}_A| = 2|\vec{\mathbf{v}}_B - \vec{\mathbf{v}}_C|$

(d) $|\vec{\mathbf{v}}_C - \vec{\mathbf{v}}_A| = 4|\vec{\mathbf{v}}_B|$

8. A student performed the experiment to measure the speed of sound in air using resonance air-column method. Two resonances in the air-column were obtained by lowering the water level. The resonance with the shorter air-column is the first resonance and that with the longer air column is the second resonance. Then,

(a) the intensity of the sound heard at the first resonance was more than that at the second resonance

(b) the prongs of the tuning fork were kept in a horizontal place above the resonance tube

(c) the amplitude of vibration of the ends of the prongs is typically around 1 cm

(d) the length of the air-column at the first resonance was somewhat shorter than 1/4th of the wavelength of the sound in air

9. Two metallic rings A and B, identical in shape and size but having different resistivities ρ_A and ρ_B, are kept on top of two identical solenoids as shown in the figure. When current I is switched on in both the solenoids in identical manner, the rings A and B jump to heights h_A and h_B, respectively, with $h_A > h_B$. The possible relation(s) between their resistivities and their masses m_A and m_B is(are)

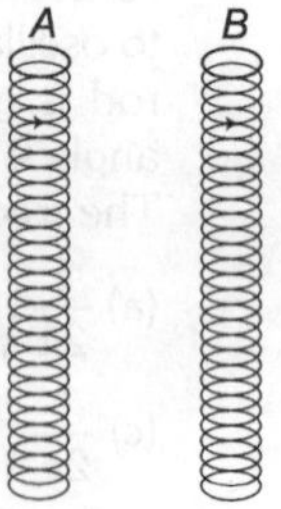

(a) $\rho_A > \rho_B$ and $m_A = m_B$

(b) $\rho_A < \rho_B$ and $m_A = m_B$

(c) $\rho_A > \rho_B$ and $m_A > m_B$

(d) $\rho_A < \rho_B$ and $m_A < m_B$

SECTION III

Matrix-Match Type

This section contains 2 questions. Each question contains statements given in two Columns, which have to be matched. The statements in Column I are labelled (A), (B), (C) and (D), while the statements in Column II are labelled p, q, r, s and t. Any given statement in Column I can have correct matching with ***ONE OR MORE*** *statement(s) in Column II. The appropriate bubbles corresponding to the answer to these questions have to be darkened as illustrated in the following example. If the correct matches are A-p, s and t; B-q and r; C-p and q; and D-s and t; then the correct darkening of bubbles will took like the following*

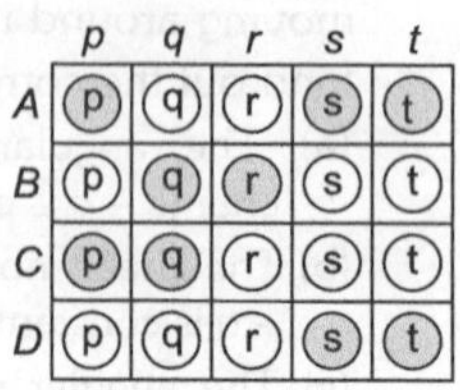

10. Column I shows four situations of standard Young's double slit arrangement with the screen placed far away from the slits S_1 and S_2. In each of these cases $S_1P_0 = S_2P_0$, $S_1P_1 - S_2P_1 = \dfrac{\lambda}{4}$ and $S_1P_2 - S_2P_2 = \dfrac{\lambda}{3}$, where λ is the wavelength of the light used. In the cases B, C and D, a transparent sheet of refractive index μ and thickness t is pasted on slit S_2. The thickness of the sheets are different in different cases. The phase difference between the light waves reaching a point P on the screen from the two slits is denoted by $\delta(P)$ and the intensity by $I(P)$. Match each situation given in Column-I with the statement(s) in Column-II valid for that situation.

Column I		Column II
A.	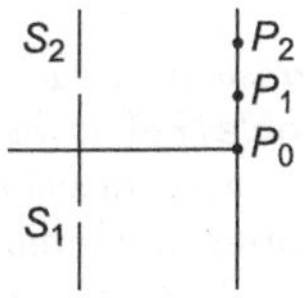	p. $\delta(P_0) = 0$
B. $(\mu - 1)t = \frac{\lambda}{4}$	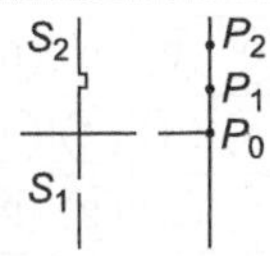	q. $\delta(P_1) = 0$
C. $(\mu - 1)t = \frac{\lambda}{2}$	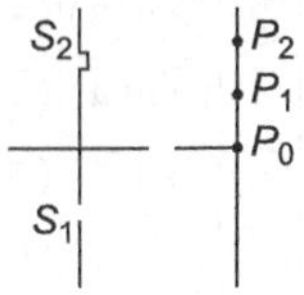	r. $I(P_1) = 0$
D. $(\mu - 1)t = \frac{3\lambda}{4}$	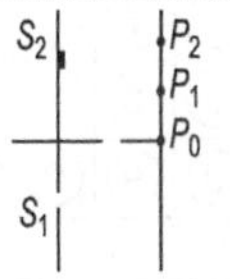	s. $I(P_0) > I(P_1)$
		t. $I(P_2) > I(P_1)$

11. Column-II gives certain systems undergoing a process. Column-I suggests changes in some of the parameters related to the system. Match the statements in Column-I to the appropriate process (es) from Column-II.

Column I	Column II
A. The energy of the system is increased	p. System : A capacitor, initially uncharged Process : It is connected to a battery
B. Mechanical energy is provided to the system, which is converted into energy of random motion of its parts	q. System : A gas in an adiabatic container fitted with an adiabatic piston
	Process : The gas is compressed by pushing the piston
C. Internal energy of the system is converted in to its mechanical energy	r. System : A gas in a rigid container
	Process : The gas gets cooled due to colder atmosphere surrounding it
D. Mass of the system is decreased	s. System : A heavy nucleus, initially at rest
	Process : The nucleus fissions into two fragments of nearly equal masses and some neutrons are emitted
	t. System : A resistive wire loop
	Process : The loop is placed in a time varying magnetic field perpendicular to its plane

SECTION IV

Integer Answer Type

This section contains 8 questions. The answer to each of the questions is a single-digit integer, ranging from 0 to 9. The appropriate bubbles below the respective question numbers in the ORS have to be darkened. For example, if the correct answer to question numbers X, Y, Z and W (say) are 6, 0, 9 and 2, respectively, then the correct darkening of bubbles will look like the following

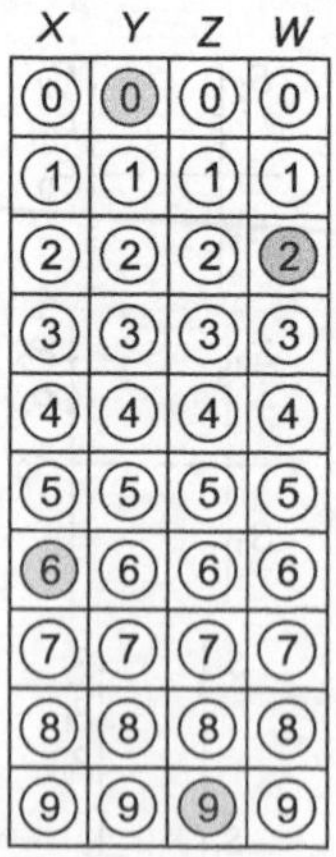

12. A steady current I goes through a wire loop PQR having shape of a right angle triangle with $PQ = 3x$, $PR = 4x$ and $QR = 5x$. If the magnitude of the magnetic field at P due to this loop is $k\left(\frac{\mu_0 I}{48\pi x}\right)$, find the value of k.

13. A light inextensible string that goes over a smooth fixed pulley as shown in the figure connects two blocks of masses 0.36 kg and 0.72 kg. Taking $g = 10\ \text{ms}^{-2}$, find the work done (in Joule) by string on the block of mass 0.36 kg during the first second after the system is released from rest.

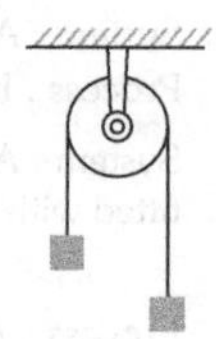

14. A solid sphere of radius R has a charge Q distributed in its volume with a charge density $\rho = kr^a$, where k and a are constants and r is the distance from its centre. If the electric field at $r = \frac{R}{2}$ is $\frac{1}{8}$ times that at $r = R$, find the value of a.

15. A metal rod AB of length $10x$ has its one end A in ice at 0°C and the other end B in water at 100°C. If a point P on the rod is maintained at 400°C, then it is found that equal amounts of water and ice evaporate and melt per unit time. The latent heat of evaporation of water is 540 calg^{-1} and latent heat of melting of ice is 80 calg^{-1}. If the point P is at a distance of λx from the ice end A, find the value of λ. [Neglect any heat loss to the surrounding.]

16. Two soap bubbles A and B are kept in a closed chamber where the air is maintained at pressure 8 Nm^{-2}. The radii of bubbles A and B are 2 cm, respectively. Surface tension of the soap-water used to make bubbles is 0.04 Nm^{-1}. Find the ratio $\frac{n_B}{n_A}$, where n_A and n_B are the number of moles of air in bubbles A and B, respectively. [Neglect the effect of gravity]

17. A 20 cm long string, having a mass of 1.0 g, is fixed at both the ends. The tension in the string is 0.5 N. The string is set into vibration using an external vibrator of frequency 100 Hz. Find the separation (in cm) between the successive nodes on the string.

18. Three objects A, B and C are kept in a straight line on a frictionless horizontal surface. These have masses m, $2m$ and m, respectively. The object A moves towards B with a speed 9 ms^{-1} and makes an elastic collision with it. Thereafter, B makes completely inelastic collision with C. All motions occur on the same straight line. Find the final speed (in ms^{-1}) of the object C.

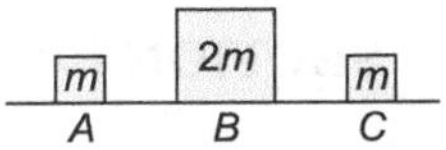

19. A cylindrical vessel of height 500 mm has an orifice (small hole) at its bottom. The orifice is initially closed and water is filled in it upto height H. Now the top is completely sealed with a cap and the orifice at the bottom is opened. Some water comes out from the orifice and the water level in the vessel becomes steady with height of water column being 200 mm. Find the fall in height (in mm) of water level due to opening of the orifice.

[Take atmospheric pressure $= 1.0 \times 10^5$ Nm^{-2}, density of water $= 1000$ kg m^{-3} and $g = 10$ ms^{-2}. Neglect any effect of surface tension.]

ANSWERS

Paper 1

1. (a) **2.** (d) **3.** (c) **4.** (a) **5.** (b) **6.** (d)
7. (c) **8.** (b) **9.** (a, d) **10.** (b, d) **11.** (c, d) **12.** (a)
13. (a) **14.** (b) **15.** (d) **16.** (d) **17.** (a) **18.** (b)
19. A→p, t B→q, s, t C→p, r, t D→r, q, t **20.** A→p, r, s B→r, s C→p, q, t D→r, s

Paper 2

1. (None) **2.** (c) **3.** (b) **4.** (d) **5.** (a) **6.** (b, d)
7. (b, c) **8.** (a, d) **9.** (b, d) **10.** A→p, s B→q C→t D→r, s, t
11. A→p, q, s, t B→q C→s D→s **12.** (7) **13.** (8) **14.** (2)
15. (9) **16.** (6) **17.** (5) **18.** (4) **19.** (6)

IIT JEE

UNSOLVED PAPER 2009

Chemistry

Paper 1

SECTION I

Single Correct Choice Type

This section contains 8 multiple choice questions numbered 1 to 8. Each question has 4 choices (a), (b), (c) and (d), out of which ***ONLY ONE*** *is correct.*

1. Among the electrolytes Na_2SO_4, $CaCl_2$ $Al_2(SO_4)_3$ and NH_4Cl, the most effective coagulating agent for Sb_2S_3 sol is

(a) Na_2SO_4 (b) $CaCl_2$

(c) $Al_2(SO_4)_3$ (d) NH_4Cl

2. The term that corrects for the attractive forces present in a real gas in the van der Waals' equation is

(a) nb (b) n^2a/V^2

(c) $-(n^2a/V^2)$ (d) $-nb$

3. The Henry's law constant for the solubility of N_2 gas in water at 298 K is 1.0×105 atm. The mole fraction of N_2 in air is 0.8. The number of moles of N_2 from air dissolved in 10 moles of water of 298 K and 5 atm pressure is

(a) 4×10^{-4} (b) 4.0×10^{-5}

(c) 5.0×10^{-4} (d) 4.0×10^{-6}

4. The reaction of P_2 with X leads selectively to P_4O_6. The X is

(a) dry O_2

(b) a mixture of O_2 and N_2

(c) moist O_2

(d) O_2 in the presence of *aq*. NaOH

5. Among cellulose, polyvinyl chloride, nylon and natural rubber, the polymer in which the intermolecular force of attraction is weakest in

(a) nylon (b) polyvinyl chloride

(c) cellulose (d) natural rubber

6. The correct acidity order of the following is

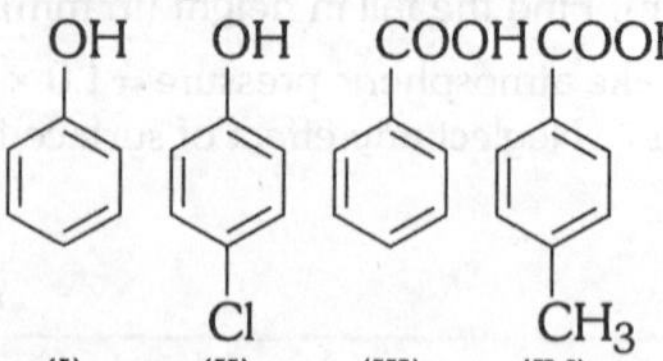

(I) (II) (III) (IV)

(a) III > IV > II > I

(b) IV > III > I > II

(c) III > II > I > IV

(d) II > III > IV > I

7. Given, that the abundances of isotopes, $_{54}Fe$, $_{56}F$ and $_{57}Fe$ are 5%, 90% and 5%, respectively, the atomic mass of Fe is

(a) 55.85 (b) 55.95

(c) 55.75 (d) 56.05

8. The IUPAC name of the following compound, is

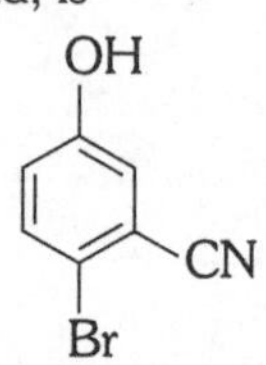

(a) 4-bromo-3-cyanohenol

(b) 2-bromo-5-hydroxybenzonitrile

(c) 2-cyano-4-hydroxybromobenzene

(d) 6-bromo-3-hydroxybenzonitrile

SECTION II

Multiple Correct Choice Type

This section contains 4 multiple correct answer(s) type questions. Each question has 4 choices (a), (b), (c) and (d), out of which of **ONE** *or* **MORE** *is/are correct.*

9. The correct statement(s) about the compound, $H_3C(HO)HC—CH=CH—CH(OH)CH_3(X)$ is (are)

(a) the total number of stereoisomers possible for X is 6

(b) the total number of diastereomers possible for X is 3

(c) if the stereochemistry about the double bond in X is *trans*, the number of enantiomers possible for X is 4

(d) if the stereochemistry about the double bond in X is *cis*, the number of enantiomers possible for X is 2

10. The compound(s) formed upon combustion of sodium metal in excess air is (are)

(a) Na_2O_2 (b) Na_2O

(c) NaO_2 (d) $NaOH$

11. The correct statement(s) regarding defects in solids is (are)

(a) Frenkel defect is usually favoured by a very small difference in the sizes of cation and anion

(b) Frenkel defect is a dislocation defect

(c) Trapping of an electron in the lattice leads to the formation of F-centre

(d) Schottky defects have no effect on the physical properties of solids

12. The compound (s) that exhibit (s) geometrical isomerism is (are)

(a) $[Pt(en)Cl_2]$ (b) $[Pt(en)_2Cl_2]$

(c) $[Pt(en)_2Cl_2]Cl_2$ (d) $[Pt(NH_3)_2Cl_2]$

SECTION III

Comprehension Type

The section contains 2 paragraphs C 13-15 and C 16-18. Based upon each paragraph, 3 multiple choice questions have to be answered. Each question has 4 choices (a), (b), (c) and (d) out of which **ONLY ONE** *is correct.*

Paragraph (for Q. Nos. 13 to 15)

A carbonyl compound *P*, which gives positive iodoform test, undergoes reaction with MeMgBr followed by dehydration to give an olefin *Q*. Ozonolysis of *Q* leads to a dicarbonyl compound *R*, which undergoes intramolecular aldol reaction to give predominantly *S*.

$$P \xrightarrow[\substack{\text{(ii) } H^+,\, H_2O \\ \text{(iii) } H_2SO_4/\text{Heat}}]{\text{(i) MeMgBr}} Q \xrightarrow{O_3/Zn\text{-}H_2O} R \xrightarrow[\text{Heat}]{OH^-} S$$

13. The structure of the carbonyl compound *P*, is

(a) 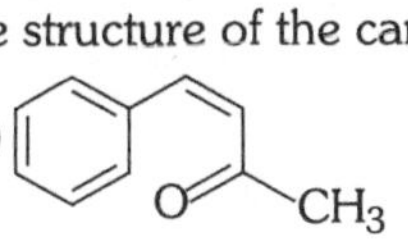

(b) 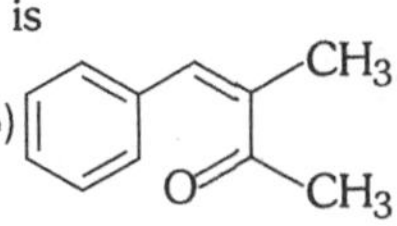

(c) 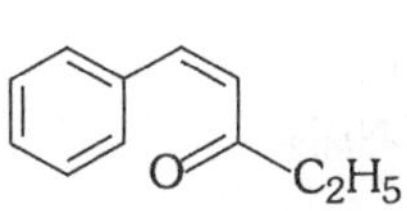

(d) 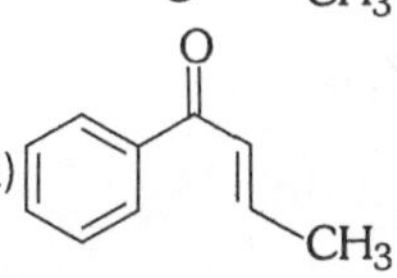

14. The structures of the products Q and R, respectively, are

(a) 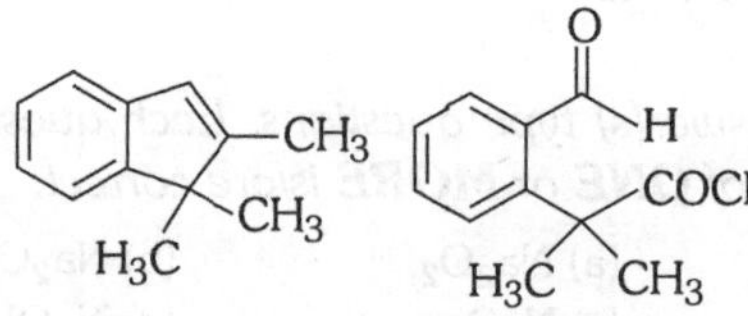

(b) 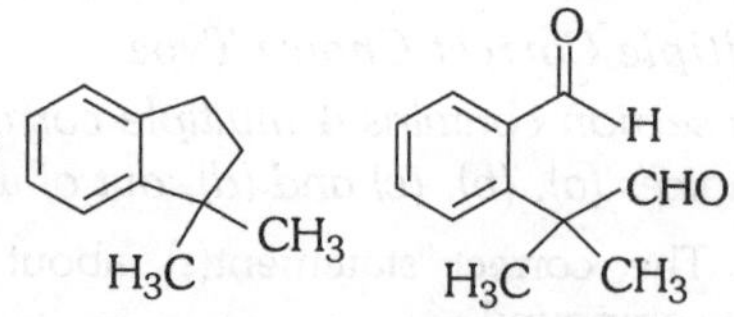

(c) 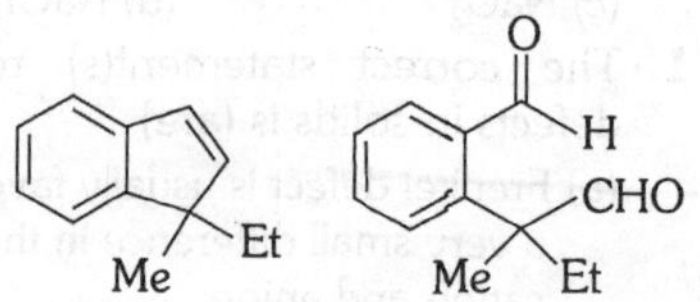

(d) 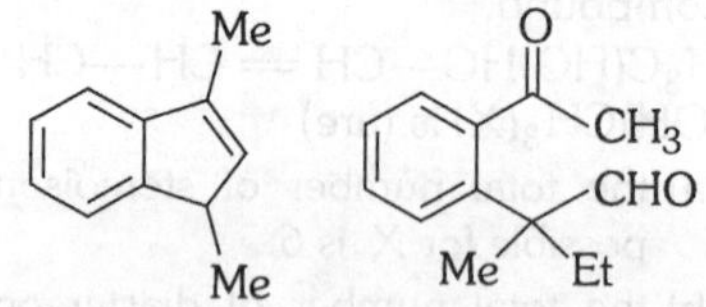

14. The structure of the product S, is

(a) 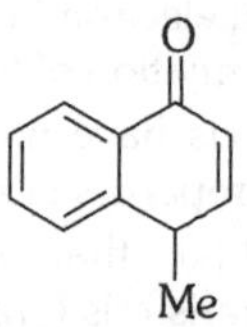

(b) 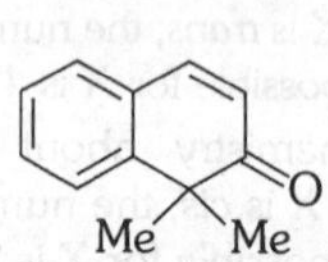

(c) 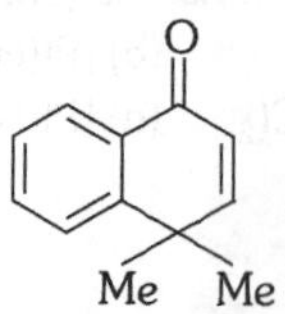

(d) 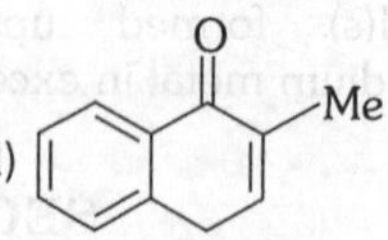

Paragraph (for Q. Nos. 16 to 18)

p-amino-N, N-dimethylaniline is added to a strongly acidic solution of X. The resulting solution is treated with a few drops of aqueous solution of Y to yield blue colouration due to the formation of methylene blue. Treatment of the aqueous solution of Y with the reagent potassium hexacyanoferrate(II) leads to the formation of an intense blue precipitate. The precipitate dissolves an excess addition of the reagent. Similarly, treatment of the solution of Y with the solution of potassium hexacyanoferrate(III) leads to a brown colouration due to the formation of Z.

16. The compound X, is

(a) $NaNO_3$ (b) $NaCl$

(c) Na_2SO_4 (d) Na_2S

17. The compound Y, is

(a) $MgCl_2$ (b) $FeCl_2$

(c) $FeCl_3$ (d) $ZnCl_2$

18. The compound Z, is

(a) $Mg_2[Fe(CN)_6]$ (b) $Fe[Fe(CN)_6]$

(c) $Fe_4[Fe(CN)_6]_3$ (d) $K_2Zn_3[Fe(CN)_6]_2$

SECTION IV

Matrix-Match Type

This section contains 2 questions. Each question contains statements given in two columns, which have to be matched. Statements in Column I are labelled as A, B, C and D, whereas statements in Column II are labelled as p, q, r,s and t. The answer to these questions have to be appropriately bubbles as illustrated in the following example.

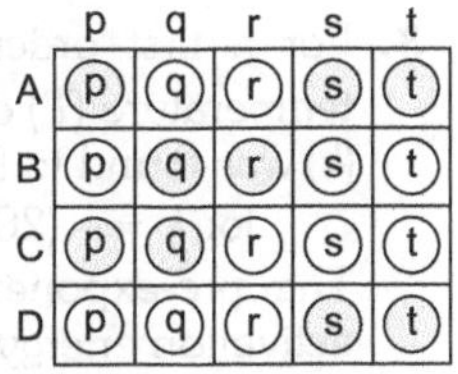

If the correct matches are A→p, s and t; B→q and r; C→p and q; and D→s and t; then the correct darkening of bubbles will look like the following.

19. Match each of the compounds is Column I with its characteristic reaction(s) is Column II.

Column I		Column II
A. $CH_3CH_2CH_2CN$	p.	Reduction with Pd-C/H_2
B. $CH_3CH_2OCOCH_3$	q.	Reduction with $SnCl_2$/HCl
C. $CH_3CH=CHCH_2OH$	r.	Development of foul smell on treatment with chloroform and alc. KOH
D. $CH_3CH_2CH_2CH_2NH_2$	s.	Reduction with disobutylaluminium hydride
	t.	Alkaline hydrolysis

20. Match each of the diatomic molecules in Column I with its property/properties in Column II.

	Column I		Column II
A.	B_2	p.	Paramagnetic
B.	N_2	q.	Undergoes oxidation
C.	O_2^-	r.	Undergoes reduction
D.	O_2	s.	Bond order ≥ 2
		t.	Mixing of 's' and 'p' orbitals

Paper 2

SECTION I

Single Correct Choice Type

This section contains 4 multiple choice questions numbered 1 to 4. Each question has 4 choices (a), (b), (c) and (d), out of which ***ONLY ONE*** *is correct.*

1. The correct stability order of the following resonance structure is

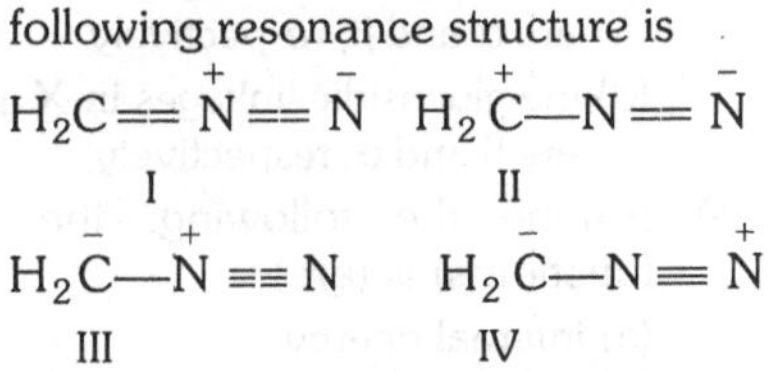

(a) I > II > III > IV

(b) I > III > II > IV

(c) II > I > III > IV

(d) III > I > IV > II

2. In the following carbocation, H / CH_3 that is most likely to migrate to the positively charged carbon is

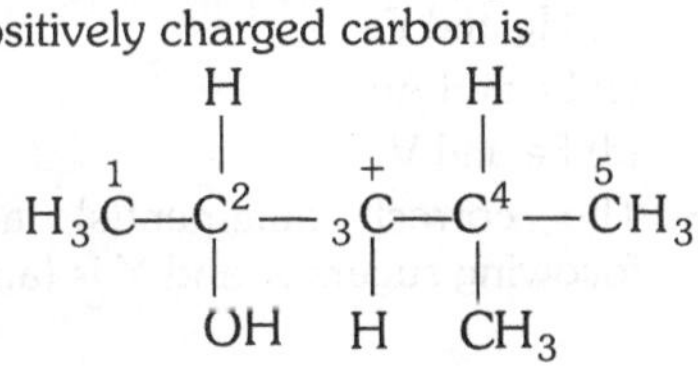

(a) CH_3 at C- 4 (b) H at C- 4
(c) CH_3 at C- 2 (d) H at C- 2

3. For a first order reaction, $A \rightarrow P$, the temperature (T) dependent rate constant (k) was found to follow the equation,

$$\log k = -(2000)/T + 6.0$$

The pre-exponential factor A and the activation energy (E_a), respectively, are

(a) $1.0 \times 10^6 s^{-1}$ and 9.2 kJ mol^{-1}
(b) $6.0 s^{-1}$ and 16.6 kJ mol^{-1}
(c) $1.0 \times 10^6 s^{-1}$ and 16.6kJ mol^{-1}
(d) $1.0 \times 10^6 s^{-1}$ and 38.3 kJ mol^{-1}

4. The spin only magnetic moment value (in Bohr magneton units) of $Cr(CO)_6$ is

(a) 0 (b) 2.84
(c) 4.90 (d) 5.92

SECTION II

Multiple Correct Choice Type

*This section contains 5 multiple correct answer(s) type questions. Each question has 4 choices (a), (b), (c) and (d), out of which **ONE** or **MORE** is/are correct.*

5. In the reaction,

$$2X + B_2H_6 \longrightarrow [BH_2(X)_2]^+[BH_4]^-$$

The amine (s) X, is (are)

(a) NH_3
(b) CH_3NH_2
(c) $(CH_3)_2NH$
(d) $(CH_3)_3N$

6. The nitrogen oxide(s) that contain(s) N.N bond(s) is (are)

(a) N_2O (b) N_2O_3
(c) N_2O_4 (d) N_2O_5

7. For the reduction of NO_3^- ion in an aqueous solution $E°$ is $+0.96$ V. Values of $E°$ for some metal ions are given below.

$V^{2+}(aq) + 2e^- \longrightarrow V; E° = -1.19$ V

$Fe^{3+}(aq) + 3e^- \longrightarrow Fe; E° = -0.04$ V

$Au^{3+}(aq) + 3e^- \longrightarrow Au; E° = +1.40$ V

$Hg^{2+}(aq) + 2e^- \longrightarrow Hg; E° = +0.86$ V

The pair(s) of metals that is (are) oxidised by NO_3^- in aqueous solution is (are)

(a) V and Hg
(b) Hg and Fe
(c) Fe and Au
(d) Fe and V

8. The correct statement(s) about the following sugars X and Y is (are)

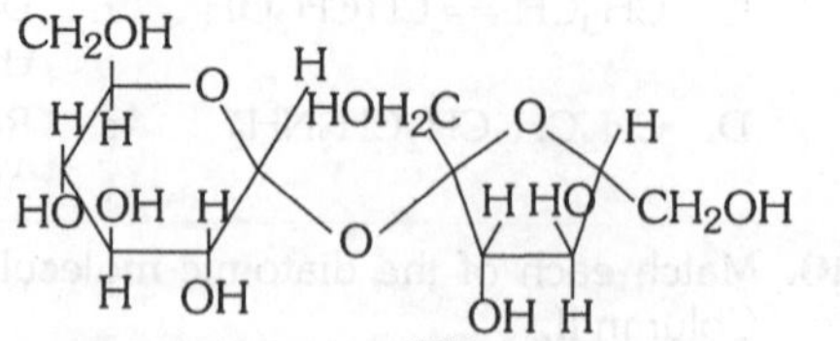

(X)

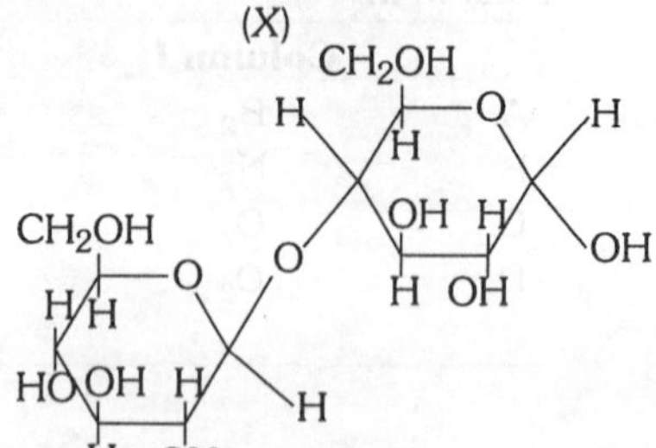

(Y)

(a) X is a reducing sugar and Y is a non-reducing sugar
(b) X is a non-reducing sugar and Y is a reducing sugar
(c) the glucosidic linkages in X and Y are α and β, respectively
(d) the glucosidic linkages in X and Y are β and α, respectively

9. Among the following, the state function(s) is (are)

(a) internal energy
(b) irreversible expansion work
(c) reversible expansion work
(d) molar enthalpy

SECTION III

Matrix-Match Type

This section contains 2 questions. Each question contains statements given in two columns, which have to be matched. Statements in Column I are labelled as A, B, C and D, whereas statements in Column II are labelled as p, q, r, s are t The answer to these questions have to be appropriately bubbles as illustrated in the following example.

If the correct matches are A→p, s and t; B→q and r; C→p and q; and D→s and t; then the correct darkening of bubbles will look like the following.

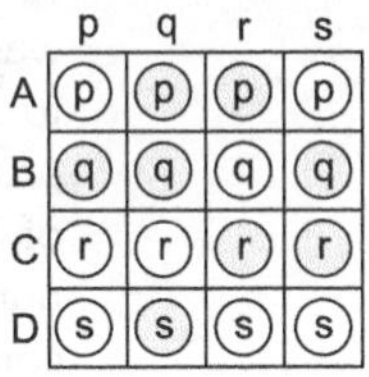

10. Match each of the compounds given in Column I with the reaction (s) that they can undergo, given in Column II.

Column I		Column II
A. 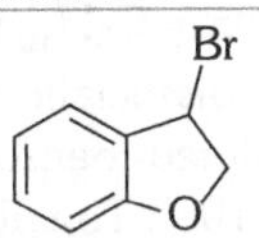	p.	Nucleophilic substitution
B. 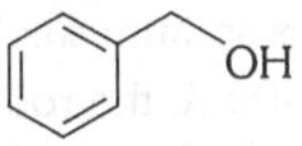	q.	Elimination
C. 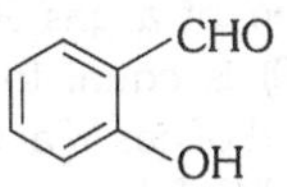	r.	Nucleophilic addition
D. 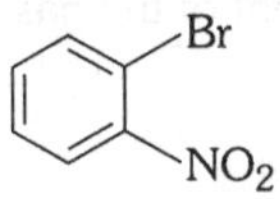	s.	Esterification with acetic anhydride
	t.	Dehydrogenation

11. Match each of the reactions given in Column I with the corresponding products(s) given in Column II.

Column I		Column II
A. Cu + dil. HNO_3	p.	NO
B. Cu+ conc. HNO_3	q.	NO_2
C. Zn + dil. HNO_3	r.	N_2O
D. Zn + conc. HNO_3	s.	$Cu(NO_3)_3$
	t.	$Zn(NO_3)_2$

SECTION IV

Integer Answer Type

This section contains 8 questions. The answer to each of the questions is a single digit integer, ranging from 0 to 9. The appropriate bubbles below the respective question numbers in the ORS have to be darkened. e.g. if the correct answers to question numbers X, Y, Z and W (say) are 6,0,9 and 2, respectively then the correct darkening of bubbles will look like the following.

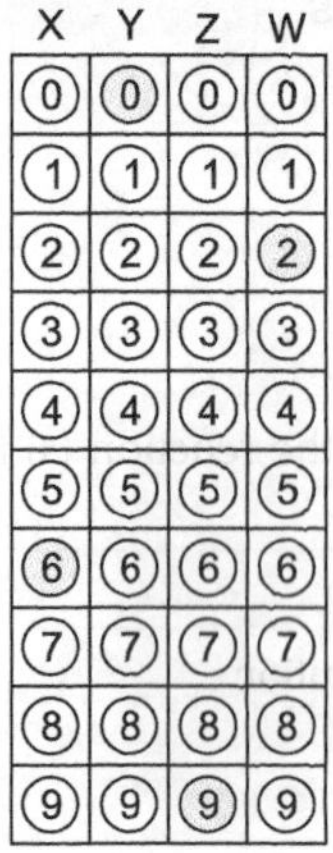

12. The total number of α and β-particles emitted in the nuclear reaction

$${}_{92}U^{238} \longrightarrow {}_{82}Pb^{214} \text{ is}$$

13. The number of water molecule(s) directly bonded to the metal centre in $CuSO_4 \cdot 5H_2O$ is

14. The oxidation number of Mn in the product of alkaline oxidative fusion of MnO_2 is

15. The coordination number of Al in the crystalline state of $AlCl_3$ is

16. In a constant volume calorimeter, 3.5 g of a gas with molecular weight = 28 was burnt in excess oxygen at 298.0 K. The temperature of the calorimeter was found to increases from 298.0 K to 298.45 K due to the combustion process. Given, that the heat capacity of the calorimeter is 2.5 kJ K^{-1}, the numerical value for the enthalpy of combustion of the gas in kJ mol^{-1} is

17. The total number of cyclic structural as well as stereoisomers possible for a compound with the molecular formula = C_5H_{10} is

18. The dissociation constant of a substituted benzoic acid at 25°C is 1.0×10^{-4}. The pH of 0.01 M solution of its sodium salt is

19. At 400 K, the root mean square (rms) speed of a gas X (molecular weight = 40) is equal to the most probable speed of gas Y at 60 K. The molecular weight of the gas Y is

ANSWERS

Paper 1

1. (c) **2.** (b) **3.** (a) **4.** (b) **5.** (d) **6.** (a)
7. (b) **8.** (b) **9.** (a, d) **10.** (a, b) **11.** (b, c) **12.** (c, d)
13. (b) **14.** (a) **15.** (b) **16.** (d) **17.** (c) **18.** (b)
19. A → p, q, s, t ; B → p, s, t; C → p; D → r **20.** A → p, r, t; B → s, t; C → p, q, r; D → p, r, s

Paper 2

1. (b) **2.** (d) **3.** (d) **4.** (a) **5.** (a, b, c) **6.** (a, b, c)
7. (a, b, d) **8.** (b, c) **9.** (a, c, d) **10.** A → p, q, t; B → p, s, t; C → r, s; D → p
11. A → p, s; B → q, s; C → r, t; D → p, r, s **12.** (8) **13.** (4) **14.** (6)
15. (5) **16.** (9) **17.** (7) **18.** (8) **19.** (4)

IIT JEE
UNSOLVED PAPER 2009

Mathematics

Paper 1

SECTION I

Single Correct Choice Type

This section contains 8 multiple choice questions. Each question has 4 choices (a), (b), (c) and (d) for its answer, out of which only one is correct.

1. Let $z = \cos\theta + i\sin\theta$. Then, the value of $\sum_{m=1}^{15} \text{Im}(z^{2m-1})$ at $\theta = 2°$ is

(a) $\frac{1}{\sin 2°}$ (b) $\frac{1}{3\sin 2°}$

(c) $\frac{1}{2\sin 2°}$ (d) $\frac{1}{4\sin 2°}$

2. The number of seven digit integers, with sum of the digits equal to 10 and formed by using the digits 1, 2 and 3 only, is

(a) 55 (b) 66 (c) 77 (d) 88

3. Let $P(3, 2, 6)$ be a point in space and Q be a point on the line $\mathbf{r} = (\hat{\mathbf{i}} - \hat{\mathbf{j}} + 2\hat{\mathbf{k}}) + \mu(-3\hat{\mathbf{i}} + \hat{\mathbf{j}} + 5\hat{\mathbf{k}})$.
Then, the value of μ for which the vector **PQ** is parallel to the plane $x - 4y + 3z = 1$ is

(a) 1/4 (b) –1/4 (c) 1/8 (d) –1/8

4. Let f be a non-negative function defined on the interval [0, 1]. If $\int_0^x \sqrt{1-\{f'(t)\}^2}\,dt = \int_0^x f(t)\,dt,\ 0 \le x \le 1$ and $f(0) = 0$, then

(a) $f\left(\frac{1}{2}\right) < \frac{1}{2}$ and $f\left(\frac{1}{3}\right) > \frac{1}{3}$

(b) $f\left(\frac{1}{2}\right) > \frac{1}{2}$ and $f\left(\frac{1}{3}\right) > \frac{1}{3}$

(c) $f\left(\frac{1}{2}\right) < \frac{1}{2}$ and $f\left(\frac{1}{3}\right) < \frac{1}{3}$

(d) $f\left(\frac{1}{2}\right) > \frac{1}{2}$ and $f\left(\frac{1}{3}\right) < \frac{1}{3}$

5. Let $z = x + iy$ be a complex number, where x and y are integers. Then, the area of the rectangle whose vertices are the roots of the equation $z\bar{z}^3 + \bar{z}z^3 = 350$ is

(a) 48 (b) 32

(c) 40 (d) 80

6. Tangents drawn from the point $P(1, 8)$ to the circle $x^2 + y^2 - 6x - 4y - 11 = 0$ touch the circle at the points A and B. The equation of the circumcircle of ΔPAB is

(a) $x^2 + y^2 + 4x - 6y + 19 = 0$

(b) $x^2 + y^2 - 4x - 10y + 19 = 0$

(c) $x^2 + y^2 - 2x + 6y - 29 = 0$

(d) $x^2 + y^2 - 6x - 4y + 19 = 0$

7. The line passing through the extremity A of the major axis and extremity B of the minor axis of the ellipse $x^2 + 9y^2 = 9$ meets its auxiliary circle at the point M. Then, the area of the triangle with vertices at A, M and the origin O is

(a) $\frac{31}{10}$ (b) $\frac{29}{10}$ (c) $\frac{21}{10}$ (d) $\frac{27}{10}$

8. If **a**, **b**, **c** and **d** are the unit vectors such that $(\mathbf{a} \times \mathbf{b}) \cdot (\mathbf{c} \times \mathbf{d}) = 1$ and $\mathbf{a} \cdot \mathbf{c} = \frac{1}{2}$, then

(a) **a**, **b**, **c** are non-coplanar

(b) **b**, **d** are non-parallel

(c) **a**, **b**, **d** are non-coplanar

(d) **a**, **d** are parallel and **b**, **c** are parallel

SECTION II

Multiple Correct Choice Type

This section contains 4 multiple choice questions. Each question has 4 choices (a), (b), (c) and (d) for its answer, out of which one or more is/are correct.

9. Let $L = \lim_{x \to 0} \dfrac{a - \sqrt{a^2 - x^2} - \dfrac{x^2}{4}}{x^4}$, $a > 0$. If L is finite, then

(a) $a = 2$ (b) $a = 1$

(c) $L = \dfrac{1}{64}$ (d) $L = \dfrac{1}{32}$

10. Area of the region bounded by the curve $y = e^x$ and lines $x = 0$ and $y = e$ is

(a) $e - 1$ (b) $\int_1^e \ln(e + 1 - y)\, dy$

(c) $e - \int_0^1 e^x dx$ (d) $\int_1^e \ln y\, dy$

11. If $\dfrac{\sin^4 x}{2} + \dfrac{\cos^4 x}{3} = \dfrac{1}{5}$, then

(a) $\tan^2 x = \dfrac{2}{3}$

(b) $\dfrac{\sin^8 x}{8} + \dfrac{\cos^8 x}{27} = \dfrac{1}{125}$

(c) $\tan^2 x = \dfrac{1}{3}$

(d) $\dfrac{\sin^8 x}{8} + \dfrac{\cos^8 x}{27} = \dfrac{2}{125}$

12. In a ΔABC with fixed base BC, the vertex A moves such that $\cos B + \cos C = 4 \sin^2 \dfrac{A}{2}$. If a, b and c denote the lengths of the sides of the triangle opposite to the angles A, B and C respectively, then

(a) $b + c = 4a$ (b) $b + c = 2a$

(c) locus of point A is an ellipse

(d) locus of point A is a pair of straight line

SECTION III

Comprehension Type

This section contains 2 groups of questions. Each group has 3 multiple choice questions based on a paragraph. Each question has 4 choices (a), (b), (c) and (d) for its answer, out of which only one is correct.

Paragraph for Question Nos. 13 to 15

Let A be the set of all 3×3 symmetric matrices all of whose entries are either 0 or 1. Five of these entries are 1 and four of them are 0.

13. The number of matrices in A is

(a) 12 (b) 6 (c) 9 (d) 3

14. The number of matrices A in A for which the system of linear equations $A\begin{bmatrix} x \\ y \\ z \end{bmatrix} = \begin{bmatrix} 1 \\ 0 \\ 0 \end{bmatrix}$ has a unique solution, is

(a) less than 4

(b) atleast 4 but less than 7

(c) atleast 7 but less than 10

(d) atleast 10

15. The number of matrices A in A for which the system of linear equations $A\begin{bmatrix} x \\ y \\ z \end{bmatrix} = \begin{bmatrix} 1 \\ 0 \\ 0 \end{bmatrix}$ is inconsistent, is

(a) 0 (b) more than 2

(c) 2 (d) 1

***Paragraph for Question Nos.* 16 *to* 18**

A fair die is tossed repeatedly until a six is obtained. Let X denotes the number of tosses required.

16. The probability that $X = 3$ equals

(a) $\frac{25}{216}$ (b) $\frac{25}{36}$ (c) $\frac{5}{36}$ (d) $\frac{125}{216}$

17. The probability that $X \geq 3$ equals

(a) $\frac{125}{216}$ (b) $\frac{25}{36}$ (c) $\frac{5}{36}$ (d) $\frac{25}{216}$

18. The conditional probability that $X \geq 6$ given $X > 3$ equals

(a) $\frac{125}{216}$ (b) $\frac{25}{216}$

(c) $\frac{5}{36}$ (d) $\frac{25}{36}$

SECTION IV

Matrix-Match

This section contains 2 questions. Each question contains statements given in two columns, which have to be matched. Statements in Column I are labelled A, B, C and D, whereas statements in Column II are labelled as p, q, r, s and t. Any given statement in Column I can have correct matching with one or more statement(s) in Column II. The appropriate bubbles corresponding to the answers to these questions have to be darkened as illustrated in the following example.

If the correct matches are A—q, A – r, B – p, B – s, C – r, C – s and D – q, then the correctly bubbled matrix will look like the following.

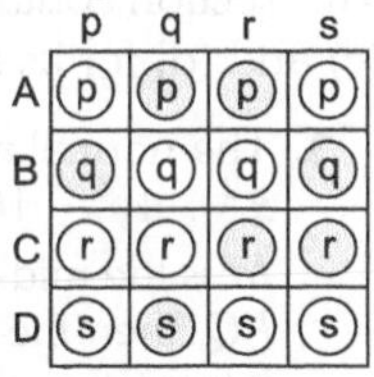

19. Match the statements/expressions in Column I with the open intervals in Column II.

Column I	Column II
A. Interval contained in the domain of definition of non-zero solutions of the differential equation $(x-3)^2 y' + y = 0$	p. $\left(-\frac{\pi}{2}, \frac{\pi}{2}\right)$
B. Interval containing the value of the integral $\int_1^5 (x-1)(x-2)(x-3)(x-4)(x-5)dx$	q. $\left(0, \frac{\pi}{2}\right)$
C. Interval in which atleast one of the points of local maximum of $\cos^2 x + \sin x$ lies	r. $\left(\frac{\pi}{8}, \frac{5\pi}{4}\right)$
D. Interval in which $\tan^{-1}(\sin x + \cos x)$ is increasing	s. $\left(0, \frac{\pi}{8}\right)$
	t. $(-\pi, \pi)$

20. Match the conics in Column I with the statements/expressions in Column II.

	Column I	Column2w II
A.	Circle	p. The locus of the point (h, k) for which the line $hx + ky = 1$ touches the circle $x^2 + y^2 = 4$
B.	Parabola	q. Points z is the complex plane satisfying $\lvert z+2\rvert - \lvert z-2\rvert = \pm 3$

Column I	Column II
C. Ellipse	r. Points of the conic have parametric representation $x=\sqrt{3}\left(\frac{1-t^2}{1+t^2}\right), y=\frac{2t}{1+t^2}$
D. Hyperbola	s. The eccentricity of the conic lies in the interval $1 \le x < \infty$
	t. Points in the complex plane satisfying $\text{Re}(z+1)^2 = \|z\|^2 + 1$

Paper 2

SECTION I

Single Correct Choice Type

This section contains 4 multiple choice questions. Each question has 4 choices (a), (b), (c) and (d) for its answer, out of which only one is correct.

1. The normal at a point P on the ellipse $x^2+4y^2=16$ meets the X-axis at Q. If M is the mid-point of the line segment PQ, then the locus of M intersects the latusrectum of the given ellipse at the points

(a) $\left(\pm\frac{3\sqrt{5}}{2}, \pm\frac{2}{7}\right)$ (b) $\left(\pm\frac{3\sqrt{5}}{2}, \pm\frac{\sqrt{19}}{4}\right)$

(c) $\left(\pm 2\sqrt{3}, \pm\frac{1}{7}\right)$ (d) $\left(\pm 2\sqrt{3}, \pm\frac{4\sqrt{3}}{7}\right)$

2. The locus of the orthocentre of the triangle formed by the lines $(1+p)x - py + p(1+p) = 0$, $(1+q)x - qy + q(1+q) = 0$ and $y = 0$, where $p \ne q$, is

(a) a hyperbola
(b) a parabola
(c) an ellipse
(d) a straight line

3. A line with positive direction cosines passes through the point $P(2, -1, 2)$ and makes equal angles with the coordinate axes. The line meets the plane $2x + y + z = 9$ at point Q. The length of the line segment PQ equals

(a) 1 (b) $\sqrt{2}$
(c) $\sqrt{3}$ (d) 2

4. If the sum of first n terms of an AP is cn^2, then the sum of squares of these n terms is

(a) $\frac{n(4n^2-1)c^2}{6}$ (b) $\frac{n(4n^2+1)c^2}{3}$

(c) $\frac{n(4n^2-1)c^2}{3}$ (d) $\frac{n(4n^2+1)c^2}{6}$

SECTION II

Multiple Correct Choice Type

This section contains 5 multiple choice question. Each questions has four choices (a), (b), (c) and (d) for its answer, out of which one or more is/are correct.

5. The tangent PT and the normal PN to the parabola $y^2 = 4ax$ at a point P on it meet its axis at points T and N, respectively. The locus of the centroid of ΔPTN is a parabola whose

(a) vertex is $\left(\frac{2a}{3}, 0\right)$

(b) directrix is $x = 0$

(c) latusrectum is $\frac{2a}{3}$

(d) focus is $(a, 0)$

6. For the function $f(x) = x \cos \frac{1}{x}, x \geq 1$,

(a) for atleast one x in the interval $[1, \infty)$, $f(x + 2) - f(x) < 2$

(b) $\lim_{x \to \infty} f'(x) = 1$

(c) for all x in the interval $[1, \infty)$, $f(x + 2) - f(x) > 2$

(d) $f'(x)$ is strictly decreasing in the interval $[1, \infty)$

7. For $0 < \theta < \frac{\pi}{2}$, the solution(s) of $\sum_{m=1}^{6} \operatorname{cosec}\left[\theta + \frac{(m-1)\pi}{4}\right] \operatorname{cosec}\left(\theta + \frac{m\pi}{4}\right) = 4\sqrt{2}$ is/are

(a) $\frac{\pi}{4}$ (b) $\frac{\pi}{6}$ (c) $\frac{\pi}{12}$ (d) $\frac{5\pi}{12}$

8. An ellipse intersects the hyperbola $2x^2 - 2y^2 = 1$ orthogonally. The eccentricity of the ellipse is reciprocal to that of the hyperbola. If the axes of the ellipse are along the coordinate axes, then

(a) equation of ellipse is $x^2 + 2y = 2$

(b) the foci of ellipse are $(\pm 1, 0)$

(c) equation of ellipse is $x^2 + 2y^2 = 4$

(d) the foci of ellipse are $(\pm \sqrt{2}, 0)$

9. If $I_n = \int_{-\pi}^{\pi} \frac{\sin nx}{(1 + \pi^x) \sin x} dx$; $n = 0, 1, 2, \ldots$, then

(a) $I_n = I_{n+2}$ (b) $\sum_{m=1}^{10} I_{2m+1} = 10\pi$

(c) $\sum_{m=1}^{10} I_{2m} = 0$ (d) $I_n = I_{n+1}$

SECTION III

Matrix-Match

This section contains 2 questions. Each question contains statements given in two columns, which have to be matched. Statements in Column I are labelled A, B, C and D, whereas statements in Column II are labelled as p, q, r, s and t. Any given statement in Column I can have correct matching with one or more statement(s) in Column II. The appropriate bubbles corresponding to the answer to these questions have to be darkened as illustrated in the following example.

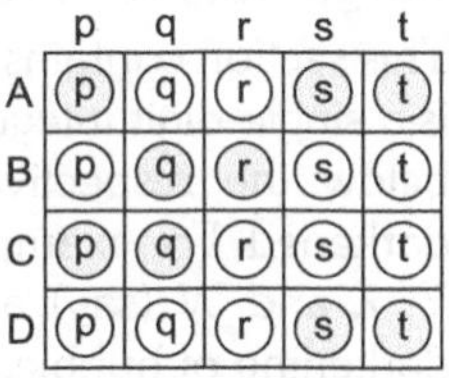

If the correct matches are A→ p, s and t; B→ q and r; C→ p and q; and D→ s and t; then the correct darkening of bubbles will look like the following.

10. Match the statements/expressions given in Column I with the values given in Column II.

Column I	Column II
A. The number of solutions of the equation $xe^{\sin x} - \cos x = 0$ in the interval $\left(0, \frac{\pi}{2}\right)$	p. 1
B. Value(s) of k for which the planes $kx + 4y + z = 0$, $4x + ky + 2z = 0$ and $2x + 2y + z = 0$ intersect in a straight line	q. 2
C. Value(s) of k for which $\|x - 1\| + \|x - 2\| + \|x + 1\| + \|x + 2\| = 4k$ has integer solution(s)	r. 3
D. If $y' = y + 1$ and $y(0) = 1$, then value(s) of $y(\ln 2)$	s. 4
	t. 5

11. Match the statements/expressions given in Column I with the values given in Column II.

Column I	Column II
A. Root(s) of the equation $2\sin^2\theta + \sin^2 2\theta = 2$	p. $\frac{\pi}{6}$
B. Points of discontinuity of the function $f(x) = \left[\frac{6x}{\pi}\right]\cos\left[\frac{3x}{\pi}\right]$, where $[y]$ denotes the largest integer less than or equal to y	q. $\frac{\pi}{4}$
C. Volume of the parallelopiped with its edges represented by the vectors $\hat{\mathbf{i}} + \hat{\mathbf{j}}, \hat{\mathbf{i}} + 2\hat{\mathbf{j}}$ and $\hat{\mathbf{i}} + \hat{\mathbf{j}} + \hat{\mathbf{k}}$	r. $\frac{\pi}{3}$
D. Angle between vectors $\mathbf{a}$ and $\mathbf{b}$, where $\mathbf{a}$, $\mathbf{b}$ and $\mathbf{c}$ are unit vectors, satisfying $\mathbf{a} + \mathbf{b} + \sqrt{3}\,\mathbf{c} = \vec{0}$	s. $\frac{\pi}{2}$
	t. π

SECTION IV

Integer Answer Type Questions

This section contains 8 questions. The answer to each of the questions is a single-digit integer, ranging from 0 to 9. The appropriate bubbles below the respective question numbers in the ORS have to be darkened. For example, if the correct answers to question numbers X, Y, Z and W (say) are 6, 0, 9 and 2 respectively, then the correct darkening of bubbles will look like the following.

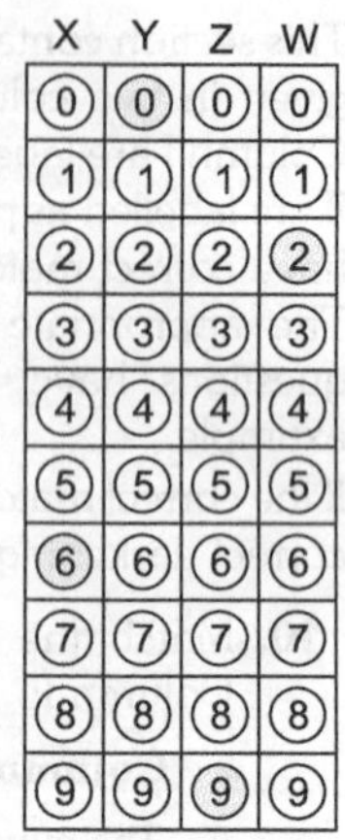

12. Let $f : R \to R$ be a continuous function, which satisfies $f(x) = \int_0^x f(t)\,dt$. Then, the value of $f(\ln 5)$ is

13. The centres of two circles C_1 and C_2 each of unit radius are at a distance of 6 units from each other. Let P be the mid-point of the line segment joining the centres of C_1 and C_2 and C be a circle touching circles C_1 and C_2 externally. If a common tangents to C_1 and C passing through P is also a common tangent to C_2 and C, then the radius of the circle C is ....

14. The smallest value of k, for which both the roots of the equation $x^2 - 8kx + 16(k^2 - k + 1) = 0$ are real, distinct and have values atleast 4, is

15. The maximum value of the function $f(x) = 2x^3 - 15x^2 + 36x - 48$ on the set $A = \{x \mid x^2 + 20 \le 9x\}$ is

16. Let ABC and ABC' be two non-congruent triangles with sides $AB = 4$, $AC = AC' = 2\sqrt{2}$ and angle $B = 30°$. The absolute value of the difference between the areas of these triangles is

17. If the function $f(x) = x^3 + e^{\frac{x}{2}}$ and $g(x) = f^{-1}(x)$, then the value of $g'(1)$ is

18. Let $p(x)$ be a polynomial of degree 4 having extremum at $x = 1, 2$ and $\lim_{x \to 0} \left[1 + \frac{p(x)}{x^2}\right] = 2$. Then, the value of $p(2)$ is

19. Let (x, y, z) be points with integer coordinates satisfying the system of homogeneous equations $3x - y - z = 0, -3x + z = 0, -3x + 2y + z = 0$. Then, the number of such points for which $x^2 + y^2 + z^2 \leq 100$ is

ANSWERS

Paper 1

1. (d)	**2.** (c)	**3.** (a)	**4.** (c)	**5.** (a)	**6.** (b)
7. (d)	**8.** (c)	**9.** (a, c)	**10.** (b, c, d)	**11.** (a, b)	**12.** (b, c)
13. (a)	**14.** (b)	**15.** (b)	**16.** (a)	**17.** (b)	**18.** (d)

19. A → p, q, s; B → p, t; C → p, q, r, t; D → s

20. A → p; B → s, t; C → r; D → q, s

Paper 2

1. (c)	**2.** (d)	**3.** (c)	**4.** (c)	**5.** (a, d)	**6.** (b, c, d)
7. (c, d)	**8.** (a, b)	**9.** (a, b, c)	**10.** A → p; B → q, s; C → q, r, s, t; D → r		
12. (0)	**13.** (8)	**14.** (2)	**15.** (5)	**16.** (4)	**17.** (2)
18. (0)	**19.** (7)				

IIT JEE

UNSOLVED PAPER 2010

Physics

Paper I

SECTION I — Single Correct Choice Type

This section contains **8 multiple choice questions**. *Each question has four choices (a), (b), (c) and (d) out of which* **ONLY ONE** *is correct.*

1. A thin uniform annular disc (*see figure*) of mass M has outer radius $4R$ and inner radius $3R$. The work required to take a unit mass from point P on its axis to infinity is

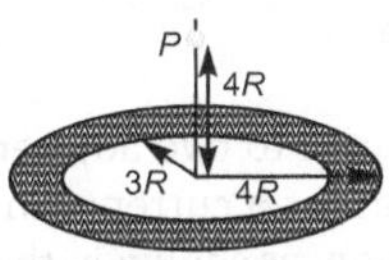

(a) $\frac{2GM}{7R}(4\sqrt{2}-5)$

(b) $-\frac{2GM}{7R}(4\sqrt{2}-5)$

(c) $\frac{GM}{4R}$

(d) $\frac{2GM}{5R}(\sqrt{2}-1)$

2. A block of mass m is on an inclined plane of angle θ. The coefficient of friction between the block and the plane is μ and $\tan\theta > \mu$. The block is held stationary by applying a force P parallel to the plane. The direction of force pointing up the plane is taken to be positive. As P is varied from $P_1 = mg(\sin\theta - \mu\cos\theta)$ to $P_2 = mg(\sin\theta + \mu\cos\theta)$, the frictional force f *versus* P graph will look like

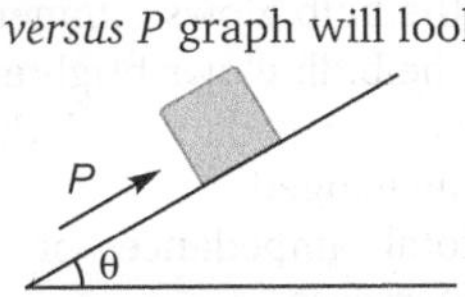

(a)
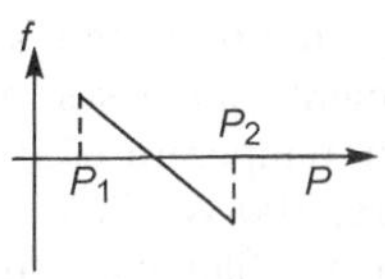

(b)
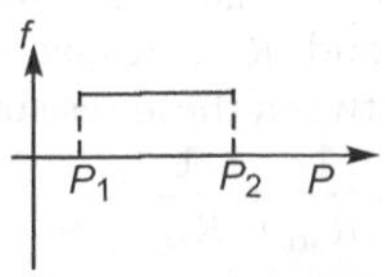

(c)
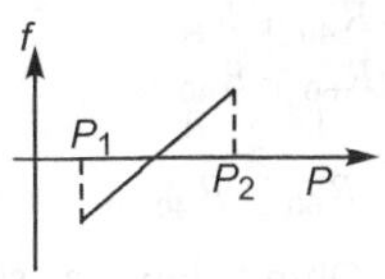

(d)
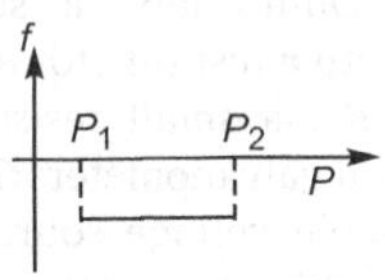

3. A real gas behaves like an ideal gas if its

(a) pressure and temperature are both high

(b) pressure and temperature are both low

(c) pressure is high and temperature is low

(d) pressure is low and temperature is high

4. Consider a thin square sheet of side L and thickness t, made of a material of resistivity ρ. The resistance between

two opposite faces, shown by the shaded areas in the figure is

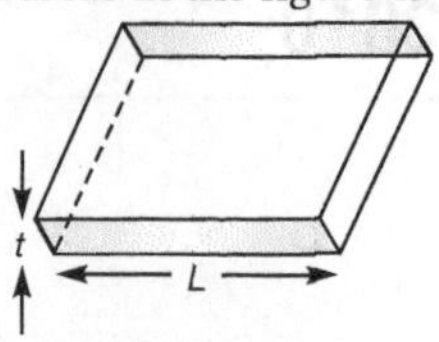

(a) directly proportional to L
(b) directly proportional to t
(c) independent of L
(d) independent of t

5. Incandescent bulbs are designed by keeping in mind that the resistance of their filament increases with the increase in temperature. If at room temperature, 100 W, 60 W and 40 W bulbs have filament resistances R_{100}, R_{60} and R_{40}, respectively, the relation between these resistances is

(a) $\dfrac{1}{R_{100}} = \dfrac{1}{R_{40}} + \dfrac{1}{R_{60}}$
(b) $R_{100} = R_{40} + R_{60}$
(c) $R_{100} > R_{60} > R_{40}$
(d) $\dfrac{1}{R_{100}} > \dfrac{1}{R_{60}} > \dfrac{1}{R_{40}}$

6. To verify Ohm's law, a student is provided with a test resistor R_T, a high resistance R_1, a small resistance R_2, two identical galvanometers G_1 and G_2, and a variable voltage source V. The correct circuit to carry out the experiment is

(a)
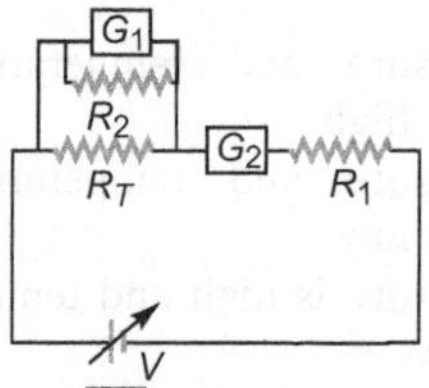

(b)
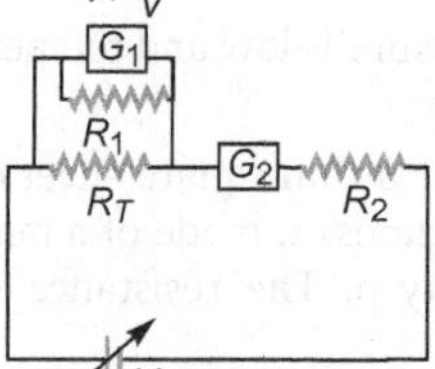

(c)
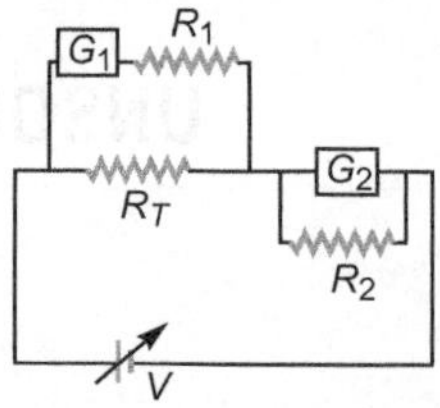

(d)
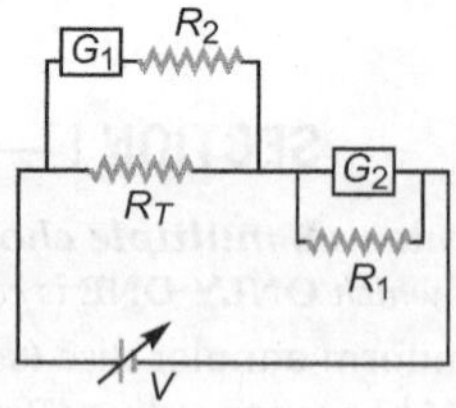

7. A thin flexible wire of length L is

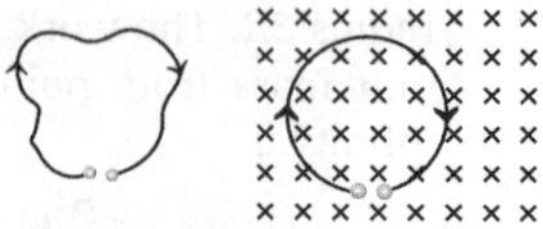

connected to two adjacent fixed points and carries a current I in the clockwise direction, as shown in the figure. When the system is put in a uniform magnetic field of strength B going into the plane of the paper, the wire takes the shape of a circle. The tension in the wire is

(a) IBL
(b) $\dfrac{IBL}{\pi}$
(c) $\dfrac{IBL}{2\pi}$
(d) $\dfrac{IBL}{4\pi}$

8. An AC voltage source of variable angular frequency ω and fixed amplitude V_0 is connected in series with a capacitance C and an electric bulb of resistance R (*inductance zero*). When ω is increased

(a) the bulb glows dimmer
(b) the bulb glows brighter
(c) total impedance of the circuit is unchanged
(d) total impedance of the circuit increases

SECTION II — Multiple Correct Choice Type

This section contains **5 *multiple choice questions***. *Each question has four choices (a), (b), (c) and (d) out of which* **ONE OR MORE** *may be correct.*

9. A student uses a simple pendulum of exactly 1m length to determine g, the acceleration due to gravity. He uses a stop watch with the least count of 1 s for this and records 40 s for 20 oscillations. For this observation, which of the following statement(s) is/are true?

(a) Error ΔT in measuring T, the time period, is 0.05 s

(b) Error ΔT in measuring T, the time period, is 1 s

(c) Percentage error in the determination of g is 5%

(d) Percentage error in the determination of g is 2.5%

10. A few electric field lines for a system of two charges Q_1 and Q_2 fixed at two different points on the x-axis are shown in the figure. These lines suggest that

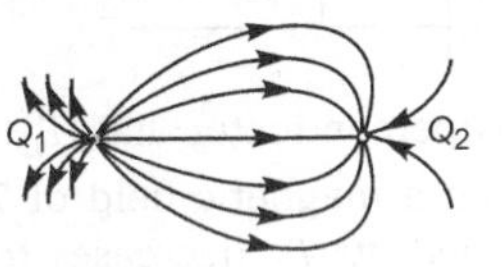

(a) $|Q_1| > |Q_2|$

(b) $|Q_1| < |Q_2|$

(c) at a finite distance to the left of Q_1 the electric field is zero

(d) at a finite distance to the right of Q_2 the electric field is zero

11. One mole of an ideal gas in initial state A undergoes a cyclic process $ABCA$, as shown in the figure. Its pressure at A is p_0. Choose the correct option(s) from the following

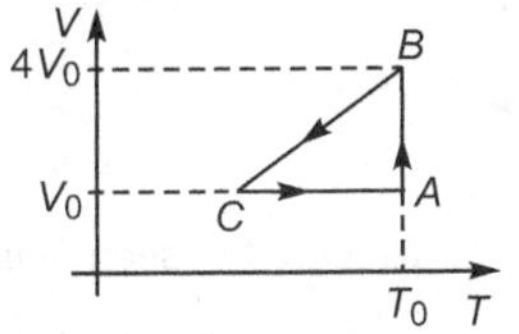

(a) Internal energies at A and B are the same

(b) Work done by the gas in process AB is $p_0V_0 \ln 4$

(c) Pressure at C is $\frac{p_0}{4}$

(d) Temperature at C is $\frac{T_0}{4}$

12. A point mass of 1 kg collides elastically with a stationary point mass of 5 kg. After their collision, the 1 kg mass reverses its direction and moves with a speed of $2\ \text{ms}^{-1}$. Which of the following statement(s) is/are correct for the system of these two masses?

(a) Total momentum of the system is $3\ \text{kg-ms}^{-1}$

(b) Momentum of 5 kg mass after collision is $4\ \text{kg-ms}^{-1}$

(c) Kinetic energy of the centre of mass is 0.75 J

(d) Total kinetic energy of the system is 4 J

13. A ray OP of monochromatic light is incident on the face AB of prism $ABCD$ near vertex B at an incident angle of 60° (*see figure*). If the refractive index of the material of the prism is $\sqrt{3}$, which of the following is (are) correct?

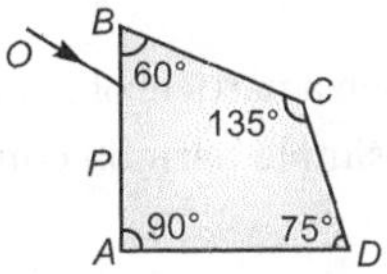

(a) The ray gets totally internally reflected at face CD

(b) The ray comes out through face AD

(c) The angle between the incident ray and the emergent ray is 90°

(d) The angle between the incident ray and the emergent ray is 120°

SECTION III — Paragraph Type

This section contains **2 *paragraphs***. *Based upon the first paragraph* 2 ***multiple choice questions*** *and based upon the second paragraph* **3 *multiple choice questions*** *have to be answered. Each of these questions has four choices (a), (b), (c) and (d) out of* ***WHICH ONLY ONE CORRECT.***

P_{14-15} Electrical resistance of certain materials, known as superconductors, changes abruptly from a non-zero value to zero as their temperature is lowered below a critical temperature $T_C(0)$. An interesting property of superconductors is that their critical temperature becomes smaller than $T_C(0)$ if they are placed in a magnetic field *i.e.*, the critical temperature $T_C(B)$ is a function of the magnetic field strength B. The dependence of $T_C(B)$ on B is shown in the figure.

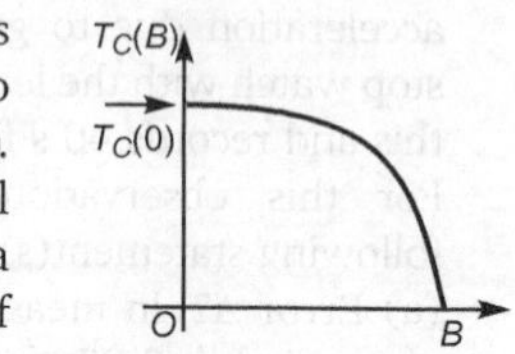

14. In the graphs below, the resistance R of a superconductor is shown as a function of its temperature T for two different magnetic fields B_1 *(solid line)* and B_2 *(dashed line)*. If B_2 is larger than B_1, which of the following graphs shows the correct variation of R with T in these fields?

(a)

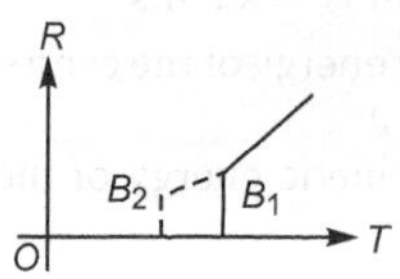

(b)

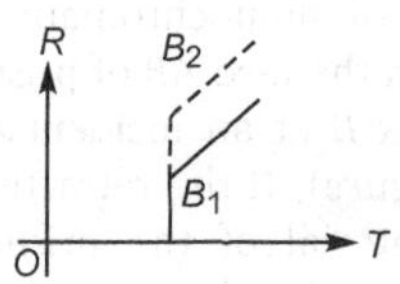

(c)

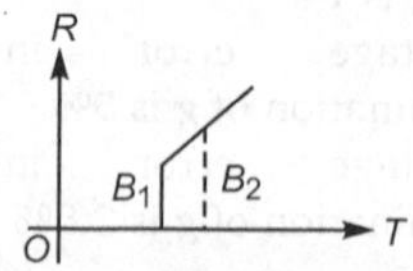

(d)

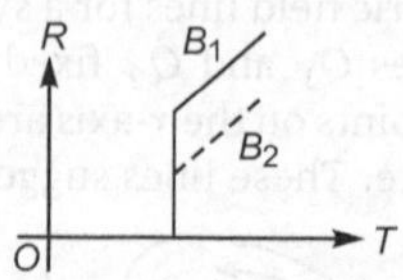

15. A superconductor has $T_C(0) = 100$ K. When a magnetic field of 7.5 Tesla is applied, its T_C decreases to 75 K. For this material one can definitely say that when (Note : T = Tesla)

(a) $B = 5\,\text{T}, T_C(B) = 80$ K

(b) $B = 5\,\text{T}, 75\,\text{K} < T_C(B) < 100$ K

(c) $B = 10\,\text{T}, 75\,\text{K} < T_C\,(B) < 100$ K

(d) $B = 10\,\text{T}, T_C\,(B) = 70$ K

P_{16-18} When a particle of mass m moves on the x-axis in a potential of the form $V(x) = kx^2$, it performs simple harmonic motion.

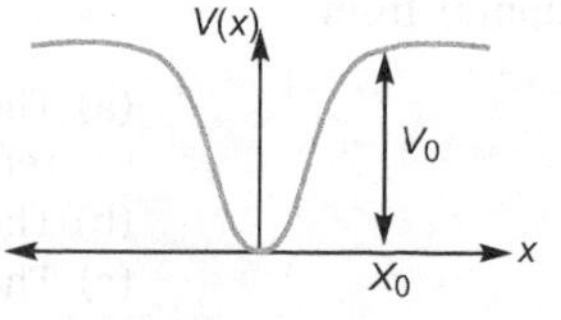

The corresponding time period is proportional to $\sqrt{\dfrac{m}{k}}$, as can be seen easily using dimensional analysis. However, the motion of a particle can be periodic even when its potential energy increases on both sides of $x = 0$ in a way different from kx^2 and its total energy is such that the particle does not escape to infinity. Consider a particle of mass m moving on the x-axis. Its potential energy is $V(x) = \alpha x^4 (\alpha > 0)$ for $|x|$ near the origin and becomes a constant equal to V_0 for $|x| \geq X_0$ *(see figure)*.

16. If the total energy of the particle is E, it will perform periodic motion only if

(a) $E < 0$ (b) $E > 0$

(c) $V_0 > E > 0$ (d) $E > V_0$

17. For periodic motion of small amplitude A, the time period T of this particle is proportional to

(a) $A\sqrt{\frac{m}{\alpha}}$ (b) $\frac{1}{A}\sqrt{\frac{m}{\alpha}}$

(c) $A\sqrt{\frac{\alpha}{m}}$ (d) $\frac{1}{A}\sqrt{\frac{\alpha}{m}}$

18. The acceleration of this particle for $|x| > X_0$ is

(a) proportional to V_0

(b) proportional to $\frac{V_0}{mX_0}$

(c) proportional to $\sqrt{\frac{V_0}{mX_0}}$

(d) Zero

SECTION IV — Integer Type

*This section contains **TEN questions**. The answer to each question is a **Single-digit integer**, ranging from 0 to 9. The correct digit below the question number in the ORS is to be bubbled.*

19. A stationary source is emitting sound at a fixed frequency f_0, which is reflected by two cars approaching the source. The difference between the frequencies of sound reflected from the cars is 1.2% of f_0. What is the difference in the speeds of the cars *(in km per hour)* to the nearest integer ? The cars are moving at constant speeds much smaller than the speed of sound which is 330 ms^{-1}.

20. The focal length of a thin biconvex lens is 20 cm. When an object is moved from a distance of 25 cm in front of it to 50 cm, the magnification of its image changes from m_{25} to m_{50}. The ratio $\frac{m_{25}}{m_{50}}$ is

21. An α-particle and a proton are accelerated from rest by a potential difference of 100 V. After this, their de Broglie wavelengths are λ_α and λ_p respectively. The ratio $\frac{\lambda_p}{\lambda_\alpha}$, to the nearest integer, is

22. When two identical batteries of internal resistance 1 Ω each are connected in series across a resistor R, the rate of heat produced in R is J_1. When the same batteries are connected in parallel across R, the rate is J_2. If $J_1 = 2.25 J_2$ then the value of R in Ω is

23. Two spherical bodies A (*radius* 6 *cm*) and B (*radius* 18 *cm*) are at temperatures T_1 and T_2, respectively. The maximum intensity in the emission spectrum of A is at 500 nm and in that of B is at 1500 nm. Considering them to be black bodies, what will be the ratio of the rate of total energy radiated by A to that of B?

24. When two progressive waves $y_1 = 4\sin(2x - 6t)$ and $y_2 = 3\sin\left(2x - 6t - \frac{\pi}{2}\right)$ are superimposed, the amplitude of the resultant wave is

25. A 0.1 kg mass is suspended from a wire of negligible mass. The length of the wire is 1 m and its cross-sectional area is 4.9×10^{-7} m^2. If the mass is pulled a little in the vertically downward direction and released, it performs simple harmonic motion of angular frequency 140 rad s^{-1}. If the Young's modulus of the material of the wire is $n \times 10^9$ Nm^{-2}, the value of n is

26. A binary star consists of two stars A (*mass* $2.2M_S$) and B (*mass* $11M_s$), where M_S is the mass of the sun. They are separated by distance d and are rotating about their centre of mass, which is stationary. The ratio of the total angular momentum of the binary star to the angular momentum of star B about the centre of mass is

27. Gravitational acceleration on the surface of a planet is $\frac{\sqrt{6}}{11}g$, where g is the gravitational acceleration on the surface of the earth. The average mass density of the planet is $\frac{2}{3}$ times that of the earth. If the escape speed on the surface of the earth is taken to be 11 kms^{-1}, the escape speed on the surface of the planet in kms^{-1} will be

28. A piece of ice (*heat capacity* $= 2100\ J\,kg^{-1}\,°C^{-1}$ *and latent heat* $= 3.36 \times 10^5\ J\,kg^{-1}$) of mass m gram is at –5°C at atmospheric pressure. It is given 420 J of heat so that the ice starts melting. Finally when the ice-water mixture is in equilibrium, it is found that 1 g of ice has melted. Assuming there is no other heat exchange in the process, the value of m is

Paper II

SECTION I — Single Correct Choice Type

This section contains 6 ***multiple choice questions****. Each question has four choices (a), (b), (c) and (d) out of which* ***ONLY ONE*** *is correct.*

1. A tiny spherical oil drop carrying a net charge q is balanced in still air with a vertical uniform electric field of strength $\frac{81\pi}{7} \times 10^5$ Vm^{-1}. When the field is switched off, the drop is observed to fall with terminal velocity 2×10^{-3} ms^{-1}. Given $g = 9.8$ ms^{-2}, viscosity of the air $= 1.8 \times 10^{-5}$ Ns m^{-2} and the density of oil $= 900$ kg m^{-3}, the magnitude of q is

(a) 1.6×10^{-19} C (b) 3.2×10^{-19} C
(c) 4.8×10^{-19} C (d) 8.0×10^{-19} C

2. A uniformly charged thin spherical shell of radius R carries uniform surface charge density of σ per unit area. It is made of two hemispherical shells, held together by pressing them with force F (*see figure*). F is proportional to

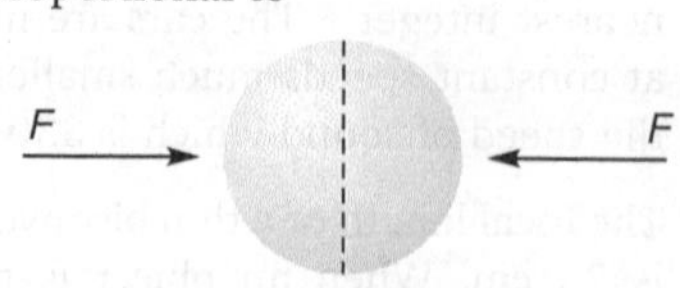

(a) $\frac{1}{\varepsilon_0}\sigma^2R^2$ (b) $\frac{1}{\varepsilon_0}\sigma^2R$
(c) $\frac{1}{\varepsilon_0}\frac{\sigma^2}{R}$ (d) $\frac{1}{\varepsilon_0}\frac{\sigma^2}{R^2}$

3. A block of mass 2 kg is free to move along the x-axis. It is at rest and from $t = 0$ onwards it is subjected to a time-dependent force $F(t)$ in the x direction. The force $F(t)$ varies with t as shown in the figure. The kinetic energy of the block after 4.5 s is

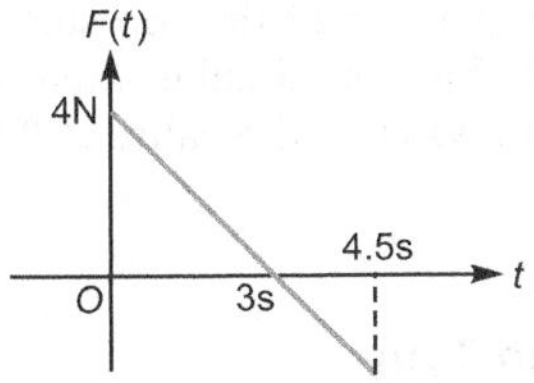

(a) 4.50 J (b) 7.50 J
(c) 5.06 J (d) 14.06 J

4. A biconvex lens of focal length 15 cm is in front of a plane mirror. The distance between the lens and the mirror is 10 cm. A small object is kept at a distance of 30 cm from the lens. The final image is

(a) virtual and at a distance of 16 cm from the mirror
(b) real and at a distance of 16 cm from the mirror
(c) virtual and at a distance of 20 cm from the mirror
(d) real and at a distance of 20 cm from the mirror

5. A vernier calipers has 1 mm marks on the main scale. It has 20 equal divisions on the vernier scale which match with 16 main scale divisions. For this vernier calipers, the least count is

(a) 0.02 mm (b) 0.05 mm
(c) 0.1 mm (d) 0.2 mm

6. A hollow pipe of length 0.8 m is closed at one end. At its open end a 0.5 m long uniform string is vibrating in its second harmonic and it resonates with the fundamental frequency of the pipe. If the tension in the wire is 50 N and the speed of sound is 320 ms^{-1}, the mass of the string is

(a) 5 g (b) 10 g
(c) 20 g (d) 40 g

SECTION II — Integer Type

This section contains **5 questions**. *The answer to each of the questions is a* ***single-digit integer****, ranging from 0 to 9. The correct digit below the question number in the* ***ORS*** *is to be bubbled.*

7. To determine the half-life of a radioactive element, a student plots a graph of $\ln\left|\frac{dN(t)}{dt}\right|$ *versus* t. Here $\frac{dN(t)}{dt}$ is the rate of radioactive decay at time t. If the number of radioactive nuclei of this element decreases by a factor of p after 4.16 yr, the value of p is

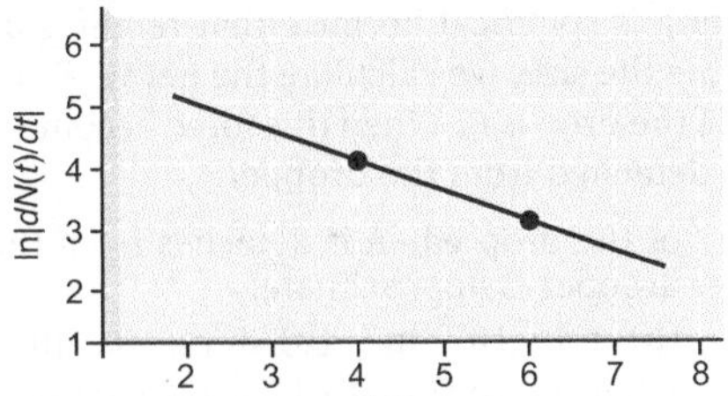

8. A diatomic ideal gas is compressed adiabatically to $\frac{1}{32}$ of its initial volume. In the initial temperature of the gas is T_i (*in kelvin*) and the final temperature is aT_i, the value of a is

9. At time $t = 0$, a battery of 10 V is connected across points A and B in the given circuit. If the capacitors have no charge initially, at what time (*in second*) does the voltage across them become 4 V?

[*Take* : $\ln 5 = 1.6$, $\ln 3 = 1.1$]

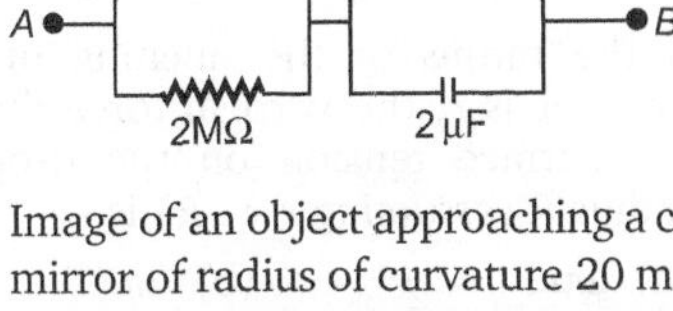

10. Image of an object approaching a convex mirror of radius of curvature 20 m along its optical axis is observed to move from $\frac{25}{3}$ m to $\frac{50}{7}$ m in 30 s. What is the speed of the object in km h^{-1} ?

11. A large glass slab $\left(\mu = \frac{5}{3}\right)$ of thickness 8 cm is placed over a point source of light on a plane surface. It is seen that light emerges out of the top surface of the slab from a circular area of radius R cm. What is the value of R?

SECTION III — Paragraph Type

This section contains 2 ***paragraphs****. Based upon each of the paragraphs* ***3 multiple choice questions*** *have to be answered. Each of these questions has four choices (a), (b), (c) and (d) out of which* ***ONLY ONE*** *is correct.*

P_{12-14} The key feature of Bohr's theory of spectrum of hydrogen atom is the quantization of angular momentum when an electron is revolving around a proton. We will extend this to a general rotational motion to find quantized rotational energy of a diatomic molecule assuming it to be rigid. The rule to be applied is Bohr's quantization condition.

12. A diatomic molecule has moment of inertia I. By Bohr's quantization condition its rotational energy in the nth level ($n = 0$ *is not allowed*) is

(a) $\frac{1}{n^2}\left(\frac{h^2}{8\pi^2 I}\right)$ (b) $\frac{1}{n}\left(\frac{h^2}{8\pi^2 I}\right)$

(c) $n\left(\frac{h^2}{8\pi^2 I}\right)$ (d) $n^2\left(\frac{h^2}{8\pi^2 I}\right)$

13. It is found that the excitation frequency from ground to the first excited state of rotation for the CO molecule is close to $\frac{4}{\pi} \times 10^{11}$ Hz. Then the moment of inertia of CO molecule about its centre of mass is close to (*Take* $h = 2\pi \times 10^{-34}$ *J-s*)

(a) 2.76×10^{-46} kg-m^2

(b) 1.87×10^{-46} kg-m^2

(c) 4.67×10^{-47} kg-m^2

(d) 1.17×10^{-47} kg-m^2

14. In a CO molecule, the distance between C (*mass* = 12 *amu*) and O (*mass* = 16 *amu*), where 1 amu $= \frac{5}{3} \times 10^{-27}$ kg, is close to

(a) 2.4×10^{-10} m (b) 1.9×10^{-10} m

(c) 1.3×10^{-10} m (d) 4.4×10^{-11} m

P_{15-17} When liquid medicine of density ρ is to be put in the eye, it is done with the help of a dropper. As the bulb on the top of the dropper is pressed, a drop forms at the opening of the dropper. We wish to estimate the size of the drop.

We first assume that the drop formed at the opening is spherical because that requires a minimum increase in its surface energy. To determine the size, we calculate the net vertical force due to the surface tension T when the radius of the drop is R. When this force becomes smaller than the weight of the drop, the drop gets detached from the dropper.

15. If the radius of the opening of the dropper is r, the vertical force due to the surface tension on the drop of radius R (*assuming* $r << R$) is

(a) $2\pi rT$ (b) $2\pi RT$

(c) $\frac{2\pi r^2 T}{R}$ (d) $\frac{2\pi R^2 T}{r}$

16. If $r = 5 \times 10^{-4}$ m, $\rho = 10^3$ kg m^{-3}, $g = 10$ ms^{-2}, $T = 0.11$ Nm^{-1}, the radius of the drop when it detaches from the dropper is approximately

(a) 1.4×10^{-3} m (b) 3.3×10^{-3} m

(c) 2.0×10^{-3} m (d) 4.1×10^{-3} m

17. After the drop detaches, its surface energy is

(a) 1.4×10^{-6} J (b) 2.7×10^{-6} J

(c) 5.4×10^{-6} J (d) 8.1×10^{-9} J

SECTION IV — Matrix Type

This section contains **2 questions**. *Each question has* **four statements** *(A, B, C and D) given in* **Column I** *and* **five statements** *(P, Q, R, S and T) in* **Column II**. *Any given statement in* **Column I** *can have correct matching with one or more statement(s) given in* **Column II**. *For example, if for a given question, statement B matches with the statements given in Q and R, then for that particular question, against statement B, darken the bubbles corresponding to Q and R in the* **ORS**.

18. You are given many resistances, capacitors and inductors. These are connected to a variable DC voltage source (*the first two circuits*) or an AC voltage source of 50 Hz frequency (*the next three circuits*) in different ways as shown in **Column II**. When a current I (*steady state for DC or rms for AC*) flows through the circuit, the corresponding voltage V_1 and V_2 (*indicated in circuits*) are related as shown in **Column I**. Match the two

	Column I		Column II
(A)	$I \neq 0$, V_1 is proportional to I	(P)	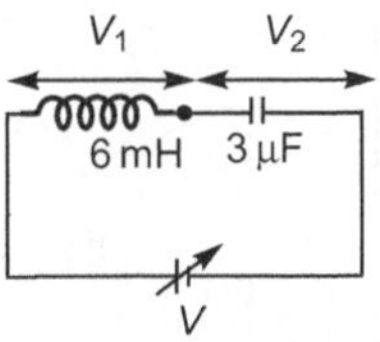
(B)	$I \neq 0$, $V_2 > V_1$	(Q)	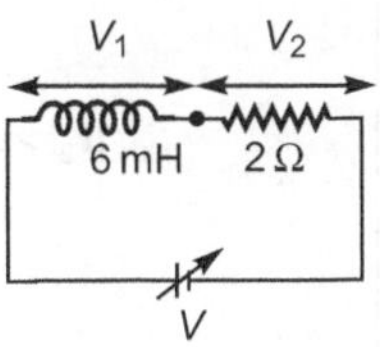
(C)	$V_1 = 0$, $V_2 = V$	(R)	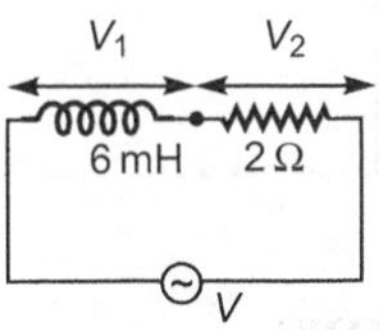
(D)	$I \neq 0$, V_2 is proportional to I	(S)	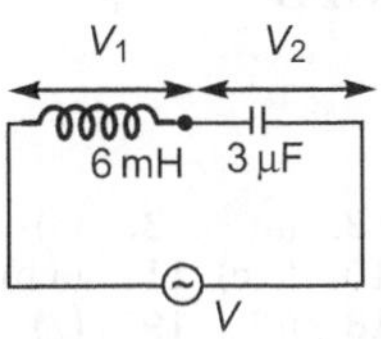
		(T)	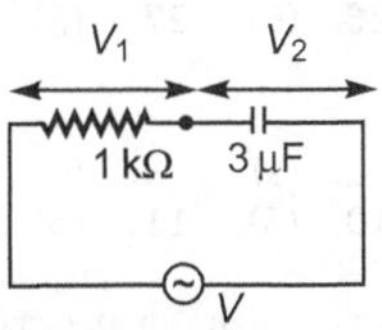

19. Two transparent media of refractive indices μ_1 and μ_3 have a solid lens shaped transparent material of refractive index μ_2 between them as shown in figures in **Column II**. A ray traversing these media is also shown in the figures. In **Column I** different relationships between μ_1, μ_2 and μ_3 are given. Match them to the ray diagram shown in **Column II**.

	Column I		Column II
(A)	$\mu_1 < \mu_2$	(P)	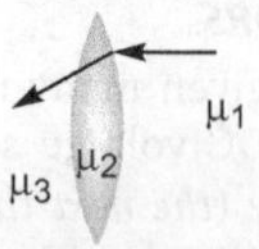
(B)	$\mu_1 > \mu_2$	(Q)	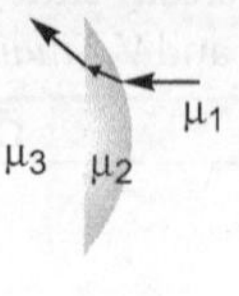
(C)	$\mu_2 = \mu_3$	(R)	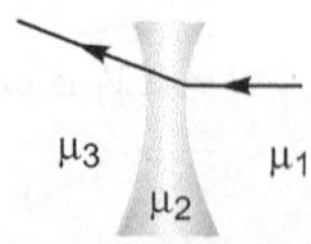
(D)	$\mu_2 > \mu_3$	(S)	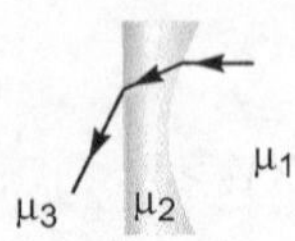
		(T)	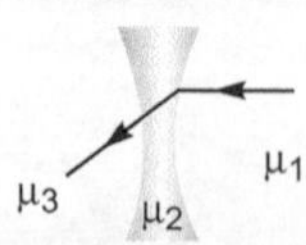

Answers

Paper I

1. (a) **2.** (a) **3.** (d) **4.** (c) **5.** (d) **6.** (c) **7.** (c) **8.** (b)
9. (a,c) **10.** (a,d) **11.** (a,b) **12.** (a,c) **13.** (a,b,c) **14.** (a) **15.** (b) **16.** (c)
17. (b) **18.** (d) **19.** (7) **20.** (6) **21.** (3) **22.** (4) **23.** (9) **24.** (5)
25. (4) **26.** (6) **27.** (3) **28.** (8)

Paper II

1. (d) **2.** (a) **3.** (c) **4.** (b) **5.** (d) **6.** (b) **7.** (8) **8.** (4)
9. (2) **10.** (3) **11.** (6) **12.** (d) **13.** (b) **14.** (c) **15.** (c) **16.** (a)
17. (b)

18. A– R, S, T; B–Q, R, S, T; C–Q, P, Q; D–Q, R, S, T

19. A– P, R; B–Q, S, T; C–P, R, T; D–Q, S

IIT JEE

UNSOLVED PAPER 2010

Chemistry

Paper I

SECTION I—Single Correct Choice Type

This section contains **8 multiple choice questions.** *Each question has 4 choices (a), (b), (c), and (d), out of which* **ONLY ONE** *is correct.*

1. The correct statement about the following disaccharide is

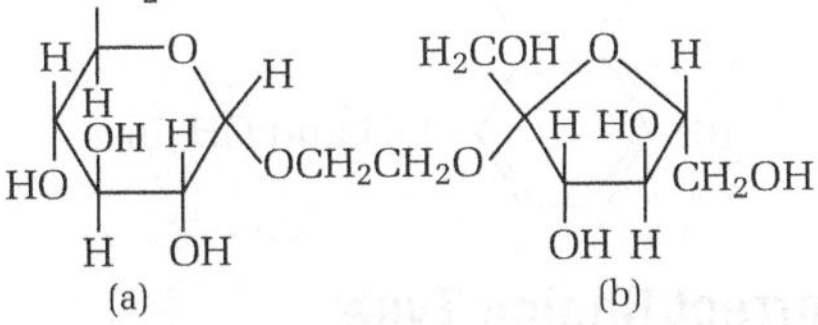

(a) Ring (a) is pyranose with α-glycosidic link

(b) Ring (a) is furanose with α-glycosidic link

(c) Ring (b) is furanose with α-glycosidic link

(d) Ring (b) is pyranose with β-glycosidic link

2. The synthesis of 3-octyne is achieved by adding a bromoalkane into a mixture of sodium amide and an alkyne. The bromoalkane and alkyne respectively are

(a) $BrCH_2CH_2CH_2CH_2CH_3$ and $CH_3CH_2C{\equiv}CH$

(b) $BrCH_2CH_2CH_3$ and $CH_3CH_2CH_2C{\equiv}CH$

(c) $BrCH_2CH_2CH_2CH_2CH_3$ and $CH_3C{\equiv}CH$

(d) $BrCH_2CH_2CH_2CH_3$ and $CH_3CH_2C{\equiv}CH$

3. The ionization isomer of $[Cr(H_2O)_4Cl(NO_2)]Cl$ is

(a) $[Cr(H_2O)_4(O_2N)]Cl_2$

(b) $[Cr(H_2O)_4Cl_2](NO_2)$

(c) $[Cr(H_2O)_4Cl(ONO)]Cl$

(d) $[Cr(H_2O)_4Cl_2(NO_2)] \cdot H_2O$

4. The correct structure of ethylenediaminetetraacetic acid (EDTA) is

(a) 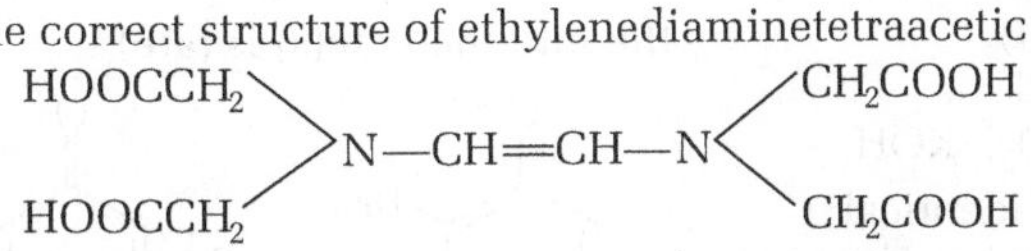

(b) 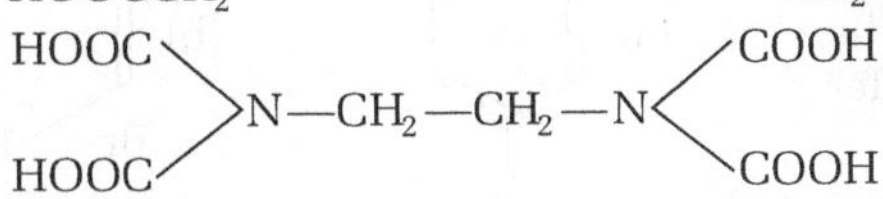

(c) 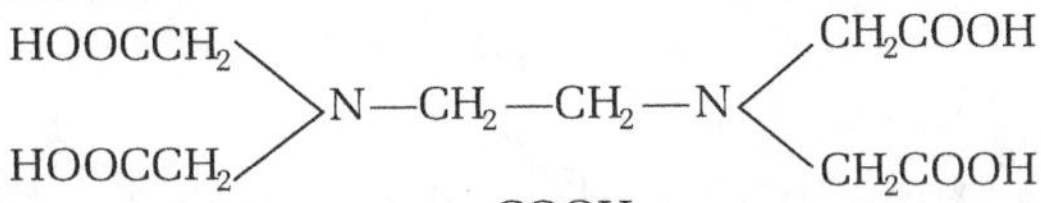

(d) 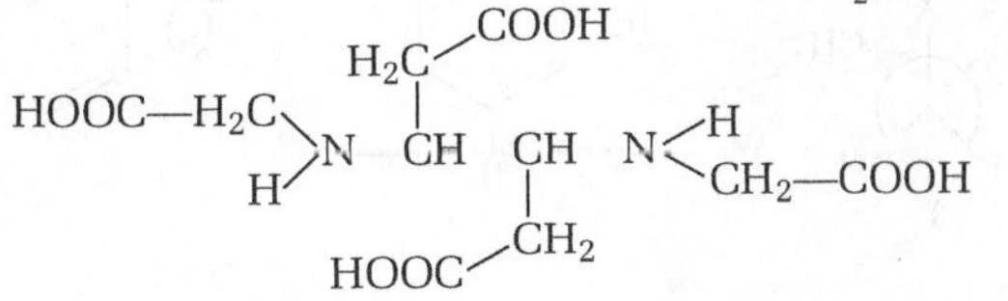

5. The bond energy (in kcal mol^{-1}) of C—C single bond is approximately

(a) 1 (b) 10

(c) 100 (d) 1000

6. The species which by definition has zero standard molar enthalpy of formation at 298 K is

(a) $Br_2(g)$

(b) $Cl_2(g)$

(c) $H_2O(g)$

(d) $CH_4(g)$

7. Plots showing the variation of the rate constant (k) with temperature (T) are given below. The plot that follows Arrhenius equation is

(a) 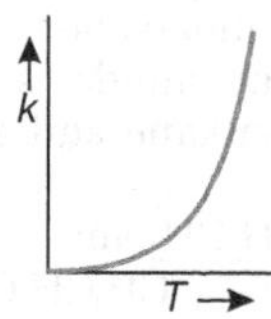

(b) 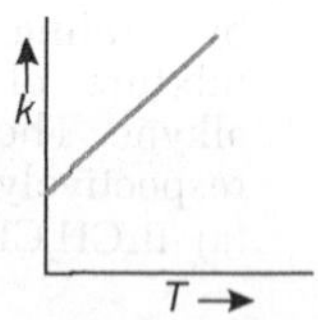

(c) 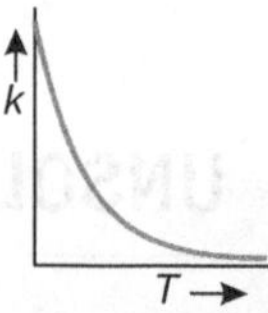

(d) 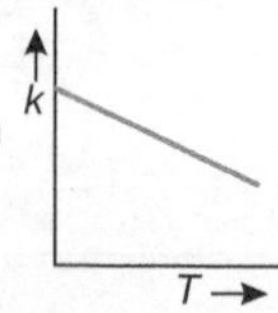

8. In the reaction, 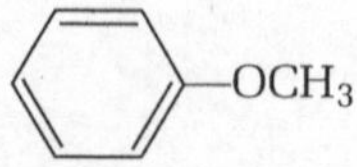

$\xrightarrow{HBr}$ the products are,

(a) 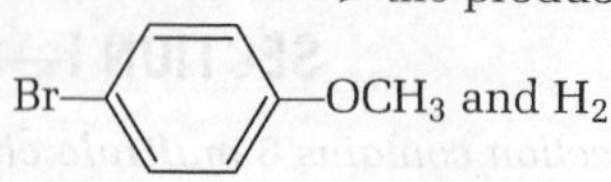

(b) 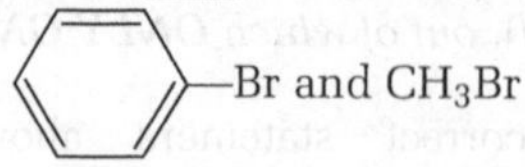

(c) 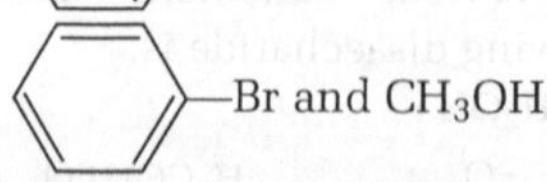

(d) 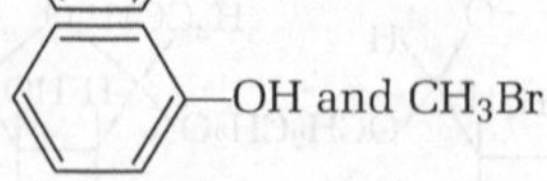

SECTION II–Multiple Correct Choice Type

This section contains **5 multiple choice questions.** *Each question has* 4 *choices* (*a*), (*b*), (*c*), *and* (*d*), *out of which* **one or more** *may be correct.*

9. Among the following, intensive property is (properties are)

(a) molar conductivity

(b) electromotive force

(c) resistance

(d) heat capacity

10. The reagent(s) used for softening the temporary hardness of water is(are)

(a) $Ca_3(PO_4)_2$ (b) $Ca(OH)_2$

(c) Na_2CO_3 (d) NaOCl

11. Aqueous solutions of HNO_3, KOH, CH_3COOH, and CH_3COONa of identical concentrations are provided. The pair(s) of solutions which form a buffer upon mixing is(are)

(a) HNO_3 and CH_3COOH

(b) KOH and CH_3COONa

(c) HNO_3 and CH_3COONa

(d) CH_3COOH and CH_3COONa

12. In the Newman projection for 2,2-dimethylbutane

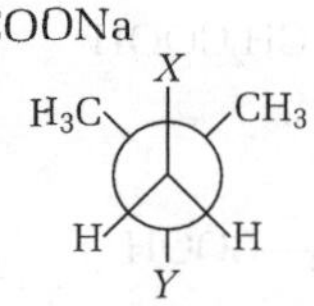

X and Y can respectively be

(a) H and H (b) H and C_2H_5

(c) C_2H_5 and H (d) CH_3 and CH_3

13. In the reaction, 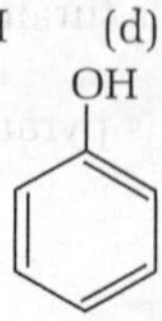

$\xrightarrow{NaOH(aq)/Br_2}$

the intermediate(s) is (are)

(a) 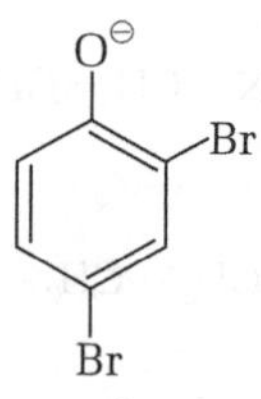

(b) 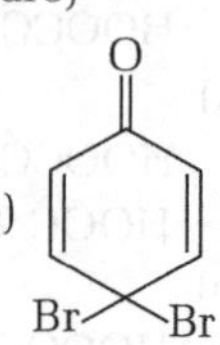

(c) 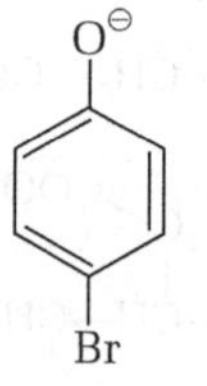

(c) 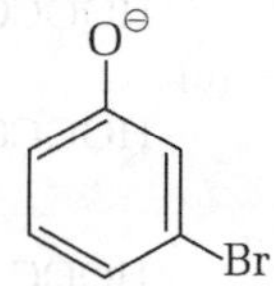

SECTION III—Paragraph Type

*This section contains **2 paragraphs.** Based upon the first paragraph 3 multiple choice questions and based upon the second paragraph **2 multiple choice questions** have to be answered. Each of these questions has 4 choices (a), (b), (c) and (d) out of **WHICH ONLY ONE CORRECT.***

C_{14-16} Copper is the most noble of the first row transition metals and occurs in small deposits in several countries. Ores of copper include chalcanthite ($CuSO_4 \cdot 5H_2O$), atacamite ($Cu_2Cl(OH)_3$), cuprite (Cu_2O), copper glance (Cu_2S) and malachite ($Cu_2(OH)_2CO_3$). However, 80% of the world copper production comes from the ore chalcopyrite ($CuFeS_2$). The extraction of copper from chalcopyrite involves partial roasting, removal of iron and self-reduction.

14. Partial roasting of chalcopyrite produces
(a) Cu_2S and FeO
(b) Cu_2O and FeO
(c) CuS and Fe_2O_3
(d) Cu_2O and Fe_2O_3

15. Iron is removed from chalcopyrite as
(a) FeO (b) FeS
(c) Fe_2O_3 (d) $FeSiO_3$

16. In self-reduction, the reducing species is
(a) S (b) O^{2-} (c) S^{2-} (d) SO_2

C_{17-18} The concentration of potassium ions inside a biological cell is at least twenty times higher than the outside. The resulting potential difference across the cell is important in several processes such as transmission of nerve impulses and maintaining the ion balance. A simple model for such a concentration cell involving a metal M is :

$M(s) | M^+(aq; 0.05 \text{ molar}) || M^+ (aq; 1 \text{ molar}) | M(s)$

For the above electrolytic cell the magnitude of the cell potential $|E_{cell}| = 70$ mV.

17. For the above cell
(a) $E_{cell} < 0; \Delta G > 0$
(b) $E_{cell} > 0; \Delta G < 0$
(c) $E_{cell} < 0; \Delta G^\circ > 0$
(d) $E_{cell} > 0; \Delta G^\circ < 0$

18. If the 0.05 molar solution of M^+ is replaced by a 0.0025 molar M^+ solution, then the magnitude of the cell potential would be
(a) 35 mV (b) 70 mV
(c) 140 mV (d) 700 mV

SECTION IV—Integer Type

*This section contains **TEN questions**. The answer to each question is a **Single digit integer** ranging from 0 to 9. The correct digit below the question number in the ORS is to be bubbled.*

19. Amongst the following, the total number of compounds whose aqueous solution turns red litmus paper blue is

KCN	K_2SO_4	$(NH_4)_2C_2O_4$	NaCl
$Zn(NO_3)_2$	$FeCl_3$	K_2CO_3	NH_4NO_3
LiCN			

20. Based on VSEPR theory, the number of 90 degree F—Br—F angles in BrF_5 is

21. The value of n in the molecular formula $Be_nAl_2Si_6O_{18}$ is

22. A student performs a titration with different burettes and finds titre values of 25.2 mL, 25.25 mL, and 25.0 mL. The number of significant figures in the average titre value is

23. The concentration of R in the reaction $R \longrightarrow P$ was measured as a function of time and the following data is obtained :

$[R]$ (molar)	1.0	0.75	0.40	0.10
t (min)	0.0	0.05	0.12	0.18

The order of the reaction is

24. The number of neutrons emitted when $^{235}_{92}U$ undergoes controlled nuclear fission to $^{142}_{54}Xe$ and $^{90}_{38}Sr$ is

25. The total number of basic groups in the following form of lysine is

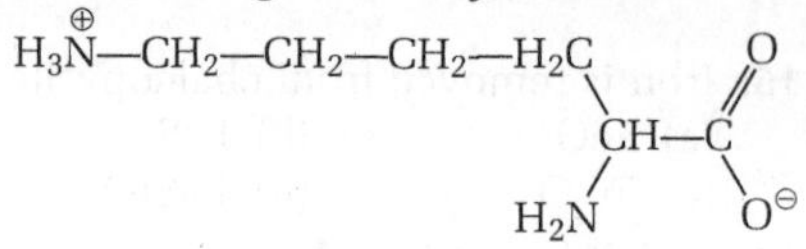

26. The total number of cyclic isomers possible for a hydrocarbon with the molecular formula C_4H_6 is

27. In the scheme given below, the total number of intramolecular aldol condensation products formed from 'Y' is

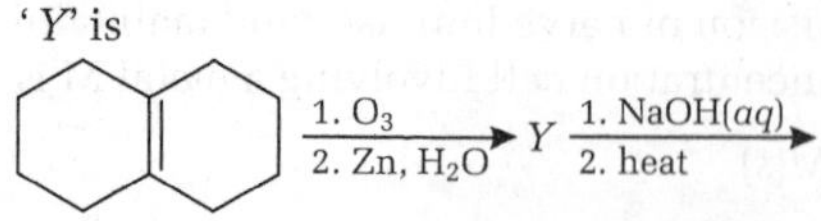

28. Amongst the following, the total number of compounds soluble in aqueous NaOH is

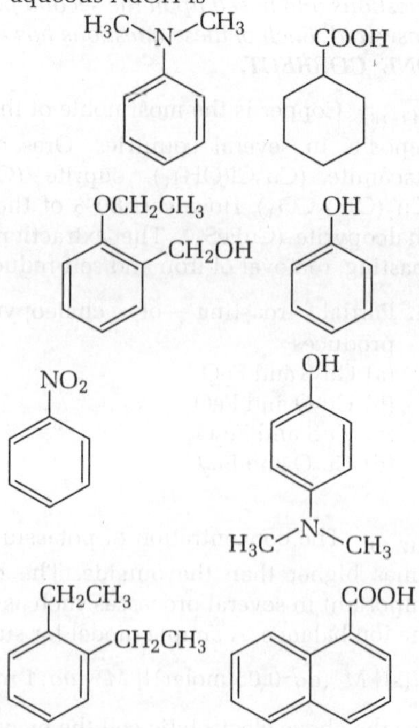

Paper II

SECTION I—Single Correct Choice Type

This section contains **6 multiple choice questions**. *Each question has* 4 *choices* (*a*), (*b*), (*c*), and (*d*), *out of which* **ONLY ONE** *is correct.*

1. The compounds P, Q and S

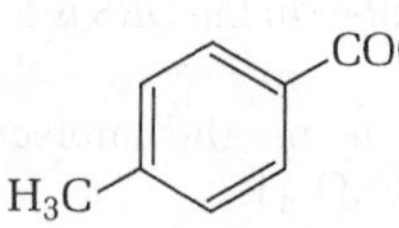

P

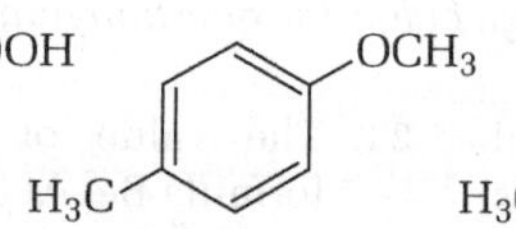

Q

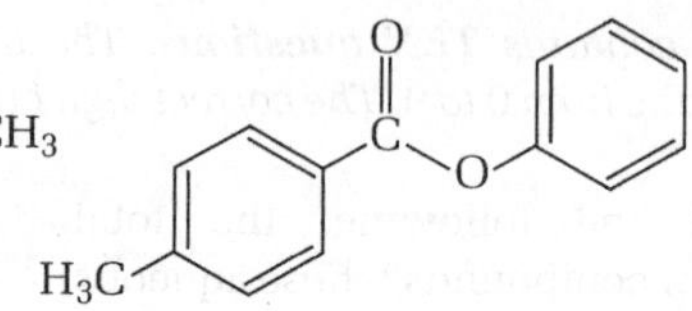

S

where separately subjected to nitration using HNO_3/H_2SO_4 mixture. The major product formed in each case respectively, is

(a) 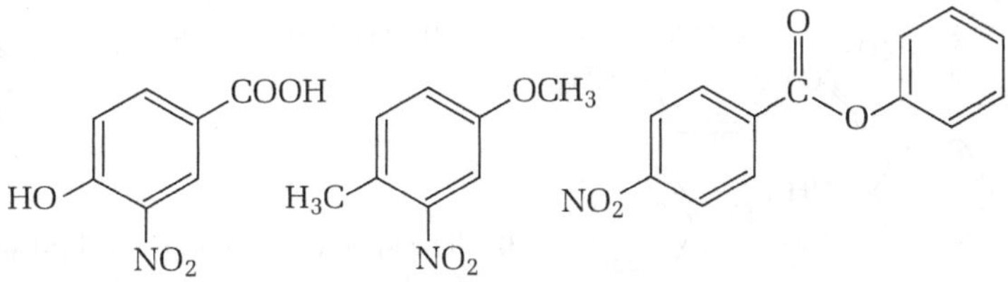

(b) 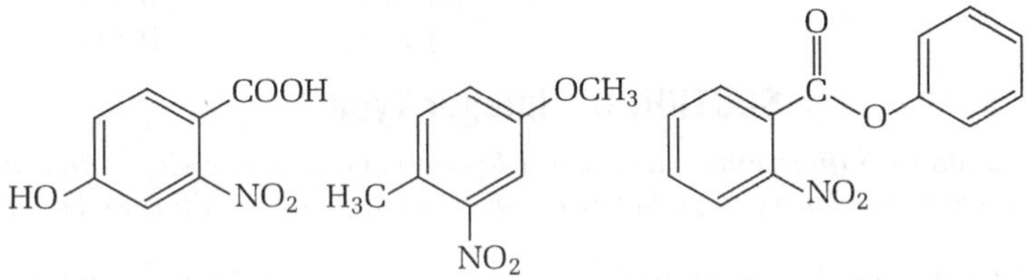

(c) 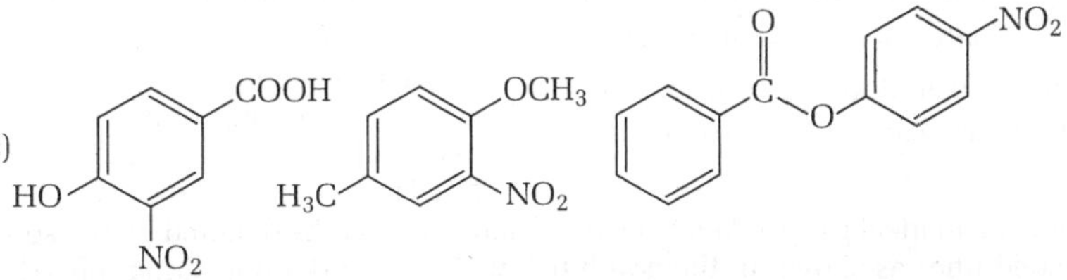

(d) 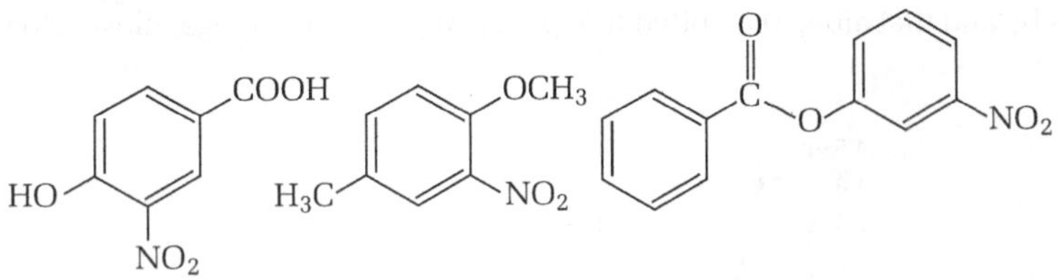

2. Assuming that Hund's rule is violated, the bond order and magnetic nature of the diatomic molecule B_2 is
   (a) 1 and diamagnetic
   (b) 0 and diamagnetic
   (c) 1 and paramagnetic
   (d) 0 and paramagnetic

3. The packing efficiency of the two-dimensional square unit cell shown below is

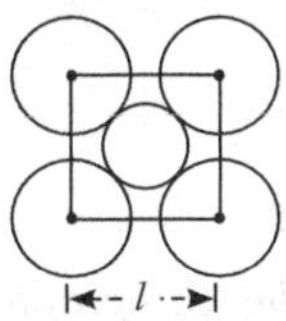

   (a) 39.27% (b) 68.02%
   (c) 74.05% (d) 78.54%

4. The complex showing a spin-only magnetic moment of 2.82 B.M. is
   (a) $Ni(CO)_4$ (b) $[NiCl_4]^{2-}$
   (c) $Ni(PPh_3)_4$ (d) $[Ni(CN)_4]^{2-}$

5. In the reaction

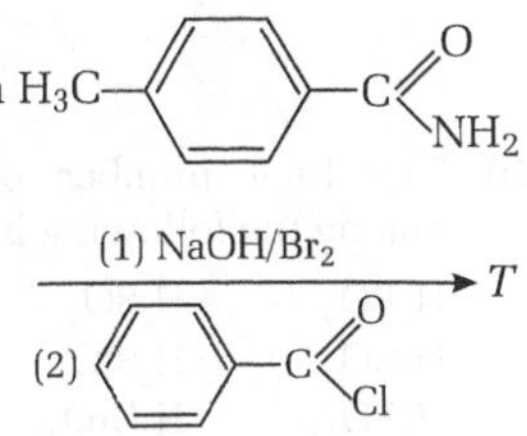

   the structure of the product T is

   (a) 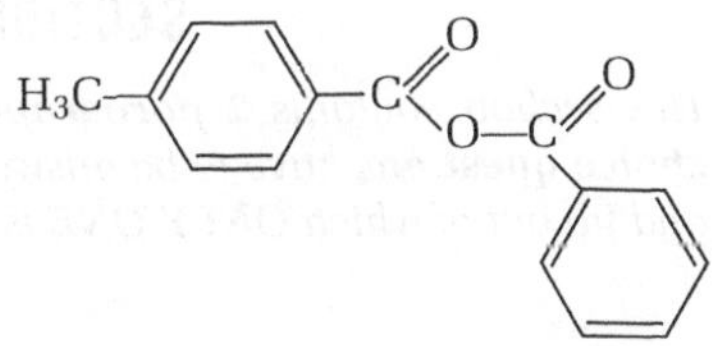

(b) 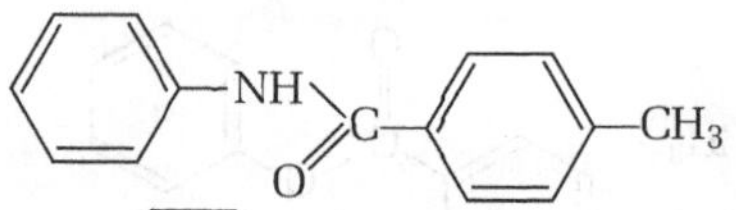

(c) 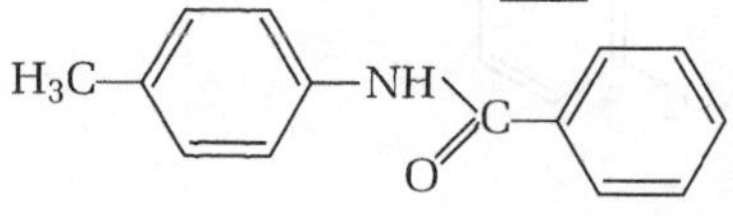

(d) 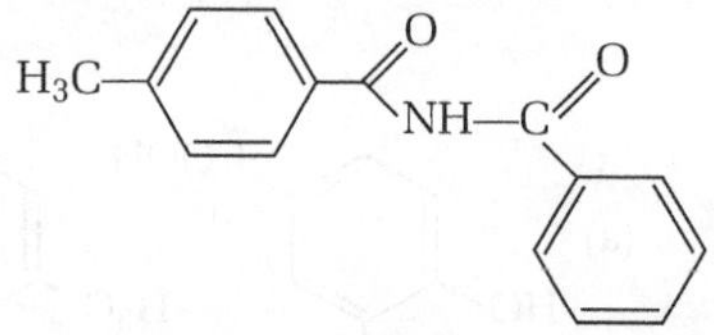

6. The species having pyramidal shape is
(a) SO_3 (b) BrF_3
(c) SiO_3^{2-} (d) OSF_2

SECTION II—Integer Type

*This section contains **5 questions**. The answer to each question is a single-digit integer, ranging from 0 to 9. The correct digit below the question no. in the ORS is to be bubbled.*

7. Silver (atomic weight $= 108 \text{ g mol}^{-1}$) has a density of 10.5 g cm^{-3}. The number of silver atoms on a surface of area 10^{-12} m^2 can be expressed in scientific notation as $y \times 10^x$. The value of x is

8. Among the following, the number of elements showing only one non-zero oxidation state is O, Cl, F, N, P, Sn, Tl, Na, Ti

9. One mole of an ideal gas is taken from a to b along two paths denoted by the solid and the dashed lines as shown in the graph below. If the work done along the solid line path is W_s and that along the dotted line path is W_d, then the integer closest to the ratio $\frac{W_d}{W_s}$ is

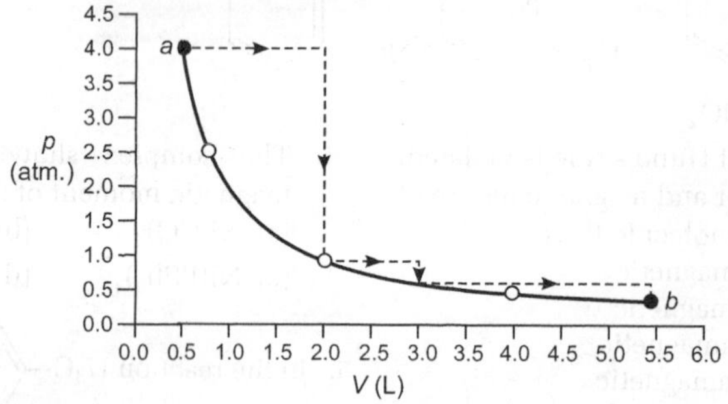

10. The total number of diprotic acids among the following is

H_3PO_4	H_2SO_4	H_3PO_3
H_2CO_3	$H_2S_2O_7$	H_3BO_3
H_3PO_2	H_2CrO_4	H_2SO_3

11. Total number of geometrical isomers for the complex $[RhCl(CO)(PPh_3)(NH_3)]$ is

SECTION III— Paragraph Type

*This section contains **2 paragraphs.** Based upon each of the paragraphs **3 multiple choice questions** have to be answered. Each of these question has 4 choices (a), (b), (c) and (d) out of which **ONLY ONE** is correct.*

C$_{12-14}$ Two aliphatic aldehydes P and Q react in the presence of aqueous K_2CO_3 to give compound R, which upon treatment with HCN provides compound S. On acidification and heating, S gives the product shown below :

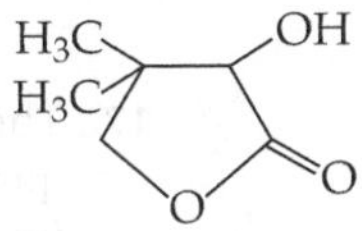

12. The compounds *P* and *Q* respectively are

(a) 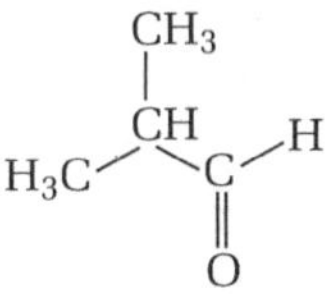 and 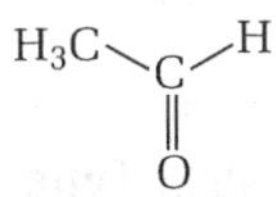

(b) 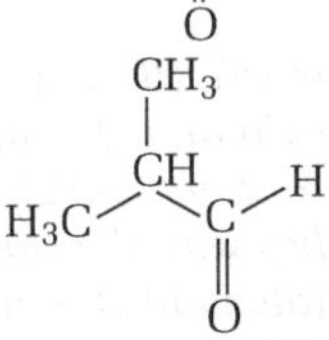 and 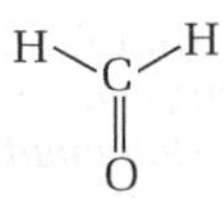

(c) 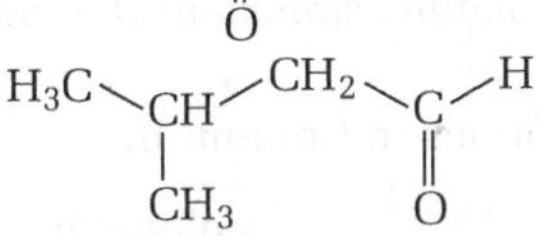 and 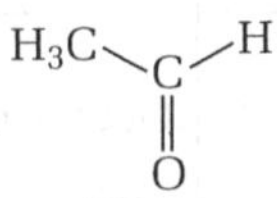

(d) 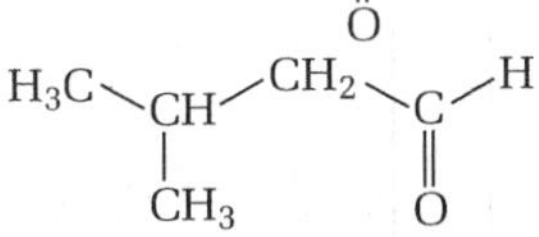 and 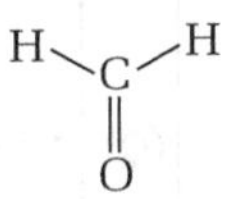

13. The compound *R* is

(a) 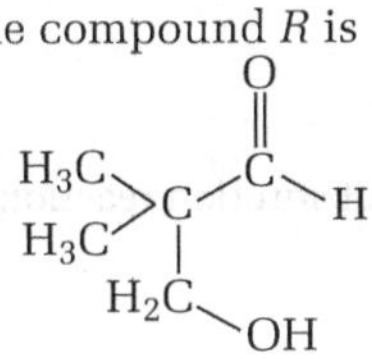

(b) 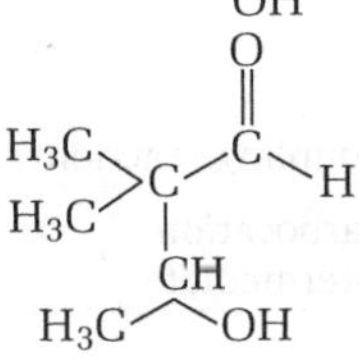

(c) 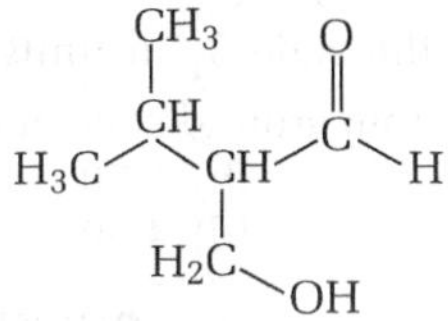

(d) 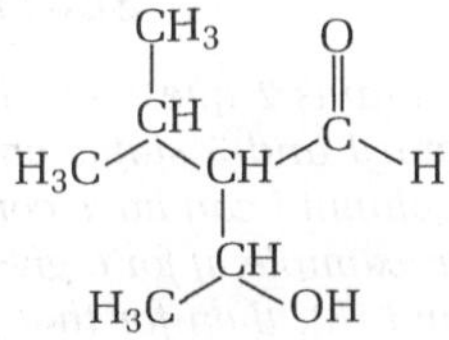

14. The compound *S* is

(a) 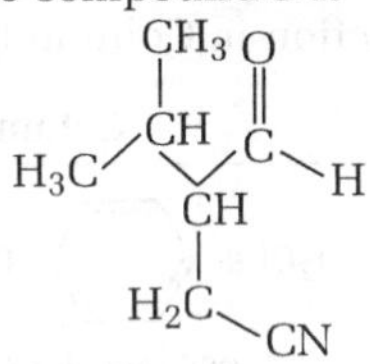

(b) 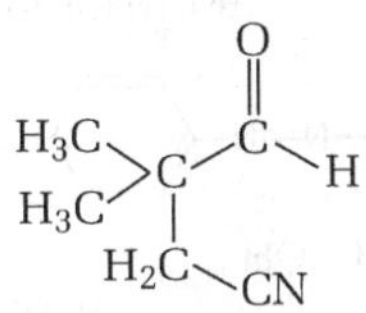

(c) 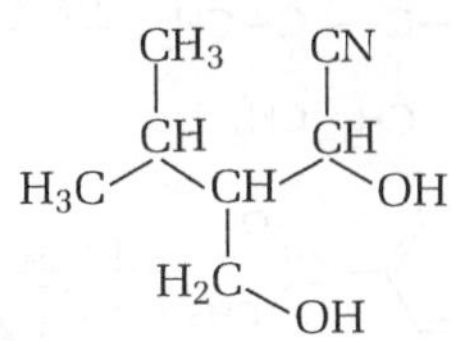

(d) 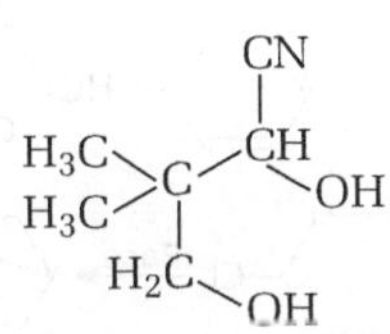

C_{15-17} The hydrogen-like species Li^{2+} is in a spherically symmetric state S_1 with one radial node. Upon absorbing light the ion undergoes transition to a state S_2. The state S_2 has one radial node and its energy is equal to the ground state energy of the hydrogen atom.

15. The state S_1 is

(a) 1*s* (b) 2*s*
(c) 2*p* (d) 3*s*

16. Energy of the state S_1 in units of the hydrogen atom ground state energy is

(a) 0.75 (b) 1.50
(c) 2.25 (d) 4.50

17. The orbtial angular momentum quantum number of the state S_2 is

(a) 0
(b) 1
(c) 2
(d) 3

SECTION IV—Matrix Type

This section contains 2 questions. Each question has 4 statements (A), (B), (C) and (D) given in Column I and 5 statements (P), (Q), (R), (S) and (T) in Column II. Any given statement in Column I can have correct matching with one or more statement(s) given in Column II. For example, if for a given question, statement B matches with the statements given in (Q) and (R), then for that particular question, against statement B, darken the bubbles corresponding to Q and R in the ***ORS****.*

18. Match the reaction in Column I with appropriate options in Column II.

	Column I		Column II
(A)	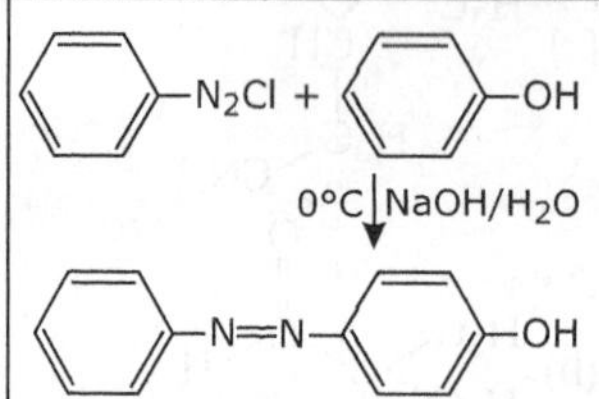	(P)	Racemic mixture
(B)	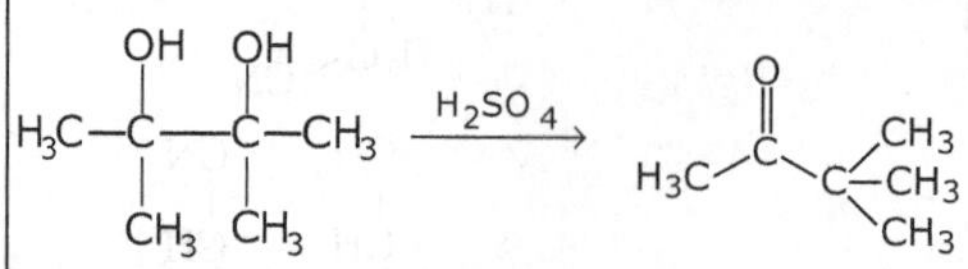	(Q)	Addition reaction
(C)	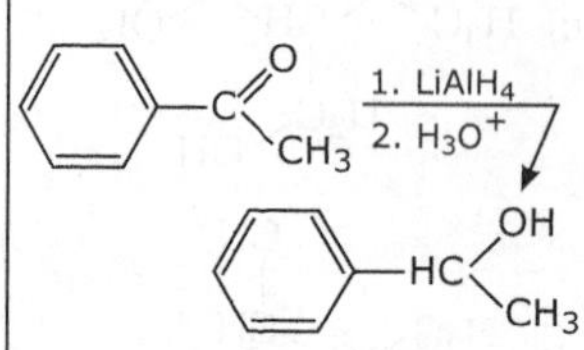	(R)	Substitution reaction
(D)	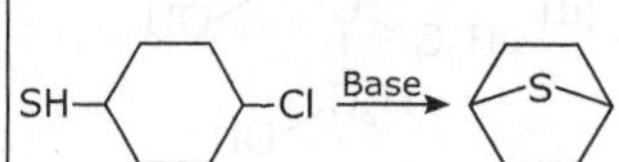	(S)	Coupling reaction
		(T)	Carbocation intermediate

19. All the compounds listed in Column I react with water. Match the result of the respective reactions with the appropriate options listed in Column II.

	Column I		Column II
(A)	$(CH_3)_2SiCl_2$	(P)	Hydrogen halide formation
(B)	XeF_4	(Q)	Redox reaction
(C)	Cl_2	(R)	Reacts with glass
(D)	VCl_5	(S)	Polymerization
		(T)	O_2 formation

Answers

Paper I

1. (a)	**2.** (d)	**3.** (b)	**4.** (c)	**5.** (c)	**6.** (b)	**7.** (a)	**8.** (d)	**9.** (a,b)	**10.** (b,c,d)
11. (c,d)	**12.** (b,d)	**13.** (a,c)	**14.** (a)	**15.** (d)	**16.** (c)	**17.** (b)	**18.** (c)	**19.** (3)	**20.** (0)
21. (3)	**22.** (3)	**23.** (0)	**24.** (3)	**25.** (2)	**26.** (5)	**27.** (1)	**28.** (4)		

Paper II

1. (c)	**2.** (a)	**3.** (d)	**4.** (b)	**5.** (b)	**6.** (d)	**7.** (7)	**8.** (2)	**9.** (2)	**10.** (6)
11. (3)	**12.** (b)	**13.** (a)	**14.** (d)	**15.** (b)	**16.** (c)	**17.** (b)			

18. A– R, S, T; B– T; C–P, Q; D– R

19. A– P, S; B–P, Q, R, T; C– P, Q, T; D–P

IIT JEE
UNSOLVED PAPER 2010

Mathematics

Paper I

SECTION I

(Straight objective Type) Only One Option Correct

This section contains **8 multiple choice questions**. *Each question has 4 choices (a), (b), (c) and (d) for its answer, out of which* **ONLY ONE** *is correct.*

1. The number of 3×3 matrices A whose entries are either 0 or 1 and for which the system $A \begin{bmatrix} x \\ y \\ z \end{bmatrix} = \begin{bmatrix} 1 \\ 0 \\ 0 \end{bmatrix}$ has exactly two distinct solutions, is

(a) 0 (b) $2^9 - 1$
(c) 168 (d) 2

2. The value of $\lim_{x \to 0} \frac{1}{x^3} \int_0^x \frac{t \log(1+t)}{t^4 + 4} dt$ is

(a) 0 (b) $\frac{1}{12}$
(c) $\frac{1}{24}$ (d) $\frac{1}{64}$

3. Let p and q be real numbers such that $p \neq 0$, $p^3 \neq q$ and $p^3 \neq -q$. If α and β are non-zero complex numbers satisfying $\alpha + \beta = -p$ and $\alpha^3 + \beta^3 = q$, then a quadratic equation having $\frac{\alpha}{\beta}$ and $\frac{\beta}{\alpha}$ as its roots is

(a) $(p^3 + q)x^2 - (p^3 + 2q)x + (p^3 + q) = 0$

(b) $(p^3 + q)x^2 - (p^3 - 2q)x + (p^3 + q) = 0$

(c) $(p^3 - q)x^2 - (5p^3 - 2q)x + (p^3 - q) = 0$

(d) $(p^3 - q)x^2 - (5p^3 + 2q)x + (p^3 - q) = 0$

4. Let f, g and h be real-valued functions defined on the interval $[0, 1]$ by $f(x) = e^{x^2} + e^{-x^2}$, $g(x) = xe^{x^2} + e^{-x^2}$ and $h(x) = x^2 e^{x^2} + e^{-x^2}$. If a, b and c denote respectively, the absolute maximum of f, g and h on $[0, 1]$, then

(a) $a = b$ and $c \neq b$ (b) $a = c$ and $a \neq b$
(c) $a \neq b$ and $c \neq b$ (d) $a = b = c$

5. If the angles A, B and C of a triangle are in an arithmetic progression and if a, b and c denote the lengths of the sides opposite to A, B and C respectively, then the value of the expression $\frac{a}{c} \sin 2C + \frac{c}{a} \sin 2A$ is

(a) $\frac{1}{2}$ (b) $\frac{\sqrt{3}}{2}$
(c) 1 (d) $\sqrt{3}$

6. Equation of the plane containing the straight line $\frac{x}{2} = \frac{y}{3} = \frac{z}{4}$ and perpendicular to the plane containing the straight lines $\frac{x}{3} = \frac{y}{4} = \frac{z}{2}$ and $\frac{x}{4} = \frac{y}{2} = \frac{z}{3}$ is

(a) $x + 2y - 2z = 0$ (b) $3x + 2y - 2z = 0$
(c) $x - 2y + z = 0$ (d) $5x + 2y - 4z = 0$

7. Let ω be a complex cube root of unity with $\omega \neq 1$. A fair die is thrown three times. If r_1, r_2 and r_3 are the numbers obtained on the die, then the probability that $\omega^{r_1} + \omega^{r_2} + \omega^{r_3} = 0$ is

(a) $\frac{1}{18}$ (b) $\frac{1}{9}$

(c) $\frac{2}{9}$ (d) $\frac{1}{36}$

8. Let P, Q, R and S be the points on the plane with position vectors $-2\hat{\mathbf{i}}-\hat{\mathbf{j}}$, $4\hat{\mathbf{i}}$, $3\hat{\mathbf{i}}+3\hat{\mathbf{j}}$ an d $-3\hat{\mathbf{i}}+2\hat{\mathbf{j}}$ respectively. The quadrilateral $PQRS$ must be a

(a) parallelogram, which is neither a rhombus nor a rectangle

(b) square

(c) rectangle, but not a square

(d) rhombus, but not a square

SECTION II

(Straight Objective Type) More than One Options are Correct

This section contains 5 **multiple choice questions.** *Each question has* 4 *choices (a), (b), (c) and (d) for its answer, out of which* ***ONE or MORE*** *is/are correct.*

9. The value(s) of $\int_0^1 \frac{x^4(1-x)^4}{1+x^2}\,dx$ is (are)

(a) $\frac{22}{7}-\pi$ (b) $\frac{2}{105}$

(c) 0 (d) $\frac{71}{15}-\frac{3\pi}{2}$

10. Let z_1 and z_2 be two distinct complex numbers and let $z=(1-t)z_1+tz_2$ for some real number t with $0<t<1$. If arg (w) denotes the principal argument of a non-zero complex number w, then

(a) $|z-z_1|+|z-z_2|=|z_1-z_2|$

(b) $\arg(z-z_1)=\arg(z-z_2)$

(c) $\begin{vmatrix} z-z_1 & \bar{z}-\bar{z}_1 \\ z_2-z_1 & \bar{z}_2-\bar{z}_1 \end{vmatrix}=0$

(d) $\arg(z-z_1)=\arg(z_2-z_1)$

11. Let f be a real-valued function defined on the interval $(0,\infty)$, by $f(x)=\ln x+\int_0^x \sqrt{1+\sin t}\,dt$. Then which of the following statement(s) is (are) true ?

(a) $f''(x)$ exists for all $x\in(0,\infty)$

(b) $f'(x)$ exists for all $x\in(0,\infty)$ and f' is continuous on $(0,\infty)$, but not differentiable on $(0,\infty)$

(c) there exists $\alpha>1$ such that $|f'(x)|<|f(x)|$ for all $x\in(\alpha,\infty)$

(d) there exists $\beta>0$ such that $|f(x)|+|f'(x)|\le\beta$ from all $x\in(0,\infty)$

12. Let ABC be a triangle such that $\angle ACB=\frac{\pi}{6}$ and let a, b and c denote the lengths of the sides opposite to A, B and C respectively. The value(s) of x for which $a=x^2+x+1$, $b=x^2-1$ and $c=2x+1$ is (are)

(a) $-(2+\sqrt{3})$ (b) $1+\sqrt{3}$

(c) $2+\sqrt{3}$ (d) $4\sqrt{3}$

13. Let A and B be two distinct points on the parabola $y^2=4x$. If the axis of the parabola touches a circle of radius r having AB as its diameter, then the slope of the line joining A and B can be

(a) $-\frac{1}{r}$ (b) $\frac{1}{r}$

(c) $\frac{2}{r}$ (d) $-\frac{2}{r}$

SECTION III

(Linked Comprehension Type)

This section contains 2 **paragraphs.** *Based upon the first paragraph* 3 **multiple choice questions** *and based upon the second paragraph* 2 **multiple choice questions** *have to answered. Each question has* 4 *choices (a), (b), (c) and (d) for its answer, out of* ***WHICH ONLY ONE.***

Passage I for Q. Nos. 14 to 16 : Let p be an odd prime number and T_p be the following set of 2×2 matrices

$$T_p = \left\{ A = \begin{bmatrix} a & b \\ c & a \end{bmatrix}; a, b, c \in \{0, 1, 2, \dots, p-1\} \right\}$$

14. The number of A in T_p such that A is either symmetric or skew-symmetric or both, and det (A) is divisible by p is

(a) $(p-1)^2$ (b) $2(p-1)$

(c) $(p-1)^2+1$ (d) $2p-1$

15. The number of A in T_p such that the trace of A is not divisible by p but det (A) is divisible by p is

[**Note :** The trace of a matrix is the sum of its diagonal entries.]

(a) $(p-1)(p^2-p+1)$

(b) $p^3-(p-1)^2$

(c) $(p-1)^2$

(d) $(p-1)(p^2-2)$

16. The number of A in T_p such that det (A) is not divisible by p, is

(a) $2p^2$ (b) p^3-5p

(c) p^3-3p (d) p^3-p^2

Passage II for Q. Nos. 17 and 18 : The circle $x^2+y^2-8x=0$ and hyperbola $\frac{x^2}{9}-\frac{y^2}{4}=1$ intersect at the points A and B.

17. Equation of a common tangent with positive slope to the circle as well as to the hyperbola is

(a) $2x-\sqrt{5}y-20=0$

(b) $2x-\sqrt{5}y+4=0$

(c) $3x-4y+8=0$

(d) $4x-3y+4=0$

18. Equation of the circle with AB as its diameter is

(a) $x^2+y^2-12x+24=0$

(b) $x^2+y^2+12x+24=0$

(c) $x^2+y^2+24x-12=0$

(d) $x^2+y^2-24x-12=0$

SECTION IV

(Numerical Value Based Questions)

*This section contains **TEN questions**. The answer to each question is a **Single-digit integer**, ranging from 0 to 9. The correct digit below the question number in the ORS is to be bubbled.*

19. Let ω be the complex number $\cos\frac{2\pi}{3}+i\sin\frac{2\pi}{3}$. Then the number of distinct complex number z satisfying

$$\begin{vmatrix} z+1 & \omega & \omega^2 \\ \omega & z+\omega^2 & 1 \\ \omega^2 & 1 & z+\omega \end{vmatrix} = 0 \text{ is equal to}$$

20. Let S_k, $k=1, 2, \dots, 100$, denote the sum of the infinite geometric series whose first term is $\frac{k-1}{k!}$ and the common ratio is $\frac{1}{k}$. Then the value of $\frac{100^2}{100!}+\sum_{k=1}^{100}|(k^2-3k+1)S_k|$ is

21. The number of all possible values of θ, where $0<\theta<\pi$, for which the system of equations

$$(y+z)\cos 3\theta = (xyz)\sin 3\theta$$

$$x\sin 3\theta = \frac{2\cos 3\theta}{y}+\frac{2\sin 3\theta}{z}$$

and $(xyz)\sin 3\theta = (y+2z)\cos 3\theta + y\sin 3\theta$

have a solution (x_0, y_0, z_0) with $y_0 z_0 \neq 0$, is

22. Let f be a real-valued differentiable function on R (the set of all real numbers) such that $f(1)=1$. If the y-intercept of the tangent at any point $P(x, y)$ on the curve $y=f(x)$ is equal to the cube of the abscissa of P, then the value of $f(-3)$ is equal to

23. The number of values of θ in the interval $\left(-\frac{\pi}{2}, \frac{\pi}{2}\right)$ such that $\theta \neq \frac{n\pi}{5}$ for $n = 0, \pm 1, \pm 2$ and $\tan\theta = \cot 5\theta$ as well as $\sin 2\theta = \cos 4\theta$ is

24. The maximum value of the expression $\frac{1}{\sin^2\theta + 3\sin\theta\cos\theta + 5\cos^2\theta}$ is

25. If $\vec{\mathbf{a}}$ and $\vec{\mathbf{b}}$ are vectors in space given by $\vec{\mathbf{a}} = \frac{\hat{\mathbf{i}} - 2\hat{\mathbf{j}}}{\sqrt{5}}$ and $\vec{\mathbf{b}} = \frac{2\hat{\mathbf{i}} + \hat{\mathbf{j}} + 3\hat{\mathbf{k}}}{\sqrt{14}}$, then the value of $(2\vec{\mathbf{a}} + \vec{\mathbf{b}}) \cdot [(\vec{\mathbf{a}} \times \vec{\mathbf{b}}) \times (\vec{\mathbf{a}} - 2\vec{\mathbf{b}})]$ is

26. The line $2x + y = 1$ is tangent to the hyperbola $\frac{x^2}{a^2} - \frac{y^2}{b^2} = 1$. If this line passes through the point of intersection of the nearest directrix and the x-axis, then the eccentricity of the hyperbola is

27. If the distance between the plane $Ax - 2y + z = d$ and the plane containing the lines $\frac{x-1}{2} = \frac{y-2}{3} = \frac{z-3}{4}$ and $\frac{x-2}{3} = \frac{y-3}{4} = \frac{z-4}{5}$ is $\sqrt{6}$, then $|d|$ is

28. For any real number x, let $[x]$ denotes the largest integer less than or equal to x. Let f be a real valued function defined on the interval $[-10, 10]$ by

$$f(x) = \begin{cases} x - [x] & \text{if } [x] \text{ is odd} \\ 1 + [x] - x & \text{if } [x] \text{ is even} \end{cases}.$$

Then the value of $\frac{\pi^2}{10}\int_{-10}^{10} f(x)\cos \pi x \, dx$ is

Paper II

(Straight Objective Type) Only One Option Correct

This section contains **6 multiple choice questions.** *Each question has 4 choices (a), (b), (c) and (d) for its answer, out of which* ***ONLY ONE*** *is correct.*

1. Let $S = \{1, 2, 3, 4\}$. The total number of unordered pairs of disjoint subsets of S is equal to

(a) 25 (b) 34
(c) 42 (d) 41

2. Let f be a real-valued function defined on the interval $(-1, 1)$ such that $e^{-x} f(x) = 2 + \int_0^x \sqrt{t^4 + 1}\, dt$, for all $x \in (-1, 1)$ and let f^{-1} be the inverse function of f. Then $(f^{-1})'(2)$ is equal to

(a) 1 (b) $\frac{1}{3}$
(c) $\frac{1}{2}$ (d) $\frac{1}{e}$

3. For $r = 0, 1, \ldots, 10$, let A_r, B_r and C_r denote, respectively, the coefficient of x^r in the expansions of $(1+x)^{10}$, $(1+x)^{20}$ and $(1+x)^{30}$. Then $\sum_{r=1}^{10} A_r (B_{10}B_r - C_{10}A_r)$ is equal to

(a) $B_{10} - C_{10}$
(b) $A_{10}(B_{10}^2 - C_{10}A_{10})$
(c) 0
(d) $C_{10} - B_{10}$

4. Two adjacent sides of a parallelogram $ABCD$ are given by $\vec{\mathbf{AB}} = 2\hat{\mathbf{i}} + 10\hat{\mathbf{j}} + 11\hat{\mathbf{k}}$ and $\vec{\mathbf{AD}} = -\hat{\mathbf{i}} + 2\hat{\mathbf{j}} + 2\hat{\mathbf{k}}$. The side AD is rotated by an acute angle α in the plane of the parallelogram so that AD becomes AD'. If AD' makes a right angle with the side AB, then the cosine of the angle α is given by

(a) $\frac{8}{9}$ (b) $\frac{\sqrt{17}}{9}$

(c) $\frac{1}{9}$ (d) $\frac{4\sqrt{5}}{9}$

5. A signal which can be green or red with probability $\frac{4}{5}$ and $\frac{1}{5}$ respectively, is received by station A and then transmitted to station B. The probability of each station receiving the signal correctly is $\frac{3}{4}$. If the signal received at station B is green, then the probability that the original signal green is

(a) $\frac{3}{5}$ (b) $\frac{6}{7}$

(c) $\frac{20}{23}$ (d) $\frac{9}{20}$

6. If the distance of the point $P(1, -2, 1)$ from the plane $x + 2y - 2z = \alpha$, where $\alpha > 0$, is 5, then the foot of the perpendicular from P to the plane is

(a) $\left(\frac{8}{3}, \frac{4}{3}, -\frac{7}{3}\right)$ (b) $\left(\frac{4}{3}, -\frac{4}{3}, \frac{1}{3}\right)$

(c) $\left(\frac{1}{3}, \frac{2}{3}, \frac{10}{3}\right)$ (d) $\left(\frac{2}{3}, -\frac{1}{3}, \frac{5}{2}\right)$

SECTION II — Integer Type

This section contains **5 *questions***. *The answer to each question is a* ***single-digit integer***, *ranging from* 0 *to* 9. The correct digit below the question number in the ORS is to be bubbled.

7. Let $a_1, a_2, a_3, \ldots, a_{11}$ be real numbers satisfying $a_1 = 15$, $27 - 2a_2 > 0$ and $a_k = 2a_{k-1} - a_{k-2}$ for $k = 3, 4, \ldots, 11$. If $\frac{a_1^2 + a_2^2 + \ldots + a_{11}^2}{11} = 90$, then the value of $\frac{a_1 + a_2 + \ldots + a_{11}}{11}$ is equal to

8. Let k be a positive real number and let

$$A = \begin{bmatrix} 2k-1 & 2\sqrt{k} & 2\sqrt{k} \\ 2\sqrt{k} & 1 & -2k \\ -2\sqrt{k} & 2k & -1 \end{bmatrix} \text{ and }$$

$$B = \begin{bmatrix} 0 & 2k-1 & \sqrt{k} \\ 1-2k & 0 & 2\sqrt{k} \\ -\sqrt{k} & -2\sqrt{k} & 0 \end{bmatrix}$$

If $\det(\text{adj}\,A) + \det(\text{adj}\,B) = 10^6$, then $[k]$ is equal to

[**Note** : adj M denotes the adjoint of a square matrix M and $[k]$ denotes the largest integer less than or equal to k].

9. Consider a triangle ABC and let a, b and c denote the lengths of the sides opposite to vertices A, B and C respectively. Suppose $a = 6$, $b = 10$ and the area of the triangle is $15\sqrt{3}$. If $\angle ACB$ is obtuse and if r denotes the radius of the incircle of the triangle, then r^2 is equal to

10. Let f be a function defined on R (the set of all real numbers) such that $f'(x) = 2010(x - 2009)(x - 2010)^2(x - 2011)^3(x - 2012)^4$, for all $x \in R$. If g is a function defined on R with values in the interval $(0, \infty)$ such that $f(x) = \ln(g(x))$, for all $x \in R$, then the number of points in R at which g has a local maximum is

11. Two parallel chords of a circle of radius 2 are at a distance $\sqrt{3} + 1$ apart. If the chords subtend at the centre, angles of $\frac{\pi}{k}$ and $\frac{2\pi}{k}$, where $k > 0$, then the value of $[k]$ is

[**Note** : $[k]$ denotes the largest integer less than or equal to k]

SECTION III

(Linked Comprehension Type)

This section contains 2 Passages. Each Passage has 3 multiple choice questions based on a paragraph. Each question has 4 choices (a), (b), (c) and (d) for its answer, out of which ONLY ONE is correct.

Passage I for Q. Nos. 12 to 14 : Consider the polynomial $f(x) = 1 + 2x + 3x^2 + 4x^3$. Let s be the sum of all distinct real roots of $f(x)$ and let $t = |s|$.

12. The real number s lies in the interval

(a) $\left(-\frac{1}{4}, 0\right)$ (b) $\left(-11, -\frac{3}{4}\right)$

(c) $\left(-\frac{3}{4}, -\frac{1}{2}\right)$ (d) $\left(0, \frac{1}{4}\right)$

13. The area bounded by the curve $y = f(x)$ and the lines $x = 0$, $y = 0$ and $x = t$, lies in the interval

(a) $\left(\frac{3}{4}, 3\right)$ (b) $\left(\frac{21}{64}, \frac{11}{16}\right)$

(c) $(9, 10)$ (d) $\left(0, \frac{21}{64}\right)$

14. The function $f'(x)$ is

(a) increasing in $\left(-t, -\frac{1}{4}\right)$ and decreasing in $\left(-\frac{1}{4}, t\right)$

(b) decreasing in $\left(-t, -\frac{1}{4}\right)$ and increasing in $\left(-\frac{1}{4}, t\right)$

(c) increasing in $(-t, t)$

(d) decreasing in $(-t, t)$

Passage II for Q. Nos. 15 to 17 : Tangents are drawn from the point $P(3, 4)$ to the ellipse $\frac{x^2}{9} + \frac{y^2}{4} = 1$ touching the ellipse at points A and B.

15. The coordinates of A and B are

(a) $(3, 0)$ and $(0, 2)$

(b) $\left(-\frac{8}{5}, \frac{2\sqrt{161}}{15}\right)$ and $\left(-\frac{9}{5}, \frac{8}{5}\right)$

(c) $\left(-\frac{8}{5}, \frac{2\sqrt{161}}{15}\right)$ and $(0, 2)$

(d) $(3, 0)$ and $\left(-\frac{9}{5}, \frac{8}{5}\right)$

16. The orthocentre of the triangle PAB is

(a) $\left(5, \frac{8}{7}\right)$ (b) $\left(\frac{7}{5}, \frac{25}{8}\right)$

(c) $\left(\frac{11}{5}, \frac{8}{5}\right)$ (d) $\left(\frac{8}{25}, \frac{7}{5}\right)$

17. The equation of the locus of the point whose distance from the point P and the line AB are equal, is

(a) $9x^2 + y^2 - 6xy - 54x - 62y + 241 = 0$

(b) $x^2 + 9y^2 + 6xy - 54x + 62y - 241 = 0$

(c) $9x^2 + 9y^2 - 6xy - 54x - 62y - 241 = 0$

(d) $x^2 + y^2 - 2xy + 27x + 31y - 120 = 0$

SECTION IV

(Matrix Matches)

This section contains 2 questions. Each question contains statements given in two columns, which have to be matched. The statements in Column I are labelled A, B, C and D, while the statements in Column II are labelled p, q, r, s and t. Any given statement in Column I can have correct matching with ONE OR MORE statement(s) in Column II.

18. Match the statements of Column I with these in Column II.
[**Note** : Here z takes values in the complex plane and Im(z) and Re(z) denote respectively, the imaginary part and the real part of z]

	Column I		Column II
(A)	The set of points z satisfying $\|z - i\|z\|\| = \|z + i\|z\|\|$ is contained in or equal to	(p)	an ellipse with eccentricity 4/5
(B)	The set of points z satisfying $\|z+4\| + \|z-4\| = 0$ is contained in or equal to	(q)	the set of points z satisfying Im $(z) = 0$
(C)	If $\|w\| = 2$, then the set of points $z = w - \frac{1}{w}$ is contained in or equal to	(r)	the set of points z satisfying $\|\text{Im}\, z\| \le 1$
(D)	If $\|w\| = 1$, then the set of points $z = w + \frac{1}{w}$ is contained in or equal to	(s)	the set of points satisfying $\|\text{Re}\, z\| \le 2$
		(t)	the set of points z satisfying $\|z\| \le 3$

19. Match the statements of Column I with values of Column II.

	Column I		Column II
(A)	A line from the origin meets the lines $\frac{x-2}{1} = \frac{y-1}{2} = \frac{z+1}{1}$ and $\frac{x-\frac{8}{3}}{2} = \frac{y+3}{-1} = \frac{z-1}{1}$ at P and Q respectively. If length $PQ = d$, then d^2 is	(p)	-4
(B)	The value of x satisfying $\tan^{-1}(x+3) - \tan^{-1}(x-3) = \sin^{-1}\left(\frac{3}{5}\right)$ are	(q)	0
(C)	Non-zero vectors $\vec{\mathbf{a}}, \vec{\mathbf{b}}$ and $\vec{\mathbf{c}}$ satisfy $\vec{\mathbf{a}} \cdot \vec{\mathbf{b}} = 0$, $(\vec{\mathbf{b}} - \vec{\mathbf{a}}) \cdot (\vec{\mathbf{b}} + \vec{\mathbf{c}}) = 0$ and $2\|\vec{\mathbf{b}} + \vec{\mathbf{c}}\| = \|\vec{\mathbf{b}} - \vec{\mathbf{a}}\|$. If $\vec{\mathbf{a}} = \mu\vec{\mathbf{b}} + 4\vec{\mathbf{c}}$, then the possible value of μ are	(r)	4
(D)	Let f be the function on $[-\pi, \pi]$ given by $f(0) = 9$ and $f(x) = \sin\left(\frac{9x}{2}\right) / \sin\left(\frac{x}{2}\right)$ for $x \neq 0$. The value of $\frac{2}{\pi}\int_{-\pi}^{\pi} f(x)dx$ is	(s)	5
		(t)	6

Answers

Paper I

1. (a)	**2.** (b)	**3.** (b)	**4.** (d)	**5.** (d)	**6.** (c)	**7.** (c)	**8.** (a)	**9.** (a)	**10.** (a,c,d)
11. (b,c)	**12.** (b)	**13.** (c,d)	**14.** (d)	**15.** (c)	**16.** (d)	**17.** (b)	**18.** (a)	**19.** 1	**20.** 4
21. 3	**22.** 9	**23.** 3	**24.** 2	**25.** 5	**26.** 2	**27.** 6	**28.** 4		

Paper II

1. (d)	**2.** (b)	**3.** (d)	**4.** (b)	**5.** (c)	**6.** (a)	**7.** 0	**8.** 4	**9.** 3	**10.** 1
11. 3	**12.** (c)	**13.** (a)	**14.** (b)	**15.** (d)	**16.** (c)	**17.** (a)			

18. $A \to q,r$; $B \to p$; $C \to p,s,t$; $D \to q,r,s,t$ **19.** $A \to t$; $B \to p,r$; $C \to p$; $D \to r$

IIT JEE

UNSOLVED PAPER 2011

Physics

Paper I

SECTION I

Directions (Q. Nos. 1-7) *This section contains 7 multiple choice questions. Each question has four choices (a), (b), (c) and (d) out of which only one is correct.*

1. Consider an electric field $\mathbf{E} = E_0\hat{\mathbf{x}}$ where $\bar{E}_0$ is a constant. The flux through the shaded area (as shown in the figure) due to this field is

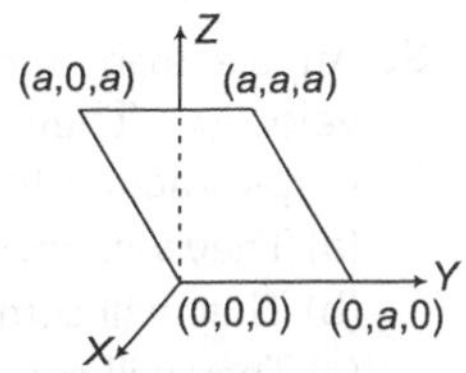

(a) $2E_0 a^2$ (b) $\sqrt{2} E_0 a^2$

(c) $E_0 a^2$ (d) $\frac{E_0 a^2}{\sqrt{2}}$

2. A ball of mass (*m*) 0.5 kg is attached to the end of a string having length (*L*) 0.5 m. The ball is rotated on a horizontal circular path about vertical axis. The maximum tension that the string can bear is 324 N. The maximum possible value of angular velocity of ball (in rad/s) is

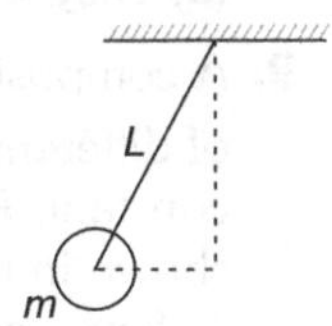

(a) 9 (b) 18 (c) 27 (d) 36

3. A 2μF capacitor is charged as shown in the figure. The percentage of its stored energy dissipated after the switch *S* is turned to position 2, is

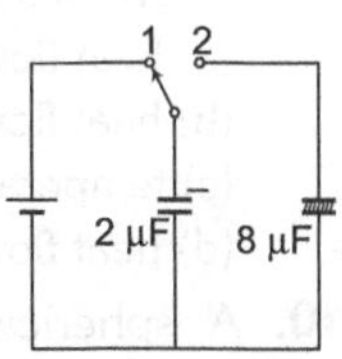

(a) 0% (b) 20%

(c) 75% (d) 80%

4. 5.6 L of helium gas at STP is adiabatically compressed to 0.7 L. Taking the initial temperature to be T_1, the work done in the process is

(a) $\frac{9}{8}RT_1$ (b) $\frac{3}{2}RT_1$ (c) $\frac{15}{8}RT_1$ (d) $\frac{9}{2}RT_1$

5. A police car with a siren of frequency 8 kHz is moving with uniform velocity 36 km/h towards a tall building which reflects the sound waves. The speed of sound in air is 320 m/s. The frequency of the siren heard by the car driver is

(a) 8.50 kHz (b) 8.25 kHz

(c) 7.75 kHz (d) 7.50 kHz

6. The wavelength of the first spectral line in the Balmer series of hydrogen atom is 6561 Å. The wavelength of the second spectral line in the Balmer series of singly ionized helium atom is

(a) 1215 Å (b) 1640 Å (c) 2430 Å (d) 4687 Å

7. A meter bridge is set up as shown in figure, to determine an unknown resistance X using a standard 10 Ω resistor. The galvanometer shows null point when tapping key is at 52 cm mark. The end-corrections are 1 cm and 2 cm respectively for the ends A and B. The determined value of X is

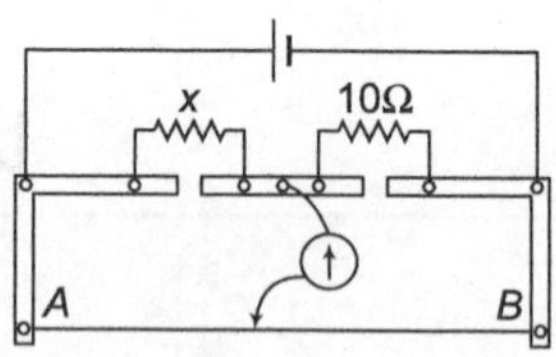

(a) 10.2 Ω (b) 10.6 Ω (c) 10.8 Ω (d) 11.1 Ω

SECTION II

Directions (Q. Nos. 8-11) *This section contains 4 multiple choice questions. Each question has four choices (a), (b), (c) and (d) out of which one or more may be correct.*

8. An electron and a proton are moving on straight parallel paths with same velocity. They enter a semi-infinite region of uniform magnetic field perpendicular to the velocity. Which of the following statement(s) is/are true?

(a) They will never come out of the magnetic field region
(b) They will come out travelling along parallel paths
(c) They will come out at the same time
(d) They will come out at different times

9. A composite block is made of slabs A, B, C, D and E of different thermal conductivities (given in terms of a constant, K) and sizes (given in terms of length, L) as shown in the figure. All slabs are of same width. Heat Q flows only from left to right through the blocks. Then, in steady state

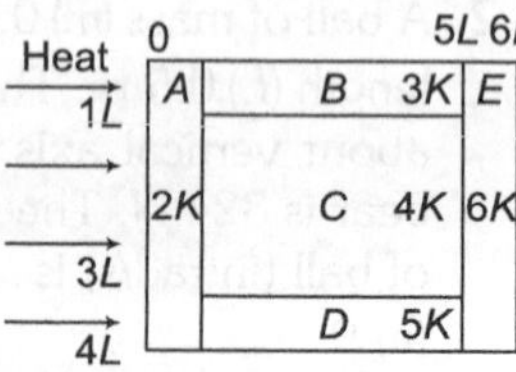

(a) heat flow through A and E slabs are same
(b) heat flow through slab E is maximum
(c) temperature difference across slab E is smallest
(d) heat flow through C = heat flow through B + heat flow through D

10. A spherical metal shell A of radius R_A and a solid metal sphere B of radius $R_B < (R_A)$ are kept far apart and each is given charge $+Q$. Now, they are connected by a thin metal wire. Then

(a) $E_A^{\text{inside}} = 0$ (b) $Q_A > Q_B$

(c) $\dfrac{\sigma_A}{\sigma_B} = \dfrac{R_B}{R_A}$ (d) $E_A^{\text{on surface}} < E_B^{\text{on surface}}$

11. A metal rod of length L and mass m is pivoted at one end. A thin disc of mass M and radius R ($< L$) is attached at its centre to the free end of the rod. Consider two ways the disc is attached. Case A—the disc is not free to rotate about its centre and case B—the disc is free to rotate about its centre. The rod-disc system performs SHM in vertical plane after being released from the same displaced position. Which of the following statement(s) is/are true?

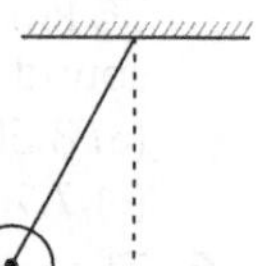

(a) Restoring torque in case $A =$ Restoring torque in case B
(b) Restoring torque in case $A <$ Restoring torque in case B
(c) Angular frequency for case $A >$ Angular frequency for case B
(d) Angular frequency for case $A <$ Angular frequency for case B

SECTION III

Directions (Q. Nos. 12-16) *This section contains 2 paragraphs. Based upon one of the paragraph 2 multiple choice questions and based on the other paragraph 3 multiple choice questions have to be answered. Each of these questions has four choices (a), (b), (c) and (d) out of which only one is correct.*

Paragraph (Q. Nos. 12-13) *A dense collection of equal number of electrons and positive ions is called neutral plasma. Certain solids containing fixed positive ions surrounded by free electrons can be treated as neutral plasma. Let N be the number density of free electrons, each of mass m. When the electrons are subjected to an electric field, they are displaced relatively away from the heavy positive ions.If the electric field becomes zero, the electrons being to oscillate about the positive ions with a natural angular frequency* ω_p, *which is called the plasma frequency. To sustain the oscillations, a time varying electric field needs to be applied that has an angular frequency* ω, *where a part of the energy is absorbed and a part of it is reflected. As* ω *approaches* ω_p, *all the free electrons are set to resonance together and all the energy is reflected. This is the explanation of high reflectivity of metals.*

12. Taking the electronic charge as e and the permittivity as ε_0, use dimensional analysis to determine the correct expression for ω_p.

(a) $\sqrt{\frac{Ne}{m\varepsilon_0}}$ (b) $\sqrt{\frac{m\varepsilon_0}{Ne}}$ (c) $\sqrt{\frac{Ne^2}{m\varepsilon_0}}$ (d) $\sqrt{\frac{m\varepsilon_0}{Ne^2}}$

13. Estimate the wavelength at which plasma reflection will occur for a metal having the density of electrons $N \approx 4 \times 10^{27}\ \text{m}^{-1}$. Take $\varepsilon_0 = 10^{-11}$ and $m = 10^{-30}$, where these quantities are in proper SI units.

(a) 800 nm (b) 600 nm (c) 300 nm (d) 200 nm

Paragraph (Q. Nos. 14-16) *Phase space diagrams are useful tools in analyzing all kinds of dynamical problems. They are especially useful in studying the changes in motion as initial position and momentum are changed. Here we consider some simple dynamical systems in one-dimension. For such systems, phase space is a plane in which position is plotted along horizontal axis and momentum is plotted along vertical axis. The phase space diagram is x(t) vs p(t) curve in this plane. The arrow on the curve indicates the time flow. For example, the phase space diagram for a particle moving with constant velocity is a straight line as shown in the figure. We use the sign convention in which position or momentum upwards (or to right) is positive and downwards (or to left) is negative.*

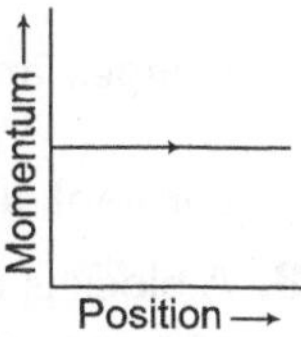

14. The phase space diagram for a ball thrown vertically up from ground is

(a)
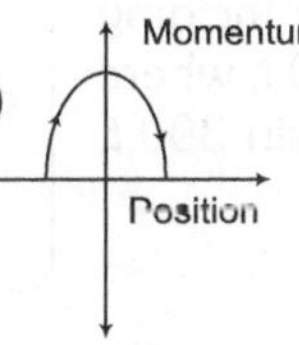

(b)
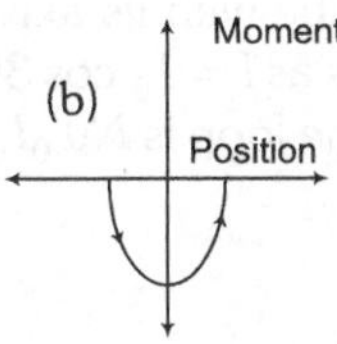

(c)
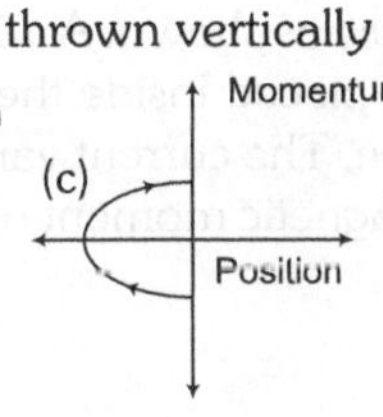

(d)
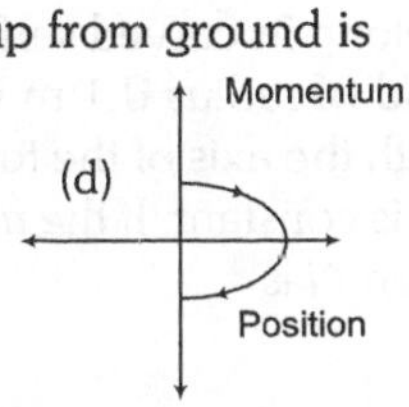

15. The phase space diagram for simple harmonic motion is a circle centered at the origin. In the figure, the two circles represent the same oscillator but for different initial conditions, and E_1 and E_2 are the total mechanical energies respectively. Then,

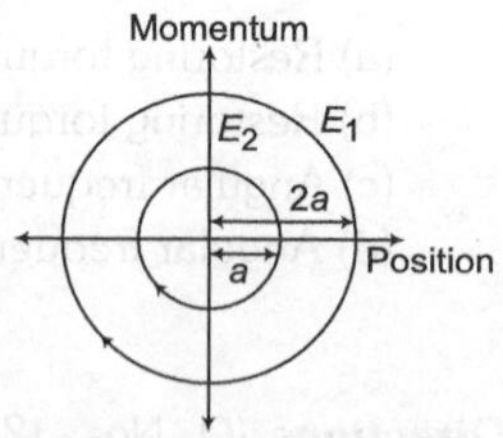

(a) $E_1 = \sqrt{2}E_2$ (b) $E_1 = 2E_2$

(c) $E_1 = 2E_2$ (d) $E_1 = 16E_2$

16. Consider the spring-mass system, with the mass submerged in water, as shown in the figure. The phase space diagram for one cycle of this system is

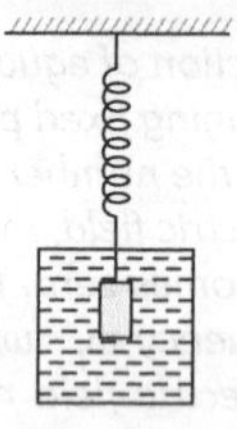

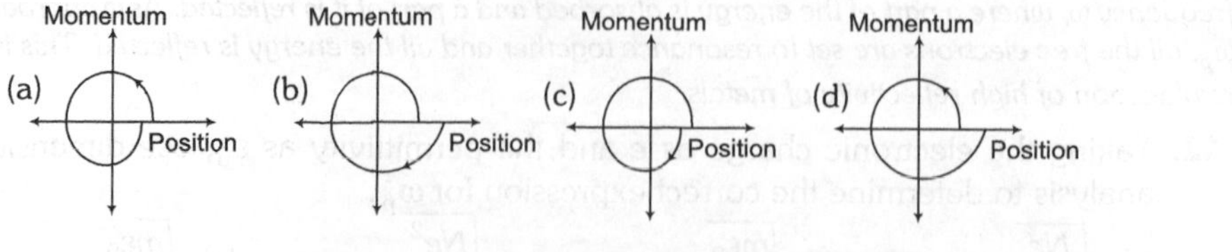

SECTION IV

Directions (Q. Nos. 17-23) *This section contains 7 questions. The answer to each question is a single digit integer, ranging from 0 to 9.*

17. Four point charges, each of $+q$, are rigidly fixed at the four corners of a square planar soap film of side a. The surface tension of the soap film is γ. The system of charges and planar film are in equilibrium, and $a = k\left[\frac{q^2}{\gamma}\right]^{1/N}$, where k is a constant. Then, N is

18. A block is moving on an inclined plane making an angle 45° with the horizontal and the coefficient of friction is μ. The force required to just push it up the inclined plane is 3 times the force required to just prevent it from sliding down. If we define $N = 10\,\mu$, then N is

19. A long circular tube of length 10 m and radius 0.3 m carries a current I along its curved surface as shown. A wire loop of resistance 0.005 Ω and of radius 0.1 m is placed inside the tube with its axis coinciding with the axis of the tube. The current varies as $I = I_0 \cos 300\,t$, where I_0 is constant. If the magnetic moment of the loop is $N\mu_0 I_0 \sin 300\,t$, then N is

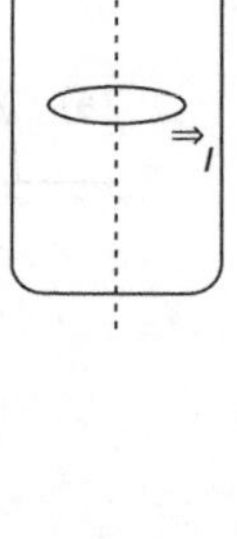

20. Steel wire of length L at 40° C is suspended from the ceiling and then a mass m is hung from its free end. The wire is cooled down from 40°C to 30°C to regain its original length L. The coefficient of linear thermal expansion of the steel is $10^{-5}/°C$, Young's modulus of steel is 10^{11} N/m^2 and radius of the wire is 1 mm. Assume that $L >>$ diameter of the wire. Then, the value of m in kg is nearly

21. Four solid spheres each of diameter $\sqrt{5}$ cm and mass 0.5 kg ar placed with their centres at the corners of a square of side 4 cm. The moment of inertia of the system about the diagonal of the square is $N \times 10^{-4}$ kg-m^2, then N is

22. A boy is pushing a ring of mass 2 kg and radius 0.5 m with a stick as shown in the figure. The stick applies a force of 2 N on the ring and rolls it without slipping with an acceleration of 0.3 m/s^2. The coefficient of friction between the ground and the ring is large enough that rolling always occurs and the coefficient of friction between the stick and the ring is $\frac{P}{10}$. The value of P is

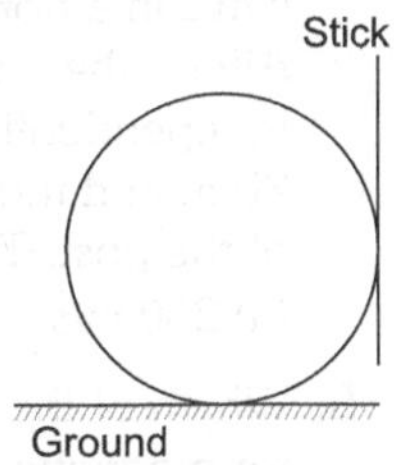

23. The activity of a freshly prepared radioactive sample is 10^{10} disintegrations per second, whose mean life is 10^{-9} s. The mass of an atom of this radioisotope is 10^{-25}kg. The mass (in mg) of the radioactive sample is

Paper II

SECTION I

Directions (Q. Nos. 1-8) *This section contains 8 multiple choice questions. Each question has four choices (a), (b), (c) and (d) out of which only one is correct.*

1. A light ray travelling in glass medium is incident on glass-air interface at an angle of incidence θ. The reflected (R) and transmitted (T) intensities, both as function of θ, are plotted. The correct sketch is

(a) 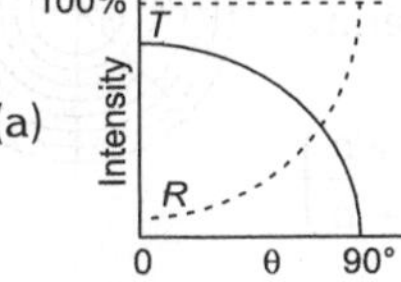

(b) 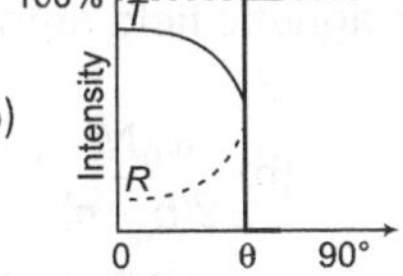

(c) 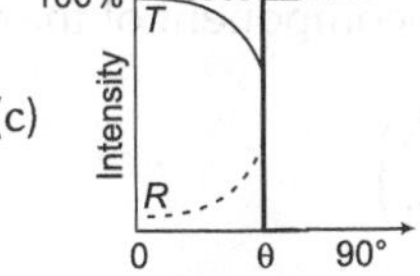

(d) 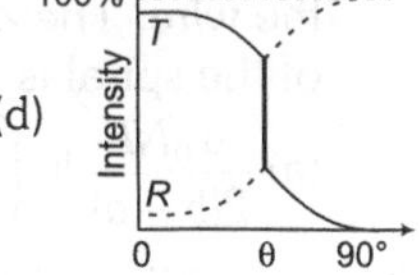

2. A wooden block performs SHM on a frictionless surface with frequency ν_0. The block carries a charge $+Q$ on its surface. If now a uniform electric field **E** is switched on as shown, then the SHM of the block will be

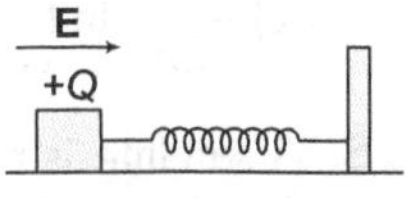

(a) of the dame frequency and with shifted mean position
(b) of the same frequency and with the same mean position
(c) of changed frequency and with shifted mean position
(d) of changed frequency and with the same mean position

3. The density of a solid ball is to be determined in an experiment. The diameter of the ball is measured with a screw gauge, whose pitch is 0.5 mm and there are 50 divisions on the circular scale. The reading on the main scale is 2.5 mm and that on the circular scale is 20 divisions. If the measured mass of the ball has a relative error of 2%, the relative percentage error in the density is

(a) 0.9% (b) 2.4% (c) 3.1% (d) 4.2%

4. A ball of mass 0.2 kg rests on a vertical post of height 5 m. A bullet of mass 0.01 kg, travelling with a velocity v m/s in a horizontal direction, hits the centre of the ball. After the collision, the ball and bullet travel independently. The ball hits the ground at a distance of 20 m and the bullet at a distance of 100 m from the foot of the post. The initial velocity v of the bullet is

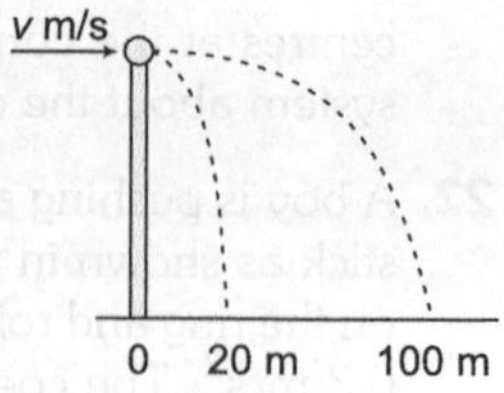

(a) 250 m/s (b) $250\sqrt{2}$ m/s (c) 400 m/s (d) 500 m/s

5. Which of the field patterns given in the figure is valid for electric field as well as for magnetic field?

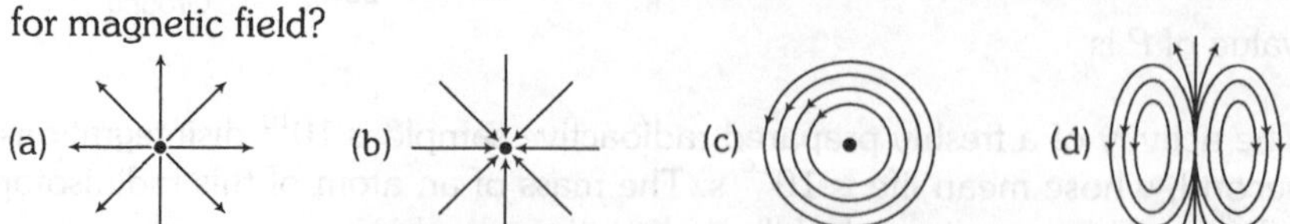

6. A point mass is subjected to two simultaneous sinusoidal displacements in x-direction, $x_1(t) = A \sin \omega t$ and $x_2(t) = A \sin\left(\omega t + \frac{2\pi}{3}\right)$. Adding a third sinusoidal displacement $x_3(t) = B \sin(\omega t + \phi)$ bring the mass to a complete rest. The values of B and ϕ are

(a) $\sqrt{2}A, \frac{3\pi}{4}$ (b) $A, \frac{4\pi}{3}$ (c) $\sqrt{3}A, \frac{5\pi}{6}$ (d) $A, \frac{\pi}{3}$

7. A long insulated copper wire is closely wound as a spiral of N turns. The spiral has inner radius a and outer radius b. The spiral lies in the XY-plane and a steady current I flows through the wire. The Z-component of the magnetic field at the centre of the spiral is

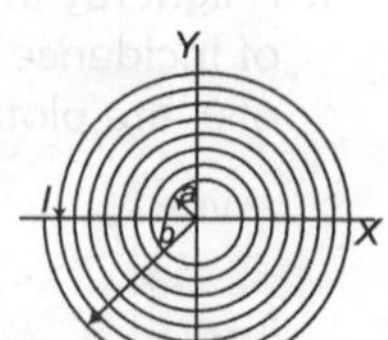

(a) $\frac{\mu_0 NI}{2(b-a)} \ln\left(\frac{b}{a}\right)$ (b) $\frac{\mu_0 NI}{2(b-a)} \ln\left(\frac{b+a}{b-a}\right)$

(c) $\frac{\mu_0 NI}{2b} \ln\left(\frac{b}{a}\right)$ (d) $\frac{\mu_0 NI}{2b} \ln\left(\frac{b+a}{b-a}\right)$

8. A satellite is moving with a constant speed v in a circular orbit about the earth. An object of mass m is ejected from the satellite such that it just escapes from the gravitational pull of the earth. At the time of its ejection, the kinetic energy of the object is

(a) $\frac{1}{2}mv^2$ (b) mv^2 (c) $\frac{3}{2}mv^2$ (d) $2mv^2$

SECTION II

Directions (Q. Nos. 9-12) *This section contains 4 multiple choice questions. Each question has four choices (a), (b), (c) and (d) out of which one or more may be correct.*

9. Two solid spheres A and B of equal volumes but of different densities d_A and d_B are connected by a string. They are fully immersed in a fluid of density d_F. They get arranged into an equilibrium state as shown in the figure with a tension in the string. The arrangement is possible only if

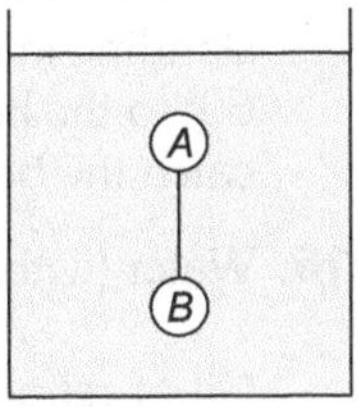

(a) $d_A < d_F$ (b) $d_B > d_F$

(c) $d_A > d_F$ (d) $d_A + d_B = 2d_F$

10. Which of the following statement(s) is/are correct?

(a) If the electric field due to a point charge varies as $r^{-2.5}$ instead of r^{-2}, then the Gauss's law will still be valid

(b) The Gauss's law can be used to calculate the field distribution around an electric dipole

(c) If the electric field between two point charges is zero somewhere, then the sign of the two charges is the same

(d) The work done by the external force in moving a unit positive charge from point A at potential V_A to point B at potential V_B is $(V_B - V_A)$

11. A series R-C circuit is connected to AC voltage source. Consider two cases; (A) when C is without a dielectric medium and (B) when C is filled with dielectric of constant 4. The current I_R through the resistor and voltage V_C across the capacitor are compared in the two cases. Which of the following is/are true?

(a) $I_R^A > I_R^B$ (b) $I_R^A < I_R^B$ (c) $V_C^A > V_C^B$ (d) $V_C^A < V_C^B$

12. A thin ring of mass 2 kg and radius 0.5 m is rolling without slipping on a horizontal plane with velocity 1 m/s. A small ball of mass 0.1 kg, moving with velocity 20 m/s in the opposite directions, hits the ring at a height of 0.75 m and goes vertically up with velocity 10 m/s. Immediately after the collision

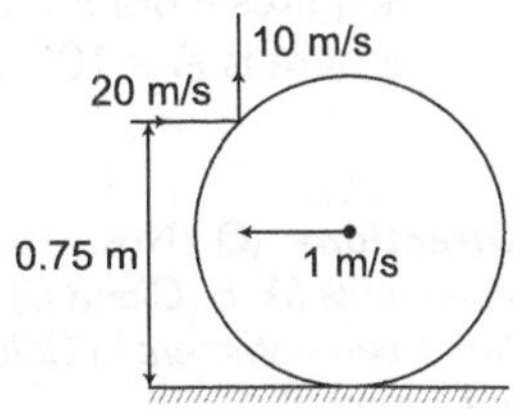

(a) the ring has pure rotation about its stationary CM

(b) the ring comes to a complete stop

(c) friction between the ring and the ground is to the left

(d) there is no friction between the ring and the ground

SECTION III

Directions (Q. Nos. 13-18) *This sections contain 6 questions. The answer to each of the question is a single-digit integer, ranging from 0 to 9.*

13. Two batteries of different emfs and different internal resistances are connected as shown. The voltage across AB in volt is

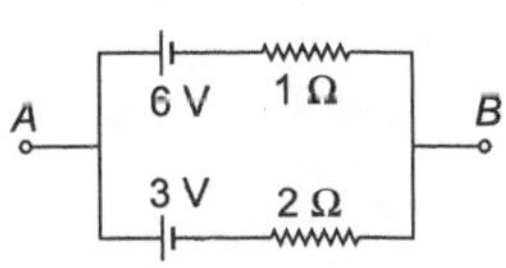

14. A series *R-C* combination is connected to an AC voltage of angular frequency $\omega = 500$ rad/s. If the impedance of the *R-C* circuit is $R\sqrt{1.25}$, the time constant (in millisecond) of the circuit is

15. A train is moving along a straight line with a constant acceleration *a*. A boy standing in the train throws a ball forward with a speed of 10 m/s, at an angle of 60° to the horizontal. The boy has to move forward by 1.15 m inside the train to catch the ball back at the initial height. The acceleration of the train in m/s^2, is

16. Water (with refractive index = $\frac{4}{3}$) in a tank is 18 cm deep. Oil of refractive index $\frac{7}{4}$ lies on water making a convex surface of radius of curvature $R = 6$ cm as shown. Consider oil to act as a thin lens. An object *S* is placed 24 cm above water surface. The location of its image is at *x* cm above the bottom of the tank. Then, *x* is

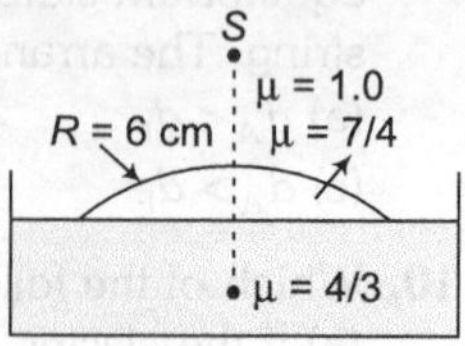

17. A block of mass 0.18 kg is attached to a spring of force constant 2 N/m. The coefficient of friction between the block and the floor is 0.1. Initially, the block is at rest and the spring is unstretched. An impulse is given to the block as shown in the figure. The block slides a distance of 0.06 m and comes to rest for the first time. The initial velocity of the block in m/s is $v = \frac{N}{10}$. Then, *N* is

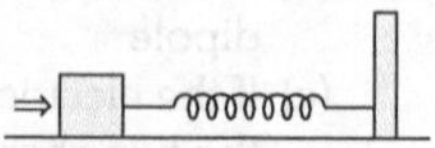

18. A silver sphere of radius 1 cm and work function 4.7 eV is suspended from an insulating thread in free-space. It is under continuous illumination of 200 nm wavelength light. As photoelectrons are emitted, the sphere gets charged and acquires a potential. The maximum number of photoelectrons emitted from the sphere is $A \times 10^Z$ (where $1 < A < 10$). The value of *Z* is

SECTION IV

Directions (Q. Nos. 19-20) *This section contains 2 questions. Each question has four statements (A, B, C and D) given in Column* I *and five statements (p,q, r, s and t) in Column* II. *Any given statement in Column* I *can have correct matching with one or more statement(s) given in Column* II.

19. One mole of a monatomic ideal gas is taken through a cycle *ABCDA* as shown in the *p-V* diagram. Column II gives the characteristics involved in the cycle. Match them with each of the processes given in Column I.

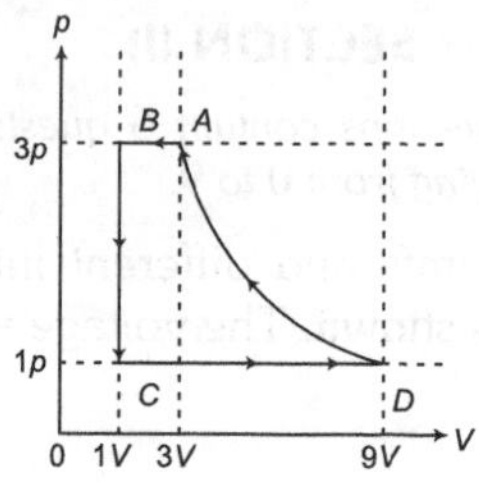

Column I	Column II
(A) Process $A \rightarrow B$	(p) Internal energy decreases
(B) Process $B \rightarrow C$	(q) Internal energy increases
(C) Process $C \rightarrow D$	(r) Heat is lost
(D) Process $D \rightarrow A$	(s) Heat is gained
	(t) Work is done on the gas

20. Column I showns four systems, each of the same length L, for producing standing waves. The lowest possible natural frequency of a system is called its fundamental frequency, whose wavelength is denoted as λ_f. Match each system with statements given in Column II describing the nature and wave length of the standing waves.

Column I	Column II
(A) Pipe closed at one end 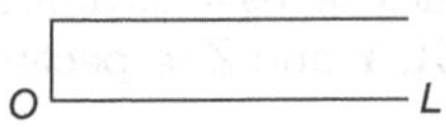	(p) Longitudinal waves
(B) Pipe open at both ends 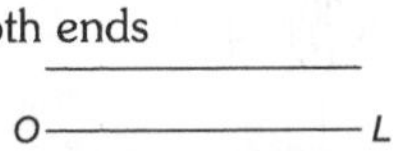	(q) Transverse waves
(C) Stretched wire clamped at both ends 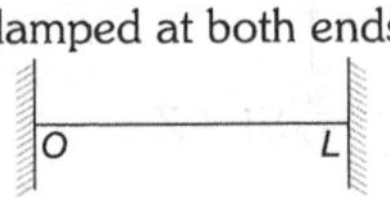	(r) $\lambda_f = L$
(D) Stretched wire clamped at both ends and at mid-point 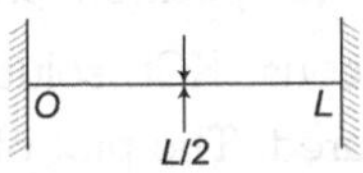	(s) $\lambda_f = 2L$
	(t) $\lambda_f = 4L$

ANSWERS

Paper I

1. (c) **2.** (d) **3.** (d) **4.** (a) **5.** (a) **6.** (a)
7. (b) **8.** (b,d) **9.** (a,c,d) **10.** (a,b,c,d) **11.** (a,d) **12.** (c)
13. (b) **14.** (d) **15.** (c) **16.** (b) **17.** (3) **18.** (5)
19. (6) **20.** (3) **21.** (9) **22.** (4) **23.** (1)

Paper II

1. (c) **2.** (a) **3.** (c) **4.** (d) **5.** (c) **6.** (b)
7. (a) **8.** (b) **9.** (a,b,d) **10.** (c,d) **11.** (b,c) **12.** (a,c)
13. (5) **14.** (4) **15.** (5) **16.** (2) **17.** (4) **18.** (7)
19. (A) $\rightarrow$ p,r,t; (B) $\rightarrow$ p,r; (C) $\rightarrow$ q, s; (D) $\rightarrow$ r **20.** (A) $\rightarrow$ p,t; (B) $\rightarrow$ p,s; (C) $\rightarrow$ q, s; (D) $\rightarrow$ q, r

IIT JEE
UNSOLVED PAPER 2011

Chemistry

Paper I

SECTION I

Directions (Q. Nos. 1 to 7) *This section contains 7 multiple choice questions. Each question has four choices (a), (b), (c), and (d), out of which only one is correct.*

1. Bombardment of aluminium by α-particle leads to its artificial disintegration in two ways, (i) and (ii) as shown. Products X, Y and Z respectively are

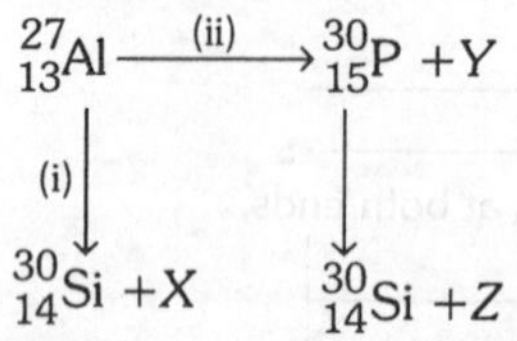

(a) proton, neutron, positron
(b) neutron, positron, proton
(c) proton, positron, neutron
(d) positron, proton, neutron

2. $AgNO_3(aq)$ was added to an aqueous KCl solution gradually and the conductivity of the solution was measured. The plot of conductance (Λ) *versus* the volume of $AgNO_3$ is

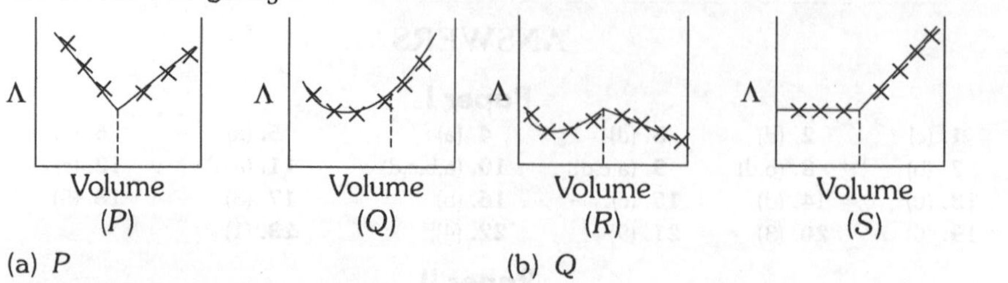

(a) P
(b) Q
(c) R
(d) S

3. The major product of the following reaction is

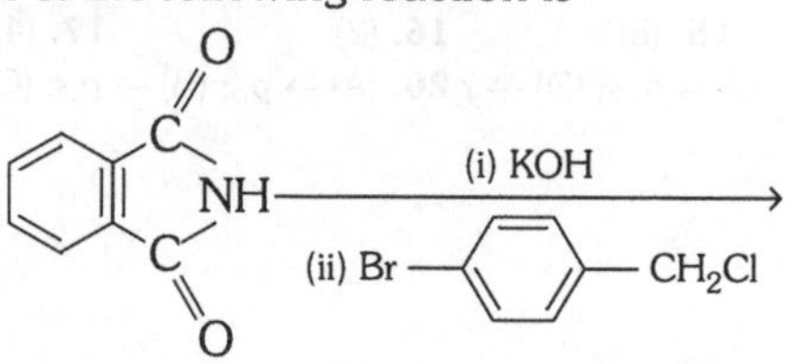

(a) 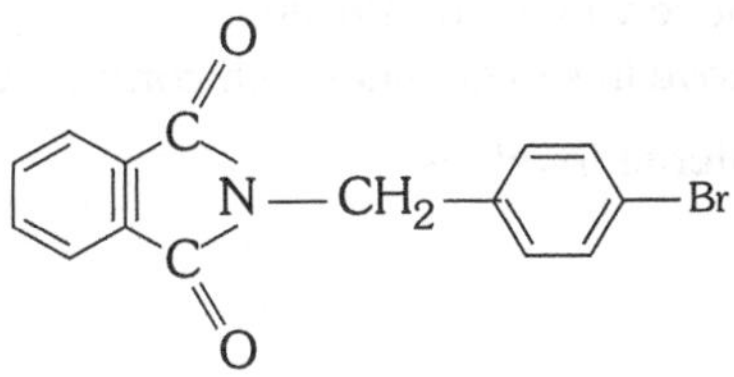 (b) 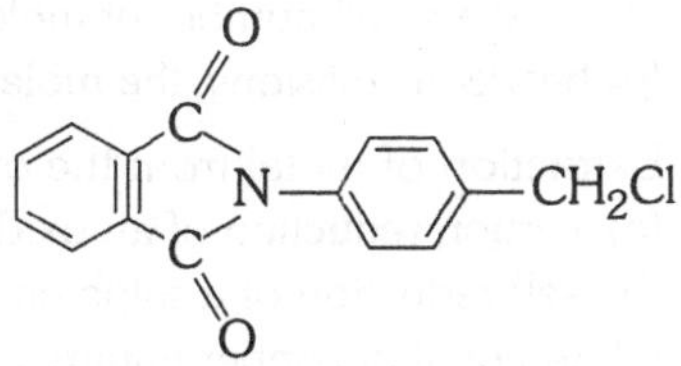

(c) 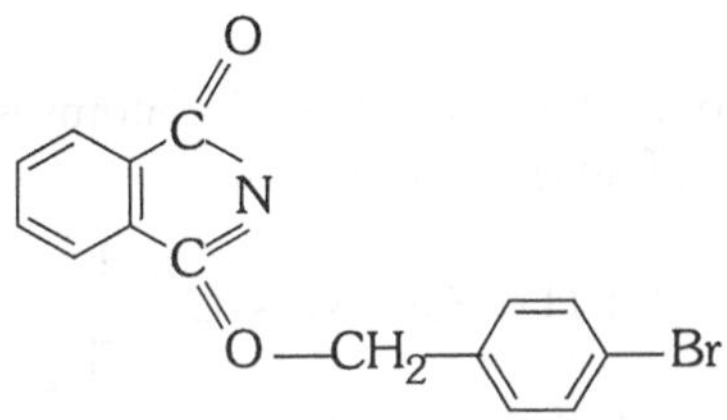 (d) 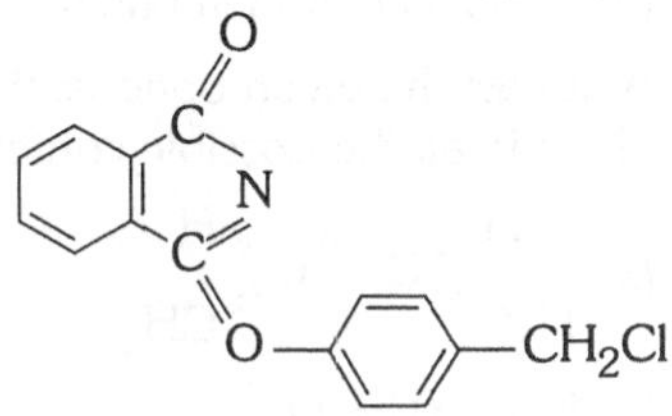

4. Geometrical shapes of the complexes formed by the reaction of Ni^{2+} with Cl^-, CN^- and H_2O, respectively, are

(a) octahedral, tetrahedral and square planar
(b) tetrahedral, square planar and octahedral
(c) square planar, tetrahedral and octahedral
(d) octahedral, square planar and octahedral

5. Extra pure N_2 can be obtained by heating

(a) NH_3 with CuO (b) NH_4NO_4 (c) $(NH_4)_2Cr_2O_7$ (d) $Ba(N_3)_2$

6. Dissolving 120 g of urea (molecular weight 60) in 1000 g of water gave a solution of density 1.15 g /mL. The molarity of the solution is

(a) 1.78 M (b) 2.00 M (c) 2.05 M (d) 2.22 M

7. Among the following compounds, the most acidic is

(a) *p*-nitrophenol (b) *p*-hydroxybenzoic acid
(c) *o*-hydroxybenzoic acid (d) *p*-toluic acid

SECTION II

Directions (Q. Nos. 8 to 11) *This section contains 4 multiple choice questions. Each question has four choices (a), (b), (c), and (d), out of which only one is correct.*

8. The correct statement(s) pertaining to the adsorption of a gas on a solid surface is(are)

(a) Adsorption is always exothermic
(b) Physisorption may transform into chemisorption at high temperature
(c) Physisorption increases with increasing temperature but chemisorption decreases with increasing temperature
(d) Chemisorption is more exothermic than physisorption, however it is very slow due to higher energy of activation

9. According to kinetic theory of gases

(a) collisions are always elastic
(b) heavier molecules transfer more momentum to the wall of the container

(c) only a small number of molecules have very high velocity
(d) between collisions, the molecules move in straight lines with constant velocities

10. Extraction of metal from the ore cassiterite involves
(a) carbon reduction of an oxide ore
(b) self-reduction of a sulphide ore
(c) removal of copper impurity
(d) removal of iron impurity

11. Amongst the given options, the compound (s) in which all the atoms are in one plane in all the possible conformations (if any), is (are)

(a) 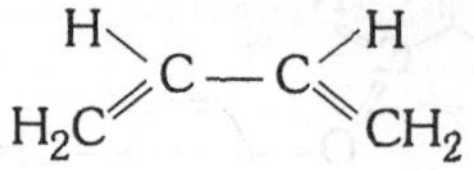

(b) 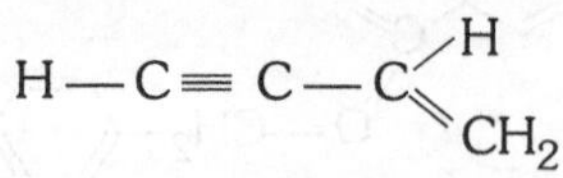

(c) $H_2C{=}C{=}O$

(d) $H_2C{=}C{=}CH_2$

SECTION III

Directions (Q. Nos. 12 to 16) *This section contains 2 paragraphs. Based on the first paragraph 2 multiple choice questions and based on the paragraph 3 multiple choice questions have to be answered. Each of these questions has four choices (a), (b), (c) and (d) out of which only one is correct.*

Paragraph 12-13 An acylic hydrocarbon P, having molecular formula C_6H_{10}, gave acetone as the only organic product through the following sequence of reactions, in which Q is an intermediate organic compounds.

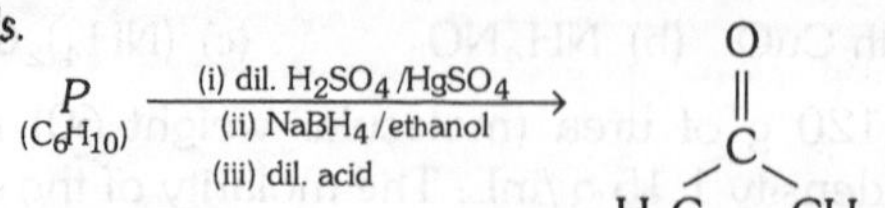

12. The structure of compound P is

(a) $CH_3CH_2CH_2CH_2{-}C{\equiv}C{-}H$

(b) $H_3CH_2C{-}C{\equiv}C{-}CH_2CH_3$

(c) 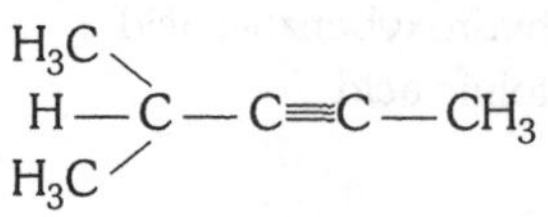

(d) 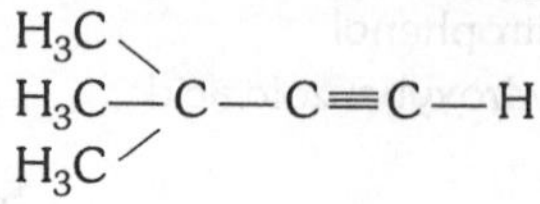

13. The structure of the compound Q is

(a) 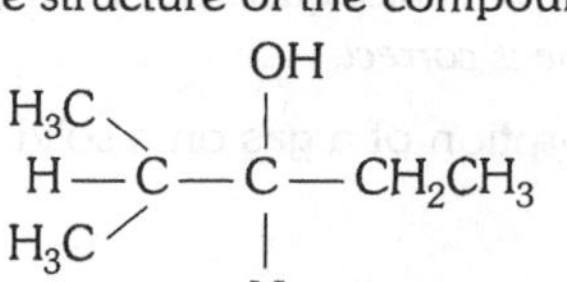

(b) 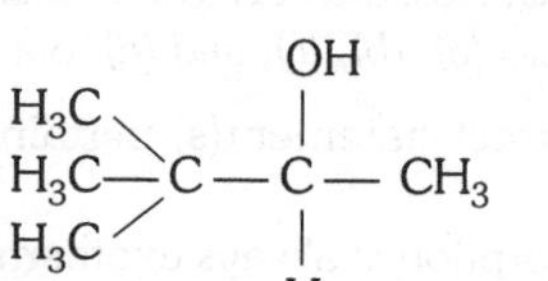

(c) 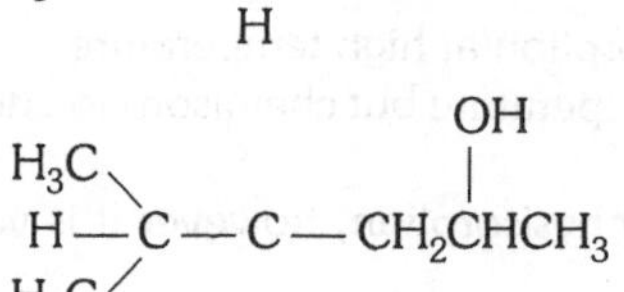

(d) 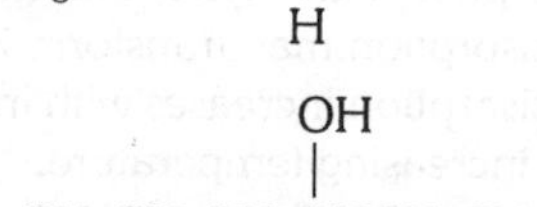

***Paragraph* 14-16** *When a metal rod M is dipped into an aqueous colourless concentrated solution of compound N, the solution turns light blue. Addition of aqueous NaCl to the blue solution gives a white precipitate O. Addition of aqueous NH_3 dissolves O and gives an intense blue solution.*

14. The metal rod *M* is

(a) Fe (b) Cu (c) Ni (d) Co

15. The compound *N* is

(a) $AgNO_3$ (b) $Zn (NO_3)_2$ (c) $Al (NO_3)_3$ (d) $Pb (NO_3)_2$

16. The final solution contains

(a) $[Pb (NH_3)_4]^{2+}$ and $[CoCl_4]^{2-}$ (b) $[Al (NH_3)_4]^{3+}$ and $[Cu(NH_3)_4]^{2+}$

(c) $[Ag (NH_3)_2]^{+}$ and $[Co(NH_3)_4]^{2+}$ (d) $[Ag (NH_3)_2]^{+}$ and $[Ni(NH_3)_6]^{2+}$

Section IV

Directions (Q. Nos. 17 to 23) *This section contains 7 questions. The answer to each of the questions is a single-digit integer, ranging from 0 to 9.*

17. The maximum number of electrons that can have principal quantum number, $n = 3$ and spin quantum number, $m_s = -\frac{1}{2}$, is

18. Reaction of Br_2 with Na_2CO_3 in aqueous solution gives sodium bromide and sodium bromate with evolution of CO_2 gas. The number of sodium bromide molecules involved in the balanced chemical equation is

19. The total number of alkenes possible by dehydrobromination of 3-bromo-3-cyclopentylhexane using alcoholic KOH is

20. The work function (Φ) of some metals is listed below. The number of metals which will show photoelectric effect when light of 300 nm wavelength falls on the metal is

Metal	Li	Na	K	Mg	Cu	Ag	Fe	Pt	W
ϕ (eV)	2.4	2.3	2.2	3.7	4.8	4.3	4.7	6.3	4.75

21. A decapeptide (molecular weight 796) on complete hydrolysis gives glycine (molecular weight 75), alanine and phenylalanine. Glycine contributes 47.0% to the total weight of the hydrolysed products. The number of glycine units present in the decapeptide is

22. The difference in the oxidation numbers of the two types of sulphur atoms in $Na_2S_4O_6$ is

23. To an evacuated vessel with movable piston under external pressure of 1 atm, 0.1 mole of He and 1.0 mole of an unknown compound (vapour pressure 0.68 atm at 0°C) are introduced. Considering the ideal gas behaviour, the total volume (in litre) of the gases at 0°C is close to

Paper II

SECTION I

Directions (Q. Nos. 1 to 8) *This section contains 8 multiple choice questions. Each question has four choices (a), (b), (c) and (d) out of which only one is correct.*

1. Oxidation states of the metal in the minerals haematite and magnetite, respectively, are

(a) II, III in haematite and III in magnetite (b) II, III in haematite and II in magnetite
(c) II in haematite and II, III in magnetite (d) III in haematite and II, III in magnetite

2. The following carbohydrate is

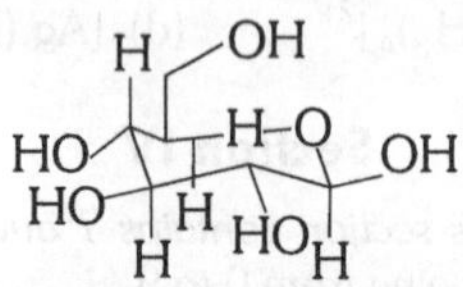

(a) a ketohexose (b) an aldohexose (c) an α-furanose (d) an α-pyranose

3. The major product of the following reaction is

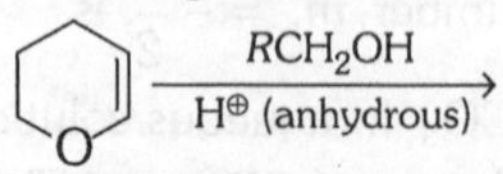

(a) a hemiacetal (b) an acetal (c) an ether (d) an ester

4. Amongst the compounds given, the one that would form a brilliant coloured dye on treatment with $NaNO_2$ in dil. HCl followed by addition to an alkaline solution of β -naphthol is

(a) 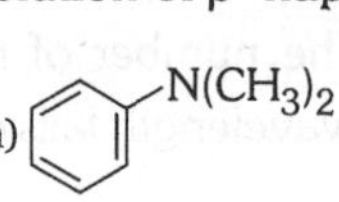 (b) 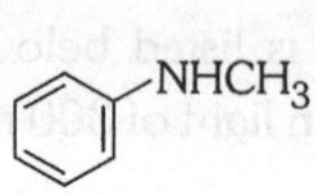 (c) 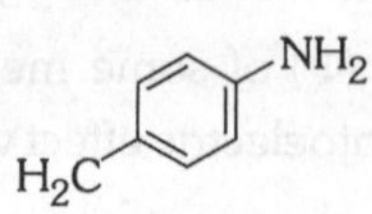 (d) 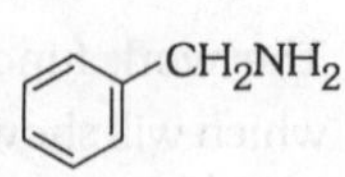

5. The freezing point (in °C) of solution containing 0.1 g of $K_3[Fe(CN)_6]$ (molecular weight 329) in 100 g of water (K_f = 1.86 K kg mol^{-1}) is

(a) -2.3×10^{-2} (b) -5.7×10^{-2} (c) -5.7×10^{-3} (d) -1.2×10^{-2}

6. Consider the following cell reaction,

$$2Fe\,(s) + O_2\,(g) + 4H^+\,(aq) \longrightarrow 2Fe^{2+}\,(aq) + 2H_2O\,(l),\ E° = 1.67\ V$$

At $[Fe^{2+}] = 10^{-3}$ M, P (O_2) = 0.1 atm and pH = 3, the cell potential at 25°C is

(a) 1.47 V (b) 1.77 V (c) 1.87 V (d) 1.57 V

7. Passing H_2S gas into a mixture of Mn^{2+}, Ni^{2+}, Cu^{2+} and Hg^{2+} ions in an acidified aqueous solution precipitates

(a) CuS and HgS (b) MnS and CuS (c) MnS and NiS (d) NiS and HgS

8. Among the following complexes ($K-P$), $K_3[Fe(CN)_6]$ (K), $[Co(NH_3)_6]Cl_3$ (L), $Na_3[Co(ox)_3]$ (M), $[Ni(H_2O)_6]Cl_2$ (N); $K_2[Pt(CN)_4]$ (O) and $[Zn(H_2O)_6](NO_3)_2$ (P)

the diamagnetic complexes are

(a) K, L, M, N (b) K, M, O, P (c) L, M, O, P (d) L, M, N, O

SECTION II

Directions (Q. Nos. 9 to 12) *This section contains 4 multiple choice questions. Each question has four choices (a), (b), (c) and (d) out of which one or more may be correct.*

9. Reduction of the metal centre in aqueous permanganate ion involves

(a) three electrons in neutral medium (b) five electrons in neutral medium

(c) three electrons in alkaline medium (d) five electrons in acidic medium

10. For the first order reaction $2N_2O_5(g) \longrightarrow 4NO_2(g) + O_2(g)$

(a) the concentration of the reactant decreases exponentially with time.

(b) the half-life of the reaction decreases with increasing temperature

(c) the half-life of the reaction depends on the initial concentration of the reactant.

(d) the reaction proceeds of 99.6% completion in eight half-life duration.

11. The equilibrium $2Cu^{I} \rightleftharpoons Cu^{0} + Cu^{II}$ in aqueous medium at 25°C shifts towards the left in the presence of

(a) NO_3^- (b) Cl^- (c) SCN^- (d) CN^-

12. The correct functional group X and the reagent/reaction conditions Y in the following schemes are $X—(CH_2)_4—X$

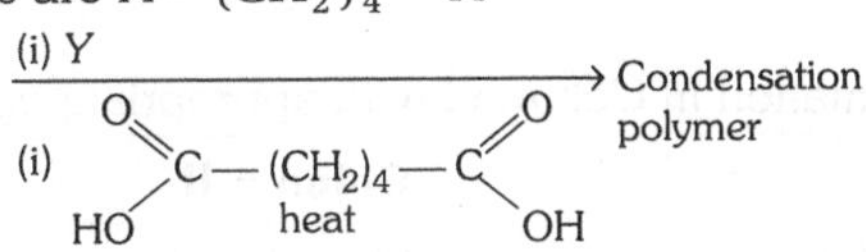

(a) $X = COOCH_3$, $Y = H_2/Ni/heat$

(b) $X = CONH_2$, $Y = H_2/Ni/heat$

(c) $X = CONH_2$, $Y = Br_2/NaOH$

(d) $X = CN$, $Y = H_2/Ni/heat$

SECTION III

Directions (Q.Nos. 13 to 18) *This section contains 6 questions. The answer to each of the question is a single digit integer, ranging from 0 to 9.*

13. In 1L saturated solution of AgCl [K_{sp} AgCl $= 1.6 \times 10^{-10}$], 0.1 mole of CuCl [K_{sp} (CuCl) $= 1.0 \times 10^{-6}$] is added. The resultant concentration of Ag^+ in the solution is 1.6×10^{-x}. The value of x is

14. The maximum number of isomers (including stereoisomers) that are possible on monochlorination of the following compounds, is

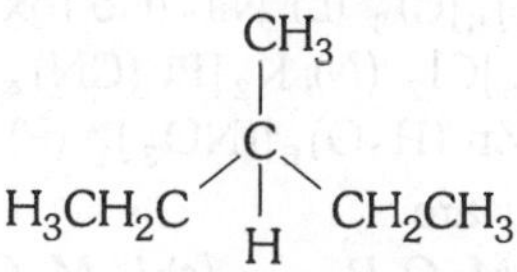

15. Among the following, the number of compounds than can react with PCl_5 to give $POCl_3$ is O_2, CO_2, SO_2, H_2O, H_2SO_4, P_4O_{10}.

16. The number of hexagonal faces that are present in a truncated octahedron is

17. The volume (in mL) of 0.1 M $AgNO_3$ required for complete precipitation of chloride ions present in 30 mL of 0.01 M solution of $[Cr(H_2O)_5Cl]Cl_2$, as silver chloride is close to

18. The total number of contributing structures showing hyperconjugation (involving C—H bonds) for the following carbocation is

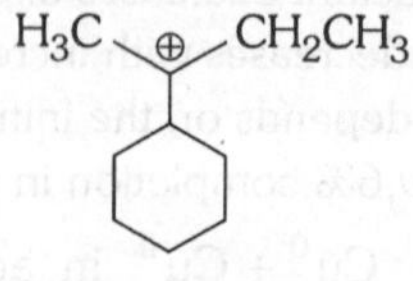

SECTION IV

Directions (Q. Nos. 19 and 20) *This section contains 2 questions. Each question has four statements (A, B, C and D) given in Column I and five statements (p, q, r, s and t) in Column II. Any given statement in Column I can have correct matching with one or more statement (s) given in Column II.*

19. Match the transformation in Column I with appropriate options in Column II.

	Column I		Column II
A.	$CO_2(s) \rightarrow CO_2(g)$	p.	Phase transition
B.	$CaCO_3(s) \rightarrow CaO(s) + CO_2(g)$	q.	Allotropic change
C.	$2H \rightarrow H_2(g)$	r.	ΔH is positive
D.	$P_{(\text{white, solid})} \rightarrow P_{(\text{red, solid})}$	s.	ΔS is positive
		t.	ΔS is negative

20. Match the reactions in Column I with appropriate types of steps/reactive intermediate involved in these reactions as given in Column II.

	Column I	Column II
A.	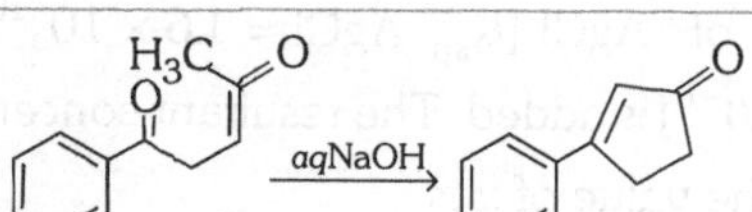	p. Nucleophilic substitution

	Column I	Column II
B.	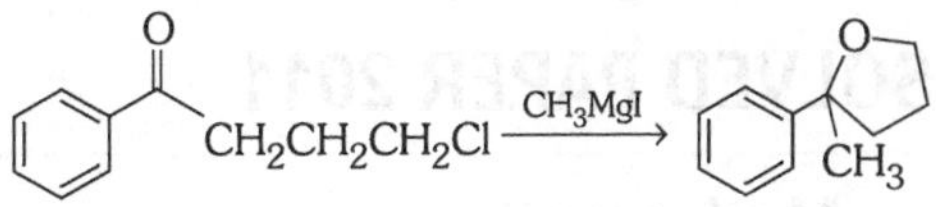	q. Electrophilic substitution
C.	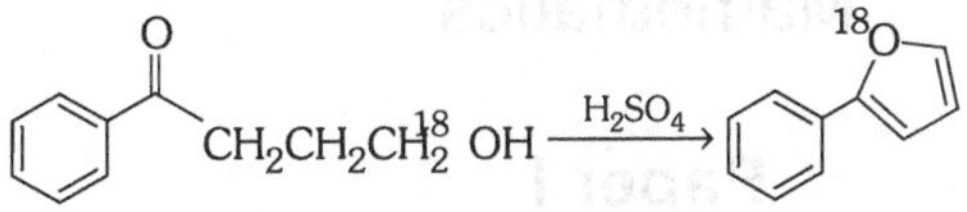	r. Dehydration
D.	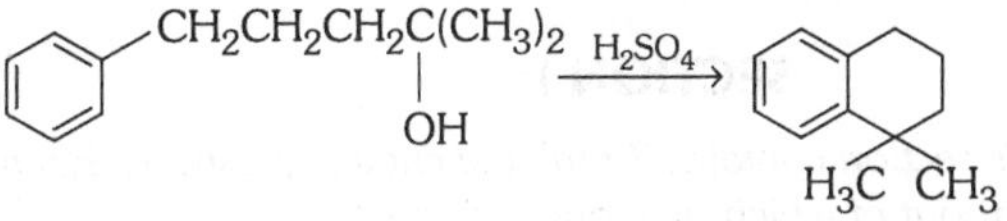	s. Nucleophilic addition
		t. Carbanion

ANSWERS

Paper I

1. (a)	**2.** (d)	**3.** (a)	**4.** (b)	**5.** (d)	**6.** (c)
7. (c)	**8.** (a,b,d)	**9.** (a,b,d)	**10.** (a,d)	**11.** (b,c)	**12.** (d)
13. (b)	**14.** (b)	**15.** (a)	**16.** (c)	**17.** (9)	**18.** (5)
19. (5)	**20.** (4)	**21.** (6)	**22.** (5)	**23.** (7)	

Paper II

1. (d)	**2.** (b)	**3.** (b)	**4.** (c)	**5.** (a)	**6.** (d)
7. (a)	**8.** (c)	**9.** (a,c,d)	**10.** (a,b,d)	**11.** (b,c,d)	**12.** (c,d)
13. (7)	**14.** (8)	**15.** (4)	**16.** (8)	**17.** (6)	**18.** (6)

19. (A) → p, r, s; (B) → r, s, (C) → t; (D) → p, q, t

20. A → r, s, t; (B) → p, s; (C) → r, s; (D) → q, r

*According to IIT official answer key.

IIT JEE

UNSOLVED PAPER 2011

Mathematics

Paper I

SECTION I

Directions (Q. Nos. 1-7) *This section contains 7 multiple choice questions. Each question has four choices (a), (b), (c) and (d) out of which only one is correct.*

1. Let $\mathbf{a} = \hat{\mathbf{i}} + \hat{\mathbf{j}} + \hat{\mathbf{k}}$, $\mathbf{b} = \hat{\mathbf{i}} - \hat{\mathbf{j}} + \hat{\mathbf{k}}$ and $\mathbf{c} = \hat{\mathbf{i}} - \hat{\mathbf{j}} - \hat{\mathbf{k}}$ be three vectors. A vector $\mathbf{v}$ in the plane of $\mathbf{a}$ and $\mathbf{b}$ whose projection of $\mathbf{c}$ is $\frac{1}{\sqrt{3}}$, is given by

(a) $\hat{\mathbf{i}} - 3\hat{\mathbf{j}} + 3\hat{\mathbf{k}}$ (b) $-3\hat{\mathbf{i}} - 3\hat{\mathbf{j}} - \hat{\mathbf{k}}$

(c) $3\hat{\mathbf{i}} - \hat{\mathbf{j}} + 3\hat{\mathbf{k}}$ (d) $\hat{\mathbf{i}} + 3\hat{\mathbf{k}} - 3\hat{\mathbf{k}}$

2. Let the straight line $x = b$ divide the area enclosed by $y = (1 - x)^2$, $y = 0$ and $x = 0$ into two parts $R_1 (0 \le x \le b)$ and $R_2 (b \le x \le 1)$ such that $R_1 - R_2 = \frac{1}{4}$. Then, b equals to

(a) $\frac{3}{4}$ (b) $\frac{1}{2}$ (c) $\frac{1}{3}$ (d) $\frac{1}{4}$

3. A straight line L through the point $(3, -2)$ is inclined at an angle $60°$ to the line $\sqrt{3}x + y = 1$. If L also intersects the X-axis, then the equation of L is

(a) $y + \sqrt{3}x + 2 - 3\sqrt{3} = 0$ (b) $y - \sqrt{3}x + 2 + 3\sqrt{3} = 0$

(c) $\sqrt{3}y - x + 3 + 2\sqrt{3} = 0$ (d) $\sqrt{3}y + x - 3 + 2\sqrt{3} = 0$

4. The value of $\int_{\sqrt{\log 2}}^{\sqrt{\log 3}} \frac{x \sin x^2}{\sin x^2 + \sin(\log 6 - x^2)} dx$ is

(a) $\frac{1}{4}\log\frac{3}{2}$ (b) $\frac{1}{2}\log\frac{3}{2}$ (c) $\log\frac{3}{2}$ (d) $\frac{1}{6}\log\frac{3}{2}$

5. Let (x_0, y_0) be the solution of the following equations $(2x)^{\log 2} = (3y)^{\log 3}$, $3^{\log x} = 2^{\log y}$, then x_0 is equal to

(a) $\frac{1}{6}$ (b) $\frac{1}{2}$ (c) $\frac{1}{2}$ (d) 6

6. Let $P = \{\theta : \sin\theta - \cos\theta = \sqrt{2}\cos\theta\}$ and $Q = \{\theta : \sin\theta + \cos\theta = \sqrt{2}\sin\theta\}$ be two sets. Then,

(a) $P \subset Q$ and $Q - P \ne \Phi$ (b) $Q \not\subset P$

(c) $P \not\subset Q$ (d) $P = Q$

7. Let α and β be the roots of $x^2 - 6x - 2 = 0$ with $\alpha > \beta$. If $a_n = \alpha^n - \beta^n$ for $n \geq 1$, then the value of $\frac{a_{10} - 2a_8}{2a_9}$ is

(a) 1 (b) 2 (c) 3 (d) 4

SECTION II

Directions (Q. Nos. 8-11) *This section contains 4 multiple choice questions. Each question has four choices (a), (b), (c) and (d) out of which one ore more may be correct.*

8. Let M and N be two 3×3 non-singular skew-symmetric matrices such that $MN = NM$. If P^T denotes the transpose of P, then $M^2N^2(M^TN)^{-1}(MN^{-1})^T$ is equal to

(a) M^2 (b) $-N^2$ (c) $-M^2$ (d) MN

9. Let the eccentricity of the hyperbola $\frac{x^2}{a^2} - \frac{y^2}{b^2} = 1$ be reciprocal to that of the ellipse $x^2 + 4y^2 = 4$. If the hyperbola passes through a focus of the ellipse, then

(a) the equation of the hyperbola is $\frac{x^2}{3^2} - \frac{y^2}{2^2} = 1$

(b) a focus of the hyperbola is (2, 0)

(c) the eccentricity of the hyperbola is $\sqrt{\frac{5}{3}}$

(d) the equation of the hyperbola is $x^2 - 3y^2 = 3$

10. The vector(s) which is/are coplanar with vectors $\hat{\mathbf{i}} + \hat{\mathbf{j}} + 2\hat{\mathbf{k}}$ and $\hat{\mathbf{i}} + 2\hat{\mathbf{j}} + \hat{\mathbf{k}}$, are perpendicular to the vector $\hat{\mathbf{i}} + \hat{\mathbf{j}} + \hat{\mathbf{k}}$ is/are

(a) $\hat{\mathbf{j}} - \hat{\mathbf{k}}$ (b) $-\hat{\mathbf{i}} + \hat{\mathbf{j}}$ (c) $\hat{\mathbf{i}} - \hat{\mathbf{j}}$ (d) $-\hat{\mathbf{j}} + \hat{\mathbf{k}}$

11. Let $f : R \to R$ be a function such that $f(x + y) = f(x) + f(y), \forall x, y \in R$. If $f(x)$ is differentiable at $x = 0$, then

(a) $f(x)$ is differentiable only in a finite interval containing zero

(b) $f(x)$ is continuous, $\forall x \in R$

(c) $f'(x)$ is constant, $\forall x \in R$

(d) $f(x)$ is differentiable except at finitely many points

SECTION III

Directions (Q. Nos. 12-16) *This section contains 2 paragraphs. Based upon one of the paragraphs 2 multiple choice questions and based on the other paragraph 3 multiple choice questions have the be answered. Each of these questions has four choices (a), (b), (c) and (d) out of which only one is correct.*

Paragraph (Q. Nos. 12-13)

Let U_1 and U_2 be two urns such that U_1 contains 3 white and 2 red balls and U_2 contains only 1 white ball. A fair coin is tossed. If head appears then 1 ball is drawn at random from U_1 and put into U_2. However, if tail appears then 2 balls are drawn at random from U_1 and put into U_2. Now, 1 ball is drawn at random from U_2.

12. The probability of the drawn ball from U_2 being white is

(a) $\frac{13}{30}$ (b) $\frac{23}{30}$ (c) $\frac{19}{30}$ (d) $\frac{11}{30}$

13. Given that the drawn ball from U_2 is white, the probability that head appeared on the coin is

(a) $\frac{17}{23}$ (b) $\frac{11}{23}$ (c) $\frac{15}{23}$ (d) $\frac{12}{23}$

Paragraph (Q. Nos. 14-16)

Let a, b and c be three real numbers satisfying $[a\ b\ c]\begin{bmatrix}1 & 9 & 7\\ 8 & 2 & 7\\ 7 & 3 & 7\end{bmatrix}=[0\ 0\ 0]$ ...(E)

14. If the point $P(a, b, c)$, with reference to Eq. (E), lies on the plane $2x + y + z = 1$, then the value of $7a + b + c$ is

(a) 0 (b) 12 (c) 7 (d) 6

15. Let ω be a solution of $x^3 - 1 = 0$ with Im $(\omega) > 0$. If $a = 2$, with b and c satisfying Eq. (E), then the value of $\frac{3}{\omega^a}+\frac{1}{\omega^b}+\frac{3}{\omega^c}$

(a) -2 (b) 2 (c) 3 (d) -3

16. Let $b = 6$, with a and c satisfying Eq. (E). If α and β are the roots of the quadratic equation $ax^2 + bx + c = 0$, then $\sum_{n=0}^{\infty}\left(\frac{1}{\alpha}+\frac{1}{\beta}\right)^n$ is

(a) 6 (b) 7 (c) $\frac{6}{7}$ (d) ∞

SECTION IV

Directions (Q. Nos. 17-23) *This section contains 7 questions. The answer to each of the questions is a single digit integer, ranging from 0 to 9.*

17. The positive integer value of $n > 3$ satisfying the equation

$$\frac{1}{\sin\left(\frac{\pi}{n}\right)}=\frac{1}{\sin\left(\frac{2\pi}{n}\right)}+\frac{1}{\sin\left(\frac{3\pi}{n}\right)} \text{ is}$$

18. Let $a_1, a_2, a_3, \ldots, a_{100}$ be an arithmetic progression with $a_1 = 3$ and $S_p = \sum_{i=1}^{p} a_i$, $1 \le p \le 100$. For any integer n with $1 \le n \le 20$, let $m = 5n$. If $\frac{S_m}{S_n}$ does not depend on n, then a_2 is

19. The minimum value of the sum of real numbers a^{-5}, a^{-4}, $3a^{-3}$, 1, a^8 and a^{10} with $a > 0$ is

20. Let $f:[1,\infty)\to[2,\infty)$ be a differentiable function such that $f(1)=2$. If $6\int_1^x f(t)dt = 3x\,f(x) - x^3$ for all $x \geq 1$, then the value of $f(2)$ is

21. Let $f(\theta) = \sin\left[\tan^{-1}\left(\dfrac{\sin\theta}{\sqrt{\cos 2\theta}}\right)\right]$, where $-\dfrac{\pi}{4} < \theta < \dfrac{\pi}{4}$. Then, the value of $\dfrac{d}{d(\tan\theta)}(f(\theta))$ is

22. Consider the parabola $y^2 = 8x$. Let Δ_1 be the area of the triangle formed by the end points of its latusrectum and the point $P\left(\dfrac{1}{2}, 2\right)$ on the parabola and Δ_2 be the area of the triangle formed by drawing tangents at P and at the end points of the latusrectum. Then, $\dfrac{\Delta_1}{\Delta_2}$ is

23. If z is any complex number satisfying $|z-3-2i| \leq 2$, then the maximum value of $|2z-6+5i|$ is

Paper II

SECTION I

Directions (Q. Nos. 1-8) *This section contains 8 multiple choice questions. Each question has four choices (a), (b), (c) and (d) out of which only one is correct.*

1. Let $P(6, 3)$ be a point on the hyperbola $\dfrac{x^2}{a^2} - \dfrac{y^2}{b^2} = 1$. If the normal at the point P intersects the X-axis at $(9, 0)$, then the eccentricity of the hyperbola is

(a) $\sqrt{\dfrac{5}{2}}$ (b) $\sqrt{\dfrac{3}{2}}$ (c) $\sqrt{2}$ (d) $\sqrt{3}$

2. Let (x, y) be any point on the parabola $y^2 = 4x$. Let P be the point that divides the line segment from $(0, 0)$ to (x, y) in the ratio $1:3$. Then, the locus of P is

(a) $x^2 = y$ (b) $y^2 = 2x$ (c) $y^2 = x$ (d) $x^2 = 2y$

3. Let $f(x) = x^2$ and $g(x) = \sin x$ for all $x \in R$. Then, the set of all x satisfying $(fogogof)(x) = (gogof)(x)$, where $(fog)(x) = f(g(x))$ is

(a) $\pm\sqrt{n\pi}, n \in \{0, 1, 2, \ldots\}$ (b) $\pm\sqrt{n\pi}, n \in \{1, 2, \ldots.\}$

(c) $\dfrac{\pi}{2} + 2n\pi, n \in \{\ldots, -2, -1, 0, 1, 2, \ldots\}$ (d) $2n\pi, n \in \{\ldots, -2, -1, 0, 1, 2, \ldots\}$

4. Let $f:[-1, 2] \to [0, \infty)$ be a continuous function such that $f(x) = f(1-x)$ for all $x \in [-1, 2]$. Let $R_1 = \int_{-1}^{2} x\,f(x)dx$ and R_2 be the area of the region bounded by $y = f(x)$, $x = -1$, $x = 2$ and the X-axis. Then,

(a) $R_1 = 2R_2$ (b) $R_1 = 3R_2$ (c) $2R_1 = R_2$ (d) $3R_1 = R_2$

5. If $\lim_{x \to 0} [1 + x \log(1 + b^2)]^{1/x} = 2b\sin^2\theta\, b > 0$ and $\theta \in (-\pi, \pi]$, then the value of θ is

(a) $\pm\frac{\pi}{4}$ (b) $\pm\frac{\pi}{3}$ (c) $\pm\frac{\pi}{6}$ (d) $\pm\frac{\pi}{2}$

6. The circle passing through the point $(-1, 0)$ and touching the Y-axis at $(0, 2)$, also passes through the point

(a) $\left(-\frac{3}{2}, 0\right)$ (b) $\left(-\frac{5}{2}, 2\right)$ (c) $\left(-\frac{3}{2}, \frac{5}{2}\right)$ (d) $(-1, -4)$

7. Let $\omega \neq 1$ be a cube root of unity and S be the set of all non-singular matrices of the form $\begin{bmatrix} 1 & a & b \\ \omega & 1 & c \\ \omega^2 & \omega & 1 \end{bmatrix}$, where each of a, b and c is either ω or ω^2. Then, the number of distinct matrices in the set S is

(a) 2 (b) 6 (c) 4 (d) 8

8. The value of b for which the equations $x^2 + bx - 1 = 0$, $x^2 + x + b = 0$ have one root in common is

(a) $-\sqrt{2}$ (b) $-i\sqrt{3}$ (c) $i\sqrt{5}$ (d) $\sqrt{2}$

SECTION II

Directions (Q. Nos. 9-12) *This section contains 4 multiple choice questions. Each question has four choices (a), (b), (c) and (d) out of which one or more may be correct.*

9. If $f(x) = \begin{cases} -x - \frac{\pi}{2}, & x \leq -\frac{\pi}{2} \\ -\cos x, & -\frac{\pi}{2} < x \leq 0 \\ x - 1, & 0 < x \leq 1 \\ \log x, & x > 1 \end{cases}$, then

(a) $f(x)$ is continuous at $x = -\frac{\pi}{2}$

(b) $f(x)$ is not differentiable at $x = 0$

(c) $f(x)$ is differentiable at $x = 1$

(d) $f(x)$ is differentiable at $x = -\frac{3}{2}$

10. Let L be a normal to the parabola $y^2 = 4x$. If L passes through the point $(9, 6)$, then L is given by

(a) $y - x + 3 = 0$ (b) $y + 3x - 33 = 0$ (c) $y + x - 15 = 0$ (d) $y - 2x + 12 = 0$

11. Let E and F be two independent events. The probability that exactly one of them occurs is 11/25 and the probability of none of them occurring is 2/25. If $P(T)$ denotes the probability of occurrence of the event T, then

(a) $P(E) = \frac{4}{5}, P(F) = \frac{3}{5}$

(b) $P(E) = \frac{1}{5}, P(F) = \frac{2}{5}$

(c) $P(E) = \frac{2}{5}, P(F) = \frac{1}{5}$

(d) $P(E) = \frac{3}{5}, P(F) = \frac{4}{5}$

12. Let $f : (0, 1) \to R$ be defined by $f(x) = \frac{b-x}{1-bx}$, where b is a constant such that $0 < b < 1$. Then,

(a) f if not invertible on (0, 1)

(b) $f \neq f^{-1}$ on (0, 1) and $f'(b) = \frac{1}{f'(0)}$

(c) $f = f^{-1}$ on (0, 1) and $f'(b) = \frac{1}{f'(0)}$

(d) f^{-1} is differentiable on (0, 1)

SECTION III

Directions (Q. Nos. 13-18) *This section contains 6 questions. The answer to each of the questions is a single digit integer, ranging from 0 to 9.*

13. Let $y'(x) + y(x)g'(x) = g(x)g'(x)\ y(0) = 0,\ x \in R$, where $f'(x)$ denotes $\frac{d\,f(x)}{dx}$ and $g(x)$ is a given non-constant differentiable function on R with $g(0) = g(2) = 0$. Then, the value of $y(2)$ is...

14. Let $\mathbf{a} = -\hat{\mathbf{i}} - \hat{\mathbf{k}}$, $\mathbf{b} = -\hat{\mathbf{i}} + \hat{\mathbf{j}}$ and $\mathbf{c} = \hat{\mathbf{i}} + 2\hat{\mathbf{j}} + 3\hat{\mathbf{k}}$ be three given vectors. If $\mathbf{r}$ is a vector such that $\mathbf{r} \times \mathbf{b} = \mathbf{c} \times \mathbf{b}$ and $\mathbf{r} \cdot \mathbf{a} = 0$, then the value of $\mathbf{r} \cdot \mathbf{b}$ is

15. Let $\omega = e^{i\pi/3}$ and a, b, c, x, y, z be non-zero complex numbers such that $a + b + c = x,\ a + b\omega + c\omega^2 = y,\ a + b\omega^2 + c\omega = z.$

Then, the value of $\frac{|x|^2 + |y|^2 + |z|^2}{|a|^2 + |b|^2 + |c|^2}$ is

16. Let M be a 3×3 matrix satisfying

$$M\begin{bmatrix}0\\1\\0\end{bmatrix} = \begin{bmatrix}-1\\2\\3\end{bmatrix},\ M\begin{bmatrix}1\\-1\\0\end{bmatrix} = \begin{bmatrix}1\\1\\-1\end{bmatrix} \text{ and } M\begin{bmatrix}1\\1\\1\end{bmatrix} = \begin{bmatrix}0\\0\\12\end{bmatrix}$$

Then, the sum of the diagonal entries of M is

17. The number of distinct real roots of $x^4 - 4x^3 + 12x^2 + x - 1 = 0$ is

18. The straight line $2x - 3y = 1$ divides the circular region $x^2 + y^2 \leq 6$ into two parts. If $S = \left\{\left(2, \frac{3}{4}\right), \left(\frac{5}{2}, \frac{3}{4}\right), \left(\frac{1}{4}, -\frac{1}{4}\right), \left(\frac{1}{8}, \frac{1}{4}\right)\right\}$, then the number of point(s) in S lying inside the smaller part is

SECTION IV

Directions (Q. Nos. 19-20) *This section contains 2 questions. Each question has four statements (A, B, C and D) given in Column I and five statements (p, q, r, s and t) in Column II. Any given statement in Column I can have correct matching with one or more statement(s) given in Column II.*

19. Match the statements given in Column I with the values given in Column II.

Column I	Column II
(A) If $\mathbf{a} = \hat{\mathbf{j}} + \sqrt{3}\hat{\mathbf{k}}$, $\mathbf{b} = -\hat{\mathbf{j}} + \sqrt{3}\hat{\mathbf{k}}$ and $\mathbf{c} = 2\sqrt{3}\hat{\mathbf{k}}$ form a triangle, then the internal angle of the triangle between $\mathbf{a}$ and $\mathbf{b}$ is	(p) $\frac{\pi}{6}$
(B) If $\int_a^b [f(x) - 3x]dx = a^2 - b^2$, then the value of $f\left(\frac{\pi}{6}\right)$ is	(q) $\frac{2\pi}{3}$
(C) The value of $\frac{\pi^2}{\log 3} \int_{7/6}^{5/6} \sec(\pi x)dx$ is	(r) $\pi/3$
(D) The maximum value of $\left|\arg\left(\frac{1}{1-z}\right)\right|$ for $|z| = 1, z \neq 1$ is given by	(s) π
	(t) $\pi/2$

20. Match the statements given in Column I with the intervals/union of intervals given in Column II.

Column I	Column II
(A) The set $\left\{\text{Re}\left(\frac{2iz}{1-z^2}\right) : z \text{ is a complex number}, |z| = 1, z \neq \pm 1\right\}$	(p) $(-\infty, -1) \cup (1, \infty)$
(B) The domain of the function $f(x) = \sin^{-1}\left[\frac{8(3)^{x-2}}{1-3^{2(x-1)}}\right]$ is	(q) $(-\infty, 0) \cup (0, \infty)$
(C) If $f(\theta) = \begin{vmatrix} 1 & \tan\theta & 1 \\ -\tan\theta & 1 & \tan\theta \\ -1 & -\tan\theta & 1 \end{vmatrix}$, then the set $\{f(\theta) : 0 \leq \theta < \frac{\pi}{2}\}$ is	(r) $[2, \infty)$
(D) If $f(x) = x^{3/2}(3x - 10)$, $x \geq 0$, then $f(x)$ is increasing in	(s) $(-\infty, -1] \cup [1, \infty)$
	(t) $(-\infty, 0] \cup [2, \infty)$

ANSWERS

Paper I

1. (c)	**2.** (b)	**3.** (b)	**4.** (a)	**5.** (c)	**6.** (d)
7. (c)	**8.** (c)	**9.** (b, d)	**10.** (a, d)	**11.** (b, c)	**12.** (b)
13. (d)	**14.** (d)	**15.** (a)	**16.** (b)	**17.** (7)	**18.** (3 or 9)
19. (8)	**20.** (8/3)	**21.** (1)	**22.** (2)	**23.** (5)	

Paper II

1. (b)	**2.** (c)	**3.** (b)	**4.** (c)	**5.** (d)	**6.** (d)
7. (a)	**8.** (b)	**9.** (a, b, c, d)	**10.** (a, b, d)	**11.** (a, d)	**12.** (a)
13. (0)	**14.** (9)	**15.** (3)	**16.** (9)	**17.** (2)	**18.** (2)

19. (A) → q; (B) → p; (C) → s; (D) → s

20. (A) → s; (B) → t; (C) → r; (D) → r

IIT JEE

UNSOLVED PAPER 2012

Physics

Paper 1

Section I

Directions (Q. Nos. 1-10) *This section contains 10 multiple choice questions. Each question has four choices (a), (b), (c) and (d) out of which only one is correct.*

1. Consider a thin spherical shell of radius R with its centre at the origin, carrying uniform positive surface charge density. The variation of the magnitude of the electric field $|\mathbf{E}(r)|$ and the electric potential $V(r)$ with the distance r from the centre, is best represented by which graph?

(a)
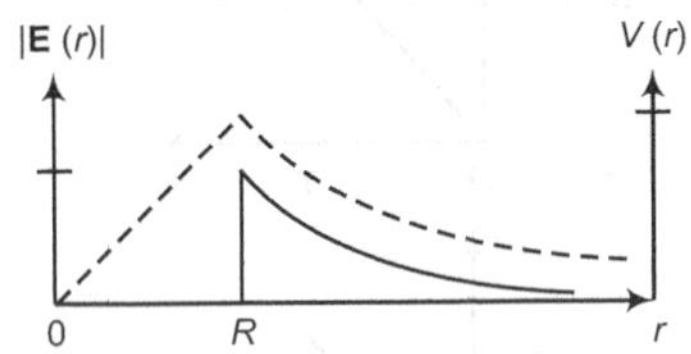

(b)
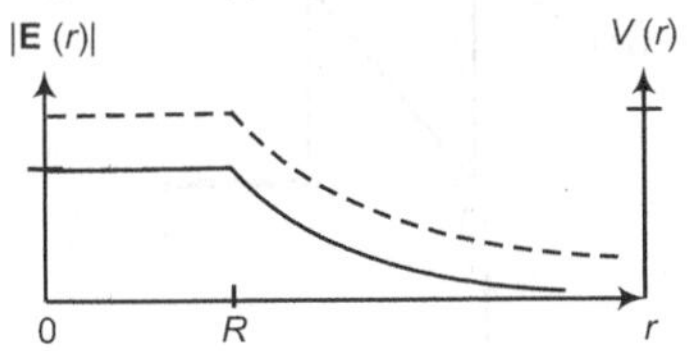

(c)
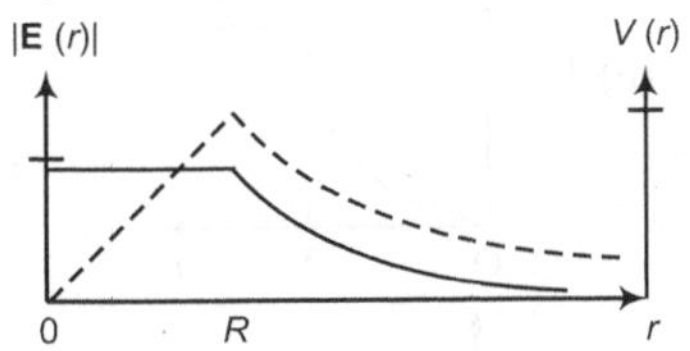

(d)
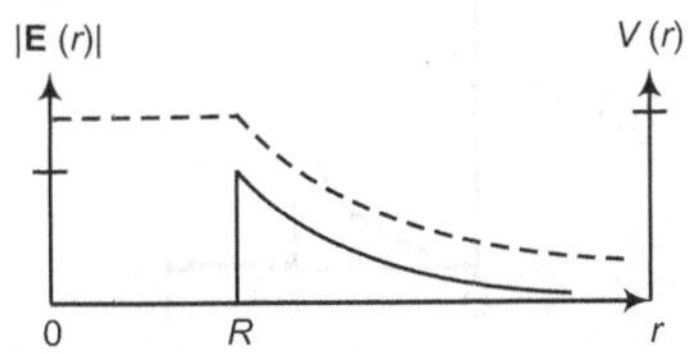

2. Young's double slit experiment is carried out by using green, red and blue light, one colour at a time. The fringe widths recorded are β_G, β_R and β_B respectively. Then,

(a) $\beta_G > \beta_B > \beta_R$
(b) $\beta_B > \beta_G > \beta_R$
(c) $\beta_R > \beta_B > \beta_G$
(d) $\beta_R > \beta_G > \beta_B$

3. Two large vertical and parallel metal plates having a separation of 1 cm are connected to a DC voltage source of potential difference X. A proton is released at rest midway between the two plates. It is found to move at 45° to the vertical just after release. Then X is nearly

(a) 1×10^{-5} V
(b) 1×10^{-7} V
(c) 1×10^{-9} V
(d) 1×10^{-10} V

4. A bi-convex lens is formed with two thin plano-convex lenses as shown in the figure. Refractive index n of the first lens is 1.5 and that of the second lens is 1.2. Both the curved surfaces are of the same radius of curvature $R = 14$ cm. For this bi-convex lens, for an object distance of 40 cm, the image distance will be

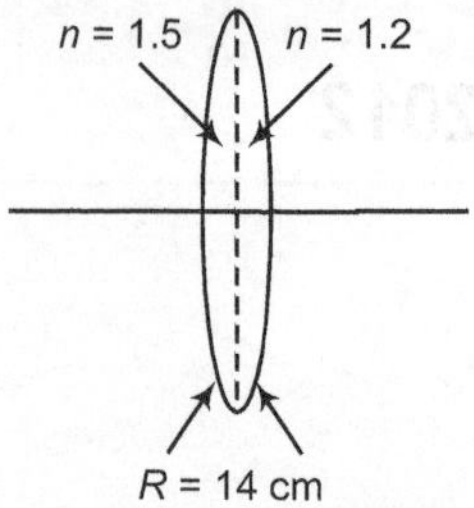

(a) – 280.0 cm (b) 40.0 cm
(c) 21.5 cm (d) 13.3 cm

5. A small mass m is attached to a massless string whose other end is fixed at P as shown in the figure. The mass is undergoing circular motion in the x-y plane with centre at O and constant angular speed ω. If the angular momentum of the system, calculated about O and P are denoted by $\mathbf{L}_O$ and $\mathbf{L}_P$ respectively, then

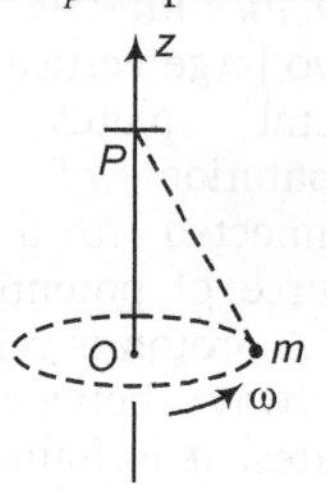

(a) $\mathbf{L}_O$ and $\mathbf{L}_P$ do not vary with time
(b) $\mathbf{L}_O$ varies with time while $\mathbf{L}_P$ remains constant
(c) $\mathbf{L}_O$ remains constant while $\mathbf{L}_P$ varies with time
(d) $\mathbf{L}_O$ and $\mathbf{L}_P$ both vary with time

6. A mixture of 2 moles of helium gas (atomic mass = 4 amu) and 1 mole of argon gas (atomic mass = 40 amu) is kept at 300 K in a container. The ratio of the rms speeds $\left(\dfrac{v_{\text{rms}}\text{ (helium)}}{v_{\text{rms}}\text{ (argon)}}\right)$ is
(a) 0.32
(b) 0.45
(c) 2.24
(d) 3.16

7. A thin uniform rod, pivoted at O, is rotating in the horizontal plane with constant angular speed ω, as shown in the figure.

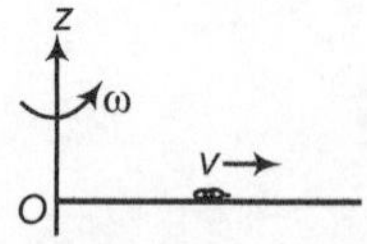

At time $t = 0$, a small insect starts from O and moves with constant speed v with respect to the rod towards the other end. It reaches the end of the rod at $t = T$ and stops. The angular speed of the system remains ω throughout. The magnitude of the torque $|\vec{\tau}|$ on the system about O, as a function of time is best represented by which plot?

(a) $|\vec{\tau}|$
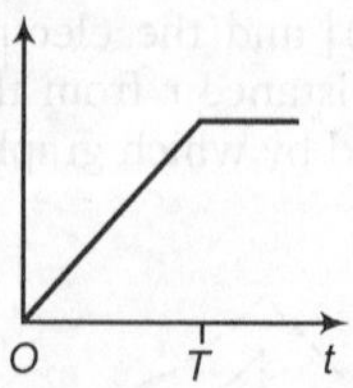

(b) $|\vec{\tau}|$
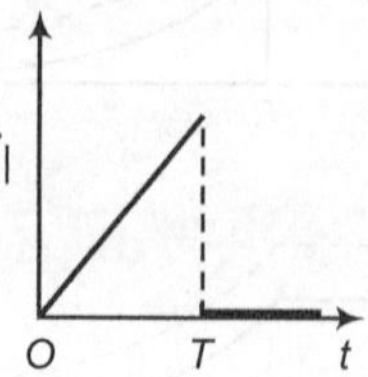

(c) $|\vec{\tau}|$
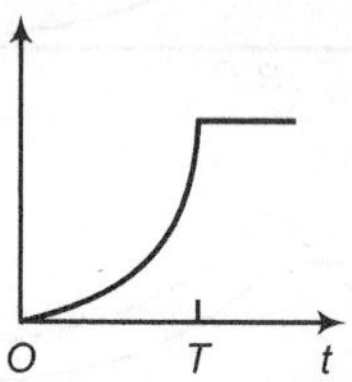

(d) $|\vec{\tau}|$
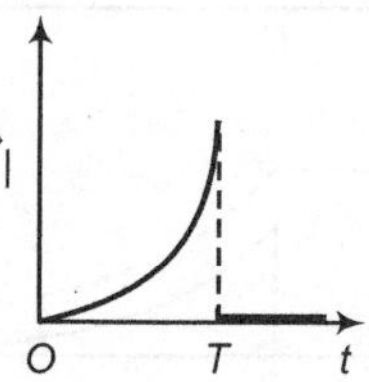

8. In the determination of Young's modulus $\left(Y = \frac{4MLg}{\pi l d^2}\right)$ by using Searle's method, a wire of length $L = 2$ m and diameter $d = 0.5$ mm is used. For a load $M = 2.5$ kg, an extension $l = 0.25$ mm in the length of the wire is observed. Quantities d and l are measured using a screw gauge and a micrometer, respectively. They have the same pitch of 0.5 mm. The number of divisions on their circular scale is 100. The contributions to the maximum probable error of the Y measurement is

(a) due to the errors in the measurements of d and l are the same

(b) due to the error in the measurement of d is twice that due to the error in the measurement of l.

(c) due to the error in the measurement of l is twice that due to the error in the measurement of d.

(d) due to the error in the measurement of d is four times that due to the error in the measurement of l.

9. A small block is connected to one end of a massless spring of unstretched length 4.9 m. The other end of the spring (see the figure) is fixed. The system lies on a horizontal frictionless surface. The block is stretched by 0.2 m and released from rest at $t = 0$. It then executes simple harmonic motion with angular frequency $\omega = \frac{\pi}{3}$ rad/s. Simultaneously at $t = 0$, a small pebble is projected with speed v from point P at an angle of 45° as shown in the figure. Point P is at a horizontal distance of 10 m from O. If the pebble hits the block at $t = 1$ s, the value of v is (take $g = 10$ m/s^2)

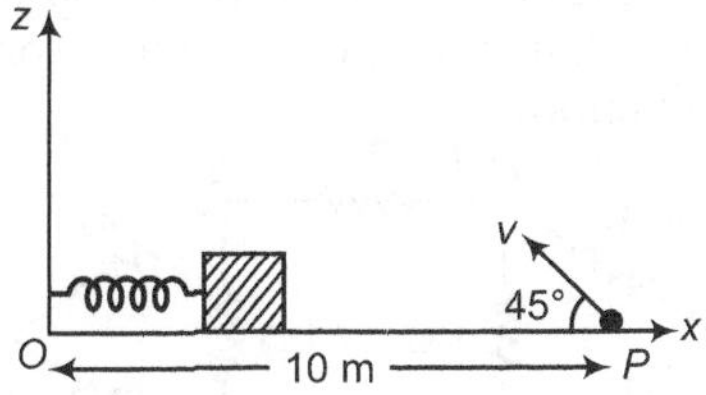

(a) $\sqrt{50}$ m/s

(b) $\sqrt{51}$ m/s

(c) $\sqrt{52}$ m/s

(d) $\sqrt{53}$ m/s

10. Three very large plates of same area are kept parallel and close to each other. They are considered as ideal black surfaces and have very high thermal conductivity. The first and third plates are maintained at temperatures $2T$ and $3T$ respectively. The temperature of the middle (*i.e.*, second) plate under steady state condition is

(a) $\left(\frac{65}{2}\right)^{\frac{1}{4}} T$

(b) $\left(\frac{97}{4}\right)^{\frac{1}{4}} T$

(c) $\left(\frac{97}{2}\right)^{\frac{1}{4}} T$

(d) $(97)^{\frac{1}{4}} T$

Section II

Directions (Q. Nos. 11-15) *This section contains 5 multiple choice questions. Each question has four choices (a), (b), (c) and (d) out of which one or more are correct.*

11. For the resistance network shown in the figure, choose the correct option(s).

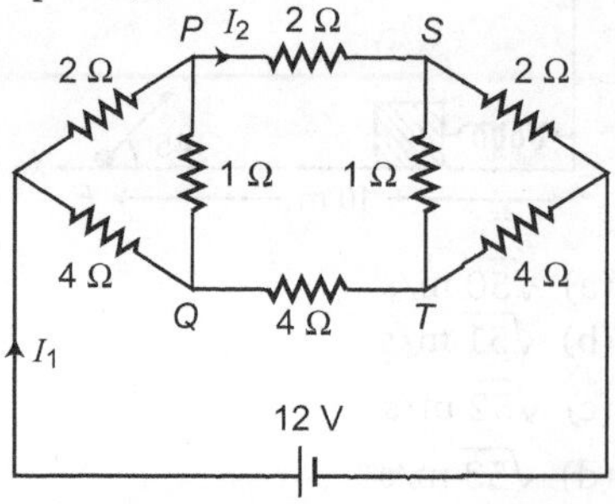

(a) The current through PQ is zero
(b) $I_1 = 3$ A
(c) The potential at S is less than that at Q
(d) $I_2 = 2$ A

12. A person blows into open-end of a long pipe. As a result, a high-pressure pulse of air travels down the pipe. When this pulse reaches the other end of the pipe,

(a) a high-pressure pulse starts travelling up the pipe, if the other end of the pipe is open
(b) a low-pressure pulse starts travelling up the pipe, if the other end of the pipe is open
(c) a low-pressure pulse starts travelling up the pipe, if the other end of the pipe is closed
(d) a high-pressure pulse starts travelling up the pipe, if the other end of the pipe is closed

13. Consider the motion of a positive point charge in a region where there are simultaneous uniform electric and magnetic fields $\mathbf{E} = E_0\,\hat{\mathbf{j}}$ and $\mathbf{B} = B_0\,\hat{\mathbf{j}}$. At time $t = 0$, this charge has velocity $\mathbf{v}$ in the x-y plane, making an angle θ with the x-axis. Which of the following option(s) is(are) correct for time $t > 0$?

(a) If $\theta = 0°$, the charge moves in a circular path in the x-z plane.
(b) If $\theta = 0°$, the charge undergoes helical motion with constant pitch along the y-axis
(c) If $\theta = 10°$, the charge undergoes helical motion with its pitch increasing with time, along the y-axis.
(d) If $\theta = 90°$, the charge undergoes linear but accelerated motion along the y-axis.

14. A cubical region of side a has its centre at the origin. It encloses three fixed point charges, $-q$ at $(0, -a/4, 0)$, $+3q$ at $(0, 0, 0)$ and $-q$ at $(0, +a/4, 0)$. Choose the correct option(s).

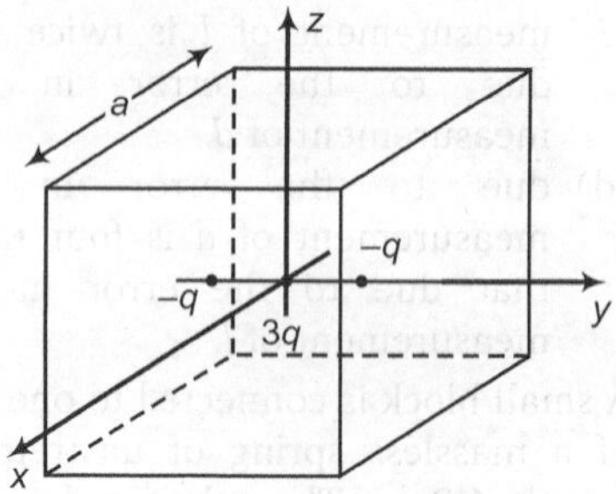

(a) The net electric flux crossing the plane $x = +a/2$ is equal to the net electric flux crossing the plane $x = -a/2$
(b) The net electric flux crossing the plane $y = +a/2$ is more than the net electric flux crossing the plane $y = -a/2$
(c) The net electric flux crossing the entire region is q/ε_0
(d) The net electric flux crossing the plane $z = +a/2$ is equal to the net electric flux crossing the plane $x = +a/2$

15. A small block of mass of 0.1 kg lies on a fixed inclined plane PQ which makes an angle θ with the horizontal. A horizontal force of 1 N acts on the block through its centre of mass as shown in the figure. The block remains stationary if (take $g = 10\ \text{m/s}^2$)

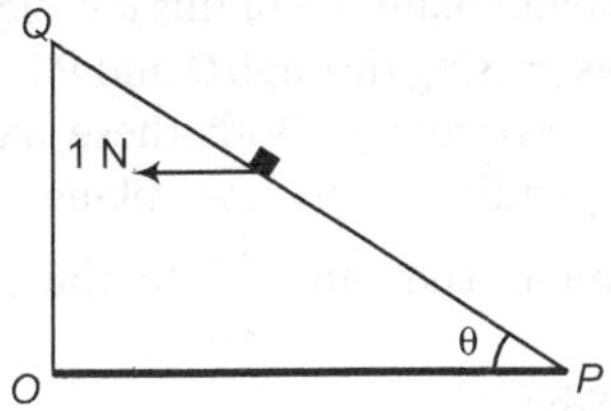

(a) $\theta = 45°$.
(b) $\theta > 45°$ and a frictional force acts on the block towards P
(c) $\theta > 45°$ and a frictional force acts on the block towards Q
(d) $\theta < 45°$ and a frictional force acts on the block towards Q

Section III

Directions (Q. Nos. 16-20) *This section contains 5 questions. The answer to each question is a single digit integer, ranging from 0 to 9 (both inclusive).*

16. A cylindrical cavity of diameter a exists inside a cylinder of diameter $2a$ as shown in the figure. Both the cylinder and the cavity are infinitely long. A uniform current density J flows along the length. If the magnitude of the magnetic field at the point P is given by $\frac{N}{12}\mu_0 aJ$, then the value of N is

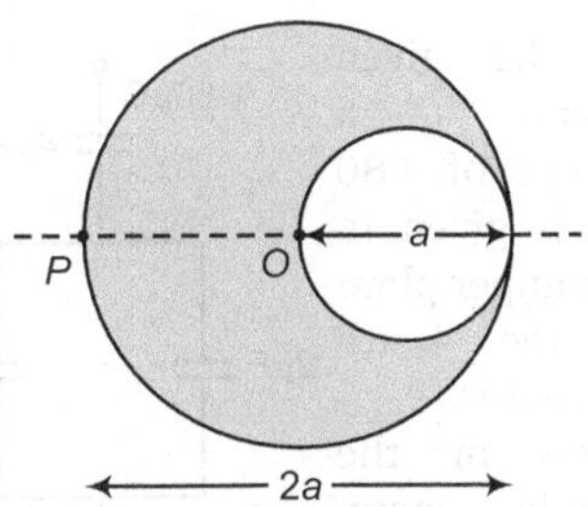

17. A proton is fired from very far away towards a nucleus with charge $Q = 120\,e$, where e is the electronic charge. It makes a closest approach of 10 fm to the nucleus. The de-Broglie wavelength (in units of fm) of the proton at its start is

[Take the proton mass, $m_p = (5/3) \times 10^{-27}$ kg; $h/e = 4.2 \times 10^{-15}$ J-s/C; $\frac{1}{4\pi\varepsilon_0} = 9 \times 10^9$ m/F; 1 fm $= 10^{-15}$ m]

18. A circular wire loop of radius R is placed in the x-y plane centred at the origin O. A square loop of side a ($a << R$) having two turns is placed with its centre at $z = \sqrt{3}\,R$ along the axis of the circular wire loop, as shown in figure. The plane of the square loop makes an angle of 45° with respect to the z-axis. If the mutual inductance between the loops is given by $\frac{\mu_0 a^2}{2^{p/2} R}$, then the value of p is

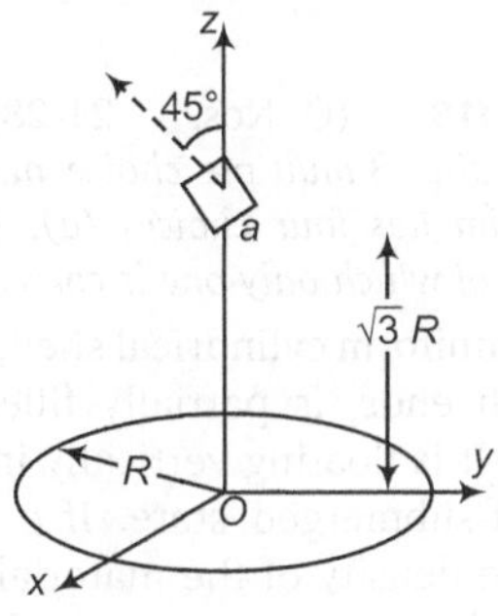

19. An infinitely long solid cylinder of radius R has a uniform volume charge density ρ. It has a spherical cavity of radius $R/2$ with its centre on the axis of the cylinder, as shown in the figure. The magnitude of the electric field at the point P, which is at a distance $2R$ from the axis of the cylinder, is given by the expression $\frac{23\rho R}{16k\varepsilon_0}$. The value of k is

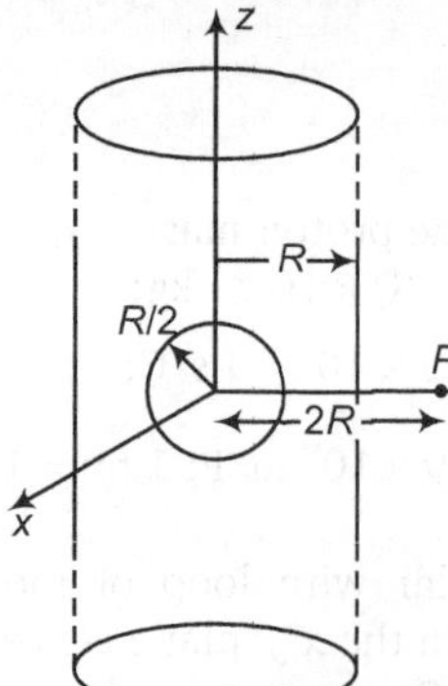

20. A lamina is made by removing a small disc of diameter $2R$ from a bigger disc of uniform mass density and radius $2R$, as shown in the figure. The moment of inertia of this lamina about axes passing through O and P is I_O and I_P, respectively. Both these axes are perpendicular to the plane of the lamina. The ratio $\frac{I_P}{I_O}$ to the nearest integer is

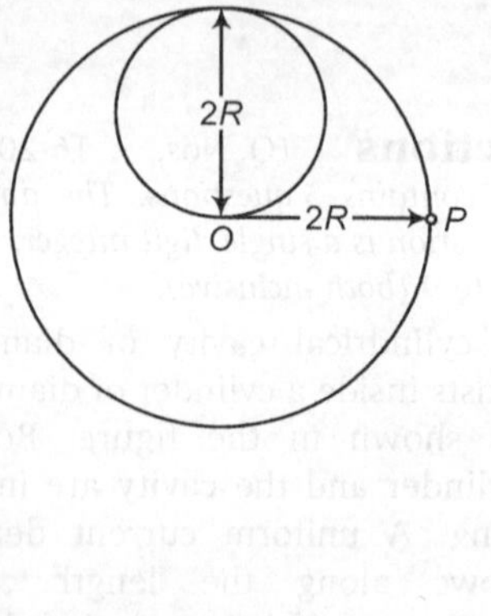

Paper 2

Section I

Directions (Q. Nos. 21-28) *This section contains 8 multiple choice questions. Each question has four choices (a), (b), (c) and (d) out of which only one is correct.*

21. A thin uniform cylindrical shell, closed at both ends, is partially filled with water. It is floating vertically in water in half-submerged state. If ρ_c is the relative density of the material of the shell with respect to water, then the correct statement is that the shell is

(a) more than half-filled if ρ_c is less than 0.5
(b) more than half-filled if ρ_c is more than 1.0
(c) half-filled if ρ_c is more than 0.5
(d) less than half-filled if ρ_c is less than 0.5

22. In the given circuit, a charge of +80 μC is given to the upper plate of the 4 μF capacitor. Then in the steady state, the charge on the upper plate of the 3 μF capacitor is

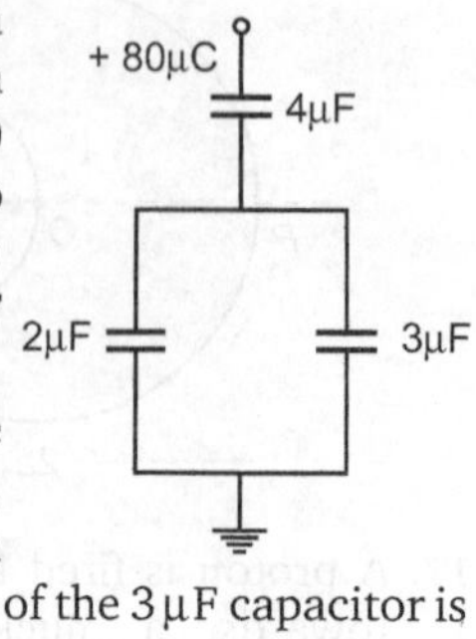

(a) + 32 μC
(b) + 40 μC
(c) + 48 μC
(d) + 80 μC

23. Two moles of ideal helium gas are in a rubber balloon at 30°C. The balloon is fully expandable and can be assumed to require no energy in its expansion. The temperature of the gas in the balloon is slowly changed to 35°C. The amount of heat required in raising the temperature is nearly (take $R = 8.31$ J/mol-K)

(a) 62 J (b) 104 J
(c) 124 J (d) 208 J

24. Consider a disc rotating in the horizontal plane with a constant angular speed ω about its centre O. The disc has a shaded region on one side of the diameter and an unshaded region on the other side as shown in the figure. When the disc is in the orientation as shown, two pebbles P and Q are simultaneously projected at an angle towards R. The velocity of projection is in the y-z plane and is same for both pebbles with respect to the disc. Assume that (i) they land back on the disc before the disc has completed $\frac{1}{8}$ rotation, (ii) their range is less than half the disc radius, and (iii) ω remains constant throughout. Then

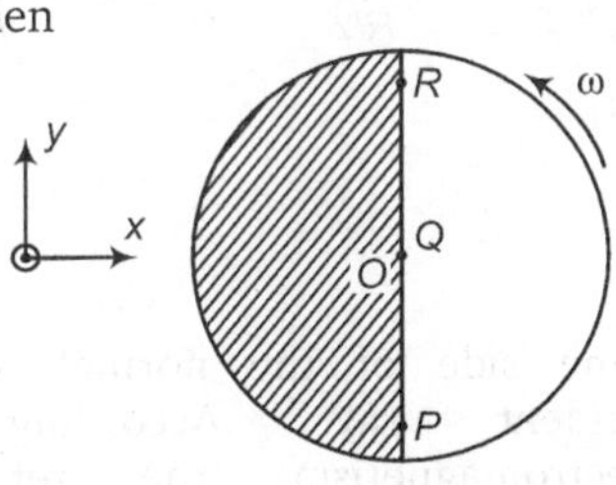

(a) P lands in the shaded region and Q in the unshaded region
(b) P lands in the unshaded region and Q in the shaded region
(c) both P and Q land in the unshaded region
(d) both P and Q land in the shaded region

25. A student is performing the experiment of resonance column. The diameter of the column tube is 4 cm. The frequency of the tuning fork is 512 Hz. The air temperature is 38° C in which the speed of sound is 336 m/s. The zero of the meter scale coincides with the top end of the resonance column tube. When the first resonance occurs, the reading of the water level in the column is

(a) 14.0 cm (b) 15.2 cm
(c) 16.4 cm (d) 17.6 cm

26. Two identical discs of same radius R are rotating about their axes in opposite directions with the same constant angular speed ω. The discs are in the same horizontal plane. At time $t = 0$, the points P and Q are facing each other as shown in the figure.

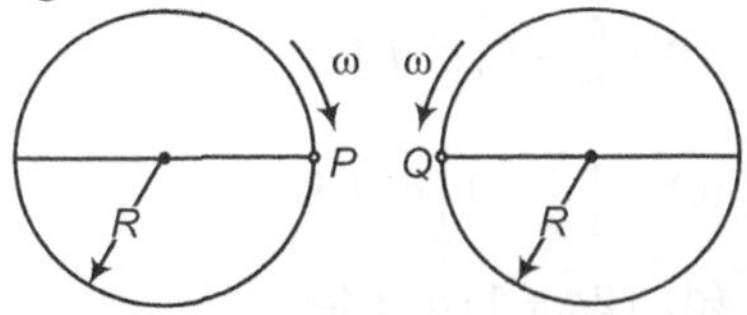

The relative speed between the two points P and Q is v_r. In one time period (T) of rotation of the discs, v_r as a function of time is best represented by

(a)
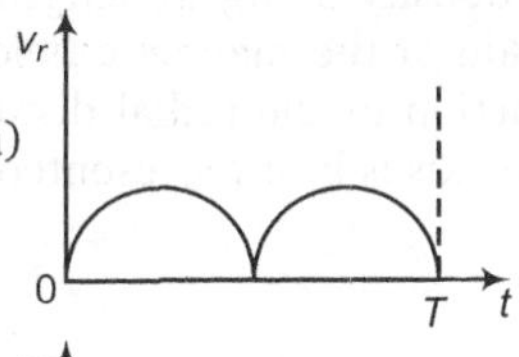

(b)
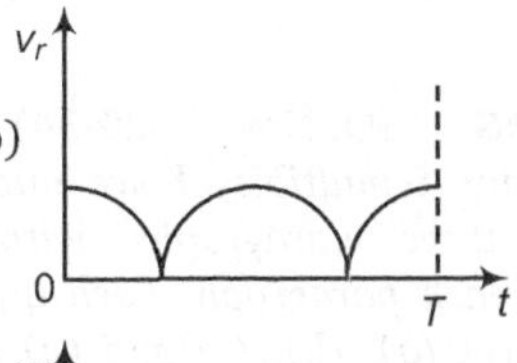

(c)
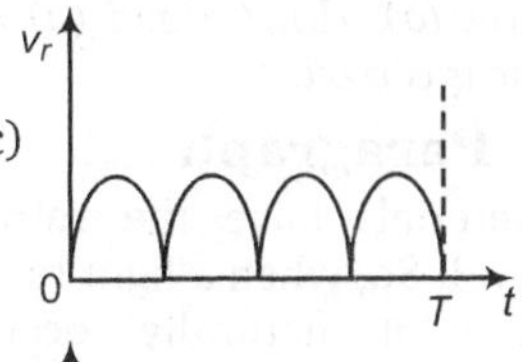

(d)
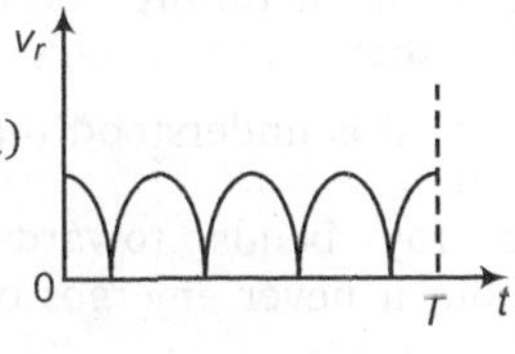

27. A loop carrying current I lies in the x-y plane as shown in the figure. The unit vector $\mathbf{k}$ is coming out of the plane of the paper. The magnetic moment of the current loop is

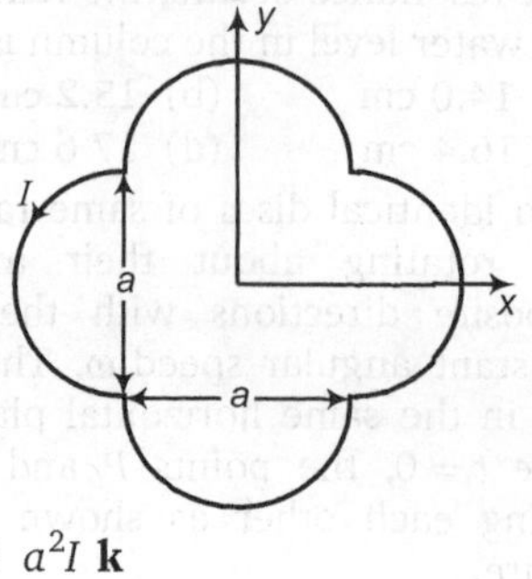

(a) $a^2 I\ \mathbf{k}$

(b) $\left(\dfrac{\pi}{2}+1\right) a^2\ I\ \mathbf{k}$

(c) $-\left(\dfrac{\pi}{2}+1\right) a^2\ I\ \mathbf{k}$

(d) $(2\pi+1)\ a^2\ I\ \mathbf{k}$

28. An infinitely long hollow conducting cylinder with inner radius $R/2$ and outer radius R carries a uniform current density along its length. The magnitude of the magnetic field, $|\mathbf{B}|$ as a function of the radial distance r from the axis is best represented by

(a) 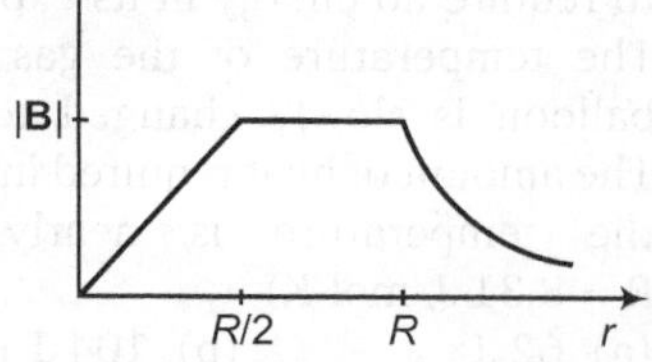

(b) 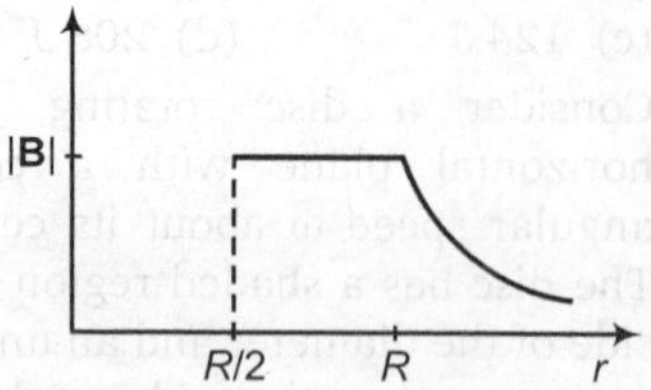

(c) 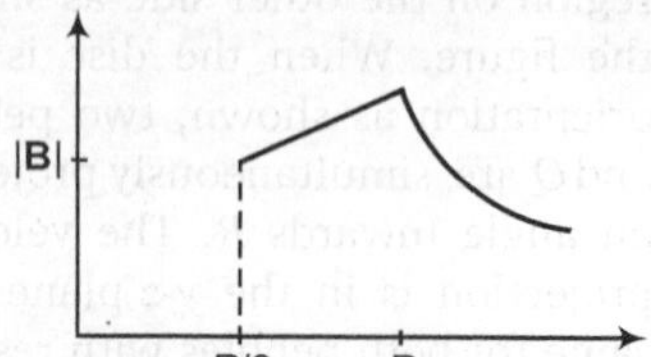

(d) 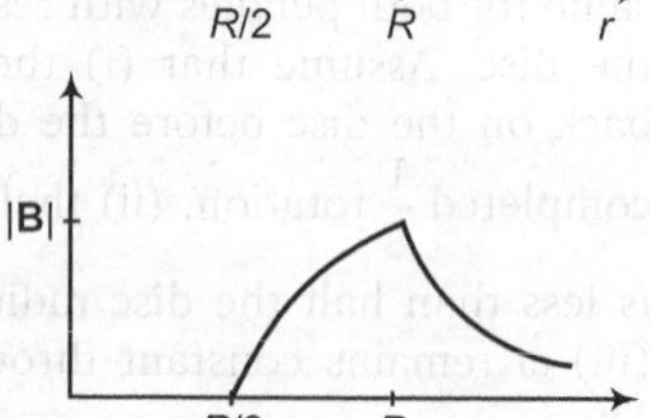

Section II

Directions (Q. Nos. 29-34) *This section contains 6 multiple choice questions relating to three paragraphs with two questions on each paragraph. Each question has four choices (a), (b), (c) and (d) out of which only one is correct.*

Paragraph

Most materials have the refractive index, $n > 1$. So, when a light ray from air enters a naturally occurring material, then by Snell's law, $\dfrac{\sin\theta_1}{\sin\theta_2} = \dfrac{n_2}{n_1}$, it is understood that the refracted ray bends towards the normal. But it never emerges on the same side of the normal as the incident ray. According to electromagnetism, the refractive index of the medium is given by the relation, $n = \left(\dfrac{c}{v}\right) = \pm\sqrt{\varepsilon_r \mu_r}$, where c is the speed of electromagnetic waves in vacuum, v its speed in the medium, ε_r and μ_r are the relative permittivity and permeability of the medium respectively.

In normal materials, both ε_r and μ_r are positive, implying positive n for the medium. When both ε_r and μ_r are negative, one must choose the negative root of n. Such negative refractive index materials can now be artificially prepared and are called meta-materials. They exhibit significantly different optical behaviour, without violating any physical laws. Since n is negative, it results in a change in the direction of propagation of the refracted light. However, similar to normal materials, the frequency of light remains unchanged upon refraction even in meta-materials.

29. Choose the correct statement.

(a) The speed of light in the meta-material is $v = c\,|n|$

(b) The speed of light in the meta-material is $v = \dfrac{c}{|n|}$

(c) The speed of light in the meta-material is $v = c$

(d) The wavelength of the light in the meta-material (λ_m) is given by $\lambda_m = \lambda_{air}\,|n|$, where λ_{air} is the wavelength of the light in air.

30. For light incident from air on a meta-material, the appropriate ray diagram is

(a)
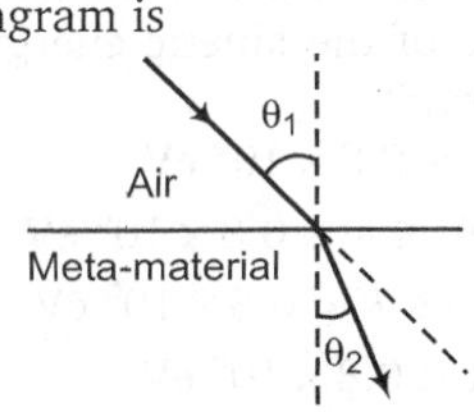

(b)
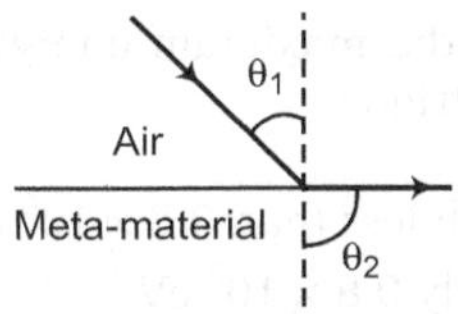

(c)
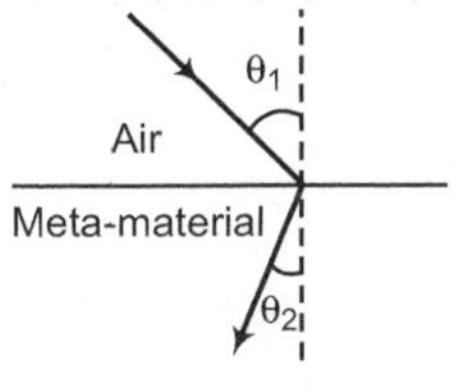

(d)
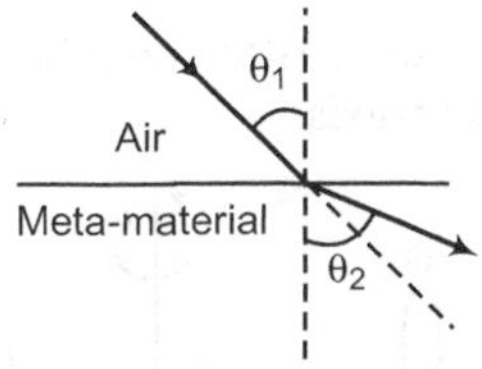

Paragraph

The general motion of a rigid body can be considered to be a combination of (i) a motion of its centre of mass about an axis, and (ii) its motion about an instantaneous axis passing through the centre of mass. These axes need not be stationary. Consider, for example, a thin uniform disc welded (rigidly fixed) horizontally at its rim to a massless stick, as shown in the figure. When the disc-stick system is rotated about the origin on a horizontal frictionless plane with angular speed ω, the motion at any instant can be taken as a combination of (i) a rotation of the centre of mass of the disc about the z-axis, and (ii) a rotation of the disc through an instantaneous vertical axis passing through its centre of mass (as is seen from the changed orientation of points P and Q). Both these motions have the same angular speed ω in this case.

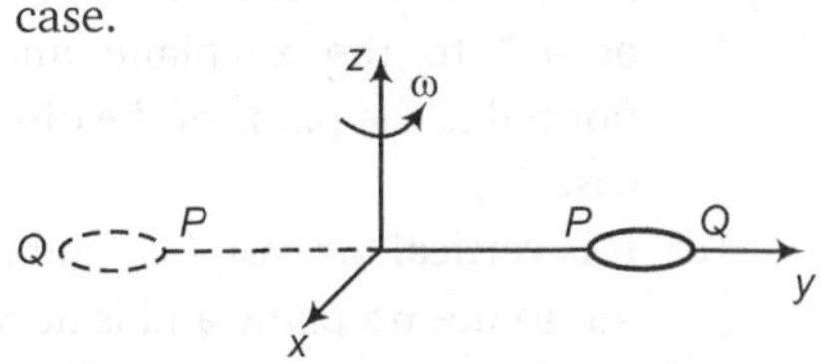

Now consider two similar systems as shown in the figure : Case (a) the disc with its face vertical and parallel to x-z plane; Case (b) the disc with its face making an angle of 45° with x-y plane and its horizontal diameter parallel to x-axis. In both the cases, the disc is welded at point P, and the systems are rotated with constant angular speed ω about the z-axis.

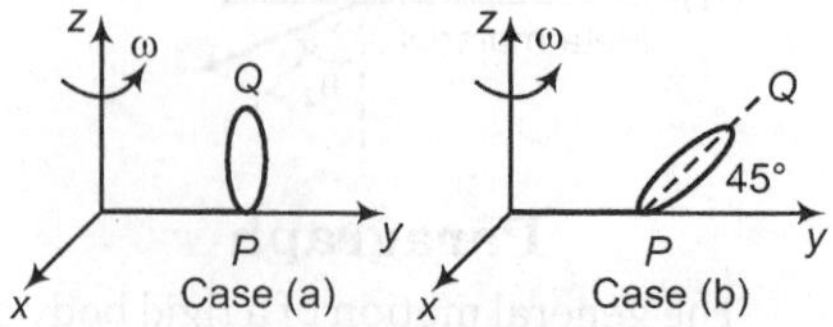

31. Which of the following statements regarding the angular speed about the instantaneous axis (passing through the centre of mass) is correct?

(a) It is $\sqrt{2}\omega$ for both the cases

(b) It is ω for case (a); and $\frac{\omega}{\sqrt{2}}$ for case (b)

(c) It is ω for case (a); and $\sqrt{2}\omega$ for case (b)

(d) It is ω for both the cases

32. Which of the following statements about the instantaneous axis (passing through the centre of mass) is correct?

(a) It is vertical for both the cases (a) and (b)

(b) It is vertical for case (a); and is at 45° to the x-z plane and lies in the plane of the disc for case (b)

(c) It is horizontal for case (a); and is at 45° to the x-z plane and is normal to the plane of the disc for case (b)

(d) It is vertical for case (a); and is at 45° to the x-z plane and is normal to the plane of the disc for case (b)

Paragraph

The β-decay process, discovered around 1900, is basically the decay of a neutron (n). In the laboratory, a proton (p) and an electron (e^-) are observed as the decay products of the neutron. Therefore, considering the decay of a neutron as a two-body decay process, it was predicted theoretically that the kinetic energy of the electron should be a constant. But experimentally, it was observed that the electron kinetic energy has a continuous spectrum. Considering a three-body decay process, *i.e.*, $n \to p + e^- + \bar{\nu}_e$, around 1930, Pauli explained the observed electron energy spectrum. Assuming the anti-neutrino ($\bar{\nu}_e$) to be massless and possessing negligible energy, and the neutron to be at rest, momentum and energy conservation principles are applied. From this calculation, the maximum kinetic energy of the electron is 0.8×10^6 eV. The kinetic energy carried by the proton is only the recoil energy.

33. If the anti-neutrino had a mass of 3 eV/c^2 (where c is the speed of light) instead of zero mass, what should be the range of the kinetic energy K, of the electron?

(a) $0 \le K \le 0.8 \times 10^6$ eV

(b) $3.0 \text{ eV} \le K \le 0.8 \times 10^6$ eV

(c) $3.0 \text{ eV} \le K < 0.8 \times 10^6$ eV

(d) $0 \le K < 0.8 \times 10^6$ eV

34. What is the maximum energy of the anti-neutrino?

(a) Zero

(b) Much less than 0.8×10^6 eV

(c) Nearly 0.8×10^6 eV

(d) Much larger than 0.8×10^6 eV

Section III

Directions (Q. Nos. 35-40) *This section contains 6 multiple choice questions. Each question has four choices (a), (b), (c) and (d) out of which one or more are correct.*

35. Six point charges are kept at the vertices of a regular hexagon of side L and centre O as shown in the figure. Given that $K = \frac{1}{4\pi\varepsilon_0}\frac{q}{L^2}$, which of the following statements(s) is(are) correct.

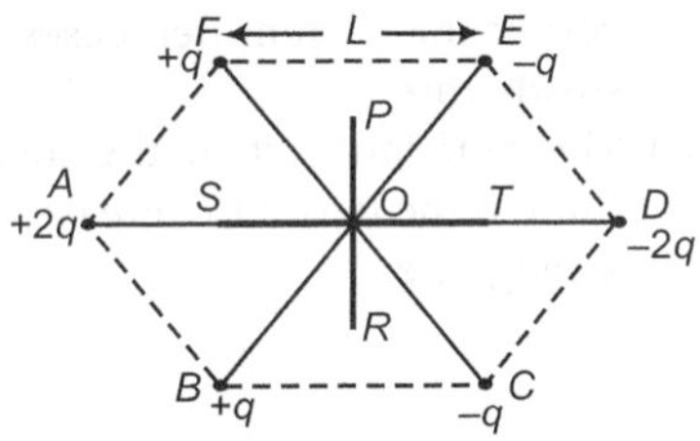

(a) The electric field at O is $6K$ along OD
(b) The potential at O is zero
(c) The potential at all points on the line PR is same
(d) The potential at all points on the line ST is same

36. Two solid cylinders P and Q of same mass and same radius start rolling down a fixed inclined plane from the same height at the same time. Cylinder P has most of its mass concentrated near its surface, while Q has most of its mass concentrated near the axis. Which statement(s) is(are) correct?

(a) Both cylinders P and Q reach the ground at the same time
(b) Cylinder P has larger linear acceleration than cylinder Q
(c) Both cylinders reach the ground with same translational kinetic energy
(d) Cylinder Q reaches the ground with larger angular speed

37. Two spherical planets P and Q have the same uniform density ρ, masses M_P and M_Q, and surface areas A and $4A$, respectively. A spherical planet R also has uniform density ρ and its mass is $(M_P + M_Q)$. The escape velocities from the planets P, Q and R, are V_P, V_Q and V_R, respectively. Then

(a) $V_Q > V_R > V_P$ (b) $V_R > V_Q > V_P$
(c) $V_R/V_P = 3$ (d) $V_P/V_Q = \frac{1}{2}$

38. The figure shows a system consisting of (i) a ring of outer radius $3R$ rolling clockwise without slipping on a horizontal surface with angular speed ω and (ii) an inner disc of radius $2R$ rotating anti-clockwise with angular speed $\omega/2$. The ring and disc are separated by frictionless ball bearings. The system is in the x-z plane. The point P on the inner disc is at a distance R from the origin, where OP makes an angle of 30° with the horizontal. Then with respect to the horizontal surface,

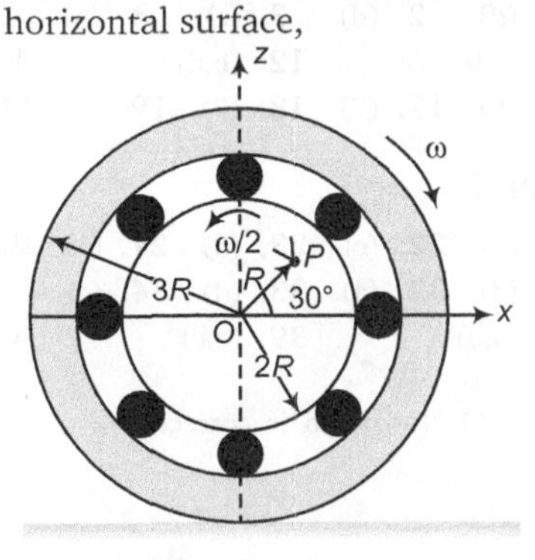

(a) the point O has a linear velocity $3R\omega\,\mathbf{i}$
(b) the point P has a linear velocity $\frac{11}{4}R\omega\,\mathbf{i} - \frac{\sqrt{3}}{4}R\omega\,\mathbf{k}$
(c) the point P has a linear velocity $\frac{13}{4}R\omega\,\mathbf{i} - \frac{\sqrt{3}}{4}R\omega\,\mathbf{k}$

(d) the point P has a linear velocity $\left(3-\frac{\sqrt{3}}{4}\right)R\omega\,\mathbf{i}+\frac{1}{4}R\omega\,\mathbf{k}$

39. In the given circuit, the AC source has $\omega = 100$ rad/s. Considering the inductor and capacitor to be ideal, the correct choice(s) is(are)

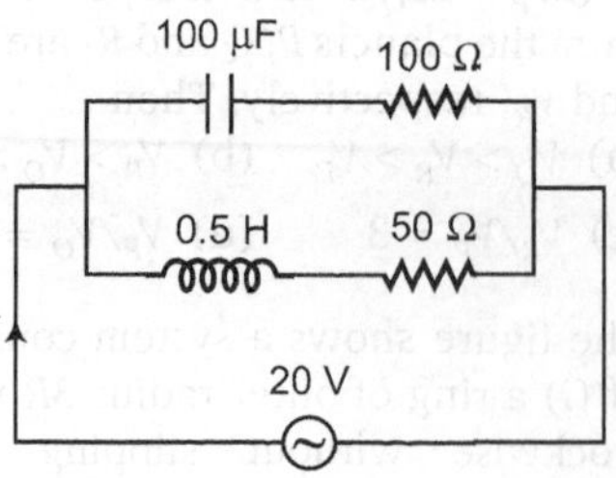

(a) The current through the circuit, I is 0.3 A

(b) The current through the circuit, I is $0.3\sqrt{2}$ A

(c) The voltage across 100 Ω resistor $= 10\sqrt{2}$ V

(d) The voltage across 50 Ω resistor $= 10$ V

40. A current carrying infinitely long wire is kept along the diameter of a circular wire loop, without touching it. The correct statement(s) is(are)

(a) The emf induced in the loop is zero if the current is constant

(b) The emf induced in the loop is finite if the current is constant

(c) The emf induced in the loop is zero if the current decreases at a steady rate

(d) The emf induced in the loop is finite if the current decreases at a steady rate

Answers

Paper 1

1. (d) **2.** (d) **3.** (c) **4.** (b) **5.** (c) **6.** (d) **7.** (b) **8.** (a) **9.** (a) **10.** (c)
11. (a,b,c,d) **12.** (b,d) **13.** (c,d) **14.** (a,c) **15.** (a,c)
16. (5) **17.** (7) **18.** (7) **19.** (6) **20.** (3)

Paper 2

21. (a) **22.** (c) **23.** (d) **24.** (*) **25.** (b) **26.** (a) **27.** (b) **28.** (d) **29.** (b) **30.** (c)
31. (d) **32.** (a) **33.** (d) **34.** (c) **35.** (a,b,c) **36.** (d) **37.** (b,d)
38. (a,b) **39.** (a,c) **40.** (a,c)

Note : (*) No option is correct.

IIT JEE
UNSOLVED PAPER 2012

Chemistry

Paper 1

Section I

Direction (Q. No. 1-10) *This section contains **10 multiple choice questions**, Each question has four choices, (a), (b), (c) and (d) out of which **ONLY ONE** is correct.*

1. Which ordering of compounds is according to the decreasing order of the oxidation state of nitrogen?
   (a) HNO_3, NO, NH_4Cl, N_2
   (b) HNO_3, NO, N_2, NH_4Cl
   (c) HNO_3, NH_4Cl, NO, N_2
   (d) NO, HNO_3, NH_4Cl, N_2

2. The kinetic energy of an electron in the second Bohr orbit of a hydrogen atom is [a_0 is Bohr radius]
   (a) $\dfrac{h^2}{4\pi^2 ma_0^2}$
   (b) $\dfrac{h^2}{16\pi^2 ma_0^2}$
   (c) $\dfrac{h^2}{32\pi^2 ma_0^2}$
   (d) $\dfrac{h^2}{64\pi^2 ma_0^2}$

3. The number of aldol reaction(s) that occurs in the given transformation is

$$CH_2CHO + 4HCHO \xrightarrow{\text{Conc. aq. NaOH}}$$

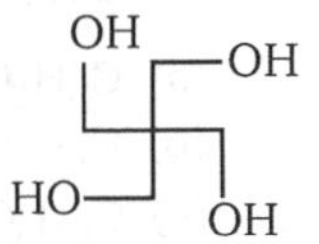

   (a) 1 (b) 2
   (c) 3 (d) 4

4. For one mole of a van der Waals' gas when $b = 0$ and $T = 300$ K, the pV *vs.* $1/V$ plot is shown below. The value of the van der Waals' constant a (atm L mol^{-2})

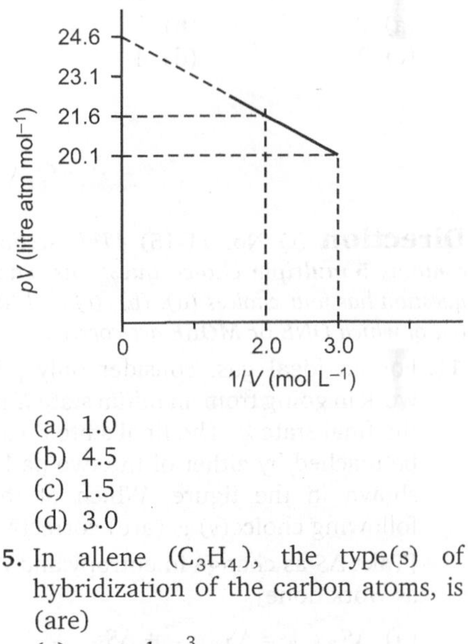

   (a) 1.0
   (b) 4.5
   (c) 1.5
   (d) 3.0

5. In allene (C_3H_4), the type(s) of hybridization of the carbon atoms, is (are)
   (a) sp and sp^3
   (b) sp and sp^2
   (c) only sp^3
   (d) sp^2 and sp^3

6. A compound M_pX_q has cubic close packing (ccp) arrangement of X. Its unit cell structure is shown below. The empirical formula of the compound, is

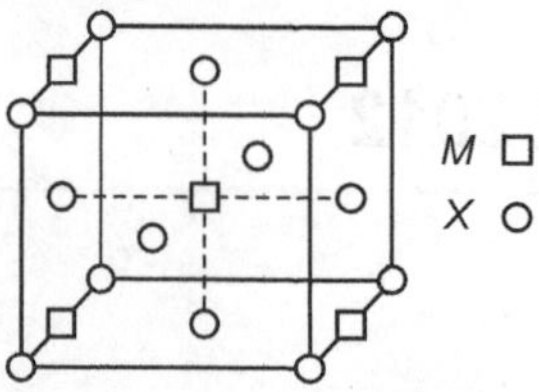

(a) MX (b) MX_2
(c) M_2X (d) M_5X_{14}

7. The number of optically active products obtained from the complete ozonolysis of the given compound, is

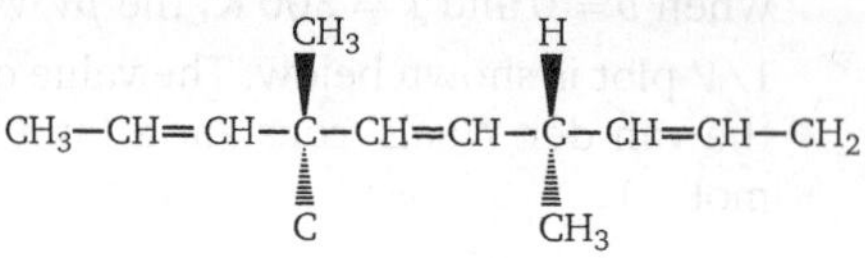

(a) 0 (b) 1
(c) 2 (d) 4

8. As per IUPAC nomenclature, the name of the complex $[Co(H_2O)_4(NH_3)_2]Cl_3$ is
(a) tetraaquadiaminecobalt (III) chloride
(b) tetraaquadiamminecobalt (III) chloride
(c) diaminetetraaquacobalt (III) chloride
(d) diamminetetraaquacobalt (III) chloride

9. The carboxyl functional group (—COOH) is present in
(a) picric acid (b) barbituric acid
(c) ascorbic acid (d) aspirin

10. The colour of light abosrbed by an aqueous solution of $CuSO_4$ is
(a) orange-red (b) blue-green
(c) yellow (d) violet

Section II

Direction (Q. No. 11-15) *This section contains* **5 multiple choice questions.** *Each question has four choices (a), (b), (c) and (d) out of which* ***ONE or MORE*** *are correct.*

11. For an ideal gas, consider only *p*-*V* work in going from an initial state *X* to the final state *Z*. The final state *Z* can be reached by either of the two paths shown in the figure. Which of the following choice(s) is (are) correct? [Take ΔS as change in entropy and *W* as work done].
(a) $\Delta S_{X \to Z} = \Delta S_{X \to Y} + \Delta S_{Y \to Z}$
(b) $W_{X \to Z} = W_{X \to Y} + W_{Y \to Z}$
(c) $W_{X \to Z \to Z} = W_{X \to Y}$
(d) $\Delta S_{X \to Y \to Z} = \Delta S_{X \to Y}$

12. Which of the following molecules, in pure form, is (are) unstable at room temperature?

(a) 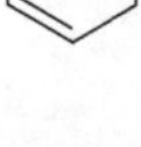 (b)

(c) 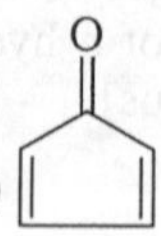 (d) 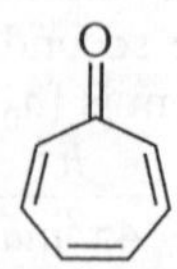

13. Identify the binary mixture(s) that can be separated into individual compounds, by differential extraction, as shown in the given scheme.

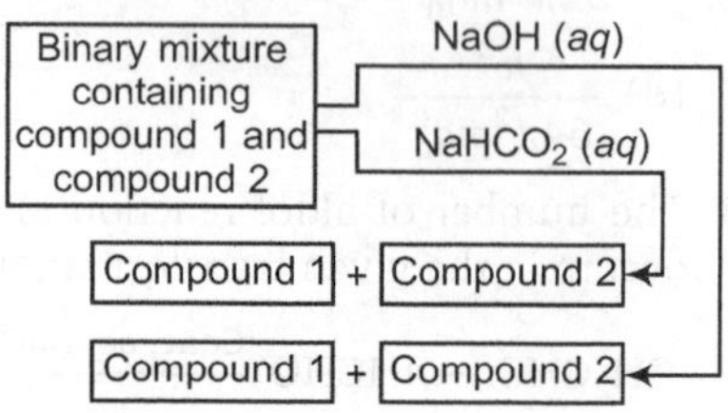

(a) C_6H_5OH and C_6H_5COOH
(b) C_6H_5COOH and $C_6H_5CH_2OH$
(c) $C_6H_5CH_2OH$ and C_6H_5OH
(d) $C_6H_5CH_2OH$ and $C_6H_5CH_2COOH$

14. Choose the correct reason(s) for the stability of the lyophobic colloidal particles.

(a) Preferential adsorption of ions on their surface from the solution.

(b) Preferential adsorption of solvent on their surface from the solution.

(c) Attraction between different particles having opposite charges on their surface.

(d) Potential difference between the fixed layer and the diffused layer of opposite charges around the colloidal particles.

15. Which of the following halides react(s) with $AgNO_3(aq)$ to give a precipitate that dissolves in $Na_2S_2O_3(aq)$?

(a) HCl (b) HF

(c) HBr (d) HI

Section III

Direction (Q. No. 16-20) *This section contains* **5 *questions.*** *The answer to each question is a single-digit integer, ranging from 0 to 9 (both inclusive).*

16. An organic compound undergoes first order decomposition. The time taken for its decomposition to 1/8 and 1/10 of its initial concentration are $t_{1/8}$ and $t_{1/10}$ respectively. What is the value of $\frac{[t_{1/8}]}{[t_{1/10}]} \times 10$? ($\log_{10} 2 = 0.3$)

17. When the following aldohexose exists in its D-configuration, the total number of stereoisomers in its pyranose form, is

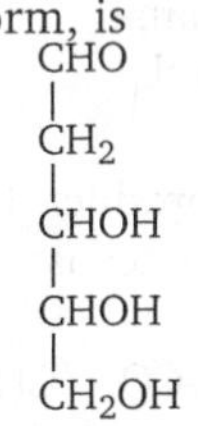

18. The substituents R_1 and R_2 for nine peptides are listed in the table given below. How many of these peptides are positively charged at pH = 7.0 ?

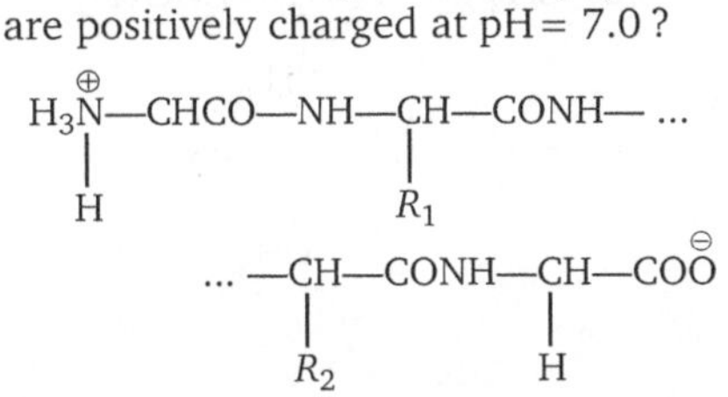

Peptide	R_1	R_2
I	H	H
II	H	CH_3
III	CH_2COOH	H
IV	CH_2CONH_2	$(CH_2)_4NH_2$
V	CH_2CONH_2	CH_2CONH_2
VI	$(CH_2)_4NH_2$	$(CH_2)_4NH_2$
VII	CH_2COOH	CH_2CONH_2
VIII	CH_2OH	$(CH_2)_4NH_2$
IX	$(CH_2)_4NH_2$	CH_3

19. The Periodic Table consists of 18 groups. An isotope of copper, on bombardment with protons, undergoes a nuclear reaction yielding element X as shown below. To which group, element X belongs in the Periodic Table?

$$^{63}_{29}Cu + ^{1}_{1}H \longrightarrow 6\,^{1}_{0}n + ^{4}_{2}\alpha + 2\,^{1}_{1}H + X$$

20. 29.2% (*w/W*) HCl stock solution has density of 1.25 g mL^{-1}. The molecular weight of HCl is 36.5 g mol^{-1}. The volume (mL) of stock solution required to prepare a 200 mL solution 0.4 M HCl is

Paper 2

Section I

Direction (Q. No. 21-28) *This section contains* ***8 multiple choice questions.*** *Each question has four choices (a), (b), (c) and (d) out of which* ***ONLY ONE*** *is correct.*

21. $NiCl_2\{P(C_2H_5)_2(C_6H_5)\}_2$ exhibits temperature dependent magnetic behaviour (paramagnetic/diamagnetic) the coordination geometries of Ni^{2+} in the paramagnetic and diamagnetic states are respectively

(a) tetrahedral and tetrahedral
(b) square planar and square planar
(c) tetrahedral and square planar
(d) square planar and tetrahedral

22. The reaction of white phosphorus with aqueous NaOH gives phosphine along with another phosphorus containing compound. The reaction type; the oxidation states of phosphorus in phosphine and the other product are respectively

(a) redox reaction; –3 and –5
(b) redox reaction; 3 and +5
(c) disproportionation reaction; –3 and +5
(d) disproportionation reaction; –3 and +3

23. In the cyanide extraction process of silver from argentite ore, the oxidizing and reducing agents used are

(a) O_2 and CO respectively
(b) O_2 and Zn dust respectively
(c) HNO_3 and Zn dust respectively
(d) HNO_3 and CO respectively

24. The compound that undergoes decarboxylation most readily under mild condition is

(a) 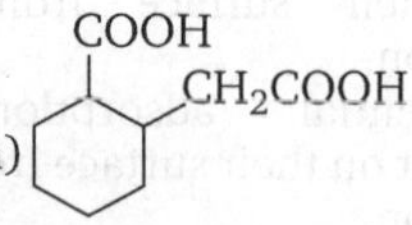

(b) 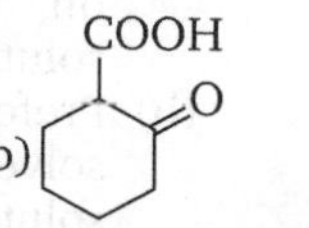

(c) 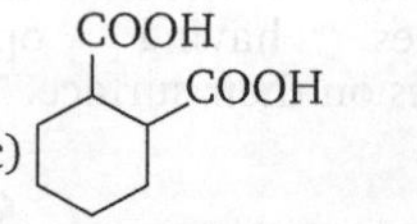

(d) 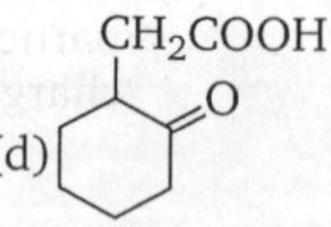

25. Using the data provided, calculate the multiple bond energy (kJ mol^{-1}) of a C≡C bond C_2H_2. That energy is (take the bond energy of a C—H bond as 350 kJ mol^{-1})

$$2C(s) + H_2(g) \longrightarrow C_2H_2(g); \quad \Delta H = 225 \text{ kJ mol}^{-1}$$

$$2C(s) \longrightarrow 2C(g); \quad \Delta H = 1410 \text{ kJ mol}^{-1}$$

$$H_2(g) \longrightarrow 2H(g); \quad \Delta H = 330 \text{ kJ mol}^{-1}$$

26. The shape of XeO_2F_2 molecule is

(a) trigonal bipyramidal
(b) square planar
(c) tetrahedral
(d) see-saw

27. The major product H in the given reaction sequence is

$$CH_3—CH_2—CO—CH_3 \xrightarrow{^{\ominus}CN} G \xrightarrow{95\%\ H_2SO_4} H$$

(a) $CH_3—CH{=}C(CH_3)—COOH$

(b) $CH_3—CH{=}C(CH_3)—CN$

(c) $CH_3—CH_2—C(OH)(CH_3)—COOH$

(d) $CH_3—CH{=}C(CH_3)—CO—NH_2$

28. For a dilute solution containing 2.5 g of a non-volatile non-electrolyte solute in 100 g of water, the elevation in boiling point at 1 atm pressure is 2°C. Assuming concentration of solute is much lower than the concentration of solvent, the vapour pressure (mm of Hg) of the solution is (take $K_b = 0.76$ K kg mol^{-1}).

(a) 724
(b) 740
(c) 736
(d) 718

Section II

Direction (Q. No. 29-34) *This section contains **6 multiple choice questions** relating to three paragraphs with two questions on each paragraph. Each question has four choices (a), (b), (c) and (d) out of which **ONLY ONE** is correct.*

Paragraph

The electrochemical cell shown below is a concentration cell.

$M|M^{2+}$ (saturated solution of a sparingly soluble salt, MX_2) $||M^{2+}$ (0.001 mol dm^{-3}) $|M$.

The emf of the cell depends on the difference in concetration of M^{2+} ions at the two electrodes. The emf of the cell at 298 is 0.059 V.

29. The solubility product (K_{sp} ; mol^{3}dm^{-9}) of MX_2 at 298 based on the information available the given concentration cell is (take $2.303 \times R \times 298/F = 0.059$ V)

(a) 1×10^{-15}
(b) 4×10^{-15}
(c) 1×10^{-12}
(d) 4×10^{-12}

30. The value of ΔG (kJ mol^{-1}) for the given cell is (take 1 F = 96500 C mol^{-1})

(a) – 5.7 (b) 5.7
(c) 11.4 (d) – 11.4

Paragraph

Bleaching powder and bleach solution are produced on a large scale and used in several house hold products. The effectiveness of bleach solution is often measured by iodometry.

31. 25 mL of household bleach solution was mixed with 30 mL of 0.50 M KI and 10 mL of 4 N acetic acid. In the titration of the liberated iodine, 48 mL of 0.25 N $Na_2S_2O_3$ was used to reach the end point. The molarity of the household bleach solution, is

(a) 0.48 M (b) 0.96 M
(c) 0.24 M (d) 0.024 M

32. Bleaching powder contains a salt of an oxoacid as one of its components. The anhydride of that oxoacid, is

(a) Cl_2O (b) Cl_2O_7
(c) ClO_2 (d) Cl_2O_6

Paragraph

In the following reactions sequence, the compound J is an intermediate.

$$I \xrightarrow[CH_3COONa]{(CH_3CO)_2O} J \xrightarrow[\text{(ii) } SOCl_2 \text{ (iii) anhyd. } AlCl_3]{\text{(i) } H_2, Pd/C} K$$

J ($C_9H_8O_2$) gives effervescence on treatment with $NaHCO_3$ and positive Baeyer's test.

33. The compound K, is

(a) 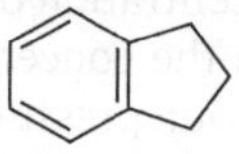 (b) 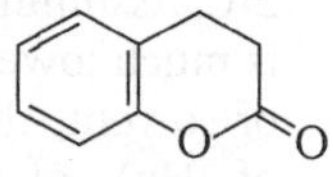

(c) 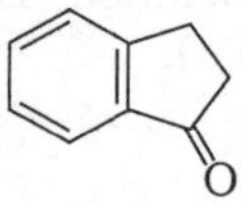

(d) 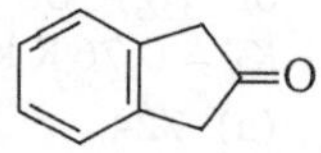

34. The compound I, is

(a) 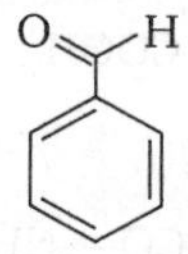

(b) 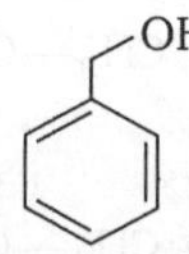

(c) 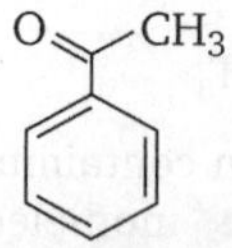

(d) 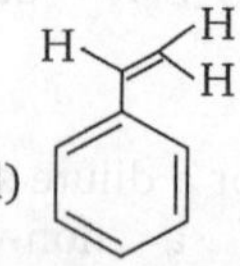

Section III

Direction (Q. No. 35-40) *This section contains* ***6 multiple choice questions.*** *Each question has four choices (a), (b), (c) and (d) out of which* ***ONE*** *or* ***MORE*** *are correct.*

35. With respect to graphite and diamond, which of the statement(s) given below is (are) correct?

(a) Graphite is harder than diamond.

(b) Graphite has higher electrical conductivity than diamond.

(c) Graphite has higher thermal conductivity than diamond.

(d) Graphite has higher C—C bond order than diamond.

36. The given graph/data I, II, III and IV represent general trends observed for different physisorption and chemisorption processes under mild conditions of temperature and pressure. Which of the following choice(s) about I, II, III and IV is (are) correct?

(I) 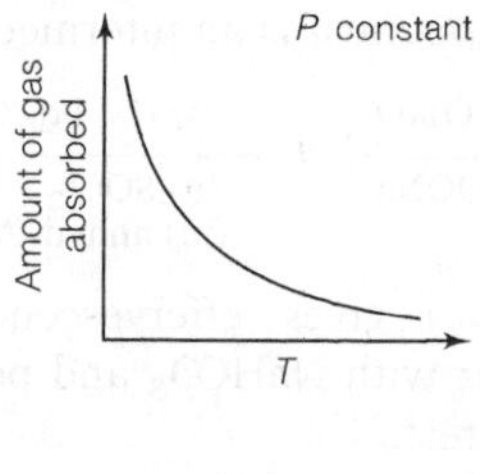

(II) 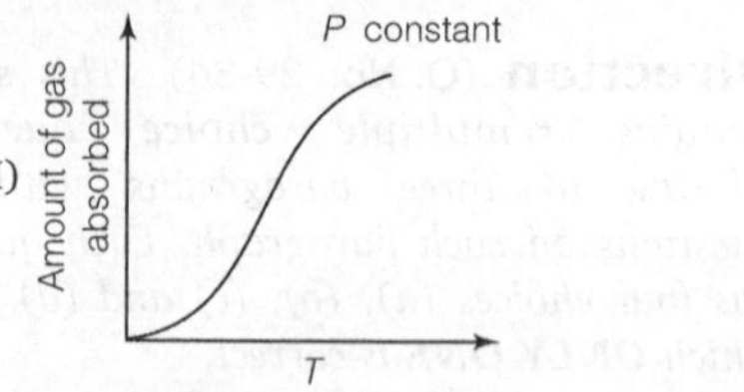

(III) 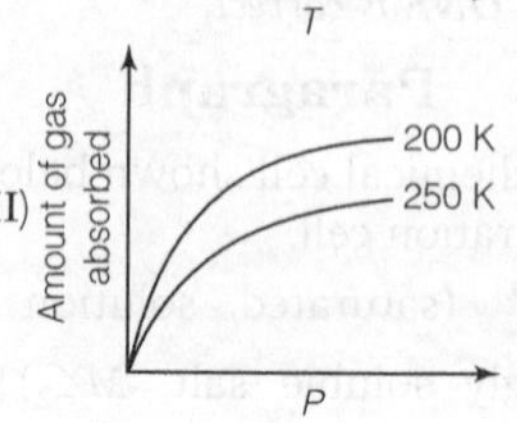

(IV) 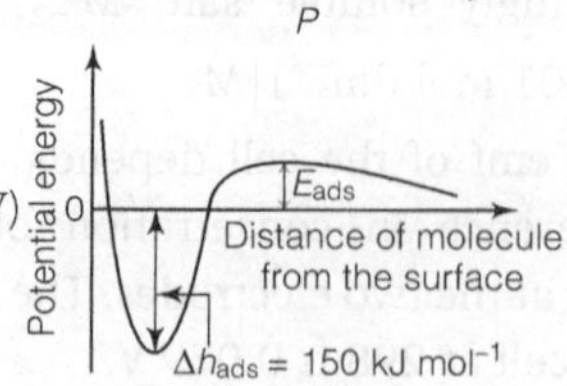

(a) I is physisorption and II is chemisorption

(b) I is physisorption and III is chemisorption

(c) IV is chemisorption and II is chemisorption

(d) IV is chemisorption and III is chemisorption

37. The reversible expansion on an ideal gas under adiabatic and isothermal conditions is shown in the figure. Which of the following statement(s) is (are) correct?

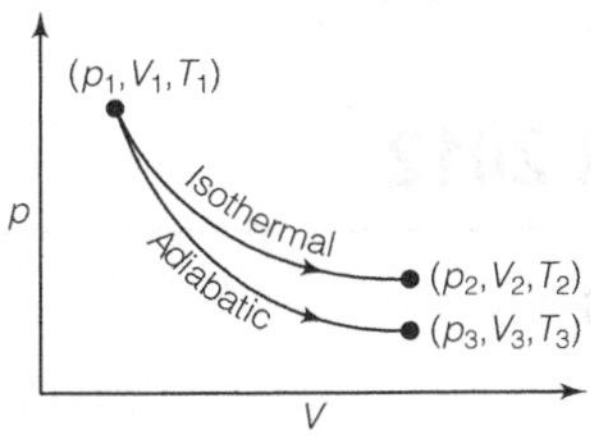

(a) $T_1 = T_2$
(b) $T_3 > T_1$
(c) $W_{\text{isothermal}} > W_{\text{adiabatic}}$
(d) $\Delta U_{\text{isothermal}} > \Delta U_{\text{adiabatic}}$

38. For the given aqueous reaction which of the statement(s) is (are) true?

Excess KI $+ K_3[Fe(CN)_6] \xrightarrow{\text{Dilute } H_2SO_4}$ Brownish-yelllow solution

$\downarrow ZnSO_4$

(White precipitate + Brownish-yellow filtrate)

$\downarrow Na_2S_2O_3$

Colourless solution

(a) The first reaction is a redox reaction
(b) White precipitate is $Zn_3[Fe(CN)_6]_2$
(c) Addition of filtrate to starch solution gives blue colour
(d) White precipitate is soluble in NaOH solution

39. With reference to the scheme given, which of the given statement (s) about T, U, V and W is (are) correct?

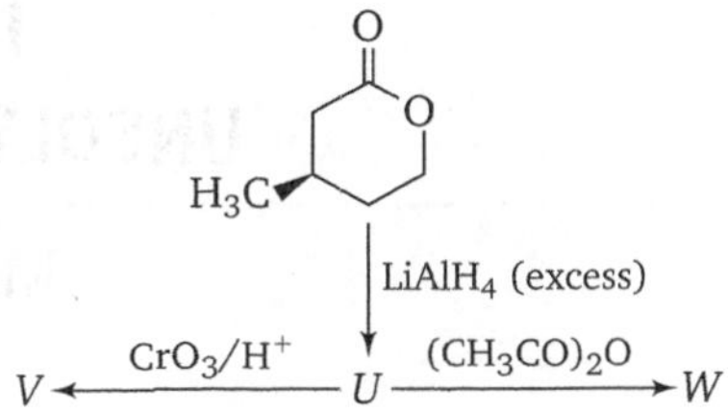

(a) T is soluble in hot aqueous NaOH.
(b) U is optically active.
(c) Molecular formula of W is $C_{10}H_{18}O_4$.
(d) V gives effervescence on treatment with aqueous $NaHCO_3$.

40. Which of the given statement(s) about N, O, P and Q with respect to M is (are) correct?

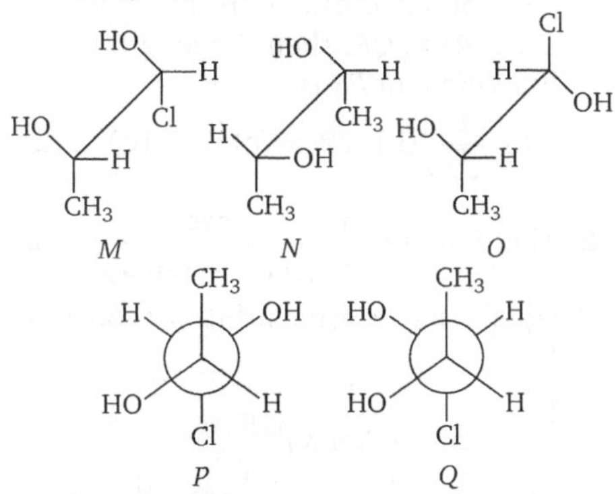

(a) M and N are non-mirror image stereoisomers
(b) M and O are identical
(c) M and P are enantiomers
(d) M and Q are identical

Answers

Paper 1

1. (b) **2.** (c) **3.** (c) **4.** (c) **5.** (b) **6.** (b) **7.** (a) **8.** (d) **9.** (d) **10.** (a)
11. (a,c) **12.** (b,c) **13.** (b,d) **14.** (a,d) **15.** (a,c,d)
16. (9) **17.** (8) **18.** (4) **19.** (8) **20.** (8)

Paper 2

21. (c) **22.** (c) **23.** (b) **24.** (b) **25.** (d) **26.** (d) **27.** (a) **28.** (a) **29.** (b) **30.** (d)
31. (c) **32.** (a) **33.** (c) **34.** (a) **35.** (b,d) **36.** (a,c) **37.** (a,c,d)
38. (a,c,d) **39.** (a,c,d) **40.** (a,b,c)

IIT JEE
UNSOLVED PAPER 2012

Mathematics

Paper 1

Section I

Direction (Q. Nos. 1-10) *This section contains 10 multiple choice questions. Each question has four choices (a), (b), (c) and (d) out of which only one is correct.*

1. The point P is the intersection of the straight line joining the points $Q(2, 3, 5)$ and $R(1, -1, 4)$ with the plane $5x - 4y - z = 1$. If S is the foot of the perpendicular drawn from the point $T(2, 1, 4)$ to QR, then the length of the line segment PS is

   (a) $\frac{1}{\sqrt{2}}$ (b) $\sqrt{2}$ (c) 2 (d) $2\sqrt{2}$

2. The integral $\int \frac{\sec^2 x}{(\sec x + \tan x)^{9/2}} dx$ equals to (for some arbitrary constant K)

   (a) $\frac{-1}{(\sec x + \tan x)^{11/2}} \left\{\frac{1}{11} - \frac{1}{7}(\sec x + \tan x)^2\right\} + K$

   (b) $\frac{1}{(\sec x + \tan x)^{11/2}} \left\{\frac{1}{11} - \frac{1}{7}(\sec x + \tan x)^2\right\} + K$

   (c) $\frac{-1}{(\sec x + \tan x)^{11/2}} \left\{\frac{1}{11} + \frac{1}{7}(\sec x + \tan x)^2\right\} + K$

   (d) $\frac{1}{(\sec x + \tan x)^{11/2}} \left\{\frac{1}{11} + \frac{1}{7}(\sec x + \tan x)^2\right\} + K$

3. Let z be a complex number such that the imaginary part of z is non-zero and $a = z^2 + z + 1$ is real. Then, a cannot take the value

   (a) -1 (b) $\frac{1}{3}$ (c) $\frac{1}{2}$ (d) $\frac{3}{4}$

4. Let $f(x) = \begin{cases} x^2 \left|\cos \frac{\pi}{x}\right|, & x \neq 0, x \in R, \\ 0, & x = 0 \end{cases}$

   then f is

   (a) differentiable both at $x = 0$ and at $x = 2$
   (b) differentiable at $x = 0$ but not differentiable at $x = 2$
   (c) not differentiable at $x = 0$ but differentiable at $x = 2$
   (d) differentiable neither at $x = 0$ nor at $x = 2$

5. The total number of ways in which 5 balls of different colours can be distributed among 3 persons so that each person gets atleast one ball is

   (a) 75 (b) 150 (c) 210 (d) 243

6. If $\lim_{x \to \infty} \left(\frac{x^2 + x + 1}{x + 1} - ax - b\right) = 4$, then

   (a) $a = 1, b = 4$ (b) $a = 1, b = -4$
   (c) $a = 2, b = -3$ (d) $a = 2, b = 3$

7. The function $f: [0, 3] \to [1, 29]$, defined by $f(x) = 2x^3 - 15x^2 + 36x + 1$, is

   (a) one-one and onto
   (b) onto but not one-one
   (c) one-one but not onto
   (d) neither one-one nor onto

8. The locus of the mid-point of the chord of contact of tangents drawn from points lying on the straight line $4x - 5y = 20$ to the circle $x^2 + y^2 = 9$ is

(a) $20(x^2 + y^2) - 36x + 45y = 0$

(b) $20(x^2 + y^2) + 36x - 45y = 0$

(c) $36(x^2 + y^2) - 20y + 45y = 0$

(d) $36(x^2 + y^2) + 20x - 45y = 0$

9. Let $P = [a_{ij}]$ be a 3×3 matrix and let $Q = [b_{ij}]$, where $b_{ij} = 2^{i+j} a_{ij}$ for $1 \le i, j \le 3$. If the determinant of P is 2, then the determinant of the matrix Q is

(a) 2^{10} (b) 2^{11} (c) 2^{12} (d) 2^{13}

10. The ellipse $E_1 : \frac{x^2}{9} + \frac{y^2}{4} = 1$ is inscribed in a rectangle R whose sides are parallel to the coordinate axes. Another ellipse E_2 passing through the point (0, 4) circumscribes the rectangle R. The eccentricity of the ellipse E_2 is

(a) $\frac{\sqrt{2}}{2}$ (b) $\frac{\sqrt{3}}{2}$

(c) $\frac{1}{2}$ (d) $\frac{3}{4}$

Section II

Direction (Q. Nos. 11-15) *This section contains 5 multiple choice questions. Each question has four choices (a), (b), (c) and (d) out of which one or more are correct.*

11. If $y(x)$ satisfies the differential equation $y' - y \tan x = 2x \sec x$ and $y(0)$, then

(a) $y\left(\frac{\pi}{4}\right) = \frac{\pi^2}{8\sqrt{2}}$ (b) $y'\left(\frac{\pi}{4}\right) = \frac{\pi^2}{18}$

(c) $y\left(\frac{\pi}{3}\right) = \frac{\pi^2}{9}$

(d) $y'\left(\frac{\pi}{3}\right) = \frac{4\pi}{3} + \frac{2\pi^2}{3\sqrt{3}}$

12. A ship is fitted with three engines E_1, E_2 and E_3. The engines function independently of each other with respective probabilities 1/2, 1/4 and 1/4. For the ship to be operational atleast two of its engines must function. Let X denote the event that the ship is operational and let X_1, X_2 and X_3 dentoes, respectively the events that the engines E_1, E_2 and E_3 are functioning. Which of the following is/are true?

(a) $P[X_1^c | X] = \frac{3}{16}$

(b) P [Exactly two engines of the ship are functioning x] = 7/8

(c) $P[X | X_2] = \frac{5}{16}$

(d) $P[X | X_1] = \frac{7}{16}$

13. Let $\theta, \phi \in [0, 2\pi]$ be such that $2\cos\theta(1 - \sin\phi) = \sin^2\theta$

$\left(\tan\frac{\theta}{2} + \cot\frac{\theta}{2}\right)\cos\phi - 1,$

$\tan(2\pi - \theta) > 0$

and $-1 < \sin\theta < -\frac{\sqrt{3}}{2}$.

Then, ϕ cannot satisfy

(a) $0 < \phi < \frac{\pi}{2}$ (b) $\frac{\pi}{2} < \phi < \frac{4\pi}{3}$

(c) $\frac{4\pi}{3} < \phi < \frac{3\pi}{2}$ (d) $\frac{3\pi}{2} < \phi < 2\pi$

14. If S be the area of the region enclosed by $y = e^{-x^2}$, $y = 0$, $x = 0$ and $x = 1$. Then,

(a) $S \ge \frac{1}{e}$

(b) $S \ge 1 - \frac{1}{e}$

(c) $S \le \frac{1}{4}\left(1 + \frac{1}{\sqrt{e}}\right)$

(d) $S \le \frac{1}{\sqrt{2}} + \frac{1}{\sqrt{e}}\left(1 - \frac{1}{\sqrt{2}}\right)$

15. Tangents are drawn to the hyperbola $\frac{x^2}{9} - \frac{y^2}{4} = 1$, parallel to the straight line $2x - y = 1$. The points of contacts of the tangents on the hyperbola are

(a) $\left(\frac{9}{2\sqrt{2}}, \frac{1}{\sqrt{2}}\right)$ (b) $\left(-\frac{9}{2\sqrt{2}}, -\frac{1}{\sqrt{2}}\right)$

(c) $(3\sqrt{3}, -2\sqrt{2})$

(d) $(-3\sqrt{3}, 2\sqrt{2})$

Section III

Direction (Q. Nos. 16-20) *This section contains 5 multiple choice questions. Each question has four choices (a), (b), (c) and (d) out of which one or more are correct.*

16. Let S be the focus of the parabola $y^2 = 8x$ and let PQ be the common chord of the circle $x^2 + y^2 - 2x - 4y = 0$ and the given parabola. The area of the ΔOPS is

17. Let $p(x)$ be a real polynomial of least degree which has a local maximum at $x = 1$ and a local minimum at $x = 3$. If $p(1) = 6$ and $p(3) = 2$, then $p'(0)$ is

18. Let $f : R \to R$ be defined as $f(x) = |x| + |x^2 - 1|$. The total number of points at which f attains either a local maximum or a local minimum is

19. The value of $6 + \log_{3/2}\left(\frac{1}{3\sqrt{2}}\sqrt{4 - \frac{1}{3\sqrt{2}}\sqrt{4 - \frac{1}{3\sqrt{2}}\sqrt{4 - \frac{1}{3\sqrt{2}}\cdots}}}\right)$ is

20. If $\mathbf{a}$, $\mathbf{b}$ and $\mathbf{c}$ are unit vectors satisfying $|\mathbf{a} - \mathbf{b}|^2 + |\mathbf{b} - \mathbf{c}|^2 + |\mathbf{c} - \mathbf{a}|^2 = 9$, then $|2\mathbf{a} + 5\mathbf{b} + 5\mathbf{c}|$, is

Paper 2

Section I

Direction (Q. Nos. 21-28) *This section contains 8 multiple choice questions. Each question has four choices (a), (b), (c) and (d) out of which only one is correct.*

21. The equation of a plane passing through the line of intersection of the planes $x + 2y + 3z = 2$ and $x - y + z = 3$ and at a distance $2/\sqrt{3}$ from the point $(3, 1, -1)$ is

(a) $5x - 11y + z = 17$

(b) $\sqrt{2}x + y = 3\sqrt{2} - 1$

(c) $x + y + z = \sqrt{3}$

(d) $x - \sqrt{2}y = 1 - \sqrt{2}$

22. If $\mathbf{a}$ and $\mathbf{b}$ are vectors such that $|\mathbf{a} + \mathbf{b}| = \sqrt{29}$ and $\mathbf{a} \times (2\mathbf{i} + 3\mathbf{j} + 4\mathbf{k}) = (2\mathbf{i} + 3\mathbf{j} + 4\mathbf{k}) \times \mathbf{b}$, then a possible value of $(\mathbf{a} + \mathbf{b}) \cdot (-7\mathbf{i} + 2\mathbf{j} + 3\mathbf{k})$ is

(a) 0 (b) 3 (c) 4 (d) 8

23. Let PQR be a triangle of area Δ with $a = 2$, $b = \frac{7}{2}$ and $c = \frac{5}{2}$, where a, b and c are the lengths of the sides of the triangle opposite to the angles at P, Q and R, respectively. Then, $\frac{2 \sin P - \sin 2P}{2 \sin P + \sin 2P}$ equals to

(a) $\frac{3}{4\Delta}$ (b) $\frac{45}{4\Delta}$

(c) $\left(\frac{3}{4\Delta}\right)^2$ (d) $\left(\frac{45}{4\Delta}\right)^5$

24. Four fair dice D_1, D_2, D_3 and D_4 each having six faces numbered 1, 2, 3, 4, 5 and 6 are rolled simultaneously. The probability that D_4 shows a number appearing on one of D_1, D_2 and D_3 is

(a) $\frac{91}{216}$ (b) $\frac{108}{216}$ (c) $\frac{125}{216}$ (d) $\frac{127}{216}$

25. The value of the integral $\int_{-\pi/2}^{\pi/2}\left(x^2 + \log\frac{\pi - x}{\pi + x}\right)\cos x\,dx$ is

(a) 0 (b) $\frac{\pi^2}{2} - 4$ (c) $\frac{\pi^2}{2} + 4$ (d) $\frac{\pi^2}{2}$

26. If P is a 3×3 matrix such that $P^T = 2P + I$, where P^T is the transpose of P and I is the 3×3 identity matrix, then there exists a column matrix, $X = \begin{bmatrix} x \\ y \\ z \end{bmatrix} \neq \begin{bmatrix} 0 \\ 0 \\ 0 \end{bmatrix}$ such that

(a) $PX = \begin{bmatrix} 0 \\ 0 \\ 0 \end{bmatrix}$ (b) $PX = X$

(c) $PX = 2X$ (d) $PX = -X$

27. Let $a_1, a_2, a_3, \ldots$ be in a harmonic progression with $a_1 = 5$ and $a_{20} = 25$. The least positive integer n for which $a_n < 0$ is

(a) 22 (b) 23 (c) 24 (d) 25

28. Let $\alpha(a)$ and $\beta(a)$ be the roots of the equation $(\sqrt[3]{1+a} - 1)x^2 - (\sqrt{1+a} - 1)x + (\sqrt[6]{1+a} - 1) = 0$, where $a > -1$. Then, $\lim_{a \to 0^+} \alpha(a)$ and $\lim_{a \to 0^+} \beta(a)$ are

(a) $-\frac{5}{2}$ and 1 (b) $-\frac{1}{2}$ and -1

(c) $-\frac{7}{2}$ and 2 (d) $-\frac{9}{2}$ and 3

Section II

Direction (Q. Nos. 29-34) *This section contains 6 multiple choice questions relating to three paragraphs with two questions on each paragraph. Each question has four choices (a), (b), (c) and (d) out of which only one is correct.*

Paragraph

Let $f(x) = (1 - x)^2 \sin^2 x + x^2$ for all $x \in R$ and

let $g(x) = \int_1^x \left(\frac{2(t-1)}{t+1} - \ln t\right) f(t)\,dt$ for all $x \in (1, \infty)$.

29. Which of the following is true?
(a) g is increasing on $(1, \infty)$
(b) g is decreasing on $(1, \infty)$
(c) g is increasing on $(1, 2)$ and decreasing on $(2, \infty)$
(d) g is decreasing on $(1, 2)$ and increasing on $(2, \infty)$

30. Consider the statements

P : There exists some $x \in R$ such that $f(x) + 2x = 2(1 + x^2)$

Q : There exists some $x \in R$ such that $2f(x) + 1 = 2x(1 + x)$

Then,

(a) Both P and Q are true
(b) P is true and Q is false
(c) P is false and Q is true
(d) Both P and Q are false

Paragraph

Let a_n denote the number of all n-digit positive integers formed by the digits 0, 1 or both such that no consecutive digits in them are 0. Let b_n = The number of such n-digit integers ending with digit 1 and c_n = The number of such n-digit integers ending with digit 0.

31. Which of the following is correct?

(a) $a_{17} = a_{16} + a_{15}$ (b) $c_{17} \neq c_{16} + c_{15}$

(c) $b_{17} \neq b_{16} + c_{16}$ (d) $a_{17} = c_{17} + b_{16}$

32. The value of b_6 is

(a) 7 (b) 8 (c) 9 (d) 11

Paragraph

A tangent PT is drawn to the circle $x^2 + y^2 = 4$ at the point $P(\sqrt{3}, 1)$. A straight line L, perpendicular to PT is a tangent to the circle $(x - 3)^2 + y^2 = 1$.

33. A common tangent of the two circles is

(a) $x = 4$ (b) $y = 2$

(c) $x + \sqrt{3}y = 4$ (d) $x + 2\sqrt{2}\, y = 6$

34. A possible equation of L is

(a) $x - \sqrt{3}y = 1$ (b) $x + \sqrt{3}y = 1$

(c) $x - \sqrt{3}y = -1$ (d) $x + \sqrt{3}y = 5$

Section III

Direction (Q. Nos. 35-40) *This section contains 6 multiple choice questions. Each question has four choices (a), (b), (c) and (d) out of which one or more are correct.*

35. Let X and Y be two events such that $P(X \mid Y) = \frac{1}{2}, P(Y \mid X) = \frac{1}{3}$ and $P(X \cap Y) \frac{1}{6}$. Which of the following is/are correct?

(a) $P(X \cup Y) = 2/3$
(b) X and Y are independent
(c) X and Y are not independent
(d) $P(X^c \cap Y) = 1/3$

36. If $f(x) = \int_0^x e^{t^2} (t-2)(t-3)\, dt$, $\forall$ $x \in (0, \infty)$, then

(a) f has a local maximum at $x = 2$
(b) f is decreasing on $(2, 3)$
(c) there exists some $c \in (0, \infty)$ such that $f''(c) = 0$
(d) f has a local minimum at $x = 3$

37. For every integer n, let a_n and b_n be real numbers. Let function $f: R \to R$ be given by

$$f(x) = \begin{cases} a_n + \sin \pi x, & \text{for } x \in [2n, 2n+1] \\ b_n + \cos \pi x, & \text{for } x \in (2n-1, 2n) \end{cases},$$

for all integers n.
If f is continuous, then which of the following hold(s) for all n?

(a) $a_{n-1} - b_{n-1} = 0$ (b) $a_n - b_n = 1$
(c) $a_n - b_{n+1} = 1$ (d) $a_{n-1} - b_n = -1$

38. If the straight lines $\frac{x-1}{2} = \frac{y+1}{K} = \frac{z}{2}$ and $\frac{x+1}{5} = \frac{y+1}{2} = \frac{z}{k}$ are coplanar, then the plane(s) containing these two lines is/are

(a) $y + 2z = -1$ (b) $y + z = -1$
(c) $y - z = -1$ (d) $y - 2z = -1$

39. If the adjoint of a 3×3 matrix P is $\begin{bmatrix} 1 & 4 & 4 \\ 2 & 1 & 7 \\ 1 & 1 & 3 \end{bmatrix}$, then the possible value(s) of the determinant of P is/are

(a) -2 (b) -1 (c) 1 (d) 2

40. Let $f: (-1, 1) \to R$ be such that $f(\cos 4\theta) = \frac{2}{2 - \sec^2 \theta}$ for $\theta \in \left(0, \frac{\pi}{4}\right) \cup \left(\frac{\pi}{4}, \frac{\pi}{2}\right)$. Then, the value(s) of $f\left(\frac{1}{3}\right)$ is/are

(a) $1 - \sqrt{\frac{3}{2}}$ (b) $1 + \sqrt{\frac{3}{2}}$

(c) $1 - \sqrt{\frac{2}{3}}$ (d) $1 + \sqrt{\frac{2}{3}}$

Answers

Paper 1

1. (a) 2. (c) 3. (d) 4. (b) 5. (b) 6. (b) 7. (b) 8. (a) 9. (d) 10. (c)
11. (a,d) 12. (b,d) 13. (a,c,d) 14. (a,b,d) 15. (a,b)
16. (4) 17. (9) 18. (5) 19. (4) 20. (3)

Paper 2

21. (a) 22. (c) 23. (c) 24. (a) 25. (b) 26. (d) 27. (d) 28. (b) 29. (b) 30. (c)
31. (a) 32. (b) 33. (d) 34. (a) 35. (a,b) 36. (a,b,c,d) 37. (b,d)
38. (b,c) 39. (a,d) 40. (a,b)

JEE Main

Unsolved Paper 2013

M.Marks 360 **Time 3 Hrs**

Instructions

- This test consists of 90 questions.
- There are three parts in the question paper A, B, C consisting of Physics, Chemistry and Mathematics having 30 questions in each part of equal weightage. Each question is allotted 4 marks for correct response.
- Candidates will be awarded marks as stated above for correct response of each question 1/4 marks will be deducted for indicating incorrect response of each question. No deduction from the total score will be made if no response is indicated for an item in the answer sheet.
- There is only one correct response for each question. Filling up more than one response in any question will be treated as wrong response and marks for wrong response will be deducted according as per instructions.

PHYSICS

1. A uniform cylinder of length L and mass M having cross-sectional area A is suspended, with its length vertical from a fixed point by a massless spring such that it is half submerged in a liquid of density σ at equilibrium position. The extension x_0 of the spring when it is in equilibrium is

(a) $\frac{Mg}{k}$ (b) $\frac{Mg}{k}\left(1-\frac{LA\sigma}{M}\right)$

(c) $\frac{Mg}{k}\left(1-\frac{LA\sigma}{2M}\right)$ (d) $\frac{Mg}{k}\left(1+\frac{LA\sigma}{M}\right)$

2. A metallic rod of length l is tied to a string of length $2l$ and made to rotate with angular speed ω on a horizontal table with one end of the string fixed. If there is a vertical magnetic field B in the region, the emf induced across the ends of the rod is

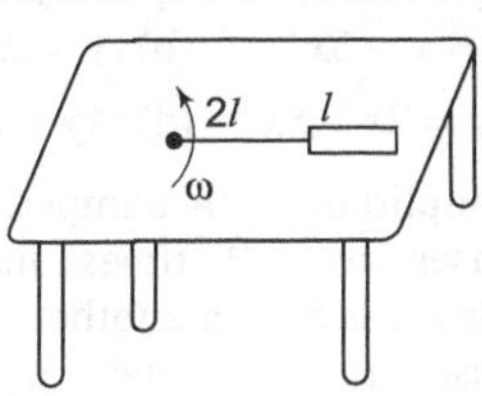

(a) $\frac{2B\omega l^3}{2}$ (b) $\frac{3B\omega l^3}{2}$

(c) $\frac{4B\omega l^2}{2}$ (d) $\frac{5B\omega l^2}{2}$

3. This question has statement I and statement II. Of the four choices given after the statements, choose the one that best describes the two statements.

Statement I A point particle of mass m moving with speed v collides with stationary point particle of mass M. If the maximum energy loss possible is given as $f\left(\frac{1}{2}mv^2\right)$, then $f = \left(\frac{m}{M+m}\right)$.

Statement II Maximum energy loss occurs when the particles get stuck together as a result of the collision.

(a) Statement I is true, Statement II is true, and Statement II is the correct explanation of Statement I
(b) Statement I is true, Statement II is true, but Statement II is not the correct explanation of Statement I
(c) Statement I is true, Statement II is false
(d) Statement I is false, Statement II is true

4. Let $[\varepsilon_0]$ denote the dimensional formula of the permittivity of vacuum. If M = mass, L = length, T = Time and A = electric current, then

(a) $[\varepsilon_0] = [M^{-1} L^{-3} T^2 A]$
(b) $[\varepsilon_0] = [M^{-1} L^{-3} T^4 A^2]$
(c) $[\varepsilon_0] = [M^{-2} L^2 T^{-1} A^{-2}]$
(d) $[\varepsilon_0] = [M^{-1} L^2 T^{-1} A^2]$

5. A projectile is given an initial velocity of $(\hat{\mathbf{i}} + 2\hat{\mathbf{j}})$ m/s, where $\hat{\mathbf{i}}$ is along the ground and $\hat{\mathbf{j}}$ is along the vertical. If $g = 10 \text{ m/s}^2$, the equation of its trajectory is

(a) $y = x - 5x^2$ (b) $y = 2x - 5x^2$
(c) $4y = 2x - 5x^2$ (d) $4y = 2x - 25x^2$

6. The amplitude of a damped oscillator decreases to 0.9 times its original magnitude is 5 s. In another 10 s it will decrease to α times its original magnitude, where α equals

(a) 0.7 (b) 0.81
(c) 0.729 (d) 0.6

7. Two capacitors C_1 and C_2 are charged to 120 V and 200 V respectively. It is found that by connecting them together the potential on each one can be made zero. Then

(a) $5C_1 = 3C_2$ (b) $3C_1 = 5C_2$
(c) $3C_1 + 5C_2 = 0$ (d) $9C_1 = 4C_2$

8. A sonometer wire of length 1.5 m is made of steel. The tension in it produces an elastic strain of 1%. What is the fundamental frequency of steel if density and elasticity of steel are 7.7×10^3 kg/m^3 and 2.2×10^{11} N/m^2 respectively?

(a) 188.5 Hz (b) 178.2 Hz
(c) 200.5 Hz (d) 770 Hz

9. A circular loop of radius 0.3 cm lies parallel to a much bigger circular loop of radius 20 cm. The centre of the smaller loop is on the axis of the bigger loop. The distance between their centres is 15 cm. If a current of 2.0 A flows through the bigger loop, then the flux linked with smaller loop is

(a) 9.1×10^{-11} Wb (b) 6×10^{-11} Wb
(c) 3.3×10^{-11} Wb (d) 6.6×10^{-9} Wb

10. Diameter of a plano-convex lens is 6 cm and thickness at the centre is 3 mm. If speed of light in material of lens is 2×10^8 m/s, the focal length of the lens is

(a) 15 cm (b) 20 cm
(c) 30 cm (d) 10 cm

11. What is the minimum energy required to launch a satellite of mass m from the surface of a planet of mass M and radius R in a circular orbit at an altitude of $2R$?

(a) $\frac{5GmM}{6R}$ (b) $\frac{2GmM}{3R}$
(c) $\frac{GmM}{2R}$ (d) $\frac{GmM}{3R}$

12. A diode detector is used to detect an amplitude modulated wave of 60% modulation by using a condenser of capacity 250 pico farad in parallel with a load resistance 100 kΩ. Find the maximum modulated frequency which could be detected by it.

(a) 10.62 MHz (b) 10.62 kHz
(c) 5.31 MHz (d) 5.31 kHz

13. A beam of unpolarized light of intensity I_0 is passed through a polaroid A and then through another polaroid B which is oriented so that its principal plane makes an angle of 45° relative to that of A. The intensity of the emergent light is

(a) I_0 (b) $I_0/2$
(c) $I_0/4$ (d) $I_0/8$

14. The supply voltage in a room is 120 V. The resistance of the lead wires is 6 Ω. A 60 W bulb is already switched on. What is the decrease of voltage across the bulb, when a 240 W heater is switched on in parallel to the bulb?

(a) zero (b) 2.9 V
(c) 13.3 V (d) 10.04 V

15. The shown *p-V* diagram represents the thermodynamic cycle of an engine, operating with an ideal monoatomic gas. The amount of heat, extracted from the source in a single cycle is

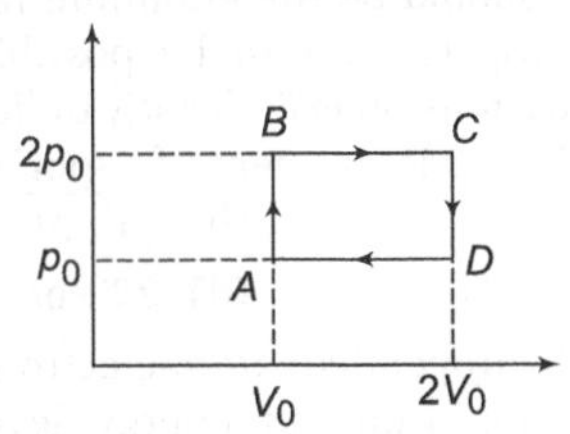

(a) p_0V_0 (b) $\left(\frac{13}{2}\right) p_0V_0$

(c) $\left(\frac{11}{2}\right) p_0V_0$ (d) $4p_0V_0$

16. A hoop of radius r and mass m rotating with an angular velocity ω_0 is placed on a rough horizontal surface. The initial velocity of the centre of the hoop is zero. What will be the velocity of the centre of the hoop when it ceases to slip?

(a) $\frac{r\omega_0}{4}$ (b) $\frac{r\omega_0}{3}$

(c) $\frac{r\omega_0}{2}$ (d) $r\omega_0$

17. An ideal gas enclosed in a vertical cylindrical container supports a freely moving piston of mass M. The piston and the cylinder have equal cross-sectional area A. When the piston is in equilibrium, the volume of the gas is V_0 and its pressure is p_0. The piston is slightly displaced from the equilibrium position and released. Assuming that the system is completely isolated from its surrounding, the piston executes a simple harmonic motion with frequency

(a) $\frac{1}{2\pi}\frac{A\gamma p_0}{V_0M}$ (b) $\frac{1}{2\pi}\frac{V_0Mp_0}{A^2\gamma}$

(c) $\frac{1}{2\pi}\sqrt{\frac{A^2\gamma p_0}{MV_0}}$ (d) $\frac{1}{2\pi}\sqrt{\frac{MV_0}{A\gamma p_0}}$

18. If a piece of metal is heated to temperature θ and then allowed to cool in a room which is at temperature θ_0. The graph between the temperature T of the metal and time t will be closed to

(a) 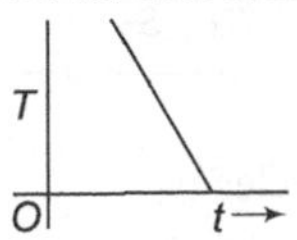

(b) 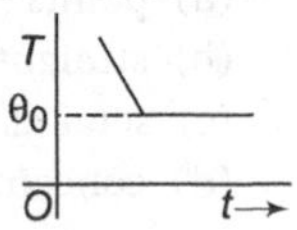

(c) 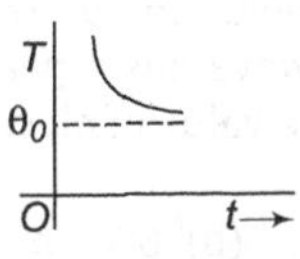

(d) 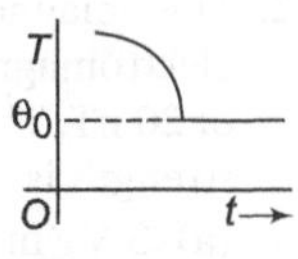

19. This question has Statement I and Statement II. Of the four choices given after the statements, choose the one that best describes the two statements.

Statement I Higher the range, greater is the resistance of ammeter.

Statement II To increase the range of ammeter, additional shunt needs to be used across it.

(a) Statement I is true, Statement II is true and Statement II is the correct explanation of Statement I
(b) Statement I is true, Statement II is true, but Statement II is not the correct explanation of Statement I
(c) Statement I is true, Statement II is false
(d) Statement I is false, Statement II is true

20. In a *L-C-R* circuit as shown below both switches are open initially. Now switch S_1 and S_2, are closed. (q is charge on the capacitor and $\tau = RC$ is capacitance time constant). Which of the following statement is correct?

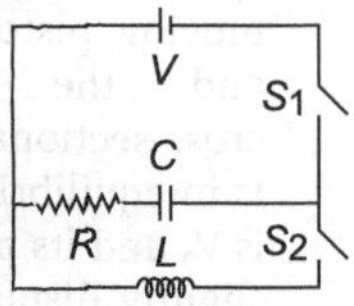

(a) Work done by the battery is half of the energy dissipated in the resistor
(b) At $t = \tau$, $q = CV/2$
(c) At $t = 2\tau$, $q = CV(1 - e^{-2})$
(d) At $t = \dfrac{\tau}{2}$, $q = CV(1 - e^{-1})$

21. Two coherent point sources S_1 and S_2 are separated by a small distance d as shown. The fringes obtained on the screen will be

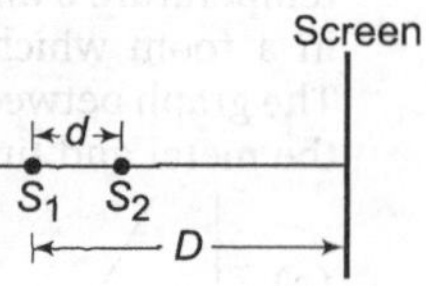

(a) points
(b) straight lines
(c) semi-circle
(d) concentric circles

22. The magnetic field in a travelling electromagnetic wave has a peak value of 20 nT. The peak value of electric field strength is

(a) 3 V/m (b) 6 V/m
(c) 9 V/m (d) 12 V/m

23. The anode voltage of a photocell is kept fixed. The wavelength λ of the light falling on the cathode is gradually changed. The plate current I of photocell varies as follows

(a) 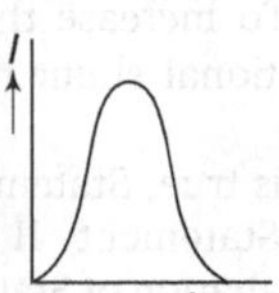
(b) 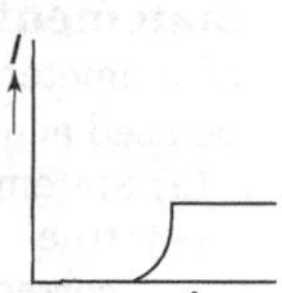
(c) 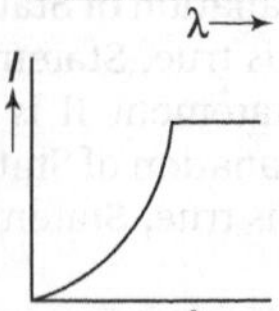
(d) 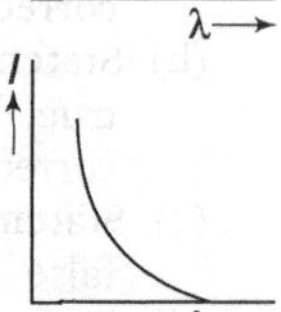

24. The *I-V* characteristic of an LED is

(a) 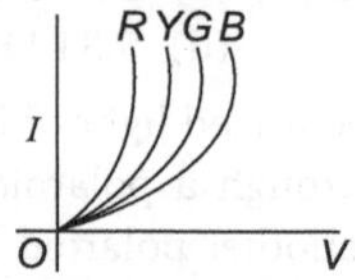

(b) 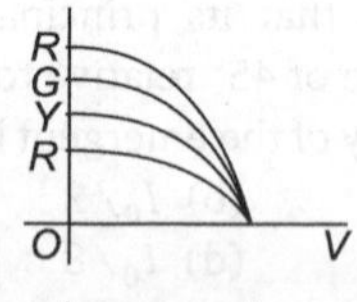

(c) 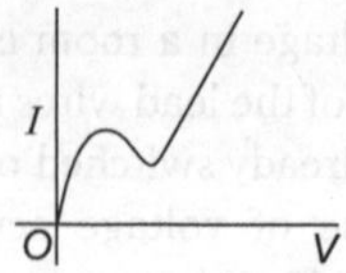

(d) 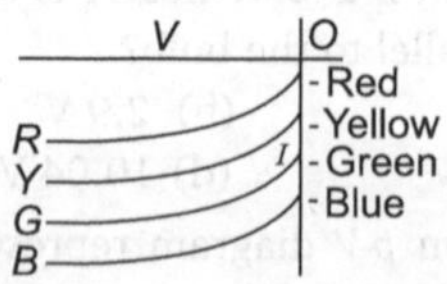

25. Assume that a drop of liquid evaporates by decrease in its surface energy, so that its temperature remains unchanged. What should be the minimum radius of the drop for this to be possible? The surface tension is T, denstiy of liquid is ρ and L is its latent heat of vaporization

(a) $\rho L/T$ (b) $\sqrt{T/\rho L}$
(c) $T/\rho L$ (d) $2T/\rho L$

26. In a hydrogen like atom electron makes transition from an energy level with quantum number n to another with quantum number $(n-1)$. If $n >> 1$, the frequency of radiation emitted is proportional to

(a) $\dfrac{1}{n}$ (b) $\dfrac{1}{n^2}$ (c) $\dfrac{1}{n^4}$ (d) $\dfrac{1}{n^3}$

27. The graph between angle of deviation (δ) and angle of incidence (i) for a triangular prism is represented by

(a) 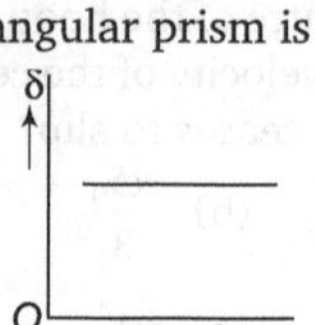
(b) 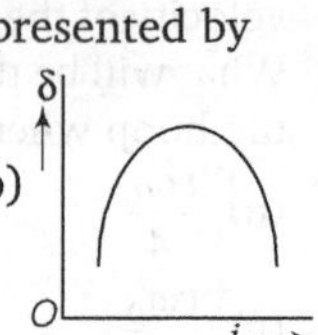

(c) 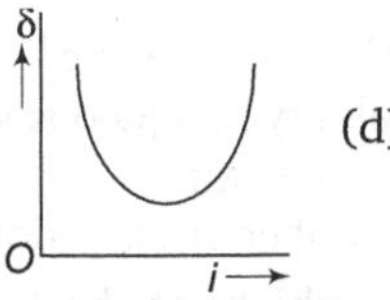 (d) 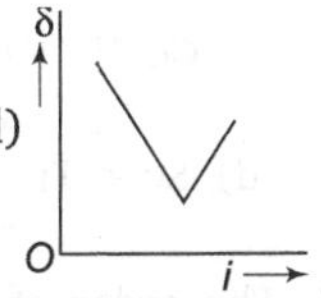

28. Two charges, each equal to q, are kept at $x = -a$ and $x = a$ on the x-axis. A particle of mass m and charge $q_0 = \frac{q}{2}$ is placed at the origin. If charge q_0 is given a small displacement $y\ (y << a)$ along the y-axis, the net force acting on the particle is proportional to

(a) y (b) $-y$ (c) $\frac{1}{y}$ (d) $-\frac{1}{y}$

29. Two short bar magnets of length 1 cm each have magnetic moments 1.20 Am^2 and 1.00 Am^2 respectively. They are placed on a hoirzontal table parallel to each other with their N poles pointing towards the south. They have a common magnetic equator and are separated by a distance of 20.0 cm. The value of the resultant horizontal magnetic induction at the mid-point O of the line joining their centres is close to (Horizontal component of the earth's magnetic induction is $3.6 \times 10^{-5} Wb/m^2$)

(a) $3.6 \times 10^{-5}\ Wb/m^2$
(b) $2.56 \times 10^{-4}\ Wb/m^2$
(c) $3.50 \times 10^{-4}\ Wb/m^2$
(d) $5.80 \times 10^{-4}\ Wb/m^2$

30. A charge Q is uniformly distributed over a long rod AB of length L as shown in the figure. The electric potential at the point O lying at distance L from the end A is

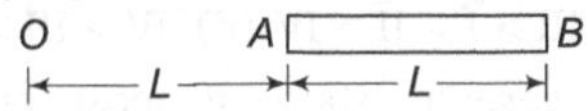

(a) $\frac{Q}{8\pi \varepsilon_0 L}$ (b) $\frac{3Q}{4\pi \varepsilon_0 L}$

(c) $\frac{Q}{4\pi \varepsilon_0 L \ln 2}$ (d) $\frac{Q \ln 2}{4\pi \varepsilon_0 L}$

CHEMISTRY

1. Which of the following complex species is not expected to exhibit optical isomerism?

(a) $[Co(en)_3]^{3+}$
(b) $[Co(en)_2 Cl_2]^+$
(c) $[Co(NH_3)_3Cl_3]$
(d) $[Co(en)(NH_3)Cl_2]^+$

2. Which one of the following molecules is expected to exhibit diamagnetic behaviour?

(a) C_2 (b) N_2
(c) O_2 (d) S_2

3. A solution of $(-l)$ 1-chloro-1-phenylethane in toluene racemises slowly in the presence of a small amount of $SbCl_5$, due to the formation of

(a) carbanion (b) carbene
(c) carbocation (d) free radical

4. Given, $E^\circ_{Cr^{3+}/Cr} = -0.74$ V;

$E^\circ_{MnO_4^-/Mn^{2+}} = 1.51$ V

$E^\circ_{Cr_2O_7^{2-}/Cr^{3+}} = 1.33$ V;

$E^\circ_{Cl/Cl^-} = 1.36$ V

Based on the data given above strongest oxidising agent will be

(a) Cl (b) Cr^{3+}
(c) Mn^{2+} (d) MnO_4^-

5. A piston filled with 0.04 mole of an ideal gas expands reversibly from 50.0 mL to 375 mL at a constant temperature of 37.0°C. As it does so, it absorbs 208 J of heat. The values of q and W for the process will be ($R = 8.314$ J/mol K, $\ln 7.5 = 2.01$)

(a) $q = +208$ J, $W = -208$ J
(b) $q = -208$ J, $W = -208$ J
(c) $q = -208$ J, $W = +208$ J
(d) $q = +208$ J, $W - +208$ J

6. The molarity of a solution obtained by mixing 750 mL of 0.5 (M) HCl with 250 mL of 2(M) HCl will be

(a) 0.875 M (b) 1.00 M
(c) 1.75 M (d) 0.0975 M

7. Arrange the following compounds in the order of decreasing acidity

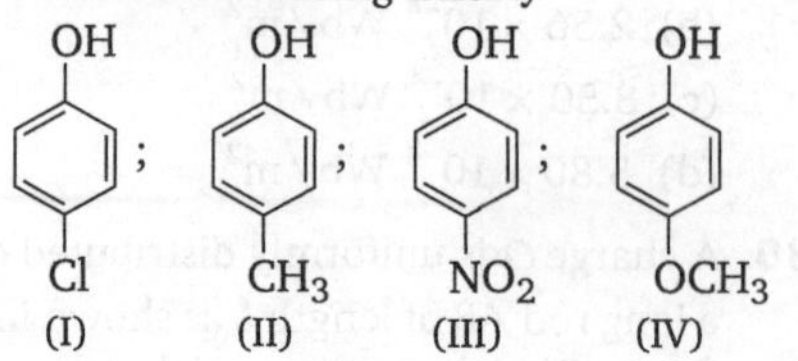

(I) (II) (III) (IV)

(a) II > IV > I > III (b) I > II > III > IV
(c) III > I > II > IV (d) IV > III > I > II

8. For gaseous state, if most probable speed is denoted by C^*, average speed by $\overline{C}$ and root square speed by C, then for a large number of molecules, the ratios of these speeds are

(a) $C^* : \overline{C} : C = 1.225 : 1.128 : 1$
(b) $C^* : \overline{C} : C = 1.128 : 1.225 : 1$
(c) $C^* : \overline{C} : C = 1 : 1.128 : 1.225$
(d) $C^* : \overline{C} : C = 1 : 1.225 : 1.128$

9. The rate of a reaction doubles when its temperature changes from 300 K to 310 K. Activation energy of such a reaction will be
($R = 8.314\ \text{JK}^{-1}\ \text{mol}^{-1}$ and $\log 2 = 0.301$)

(a) 53.6 kJ mol^{-1} (b) 48.6 kJ mol^{-1}
(c) 58.5 kJ mol^{-1} (d) 60.5 kJ mol^{-1}

10. A compound with molecular mass 180 is acylated with CH_3COCl to get a compound with molecular mass 390. The number of amino groups present per molecule of the former compound is

(a) 2 (b) 5
(c) 4 (d) 6

11. Which of the following arrangements does not represent the correct order of the property stated against it?

(a) $V^{2+} < Cr^{2+} < Mn^{2+} < Fe^{2+}$: paramagnetic behaviour
(b) $Ni^{2+} < Co^{2+} < Fe^{2+} < Mn^{2+}$: ionic size
(c) $Co^{3+} < Fe^{3+} < Cr^{3+} < Sc^{3+}$: stability in aqueous solution
(d) Sc < Ti < Cr < Mn : number of oxidation states

12. The order of stability of the following carbocations

$\overset{}{CH_2}{=}CH{-}\overset{\oplus}{C}H_2$; $CH_3{-}CH_2{-}\overset{\oplus}{C}H_2$;
(I) (II)

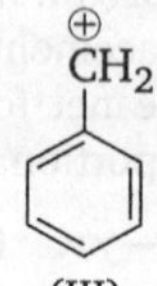 is

(III)

(a) III > II > I (b) II > III > I
(c) I > II > III (d) III > I > II

13. Consider the following reaction,

$$xMnO_4^- + yC_2O_4^{2-} + zH^+ \longrightarrow xMn^{2+} + 2yCO_2 + \frac{z}{2}H_2O$$

The values of x, y and z in the reaction are, respectively

(a) 5, 2 and 16 (b) 2, 5 and 8
(c) 2, 5 and 16 (d) 5, 2 and 8

14. Which of the following is the wrong statement?

(a) ONCl and ONO^- are not isoelectronic
(b) O_3 molecule is bent
(c) Ozone is violet-black in solid state
(d) Ozone is diamagnetic gas

15. A gaseous hydrocarbon gives upon combustion 0.72 g of water and 3.08 g of CO_2. The empirical formula of the hydrocarbon is

(a) C_2H_4 (b) C_3H_4 (c) C_6H_5 (d) C_7H_8

16. In which of the following pairs of molecules/ions both the species are not likely to exist?

(a) H_2^+, He_2^{2-} (b) H_2^-, He_2^{2-}
(c) H_2^{2+}, He_2 (d) H_2^-, He_2^{2+}

17. Which of the following exists as covalent crystals in the solid state?

(a) Iodine (b) Silicon
(c) Sulphur (d) Phosphorus

18. Synthesis of each molecule of glucose in photosynthesis involves
(a) 18 molecules of ATP
(b) 10 molecules of ATP
(c) 8 molecules of ATP
(d) 6 molecules of ATP

19. The coagulating power of electrolytes having ions Na^+, Al^{3+} and Ba^{2+} for arsenic sulphide sol increases in the order
(a) $Al^{3+} < Ba^{2+} < Na^+$
(b) $Na^+ < Ba^{2+} < Al^{3+}$
(c) $Ba^{2+} < Na^{2+} < Al^{3+}$
(d) $Al^{3+} < Na^+ < Ba^{2+}$

20. Which of the following represents the correct order of increasing first ionization enthalpy for Ca, Ba, S, Se and Ar?
(a) Ca < S < Ba < Se < Ar
(b) S < Se < Ca < Ba < Ar
(c) Ba < Ca < Se < S < Ar
(d) Ca < Ba < S < Se < Ar

21. Energy of an electron is given by

$$E = -2.178 \times 10^{-18}\ J\left(\frac{Z^2}{n^2}\right)$$

Wavelength of light required to excite an electron in an hydrogen atom from level $n = 1$ to $n = 2$ will be
($h = 6.62 \times 10^{-34}$ Js and $c = 3.0 \times 10^8\ ms^{-1}$)
(a) 1.214×10^{-7} m
(b) 2.816×10^{-7} m
(c) 6.500×10^{-7} m
(d) 8.500×10^{-7} m

22. Compound (*A*), C_8H_9Br gives a white precipitate when warmed with alcoholic $AgNO_3$. Oxidation of (*A*) gives an acid (*B*), $C_8H_6O_4$. (*B*) easily forms anyhydride on heating. Identify the compound (*A*).

(a) 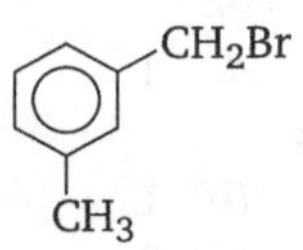 (b) 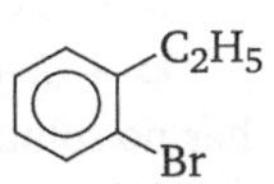

(c) 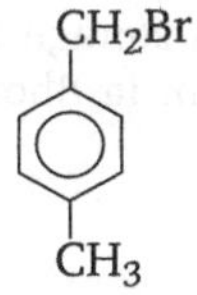 (d) 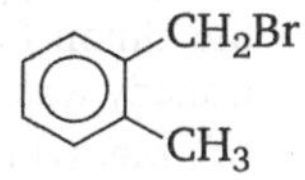

23. Four successive members of the first row transition elements listed below with atomic numbers. Which one of them is expected to have the highest $E^\circ_{M^{3+}/M^{2+}}$ value?
(a) Cr (Z = 24) (b) Mn (Z = 25)
(c) Fe (Z = 26) (d) Co (Z = 27)

24. How many litres of water must be added to 1 L of an aqueous solution of HCl with a pH of 1 to create an aqueous solution with pH of 2 ?
(a) 0.1 L (b) 0.9 L (c) 2.0 L (d) 9.0 L

25. The first ionisation potential of Na is 5.1 eV. The value of electron gain enthalpy of Na^+ will be
(a) – 2.55 eV (b) – 5.1 eV
(c) – 10.2 eV (d) + 2.55 eV

26. An organic compound *A* upon reacting with NH_3 gives *B*. On heating, *B* gives *C*. *C* in the presence of KOH reacts with Br_2 to give $CH_3CH_2NH_2$. *A* is
(a) CH_3COOH
(b) $CH_3CH_2CH_2COOH$
(c) $CH_3—\underset{\underset{CH_3}{|}}{C}H—COOH$
(d) CH_3CH_2COOH

27. Stability of the species Li_2, Li_2^- and Li_2^+ increases in the order of
(a) $Li_2 < Li_2^+ < Li_2^-$ (b) $Li_2^- < Li_2^+ < Li_2$
(c) $Li_2 < Li_2^- < Li_2^+$ (d) $Li_2^- < Li_2 < Li_2^+$

28. An unknown alcohol is treated with the "Lucas reagent" to determine whether the alcohol is primary, secondary or tertiary. Which alcohol reacts fastest and by what mechanism?
(a) Secondary alcohol by S_N1
(b) Tertiary alcohol by S_N1
(c) Secondary alcohol by S_N2
(d) Tertiary alcohol by S_N2

29. The gas leaked from a storage tank of the Union Carbide plant in Bhopal gas tragedy was

(a) Methyl isocyanate
(b) Methylamine
(c) Ammonia
(d) Phosgene

30. Experimentally it was found that a metal oxide has formula $M_{0.98}O$. Metal M, present as M^{2+} and M^{3+} in its oxide. Fraction of the metal which exists as M^{3+} would be

(a) 7.01% (b) 4.08%
(c) 6.05% (d) 5.08%

MATHEMATICS

1. Distance between two parallel planes $2x + y + 2z = 8$ and $4x + 2y + 4z + 5 = 0$ is

(a) $\frac{3}{2}$ (b) $\frac{5}{2}$ (c) $\frac{7}{2}$ (d) $\frac{9}{2}$

2. At present, a firm is manufacturing 2000 items. It is estimated that the rate of change of production P with respect to additional number of workers x is given by

$\frac{dP}{dx} = 100 - 12\sqrt{x}$. If the firm employees 25 more workers, then the new level of production of items is

(a) 2500 (b) 3000 (c) 3500 (d) 4500

3. Let A and B two sets containing 2 elements and 4 elements respectively. The number of subsets of $A \times B$ having 3 or more elements is

(a) 256 (b) 220 (c) 219 (d) 211

4. If the lines

$$\frac{x-2}{1} = \frac{y-3}{1} = \frac{z-4}{-k} \text{ and}$$

$$\frac{x-1}{k} = \frac{y-4}{2} = \frac{z-5}{1}$$

are coplanar, then k can have

(a) any value
(b) exactly one value
(c) exactly two values
(d) exactly three values

5. If the vectors $\mathbf{AB} = 3\hat{\mathbf{i}} + 4\hat{\mathbf{k}}$ and $\mathbf{AC} = 5\hat{\mathbf{i}} - 2\hat{\mathbf{j}} + 4\hat{\mathbf{k}}$ are the sides of a ΔABC, then the length of the median through A is

(a) $\sqrt{18}$ (b) $\sqrt{72}$
(c) $\sqrt{33}$ (d) $\sqrt{45}$

6. The real number k for which the equation, $2x^3 + 3x + k = 0$ has two distinct real roots in $[0, 1]$

(a) lies between 1 and 2
(b) lies between 2 and 3
(c) lies between -1 and 0
(d) does not exist

7. The sum of first 20 terms of the sequence 0.7, 0.77, 0.777,..., is

(a) $\frac{7}{81}(179 - 10^{-20})$

(b) $\frac{7}{9}(99 - 10^{-20})$

(c) $\frac{7}{81}(179 + 10^{-20})$

(d) $\frac{7}{9}(99 + 10^{-20})$

8. A ray of light along $x + \sqrt{3}y = \sqrt{3}$ gets reflected upon reaching x-axis, the equation of the reflected ray is

(a) $y = x + \sqrt{3}$
(b) $\sqrt{3}y = x - \sqrt{3}$
(c) $y = \sqrt{3}x - \sqrt{3}$
(d) $\sqrt{3}y = x - 1$

9. The number of values of k, for which the system of equations

$$(k+1)x + 8y = 4k$$
$$kx + (k+3)y = 3k - 1$$

has no solution, is

(a) infinite (b) 1
(c) 2 (d) 3

10. If the equations $x^2 + 2x + 3 = 0$ and $ax^2 + bx + c = 0$, $a, b, c \in R$, have a common root, then $a : b : c$ is

(a) 1 : 2 : 3 (b) 3 : 2 : 1

(c) 1 : 3 : 2 (d) 3 : 1 : 2

11. The circle passing through (1, – 2) and touching the axis of x at (3, 0) also passes through the point

(a) (– 5, 2) (b) (2, – 5)

(c) (5, – 2) (d) (– 2, 5)

12. If x, y and z are in AP and $\tan^{-1} x$, $\tan^{-1} y$ and $\tan^{-1} z$ are also in AP, then

(a) $x = y = z$ (b) $2x = 3y = 6z$

(c) $6x = 3y = 2z$ (d) $6x = 4y = 3z$

13. Consider

Statement I
$(p \wedge \sim q) \wedge (\sim p \wedge q)$ is a fallacy.

Statement II
$(p \rightarrow q) \leftrightarrow (\sim q \rightarrow \sim p)$ is a tautology.

(a) Statement I is true; Statement II is true; Statement II is a correct explanation for Statement I.

(b) Statement I is true; Statement II is true; Statement II is not a correct explanation for Statement I.

(c) Statement I is true; Statement II is false.

(d) Statement I is false; Statement II is true.

14. If $\int f(x)\,dx = \psi(x)$, then $\int x^5 f(x^3) dx$ is equal to

(a) $\frac{1}{3}[x^3\psi(x^3) - \int x^2\psi(x^3)dx] + C$

(b) $\frac{1}{3}x^3\psi(x^3) - 3\int x^3\psi(x^3)\,dx + C$

(c) $\frac{1}{3}x^3\psi(x^3) - \int x^2\psi(x^3)\,dx + C$

(d) $\frac{1}{3}[x^3\psi(x^3) - \int x^3\psi(x^3)\,dx] + C$

15. $\lim_{x \to 0} \frac{(1 - \cos 2x)(3 + \cos x)}{x \tan 4x}$ is equal to

(a) $-\frac{1}{4}$ (b) $\frac{1}{2}$ (c) 1 (d) 2

16. **Statement I** The value of the integral

$$\int_{\pi/6}^{\pi/3} \frac{dx}{1 + \sqrt{\tan x}}$$ is equal to $\pi/6$.

Statement II

$$\int_a^b f(x)\,dx = \int_a^b f(a + b - x)\,dx$$

(a) Statement I is true; Statement II is true; Statement II is a correct explanation for Statement I.

(b) Statement I is true; Statement II is true; Statement II is not a correct explanation for Statement I.

(c) Statement I is true; Statement II is false.

(d) Statement I is false; Statement II is true.

17. The equation of the circle passing through the foci of the ellipse $\frac{x^2}{16} + \frac{y^2}{9} = 1$ and having centre at (0, 3) is

(a) $x^2 + y^2 - 6y - 7 = 0$

(b) $x^2 + y^2 - 6y + 7 = 0$

(c) $x^2 + y^2 - 6y - 5 = 0$

(d) $x^2 + y^2 - 6y + 5 = 0$

18. A multiple choice examination has 5 questions. Each question has three alternative answers of which exactly one is correct. The probability that a student will get 4 or more correct answers just by guessing is

(a) $\frac{17}{3^5}$ (b) $\frac{13}{3^5}$

(c) $\frac{11}{3^5}$ (d) $\frac{10}{3^5}$

19. The x-coordinate of the incentre of the triangle that has the coordinates of mid-points of its sides as (0, 1), (1, 1) and (1, 0) is

(a) $2 + \sqrt{2}$

(b) $2 - \sqrt{2}$

(c) $1 + \sqrt{2}$

(d) $1 - \sqrt{2}$

20. The term independent of x in expansion of $\left(\frac{x+1}{x^{2/3}-x^{1/3}+1}-\frac{x-1}{x-x^{1/2}}\right)^{10}$ is
(a) 4 (b) 120
(c) 210 (d) 310

21. The area (in square units) bounded by the curves $y=\sqrt{x}$, $2y-x+3=0$, x-axis and lying in the first quadrant is
(a) 9 (b) 36
(c) 18 (d) $\frac{27}{4}$

22. Let T_n be the number of all possible triangles formed by joining vertices of an n-sided regular polygon. If $T_{n+1}-T_n=10$, then the value of n is
(a) 7 (b) 5
(c) 10 (d) 8

23. If z is a complex number of unit modulus and argument θ, then $\arg\left(\frac{1+z}{1+\bar{z}}\right)$ is equal to
(a) $-\theta$ (b) $\frac{\pi}{2}-\theta$
(c) θ (d) $\pi-\theta$

24. $ABCD$ is a trapezium such that AB and CD are parallel and $BC \perp CD$. If $\angle ADB=\theta$, $BC=p$ and $CD=q$, then AB is equal to
(a) $\frac{(p^2+q^2)\sin\theta}{p\cos\theta+q\sin\theta}$
(b) $\frac{p^2+q^2\cos\theta}{p\cos\theta+q\sin\theta}$
(c) $\frac{p^2+q^2}{p^2\cos\theta+q^2\sin\theta}$
(d) $\frac{(p^2+q^2)\sin\theta}{(p\cos\theta+q\sin\theta)^2}$

25. If $P=\begin{bmatrix}1 & \alpha & 3\\ 1 & 3 & 3\\ 2 & 4 & 4\end{bmatrix}$ is the adjoint of a 3×3 matrix A and $|A|=4$, then α is equal to
(a) 4 (b) 11
(c) 5 (d) 0

26. The intercepts on x-axis made by tangents to the curve, $y=\int_0^x |t|\,dt$, $x\in R$, which are parallel to the line $y=2x$, are equal to
(a) ±1 (b) ±2 (c) ±3 (d) ±4

27. Given A circle, $2x^2+2y^2=5$ and a parabola, $y^2=4\sqrt{5}x$.

Statement I An equation of a common tangent to these curves is $y=x+\sqrt{5}$.

Statement II If the line, $y=mx+\frac{\sqrt{5}}{m}$ $(m\neq0)$ is the common tangent, then m satisfies $m^4-3m^2+2=0$.

(a) Statement I is true; Statement II is true; Statement II is a correct explanation for Statement I.
(b) Statement I is true; Statement II is true; Statement II is not a correct explanation for Statement I.
(c) Statement I is true; Statement II is false.
(d) Statement I is false; Statement II is true.

28. If $y=\sec(\tan^{-1}x)$, then $\frac{dy}{dx}$ at $x=1$ is equal to
(a) $\frac{1}{\sqrt{2}}$ (b) $\frac{1}{2}$ (c) 1 (d) $\sqrt{2}$

29. The expression $\frac{\tan A}{1-\cot A}+\frac{\cot A}{1-\tan A}$ can be written as
(a) $\sin A\cos A+1$
(b) $\sec A\,\mathrm{cosec}\,A+1$
(c) $\tan A+\cot A$
(d) $\sec A+\mathrm{cosec}\,A$

30. All the students of a class performed poorly in Mathematics. The teacher decided to give grace marks of 10 to each of the students. Which of the following statistical measures will not change even after the grace marks were given?
(a) Mean (b) Median
(c) Mode (d) Variance

Answers

Physics

1. (c)	2. (d)	3. (d)	4. (b)	5. (b)	6. (c)	7. (b,c)	8. (b)	9. (a)	10. (c)
11. (a)	12. (b)	13. (c)	14. (d)	15. (b)	16. (c)	17. (c)	18. (c)	19. (d)	20. (c)
21. (d)	22. (b)	23. (d)	24. (a)	25. (d)	26. (d)	27. (c)	28. (a)	29. (b)	30. (d)

Chemistry

1. (c)	2. (a.b)	3. (c)	4. (d)	5. (a)	6. (a)	7. (c)	8. (c)	9. (a)	10. (b)
11. (a)	12. (d)	13. (c)	14. (*)	15. (d)	16. (c)	17. (b)	18. (a)	19. (b)	20. (c)
21. (a)	22. (d)	23. (d)	24. (d)	25. (b)	26. (d)	27. (b)	28. (b)	29. (a)	30. (b)

Mathematics

1. (c)	2. (c)	3. (c)	4. (c)	5. (c)	6. (d)	7. (c)	8. (b)	9. (b)	10. (a)
11. (c)	12. (a)	13. (b)	14. (c)	15. (d)	16. (d)	17. (a)	18. (c)	19. (b)	20. (c)
21. (a)	22. (b)	23. (c)	24. (a)	25. (b)	26. (a)	27. (b)	28. (a)	29. (b)	30. (d)

JEE Advanced

Unsolved Paper 2013

PAPER 1

- **M. Marks 180** **Time 3 Hrs**

Instructions

- The question paper consists of three sections.
- Section 1 contains 10 multiple choice questions each of 2 marks in this section there is no negative marking. Each question has four choices (a), (b), (c) and (d) out of which only one is correct.
- Section 2 contains 5 multiple choice questions each of 4 marks with negative marking of 1 mark. Each question has four choices (a), (b), (c) and (d) out of which one or more are correct.
- Section 3 contains 5 questions each of 4 marks with negative marking of 1 mark. The answer to each question is a single-digit integer, ranging from 0 to 9 (both inclusive)

PHYSICS

Section I

Direction (Q. No. 1-10) *This section contains* 10 *multiple choice questions. Each question has four choices (a), (b), (c) and (d) out of which only one is correct.*

1. A particle of mass m is projected from the ground with an initial speed u_0 at an angle α with the horizontal. At the highest point of its trajectory, it makes a completely inelastic collision with another identical particle, which was thrown vertically upward from the ground with the same initial speed u_0. The angle that the composite system makes with the horizontal immediately after the collision is

(a) $\frac{\pi}{4}$ (b) $\frac{\pi}{4} + \alpha$

(c) $\frac{\pi}{4} - \alpha$ (d) $\frac{\pi}{2}$

2. The image of an object, formed by a plano-convex lens at a distance of 8 m behind the lens, is real and is one-third the size of the object. The wavelength of light inside the lens is $\frac{2}{3}$ times the wavelength in free space. The radius of the curved surface of the lens is

(a) 1 m (b) 2 m (c) 3 m (d) 6 m

3. The diameter of a cylinder is measured using a vernier callipers with no zero error. It is found that the zero of the vernier scale lies between 5.10 cm and 5.15 cm of the main scale. The vernier scale has 50 division equivalent to 2.45 cm. The 24th division of the vernier scale exactly coincides with one of the main scale divisions. The diameter of the cylinder is

(a) 5.112 cm (b) 5.124 cm
(c) 5.136 cm (d) 5.148 cm

4. The work done on a particle of mass m by a force,

$$K\left[\frac{x}{(x^2+y^2)^{3/2}}\,\hat{\mathbf{i}}+\frac{y}{(x^2+y^2)^{3/2}}\,\hat{\mathbf{j}}\right]$$

(K being a constant of appropriate dimensions), when the particle is taken from the point $(a, 0)$ to the point $(0, a)$ along a circular path of radius a about the origin in the x-y plane is

(a) $\frac{2K\pi}{a}$ (b) $\frac{K\pi}{a}$ (c) $\frac{K\pi}{2a}$ (d) 0

5. One end of a horizontal thick copper wire of length $2L$ and radius $2R$ is welded to an end of another horizontal thin copper wire of length L and radius R. When the arrangement is stretched by applying forces at two ends, the ratio of the elongation in the thin wire to that in the thick wire is

(a) 0.25 (b) 0.50
(c) 2.00 (d) 4.00

6. A ray of light travelling in the direction $\frac{1}{2}(\hat{\mathbf{i}}+\sqrt{3}\,\hat{\mathbf{j}})$ is incident on a plane mirror. After reflection, it travels along the direction $\frac{1}{2}(\hat{\mathbf{i}}-\sqrt{3}\,\hat{\mathbf{j}})$. The angle of incidence is

(a) 30° (b) 45°
(c) 60° (d) 75°

7. Two rectangular blocks, having indentical dimensions, can be arranged either in configuration I or in configuration II as shown in the figure. One of the blocks has thermal conductivity K and the other $2K$. The temperature difference between the ends along the x-axis is the same in both the configurations. It takes 9 s to transport a certain amount of heat from the hot end to the cold end in the configuration I. The time to transport the same amount of heat in the configuration II is

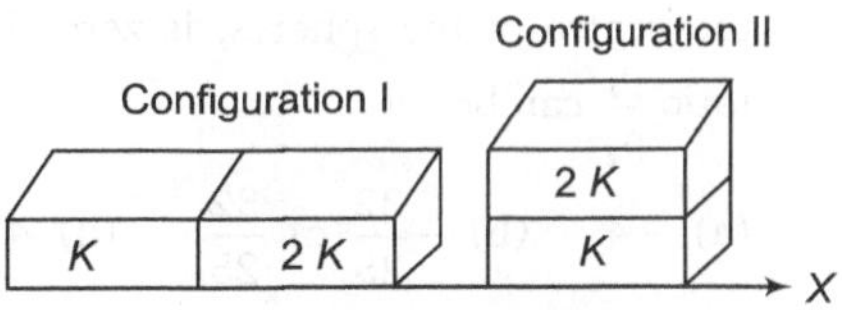

(a) 2.0 s (b) 3.0 s
(c) 4.5 s (d) 6.0 s

8. A pulse of light of duration 100 ns is absorbed completely by a small object initially at rest. Power of the pulse is 30 mV and the speed of light is 3×10^8 ms^{-1}. The final momentum of the object is

(a) 0.3×10^{-17} kg-ms^{-1}
(b) 1.0×10^{-17} kg-ms^{-1}
(c) 3.0×10^{-17} kg-ms^{-1}
(d) 9.0×10^{-17} kg-ms^{-1}

9. In the Young's double slit experiment using a monochromatic light of wavelength λ the path difference (in terms of an integer n) corresponding to any point having half the peak intensity is

(a) $(2n+1)\frac{\lambda}{2}$ (b) $(2n+1)\frac{\lambda}{4}$
(c) $(2n+1)\frac{\lambda}{8}$ (d) $(2n+1)\frac{\lambda}{16}$

10. Two non-reactive monoatomic ideal gases have their atomic masses in the ratio 2 : 3. The ratio of their partial pressures, when enclosed in a vessel kept at a constant temperature, is 4 : 3. The ratio of their densities is

(a) 1 : 4 (b) 1 : 2

(c) 6 : 9 (d) 8 : 9

Section II

Direction (Q. No. 11-15) *This section contains 5 multiple choice questions. Each question has four choices (a), (b), (c) and (d) out of which one or more are correct.*

11. Two non-conducting solid spheres of radii R and $2R$, having uniform volume charge densities ρ_1 and ρ_2 respectively, touch each other. The net electric field at a distance $2R$ from the centre of the smaller sphere, along the line joining the centre of the spheres, is zero. The ratio $\frac{\rho_1}{\rho_2}$ can be

(a) -4 (b) $-\frac{32}{25}$ (c) $\frac{32}{25}$ (d) 4

12. A horizontal stretched string, fixed at two ends, is vibrating in its fifth harmonic according to the equation, $y(x, t) = (0.01\text{ m})\,[\sin(62.8\text{ m}^{-1})x]\,\cos[(628\text{ s}^{-1})t]$.

Assuming $\pi = 3.14$, the correct statement(s) is (are)

(a) the number of nodes is 5

(b) the length of the string is 0.25 m

(c) the maximum displacement of the mid-point of the string from its equilibrium position is 0.01 m

(d) the fundamental frequency is 100 Hz

13. In the circuit shown in the figure, there are two parallel plate capacitors each of capacitance C. The switch S_1 is pressed first to fully charge the capacitor C_1 and then released. The switch S_2 is then pressed to charge the capacitor C_2. After some time, S_2 is released and then S_3 is pressed. After some time

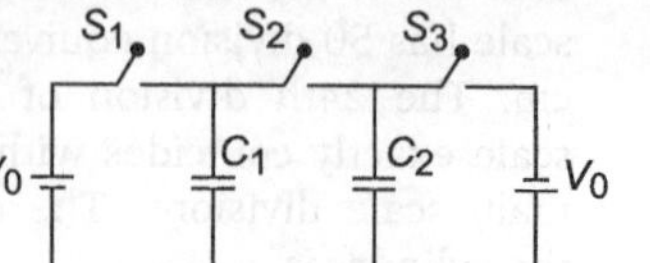

(a) the charge on the upper plate of C_1 is $2CV_0$

(b) the charge on the upper plate of C_1 is CV_0

(c) the charge on the upper plate of C_2 is 0

(d) the charge on the upper plate of C_2 is $-CV_0$

14. A particle of mass M and positive charge Q, moving with a constant velocity $\mathbf{u}_1 = 4\hat{\mathbf{i}}\text{ ms}^{-1}$, enters a region of uniform static magnetic field normal to the x-y plane. The region of the magnetic field extends from $x = 0$ to $x = L$ for all values of y. After passing through this region, the particle emerges on the other side after 10 milliseconds with a velocity $\mathbf{u}_2 = 2(\sqrt{3}\,\hat{\mathbf{i}} + \hat{\mathbf{j}})\text{ ms}^{-1}$. The correct statement(s) is (are)

(a) the direction of the magnetic field is $-z$ direction

(b) the direction of the magnetic field is $+z$ direction

(c) the magnitude of the magnetic field is $\frac{50\pi M}{3Q}$ units

(d) the magnitude of the magnetic field is $\frac{100\pi M}{3Q}$ units

15. A solid sphere of radius R and density ρ is attached to one end of a massless spring of force constant k. The other end of the spring is connected to another solid sphere of radius R and density 3ρ. The complete arrangement is placed in a liquid of density 2ρ and is allowed to reach equilibrium. The correct statement(s) is (are)

(a) the net elongation of the spring is $\frac{4\pi R^3 \rho g}{3k}$

(b) the net elongation of the spring is $\frac{8\pi R^3 \rho g}{3k}$

(c) the light sphere is partially submerged

(d) the light sphere is completely submerged

Section III

Direction (Q. No. 16-20) *This section contains* 5 *questions. The answer to each question is a single digit integer, ranging from* 0 *to* 9 (*both inclusive*)

16. A uniform circular disc of mass 50 kg and radius 0.4 m is rotating with an angular velocity of 10 rad/s about its own axis, which is vertical. Two uniform circular rings, each of mass 6.25 kg and radius 0.2 m, are gently placed symmetrically on the disc in such a manner that they are touching each other along the axis of the disc and are horizontal. Assume that the friction is large enough such that the rings are at rest relative to the disc and the system rotates about the original axis. The new angular velocity (in rad s^{-1}) of the system is

17. The work functions of silver and sodium are 4.6 and 2.3 eV, respectively. The ratio of the slope of the stopping potential *versus* frequency plot for silver to that of sodium is

18. A bob of mass m, suspended by a string of length l_1, is given a minimum velocity required to complete a full circle in the vertical plane, At the highest point, it collides elastically with another bob of mass m suspended by a string of length l_2, which is initially at rest. Both the strings are massless and inextensible. If the second bob, after collision acquires the minimum speed required to complete a full circle in the vertical plane, the ratio $\frac{l_1}{l_2}$ is

19. A particle of mass 0.2 kg is moving in one dimension under a force that delivers a constant power 0.5 W to the particle. If the initial speed (in ms^{-1}) of the particle is zero, the speed (in ms^{-1}) after 5s is

20. A freshly prepared sample of a radioisotope of half-life 1386 s has activity 10^3 disintegrations per second. Given that $\ln 2 = 0.693$, the fraction of the initial number of nuclei (expressed in nearest integer percentage) that will decay in the first 80 s after preparation of the sample is

CHEMISTRY

Section I

Direction (Q. No. 1-10) *This section contains* 10 *multiple choice questions. Each question has four choices (a), (b), (c) and (d) out of which only one is correct.*

1. The standard enthalpies of formation of $CO_2(g)$, $H_2O(l)$ and glucose(s) at 25°C are – 400 kJ/mol, – 300 kJ/mol and – 1300 kJ/mol, respectively. The standard enthalpy of combustion per gram of glucose at 25°C is

(a) + 2900 kJ (b) – 2900 kJ
(c) – 16.11 kJ (d) + 16.11 kJ

2. KI in acetone, undergoes S_N2 reaction with each *P*, *Q*, *R* and *S*. The rates of the reaction vary as

$H_3C—Cl$
P

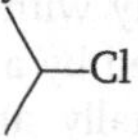
Q

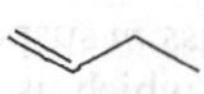
R

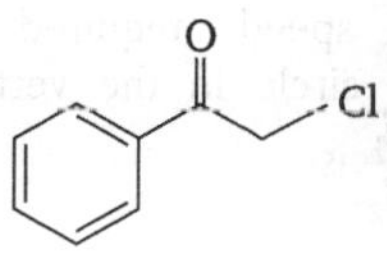
S

(a) $P > Q > R > S$
(b) $S > P > R > Q$
(c) $P > R > Q > S$
(d) $R > P > S > Q$

3. The compound that does not liberate CO_2, on treatment with aqueous sodium bicarbonate solution, is

(a) benzoic acid
(b) benzenesulphonic acid
(c) salicylic acid
(d) carbolic acid (Phenol)

4. Consider the following complex ions, *P*, *Q* and *R*.

$P = [FeF_6]^{3-}$, $Q = [V(H_2O)_6]^{2+}$ and $R = [Fe(H_2O)_6]^{2+}$

The correct order of the complex ions, according to their spin-only magnetic moment values (in BM) is

(a) $R < Q < P$ (b) $Q < R < P$
(c) $R < P < Q$ (d) $Q < P < R$

5. In the reaction,

$$P + Q \longrightarrow R + S$$

the time taken for 75% reaction of *P* is twice the time taken for 50% reaction of *P*. The concentration of *Q* varies with reaction time as shown in the figure. The overall order of the reaction is

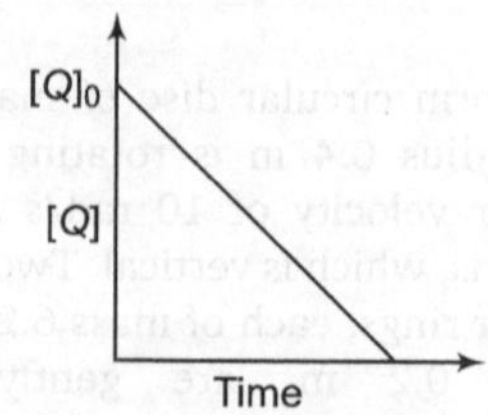

(a) 2 (b) 3 (c) 0 (d) 1

6. Concentrated nitric acid, upon long standing, turns yellow-brown due to the formation of

(a) NO (b) NO_2
(c) N_2O (d) N_2O_4

7. The arrangement of X^- ions around A^+ ion in solid *AX* is given in the figure (not drawn to scale). If the radius of X^- is 250 pm, the radius of A^+ is

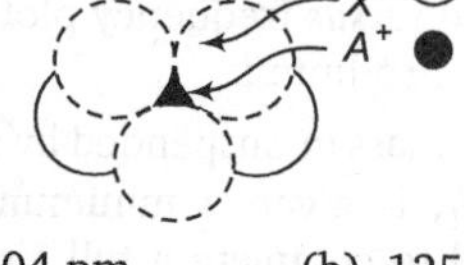

(a) 104 pm (b) 125 pm
(c) 183 pm (d) 57 pm

8. Upon treatment with ammoniacal H_2S, the metal ion that precipitates as a sulphide is

(a) Fe (III) (b) Al (III)
(c) Mg (II) (d) Zn (II)

9. Methylene blue, from its aqueous solution, is adsorbed on activated charcoal at 25°C. For this process, the correct statement is

(a) the adsorption requires activation at 25°C

(b) the adsorption is accompanied by a decreases in enthalpy

(c) the adsorption increases with increase of temperature

(d) the adsorption is irreversible

10. Sulphide ores are common for the metals

(a) Ag, Cu and Pb (b) Ag, Cu and Sn

(c) Ag, Mg and Pb (d) Al, Cu and Pb

Section II

Direction (Q. No. 11-15) *This section contains 5 multiple choice questions. Each question has four choices (a), (b), (c) and (d) out of which one or more are correct.*

11. Benzene and naphthalene form an ideal solution at room temperature. For this process, the true statement(s) is (are)

(a) ΔG is positive

(b) ΔS_{system} is positive

(c) $\Delta S_{\text{surroundings}} = 0$

(d) $\Delta H = 0$

12. The pair(s) of coordination complexes/ions exhibiting the same kind of isomerism is(are)

(a) $[Cr(NH_3)_5Cl]Cl_2$ and $[Cr(NH_3)_4Cl_2]Cl$

(b) $[Co(NH_3)_4Cl_2]^+$ and $[Pt(NH_3)_2(H_2O)Cl]^+$

(c) $[CoBr_2Cl_2]^{2-}$ and $[PtBr_2Cl_2]^{2-}$

(d) $[Pt(NH_3)_3(NO_3)]Cl$ and $[Pt(NH_3)_3Cl]Br$

13. The initial rate of hydrolysis of methyl acetate (1 M) by a weak acid (HA, 1M) is 1/100th of that of a strong acid (HX, 1M), at 25°C. The K_a (HA) is

(a) 1×10^{-4} (b) 1×10^{-5}

(c) 1×10^{-6} (d) 1×10^{-3}

14. The hyperconjugative stabilities of *tert*-butyl cation and 2-butene, respectively, are due to

(a) $\sigma \to p$ (empty) and $\sigma \to \pi^*$ electron delocalisations

(b) $\sigma \to \sigma^*$ and $\sigma \to \pi$ electron delocalisations

(c) $\sigma \to p$ (filled) and $\sigma \to \pi$ electron delocalisations

(d) p (filled) $\to \sigma^*$ and $\sigma \to \pi^*$ electrons delocalisations

15. Among *P*, *Q*, *R* and *S*, the aromatic compounds(s) is/are

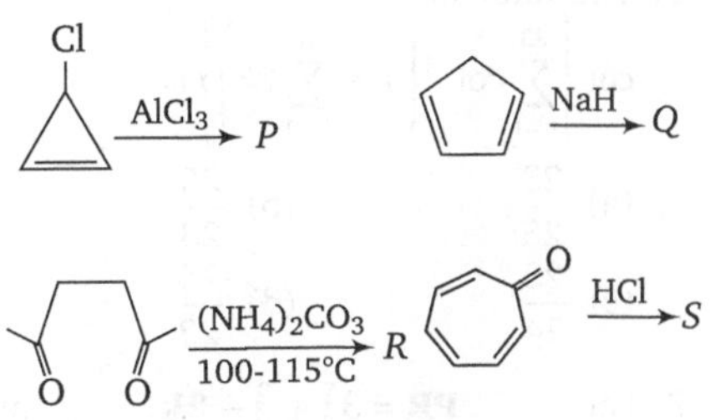

(a) *P* (b) *Q*

(c) *R* (d) *S*

Section III

Direction (Q. No. 16-20) *This section contains 5 questions. The answer to each question is a single digit integer, ranging from 0 to 9 (both inclusive).*

16. The total number of lone-pairs of electrons in melamine is

17. The total number of carboxylic acid groups in the product P is

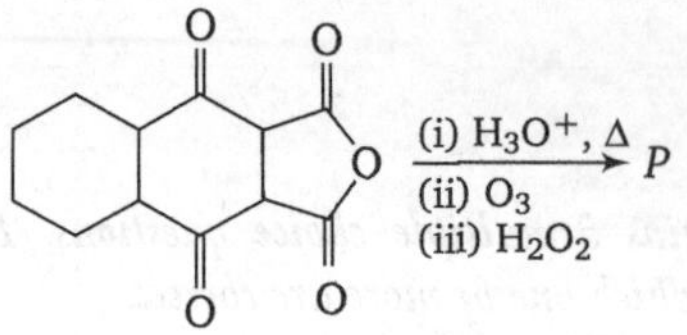

18. The atomic masses of He and Ne are 4 and 20 amu, respectively. The value of the de-Broglie wavelength of He gas at −73°C is "M" times that of the de-Broglie wavelength of Ne at 727°C. M is

19. $EDTA^{4-}$ is ethylenediaminetetraacetate ion. The total number of N—Co—O bond angles in $[Co(EDTA)]^-$ complex ion is

20. A tetrapeptide has —COOH group on alanine. This produces glycine (Gly), valine (Val), phenyl alanine (Phe) and alanine (Ala), on complete hydrolysis. For this tetrapeptide, the number of possible sequences (primary structures) with $—NH_2$ group attached to a chiral center is

MATHEMATICS

Section I

Directions (Q. Nos. 1-10) *This section contains 10 multiple choice questions. Each question has four choices (a), (b), (c) and (d) out of which only one is correct.*

1. The value of

$$\cot\left\{\sum_{n=1}^{23}\cot^{-1}\left(1+\sum_{k=1}^{n}2k\right)\right\}$$ is

(a) $\frac{23}{25}$ (b) $\frac{25}{23}$

(c) $\frac{23}{24}$ (d) $\frac{24}{23}$

2. Let $\mathbf{PR} = 3\hat{\mathbf{i}} + \hat{\mathbf{j}} - 2\hat{\mathbf{k}}$ and $\mathbf{SQ} = \hat{\mathbf{i}} - 3\hat{\mathbf{j}} - 4\hat{\mathbf{k}}$ determine diagonals of a parallelogram $PQRS$ and $\mathbf{PT} = \hat{\mathbf{i}} + 2\hat{\mathbf{j}} + 3\hat{\mathbf{k}}$ be another vector. Then, the volume of the parallelopiped determined by the vectors **PT**, **PQ** and **PS** is

(a) 5 (b) 20

(c) 10 (d) 30

3. Let complex numbers α and $\frac{1}{\alpha}$ lies on circles $(x - x_0)^2 + (y - y_0)^2 = r^2$ and $(x - x_0)^2 + (y - y_0)^2 = 4r^2$, respectively.
If $z_0 = x_0 + iy_0$ satisfies the equation $2|z_0|^2 = r^2 + 2$, then $|\alpha|$ is equal to

(a) $\frac{1}{\sqrt{2}}$ (b) $\frac{1}{2}$

(c) $\frac{1}{\sqrt{7}}$ (d) $\frac{1}{3}$

4. For $a > b > c > 0$, the distance between (1, 1) and the point of intersection of the lines $ax + by + c = 0$ and $bx + ay + c = 0$ is less than $2\sqrt{2}$. Then,

(a) $a + b - c > 0$ (b) $a - b + c < 0$

(c) $a - b + c > 0$ (d) $a + b - c < 0$

5. Perpendicular are drawn from points on the line $\frac{x+2}{2}=\frac{y+1}{-1}=\frac{z}{3}$ to the plane $x+y+z=3$. The feet of perpendiculars lie on the line

(a) $\frac{x}{5}=\frac{y-1}{8}=\frac{z-2}{-13}$

(b) $\frac{x}{2}=\frac{y-1}{3}=\frac{z-2}{-5}$

(c) $\frac{x}{4}=\frac{y-1}{3}=\frac{z-2}{-7}$

(d) $\frac{x}{2}=\frac{y-1}{-7}=\frac{z-2}{5}$

6. Four persons independently solve a certain problem correctly with probabilities $\frac{1}{2}, \frac{3}{4}, \frac{1}{4}, \frac{1}{8}$. Then, the probability that the problem is solved correctly by atleast one of them is

(a) $\frac{235}{256}$ (b) $\frac{21}{256}$ (c) $\frac{3}{256}$ (d) $\frac{253}{256}$

7. The area enclosed by the curves $y=\sin x+\cos x$ and $y=|\cos x-\sin x|$ over the interval $\left[0, \frac{\pi}{2}\right]$ is

(a) $4(\sqrt{2}-1)$ (b) $2\sqrt{2}(\sqrt{2}-1)$

(c) $2(\sqrt{2}+1)$ (d) $2\sqrt{2}(\sqrt{2}+1)$

8. A curve passes through the point $\left(1, \frac{\pi}{6}\right)$. Let the slope of the curve at each point (x, y) be $\frac{y}{x}+\sec\left(\frac{y}{x}\right), x>0$.

Then, the equation of the curve is

(a) $\sin\left(\frac{y}{x}\right)=\log x+\frac{1}{2}$

(b) $\text{cosec}\left(\frac{y}{x}\right)=\log x+2$

(c) $\sec\left(\frac{2y}{x}\right)=\log x+2$

(d) $\cos\left(\frac{2y}{x}\right)=\log x+\frac{1}{2}$

9. Let $f:\left[\frac{1}{2}, 1\right]\to R$ (the set of all real numbers) be a positive, non-constant and differentiable function such that $f'(x)<2f(x)$ and $f\left(\frac{1}{2}\right)=1$. Then, the value of $\int_{1/2}^{1} f(x)\,dx$ lies in the interval

(a) $(2e-1, 2e)$

(b) $(e-1, 2e-1)$

(c) $\left(\frac{e-1}{2}, e-1\right)$

(d) $\left(0, \frac{e-1}{2}\right)$

10. The number of points in $(-\infty, \infty)$, for which $x^2-x\sin x-\cos x=0$, is

(a) 6 (b) 4 (c) 2 (d) 0

Section II

Directions (Q. Nos. 11-15) *This section contains 5 multiple choice questions. Each question has four choices (a), (b), (c) and (d) out of which one or more are correct.*

11. A rectangular sheet of fixed perimeter with sides having their lengths in the ratio 8 : 15 is converted into an open rectangular box by folding after removing squares of equal area from all four corners. If the total area of removed squares is 100, the resulting box has maximum volume. The lengths of the sides of the rectangular sheet are

(a) 24 (b) 32

(c) 45 (d) 60

12. Let $S_n=\sum_{2}^{4n}(-1)^{\frac{k(k+1)}{2}}k^2$. Then, S_n can take value(s)

(a) 1056 (b) 1088

(c) 1120 (d) 1332

13. A line l passing through the origin is perpendicular to the lines

$$l_1 : (3+t)\hat{\mathbf{i}} + (-1+2t)\hat{\mathbf{j}} + (4+2t)\hat{\mathbf{k}}, -\infty < t < \infty$$

$$l_2 : (3+2s)\hat{\mathbf{i}} + (3+2s)\hat{\mathbf{j}} + (2+s)\hat{\mathbf{k}}, -\infty < s < \infty$$

Then, the coordinate(s) of the point(s) on l_2 at a distance of $\sqrt{17}$ from the point of intersection of l and l_1 is (are)

(a) $\left(\frac{7}{3}, \frac{7}{3}, \frac{5}{3}\right)$ (b) $(-1, -1, 0)$

(c) $(1, 1, 1)$ (d) $\left(\frac{7}{9}, \frac{7}{9}, \frac{8}{9}\right)$

14. Let $f(x) = x \sin \pi x$, $x > 0$. Then, for all natural numbers n, $f'(x)$ vanishes at

(a) a unique point in the interval $\left(n, n+\frac{1}{2}\right)$

(b) a unique point in the interval $\left(n+\frac{1}{2}, n+1\right)$

(c) a unique point in the interval $(n, n+1)$

(d) two points in the interval $(n, n+1)$

15. For 3×3 matrices M and N, which of the following statement(s) is (are) not correct?

(a) $N^T M N$ is symmetric or skew-symmetric, according as M is symmetric or skew-symmetric

(b) $MN - NM$ is symmetric for all symmetric matrices M and N

(c) $M N$ is symmetric for all symmetric matrices M and N

(d) $(\text{adj } M)$ $(\text{adj } N) = \text{adj}(MN)$ for all invertible matrices M and N

Section III

Directions (Q. Nos. 16-20) *This section contains 5 questions. The answer to each question is a single digit integer, ranging from 0 to 9 (both inclusive.)*

16. A vertical line passing through the point $(h, 0)$ intersects the ellipse $\frac{x^2}{4} + \frac{y^2}{3} = 1$ at the points P and Q. Let the tangents to the ellipse at P and Q meet at the point R. If $\Delta(h) =$ area of the ΔPQR, $\Delta_1 = \max\limits_{1/2 \le h \le 1} \Delta(h)$ and $\Delta_2 = \min\limits_{1/2 \le h \le 1} \Delta(h)$, then $\frac{8}{\sqrt{5}} \Delta_1 - 8\Delta_2$ is equal to

17. The coefficients of three consecutive terms of $(1+x)^{n+5}$ are in the ratio $5:10:14$. Then, n is equal to

18. Consider the set of eight vectors $V = [a\hat{\mathbf{i}} + b\hat{\mathbf{j}} + c\hat{\mathbf{k}} : a, b, c \in \{-1, 1\}]$. Three non-coplanar vectors can be chosen from V in 2^p ways. Then, p is

19. Of the three independent events E_1, E_2 and E_3, the probability that only E_1 occurs is α, only E_2 occurs is β and only E_3 occurs is γ. Let the probability p that none of events E_1, E_2 or E_3 occurs satisfy the equations $(\alpha - 2\beta)$, $p = \alpha\beta$ and $(\beta - 3\gamma)$ $p = 2\beta\gamma$. All the given probabilities are assumed to lie in the interval $(0, 1)$.

Then, $\frac{\text{probability of occurrence of } E_1}{\text{probability of occurrence of } E_3}$ is equal to

20. A pack contains n card numbered from 1 to n. Two consecutive numbered card are removed from the pack and the sum of the numbers on the remaining cards is 1224. If the smaller of the numbers on the removed cards is k, then $k - 20$ is equal to

PAPER 2

Instructions

- The question paper consists of three sections.
- Section 1 contains 8 multiple choice questions each of 3 marks with negative marking of 1 mark. Each question has four choices (a), (b), (c) and (d) out of which one or more are correct.
- Section 2 contains 4 paragraphs each describing theory, experiment, data etc. 8 question sections to four paragraphs with two questions on each paragraph, each of 3 marks with negative marking of 1 mark. Each question of paragraph has only one correct answer among the four choice (a), (b), (c) and (d).
- Section 3 contains 4 multiple choice questions each of 3 marks with negative marking of 1 mark. Each question has matching lists. The codes for the lists have choices (a), (b), (c) and (d) out of which only one is correct.

PHYSICS

Section I

Direction (Q. No. 1-8) *This section contains* 8 *multiple choice questions. Each question has four choices (a), (b), (c) and (d) out of which one or more than one are correct.*

1. Two bodies, each of mass M, are kept fixed with a separation $2L$. A particle of mass m is projected from the mid-point of the line joining their centres, perpendicular to the line. The gravitational constant is G. The correct statement(s) is (are)

(a) The minimum initial velocity of the mass m to escape the gravitational field of the two bodies is $4\sqrt{\frac{GM}{L}}$

(b) The minimum initial velocity of the mass m to escape the gravitational field of the two bodies is $2\sqrt{\frac{GM}{L}}$

(c) The minimum initial velocity of the mass m to escape the gravitational field of the two bodies is $\sqrt{\frac{2GM}{L}}$

(d) The energy of the mass m remains constant

2. A particle of mass m is attached to one end of a mass-less spring of force constant k, lying on a frictionless horizontal plane. The other end of the spring is fixed. The particle starts moving horizontally from its equilibrium position at time $t = 0$ with an initial velocity u_0. When the speed of the particle is $0.5\,u_0$, it collides elastically with a rigid wall. After this collision

(a) the speed of the particle when it returns to its equilibrium position is u_0

(b) the time at which the particle passes through the equilibrium position for the first time is $t = \pi\sqrt{\frac{m}{k}}$

(c) the time at which the maximum compression of the spring occurs is $t = \frac{4\pi}{3}\sqrt{\frac{m}{k}}$

(d) the time at which the particle passes through the equilibrium position for the second time is $t = \frac{5\pi}{3}\sqrt{\frac{m}{k}}$

3. A steady current I flows along an infinitely long hollow cylindrical conductor of radius R. This cylinder is placed coaxially inside an infinite solenoid of radius $2R$. The solenoid has n turns per unit length and carries a steady current I. Consider a point P at a distance r from the common axis. The correct statement(s) is (are)

(a) In the region $0 < r < R$, the magnetic field is non-zero

(b) In the region $R < r < 2R$, the magnetic field is along the common axis

(c) In the region $R < r < 2R$, the magnetic field is tangential to the circle of radius r, centered on the axis

(d) In the region $r > 2R$, the magnetic field is non-zero

4. Two vehicles, each moving with speed u on the same horizontal straight road, are approaching each other. Wind blows along the road with velocity w. One of these vehicles blows a whistle of frequency f_1. An observer in the other vehicle hears the frequency of the whistle to be f_2. The speed of sound in still air is v. The correct statement(s) is (are)

(a) If the wind blows from the observer to the source, $f_2 > f_1$.

(b) If the wind blows from the source to the observer, $f_2 > f_1$

(c) If the wind blows from the observer to the source, $f_2 < f_1$

(d) If the wind blows from the source to the observer, $f_2 < f_1$

5. Using the expression $2d \sin\theta = \lambda$, one calculates the values of d by measuring the corresponding angles θ in the range 0 to 90°. The wavelength λ is exactly known and the error in θ is constant for all values of θ. As θ increases from 0°

(a) the absolute error in d remains constant

(b) the absolute error in d increases

(c) the fractional error in d remains constant

(d) the fractional error in d decreases

6. Two non-conducting spheres of radii R_1 and R_2 and carrying uniform volume charge densities $+\rho$ and $-\rho$, respectively, are placed such that they partially overlap, as shown in the figure. At all points in the overlapping region

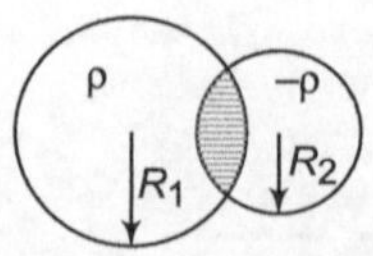

(a) the electrostatic field is zero

(b) the electrostatic potential is constant

(c) the electrostatic field is constant in magnitude

(d) the electrostatic field has same direction

7. The figure below shows the variation of specific heat capacity (C) of a solid as a function of temperature (T). The temperature is increased continuously from 0 to 500 K at a constant rate. Ignoring any volume change, the following statement(s) is (are) correct to reasonable approximation.

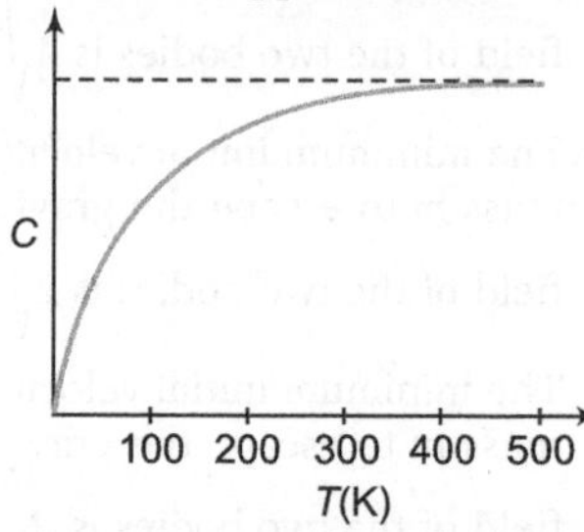

(a) the rate at which heat is absorbed in the range 0-100 K varies linearly with temperature T

(b) heat absorbed in increasing the temperature from 0-100 K is less than the heat required for increasing the temperature from 400-500 K

(c) there is no change in the rate of heat absorbtion in the range 400-500 K

(d) the rate of heat absorption increases in the range 200-300 K

8. The radius of the orbit of an electron in a Hydrogen-like atom is $4.5\, a_0$ where a_0 is the Bohr radius. Its orbital angular momentum is $\frac{3h}{2\pi}$. It is given that h is Planck constant and R is Rydberg constant. The possible wavelength(s), when the atom de-excites, is (are)

(a) $\frac{9}{32R}$ (b) $\frac{9}{16R}$ (c) $\frac{9}{5R}$ (d) $\frac{4}{3R}$

Section II

Direction (Q. No. 9-16) *This section contains 8 multiple choice questions relating to four paragraphs with two questions on each paragraph. Each question has four choices (a), (b), (c) and (d) out of which only one is correct.*

Paragraph I

A small block of mass 1 kg is released from rest at the top of a rough track. The track is a circular arc of radius 40 m. The block slides along the track without toppling and a frictional force acts on it in the direction opposite to the instantaneous velocity. The work done in overcoming the friction up to the point Q, as shown in the figure, is 150 J. (Take the acceleration due to gravity, $g = 10\ ms^{-2}$)

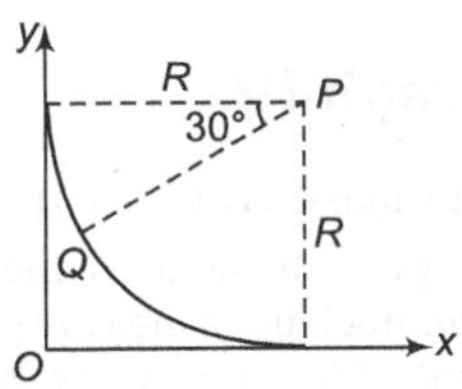

9. The speed of the block when it reaches the point Q is

(a) $5\ ms^{-1}$ (b) $10\ ms^{-1}$

(c) $10\sqrt{3}\ ms^{-1}$ (d) $20\ ms^{-1}$

10. The magnitude of the normal reaction that acts on the block at the point Q is

(a) 7.5 N

(b) 8.6 N

(c) 11.5 N

(d) 22.5 N

Paragraph II

A thermal power plant produces electric power of 600 kW at 4000 V, which is to be transported to a place 20 km away from the power plant for consumers' usage. It can be transported either directly with a cable of large current carrying capacity or by using a combination of step-up and step-down transformers at the two ends. The drawback of the direct transmission is the large energy dissipation. In the method using transformers, the dissipation is much smaller. In this method, a step-up transformer is used at the plant side so that the current is reduced to a smaller value. At the consumers' end a step-down transformer is used to supply power to the consumers at the specified lower voltage. It is reasonable to assume that the power cable is purely resistive and the transformers are ideal with a power factor unity. All the current and voltages mentioned are rms values.

11. If the direct transmission method with a cable of resistance $0.4\ \Omega\ km^{-1}$ is used, the power dissipation (in %) during transmission is

(a) 20 (b) 30

(c) 40 (d) 50

12. In the method using the transformers, assume that the ratio of the number of turns in the primary to that in the secondary in the step-up transformer is 1 : 10. If the power to the consumers has to be supplied at 200V, the ratio of the number of turns in the primary to that in the secondary in the step-down transformer is
(a) 200 : 1 (b) 150 : 1
(c) 100 : 1 (d) 50 : 1

Paragraph III

A point charge Q is moving in a circular orbit of radius R in the x-y plane with an angular velocity ω. This can be considered as equivalent to a loop carrying a steady current $\frac{Q\omega}{2\pi}$. A uniform magnetic field along the positive z-axis is now switched on, which increases at a constant rate from 0 to B in one second. Assume that the radius of the orbit remains constant. The applications of the magnetic field induces an emf in the orbit. The induced emf is defined as the work done by an induced electric field in moving a unit positive charge around a closed loop. It is known that, for an orbiting charge, the magnetic dipole moment is proportional to the angular momentum with a proportionality constant γ.

13. The magnitude of the induced electric field in the orbit at any instant of time during the time interval of the magnetic field change is
(a) $\frac{BR}{4}$ (b) $\frac{-BR}{2}$ (c) BR (d) $2BR$

14. The change in the magnetic dipole moment associated with the orbit, at the end of the time interval of the magnetic field change, is
(a) γBQR^2 (b) $-\gamma\frac{BQR^2}{2}$
(c) $\gamma\frac{BQR^2}{2}$ (d) γBQR^2

Paragraph IV

The mass of a nucleus ${}_Z^AX$ is less that the sum of the masses of $(A - Z)$ number of neutrons and Z number of protons in the nucleus. The energy equivalent to the corresponding mass difference is known as the binding energy of the nucleus. A heavy nucleus of mass M can break into two light nuclei of masses m_1 and m_2 only if $(m_1 + m_2) < M$. Also two light nuclei of masses m_3 and m_4 can undergo complete fusion and form a heavy nucleus of mass M' only if $(m_3 + m_4) > M'$. The masses of some neutral atoms are given in the table below :

${}_1^1H$	1.007825u	${}_1^2H$	2.014102u	${}_1^3H$	3.016050u
${}_3^6Li$	6.01513u	${}_3^7Li$	7.016004u	${}_2^4He$	4.002603u
${}_{64}^{152}Gd$	151.919803u	${}_{82}^{206}Pb$	205.974455u	${}_{30}^{70}Zn$	69.925325u
${}_{34}^{82}Se$	81.916709u	${}_{83}^{209}Bi$	208.980388u	${}_{84}^{210}Po$	209.982876u

15. The correct statement is
(a) The nucleus ${}_3^6Li$ can emit an alpha particle
(b) The nucleus ${}_{84}^{210}Po$ can emit a proton
(c) Deuteron and alpha particle can undergo complete fusion
(d) The nuclei ${}_{30}^{70}Zn$ and ${}_{34}^{82}Se$ can undergo complete fusion

16. The Kinetic energy (in keV) of the alpha particle, when the nucleus ${}_{84}^{210}Po$ at rest undergoes alpha decay, is
(a) 5319 (b) 5422
(c) 5707 (d) 5818

Section III

Direction (Q. No. 17-20) *This section contains 4 multiple choice questions. Each question has matching lists. The codes for the lists have choices (a), (b), (c) and (d) out of which only one is correct.*

17. A right angled prism of refractive index μ_1 is placed in a rectangular block of refractive index μ_2, which is surrounded by a medium of refractive index μ_3, as shown in the figure, *A* ray of light '*e*' enters the rectangular block at normal incidence. Depending upon the relationships between μ_1, μ_2 and μ_3, it takes one of the four possible paths '*ef*', '*eg*', '*eh*' or '*ei*'.

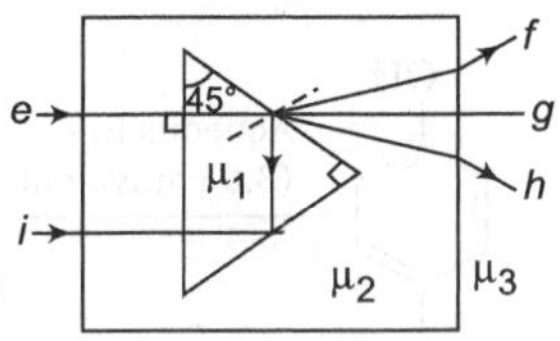

Match the paths in List I with conditions of refractive indices in List II and select the correct answer using the codes given below the lists.

	List I		List II
P.	$e \to f$	1.	$\mu_1 > \sqrt{2}\mu_2$
Q.	$e \to g$	2.	$\mu_1 > \mu_1$ and $\mu_2 > \mu_3$
R.	$e \to h$	3.	$\mu_1 = \mu_2$
S.	$e \to i$	4.	$\mu_2 < \mu_1 < \sqrt{2}\mu_2$ and $\mu_2 > \mu_3$

Codes

	P	Q	R	S		P	Q	R	S
(a)	2	3	1	4	(b)	1	2	4	3
(c)	4	1	2	3	(d)	2	3	4	1

18. Match List I with List II and select the correct answer using the codes given below the lists.

	List I		List II
P.	Boltzmann constant	1.	$[ML^2T^{-1}]$
Q.	Coefficient of viscosity	2.	$[ML^{-1}T^{-1}]$
R.	Planck constant	3.	$[MLT^{-3}K^{-1}]$
S.	Thermal conductivity	4.	$[ML^2T^{-2}K^{-1}]$

Codes

	P	Q	R	S		P	Q	R	S
(a)	3	1	2	4	(b)	3	2	1	4
(c)	4	2	1	3	(d)	4	1	2	3

19. One mole of a monatomic ideal gas is taken along two cyclic processes $E \to F \to G \to E$ and $E \to F \to H \to E$ as shown in the *p-V* diagram.

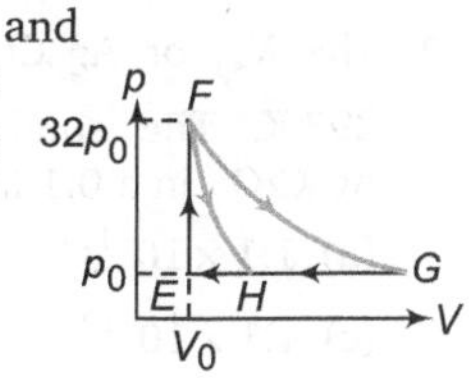

The processes involved are purely isochoric, isobaric, isothermal or adiabatic.

Match the paths in List I with the magnitudes of the work done in List II and select the correct answer using the codes given below the lists.

	List I		List II
P.	$G \to E$	1.	$160\ p_0V_0 \ln 2$
Q.	$G \to H$	2.	$36\ p_0V_0$
R.	$F \to H$	3.	$24\ p_0V_0$
S.	$F \to G$	4.	$31\ p_0V_0$

Codes

	P	Q	R	S		P	Q	R	S
(a)	4	3	2	1	(b)	4	3	1	2
(c)	3	1	2	4	(d)	1	3	2	4

20. Match List I of the nuclear process with List II containing parent nucleus and one of the end products of each process and then select the correct answer using the codes given below the lists.

	List I		List II
P.	Alpha decay	1.	${}^{15}_{8}O \to {}^{17}_{7}N + \dots$
Q.	β^+ decay	2.	${}^{238}_{92}U \to {}^{234}_{90}Th + \dots$
R.	Fission	3.	${}^{185}_{83}Bi \to {}^{184}_{82}Pb + \dots$
S.	Proton emission	4.	${}^{239}_{94}Pu \to {}^{140}_{57}La + \dots$

Codes

	P	Q	R	S		P	Q	R	S
(a)	4	2	1	3	(b)	1	3	2	4
(c)	2	1	4	3	(d)	4	3	2	1

CHEMISTRY

Section I

Direction (Q. No. 1-8) *This section contains 8 multiple choice questions. Each question has four choices (a), (b), (c) and (d) out of which one or more than one is correct.*

1. The K_{sp} of Ag_2CrO_4 is 1.1×10^{-12} at 298 K. The solubility (in mol/L) of Ag_2CrO_4 in a 0.1 M $AgNO_3$ solution is

(a) 1.1×10^{-11} (b) 1.1×10^{-10}

(c) 1.1×10^{-12} (d) 1.1×10^{-9}

2. In the following reaction, the product(s) formed is (are)

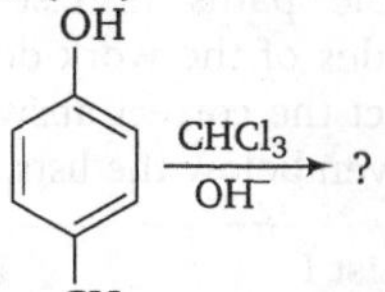

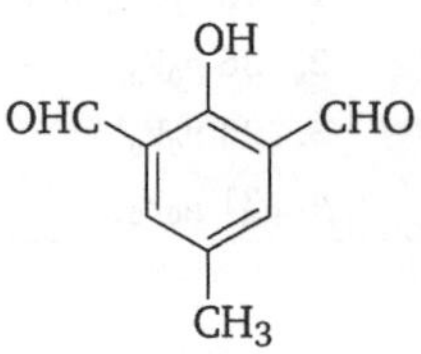

(P)

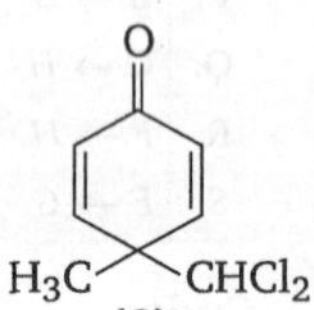

(Q)

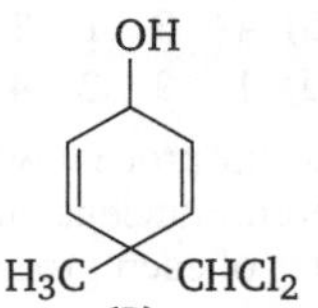

(R)

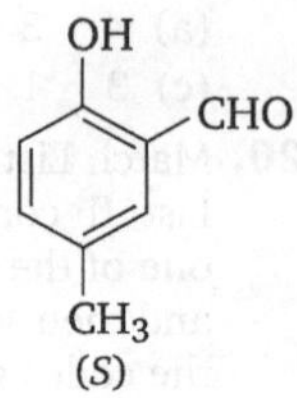

(S)

(a) *P* (major)

(b) *Q* (minor)

(c) *R* (minor)

(d) *S* (major)

3. The major product(s) of the following reaction is (are)

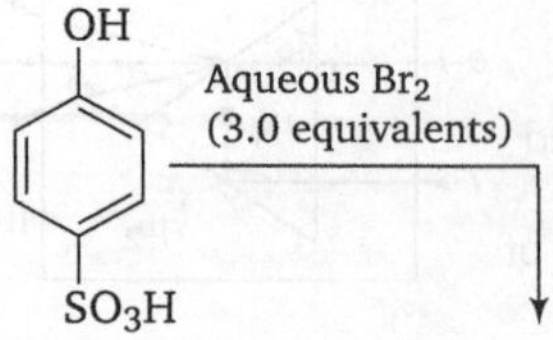

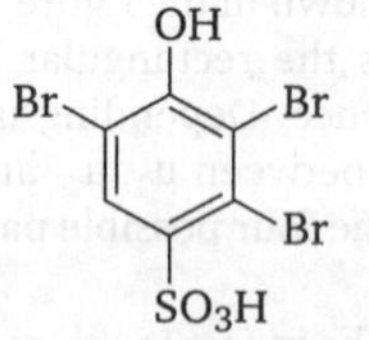

(P)

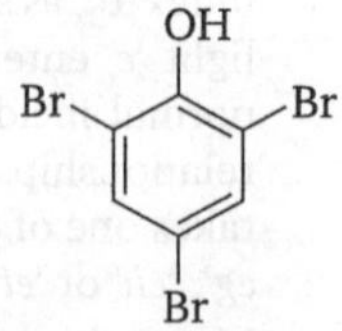

(Q)

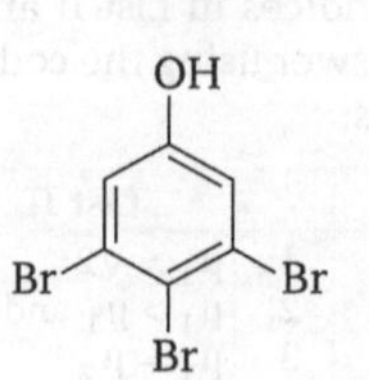

(R)

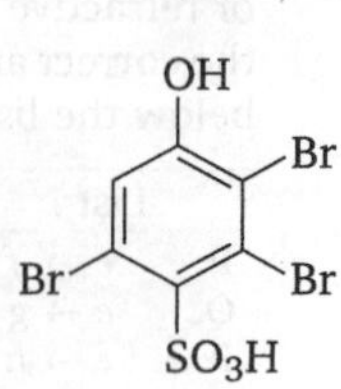

(S)

(a) *P* (b) *Q*

(c) *R* (d) *S*

4. After completion of the reactions (I and II), the organic compound(s) in the reaction mixtures is (are)

Reaction I 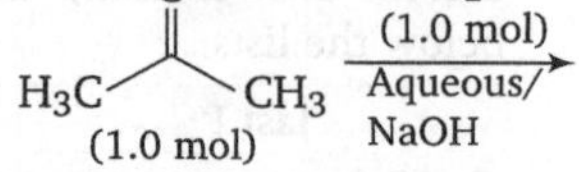

Reaction II 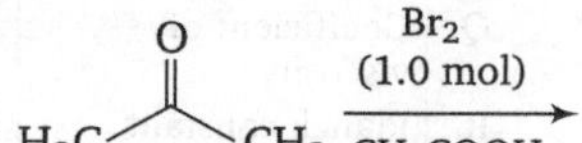

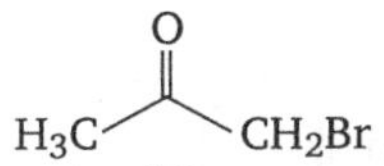
(P)

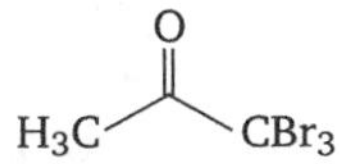
(Q)

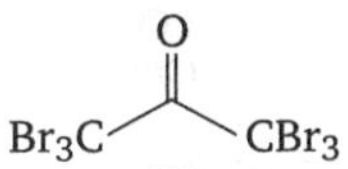
(R)

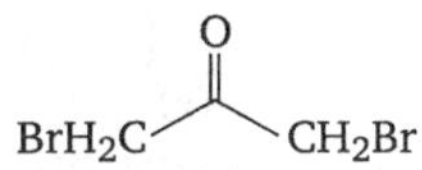
(S)

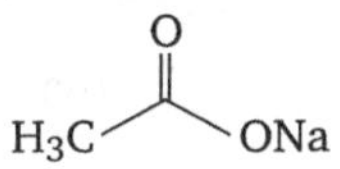
(T)

$CHBr_3$
(U)

(a) Reaction I : *P* and Reaction II : *P*
(b) Reaction I : *U*, acetone and Reaction II : *Q*, acetone
(c) Reaction I : *T*, *U*, acetone and Reaction II : *P*
(d) Reaction I : *R*, acetone and Reaction II : *S*, acetone

5. The correct statement(s) about O_3 is (are)
(a) O—O bond lengths are equal
(b) thermal decomposition of O_3 is endothermic
(c) O_3 is diamagnetic in nature
(d) O_3 has a bent structure

6. In the nuclear transmutation
$${}^{9}_{4}Be + X \longrightarrow {}^{8}_{4}Be + Y$$
(*X*, *Y*) is (are)
(a) (γ, *n*) (b) (*p*, *D*)
(c) (*n*, *D*) (d) (γ, *p*)

7. The carbon-based reduction method is not used for the extraction of
(a) tin from SnO_2
(b) iron from Fe_2O_3
(c) aluminium from Al_2O_3
(d) magnesium from $MgCO_3$, $CaCO_3$

8. The thermal dissociation of equilibrium of $CaCO_3(s)$ is studied under different conditions.
$$CaCO_3(s) \rightleftharpoons CaO(s) + CO_2(g)$$
For this equilibrium, the correct statement(s) is (are)
(a) ΔH is dependent on *T*
(b) *K* is independent of the initial amount of $CaCO_3$
(c) *K* is dependent on the pressure of CO_2 at a given *T*
(d) ΔH is independent of the catalyst, if any

Section II

Direction (Q. No. 9-16) *This section contains 8 multiple choice questions relating to four paragraphs with two questions on each paragraph. Each question has four choices (a), (b), (c) and (d) out of which only one is correct.*

Paragraph I

An aqueous solution of a mixture of two inorganic salts,when treated with dilute HCl, gave a precipitate (P) and filtrate (Q). The precipitate P was found to dissolve in hot water. The filtrate (Q) remained unchanged, when treated with H_2S in a dilute mineral acid medium. However, it gave a precipitate (R) with H_2S in an ammoniacal medium. The precipitate R gave a coloured solution (S), when treated with H_2O_2 in an aqueous NaOH medium.

9. The precipitate *P* contains
(a) Pb^{2+} (b) Hg_2^{2+}
(c) Ag^+ (d) Hg^{2+}

10. The coloured solution *S* contains
(a) $Fe_2(SO_4)_3$
(b) $CuSO_4$
(c) $ZnSO_4$
(d) Na_2CrO_4

Paragraph II

P and Q are isomers of dicarboxylic acid $C_4H_4O_4$. Both decolourize Br_2/H_2O. On heating, P forms the cyclic anhydride.
Upon treatment with dilute alkaline $KMnO_4$. P as well as Q could produce one or more than one form S, T and U.

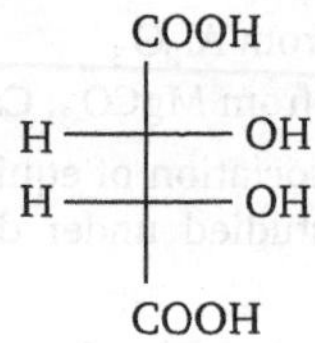

S

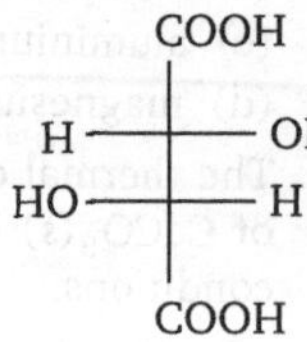

T

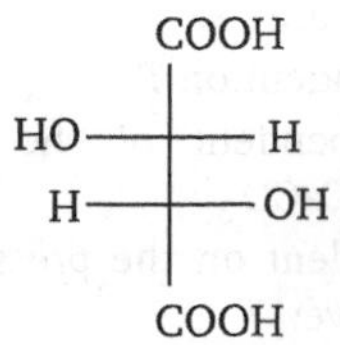

U

11. Compounds formed from P and Q are, respectively

(a) Optically active S and optically active pair (T, U)
(b) Optically inactive S and optically inactive pair (T, U)
(c) Optically active pair (T, U) and optically active S
(d) Optically inactive pair (T, U) and optically inactive S

12. In the following reaction sequences V and W are respectively

$$Q \xrightarrow[\Delta]{H_2/Ni} V$$

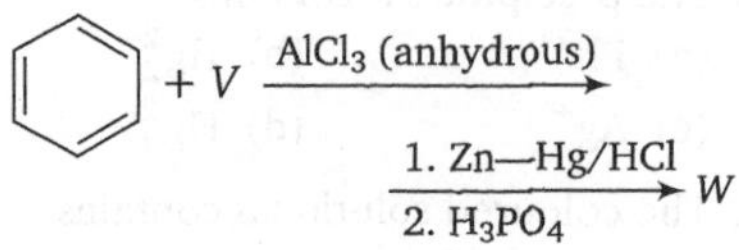

(a)  and 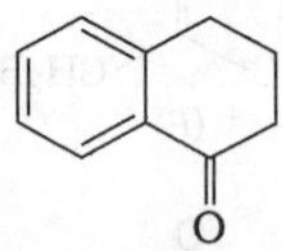
(V) (W)

(b) 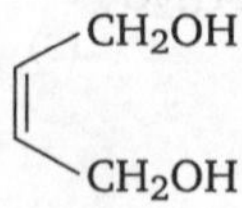 and 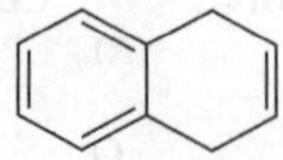
(V) (W)

(c)  and 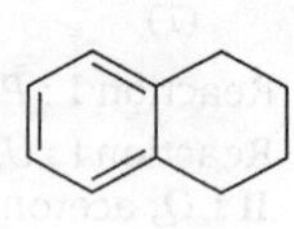
(V) (W)

(d) 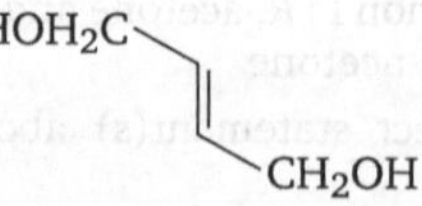
(V)
and 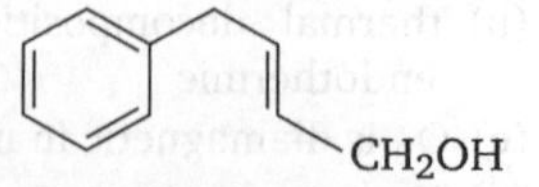
(W)

Paragraph III

A fixed mass m of a gas is subjected to transformation of states from K to L to M to N and back to K as shown in the figure.

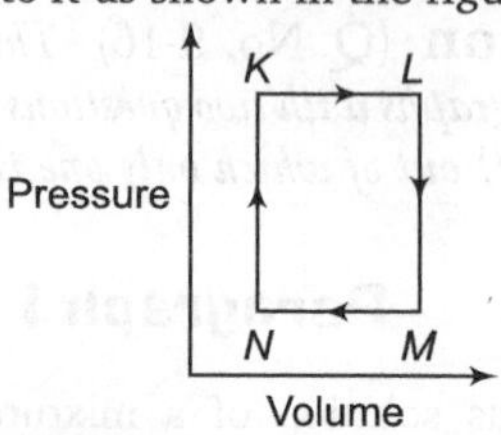

13. The succeeding operations that enable this transformation of states are

(a) heating, cooling, heating, cooling
(b) cooling, heating, cooling, heating
(c) heating, cooling, cooling, heating
(d) cooling, heating, heating, cooling

14. The pair of isochoric processes among the transormation of states is

(a) *K* to *L* and *L* to *M*
(b) *L* to *M* and *N* to *K*
(c) *L* to *M* and *M* to *N*
(d) *M* to *N* and *N* to *K*

Paragraph IV

The reactions of Cl_2 gas with cold-dilute and hot-concentrated NaOH in water give sodium salts of two (different) oxoacids of chlorine, *P* and *Q*, respectively. The Cl_2 gas reacts with SO_2 gas, in the presence of charcoal, to give a product *R*. *R* reacts with white phosphorus to give a compound *S*. On hydrolysis, *S* gives an oxoacid of phosphorous *T*.

15. *P* and *Q* respectively, are the sodium salts of

(a) hypochlorous and chloric acids
(b) hypochlorous and chlorous acids
(c) chloric and perchloric acids
(d) chloric and hypochlorous acids

16. *R*, *S* and *T*, respectively, are

(a) SO_2Cl_2, PCl_5 and H_3PO_4
(b) SO_2Cl_2, PCl_3 and H_3PO_3
(c) $SOCl_2$, PCl_3 and H_3PO_2
(d) $SOCl_2$, PCl_5 and H_3PO_4

Section III

Direction (Q. No. 17-20) *This section contains 4 multiple choice questions. Each question has four choices (a), (b), (c) and (d) out of which only one is correct.*

17. The unbalanced chemical reactions given in List I show missing reagent or condition (?) which are provided in List II. Match List I with List II and select the correct answer using the code given below the lists.

	List I		List II
P.	$PbO_2 + H_2SO_4 \xrightarrow{?} PbSO_4 + O_2 +$ other product	1.	NO
Q.	$Na_2S_2O_3 + H_2O \xrightarrow{?} NaHSO_4 +$ other product	2.	I_2
R.	$N_2H_4 \xrightarrow{?} N_2 +$ other product	3.	Warm
S.	$XeF_2 \xrightarrow{?} Xe +$ other product	4.	Cl_2

Codes

	P	Q	R	S		P	Q	R	S
(a)	4	2	3	1	(b)	3	2	1	4
(c)	1	4	2	3	(d)	3	4	2	1

18. Match the chemical conversion in List I with the appropriate reagents in List II and select the correct answer using the code given below the lists.

	List I		List II
P.	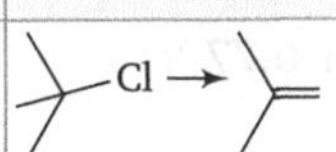	1.	(i) $Hg(OAc)_2$; (ii) $NaBH_4$
Q.	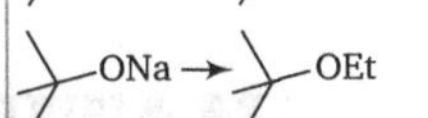	2.	NaOEt
R.	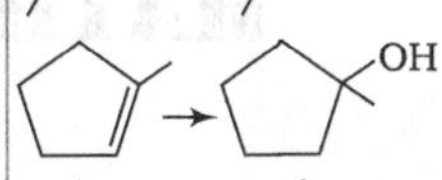	3.	Et-Br
S.	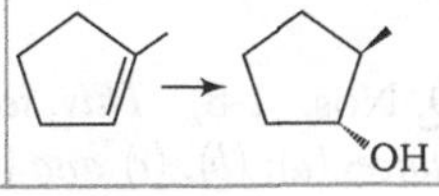	4.	(i) BH_3 ; (ii) H_2O_2/ NaOH

Codes

	P	Q	R	S
(a)	2	3	1	4
(b)	3	2	1	4
(c)	2	3	4	1
(d)	3	2	4	1

19. An aqueous solution of X is added slowly to an aqueous solution of Y as shown in List I. The variation in conductivity of these reactions is given in List II. Match List I with List II and select the correct answer using the code given below the lists.

	List I		List II
P.	$(C_2H_5)_3N$ (X) + CH_3COOH (Y)	1.	Conductivity decreases and then increases
Q.	KI(0.1 M) (X) + $AgNO_3$(0.01 M) (Y)	2.	Conductivity decreases and then does not change much
R.	CH_3COOH (X) + KOH (Y)	3.	Conductivity increases and then does not change much
S.	NaOH (X) + HI (Y)	4.	Conductivity does not change much and then increases

Codes

	P	Q	R	S		P	Q	R	S
(a)	3	4	2	1	(b)	4	3	2	1
(c)	2	3	4	1	(d)	1	4	3	2

20. The standard reduction potential data at 25°C is given below.

$E°(Fe^{3+}/Fe^{2+}) = +0.77$ V;

$E°(Fe^{2}/Fe) = -0.44$ V;

$E°(Cu^{2+}/Cu) = +0.34$ V;

$E°(Cu^{+}/Cu) = +0.52$ V;

$E°(O_2(g) + 4H^+ + 4e^- \longrightarrow 2H_2O) = +1.23$ V;

$E°(O_2(g) + 2H_2O + 4e^- \longrightarrow 4OH) = +0.40$ V

$E°(Cr^{3+}/Cr) = -0.74$ V;

$E°(Cr^{2+}/Cr) = +0.91$ V

Match $E°$ of the rebox pair in List I with the values given in List II and select the correct answer using the code given below the lists.

	List I		List II
P.	$E°(Fe^{3+}/Fe)$	1.	−0.18 V
Q.	$E°(4H_2O \rightleftharpoons 4H^+ + 4OH^-)$	2.	−0.4 V
R.	$E°(Cu^{2+} + Cu \longrightarrow 2Cu^+)$	3.	−0.04 V
S.	$E°(Cr^{3+}, Cr^{2+})$	4.	−0.83 V

Codes

	P	Q	R	S		P	Q	R	S
(a)	4	1	2	3	(b)	2	3	4	1
(c)	1	2	3	4	(d)	3	4	1	2

MATHEMATICS

Section I

Directions (Q. Nos. 1-8) *This section contains 8 multiple choice questions. Each question has four choices (a), (b), (c) and (d) out of which one or more are correct.*

1. For $a \in R$ (the set of all real numbers), $a \neq -1$,

$$\lim_{n\to\infty} \frac{(1^a + 2^a + \ldots + n^a)}{(n+1)^{a-1}[(na+1) + (na+2) + \ldots + (na+n)]} = \frac{1}{60}.$$ Then, a is equal to

(a) 5 (b) 7

(c) $\frac{-15}{2}$ (d) $\frac{-17}{2}$

2. Circle(s) touching x-axis at a distance 3 from the origin and having an intercept of length $2\sqrt{7}$ on y-axis is (are)

(a) $x^2 + y^2 - 6x + 8y + 9 = 0$

(b) $x^2 + y^2 - 6x + 7y + 9 = 0$

(c) $x^2 + y^2 - 6x - 8y + 9 = 0$

(d) $x^2 + y^2 - 6x - 7y + 9 = 0$

3. Two lines $L_1 : x = 5, \frac{y}{3-\alpha} = \frac{z}{-2}$ and $L_2 : x = \alpha, \frac{y}{-1} = \frac{z}{2-\alpha}$ are coplanar.

Then, α can take value(s)

(a) 1 (b) 2 (c) 3 (d) 4

4. In a ΔPQR, P is the largest angle and $\cos P = \frac{1}{3}$. Further in circle of the triangle touches the sides PQ, QR and RP at N, L and M respectively, such that the lengths of PN, QL and RM are consecutive even integers. Then, possible length(s) of the side(s) of the triangle is (are)

(a) 16 (b) 18

(c) 24 (d) 22

5. Let $w = \frac{\sqrt{3}+i}{2}$ and $P = \{w^n : n = 1, 2, 3, \ldots\}$. Further

$H_1 = \left\{z \in C : \text{Re } z > \frac{1}{2}\right\}$ and

$H_2 = \left\{z \in C : \text{Re } z < -\frac{1}{2}\right\}$, where C is the set of all complex numbers,

if $z_1 \in P \cap H_1$, $z_2 \in P \cap H_2$ and O represents the origin, then $\angle z_1 O z_2$ is equal to

(a) $\frac{\pi}{2}$ (b) $\frac{\pi}{6}$ (c) $\frac{2\pi}{3}$ (d) $\frac{5\pi}{6}$

6. If $3^x = 4^{x-1}$, then x is equal to

(a) $\frac{2\log_3 2}{2\log_3 2 - 1}$ (b) $\frac{2}{2 - \log_2 3}$

(c) $\frac{1}{1 - \log_4 3}$ (d) $\frac{2\log_2 3}{2\log_2 3 - 1}$

7. Let ω be a complex cube root of unity with $\omega \neq 1$ and $P = [p_{ij}]$ be a $n \times n$ matrix with $p_{ij} = \omega^{i+j}$. Then, $P^2 \neq 0$, when n is equal to

(a) 57 (b) 55 (c) 58 (d) 56

8. The function $f(x) = 2|x| + |x+2| - ||x+2| - 2|x||$ has a local minimum or a local maximum at x is equal to

(a) -2 (b) $\frac{-2}{3}$ (c) 2 (d) $\frac{2}{3}$

Section II

Directions (Q. Nos. 9-16) *This section contains 4 paragraphs each describing theory, experiment, data etc., eight questions relate to four paragraphs with two questions on each paragraph. Each question of a paragraph has only one correct answer among the four choices (a), (b), (c) and (d).*

Paragraph I

Let $f : [0, 1] \to R$ (the set of all real numbers) be a function. Suppose the function f is twice differentiable, $f(0) = f(1) = 0$ and satisfies $f''(x) - 2f'(x) + f(x) \geq e^x$, $x \in [0, 1]$.

9. Which of the following is true for $0 < x < 1$?

(a) $0 < f(x) < \infty$ (b) $-\frac{1}{2} < f(x) < \frac{1}{2}$

(c) $-\frac{1}{4} < f(x) < 1$ (d) $-\infty < f(x) < 0$

10. If the function $e^{-x}f(x)$ assumes its minimum in the interval [0, 1] a $x = \frac{1}{4},$ which of the following is true?

(a) $f'(x) < f(x), \frac{1}{4} < x < \frac{3}{4}$

(b) $f'(x) > f(x), 0 < x < \frac{1}{4}$

(c) $f'(x) < f(x), 0 < x < \frac{1}{4}$

(d) $f'(x) < f(x), \frac{3}{4} < x < 1$

Paragraph II

Let PQ be a focal chord of the parabola $y^2 = 4ax$. The tangents to the parabola at P and Q meet at a point lying on the line $y = 2x + a, a > 0$.

11. Length of chord PQ is

(a) $7a$ (b) $5a$

(c) $2a$ (d) $3a$

12. If chord PQ subtends an angle θ at the vertex of $y^2 = 4ax$, then $\tan\theta$ is equal to

(a) $\frac{2}{3}\sqrt{7}$ (b) $\frac{-2}{3}\sqrt{7}$

(c) $\frac{2}{3}\sqrt{5}$ (d) $\frac{-2}{3}\sqrt{5}$

Paragraph III

Let $S = S_1 \cap S_2 \cap S_3$, where

$S_1 = \{z \in C : |z| < 4\}, S_2$

$= \left\{z \in C : \text{Im}\left[\frac{z - 1 + \sqrt{3}i}{1 - \sqrt{3}i}\right] > 0\right\}$

and $S_3 : \{z \in C : \text{Re } z > 0\}$

13. Area of S is equal to

(a) $\frac{10\pi}{3}$ (b) $\frac{20\pi}{3}$

(c) $\frac{16\pi}{3}$ (d) $\frac{32\pi}{3}$

14. $\min_{z \in S} |1 - 3i - z|$ is equal to

(a) $\frac{2 - \sqrt{3}}{2}$

(b) $\frac{2 + \sqrt{3}}{2}$

(c) $\frac{3 - \sqrt{3}}{2}$

(d) $\frac{3 + \sqrt{3}}{2}$

Paragraph IV

A box B_1 contains 1 white ball, 3 red balls and 2 black balls. Another box B_2 contains 2 white balls, 3 red balls and 4 black balls. A third box B_3 contains 3 white balls, 4 red balls and 5 black balls.

15. If 1 ball is drawn from each of the boxes B_1, B_2 and B_3, the probability that all 3 drawn balls are of the same colour is

(a) $\frac{82}{648}$ (b) $\frac{90}{648}$

(c) $\frac{558}{648}$ (d) $\frac{566}{648}$

16. If 2 balls are drawn (without replacement) from a randomly selected box and one of the balls is white and the other ball is red, the probability that these 2 balls are drawn from box B_2 is

(a) $\frac{116}{181}$ (b) $\frac{126}{181}$

(c) $\frac{65}{181}$ (d) $\frac{55}{181}$

Section III

Directions (Q. Nos. 17-20) *This section contains 4 multiple choice questions. Each question has four choices (a), (b), (c) and (d) out of which only one is correct.*

17. Match List I with List II and select the correct answer using the code given below the lists.

	List I		List II
P.	$\left[\frac{1}{y^2}\left\{\frac{\cos(\tan^{-1} y) + y\sin(\tan^{-1} y)}{\cot(\sin^{-1} y) + \tan(\sin^{-1} y)}\right\}^2 + y^4\right]^{1/2}$ takes value	1.	$\frac{1}{2}\sqrt{\frac{5}{3}}$
Q.	If $\cos x + \cos y + \cos z = 0 = \sin x + \sin y + \sin z$, then possible value of $\cos\frac{x-y}{2}$ is	2.	$\sqrt{2}$
R.	If $\cos\left(\frac{\pi}{4} - x\right)\cos 2x + \sin x \sin 2x \sec x = \cos x \sin 2x \sec x + \cos\left(\frac{\pi}{4} + x\right)\cos 2x$, then possible value of $\sec x$ is	3.	$\frac{1}{2}$
S.	If $\cot\left(\sin^{-1}\sqrt{1-x^2}\right) = \sin[\tan^{-1}(x\sqrt{6})]$, $x \neq 0$. Then, possible value of x is	4.	1

Codes

	P	Q	R	S		P	Q	R	S
(a)	4	3	2	1	(b)	4	3	2	1
(c)	3	4	2	1	(d)	3	4	1	2

18. A line $L : y = mx + 3$ meets y-axis at $E(0, 3)$ and the arc of the parabola $y^2 = 16x, 0 \le y \le 6$ at the point $F(x_0, y_0)$. The tangent to the parabola at $F(x_0, y_0)$ intersects the y-axis at $G(0, y_1)$. The slope m of the line L is chosen such that the area of the ΔEFG has a local maximum

Match List I with List II and select the correct answer using the code given below the lists

	List I		List II
P.	$m =$	1.	1/2
Q.	Maximum area of ΔEFG is	2.	4
R.	$y_0 =$	3.	2
S.	$y_1 =$	4.	1

Codes

	P	Q	R	S		P	Q	R	S
(a)	4	1	2	3	(b)	3	4	1	2
(c)	1	3	2	4	(d)	1	3	4	2

19. Match List I with List II and select the correct answer using the code given below the lists

	List I		List II
P.	Volume of parallelopiped determined by vectors **a**, **b** and **c** is 2. Then, the volume of the parallelopiped determined by vectors 2(**a** × **b**), 3(**b** × **c**) and (**c** × **a**) is	1.	100
Q.	Volume of parallelopiped determined by vectors **a**, **b** and **c** is 5. Then, the volume of the parallelopiped determined by vectors 3(**a** + **b**), (**b** + **c**) and 2(**c** + **a**) is	2.	30
R.	Area of a triangle with adjacent sides determined by vector **a** and **b** is 20. Then, the area of the triangle with adjacent sides determined by vectors (2**a** + 3**b**) and (**a** − **b**) is	3.	24
S.	Area of a parallelogram with adjacent sides determined by vectors **a** and **b** is 30. Then, the area of the parallelogram with adjacent sides determined by vectors (**a** + **b**) and **a** is	4.	60

Codes

	P	Q	R	S		P	Q	R	S
(a)	4	2	3	1	(b)	2	3	1	4
(c)	3	4	1	2	(d)	1	4	3	2

20. Consider the lines

$$L_1 : \frac{x-1}{2} = \frac{y}{-1} = \frac{z+3}{1},$$

$$L_2 : \frac{x-4}{1} = \frac{y+3}{1} = \frac{z+3}{2}$$

and the planes $P_1 : 7x + y + 2z = 3$, $P_2 : 3x + 5y - 6z = 4$. Let $ax + by + cz = d$ the equation of the plane passing through the point of intersection of lines L_1 and L_2 and perpendicular to planes P_1 and P_2.

Match List I with List II and select the correct answer using the code given below the lists.

List I		List II	
P.	$a =$	1.	13
Q.	$b =$	2.	-13
R.	$c =$	3.	1
S.	$d =$	4.	-2

Codes

	P	Q	R	S		P	Q	R	S
(a)	3	2	4	1	(b)	1	3	4	2
(c)	3	2	1	4	(d)	2	4	1	3

Answers

Paper 1

Physics

1. (a) 2. (c) 3. (b) 4. (d) 5. (c) 6. (a) 7. (a) 8. (b) 9. (b) 10. (d)
11. (b,d) 12. (b,c) 13. (b,d) 14. (a,c) 15. (a,d) 16. (8) 17. (1) 18. (5) 19. (8) 20. (4)

Chemistry

1. (c) 2. (b) 3. (d) 4. (b) 5. (d) 6. (b) 7. (a) 8. (d) 9. (b) 10. (a)
11. (b,c,d) 12. (b,d) 13. (a) 14. (a) 15. (a,b,c,d) 16. (6) 17. (2) 18. (5) 19. (8) 20. (4)

Mathematics

1. (b) 2. (c) 3. (c) 4. (a) 5. (d) 6. (a) 7. (b) 8. (a) 9. (d) 10. (c)
11. (a,c) 12. (a,d) 13. (b,d) 14. (b,c) 15. (c,d) 16. (9) 17. (6) 18. (5) 19. (6) 20. (5)

Paper 2

Physics

1. (b,d) 2. (a,d) 3. (a,d) 4. (a,b) 5. (d) 6. (c,d) 7. (b,c,d) 8. (a,c) 9. (b) 10. (a)
11. (b) 12. (a) 13. (b) 14. (b) 15. (c) 16. (a) 17. (d) 18. (c) 19. (a) 20. (c)

Chemistry

1. (b) 2. (b,d) 3. (b) 4. (c) 5. (a,c,d) 6. (a,b) 7. (c,d) 8. (a,b,c,d) 9. (a) 10. (d)
11. (b) 12. (a) 13. (c) 14. (b) 15. (a) 16. (a) 17. (d) 18. (a) 19. (a) 20. (d)

Mathematics

1. (b,d) 2. (a,c) 3. (a,d) 4. (b,d) 5. (c,d) 6. (a,b,c) 7. (b,c,d) 8. (a,b) 9. (b) 10. (c)
11. (b) 12. (d) 13. (b) 14. (c) 15. (a) 16. (d) 17. (b) 18. (a) 19. (c) 20. (a)

JEE Main
Unsolved Paper 2014

M. Marks : 360 Time : 3 Hrs

Instructions

- This test consists of 90 questions.
- There are three parts in the question paper A, B, C consisting of Physics, Chemistry and Mathematics having 30 questions in each part of equal weightage. Each question is allotetted 4 marks for correct response.
- Candidates will be awarded marks as stated above for correct response of each question. 1/4 marks will be deducted for indicating incorrect response of each question. No deduction from the total score will be made if no response is indicated for an item in the answer sheet.
- There is only one correct response for each question. Filling up more than one response in any question will be treated as wrong response and marks for wrong response will be deducted according as per instructions.

Physics

1. The current voltage relation of diode is given by $I = (e^{1000V/T} - 1)$ mA, where the applied voltage V is in volt and the temperature T is in kelvin. If a student makes an error measuring ± 0.01 V while measuring the current of 5 mA at 300K, what will be the error in the value of current in mA?

(a) 0.2 mA (b) 0.02 mA
(c) 0.5 mA (d) 0.05 mA

2. From a tower of height H, a particle is thrown vertically upwards with a speed u. The time taken by the particle to hit the ground, is n times that taken by it to reach the highest point of its path. The relation between H, u and n is

(a) $2gH = n^2u^2$
(b) $gH = (n-2)^2u^2$
(c) $2gH = nu^2(n-2)$
(d) $gH = (n-2)^2u^2$

3. A mass m supported by a massless string wound around a uniform hollow cylinder of mass m and radius R. If the string does not slip on the cylinder, with what acceleration will the mass fall on release?

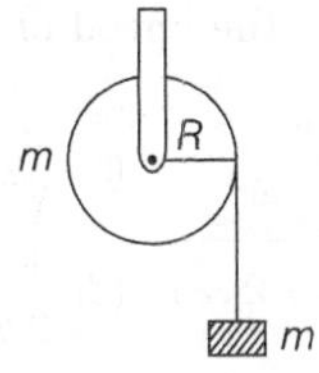

(a) $2g/3$ (b) $g/2$ (c) $5g/6$ (d) g

4. A block of mass m is placed on a surface with a vertical cross-section given by $y = x^3/6$. If the coefficient of friction is 0.5, the maximum height above the ground at which the block can be placed without slipping is

(a) $\frac{1}{6}$ m (b) $\frac{2}{3}$ m (c) $\frac{1}{3}$ m (d) $\frac{1}{2}$ m

5. When a rubber band is stretched by a distance x, it exerts a restoring force of magnitude $F = ax + bx^2$, where a and b are constants. The work done in stretching the unstretched rubber-band by L is

(a) $aL^2 + bL^3$

(b) $\frac{1}{2}(aL^2 + bL^3)$

(c) $\frac{aL^2}{2} + \frac{bL^3}{3}$

(d) $\frac{1}{2}\left(\frac{aL^2}{2} + \frac{bL^3}{3}\right)$

6. A bob of mass m attached to an inextensible string of length l is suspended from a vertical support. The bob rotates in a horizontal circle with an angular speed ω rad/s about the vertical support. About the point of suspension

(a) angular momentum is conserved

(b) angular momentum changes in magnitude but not in direction

(c) angular momentum changes in direction but not in magnitude

(d) angular momentum changes both in direction and magnitude

7. Four particles, each of mass M and equidistant from each other, move along a circle of radius R under the action of their mutual gravitational attraction, the speed of each particle is

(a) $\sqrt{\frac{GM}{R}}$ (b) $\sqrt{2\sqrt{2}\frac{GM}{R}}$

(c) $\sqrt{\frac{GM}{R}(1 + 2\sqrt{2})}$ (d) $\frac{1}{2}\sqrt{\frac{GM}{R}(1 + 2\sqrt{2})}$

8. The pressure that has to be applied to the ends of a steel wire of length 10 cm to keep its length constant when its temperature is raised by 100°C is (For steel, Young's modulus is $2 \times 10^{11}\,\text{Nm}^{-2}$ and coefficient of thermal expansion is $1.1 \times 10^{-5}\,\text{K}^{-1}$)

(a) 2.2×10^8 Pa (b) 2.2×10^9 Pa

(c) 2.2×10^7 Pa (d) 2.2×10^6 Pa

9. There is a circular tube in a vertical plane. Two liquids which do not mix and of densities d_1 and d_2 are filled in the tube. Each liquid subtends 90° angle at centre. Radius joining their interface makes an angle α with vertical. Ratio d_1/d_2 is

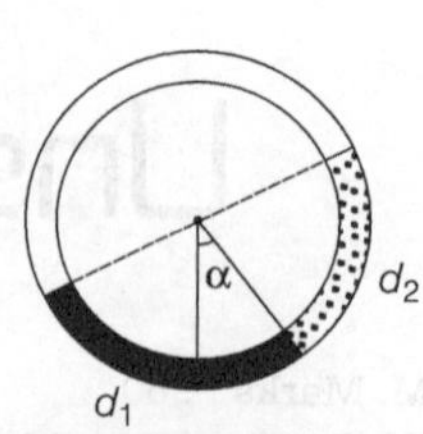

(a) $\frac{1 + \sin\alpha}{1 - \sin\alpha}$ (b) $\frac{1 + \cos\alpha}{1 - \cos\alpha}$

(c) $\frac{1 + \tan\alpha}{1 - \tan\alpha}$ (d) $\frac{1 + \sin\alpha}{1 - \cos\alpha}$

10. On heating water, bubbles beings formed at the bottom of the vessel detach and rise. Take the bubbles to be spheres of radius R and making a circular contact of radius r with the bottom of the vessel. If $r << R$ and the surface tension of water is T, value of r just before bubbles detach is (density of water is ρ)

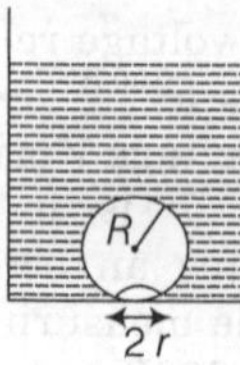

(a) $R^2\sqrt{\frac{\rho_w g}{3T}}$ (b) $R^2\sqrt{\frac{\rho_w g}{6T}}$

(c) $R^2\sqrt{\frac{\rho_w g}{T}}$ (d) $R^2\sqrt{\frac{3\rho_w g}{T}}$

11. Three rods of copper, brass and steel are welded together to form a Y-shaped structure. Area of cross-section of each rod is $4\,\text{cm}^2$. End of copper rod is maintained at 100°C whereas ends of brass and steel are kept at 0°C. Lengths of the copper, brass and steel rods are 46, 13 and 12 cm respectively.

The rods are thermally insulated from surroundings except at ends. Thermal conductivities of copper, brass and steel are 0.92, 0.26 and 0.12 in CGS units, respectively. Rate of heat flow through copper rod is

(a) 1.2 cal/s
(b) 2.4 cal/s
(c) 4.8 cal/s
(d) 6.0 cal/s

12. One mole of diatomic ideal gas undergoes a cyclic process ABC as shown in figure. The process BC is adiabatic. The temperatures at A, B and C are 400 K, 800 K and 600 K, respectively. Choose the correct statement.

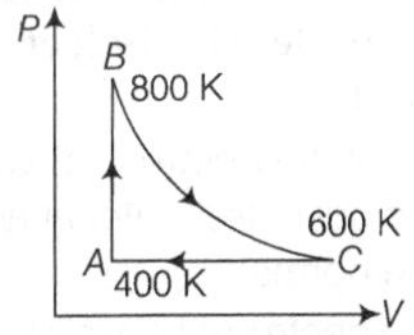

(a) The change in internal energy in whole cyclic process is 250 R
(b) The change in internal energy in the process CA is 700 R
(c) The change in internal energy in the process AB is $-350\,R$
(d) The change in internal energy in the process BC is $-500\,R$

13. An open glass tube is immersed in mercury in such a way that a length of 8 cm extends above the mercury level. The open end of the tube is then closed and sealed and the tube is raised vertically up by additional 46 cm. What will be length of the air column above mercury in the tube now? (Atmospheric pressure = 76 cm of Hg)

(a) 16 cm (b) 22 cm
(c) 38 cm (d) 6 cm

14. A particle moves with simple harmonic motion in a straight line. In first τ sec, after starting from rest it travels a distance a and in next τ sec, it travels $2a$, in same direction, then

(a) amplitude of motion is $3a$
(b) time period of oscillations is 8π
(c) amplitude of motion is $4a$
(d) time period of oscillations is 6π

15. A pipe of length 85 cm is closed from one end. Find the number of possible natural oscillations of air column in the pipe whose frequencies lie below 1250 Hz. The velocity of sound in air is 340 m/s.

(a) 12 (b) 8 (c) 6 (d) 4

16. Assume that an electric field $\mathbf{E} = 30x^2\,\hat{\mathbf{i}}$ exists in space. Then, the potential difference $V_A - V_O$, where V_O is the potential at the origin and V_A the potential at $x = 2$ m is

(a) 120 J (b) -120 J
(c) -80 J (d) 80 J

17. A parallel plate capacitor is made of two circular plates separated by a distance of 5 mm and with a dielectric of dielectric constant 2.2 between them. When the electric field in the dielectric is 3×10^4 V/m, the charge density of the positive plate will be close to

(a) 6×10^{-7} C/m^2
(b) 3×10^{-7} C/m^2
(c) 3×10^{4} C/m^2
(d) 6×10^{4} C/m^2

18. In a large building, there are 15 bulbs of 40 W, 5 bulbs of 100 W, 5 fans of 80 W and 1 heater of 1 kW. The voltage of the electric mains is 220 V. The minimum capacity of the main fuse of the building will be

(a) 8 A (b) 10 A (c) 12 A (d) 14 A

19. A conductor lies along the z-axis at $-1.5 \le z < 1.5$ m and carries a fixed current of 10.0 A in $-a_z$ direction (see figure). For a field $\mathbf{B} = 3.0 \times 10^{-4} e^{-0.2x} a_y$ T, find the power required to move the conductor at constant speed to $x = 2.0$ m, $y = 0$ in 5×10^{-3} s. Assume parallel motion along the x-axis.

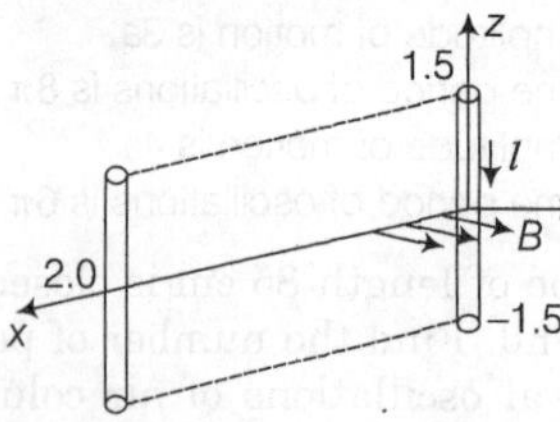

(a) 1.57 W (b) 2.97 W
(c) 14.85 W (d) 29.7 W

20. The coercivity of a small magnet where the ferromagnet gets demagnetised is 3×10^3 Am^{-1}. The current required to be passed in a solenoid of length 10 cm and number of turns 100, so that the magnet gets demagnetised when inside the solenoid is

(a) 30 mA (b) 60 mA
(c) 3 A (d) 6 A

21. In the circuit shown here, the point C is kept connected to point A till the current flowing through the circuit becomes constant. Afterward, suddenly point C is disconnected from point A and connected to point B at time $t = 0$. Ratio of the voltage across resistance and the inductor at $t = L/R$ will be equal to

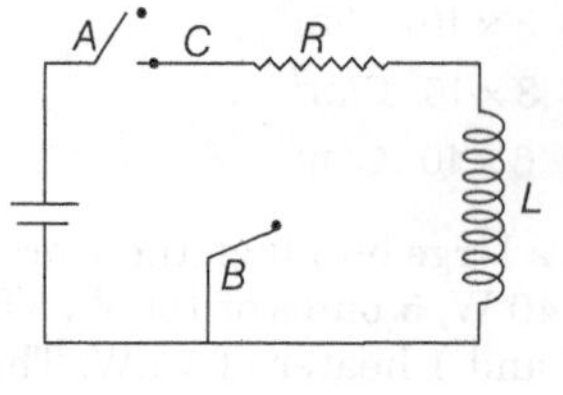

(a) $\frac{e}{1-e}$ (b) 1
(c) −1 (d) $\frac{1-e}{e}$

22. During the propagation of electromagnetic waves in a medium,

(a) electric energy density is double of the magnetic energy density
(b) electric energy density is half of the magnetic energy density
(c) electric energy density is equal to the magnetic energy density
(d) Both electric and magnetic energy densities are zero

23. A thin convex lens made from crown glass ($\mu = 3/2$) has focal length f. When it is measured in two different liquids having refractive indices 4/3 and 5/3. It has the focal lengths f_1 and f_2, respectively. The correct relation between the focal length is

(a) $f_1 = f_2 < f$
(b) $f_1 > f$ and f_2 becomes negative
(c) $f_2 > f$ and f_1 becomes negative
(d) f_1 and f_2 both become negative

24. A green light is incident from the water to the air-water interface at the critical angle (θ). Select the correct statement.

(a) The entire spectrum of visible light will come out of the water at an angle of 90° to the normal
(b) The spectrum of visible light whose frequency is less than that of green light will come out of the air medium.
(c) The spectrum of visible light whose frequency is more than that of green light will come out to the air medium.
(d) The entire spectrum of visible light will come out of the water at various angles to the normal.

25. Two beams, A and B, of plane polarised light with mutually perpendicular planes of polarisation are seen through a polaroid. From the position when the beam A has maximum intensity (and beam B has zero intensity), a rotation of polaroid through 30° makes the two beams appear equally bright. If the initial intensities of the two beams are I_A and I_B respectively, then I_A / I_B equals

(a) 3 (b) 3/2
(c) 1 (d) 1/3

26. The radiation corresponding to $3 \rightarrow 2$ transition of hydrogen atom falls on a metal surface to produce photoelectrons. These electrons are made to enter a magnetic field of 3×10^{-4} T. If the radius of the largest circular path followed by these electrons is 10.0 mm, the work function of the metal is close to

(a) 1.8 eV (b) 1.1 eV
(c) 0.8 eV (d) 1.6 eV

27. Hydrogen $({}_1H^1)$, deuterium $({}_1H^2)$, singly ionised helium $({}_2He^4)^+$ and doubly ionised lithium $({}_3Li^8)^{++}$ all have one electron around the nucleus. Consider an electron transition from $n = 2$ to $n = 1$. If the wavelengths of emitted radiation are $\lambda_1, \lambda_2, \lambda_3$ and λ_4, respectively for four elements, then approximately which one of the following is correct?

(a) $4\lambda_1 = 2\lambda_2 = 2\lambda_3 = \lambda_4$
(b) $\lambda_1 = 2\lambda_2 = 2\lambda_3 = \lambda_4$
(c) $\lambda_1 = \lambda_2 = 4\lambda_3 = 9\lambda_4$
(d) $\lambda_1 = 2\lambda_2 = 3\lambda_3 = 4\lambda_4$

28. The forward biased diode connection is

(a) 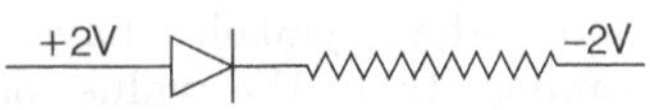

(b) 

(c) 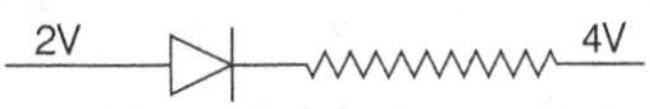

(d) 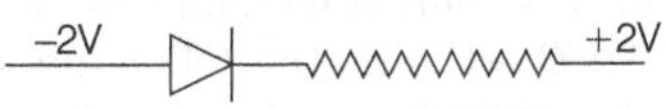

29. A student measured the length of a rod and wrote it as 3.50 cm. Which instrument did he use to measure it?

(a) A meter scale
(b) A vernier calliper where the 10 divisions in vernier scale matches with 9 divisions in main scale and main scale has 10 divisions in 1 cm
(c) A screw gauge having 100 divisions in the circular scale and pitch as 1 mm
(d) A screw gauge having 50 divisions in the circular scale and pitch as 1 mm

30. Match List I (Electromagnetic wave type) with List II (Its association/application) and select the correct option from the choices given below the lists.

List I	List II
A. Infrared waves	1. To treat muscular strain
B. Radio waves	2. For broadcasting
C. X-rays	3. To detect fracture of bones
D. Ultraviolet	4. Absorbed by the ozone layer of the atmosphere

Codes

	A	B	C	D
(a)	4	3	2	1
(b)	1	2	4	3
(c)	3	2	1	4
(d)	1	2	3	4

Chemistry

1. The correct set of four quantum numbers for the valence electrons of rubidium atom $(Z = 37)$ is

(a) $5, 0, 0, +\frac{1}{2}$ (b) $5, 1, 0, +\frac{1}{2}$
(c) $5, 1, ,1, +\frac{1}{2}$ (d) $5, 0, 1, +\frac{1}{2}$

2. If Z is a compressibility factor, van der Waals' equation at low pressure can be written as

(a) $Z = 1 + \frac{RT}{pb}$ (b) $Z = 1 - \frac{a}{VRT}$
(c) $Z = 1 - \frac{pb}{RT}$ (d) $Z = 1 + \frac{pb}{RT}$

3. CsCl crystallises in body centred cubic lattice. If '*a*' its edge length, then which of the following expressions is correct?

(a) $r_{Cs^+} + r_{Cl^-} = 3a$
(b) $r_{Cs^+} + r_{Cl^-} = \frac{3a}{2}$
(c) $r_{Cs^+} + r_{Cl^-} = \frac{\sqrt{3}}{2}a$
(d) $r_{Cs^+} + r_{Cl^-} = \sqrt{3}a$

4. For the estimation of nitrogen, 1.4 g of an organic compound was digested by Kjeldahl's method and the evolved ammonia was absorbed in 60 mL of $\frac{M}{10}$ sulphuric acid. The unreacted acid required 20 mL of $\frac{M}{10}$ sodium hydroxide for complete neutralisation. The percentage of nitrogen in the compound is

(a) 6% (b) 10% (c) 3% (d) 5%

5. Resistance of 0.2 M solution of an electrolyte is 50 Ω. The specific conductance of the solution of 0.5 M solution of same electrolyte is 1.4 S m^{-1} and resistance of same solution of the same electrolyte is 280 Ω. The molar conductivity of 0.5 M solution of the electrolyte in S $m^2 mol^{-1}$ is

(a) 5×10^{-4} (b) 5×10^{-3}
(c) 5×10^{3} (d) 5×10^{2}

6. For the complete combustion of ethanol, $C_2H_5OH(l) + 3O_2(g) \longrightarrow 2CO_2(g) + 3H_2O(l)$, the amount of heat produced as measured in bomb calorimeter, is 1364.47 kJ mol^{-1} at 25°C. Assuming ideality the enthalpy of combustion, $\Delta_C H$, for the reaction will be ($R = 8.314 \text{ J K}^{-1}\text{mol}^{-1}$)

(a) -1366.95 kJ mol^{-1}
(b) -1361.95 kJ mol^{-1}
(c) -1460.50 kJ mol^{-1}
(d) -1350.50 kJ mol^{-1}

7. The equivalent conductance of NaCl at concentration C and at infinite dilution are λ_C and λ_∞, respectively. The correct relationship between λ_C and λ_∞ is given as (where, the constant B is positive)

(a) $\lambda_C = \lambda_\infty + (B)C$
(b) $\lambda_C = \lambda_\infty - (B)C$
(c) $\lambda_C = \lambda_\infty - (B)\sqrt{C}$
(d) $\lambda_C = \lambda_\infty + (B)\sqrt{C}$

8. Consider separate solution of 0.500 M $C_2H_5OH(aq)$, 0.100 M $Mg_3(PO_4)_2(aq)$, 0.250 M $KBr(aq)$ and 0.125 M $Na_3PO_4(aq)$ at 25°C. Which statement is true about these solution, assuming all salts to be strong electrolytes?

(a) They all have the same osmotic pressure
(b) 0.100 M $Mg_3(PO_4)_2(aq)$ has the highest osmotic pressure
(c) 0.125 M $Na_3PO_4(aq)$ has the highest osmotic pressure
(d) 0.500 M C_2H_5OH (aq) has the highest osmotic pressure

9. For the reaction,

$$SO_2(g) + \frac{1}{2}O_2(g) \rightleftharpoons SO_3(g)$$

if $K_p = K_C(RT)^x$

where, the symbols have usual meaning, then the value of x is (assuming ideality)

(a) -1 (b) $-\frac{1}{2}$ (c) $\frac{1}{2}$ (d) 1

10. For the non-stoichiometric reaction $2A + B \rightarrow C + D$, the following kinetic data were obtained in three separate experiments, all at 298 K.

	Initial concentration (*A*)	Initial concentration (*B*)	Initial rate of formation of *C* (mol $L^{-1}S^{-1}$)
(i)	0.1 M	0.1 M	1.2×10^{-3}
(ii)	0.1 M	0.2 M	1.2×10^{-3}
(iii)	0.2 M	0.1 M	2.4×10^{-3}

The rate law for the formation of C is

(a) $\frac{dC}{dt} = k[A][B]$ (b) $\frac{dC}{dt} = k[A]^2[B]$
(c) $\frac{dC}{dt} = k[A][B]^2$ (d) $\frac{dC}{dt} = k[A]$

11. Among the following oxoacids, the correct decreasing order of acid strength is
(a) $HOCl > HClO_2 > HClO_3 > HClO_4$
(b) $HClO_4 > HOCl > HClO_2 > HClO_3$
(c) $HClO_4 > HClO_3 > HClO_2 > HOCl$
(d) $HClO_2 > HClO_4 > HClO_3 > HOCl$

12. The metal that cannot be obtained by electrolysis of an aqueous solution of its salts is
(a) Ag (b) Ca (c) Cu (d) Cr

13. The octahedral complex of a metal ion M^{3+} with four monodentate ligands L_1, L_2, L_3 and L_4 absorb wavelengths in the region of red, green, yellow and blue, respectively. The increasing order of ligand strength of the four ligands is
(a) $L_4 < L_3, L_2 < L_1$
(b) $L_1 < L_3 < L_2 < L_4$
(c) $L_3 < L_2 < L_4 < L_1$
(d) $L_1 < L_2 < L_4 < L_3$

14. Which of the following properties is not shown by NO?
(a) It is diamagnetic in gaseous state
(b) It is a neutral oxide
(c) It combines with oxygen to form nitrogen dioxide
(d) Its bond order is 2.5

15. In which of the following reactions H_2O_2 acts as a reducing agent?
I. $H_2O_2 + 2H^+ + 2e^- \longrightarrow 2H_2O$
II. $H_2O_2 - 2e^- \longrightarrow O_2 + 2H^+$
III. $H_2O_2 + 2e^- \longrightarrow 2OH^-$
IV. $H_2O_2 + 2OH^- - 2e^- \rightarrow O_2 + 2H_2O$
(a) I and II (b) III and IV
(c) I and III (d) II and IV

16. The correct statement for the molecule, CsI_3 is
(a) it is a covalent molecule
(b) it contains Cs^+ and I_3^-
(c) it contains Cs^{3+} and I^- ions
(d) it contains Cs^+, I^- and lattice I_2 molecule

17. The ratio of masses of oxygen and nitrogen of a particular gaseous mixture is 1 : 4. The ratio of number of their molecule is
(a) 1 : 4
(b) 7 : 32
(c) 1 : 8
(d) 3 : 16

18. Given below are the half-cell reactions
$$Mn^{2+} + 2e^- \longrightarrow Mn \,; E° = -1.18 \text{ eV}$$
$$2(Mn^{3+} + e^- \longrightarrow Mn^{2+});$$
$$E° = +1.51 \text{ eV}$$
The $E°$ for $3Mn^{2+} \rightarrow Mn + 2Mn^{3+}$ will be
(a) -2.69 V; the reaction will not occur
(b) -2.69 V; the reaction will occur
(c) -0.33 V; the reaction will not occur
(d) -0.33 V; the reaction will occur

19. Which series of reactions correctly represents chemical relations related to iron and its compound?
(a) $Fe \xrightarrow{\text{Dil. } H_2SO_4} FeSO_4 \xrightarrow{H_2SO_4, O_2} Fe_2(SO_4)_3 \xrightarrow{\text{Heat}} Fe$
(b) $Fe \xrightarrow{O_2, \text{ Heat}} FeO \xrightarrow{\text{Dil. } H_2SO_4} FeSO_4 \xrightarrow{\text{Heat}} Fe$
(c) $Fe \xrightarrow{Cl_2, \text{ Heat}} FeCl_3 \xrightarrow{\text{Heat, air}} FeCl_2 \xrightarrow{Zn} Fe$
(d) $Fe \xrightarrow{O_2, \text{ Heat}} Fe_3O_4 \xrightarrow{CO, 600°C} FeO \xrightarrow{CO, 700°C} Fe$

20. The equation which is balanced and represents the correct product(s) is
(a) $Li_2O + 2KCl \longrightarrow 2LiCl + K_2O$
(b) $[CoCl(NH_3)_5]^+ + 5H^+ \longrightarrow Co^{2+} + 5NH_4^+ + Cl^-$
(c) $[Mg(H_2O)_6]^{2+} + (EDTA)^{4-} \xrightarrow{\text{Excess NaOH}} [Mg(EDTA)]^{2+} + 6H_2O$
(d) $CuSO_4 + 4KCN \longrightarrow K_2[Cu(CN)_4] + K_2SO_4$

21. In S_N2 reactions, the correct order of reactivity for the following compounds CH_3Cl, CH_3CH_2Cl, $(CH_3)_2CHCl$ and $(CH_3)_3CCl$ is

(a) $CH_3Cl > (CH_3)_2CHCl > CH_3CH_2Cl > (CH_3)_3CCl$

(b) $CH_3Cl > CH_3CH_2Cl > (CH_3)_2CHCl > (CH_3)_3CCl$

(c) $CH_3CH_2Cl > CH_3Cl > (CH_3)_2CHCl > (CH_3)_3CCl$

(d) $(CH_3)_2CHCl > CH_3CH_2Cl > CH_3Cl > (CH_3)_3CCl$

22. On heating an aliphatic primary amine with chloroform and ethanolic potassium hydroxide, the organic compound formed is

(a) an alkanol
(b) an alkanediol
(c) an alkyl cyanide
(d) an alkyl isocyanide

23. The most suitable reagent for the conversion of $R—CH_2—OH \longrightarrow R—CHO$ is

(a) $KMnO_4$
(b) $K_2Cr_2O_7$
(c) CrO_3
(d) PCC (pyridinium chlorochromate)

24. The major organic compound formed by the reaction of 1,1,1-trichloroethane with silver powder is

(a) acetylene
(b) ethene
(c) 2-butyne
(d) 2-butene

25. Sodium phenoxide when heated with CO_2 under pressure at 125°C yields a product which on acetylation produces C.

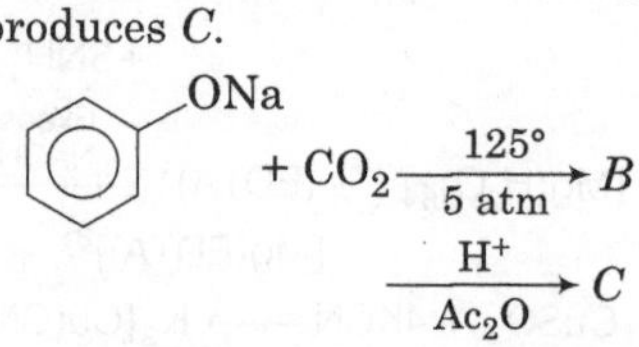

The major product C would be

(a) 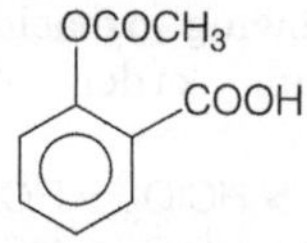

(b) 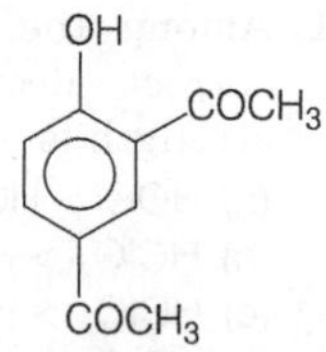

(c) 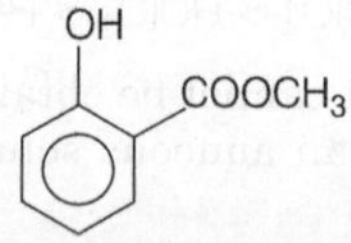

(d) 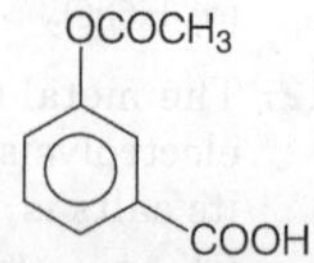

26. Considering the basic strength of amines in aqueous solution, which one has the smallest pK_b value?

(a) $(CH_3)_2NH$ (b) CH_3NH_2
(c) $(CH_3)_3N$ (d) $C_6H_5NH_2$

27. For which of the following molecule significant $\mu \neq 0$?

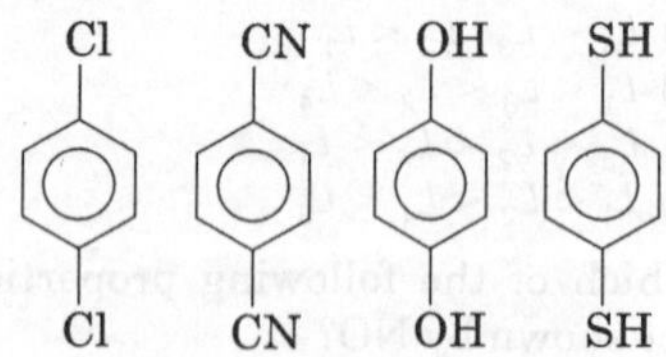

I II III IV

(a) Only I (b) I and II
(c) Only III (d) III and IV

28. Which one is classified as a condensation polymer?

(a) Dacron (b) Neoprene
(c) Teflon (d) Acrylonitrile

29. Which one of the following bases is not present in DNA?

(a) Quinoline (b) Adenine
(c) Cytosine (d) Thymine

30. In the reaction,

$$CH_3COOH \xrightarrow{LiAlH_4} A \xrightarrow{PCl_5} B \xrightarrow{\text{Alc. KOH}} C$$

The product C is

(a) acetaldehyde
(b) acetylene
(c) ethylene
(d) acetyl chloride

Mathematics

1. If $X = \{4^n - 3n - 1 : n \in N\}$ and $Y = \{9(n-1) : n \in N\}$; where N is the set of natural numbers, then $X \cup Y$ is equal to

(a) N (b) $Y - X$
(c) X (d) Y

2. If z is a complex number such that $|z| \geq 2$, then the minimum value of $\left|z + \frac{1}{2}\right|$

(a) is equal to 5/2
(b) lies in the interval (1, 2)
(c) is strictly greater than 5/2
(d) is strictly greater than 3/2 but less than 5/2

3. If $a \in R$ and the equation $-3(x-[x])^2 + 2(x-[x]) + a^2 = 0$ (where, $[x]$ denotes the greatest integer $\leq x$) has no integral solution, then all possible values of a lie in the interval

(a) $(-1, 0) \cup (0, 1)$ (b) $(1, 2)$
(c) $(-2, -1)$ (d) $(-\infty, -2) \cup (2, \infty)$

4. Let α and β be the roots of equation $px^2 + qx + r = 0$, $p \neq 0$. If p, q and r are in AP and $\frac{1}{\alpha} + \frac{1}{\beta} = 4$, then the value of $|\alpha - \beta|$ is

(a) $\frac{\sqrt{61}}{9}$ (b) $\frac{2\sqrt{17}}{9}$
(c) $\frac{\sqrt{34}}{9}$ (d) $\frac{2\sqrt{13}}{9}$

5. If $\alpha, \beta \neq 0$ and $f(n) = \alpha^n + \beta^n$ and

$$\begin{vmatrix} 3 & 1+f(1) & 1+f(2) \\ 1+f(1) & 1+f(2) & 1+f(3) \\ 1+f(2) & 1+f(3) & 1+f(4) \end{vmatrix} = K(1-\alpha)^2(1-\beta)^2(\alpha-\beta)^2,$$ then K is equal to

(a) $\alpha\beta$ (b) $\frac{1}{\alpha\beta}$
(c) 1 (d) -1

6. If A is a 3×3 non-singular matrix such that $AA^T = A^T A$ and $B = A^{-1}A^T$, then BB^T is equal to

(a) $I + B$ (b) I (c) B^{-1} (d) $(B^{-1})^T$

7. If the coefficients of x^3 and x^4 in the expansion of $(1 + ax + bx^2)(1 - 2x)^{18}$ in powers of x are both zero, then (a, b) is equal to

(a) $\left(16, \frac{251}{3}\right)$ (b) $\left(14, \frac{251}{3}\right)$
(c) $\left(14, \frac{272}{3}\right)$ (d) $\left(16, \frac{272}{3}\right)$

8. The angle between the lines whose direction cosines satisfy the equations $l + m + n = 0$ and $l^2 = m^2 + n^2$ is

(a) $\frac{\pi}{3}$ (b) $\frac{\pi}{4}$
(c) $\frac{\pi}{6}$ (d) $\frac{\pi}{2}$

9. If $(10)^9 + 2(11)^1(10)^8 + 3(11)^2(10)^7 + \ldots + 10(11)^9 = k(10)^9$, then k is equal to

(a) $\frac{121}{10}$ (b) $\frac{441}{100}$ (c) 100 (d) 110

10. Three positive numbers form an increasing GP. If the middle term in this GP is doubled, then new numbers are in AP. Then, the common ratio of the GP is

(a) $\sqrt{2} + \sqrt{3}$ (b) $3 + \sqrt{2}$
(c) $2 - \sqrt{3}$ (d) $2 + \sqrt{3}$

11. $\lim_{x \to 0} \frac{\sin(\pi \cos^2 x)}{x^2}$ is equal to

(a) $\frac{\pi}{2}$ (b) 1
(c) $-\pi$ (d) π

12. If g is the inverse of a function f and $f'(x) = \frac{1}{1 + x^5}$, then $g'(x)$ is equal to

(a) $1 + x^5$ (b) $5x^4$
(c) $\frac{1}{1 + \{g(x)\}^5}$ (d) $1 + \{g(x)\}^5$

13. If f and g are differentiable functions in $(0, 1)$ satisfying $f(0) = 2 = g(1)$, $g(0) = 0$ and $f(1) = 6$, then for some $c \in]0, 1[$

(a) $2f'(c) = g'(c)$ (b) $2f'(c) = 3g'(c)$

(c) $f'(c) = g'(c)$ (d) $f'(c) = 2g'(c)$

14. If $x = -1$ and $x = 2$ are extreme points of $f(x) = \alpha \log|x| + \beta x^2 + x$, then

(a) $\alpha = -6, \beta = \frac{1}{2}$ (b) $\alpha = -6, \beta = -\frac{1}{2}$

(c) $\alpha = 2, \beta = -\frac{1}{2}$ (d) $\alpha = 2, \beta = \frac{1}{2}$

15. The integral $\int \left(1 + x - \frac{1}{x}\right) e^{x + \frac{1}{x}} dx$ is equal to

(a) $(x - 1) e^{x + \frac{1}{x}} + C$

(b) $x e^{x + \frac{1}{x}} + C$

(c) $(x + 1) e^{x + \frac{1}{x}} + C$

(d) $-x e^{x + \frac{1}{x}} + C$

16. The integral $\int_0^{\pi} \sqrt{1 + 4\sin^2 \frac{x}{2} - 4\sin \frac{x}{2}}\, dx$ is equal to

(a) $\pi - 4$ (b) $\frac{2\pi}{3} - 4 - 4\sqrt{3}$

(c) $4\sqrt{3} - 4$ (d) $4\sqrt{3} - 4 - \frac{\pi}{3}$

17. The area of the region described by $A = \{(x, y) : x^2 + y^2 \le 1 \text{ and } y^2 \le 1 - x\}$ is

(a) $\frac{\pi}{2} + \frac{4}{3}$ (b) $\frac{\pi}{2} - \frac{4}{3}$

(c) $\frac{\pi}{2} - \frac{2}{3}$ (d) $\frac{\pi}{2} + \frac{2}{3}$

18. Let the population of rabbits surviving at a time t be governed by the differential equation $\frac{dp(t)}{dt} = \frac{1}{2} p(t) - 200$. If $p(0) = 100$, then $p(t)$ is equal to

(a) $400 - 300e^{\frac{t}{2}}$ (b) $300 - 200e^{-\frac{t}{2}}$

(c) $600 - 500e^{\frac{t}{2}}$ (d) $400 - 300e^{-\frac{t}{2}}$

19. If PS is the median of the triangle with vertices $P(2, 2)$, $Q(6, -1)$ and $R(7, 3)$, then equation of the line passing through $(1, -1)$ and parallel to PS is

(a) $4x - 7y - 11 = 0$

(b) $2x + 9y + 7 = 0$

(c) $4x + 7y + 3 = 0$

(d) $2x - 9y - 11 = 0$

20. Let a, b, c and d be non-zero numbers. If the point of intersection of the lines $4ax + 2ay + c = 0$ and $5bx + 2by + d = 0$ lies in the fourth quadrant and is equidistant from the two axes, then

(a) $2bc - 3ad = 0$ (b) $2bc + 3ad = 0$

(c) $2ad - 3bc = 0$ (d) $3bc + 2ad = 0$

21. The locus of the foot of perpendicular drawn from the centre of the ellipse $x^2 + 3y^2 = 6$ on any tangent to it is

(a) $(x^2 - y^2)^2 = 6x^2 + 2y^2$

(b) $(x^2 - y^2)^2 = 6x^2 - 2y^2$

(c) $(x^2 + y^2)^2 = 6x^2 + 2y^2$

(d) $(x^2 + y^2)^2 = 6x^2 - 2y^2$

22. Let C be the circle with centre at $(1, 1)$ and radius 1. If T is the circle centred at $(0, y)$ passing through origin and touching the circle C externally, then the radius of T is equal to

(a) $\frac{\sqrt{3}}{\sqrt{2}}$ (b) $\frac{\sqrt{3}}{2}$

(c) $\frac{1}{2}$ (d) $\frac{1}{4}$

23. The slope of the line touching both the parabolas $y^2 = 4x$ and $x^2 = -32y$ is

(a) $\frac{1}{2}$ (b) $\frac{3}{2}$ (c) $\frac{1}{8}$ (d) $\frac{2}{3}$

24. The image of the line $\frac{x-1}{3} = \frac{y-3}{1} = \frac{z-4}{-5}$ in the plane $2x - y + z + 3 = 0$ is the line

(a) $\frac{x+3}{3} = \frac{y-5}{1} = \frac{z-2}{-5}$

(b) $\frac{x+3}{-3} = \frac{y-5}{-1} = \frac{z+2}{5}$

(c) $\frac{x-3}{3} = \frac{y+5}{1} = \frac{z-2}{-5}$

(d) $\frac{x-3}{-3} = \frac{y+5}{-1} = \frac{z-2}{5}$

25. If $[\mathbf{a}\times\mathbf{b}\ \mathbf{b}\times\mathbf{c}\ \mathbf{c}\times\mathbf{a}] = \lambda\,[\mathbf{a}\,\mathbf{b}\,\mathbf{c}]^2$, then λ is equal to

(a) 0 (b) 1
(c) 2 (d) 3

26. Let A and B be two events such that $P(\overline{A \cup B}) = \frac{1}{6}$, $P(A \cap B) = \frac{1}{4}$ and $P(\overline{A}) = \frac{1}{4}$, where $\overline{A}$ stands for the complement of the event A. Then , the events A and B are

(a) independent but not equally likely
(b) independent and equally likely
(c) mutually exclusive and independent
(d) equally likely but not independent

27. The variance of first 50 even natural numbers is

(a) $\frac{833}{4}$ (b) 833 (c) 437 (d) $\frac{437}{4}$

28. If $f_k(x) = 1/k\,(\sin^k x + \cos^k x)$, where $x \in R$ and $k \geq 1$, then $f_4(x) - f_6(x)$ is equal to

(a) 1/6 (b) 1/3 (c) 1/4 (d) 1/12

29. A bird is sitting on the top of a vertical pole 20 m high and its elevation from a point O on the ground is 45°. It flies off horizontally straight away from the point O. After 1s, the elevation of the bird from O is reduced to 30°. Then, the speed (in m/s) of the bird is

(a) $40(\sqrt{2}-1)$ (b) $40(\sqrt{3}-\sqrt{2})$
(c) $20\sqrt{2}$ (d) $20(\sqrt{3}-1)$

30. The statement $\sim (p \leftrightarrow \sim q)$ is

(a) equivalent to $p \leftrightarrow q$
(b) equivalent to $\sim p \leftrightarrow q$
(c) a tautology
(d) a fallacy

Answers

Physics

1. (a)	**2.** (c)	**3.** (b)	**4.** (a)	**5.** (c)	**6.** (c)	**7.** (d)	**8.** (a)	**9.** (c)	**10.** (*)
11. (c)	**12.** (d)	**13.** (a)	**14.** (d)	**15.** (c)	**16.** (c)	**17.** (a)	**18.** (c)	**19.** (b)	**20.** (c)
21. (c)	**22.** (c)	**23.** (b)	**24.** (d)	**25.** (d)	**26.** (b)	**27.** (c)	**28.** (a)	**29.** (b)	**30.** (d)

Note '*' Means, none options are not correct.

Chemistry

1. (a)	**2.** (b)	**3.** (c)	**4.** (b)	**5.** (a)	**6.** (a)	**7.** (c)	**8.** (a)	**9.** (b)	**10.** (d)
11. (c)	**12.** (b)	**13.** (b)	**14.** (a)	**15.** (d)	**16.** (b)	**17.** (b)	**18.** (a)	**19.** (d)	**20.** (b)
21. (b)	**22.** (d)	**23.** (d)	**24.** (c)	**25.** (a)	**26.** (a)	**27.** (d)	**28.** (a)	**29.** (a)	**30.** (c)

Mathematics

1. (d)	**2.** (b)	**3.** (a)	**4.** (d)	**5.** (c)	**6.** (b)	**7.** (d)	**8.** (a)	**9.** (c)	**10.** (d)
11. (d)	**12.** (d)	**13.** (d)	**14.** (c)	**15.** (b)	**16.** (d)	**17.** (a)	**18.** (a)	**19.** (b)	**20.** (c)
21. (c)	**22.** (d)	**23.** (a)	**24.** (a)	**25.** (b)	**26.** (a)	**27.** (b)	**28.** (d)	**29.** (d)	**30.** (a)

JEE Advanced
Unsolved Paper 2014

Time : 3 Hrs **M. Marks : 180**

Paper 1

Instructions

- The question paper consists of two sections.
- Section 1 contains 10 multiple choice questions. Each question has four choices (a), (b), (c) and (d) out of which one or more than one are correct.
- Section 2 contains 10 questions. The answer to each of the question is a single-digit integer, ranging from 0 to 9 (both inclusive)
- For each question in Section 1, you will be awarded 3 marks if you darken all the bubble(s) corresponding to the correct answer(s) and zero mark if no bubbles are darkened. No negative marks will be awarded for incorrect answers in this section.
- For each question in Section 2, you will be awarded 3 marks if you darken only the bubble corresponding to the correct answer and zero mark if no bubble is darkened. No negative marks will be awarded for incorrect answer in this section.

Physics

Section 1

Direction (Q. No. 1-10) *This section contains 10 multiple choice questions. Each question has four choices (a), (b), (c) and (d) out of which one or more than one are correct.*

1. At time $t = 0$, terminal A in the circuit shown in the figure is connected to B by a key and an alternating current $I(t) = I_0 \cos(\omega t)$, with $I_0 = 1$ A and $\omega = 500$ rad s^{-1} starts flowing in it with the initial direction shown in the figure. At $t = \frac{7\pi}{6\omega}$, the key is switched from B to D. Now onwards only A and D are connected. A total charge Q flows from the battery to charge the capacitor fully. If $C = 20\,\mu F$, $R = 10\,\Omega$ and the battery is ideal with emf of 50 V, identify the correct statement(s).

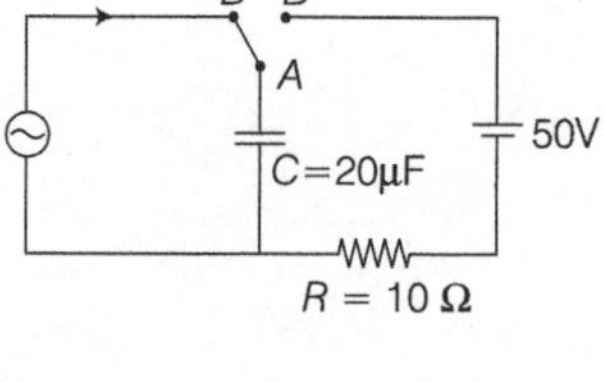

(a) Magnitude of the maximum charge on the capacitor before $t = \frac{7\pi}{6\omega}$ is 1×10^{-3} C

(b) The current in the left part of the circuit just before $t = \frac{7\pi}{6\omega}$ is clockwise

(c) Immediately after A is connected to D, the current in R is 10 A

(d) $Q = 2 \times 10^{-3}$ C

2. A light source, which emits two wavelengths $\lambda_1 = 400$ nm and $\lambda_2 = 600$ nm, is used in a Young's double-slit experiment. If recorded fringe widths for λ_1 and λ_2 are β_1 and β_2 and the number of fringes for them within a distance y on one side of the central maximum are m_1 and m_2, respectively, then

(a) $\beta_2 > \beta_1$

(b) $m_1 > m_2$

(c) from the central maximum, 3rd maximum of λ_2 overlaps with 5th minimum of λ_1

(d) the angular separation of fringes of λ_1 is greater than λ_2

3. One end of a taut string of length 3 m along the x-axis is fixed at $x = 0$. The speed of the waves in the string is 100 ms^{-1}. The other end of the string is vibrating in the y-direction so that stationary waves are set up in the string. The possible waveform(s) of these stationary wave is (are)

(a) $y(t) = A \sin \frac{\pi x}{6} \cos \frac{50\pi t}{3}$

(b) $y(t) = A \sin \frac{\pi x}{3} \cos \frac{100\pi t}{3}$

(c) $y(t) = A \sin \frac{5\pi x}{6} \cos \frac{250\pi t}{3}$

(d) $y(t) = A \sin \frac{5\pi x}{2} \cos 250\pi t$

4. A parallel plate capacitor has a dielectric slab of dielectric constant K between its plates that covers 1/3 of the area of its plates, as shown in the figure. The total capacitance of the capacitor is C while that of the portion with dielectric in between is C_1. When the capacitor is charged, the plate area covered by the dielectric gets charge Q_1 and the rest of the area gets charge Q_2. The electric field in the dielectric is E_1 and that in the other portion is E_2. Choose the correct option/options, ignoring edge effects.

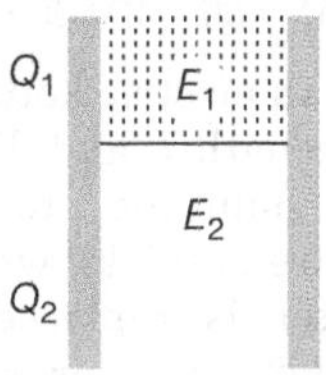

(a) $\frac{E_1}{E_2} = 1$

(b) $\frac{E_1}{E_2} = \frac{1}{K}$

(c) $\frac{Q_1}{Q_2} = \frac{3}{K}$

(d) $\frac{C}{C_1} = \frac{2 + K}{K}$

5. Let $E_1(r), E_2(r)$ and $E_3(r)$ be the respective electric fields at a distance r from a point charge Q, an infinitely long wire with constant linear charge density λ, and an infinite plane with uniform surface charge density σ.

If $E_1(r_0) = E_2(r_0) = E_3(r_0)$ at a given distance r_0, then

(a) $Q = 4\sigma\pi r_0^2$

(b) $r_0 = \frac{\lambda}{2\pi\sigma}$

(c) $E_1\left(\frac{r_0}{2}\right) = 2E_2\left(\frac{r_0}{2}\right)$

(d) $E_2\left(\frac{r_0}{2}\right) = 4E_3\left(\frac{r_0}{2}\right)$

6. A student is performing an experiment using a resonance column and a tuning fork of frequency 244 s^{-1}. He is told that the air in the tube has been replaced by another gas (assume that the column remains filled with the gas). If the minimum height at which resonance occurs is (0.350 ± 0.005) m, the gas in the tube is

(Useful information
$\sqrt{167RT} = 640\ \text{J}^{1/2}\ \text{mole}^{-1/2}$;
$\sqrt{140RT} = 590\ \text{J}^{1/2}\ \text{mole}^{-1/2}$. The molar masses M in grams are given in the options. Take the value of $\sqrt{10/M}$ for each gas as given there.)

(a) Neon ($M = 20$, $\sqrt{10/20} = 7/10$)
(b) Nitrogen ($M = 28$, $\sqrt{10/28} = 3/5$)
(c) Oxygen ($M = 32$, $\sqrt{10/32} = 9/16$)
(d) Argon ($M = 36$, $\sqrt{10/36} = 17/32$)

7. Heater of an electric kettle is made of a wire of length L and diameter d. It takes 4 minutes to raise the temperature of 0.5 kg water by 40 K. This heater is replaced by a new heater having two wires of the same material, each of length L and diameter $2d$.

The way these wires are connected is given in the options. How much time in minutes will it take to raise the temperature of the same amount of water by 40 K?

(a) 4, if wires are in parallel
(b) 2, if wires are in series
(c) 1, if wires are in series
(d) 0.5, if wires are in parallel

8. In the figure, a ladder of mass m is shown leaning against a wall. It is in static equilibrium making an angle θ with the horizontal floor. The coefficient of friction between the wall and the ladder is μ_1 and that between the floor and the ladder is μ_2. The normal reaction of the wall on the ladder is N_1 and that of the floor is N_2. If the ladder is about to slip, then

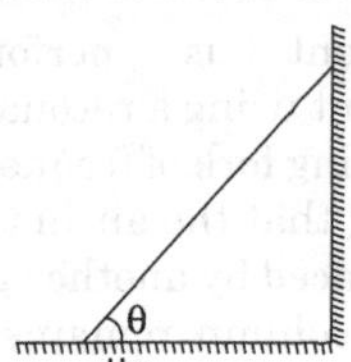

(a) $\mu_1 = 0, \mu_2 \neq 0$ and $N_2 \tan\theta = \dfrac{mg}{2}$
(b) $\mu_1 \neq 0, \mu_2 = 0$ and $N_1 \tan\theta = \dfrac{mg}{2}$
(c) $\mu_1 \neq 0, \mu_2 \neq 0$ and $N_2 = \dfrac{mg}{1 + \mu_1\mu_2}$
(d) $\mu_1 = 0, \mu_2 \neq 0$ and $N_1 \tan\theta = \dfrac{mg}{2}$

9. A transparent thin film of uniform thickness and refractive index $n_1 = 1.4$ is coated on the convex spherical surface of radius R at one end of a long solid glass cylinder of refractive index $n_2 = 1.5$, as shown in the figure. Rays of light parallel to the axis of the cylinder traversing through the film from air to glass get focused at distance f_1 from the film, while rays of light traversing from glass to air get focused at distance f_2 from the film. Then

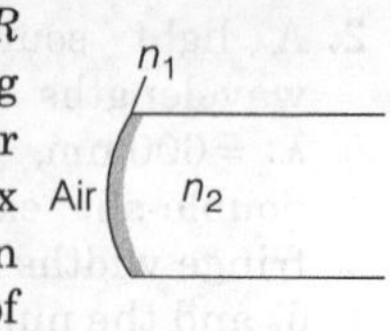

(a) $|f_1| = 3R$
(b) $|f_1| = 2.8R$
(c) $|f_2| = 2R$
(d) $|f_2| = 1.4R$

10. Two ideal batteries of emf V_1 and V_2 and three resistances R_1, R_2 and R_3 are connected as shown in the figure. The current in resistance R_2 would be zero if

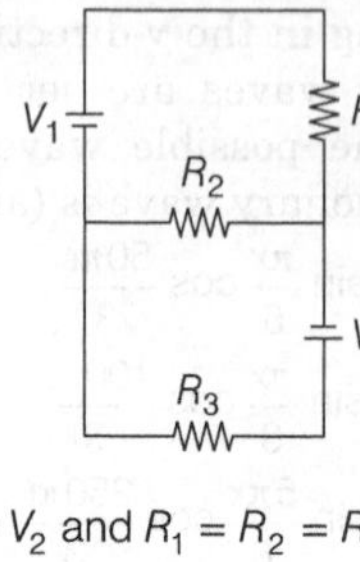

(a) $V_1 = V_2$ and $R_1 = R_2 = R_3$
(b) $V_1 = V_2$ and $R_1 = 2R_2 = R_3$
(c) $V_1 = 2V_2$ and $2R_1 = 2R_2 = R_3$
(d) $2V_1 = V_2$ and $2R_1 = R_2 = R_3$

Section 2

Direction (Q. No. 11-20) *This section contains 10 questions. Each question, when worked out will result in one integer from 0 to 9 (both inclusive)*

11. Airplanes A and B are flying with constant velocity in the same vertical plane at angles 30° and 60° with respect to the horizontal respectively as shown in figure. The speed of A is $100\sqrt{3}\ \text{ms}^{-1}$. At time $t = 0$ s, an observer in A finds B at a distance of 500 m. This observer sees B moving with a constant velocity perpendicular to the line of motion of A. If at $t = t_0$, A just escapes being hit by B, t_0 in seconds is

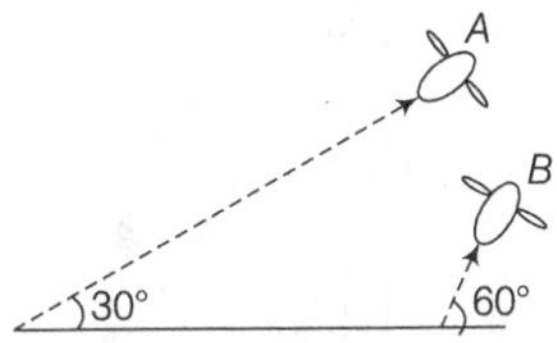

12. During Searle's experiment, zero of the vernier scale lies between 3.20×10^{-2} m and 3.25×10^{-2} m of the main scale. The 20th division of the vernier scale exactly coincides with one of the main scale divisions. When an additional load of 2 kg is applied to the wire, the zero of the vernier scale still lies between 3.20×10^{-2} m and 3.25×10^{-2} m of the main scale but now the 45th division of vernier scale coincides with one of the main scale divisions. The length of the thin metallic wire is 2 m and its cross-sectional area is $8 \times 10^{-7}\ \text{m}^2$. The least count of the vernier scale is 1.0×10^{-5} m. The maximum percentage error in the Young's modulus of the wire is

13. A uniform circular disc of mass 1.5 kg and radius 0.5 m is initially at rest on a horizontal frictionless surface. Three forces of equal magnitude $F = 0.5$ N are applied simultaneously along the three sides of an equilateral triangle XYZ with its vertices on the perimeter of the disc (see figure). One second after applying the forces, the angular speed of the disc in rad s^{-1} is

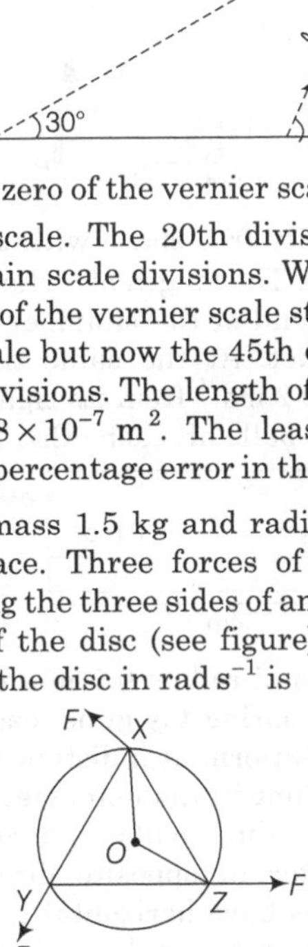

14. Two parallel wires in the plane of the paper are distance X_0 apart. A point charge is moving with speed u between the wires in the same plane at a distance X_1 from one of the wires. When the wires carry current of magnitude I in the same direction, the radius of curvature of the path of the point charge is R_1. In contrast, if the currents I in the two wires have directions opposite to each other, the radius of curvature of the path is R_2. If $\dfrac{X_0}{X_1} = 3$, and value of R_1/R_2 is

15. To find the distance d over which a signal can be seen clearly in foggy conditions, a railways engineer uses dimensional analysis and assumes that the distance depends on the mass density ρ of the fog, intensity (power/area) S of the light from the signal and its frequency f. The engineer finds that d is proportional to $S^{1/n}$. The value of n is

16. A galvanometer gives full scale deflection with 0.006 A current. By connecting it to a 4990 Ω resistance, it can be converted into a voltmeter of range 0-30 V. If connected to a $\frac{2n}{249}$ Ω resistance, it becomes an ammeter of range 0-1.5 A. The value of n is

17. Consider an elliptically shaped rail PQ in the vertical plane with $OP = 3$ m and $OQ = 4$ m. A block of mass 1 kg is pulled along the rail from P to Q with a force of 18 N, which is always parallel to line PQ (see figure). Assuming no frictional losses, the kinetic energy of the block when it reaches Q is $(n \times 10)$ J. The value of n is (take acceleration due to gravity = 10 ms^{-2})

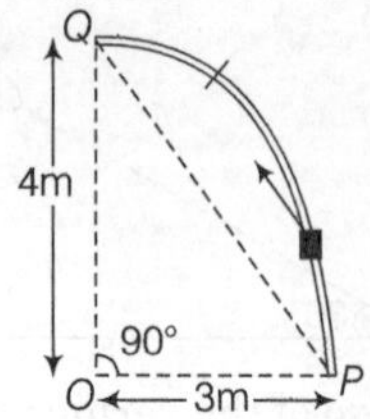

18. A rocket is moving in a gravity free space with a constant acceleration of 2 ms^{-2} along $+x$ direction (see figure). The length of a chamber inside the rocket is 4 m. A ball is thrown from the left end of the chamber in $+x$ direction with a speed of 0.3 ms^{-1} relative to the rocket. At the same time, another ball is thrown in $-x$ direction with a speed of 0.2 ms^{-1} from its right end relative to the rocket. The time in seconds when the two balls hit each other is

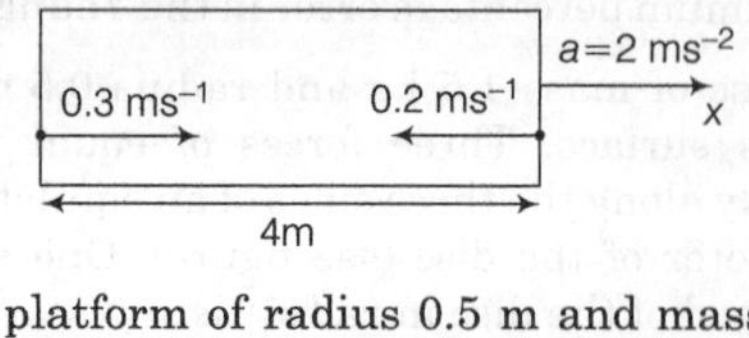

19. A horizontal circular platform of radius 0.5 m and mass 0.45 kg is free to rotate about its axis. Two massless spring toy-guns, each carrying a steel ball of mass 0.05 kg are attached to the platform at a distance 0.25 m from the centre on its either sides along its diameter (see figure). Each gun simultaneously fires the balls horizontally and perpendicular to the diameter in opposite directions. After leaving the platform, the balls have horizontal speed of 9 ms^{-1} with respect to the ground. The rotational speed of the platform in rad s^{-1} after the balls leave the platform is

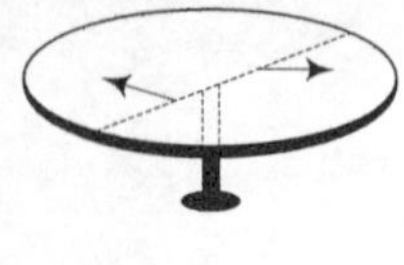

20. A thermodynamic system is taken from an initial state i with internal energy $U_i = 100$ J to the final state f along two different paths iaf and ibf, as schematically shown in the figure. The work done by the system along the paths af, ib and bf are $W_{af} = 200$ J, $W_{ib} = 50$ J and $W_{bf} = 100$ J respectively. The heat supplied to the system along the path iaf, ib and bf are Q_{iaf}, Q_{ib} and Q_{bf} respectively. If the internal energy of the system in the state b is $U_b = 200$ J and $Q_{iaf} = 500$ J, the ratio Q_{bf}/Q_{ib} is

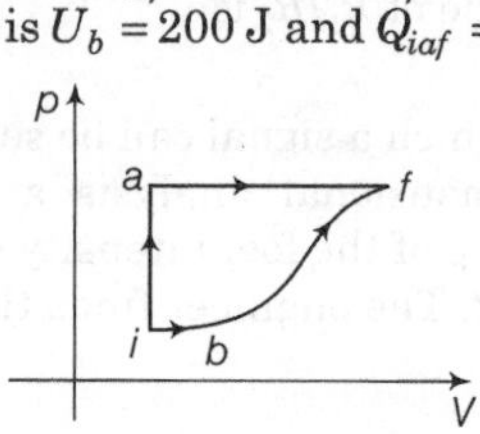

Chemistry

Section I

Directions (Q. Nos. 1-10) *This section contains 10 multiple choice questions. Each question has four choices (a), (b), (c) and (d) out of which one or more than one are correct.*

1. For the reaction,

$I^- + CiO_3^- + H_2SO_4 \rightarrow Cl^- + HSO_4^- + I_2$

the correct statement(s) in the balanced equation is/are

(a) stoichiometric coefficient of HSO_4^- is 6
(b) iodide is oxidised
(c) sulphur is reduced
(d) H_2O is one of the products

2. The pair(s) of reagents that yield paramagnetic species is/are

(a) Na and excess of NH_3
(b) K and excess of O_2
(c) Cu and dilute HNO_3
(d) O_2 and 2-ethylanthraquinol

3. In the reaction shown below, the major product(s) formed is/are

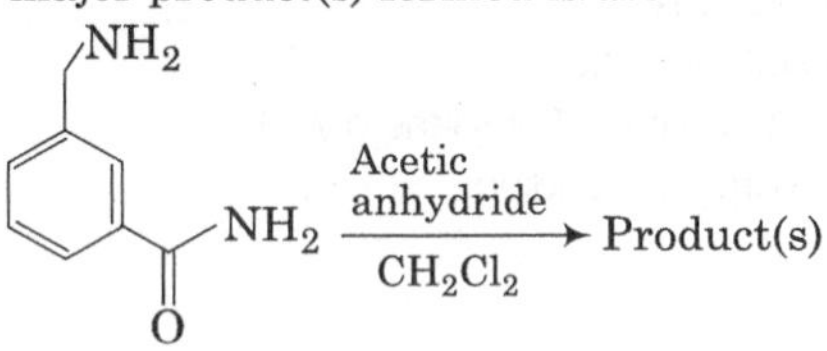

(a) 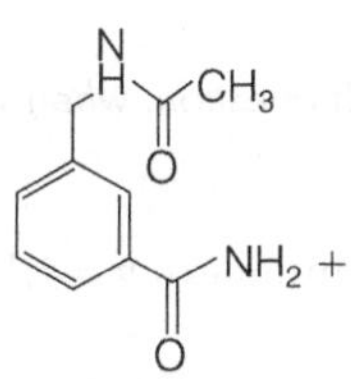 $+ CH_3COOH$

(b) 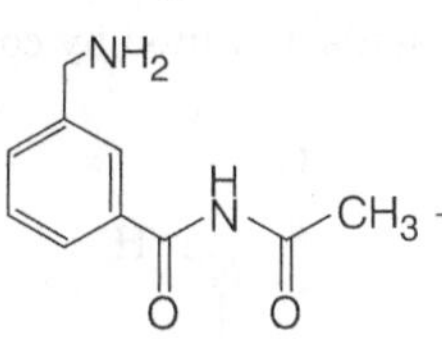 $+ CH_3COOH$

(c) 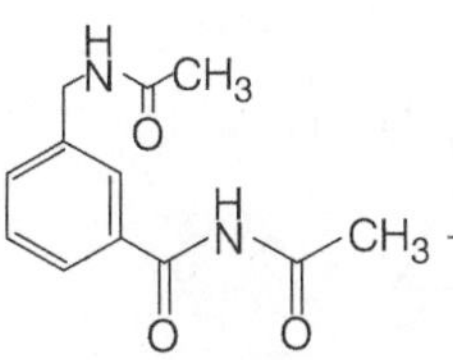 $+ H_2O$

(d) 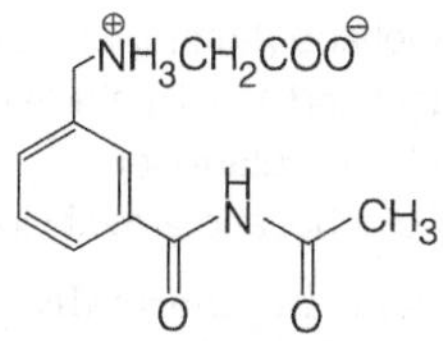

4. In a galvanic cell, the salt bridge

(a) does not participate chemically in the cell reaction
(b) stops the diffusion of ions from one electrode to another
(c) is necessary for the occurrence of the cell reaction
(d) ensures mixing of the two electrolytic solutions

5. Upon heating with Cu_2S, the reagent(s) that give copper metal is/are

(a) $CuFeS_2$ (b) CuO (c) Cu_2O (d) $CuSO_4$

6. Hydrogen bonding plays a central role in the following phenomena

(a) ice floats in water
(b) higher Lewis basicity of primary amines than tertiary amines in aqueous solutions
(c) formic acid is more acidic than acetic acid
(d) dimerisation of acetic acid in benzene

7. The reactivity of compound Z with different halogens under appropriate conditions is given below

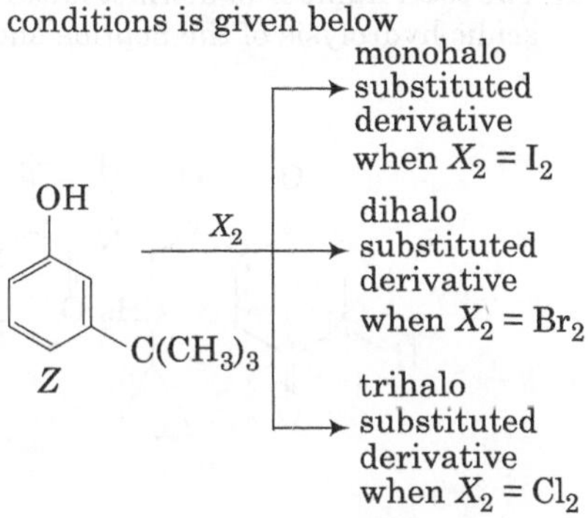

The observed pattern of electrophilic substitution can be explained by

(a) the steric effect on the halogen

(b) the steric effect of the *tert*-butyl group

(c) the electronic effect of the phenolic group

(d) the electronic effect of the *tert*-butyl group

8. The correct combination of names for isomeric alcohols with molecular formula $C_4H_{10}O$ is/are

(a) *tert*-butanol and 2-methylpropan-2-ol

(b) *tert*-butanol and 1,1-dimethylethan-1-ol

(c) *n*-butanol and butan-1-ol

(d) *iso*-butyl alcohol and 2-methylpropan-1-ol

9. An ideal gas in thermally insulated vessel at internal pressure $= p_1$, volume $= V_1$ and absolute temperature $= T_1$ expands irreversibly against zero external pressure, as shown in the diagram.

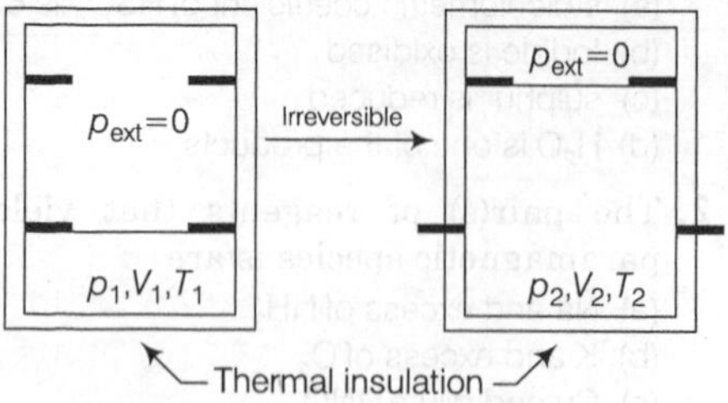

The final internal pressure, volume and absolute temperature of the gas are p_2, V_2 and T_2, respectively. For this expansion

(a) $q = 0$ (b) $T_2 = T_1$ (c) $p_2V_2 = p_1V_1$ (d) $p_2V_2^{\gamma} = p_1V_1^{\gamma}$

10. The correct statement(s) for orthoboric acid is/are

(a) it behaves as a weak acid in water due to self ionisation

(b) acidity of its aqueous solution increases upon addition of ethylene glycol

(c) it has a three dimensional structure due to hydrogen bonding

(d) it is a weak electrolyte in water

Section II

Directions (Q. Nos. 11-20) *This section contains 10 question. Each question, when worked out will result in one integer from 0 to 9 (both inclusive)*

11. In an atom, the total number of electrons having quantum numbers $n = 4, |m_l| = 1$ and $m_s = -1/2$ is

12. The total number of distinct naturally occurring amino acids obtained by complete acidic hydrolysis of the peptide shown below is

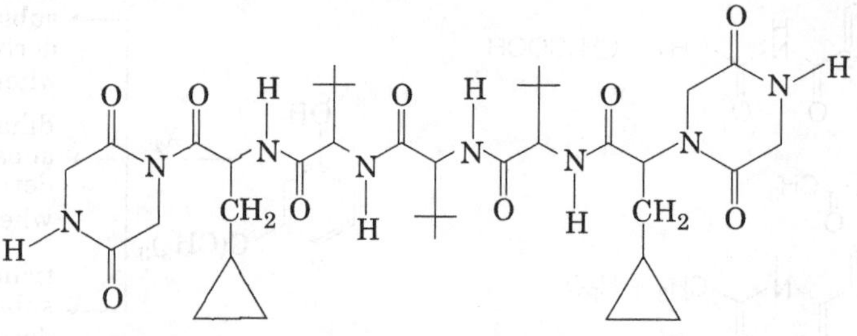

13. If the value of Avogadro number is 6.023×10^{23} mol^{-1} and the value of Boltzmann constant is 1.380×10^{-23} JK^{-1}, then the number of significant digits in the calculated value of the universal gas constant is

14. A compound H_2X with molar weight of 80 g is dissolved in a solvent having density of 0.4 g mL^{-1}. Assuming no change in volume upon dissolution, the molality of a 3.2 molar solution is

15. MX_2 dissociates into M^{2+} and X^- ions in an aqueous solution, with a degree of dissociation (α) of 0.5. The ratio of the observed depression of freezing point of the aqueous solution to the value of the depression of freezing point in the absence of ionic dissociation is

16. Consider the following list of reagents, acidified $K_2Cr_2O_7$, alkaline $KMnO_4$, $CuSO_4$, H_2O_2, Cl_2, O_3, $FeCl_3$, HNO_3 and $Na_2S_2O_3$. The total number of reagents that can oxidise aqueous iodide to iodine is

17. The total number(s) of stable conformers with non-zero dipole moment for the following compound is(are)

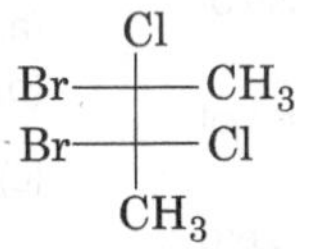

18. Among PbS, CuS, HgS, MnS, Ag_2S, NiS, CoS, Bi_2S_3 and SnS_2 the total number of black coloured sulphides is

19. Consider all possible isomeric ketones including stereoisomers of MW = 100. All these isomers are independently reacted with $NaBH_4$ (**Note** stereoisomers are also reacted separately). The total number of ketones that give a racemic product(s) is/are

20. A list of species having the formula XZ_4 is given below

XeF_4, SF_4, SiF_4, BF_4^-, BrF_4^-, $[Cu(NH_3)_4]^{2+}$, $[FeCl_4]^{2-}$, $[CoCl_4]^{2-}$ and $[PtCl_4]^{2-}$

Defining shape on the basis of the location of X and Z atoms, the total number of species having a square planar shape is

Mathematics

Section I

Directions (Q. Nos. 1-10) *This section contains 10 multiple choice questions. Each question has four choices (a), (b), (c) and (d) out of which one or more than one are correct.*

1. Let $f : [a, b] \to [1, \infty)$ be a continuous function and $g : R \to R$ be defined as

$$g(x) = \begin{cases} 0, & \text{if } x < a \\ \int_a^x f(t)dt, & \text{if } a \le x \le b. \\ \int_a^b f(t)dt, & \text{if } x > b \end{cases}$$

Then,

(a) $g(x)$ is continuous but not differentiable at a

(b) $g(x)$ is differentiable on R

(c) $g(x)$ is continuous but not differentiable at b

(d) $g(x)$ is continuous and differentiable at either a or b but not both

2. For every pair of continuous function $f, g: [0,1] \to R$ such that max $\{f(x): x \in [0,1]\} = \max\{g(x): x \in [0,1]\}$.

The correct statement(s) is (are)

(a) $[f(c)]^2 + 3f(c) = [g(c)]^2 + 3g(c)$ for some $c \in [0,1]$

(b) $[f(c)]^2 + f(c) = [g(c)]^2 + 3g(c)$ for some $c \in [0,1]$

(c) $[f(c)]^2 + 3f(c) = [g(c)]^2 + g(c)$ for some $c \in [0,1]$

(d) $[f(c)]^2 = [g(c)]^2$ for some $c \in [0,1]$

3. Let M be a 2×2 symmetric matrix with integer entries. Then, M is invertible, if

(a) the first column of M is the transpose of the second row of M

(b) the second row of M is the transpose of the first column of M

(c) M is a diagonal matrix with non-zero entries in the main diagonal

(d) the product of entries in the main diagonal of M is not the square of an integer

4. Let $\mathbf{x}$, $\mathbf{y}$ and $\mathbf{z}$ be three vectors each of magnitude $\sqrt{2}$ and the angle between each pair of them is $\frac{\pi}{3}$. If $\mathbf{a}$ is a non-zero vector perpendicular to $\mathbf{x}$ and $\mathbf{y}\times\mathbf{z}$ and $\mathbf{b}$ is non-zero vector perpendicular to $\mathbf{y}$ and $\mathbf{z}\times\mathbf{x}$, then

(a) $\mathbf{b} = (\mathbf{b}\cdot\mathbf{z})(\mathbf{z}-\mathbf{x})$

(b) $\mathbf{a} = (\mathbf{a}\cdot\mathbf{y})(\mathbf{y}-\mathbf{z})$

(c) $\mathbf{a}\cdot\mathbf{b} = -(\mathbf{a}\cdot\mathbf{y})(\mathbf{b}\cdot\mathbf{z})$

(d) $\mathbf{a} = (\mathbf{a}\cdot\mathbf{y})(\mathbf{z}-\mathbf{y})$

5. From a point $P(\lambda, \lambda, \lambda)$, perpendiculars PQ and PR are drawn respectively on the lines $y = x, z = 1$ and $y = -x, z = -1$. If P is such that $\angle QPR$ is a right angle, then the possible value(s) of λ is (are)

(a) $\sqrt{2}$ (b) 1

(c) -1 (d) $-\sqrt{2}$

6. Let M and N be two 3×3 matrices such that $MN = NM$. Further, if $M \neq N^2$ and $M^2 = N^4$, then

(a) determinant of $(M^2 + MN^2)$ is 0

(b) there is a 3×3 non-zero matrix U such that $(M^2 + MN^2)U$ is zero matrix

(c) determinant of $(M^2 + MN^2) \geq 1$

(d) for a 3×3 matrix U, if $(M^2 + MN^2)U$ equals the zero matrix, then U is the zero matrix

7. Let $f: (0, \infty) \to R$ be given by

$$f(x) = \int_{1/x}^{x} e^{-\left(t+\frac{1}{t}\right)} \frac{dt}{t}.$$

Then,

(a) $f(x)$ is monotonically increasing on $[1, \infty)$

(b) $f(x)$ is monotonically decreasing on $[0, 1)$

(c) $f(x) + f\left(\frac{1}{x}\right) = 0$, for all $x \in (0, \infty)$

(d) $f(2^x)$ is an odd function of x on R

8. Let $f: \left(-\frac{\pi}{2}, \frac{\pi}{2}\right) \to R$ be given by

$$f(x) = [\log(\sec x + \tan x)]^3.$$

Then,

(a) $f(x)$ is an odd function

(b) $f(x)$ is a one-one function

(c) $f(x)$ is an onto function

(d) $f(x)$ is an even function

9. A circle S passes through the point $(0, 1)$ and is orthogonal to the circles $(x-1)^2 + y^2 = 16$ and $x^2 + y^2 = 1$. Then,

(a) radius of S is 8

(b) radius of S is 7

(c) centre of S is $(-7, 1)$

(d) centre of S is $(-8, 1)$

10. Let $a \in R$ and $f: R \to R$ be given by $f(x) = x^5 - 5x + a$.

Then,

(a) $f(x)$ has three real roots, if $a > 4$

(b) $f(x)$ has only one real root, if $a > 4$

(c) $f(x)$ has three real roots, if $a < -4$

(d) $f(x)$ has three real roots, if $-4 < a < 4$

Section II

Directions (Q. Nos. 11-20) *This section contains 10 questions. Each question, when worked out will result in one integer from 0 to 9 (both inclusive).*

11. The slope of the tangent to the curve $(y-x^5)^2 = x(1+x^2)^2$ at the point (1, 3) is

12. Let $f: [0,4\pi] \to [0,\pi]$ be defined by $f(x) = \cos^{-1}(\cos x)$. The number of points $x \in [0,4\pi]$ satisfying the equation $f(x) = \dfrac{10-x}{10}$ is

13. The largest value of the non-negative integer a for which $\lim\limits_{x \to 1}\left\{\dfrac{-ax+\sin(x-1)+a}{x+\sin(x-1)-1}\right\}^{\frac{1-x}{1-\sqrt{x}}} = \dfrac{1}{4}$ is

14. Let $f: R \to R$ and $g: R \to R$ be respectively given by $f(x) = |x|+1$ and $g(x) = x^2+1$. Define $h: R \to R$ by

$$h(x) = \begin{cases} \max\{f(x), g(x)\}, & \text{if } x \le 0. \\ \min\{f(x), g(x)\}, & \text{if } x > 0. \end{cases}$$

The number of points at which $h(x)$ is not differentiable is

15. For a point P in the plane, let $d_1(P)$ and $d_2(P)$ be the distances of the point P from the lines $x-y=0$ and $x+y=0$, respectively. The area of the region R consisting of all points P lying in the first quadrant of the plane and satisfying $2 \le d_1(P)+d_2(P) \le 4$, is

16. Let $n_1 < n_2 < n_3 < n_4 < n_5$ be positive integers such that $n_1+n_2+n_3+n_4+n_5 = 20$. The number of such distinct arrangements $(n_1, n_2, n_3, n_4, n_5)$ is

17. The value of $\int_0^1 4x^3 \left\{\dfrac{d^2}{dx^2}(1-x^2)^5\right\} dx$ is

18. Let **a, b** and **c** be three non-coplanar unit vectors such that the angle between every pair of them is $\pi/3$. If $\mathbf{a} \times \mathbf{b} + \mathbf{b} \times \mathbf{c} = p\mathbf{a} + q\mathbf{b} + r\mathbf{c}$, where p, q and r are scalars, then the value of $\dfrac{p^2+2q^2+r^2}{q^2}$ is

19. Let a, b, c be positive integers such that b/a is an integer. If a, b, c are in geometric progression and the arithmetic mean of a, b, c is $b+2$, then the value of $\dfrac{a^2+a-14}{a+1}$ is

20. Let $n \ge 2$ be an integer. Take n distinct points on a circle and join each pair of points by a line segment. Colour the line segment joining every pair of adjacent points by blue and the rest by red. If the number of red and blue line segments are equal, then the value of n is

Paper 2

Instructions

- Section 1 contains 10 multiple choice questions. Each question has four choices (a), (b), (c) and (d) out of which one is correct.
- Section 2 contains 3 paragraphs each describing theory, experiment and data etc. Six questions relate to three paragraphs with two questions on each paragraph. Each question pertaining to a particular passage should have only one correct answer among the four given choices (a), (b), (c) and (d).
- Section 3 contains 4 multiple choice questions. Each question has two lists (**List I** P, Q, R and S; **List II** 1, 2, 3 and 4). The options for the correct match are provided as (a), (b), (c) and (d) out of which only one is correct.
- For each question in Section 1, 2 and 3 you will be awarded 3 marks if you darken the bubble corresponding to the correct answer and zero mark if no bubble is darkened. In all other cases, minus one (–1) mark will be awarded.

Physics

Section 1

Direction (Q. No. 1-10) *This section contains 10 multiple choice questions. Each question has four choices (a), (b), (c) and (d) out of which only one option is correct.*

1. A tennis ball is dropped on a horizontal smooth surface. It bounces back to its original position after hitting the surface. The force on the ball during the collision is proportional to the length of compression of the ball.

Which one of the following sketches describes the variation of its kinetic energy K with time t most appropriately? The figures are only illustrative and not to the scale.

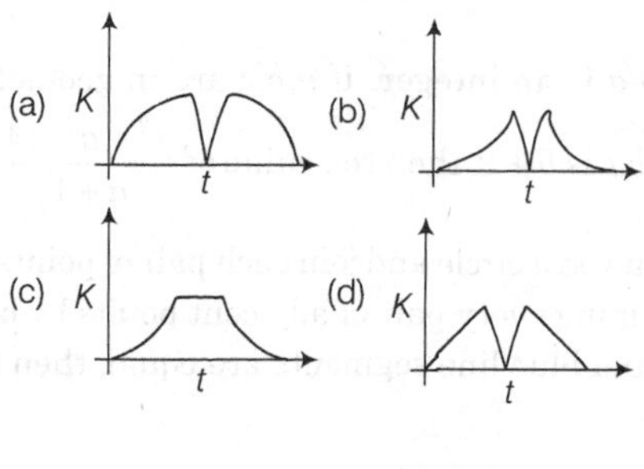

2. A wire, which passes through the hole in a small bead, is bent in the form of quarter of a circle. The wire is fixed vertically on ground as shown in the figure. The bead is released from near the top of the wire and it slides along the wire without friction. As the bead moves from A to B, the force it applies on the wire is

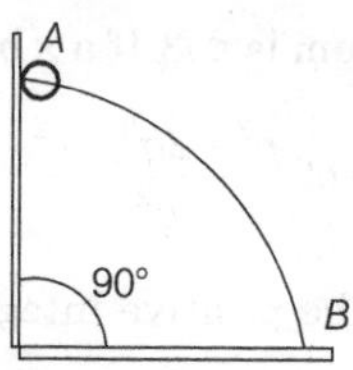

(a) always radially outwards
(b) always radially inwards
(c) radially outwards initially and radially inwards later
(d) radially inwards initially and radially outwards later

3. During an experiment with a metre bridge, the galvanometer shows a null point when the jockey is pressed at 40.0 cm using a standard resistance of 90 Ω, as shown in the figure. The least count of the scale used in the metre bridge is 1 mm. The unknown resistance is

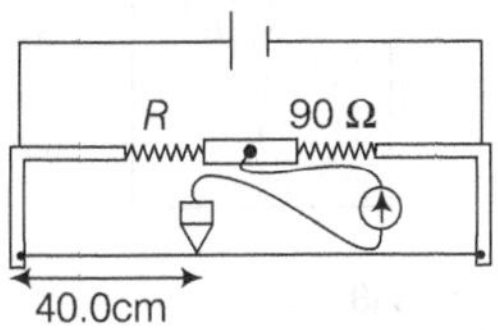

(a) 60 ± 0.15 Ω
(b) 135 ± 0.56 Ω
(c) 60 ± 0.25 Ω
(d) 135 ± 0.23 Ω

4. Charges $Q, 2Q$ and $4Q$ are uniformly distributed in three dielectric solid spheres 1, 2 and 3 of radii $R/2, R$ and $2R$ respectively, as shown in figure. If magnitudes of the electric fields at point P at a distance R from the centre of spheres 1, 2 and 3 are E_1, E_2 and E_3 respectively, then

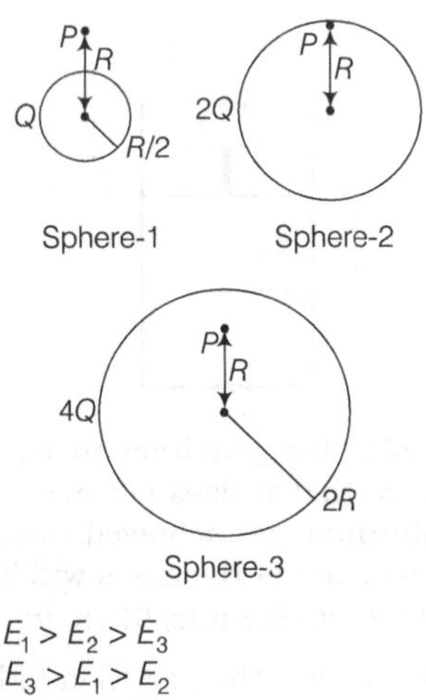

(a) $E_1 > E_2 > E_3$
(b) $E_3 > E_1 > E_2$
(c) $E_2 > E_1 > E_3$
(d) $E_3 > E_2 > E_1$

5. A point source S is placed at the bottom of a transparent block of height 10 mm and refractive index 2.72. It is immersed in a lower refractive index liquid as shown in the figure. It is found that the light emerging from the block to the liquid forms a circular bright spot of diameter 11.54 mm on the top of the block. The refractive index of the liquid is

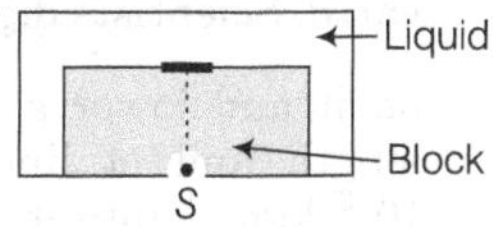

(a) 1.21 (b) 1.30 (c) 1.36 (d) 1.42

6. Parallel rays of light of intensity $I = 912 \text{ Wm}^{-2}$ are incident on a spherical black body kept in surroundings of temperature 300 K. Take Stefan constant $\sigma = 5.7 \times 10^{-8} \text{ Wm}^{-2} \text{ K}^{-4}$ and assume that the energy exchange with the surroundings is only through radiation. The final steady state temperature of the black body is close to

(a) 330 K (b) 660 K (c) 990 K (d) 1550 K

7. A metal surface is illuminated by light of two different wavelengths 248 nm and 310 nm. The maximum speeds of the photoelectrons corresponding to these wavelengths are u_1 and u_2, respectively. If the ratio $u_1 : u_2 = 2:1$ and $hc = 1240$ eV nm, the work function of the metal is nearly

(a) 3.7 eV (b) 3.2 eV
(c) 2.8 eV (d) 2.5 eV

8. If λ_{Cu} is the wavelength of K_α, X-ray line of copper (atomic number 29) and λ_{MO} is the wavelength of the K_α, X-ray line of molybdenum (atomic number 42), then the ratio $\lambda_{Cu}/\lambda_{Mo}$ is close to

(a) 1.99 (b) 2.14 (c) 0.50 (d) 0.48

9. A planet of radius $R = \frac{1}{10} \times$ (radius of earth) has the same mass density as earth. Scientists dig a well of depth $\frac{R}{5}$ on it and lower a wire of the same length and of linear mass density 10^{-3} kgm^{-1} into it. If the wire is not touching anywhere, the force applied at the top of the wire by a person holding it in place is (take the radius of earth $= 6 \times 10^6$ m and the acceleration due to gravity of earth is 10 ms^{-2})

(a) 96 N (b) 108 N
(c) 120 N (d) 150 N

10. A glass capillary tube is of the shape of truncated cone with an apex angle α so that its two ends have cross-sections of different radii. When dipped in water vertically, water rises in it to a height h, where the radius of its cross-section is b. If the surface tension of water is S, its density is ρ, and its contact angle with glass is θ, the value of h will be (g is the acceleration due to gravity)

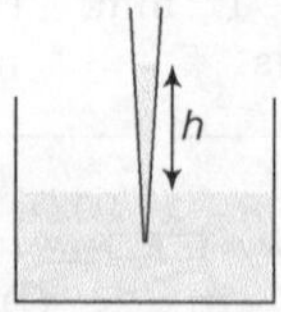

(a) $\frac{2S}{b\rho g}\cos(\theta - \alpha)$

(b) $\frac{2S}{b\rho g}\cos(\theta + \alpha)$

(c) $\frac{2S}{b\rho g}\cos(\theta - \alpha/2)$

(d) $\frac{2S}{b\rho g}\cos(\theta + \alpha/2)$

Section 2

Direction (Q. No. 11-16) *This section contains 6 multiple choice questions relating to three paragraphs with two questions on each paragraph. Each question has four choices (a), (b), (c) and (d) out of which only one is correct.*

Paragraph 1

In the figure a container is shown to have a movable (without friction) piston on top. The container and the piston are all made of perfectly insulating material allowing no heat transfer between outside and inside the container. The container is divided into two compartments by a rigid partition made of a thermally conducting material that allows slow transfer of heat. The lower compartment of the container is filled with 2 moles of an ideal monoatomic gas at 700 K and the upper compartment is filled with 2 moles of an ideal diatomic gas at 400 K. The heat capacities per mole of an ideal monoatomic gas are $C_V = \frac{3}{2}R$, $C_p = \frac{5}{2}R$, and those for an ideal diatomic gas are $C_V = \frac{5}{2}R, C_p = \frac{7}{2}R$.

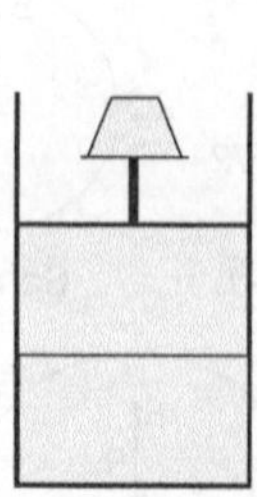

11. Consider the partition to be rigidly fixed so that it does not move. When equilibrium is achieved, the final temperature of the gases will be

(a) 550 K (b) 525 K (c) 513 K (d) 490 K

12. Now consider the partition to be free to move without friction so that the pressure of gases in both compartments is the same. Then total work done by the gases till the time they achieve equilibrium will be

(a) 250R (b) 200R (c) 100R (d) $-100R$

Paragraph 2

A spray gun is shown in the figure where a piston pushes air out of nozzle. A thin tube of uniform cross-section is connected to the nozzle. The other end of the tube is in a small liquid container. As the piston pushes air through the nozzle, the liquid from the container rises into the nozzle and is sprayed out.

For the spray gun shown, the radii of the piston and the nozzle are 20 mm and 1 mm respectively. The upper end of the container is open to the atmosphere.

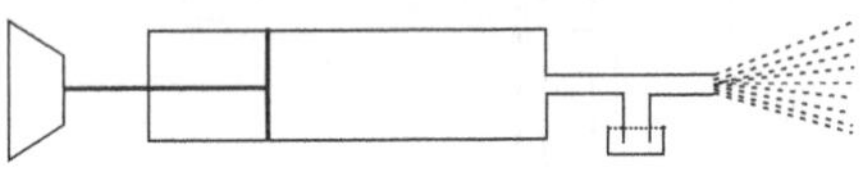

13. If the piston is pushed at a speed of $5\ mms^{-1}$, the air comes out of the nozzle with a speed of

(a) $0.1\ ms^{-1}$ (b) $1\ ms^{-1}$
(c) $2\ ms^{-1}$ (d) $8\ ms^{-1}$

14. If the density of air is ρ_a and that of the liquid ρ_l, then for a given piston speed the rate (volume per unit time) at which the liquid is sprayed will be proportional to

(a) $\sqrt{\frac{\rho_a}{\rho_l}}$ (b) $\sqrt{\rho_a \rho_l}$
(c) $\sqrt{\frac{\rho_l}{\rho_a}}$ (d) ρ_l

Paragraph 3

The figure shows a circular loop of radius a with two long parallel wires (numbered 1 and 2) all in the plane of the paper. The distance of each wire from the centre of the loop is d. The loop and the wires are carrying the same current I. The current in the loop is in the counter-clockwise direction if seen from above.

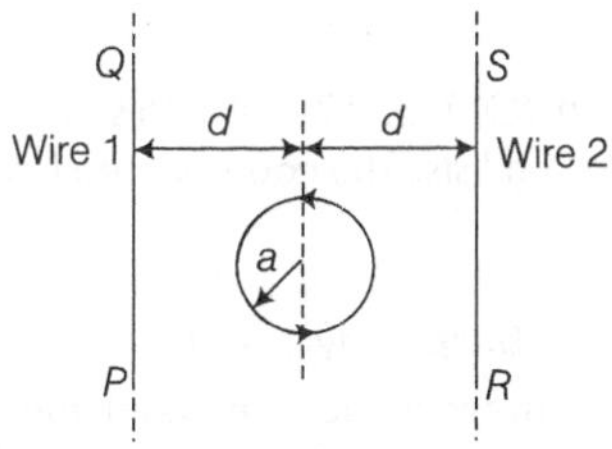

15. When $d \approx a$ but wires are not touching the loop, it is found that the net magnetic field on the axis of the loop is zero at a height h above the loop. In that case

(a) current in wire 1 and wire 2 is the direction PQ and RS, respectively and $h \approx a$
(b) current in wire 1 and wire 2 is the direction PQ and SR, respectively and $h \approx a$
(c) current in wire 1 and wire 2 is the direction PQ and SR, respectively and $h \approx 1.2a$
(d) current in wire 1 and wire 2 is the direction PQ and RS, respectively and $h \approx 1.2a$

16. Consider $d >> a$, and the loop is rotated about its diameter parallel to the wires by 30° from the position shown in the figure. If the currents in the wires are in the opposite directions, the torque on the loop at its new position will be (assume that the net field due to the wires is constant over the loop)

(a) $\frac{\mu_0 I^2 a^2}{d}$
(b) $\frac{\mu_0 I^2 a^2}{2d}$
(c) $\frac{\sqrt{3}\mu_0 I^2 a^2}{d}$
(d) $\frac{\sqrt{3}\mu_0 I^2 a^2}{2d}$

Section 3

Direction (Q. No. 17-20) *This section contains 4 multiple choice questions. Each question has matching lists. The codes for the lists have choices (a), (b), (c) and (d) out of which only one is correct.*

17. Four charges Q_1, Q_2, Q_3 and Q_4 of same magnitude are fixed along the x-axis at $x = -2a, -a, +a$ and $+2a$ respectively. A positive charge q is placed on the positive y-axis at a distance $b > 0$. Four options of the signs of these charges are given in List I. The direction of the forces on the charge q is given in List II. Match List I with List II and select the correct answer using the code given below the lists.

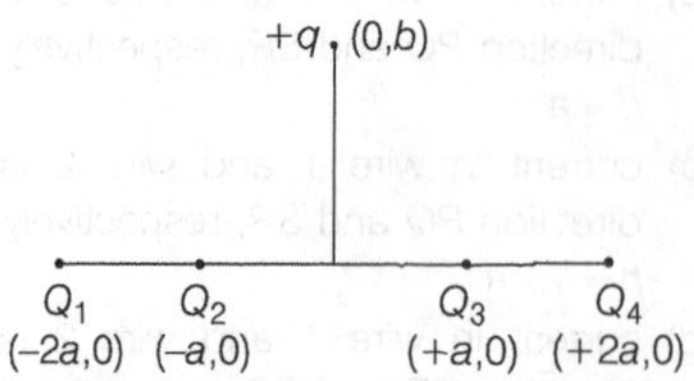

	List I		List II
P.	Q_1, Q_2, Q_3, Q_4 all positive	1.	$+x$
Q.	Q_1, Q_2 positive; Q_3, Q_4 negative	2.	$-x$
R.	Q_1, Q_4 positive; Q_2, Q_3 negative	3.	$+y$
S.	Q_1, Q_3 positive; Q_2, Q_4 negative	4.	$-y$

Codes

	P	Q	R	S
(a)	3	1	4	2
(b)	4	2	3	1
(c)	3	1	2	4
(d)	4	2	1	3

18. Four combinations of two thin lenses are given in List I.The radius of curvature of all curved surfaces is r and the refractive index of all the lenses is 1.5. Match lens combinations in List I with their focal length in List II and select the correct answer using the code given below the lists.

	List I		List II
P.	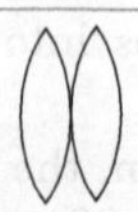	1.	$2r$
Q.	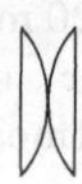	2.	$r/2$
R.		3.	$-r$
S.		4.	r

Codes

	P	Q	R	S
(a)	1	2	3	4
(b)	2	4	3	1
(c)	4	1	2	3
(d)	2	1	3	4

19. A block of mass $m_1 = 1$ kg another mass $m_2 = 2$ kg are placed together (see figure) on an inclined plane with angle of inclination θ. Various values of θ are given in List I.

The coefficient of friction between the block m_1 and the plane is always zero. The coefficient of static and dynamic friction between the block m_2 and the plane are equal to $\mu = 0.3$.

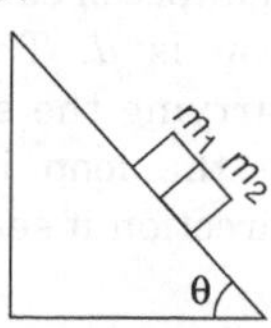

In List II expressions for the friction on the block m_2 are given. Match the correct expression of the friction in List II with the angles given in List I, and choose the correct option.

The acceleration due to gravity is denoted by g. **[Useful information** $\tan(5.5°) \approx 0.1$; **$\tan(11.5°) \approx 0.2$; $\tan(16.5°) \approx 0.3$]**

	List I		List II
P.	$\theta = 5°$	1.	$m_2 g \sin\theta$
Q.	$\theta = 10°$	2.	$(m_1 + m_2) g \sin\theta$
R.	$\theta = 15°$	3.	$\mu m_2 g \cos\theta$
S.	$\theta = 20°$	4.	$\mu (m_1 + m_2) g \cos\theta$

Codes

	P	Q	R	S
(a)	1	1	1	3
(b)	2	2	2	3
(c)	2	2	2	4
(d)	2	2	3	3

20. A person in a lift is holding a water jar, which has a small hole at the lower end of its side. When the lift is at rest, the water jet coming out of the hole hits the floor of the lift at a distance d of 1.2 m from the person.

In the following, state of the lift's motion is given in List I and the distance where the water jet hits the floor of the lift is given in List II. Match the statements from List I with those in List II and select the correct answer using the code given below the lists.

	List I		List II
P.	Lift is accelerating vertically up.	1.	$d = 1.2$ m
Q.	Lift is accelerating vertically down with an acceleration less than the gravitational acceleration.	2.	$d > 1.2$ m
R.	Lift is moving vertically up with constant speed.	3.	$d < 1.2$ m
S.	Lift is falling freely.	4.	No water leaks out of the jar

Codes

	P	Q	R	S
(a)	2	3	2	4
(b)	2	3	1	4
(c)	1	1	1	4
(d)	2	3	1	1

Chemistry

Section I

Directions (Q. Nos. 1-10) *This section contains 10 multiple choice questions. Each question has four choices (a), (b), (c) and (d) out of which only one is correct.*

1. Assuming $2s$-$2p$ mixing is not operative, the paramagnetic species among the following is

(a) Be_2 (b) B_2
(c) C_2 (d) N_2

2. For the process,

$$H_2O\,(l) \longrightarrow H_2O\,(g)$$

at $T = 100°C$ and 1 atmosphere pressure, the correct choice is

(a) $\Delta S_{system} > 0$ and $\Delta S_{surrounding} > 0$
(b) $\Delta S_{system} > 0$ and $\Delta S_{surrounding} < 0$
(c) $\Delta S_{system} < 0$ and $\Delta S_{surrounding} > 0$
(d) $\Delta S_{system} < 0$ and $\Delta S_{surrounding} < 0$

3. For the elementary reaction $M \rightarrow N$, the rate of disappearance of M increases by a factor of 8 upon doubling the concentration of M. The order of the reaction with respect to M is

(a) 4 (b) 3 (c) 2 (d) 1

4. For the identification of β-naphthol using dye test, it is necessary to use

(a) dichloromethane solution of β-naphthol
(b) acidic solution of β-naphthol
(c) neutral solution of β-naphthol
(d) alkaline solution of β-naphthol

5. Isomers of hexane, based on their branching, can be divided into three distinct classes as shown in the figure.

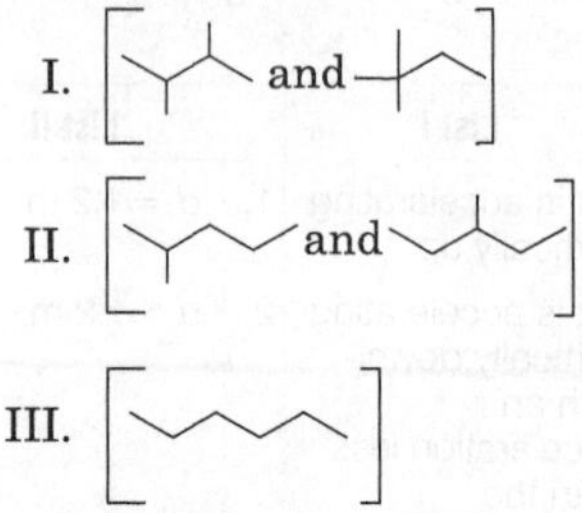

The correct order of their boiling point is

(a) I > II > III
(b) III > II > I
(c) II > III > I
(d) III > I > II

6. The major product in the following reaction is

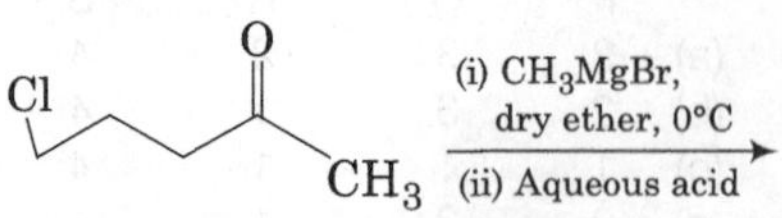

(a) 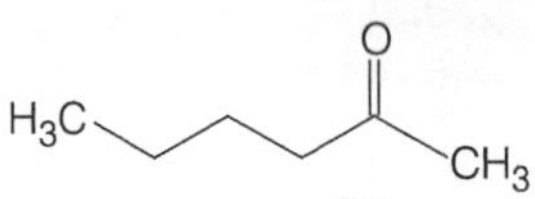

(b) 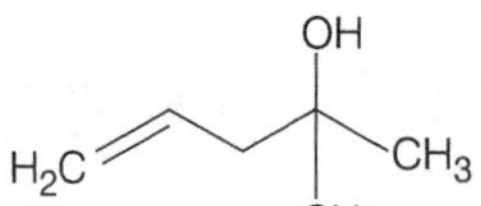

(c) 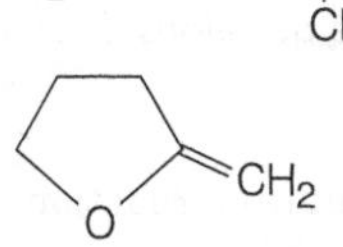

(d) 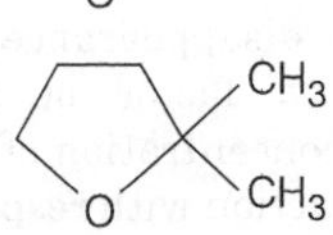

7. Under ambient conditions, the total number of gases released as products in the final step of the reaction scheme shown below is

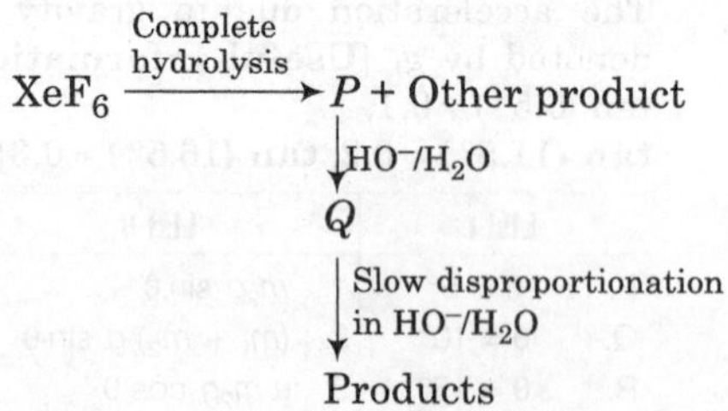

(a) 0 (b) 1 (c) 2 (d) 3

8. The product formed in the reaction of $SOCl_2$ with white phosphorus is

(a) PCl_3 (b) SO_2Cl_2 (c) SCl_2 (d) $POCl_3$

9. Hydrogen peroxide in its reaction with KIO_4 and NH_2OH respectively, is acting as a

(a) reducing agent, oxidising agent
(b) reducing agent, reducing agent
(c) oxidising agent, oxidising agent
(d) oxidising agent, reducing agent

10. The acidic hydrolysis of ether (X) shown below is fastest when

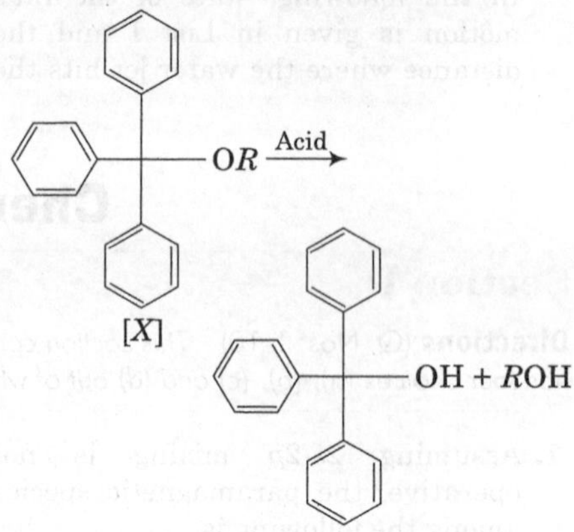

(a) one phenyl group is replaced by a methyl group
(b) one phenyl group is replaced by a *para*-methoxyphenyl group
(c) two phenyl groups are replaced by two *para*-methoxyphenyl groups
(d) no structural change is made to X

Section II

Directions (Q. Nos. 11-16) *This section contains 3 paragraphs, each describing theory, experiments, data etc. Six questions relate to the three paragraphs with two questions on each paragraph. Each question has only one correct answer among the four given options (a), (b), (c) and (d).*

Paragraph 1

X and Y are two volatile liquids with molar weights of 10 g mol^{-1} and 40 g mol^{-1} respectively. Two cotton plugs, one soaked in X and the other soaked in Y, are simultaneously placed at the ends of a tube of length $L = 24$ cm, as shown in the figure. The tube is filled with an inert gas at 1 atm pressure and a temperature of 300 K. Vapours of X and Y react to form a product which is first observed at a distance d cm from the plug soaked in X. Take X and Y to have equal molecular diameters and assume ideal behaviour for the inert gas and the two vapours.

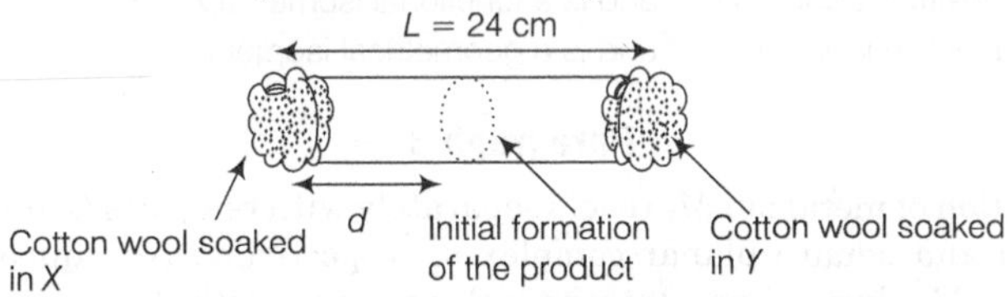

11. The value of d in cm (shown in the figure), as estimated from Graham's law, is

(a) 8 (b) 12 (c) 16 (d) 20

12. The experimental value of d is found to be smaller than the estimate obtained using Graham's law. This due to

(a) larger mean free path for X as a compared of that of Y

(b) larger mean free path for Y as compared to that of X

(c) increased collision frequency of Y with the inert gas as compared to that of X with the inert gas

(d) increased collision frequency of X with the inert gas as compared to that of Y with the inert gas

Paragraph 2

Schemes 1 and 2 describe sequential transformation of alkynes M and N. Consider only the major products formed in each step for both schemes.

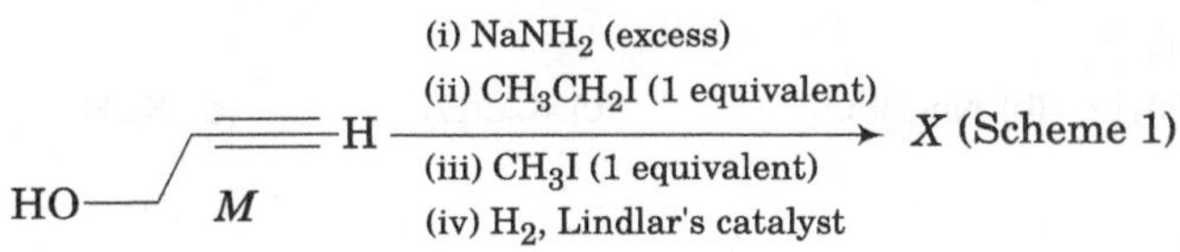

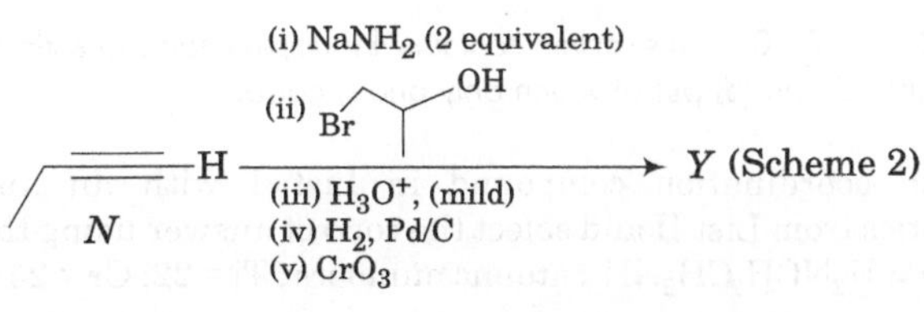

13. The product X is

(a) 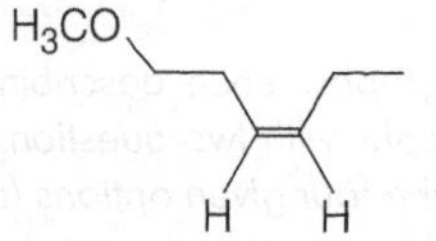

(b) 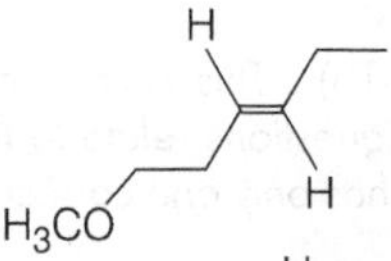

(c) 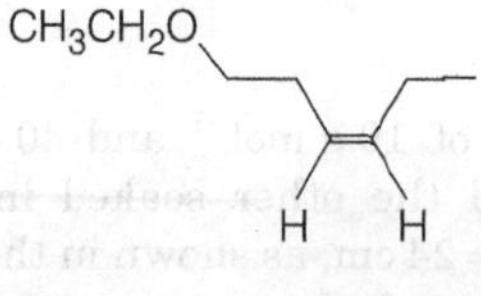

(d) 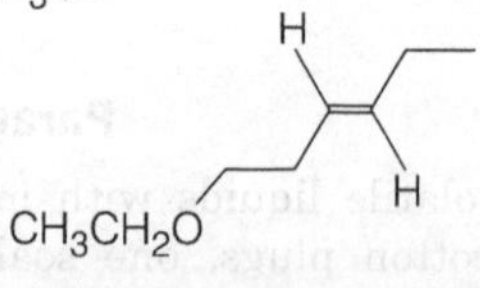

14. The correct statement with respect to product Y is

(a) it gives a positive Tollen's test and is a functional isomer of X
(b) it gives a positive Tollen's test and is a geometrical isomer of X
(c) it gives a positive iodoform test and is a functional isomer of X
(d) it gives a positive iodoform test and is a geometrical isomer of X

Paragraph 3

An aqueous solution of metal ion M_1 reacts separately with reagents Q and R in excess to give tetrahedral and square planar complexes, respectively. An aqueous solution of another metal ion M_2 always forms tetrahedral complexes with these reagents. Aqueous solution of M_2 on reaction with reagent S gives white precipitate which dissolves in excess of S. The reactions are summarized in the scheme given below

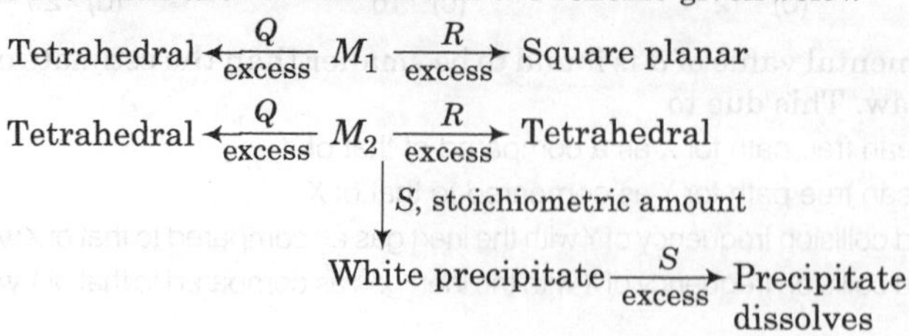

15. M_1, Q and R, respectively are

(a) Zn^{2+}, KCN and HCl
(b) Ni^{2+}, HCl and KCN
(c) Cd^{2+}, KCN and HCl
(d) Co^{2+}, HCl and KCN

16. Reagent S is

(a) $K_4[Fe(CN)_6]$ (b) Na_2HPO_4 (c) K_2CrO_4 (d) KOH

Section III

Directions (Q. Nos. 17-20) *This section contains 4 multiple choice questions. Each question has four choices (a), (b), (c) and (d) out of which only one is correct.*

17. Match each coordination compound in List I with an appropriate pair of characteristics from List II and select the correct answer using the code given below the lists (en = $H_2NCH_2CH_2NH_2$; atomic numbers : Ti = 22; Cr = 24; Co = 27; Pt = 78)

	List I		List II
P.	$[Cr(NH_3)_4Cl_2]Cl$	1.	Paramagnetic and exhibits ionisation isomerism
Q.	$[Ti(H_2O)_5Cl](NO_3)_2$	2.	Diamagnetic and exhibits *cis-trans* isomerism
R.	$[Pt(en)(NH_3)Cl]NO_3$	3.	Paramagnetic and exhibits *cis-trans* isomerism
S.	$[Co(NH_3)_4(NO_3)_2]NO_3$	4.	Diamagnetic and exhibits ionisation isomerism

Codes

	P	Q	R	S		P	Q	R	S
(a)	4	2	3	1	(b)	3	1	4	2
(c)	2	1	3	4	(d)	1	3	4	2

18. Match the orbital overlap figures shown in List I with the description given in List II and select the correct answer using the code given below the lists.

	List I		List II
P.	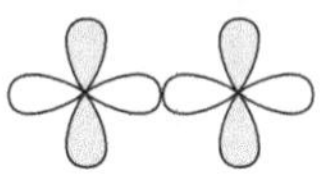	1.	*p-d* π antibonding
Q.	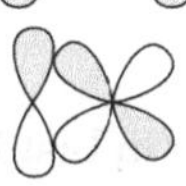	2.	*d-d* σ bonding
R.	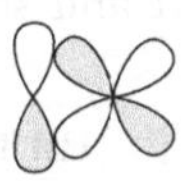	3.	*p-d* π bonding
S.	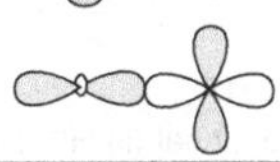	4.	*d-d* σ antibonding

Codes

	P	Q	R	S		P	Q	R	S
(a)	2	1	3	4	(b)	4	3	1	2
(c)	2	3	1	4	(d)	4	1	3	2

19. Different possible thermal decomposition pathways for peroxyesters are shown below. Match each pathway from List I with an appropriate structure from List II and select the correct answer using the code given below the lists.

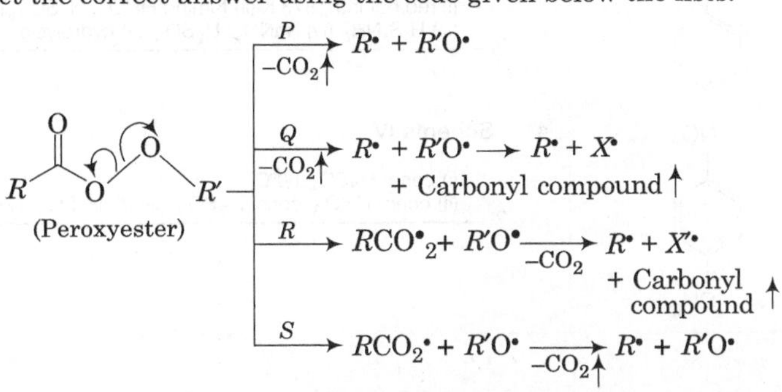

	List I		List II
P.	Pathway *P*	1.	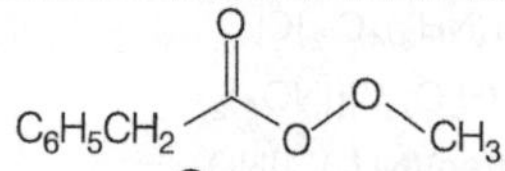
Q.	Pathway *Q*	2.	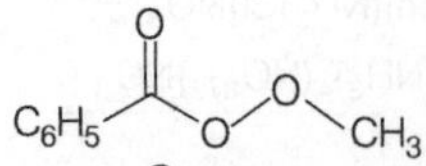
R.	Pathway *R*	3.	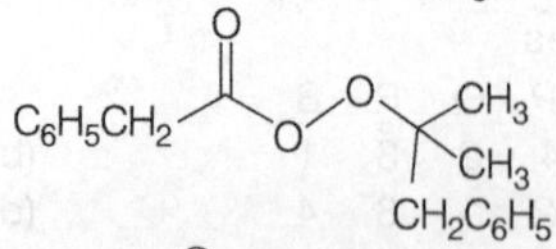
S.	Pathway *S*	4.	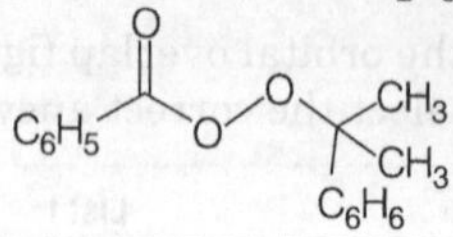

Codes

	P	Q	R	S		P	Q	R	S
(a)	1	3	4	2	(b)	2	4	3	1
(c)	4	1	2	3	(d)	3	2	1	4

20. **Match the four starting materials (P, Q, R, S) given in List I with the corresponding reaction schemes (I, II, III, IV) provided in List II and select the correct answer using the code given below the lists.**

	List I		List II
P.	$H-\equiv-H$	1.	**Scheme I** $? \xrightarrow[\text{(iii) } SOCl_2 \text{ (iv) } NH_3]{\text{(i) } KMnO_4,\ HO^-, \text{ heat (ii) } H^+, H_2O} C_7H_6N_2O_3$
Q.	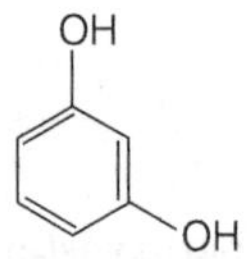	2.	Scheme II $? \xrightarrow[\text{(iv) } HNO_3 \text{ (v) dil. } H_2SO_4, \text{ heat (vi) } HO^-]{\text{(i) Sn/HCl (ii) } CH_3COCl \text{ (iii) conc. } H_2SO_4} C_6H_6N_2O_2$
R.	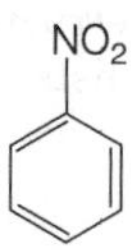	3.	**Scheme III** $? \xrightarrow[\text{(iii) } H_2S.NH_3 \text{ (iv) } NaNO_2, H_2SO_4 \text{ (v) hydrolysis}]{\text{(i) red hot iron, 873 K (ii) fuming } HNO_3, H_2SO_4, \text{ heat}} C_6H_5NO_3$
S.		4.	**Scheme IV** $? \xrightarrow[\text{(ii) conc. } HNO_3, \text{ conc. } H_2SO_4 \text{ (iii) dil.} H_2SO_4, \text{ heat}]{\text{(i) conc. } H_2SO_4, 60°C} C_6H_5NO_4$

Codes

	P	Q	R	S		P	Q	R	S
(a)	1	4	2	3	(b)	3	1	4	2
(c)	3	4	2	1	(d)	4	1	3	2

Mathematics

Section I

Directions (Q. Nos. 1-10) *This section contains 10 multiple choice questions. Each question has four choices (a), (b), (c) and (d) out of which one is correct.*

1. Three boys and two girls stand in a queue. The probability that the number of boys ahead of every girl is atleast one more than the number of girls ahead of her, is

(a) 1/2 (b) 1/3 (c) 2/3 (d) 3/4

2. In a triangle, the sum of two sides is x and the product of the same two sides is y. If $x^2 - c^2 = y$, where c is the third side of the triangle, then the ratio of the inradius to the circumradius of the triangle is

(a) $\frac{3y}{2x(x+c)}$ (b) $\frac{3y}{2c(x+c)}$

(c) $\frac{3y}{4x(x+c)}$ (d) $\frac{3y}{4c(x+c)}$

3. Six cards and six envelopes are numbered 1, 2, 3, 4, 5, 6 and cards are to be placed in envelopes so that each envelope contains exactly one card and no card is placed in the envelope bearing the same number and moreover the card numbered 1 is always placed in envelope numbered 2. Then, the number of ways it can be done is

(a) 264 (b) 265 (c) 53 (d) 67

4. The common tangents to the circle $x^2 + y^2 = 2$ and the parabola $y^2 = 8x$ touch the circle at the points P, Q and the parabola at the points R, S. Then, the area of the quadrilateral $PQRS$ is

(a) 3 (b) 6 (c) 9 (d) 15

5. The quadratic equation $p(x) = 0$ with real coefficients has purely imaginary roots. Then, the equation $p[p(x)] = 0$ has

(a) only purely imaginary roots

(b) all real roots

(c) two real and two purely imaginary roots

(d) neither real nor purely imaginary roots

6. The following integral $\int_{\pi/4}^{\pi/2} (2\operatorname{cosec} x)^{17} dx$ is equal to

(a) $\int_0^{\log(1+\sqrt{2})} 2(e^u + e^{-u})^{16} du$

(b) $\int_0^{\log(1+\sqrt{2})} (e^u + e^{-u})^{17} du$

(c) $\int_0^{\log(1+\sqrt{2})} (e^u - e^{-u})^{17} du$

(d) $\int_0^{\log(1+\sqrt{2})} 2(e^u - e^{-u})^{16} du$

7. The function $y = f(x)$ is the solution of the differential equation $\frac{dy}{dx} + \frac{xy}{x^2-1} = \frac{x^4+2x}{\sqrt{1-x^2}}$ in $(-1,1)$ satisfying $f(0) = 0$. Then, $\int_{-\frac{\sqrt{3}}{2}}^{\sqrt{3}/2} f(x)\, dx$ is

(a) $\frac{\pi}{3} - \frac{\sqrt{3}}{2}$ (b) $\frac{\pi}{3} - \frac{\sqrt{3}}{4}$

(c) $\frac{\pi}{6} - \frac{\sqrt{3}}{4}$ (d) $\frac{\pi}{6} - \frac{\sqrt{3}}{2}$

8. Let $f: [0,2] \to R$ be a function which is continuous on $[0,2]$ and is differentiable on $(0,2)$ with $f(0) = 1$.

Let $F(x) = \int_0^{x^2} f(\sqrt{t})\, dt$, for $x \in [0,2]$. If $F'(x) = f'(x)$, for all $x \in (0,2)$, then $F(2)$ equals to

(a) $e^2 - 1$ (b) $e^4 - 1$

(c) $e - 1$ (d) e^4

9. Coefficient of x^{11} in the expansion of $(1+x^2)^4 (1+x^3)^7 (1+x^4)^{12}$ is

(a) 1051 (b) 1106

(c) 1113 (d) 1120

10. For $x \in (0, \pi)$, the equation $\sin x + 2\sin 2x - \sin 3x = 3$ has

(a) infinitely many solutions

(b) three solutions

(c) one solution

(d) no solution

Section II

Directions (Q. Nos. 11-16) *This section contains 3 paragraph, each describing theory, experiments, data etc. Six questions relate to the three paragraphs with two questions on each paragraph. Each question has only one correct answer among the four given options (a), (b), (c) and (d).*

Paragraph 1

Box I contains three cards bearing numbers 1, 2, 3 ; box II contains five cards bearing numbers 1, 2, 3, 4, 5; and box III contains seven cards bearing numbers 1, 2, 3, 4, 5, 6, 7. A card is drawn from each of the boxes. Let x_i be the number on the card drawn from the ith box $i = 1, 2, 3$.

11. The probability that $x_1 + x_2 + x_3$ is odd, is

(a) $\frac{29}{105}$ (b) $\frac{53}{105}$

(c) $\frac{57}{105}$ (d) $\frac{1}{2}$

12. The probability that x_1, x_2, x_3 are in an arithmetic progression, is

(a) $\frac{9}{105}$ (b) $\frac{10}{105}$

(c) $\frac{11}{105}$ (d) $\frac{7}{105}$

Paragraph 2

Let a, r, s, t be non-zero real numbers. Let $P(at^2, 2at), Q, R(ar^2, 2ar)$ and $S(as^2, 2as)$ be distinct point on the parabola $y^2 = 4ax$. Suppose that PQ is the focal chord and lines QR and PK are parallel, where K is the point $(2a, 0)$.

13. The value of r is

(a) $-\frac{1}{t}$ (b) $\frac{t^2+1}{t}$

(c) $\frac{1}{t}$ (d) $\frac{t^2-1}{t}$

14. If $st = 1$, then the tangent at P and the normal at S to the parabola meet at a point whose ordinate is

(a) $\frac{(t^2+1)^2}{2t^3}$ (b) $\frac{a(t^2+1)^2}{2t^3}$

(c) $\frac{a(t^2+1)^2}{t^3}$ (d) $\frac{a(t^2+2)^2}{t^3}$

Paragraph 3

Given that for each $a \in (0, 1)$, $\lim_{h \to 0^+} \int_h^{1-h} t^{-a}(1-t)^{a-1} dt$ exists. Let this limit be $g(a)$. In addition, it is given that the function $g(a)$ is differentiable on $(0, 1)$.

15. The value of $g\left(\frac{1}{2}\right)$ is

(a) π (b) 2π

(c) $\pi/2$ (d) $\pi/4$

16. The value of $g'(1/2)$ is

(a) $\frac{\pi}{2}$ (b) π

(c) $-\frac{\pi}{2}$ (d) 0

Section III

Directions (Q. Nos. 17-20) *This section contains 4 multiple choice questions. Each question has four choices (a), (b), (c) and (d) out of which only one is correct.*

17. Match the following

	List I		List II
P.	The number of polynomials $f(x)$ with non-negative integer coefficients of degree ≤ 2, satisfying $f(0) = 0$ and $\int_0^1 f(x)dx = 1$, is	(i)	8
Q.	The number of points in the interval $[-\sqrt{13}, \sqrt{13}]$ at which $f(x) = \sin(x^2) + \cos(x^2)$ attains its maximum value, is	(ii)	2
R.	$\int_{-2}^{2} \frac{3x^2}{(1+e^x)} dx$ equals	(iii)	4
S.	$\frac{\left(\int_{-1/2}^{1/2} \cos 2x \log\left(\frac{1+x}{1-x}\right) dx\right)}{\left(\int_0^{1/2} \cos 2x \log\left(\frac{1+x}{1-x}\right) dx\right)}$ equals	(iv)	0

Codes

	P	Q	R	S		P	Q	R	S
(a)	(iii)	(ii)	(iv)	(i)	(b)	(ii)	(iii)	(iv)	(i)
(c)	(iii)	(ii)	(i)	(iv)	(d)	(ii)	(iii)	(i)	(iv)

18. Match List I with List II and select the correct answer using the codes given below the lists.

	List I		List II
P.	Let $y(x) = \cos(3\cos^{-1} x)$, $x \in [-1, 1]$, $x \neq \pm \frac{\sqrt{3}}{2}$. Then, $\frac{1}{y(x)}\left\{(x^2 - 1)\frac{d^2y(x)}{dx^2} + \frac{xdy(x)}{dx}\right\}$ equals	(i)	1
Q.	Let $A_1, A_2, \ldots A_n, (n > 2)$ be the vertices of a regular polygon of n sides with its centre at the origin. Let a_k be the position vector of the point $A_k, k = 1, 2, \ldots, n$. If $\left\|\sum_{k=1}^{n-1} (\mathbf{a}_k \cdot \mathbf{a}_{k+1})\right\| = \left\|\sum_{k=1}^{n-1} (\mathbf{a}_k \cdot \mathbf{a}_{k+1})\right\|$, then the minimum value of n is	(ii)	2
R.	If the normal from the point $P(h, 1)$ on the ellipse $\frac{x^2}{6} + \frac{y^2}{3} = 1$ is perpendicular to the line $x + y = 8$, then the value of h is	(iii)	8
S.	Number of positive solutions satisfying the equation $\tan^{-1}\left(\frac{1}{2x+1}\right) + \tan^{-1}\left(\frac{1}{4x+1}\right) = \tan^{-1}(2/x^2)$ is	(iv)	9

Codes

	P	Q	R	S		P	Q	R	S
(a)	(iv)	(iii)	(ii)	(i)	(b)	(ii)	(iv)	(iii)	(i)
(c)	(iv)	(iii)	(i)	(ii)	(d)	(ii)	(iv)	(i)	(iii)

19. Let $f_1 : R \to R, f_2 : [0, \infty] \to R, f_3 : R \to R$ and $f_4 : R \to [0, \infty)$ be defined by

$$f_1(x) = \begin{cases} |x|, & \text{if } x < 0 \\ e^x, & \text{if } x \geq 0 \end{cases}; \; f_2(x) = x^2;$$

$$f_3(x) = \begin{cases} \sin x, & \text{if } x < 0 \\ x, & \text{if } x \geq 0 \end{cases}$$

$$\text{and } f_4(x) = \begin{cases} f_2[f_1(x)], & \text{if } x < 0 \\ f_2[f_1(x)] - 1, & \text{if } x \geq 0 \end{cases}$$

	List I		List II
P.	f_4 is	(i)	onto but not one-one
Q.	f_3 is	(ii)	neither continuous nor one-one
R.	$f_2 \circ f_1$ is	(iii)	differentiable but not one-one
S.	f_2 is	(iv)	continuous and one-one

	P	Q	R	S
(a)	(iii)	(i)	(iv)	(ii)
(b)	(i)	(iii)	(iv)	(ii)
(c)	(iii)	(i)	(ii)	(iv)
(d)	(i)	(iii)	(ii)	(iv)

20. Let $z_k = \cos\left(\frac{2k\pi}{10}\right) + i \sin\left(\frac{2k\pi}{10}\right)$; $k = 1, 2, \ldots, 9$.

	List I		List II
P.	For each z_k, there exists a z_j such that $z_k \cdot z_j = 1$	(i)	True
Q.	There exists a $k \in \{1, 2, \ldots, 9\}$ such that $z_1 \cdot z = z_k$ has no solution z in the set of complex numbers	(ii)	False
R.	$\frac{\|1 - z_1\|\|1 - z_2\| \ldots \|1 - z_9\|}{10}$ equals	(iii)	1
S.	$1 - \sum_{k=1}^{9} \cos\left(\frac{2k\pi}{10}\right)$ equals	(iv)	2

Codes

	P	Q	R	S
(a)	(i)	(ii)	(iv)	(iii)
(b)	(ii)	(i)	(iii)	(iv)
(c)	(i)	(ii)	(iv)	(iii)
(d)	(ii)	(i)	(iv)	(iii)

Answers

Paper 1

Physics

1. (c,d) **2.** (a,b,c) **3.** (a,c,d) **4.** (a,d) **5.** (c) **6.** (d) **7.** (b,d) **8.** (c,d) **9.** (a,c) **10.** (a,b,d)
11. (5) **12.** (4) **13.** (2) **14.** (3) **15.** (3) **16.** (5) **17.** (5) **18.** (2 or 8) **19.** (4) **20.** (2)

Chemistry

1. (a,b,d) **2.** (a,b,c) **3.** (a) **4.** (a) **5.** (b,c,d) **6.** (a,b,d) **7.** (a,b,c) **8.** (a,c,d) **9.** (a,b,c) **10.** (b,d)
11. (6) **12.** (1) **13.** (4) **14.** (8) **15.** (2) **16.** (7) **17.** (3) **18.** (6/7) **19.** (5) **20.** (4)

Mathematics

1. (a,c) **2.** (a,d) **3.** (c,d) **4.** (a,b,c) **5.** (c) **6.** (a,b) **7.** (a,c,d) **8.** (a,b,c) **9.** (b,c) **10.** (b,d)
11. (8) **12.** (3) **13.** (0) **14.** (3) **15.** (6) **16.** (7) **17.** (2) **18.** (4) **19.** (4) **20.** (5)

Paper 2

Physics

1. (b) **2.** (d) **3.** (c) **4.** (c) **5.** (c) **6.** (a) **7.** (a) **8.** (b) **9.** (b) **10.** (d)
11. (d) **12.** (d) **13.** (c) **14.** (a) **15.** (c) **16.** (b) **17.** (a) **18.** (b) **19.** (d) **20.** (c)

Chemistry

1. (c) **2.** (b) **3.** (b) **4.** (d) **5.** (b) **6.** (d) **7.** (c) **8.** (a) **9.** (a) **10.** (c)
11. (c) **12.** (d) **13.** (a) **14.** (c) **15.** (b) **16.** (d) **17.** (b) **18.** (c) **19.** (a) **20.** (c)

Mathematics

1. (a) **2.** (b) **3.** (c) **4.** (d) **5.** (d) **6.** (a) **7.** (b) **8.** (b) **9.** (c) **10.** (d)
11. (b) **12.** (c) **13.** (d) **14.** (b) **15.** (a) **16.** (d) **17.** (d) **18.** (a) **19.** (d) **20.** (c)

Unsolved Paper 2015
JEE Main
Joint Entrance Examination

Instructions

1. This test consists of 90 questions.
2. There are three parts in the question paper A, B, C consisting of Physics, Chemistry & Mathematics having 30 questions in each part of equal weightage. Each question is allotted 4 marks for correct response.
3. Candidates will be awarded marks as stated above in instruction no. 2 for correct response of each question. 1 marks will be deducted for indicating incorrect response of each question. No deduction from the total score will be made if no response is indicated for an item in the answer sheet.
4. There is only one correct response for each question. Filling up more than one response in any question will be treated as wrong response and marks for wrong response will be deducted according as per instructions.

Physics

1. Two stones are thrown up simultaneously from the edge of a cliff 240 m high with initial speed of 10 m/s and 40 m/s respectively. Which of the following graph best represents the time variation of relative position of the second stone with respect to the first? Assume stones do not rebound after hitting the ground and neglect air resistance, take $g = 10\,\text{m/s}^2$)

(a)
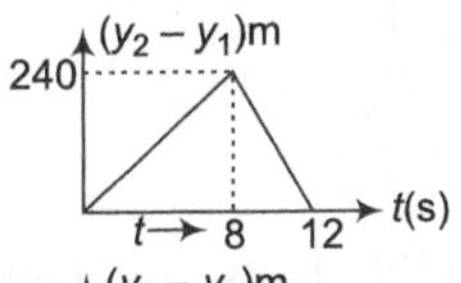

(c)
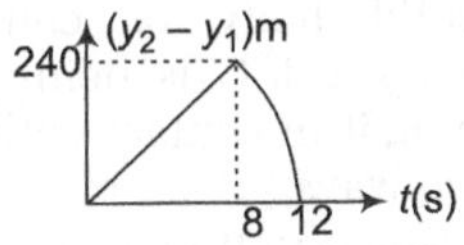

(b)
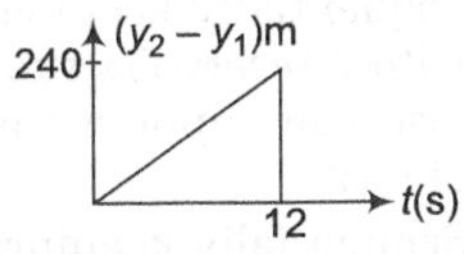

(d)
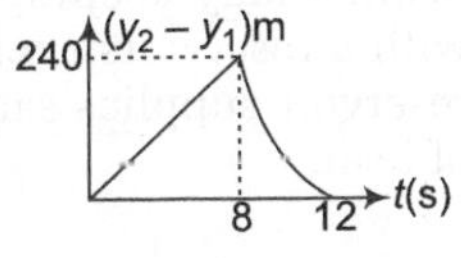

2. The period of oscillation of a simple pendulum is $T = 2\pi\sqrt{\frac{L}{g}}$. Measured value of L is 20.0 cm known to 1 mm accuracy and time for 100 oscillations of the pendulum is found to be 90 s using a wrist watch of 1s resolution. The accuracy in the determination of g is

(a) 2% (b) 3% (c) 1% (d) 5%

3. Given in the figure are two blocks A and B of weight 20 N and 100 N respectively. These are being pressed against a wall by a force F as shown in figure. If the coefficient of friction between the blocks is 0.1 and between block B and the wall is 0.15, the frictional force applied by the wall in block B is

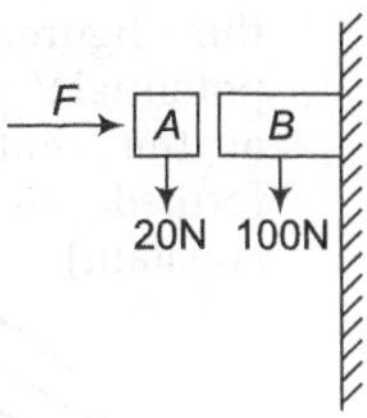

(a) 100 N
(b) 80 N
(c) 120 N
(d) 150 N

4. A particle of mass m moving in the x-direction with speed $2v$ is hit by another particle of mass $2m$ moving in the y-direction with speed v. If the collision is perfectly inelastic, the percentage loss in the energy during the collision is close to

(a) 44% (b) 50% (c) 56% (d) 62%

5. Distance of the centre of mass of a solid uniform cone from its vertex is z_0. If the radius of its base is R and its height is h, then z_0 is equal to

(a) $\frac{h^2}{4R}$ (b) $\frac{3h}{4}$

(c) $\frac{5h}{8}$ (d) $\frac{3h^2}{8R}$

6. From a solid sphere of mass M and radius R, a cube of maximum possible volume is cut. Moment of inertia of cube about an axis passing through its centre and perpendicular to one of its faces is

(a) $\frac{MR^2}{32\sqrt{2}\pi}$ (b) $\frac{MR^2}{16\sqrt{2}\pi}$

(c) $\frac{4MR^2}{9\sqrt{3}\pi}$ (d) $\frac{4MR^2}{3\sqrt{3}\pi}$

7. From a solid sphere of mass M and radius R, a spherical portion of radius $\left(\frac{R}{2}\right)$ is removed as shown in the figure. Taking gravitational potential $V = 0$ at $r = \infty$, the potential at the centre of the cavity thus formed is (G = gravitational constant)

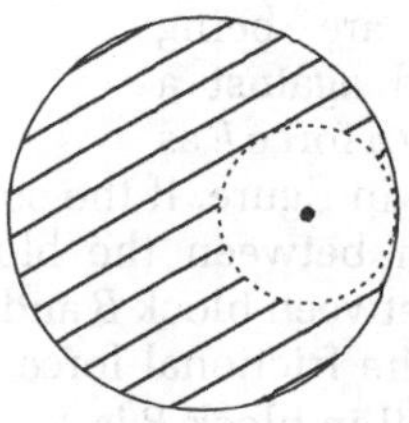

(a) $\frac{-GM}{2R}$ (b) $\frac{-GM}{R}$

(c) $\frac{-2GM}{3R}$ (d) $\frac{-2GM}{R}$

8. A pendulum made of a uniform wire of cross-sectional area A has time period T. When an additional mass M is added to its bob, the time period changes T_M. If the Young's modulus of the material of the wire is Y, then $\frac{1}{Y}$ is equal to (g = gravitational acceleration)

(a) $\left[\left(\frac{T_M}{T}\right)^2 - 1\right]\frac{A}{Mg}$

(b) $\left[\left(\frac{T_M}{T}\right)^2 - 1\right]\frac{Mg}{A}$

(c) $\left[1-\left(\frac{T_M}{T}\right)^2\right]\frac{A}{Mg}$

(d) $\left[1-\left(\frac{T}{T_M}\right)^2\right]\frac{A}{Mg}$

9. Consider a spherical shell of radius R at temperature T. The black body radiation inside it can be considered as an ideal gas of photons with internal energy per unit volume $u = \frac{U}{V} \propto T^4$ and pressure $p = \frac{1}{3}\left(\frac{U}{V}\right)$. If the shell now undergoes an adiabatic expansion, the relation between T and R is

(a) $T \propto e^{-R}$

(b) $T \propto e^{-3R}$

(c) $T \propto \frac{1}{R}$

(d) $T \propto \frac{1}{R^3}$

10. A solid body of constant heat capacity 1 J/°C is being heated by keeping it in contact with reservoirs in two ways

(i) Sequentially keeping in contact with 2 reservoirs such that each reservoir supplies same amount of heat.

(ii) Sequentially keeping in contact with 8 reservoirs such that each reservoir supplies same amount of heat.

22. An inductor ($L = 0.03$ H) and a resistor ($R = 0.15$ kΩ) are connected in series to a battery of 15V EMF in a circuit shown below. The key K_1 has been kept closed for a long time. Then at $t = 0$, K_1 is opened and key K_2 is closed simultaneously. At $t = 1$ms, the current in the circuit will be ($e^5 \cong 150$)

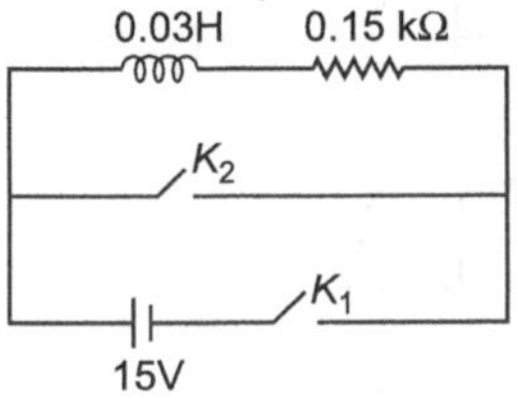

(a) 100 mA (b) 67 mA
(c) 6.7 mA (d) 0.67mA

23. A red LED emits light at 0.1 W uniformly around it. The amplitude of the electric field of the light at a distance of 1 m from the diode is

(a) 1.73 V/m
(b) 2.45 V/m
(c) 5.48 V/m
(d) 7.75 V/m

24. Monochromatic light is incident on a glass prism of angle A. If the refractive index of the material of the prism is μ, a ray incident at an angle θ, on the face AB would get transmitted through the face AC of the prism provided

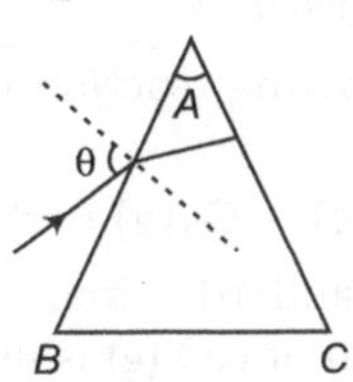

(a) $\theta > \sin^{-1}\left[\mu \sin\left(A - \sin^{-1}\left(\frac{1}{\mu}\right)\right)\right]$

(b) $\theta < \sin^{-1}\left[\mu \sin\left(A - \sin^{-1}\left(\frac{1}{\mu}\right)\right)\right]$

(c) $\theta > \cos^{-1}\left[\mu \sin\left(A + \sin^{-1}\left(\frac{1}{\mu}\right)\right)\right]$

(d) $\theta < \cos^{-1}\left[\mu \sin\left(A + \sin^{-1}\left(\frac{1}{\mu}\right)\right)\right]$

25. On a hot summer night, the refractive index of air is smallest near the ground and increases with height from the ground. When a light beam is directed horizontally, the Huygens principle leads us to conclude that as it travels, the light beam

(a) becomes narrower
(b) goes horizontally without any deflection
(c) bends downwards
(d) bends upwards

26. Assuming human pupil to have a radius of 0.25 cm and a comfortable viewing distance of 25 cm, the minimum separation between two objects that human eye can resolve at 500 nm wavelength is

(a) 1 μm (b) 30 μm (c) 100 μm(d) 300 μm

27. As an electron makes a transition from an excited state to the ground state of a hydrogen like atom/ion

(a) its kinetic energy increases but potential energy and total energy decrease
(b) kinetic energy, potential energy and total energy decrease
(c) kinetic energy decreases, potential energy increases but total energy remains same
(d) kinetic energy and total energy decrease but potential energy increases

28. Match List I (fundamental experiment) with List II (its conclusion) and select the correct option from the choices given below the list.

List I	List II
A. Franck-Hertz experiment	1. Particle nature of light
B. Photo-electric experiment	2. Discrete energy levels of atom
C. Davisson-Germer experiment	3. Wave nature of electron
	4. Structure of atom

	A	B	C		A	B	C
(a)	1	4	3	(b)	2	4	3
(c)	2	1	3	(d)	4	3	2

29. A signal of 5 kHz frequency is amplitude modulated on a carrier wave of frequency 2MHz. The frequencies of the resultant signal is/are

(a) 2 MHz only
(b) 2005 kHz and 1995 kHz
(c) 2005 kHz 2000 kHz and 1995 kHz
(d) 2000 kHz and 1995 kHz

30. An LCR circuit is equivalent to a damped pendulum. In an LCR circuit, the capacitor is charged to Q_0 and then connected to the L and R as shown below.

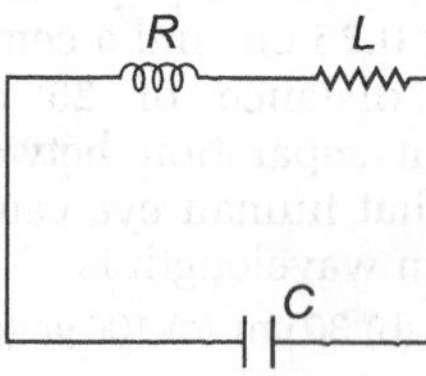

If a student plots graphs of the square of maximum charge (Q^2_{Max}) on the capacitor with time (t) for two different values L_1 and L_2 ($L_1 > L_2$) of L, then which of the following represents this graph correctly? (plots are schematic and not drawn to scale)

(a) 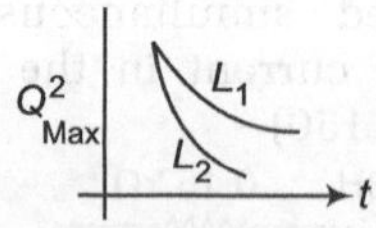

(c) 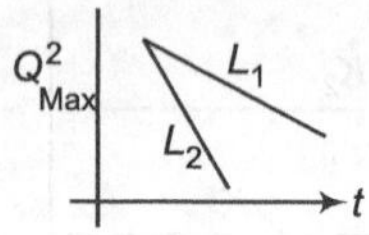

(b) 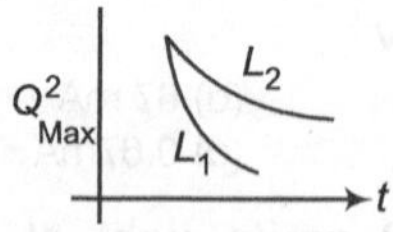

(d) 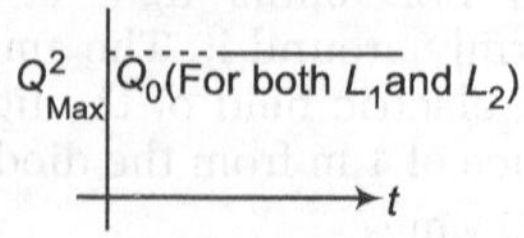

Chemistry

1. The molecular formula of a commercial resin used for exchanging ions in water softening is $C_8H_7SO_3Na$ (mol. wt. = 206). What would be the maximum uptake of Ca^{2+} ions by the resin when expressed in mole per gram resin?

(a) $\frac{1}{103}$ (b) $\frac{1}{206}$
(c) $\frac{2}{309}$ (d) $\frac{1}{412}$

2. Sodium metal crystallises in a body centred cubic lattice with a unit cell edge of 4.29 Å. The radius of sodium atom is approximately

(a) 1.86 Å (b) 3.22 Å (c) 5.72 Å (d) 0.93 Å

3. Which of the following is the energy of a possible excited state of hydrogen?

(a) +13.6 eV (b) −6.8 eV
(c) −3.4 eV (d) +6.8 eV

4. The intermolecular interaction that is dependent on the inverse cube of distance between the molecules is

(a) ion-ion interaction
(b) ion-dipole interaction
(c) London force
(d) hydrogen bond

5. The following reaction is performed at 298K

$$2NO(g) + O_2(g) \rightleftharpoons 2NO_2(g)$$

The standard free energy of formation of NO (g) is 86.6 kJ/mol at 298 K. What is the standard free energy of formation of $NO_2(g)$ at 298 K? ($K_p = 1.6 \times 10^{12}$)

(a) $R(298)\ \ln(1.6\times 10^{12}) - 86600$
(b) $86600 + R(298)\ \ln(1.6\times 10^{12})$
(c) $86600 - \frac{\ln(1.6\times 10^{12})}{R(298)}$
(d) $0.5\ [2\times 86600 - R(298)\ \ln(1.6\times 10^{12})]$

6. The vapour pressure of acetone at 20°C is 185 torr. When 1.2 g of a non-volatile substance was dissolved in 100 g of acetone at 20°C, its vapour pressure was 183 torr. The molar mass $(g\,mol^{-1})$ of the substance is
(a) 32 (b) 64 (c) 128 (d) 488

7. The standard Gibbs energy change at 300K for the reaction, $2A \rightleftharpoons B + C$ is 2494. 2 J. At a given time, the composition of the reaction mixture is $[A] = \frac{1}{2}$, $[B] = 2$ and $[C] = \frac{1}{2}$. The reaction proceeds in the $[R = 8.314\ JK/mol, e = 2.718]$
(a) forward direction because $Q > K_c$
(b) reverse direction because $Q > K_c$
(c) forward direction because $Q < K_c$
(d) reverse direction because $Q < K_c$

8. Two Faraday of electricity is passed through a solution of $CuSO_4$. The mass of copper deposited at the cathode is (at. mass of Cu = 63.5 u)
(a) 0g (b) 63.5g (c) 2g (d) 127g

9. Higher order (>3) reactions are rare due to
(a) low probability of simultaneous collision of all the reacting species
(b) increase in entropy and activation energy as more molecules are involved
(c) shifting of equilibrium towards reactants due to elastic collisions
(d) loss of active species on collision

10. 3 g of activated charcoal was added to 50 mL of acetic acid solution (0.06 N) in a flask. After an hour it was filtered and the strength of the filtrate was found to be 0.042 N. The amount of acetic acid adsorbed (per gram of charcoal) is
(a) 18 mg (b) 36 mg
(c) 42 mg (d) 54 mg

11. The ionic radii (in Å) of N^{3-}, O^{2-} and F^- respectively are
(a) 1.36, 1.40 and 1.71
(b) 1.36, 1.71 and 1.40
(c) 1.71, 1.40 and 1.36
(d) 1.71, 1.36 and 1.40

12. In the context of the Hall-Heroult process for the extraction of Al, which of the following statements is false?
(a) CO and CO_2 are produced in this process
(b) Al_2O_3 is mixed with CaF_2 which lowers the melting point of the mixture and brings conductivity
(c) Al^{3+} is reduced at the cathode to form Al
(d) Na_3AlF_6 serves as the electrolyte

13. From the following statements regarding H_2O_2, choose the incorrect statement.
(a) It can act only as an oxidising agent
(b) It decomposed on exposure to light
(c) It has to be stored in plastic or wax lined glass bottles in dark
(d) It has to be kept away from dust

14. Which one of the following alkaline earth metal sulphates has its hydration enthalpy greater than its lattice enthalpy?
(a) $CaSO_4$ (b) $BeSO_4$ (c) $BaSO_4$ (d) $SrSO_4$

15. Which among the following is the most reactive?
(a) Cl_2 (b) Br_2 (c) I_2 (d) ICl

16. Match the catalysts to the correct processes.

Catalyst	Process
(A) $TiCl_3$	(i) Wacker process
(B) $PdCl_2$	(ii) Ziegler- Natta polymerisation
(C) $CuCl_2$	(iii) Contact process
(D) V_2O_5	(iv) Deacon's process

(a) (A)- (iii), (B) - (ii), (C) - (iv), (D) - (i)
(b) (A)- (ii), (B) - (i), (C) - (iv), (D) - (iii)
(c) (A)- (ii), (B) - (iii), (C) - (iv), (D) - (i)
(d) (A)- (iii), (B) - (i), (C) - (ii), (D) - (iv)

17. Which one has the highest boiling point?
(a) He (b) Ne (c) Kr (d) Xe

18. The number of geometric isomers that can exist for square planar $[Pt(Cl)(py)(NH_3)(NH_2OH)]^+$ is (py = pyridine)
(a) 2 (b) 3 (c) 4 (d) 6

19. The colour of $KMnO_4$ is due to
(a) $M \rightarrow L$ charge transfer transition
(b) $d - d$ transition
(c) $L \rightarrow M$ charge transfer transition
(d) $\sigma - \sigma^*$ transition

20. Assertion (A) Nitrogen and oxygen are the main components in the atmosphere but these do not react to form oxides of nitrogen.

Reason (R) The reaction between nitrogen and oxygen requires high temperature.
(a) Both Assertion and Reason are correct and the reason is the correct explanation for the Assertion.
(b) Both Assertion and Reason are correct but the reason is not the correct explanation for the Assertion.
(c) The Assertion is incorrect but the Reason is correct.
(d) Both the Assertion and Reason are incorrect.

21. In Carius method of estimation of halogens, 250 mg of an organic compound gave 141 mg of AgBr. The percentage of bromine in the compound is
(at. mass Ag = 108, Br = 80)
(a) 24 (b) 36 (c) 48 (d) 60

22. Which of the following compound will exhibit geometrical isomerism?
(a) 1-phenyl-2-butene
(b) 3-phenyl-1-butene
(c) 2-phenyl-1-butene
(d) 1, 1-diphenyl-1-propane

23. Which compound would give 5-keto-2-methyl hexanal upon ozonolysis?

(a) 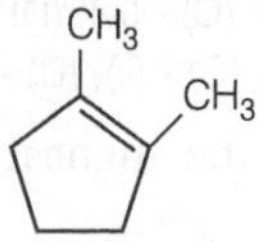

(b) 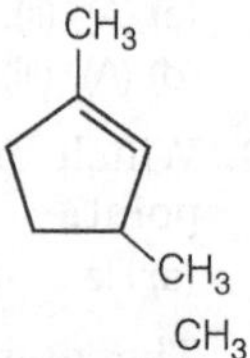

(c) 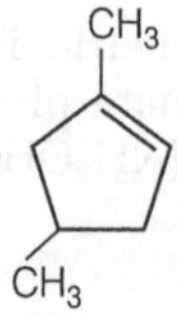

(d) 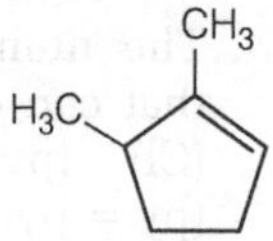

24. The synthesis of alkyl fluorides is best accomplished by
(a) free radical fluorination
(b) Sandmeyer's reaction
(c) Finkelstein reaction
(d) Swarts reaction

25. In the following sequence of reaction,
Toluene $\xrightarrow{KMnO_4} A \xrightarrow{SOCl_2} B \xrightarrow[BaSO_4]{H_2/Pd} C$
The product C is
(a) C_6H_5COOH
(b) $C_6H_5CH_3$
(c) $C_6H_5CH_2OH$
(d) C_6H_5CHO

26. In the reaction,

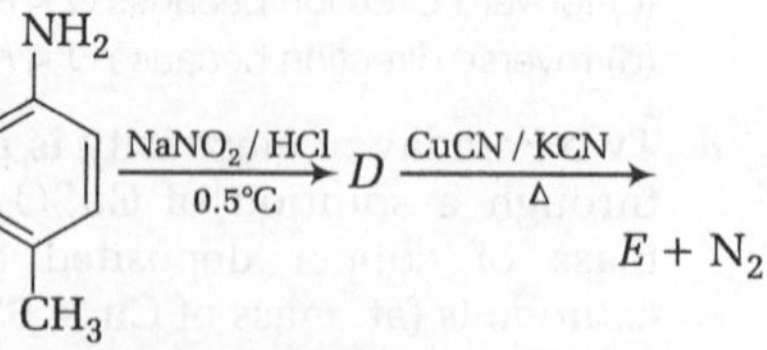

The product E is

(a) 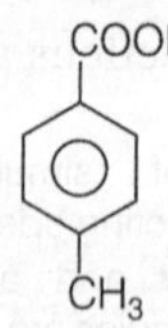

(b) 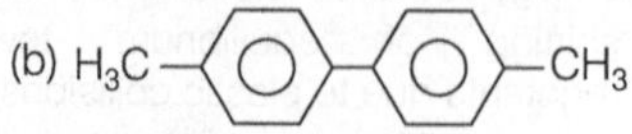

(c) 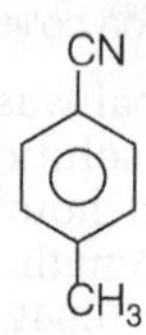

(d) 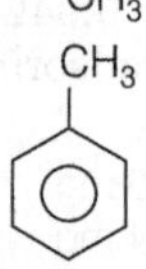

27. Which polymer is used in the manufacture of paints and lacquers?
(a) Bakelite
(b) Glyptal
(c) Polypropene
(d) Polyvinyl chloride

28. Which of the vitamins given below is water soluble?
(a) Vitamin C (b) Vitamin D
(c) Vitamin E (d) Vitamin K

29. Which of the following compounds is not an antacid?
(a) Aluminium hydroxide
(b) Cimetidine
(c) Phenelzine
(d) Ranitidine

30. Which of the following compounds is not coloured yellow?
(a) $Zn_2[Fe(CN)_6]$
(b) $K_3[Co(NO_2)_6]$
(c) $(NH_4)_3[As(Mo_3O_{10})_4]$
(d) $BaCrO_4$

Mathematics

1. Let A and B be two sets containing four and two elements respectively. Then, the number of subsets of the set $A \times B$, each having atleast three elements are
(a) 219
(b) 256
(c) 275
(d) 510

2. A complex number z is said to be unimodular, if $|z| = 1$. Suppose z_1 and z_2 are complex numbers such that $\frac{z_1 - 2z_2}{2 - z_1\bar{z}_2}$ is unimodular and z_2 is not unimodular. Then, the point z_1 lies on a
(a) straight line parallel to X-axis
(b) straight line parallel to Y-axis
(c) circle of radius 2
(d) circle of radius $\sqrt{2}$

3. Let α and β be the roots of equation $x^2 - 6x - 2 = 0$. If $a_n = \alpha^n - \beta^n$, for $n \ge 1$, then the value of $\frac{a_{10} - 2a_8}{2a_9}$ is equal to
(a) 6 (b) −6 (c) 3 (d) −3

4. If $A = \begin{bmatrix} 1 & 2 & 2 \\ 2 & 1 & -2 \\ a & 2 & b \end{bmatrix}$ is a matrix satisfying the equation $AA^T = 9I$, where I is 3×3 identity matrix, then the ordered pair (a, b) is equal to
(a) (2, −1) (b) (−2, 1)
(c) (2, 1) (d) (−2, −1)

5. The set of all values of λ for which the system of linear equations $2x_1 - 2x_2 + x_3 = \lambda x_1$, $2x_1 - 3x_2 + 2x_3 = \lambda x_2$ and $-x_1 + 2x_2 = \lambda x_3$ has a non-trivial solution,
(a) is an empty set
(b) is a singleton set
(c) contains two elements
(d) contains more than two elements

6. The number of integers greater than 6000 that can be formed, using the digits 3, 5, 6, 7 and 8 without repetition, is
(a) 216 (b) 192 (c) 120 (d) 72

7. The sum of coefficients of integral powers of x in the binomial expansion of $(1 - 2\sqrt{x})^{50}$ is
(a) $\frac{1}{2}(3^{50} + 1)$
(b) $\frac{1}{2}(3^{50})$
(c) $\frac{1}{2}(3^{50} - 1)$
(d) $\frac{1}{2}(2^{50} + 1)$

8. If m is the AM of two distinct real numbers l and $n (l, n > 1)$ and G_1, G_2 and G_3 are three geometric means between l and n, then $G_1^4 + 2G_2^4 + G_3^4$ equals
(a) $4 l^2mn$
(b) $4 lm^2n$
(c) $4 lmn^2$
(d) $4l^2m^2n^2$

9. The sum of first 9 terms of the series $\frac{1^3}{1}+\frac{1^3+2^3}{1+3}+\frac{1^3+2^3+3^3}{1+3+5}+\ldots$ is

(a) 71 (b) 96
(c) 142 (d) 192

10. $\lim_{x\to 0}\frac{(1-\cos 2x)(3+\cos x)}{x\tan 4x}$ is equal to

(a) 4 (b) 3
(c) 2 (d) $\frac{1}{2}$

11. If the function $g(x)=\begin{cases}k\sqrt{x+1}, & 0\le x\le 3\\ mx+2, & 3<x\le 5\end{cases}$ is differentiable, then the value of $k+m$ is

(a) 2 (b) $\frac{16}{5}$
(c) $\frac{10}{3}$ (d) 4

12. The normal to the curve $x^2+2xy-3y^2=0$ at $(1,1)$

(a) does not meet the curve again
(b) meets the curve again in the second quadrant
(c) meets the curve again in the third quadrant
(d) meets the curve again in the fourth quadrant

13. Let $f(x)$ be a polynomial of degree four having extreme values at $x=1$ and $x=2$. If $\lim_{x\to 0}\left[1+\frac{f(x)}{x^2}\right]=3$, then $f(2)$ is equal to

(a) −8
(b) −4
(c) 0
(d) 4

14. The integral $\int\frac{dx}{x^2(x^4+1)^{\frac{3}{4}}}$ equals

(a) $\left(\frac{x^4+1}{x^4}\right)^{\frac{1}{4}}+c$ (b) $(x^4+1)^{\frac{1}{4}}+c$
(c) $-(x^4+1)^{\frac{1}{4}}+c$ (d) $-\left(\frac{x^4+1}{x^4}\right)^{\frac{1}{4}}+c$

15. The integral $\int_2^4\frac{\log x^2}{\log x^2+\log(36-12x+x^2)}dx$ is equal to

(a) 2 (b) 4
(c) 1 (d) 6

16. The area (in sq units) of the region described by $\{(x,y): y^2\le 2x$ and $y\ge 4x-1\}$ is

(a) $\frac{7}{32}$ (b) $\frac{5}{64}$ (c) $\frac{15}{64}$ (d) $\frac{9}{32}$

17. Let $y(x)$ be the solution of the differential equation $(x\log x)\frac{dy}{dx}+y=2x\log x,\quad (x\ge 1)$. Then, $y(e)$ is equal to

(a) e (b) 0
(c) 2 (d) $2e$

18. The number of points having both coordinates as integers that lie in the interior of the triangle with vertices $(0,0)$, $(0,41)$ and $(41,0)$ is

(a) 901 (b) 861
(c) 820 (d) 780

19. Locus of the image of the point $(2,3)$ in the line $(2x-3y+4)+k(x-2y+3)=0$, $k\in R$, is a

(a) straight line parallel to X-axis
(b) straight line parallel to Y-axis
(c) circle of radius $\sqrt{2}$
(d) circle of radius $\sqrt{3}$

20. The number of common tangents to the circles $x^2+y^2-4x-6y-12=0$ and $x^2+y^2+6x+18y+26=0$ is

(a) 1 (b) 2 (c) 3 (d) 4

21. The area (in sq units) of the quadrilateral formed by the tangents at the end points of the latusrectum to the ellipse $\frac{x^2}{9}+\frac{y^2}{5}=1$ is

(a) $\frac{27}{4}$ (b) 18
(c) $\frac{27}{2}$ (d) 27

22. Let O be the vertex and Q be any point on the parabola $x^2 = 8y$. If the point P divides the line segment OQ internally in the ratio 1 : 3, then the locus of P is

(a) $x^2 = y$
(b) $y^2 = x$
(c) $y^2 = 2x$
(d) $x^2 = 2y$

23. The distance of the point (1, 0, 2) from the point of intersection of the line $\frac{x-2}{3} = \frac{y+1}{4} = \frac{z-2}{12}$ and the plane $x - y + z = 16$ is

(a) $2\sqrt{14}$ (b) 8
(c) $3\sqrt{21}$ (d) 13

24. The equation of the plane containing the line $2x - 5y + z = 3$, $x + y + 4z = 5$ and parallel to the plane $x + 3y + 6z = 1$ is

(a) $2x + 6y + 12z = 13$
(b) $x + 3y + 6z = -7$
(c) $x + 3y + 6z = 7$
(d) $2x + 6y + 12z = -13$

25. Let **a**, **b** and **c** be three non-zero vectors such that no two of them are collinear and $(\mathbf{a} \times \mathbf{b}) \times \mathbf{c} = \frac{1}{3}|\mathbf{b}||\mathbf{c}|\mathbf{a}$. If θ is the angle between vectors **b** and **c**, then a value of $\sin\theta$ is

(a) $\frac{2\sqrt{2}}{3}$
(b) $\frac{-\sqrt{2}}{3}$
(c) $\frac{2}{3}$
(d) $\frac{-2\sqrt{3}}{3}$

26. If 12 identical balls are to be placed in 3 identical boxes, then the probability that one of the boxes contains exactly 3 balls, is

(a) $\frac{55}{3}\left(\frac{2}{3}\right)^{11}$ (b) $55\left(\frac{2}{3}\right)^{10}$
(c) $220\left(\frac{1}{3}\right)^{12}$ (d) $22\left(\frac{1}{3}\right)^{11}$

27. The mean of the data set comprising of 16 observations is 16. If one of the observation valued 16 is deleted and three new observations valued 3, 4 and 5 are added to the data, then the mean of the resultant data is

(a) 16.8 (b) 16.0
(c) 15.8 (d) 14.0

28. If the angles of elevation of the top of a tower from three collinear points A, B and C on a line leading to the foot of the tower are 30°, 45° and 60° respectively, then the ratio $AB : BC$ is

(a) $\sqrt{3} : 1$ (b) $\sqrt{3} : \sqrt{2}$
(c) $1 : \sqrt{3}$ (d) 2 : 3

29. Let

$$\tan^{-1} y = \tan^{-1} x + \tan^{-1}\left(\frac{2x}{1-x^2}\right),$$

where $|x| < \frac{1}{\sqrt{3}}$. Then, a value of y is

(a) $\frac{3x - x^3}{1 - 3x^2}$ (b) $\frac{3x + x^3}{1 - 3x^2}$
(c) $\frac{3x - x^3}{1 + 3x^2}$ (d) $\frac{3x + x^3}{1 + 3x^2}$

30. The negation of $\sim s \vee (\sim r \wedge s)$ is equivalent to

(a) $s \wedge \sim r$ (b) $s \wedge (r \wedge \sim s)$
(c) $s \vee (r \vee \sim s)$ (d) $s \wedge r$

Answers

Physics

1. (c)	**2.** (b)	**3.** (c)	**4.** (c)	**5.** (b)	**6.** (c)	**7.** (b)	**8.** (a)	**9.** (c)	**10.** (b)
11. (c)	**12.** (b)	**13.** (b)	**14.** (a)	**15.** (c,d)	**16.** (b)	**17.** (d)	**18.** (c)	**19.** (a)	**20.** (b)
21. (c)	**22.** (d)	**23.** (b)	**24.** (a)	**25.** (d)	**26.** (b)	**27.** (a)	**28.** (c)	**29.** (c)	**30.** (a)

Chemistry

1. (d)	**2.** (a)	**3.** (c)	**4.** (b)	**5.** (d)	**6.** (b)	**7.** (b)	**8.** (b)	**9.** (a)	**10.** (a)
11. (c)	**12.** (d)	**13.** (a)	**14.** (b)	**15.** (d)	**16.** (b)	**17.** (d)	**18.** (b)	**19.** (c)	**20.** (a)
21. (a)	**22.** (a)	**23.** (b)	**24.** (d)	**25.** (d)	**26.** (c)	**27.** (b)	**28.** (a)	**29.** (c)	**30.** (a)

Mathematics

1. (a)	**2.** (c)	**3.** (c)	**4.** (d)	**5.** (c)	**6.** (b)	**7.** (a)	**8.** (b)	**9.** (b)	**10.** (c)
11. (a)	**12.** (d)	**13.** (c)	**14.** (d)	**15.** (c)	**16.** (d)	**17.** (c)	**18.** (d)	**19.** (c)	**20.** (c)
21. (d)	**22.** (d)	**23.** (d)	**24.** (c)	**25.** (a)	**26.** (a)	**27.** (d)	**28.** (a)	**29.** (a)	**30.** (d)

Unsolved Paper 2015
JEE Advanced

Paper 1

Physics

Section 1 (Maximum Marks : 32)

- This section contains EIGHT questions.
- The answer to each question is a SINGLE DIGIT INTEGER ranging from 0 to 9, both inclusive.
- For each question, darken the bubble corresponding to the correct integer in ORS.
- Marking scheme :
  + 4 If the bubble corresponding to the answer is darkened
  0 in all other cases

1. A Young's double slit interference arrangement with slits S_1 and S_2 is immersed in water (refractive index $= 4/3$) as shown in the figure. The positions of maxima on the surface of water are given by $x^2 = p^2m^2\lambda^2 - d^2$, where λ is the wavelength of light in air (refractive index $= 1$), $2d$ is the separation between the slits and m is an integer. The value of p is

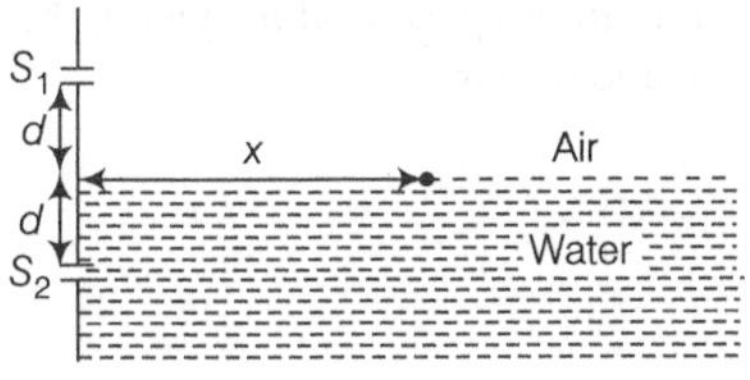

2. Consider a concave mirror and a convex lens (refractive index $= 1.5$) of focal length 10 cm each, separated by a distance of 50 cm in air (refractive index $= 1$) as shown in the figure. An object is placed at a distance of 15 cm from the mirror. Its erect image formed by this combination has magnification M_1. When the set-up is kept in a medium of refractive index $\frac{7}{6}$, the magnification becomes M_2. The magnitude $\left|\frac{M_2}{M_1}\right|$ is

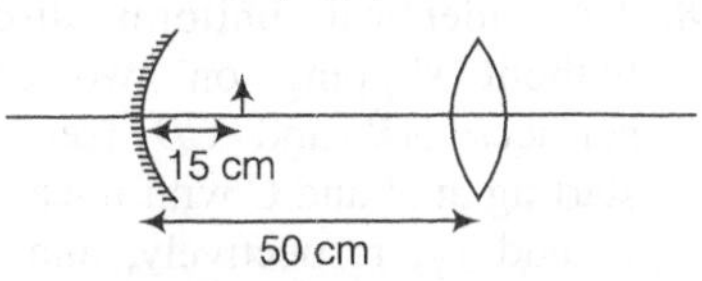

3. An infinitely long uniform line charge distribution of charge per unit length λ lies parallel to the y-axis in the y-z plane at $z = \frac{\sqrt{3}}{2}a$ (see figure). If the magnitude of the flux of the electric field through the rectangular surface $ABCD$ lying in the x-y plane with its centre at the origin is $\frac{\lambda L}{n\varepsilon_0}$ (ε_0 = permittivity of free space), then the value of n is

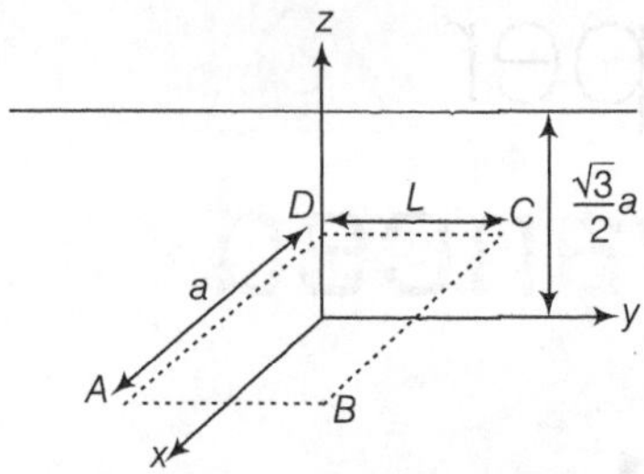

4. Consider a hydrogen atom with its electro in the nth orbital. An electromagnetic radiation of wavelength 90 nm is used to ionize the atom. If the kinetic energy of the ejected electron is 10.4 eV, then the value of n is ($hc = 1242$ eV nm)

5. A bullet is fired vertically upwards with velocity v from the surface of a spherical planet. When it reaches its maximum height, its acceleration due to the planet's gravity is $\frac{1}{4}$th of its value at the surface of the planet. If the escape velocity from the planet is $v_{sec} = v\sqrt{N}$, then the value of N is (ignore energy loss due to atmosphere)

6. Two identical uniform discs roll without slipping on two different surfaces AB and CD (see figure) starting at A and C with linear speeds v_1 and v_2, respectively, and always remain in contact with the surfaces. If they reach B and D with the same linear speed and $v_1 = 3$ m/s, then v_2 in m/s is ($g = 10\,m/s^2$)

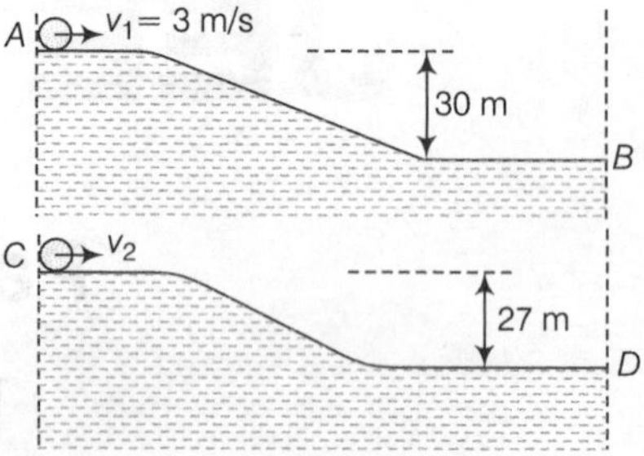

7. Two spherical stars A and B emit blackbody radiation. The radius of A is 400 times that of B and A emits 10^4 times the power emitted from B. The ratio $\left(\frac{\lambda_A}{\lambda_B}\right)$ of their wavelengths λ_A and λ_B at which the peaks occur in their respective radiation curves is

8. A nuclear power plant supplying electrical power to a village uses a radioactive material of half life T years as the fuel.

The amount of fuel at the beginning is such that the total power requirement of the village is 12.5% of the electrical power available from the plant at that time. If the plant is able to meet the total power needs of the village for a maximum period of nT years, then the value of n is

Section 2 (Maximum Marks : 40)

- This section contains TEN questions.
- Each question has FOUR options (a), (b), (c) and (d), ONE OR MORE THAN ONE of these four option(s) is(are) correct.
- For each question, darken the bubble(s) corresponding to all the correct option(s) in the ORS.
- Marking scheme :

  + 4 If only the bubble(s) corresponding to all the correct option(s) is(are) darkened.

  0 If none of the bubbles is darkened.

  − 2 In all other cases.

9. A ring of mass M and radius R is rotating with angular speed ω about a fixed vertical axis passing through its centre O with two point masses each of mass $\frac{M}{8}$ at rest at O. These masses can move radially outwards along two massless rods fixed on the ring as shown in the figure. At some instant, the angular speed of the system is $\frac{8}{9}\omega$ and one of the masses is at a distance of $\frac{3}{5}R$ from O. At this instant, the distance of the other mass from O is

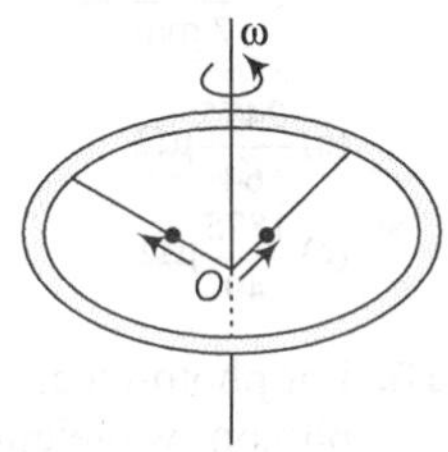

(a) $\frac{2}{3}R$ (b) $\frac{1}{3}R$

(c) $\frac{3}{5}R$ (d) $\frac{4}{5}R$

10. The figures below depict two situations in which two infinitely long static line charges of constant positive line charge density λ are kept parallel to each other.

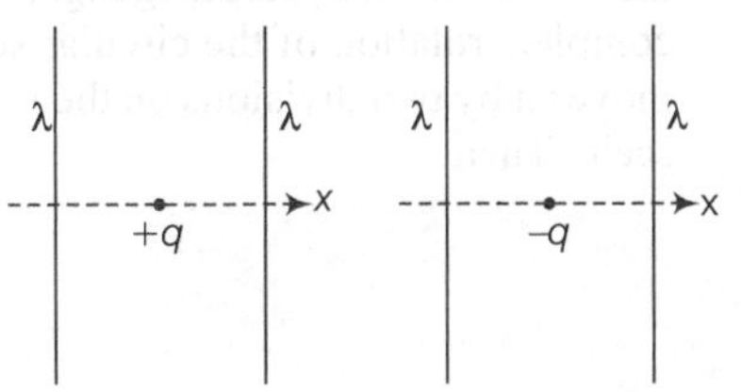

In their resulting electric field, point charges q and $-q$ are kept in equilibrium between them. The point charges are confined to move in the x direction only. If they are given a small displacement about their equilibrium positions, then the correct statements is/are

(a) both charges execute simple harmonic motion

(b) both charges will continue moving in the direction of their displacement

(c) charge $+q$ executes simple harmonic motion while charge $-q$ continues moving in the direction of its displacement

(d) charge $-q$ executes simple harmonic motion while charge $+q$ continues moving in the direction of its displacement

11. Two identical glass rods S_1 and S_2 (refractive index $= 1.5$) have one convex end of radius of curvature 10 cm. They are placed with the curved surfaces at a distance d as shown in the figure, with their axes (shown by the dashed line) aligned. When a point source of light P is placed inside rod S_1 on its axis at a distance of 50 cm from the curved face, the light rays emanating from it are found to be parallel to the axis inside S_2. The distance d is

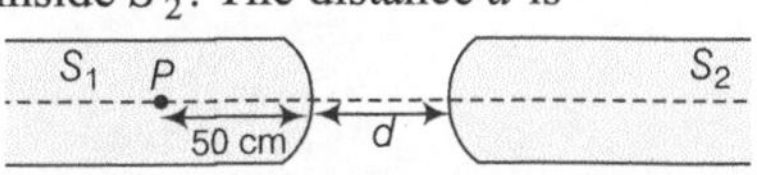

(a) 60 cm (b) 70 cm (c) 80 cm (d) 90 cm

12. A conductor (shown in the figure) carrying constant current I is kept in the x-y plane in a uniform magnetic field **B**. If F is the magnitude of the total magnetic force acting on the conductor, then the correct statements is/are

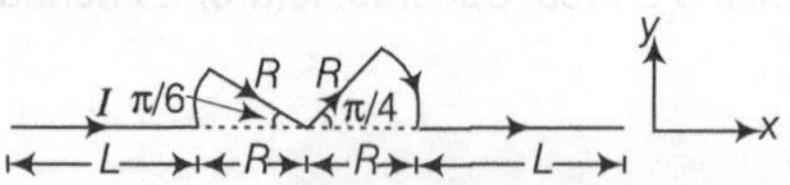

(a) if **B** is along $\hat{\mathbf{z}}$, $F \propto (L + R)$
(b) if **B** is along $\hat{\mathbf{x}}$, $F = 0$
(c) if **B** is along $\hat{\mathbf{y}}$, $F \propto (L + R)$
(d) if **B** is along $\hat{\mathbf{z}}$, $F = 0$

13. A container of fixed volume has a mixture of one mole of hydrogen and one mole of helium in equilibrium at temperature T. Assuming the gases are ideal, the correct statements is/are

(a) The average energy per mole of the gas mixture is $2RT$
(b) The ratio of speed of sound in the gas mixture to that in helium gas is $\sqrt{\frac{6}{5}}$
(c) The ratio of the rms speed of helium atoms to that of hydrogen molecules is $\frac{1}{2}$
(d) The ratio of the rms speed of helium atoms to that of hydrogen molecules is $\frac{1}{\sqrt{2}}$

14. In an aluminium (Al) bar of square cross section, a square hole is drilled and is filled with iron (Fe) as shown in the figure. The electrical resistivities of Al and Fe are $2.7 \times 10^{-8}\ \Omega$ m and $1.0 \times 10^{-7}\ \Omega$ m, respectively. The electrical resistance between the two faces P and Q of the composite bar is

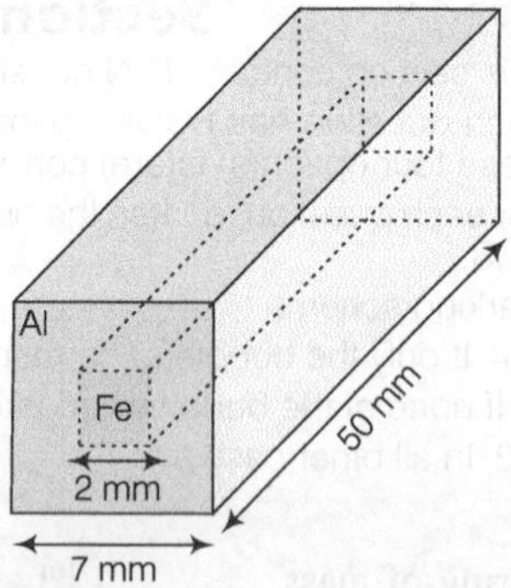

(a) $\frac{2475}{64}\ \mu\Omega$
(b) $\frac{1875}{64}\ \mu\Omega$
(c) $\frac{1875}{49}\ \mu\Omega$
(d) $\frac{2475}{132}\ \mu\Omega$

15. For photo-electric effect with incident photon wavelength λ, the stopping potential is V_0. Identify the correct variation(s) of V_0 with λ and $\frac{1}{\lambda}$.

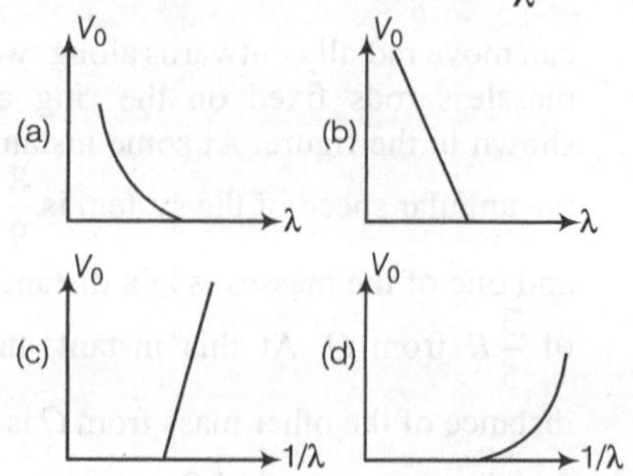

16. Consider a Vernier callipers in which each 1 cm on the main scale is divided into 8 equal divisions and a screw gauge with 100 divisions on its circular scale. In the Vernier callipers, 5 divisions of the Vernier scale coincide with 4 divisions on the main scale and in the screw gauge, one complete rotation of the circular scale moves it by two divisions on the linear scale. Then

(a) If the pitch of the screw gauge is twice the least count of the Vernier callipers, the least count of the screw gauge is 0.01 mm
(b) If the pitch of the screw gauge is twice the least count of the Vernier callipers, the least count of the screw gauge is 0.05 mm
(c) If the least count of the linear scale of the screw gauge is twice the least count of the Vernier callipers, the least count of the screw gauge is 0.01 mm
(d) If the least count of the linear scale of the screw gauge is twice the least count of the Vernier callipers, the least count of the screw gauge is 0.005 mm

17. Planck's constant h, speed of light c and gravitational constant G are used to form a unit of length L and a unit of mass M. Then, the correct options is/are

(a) $M \propto \sqrt{c}$ (b) $M \propto \sqrt{G}$
(c) $L \propto \sqrt{h}$ (d) $L \propto \sqrt{G}$

18. Two independent harmonic oscillators of equal masses are oscillating about the origin with angular frequencies ω_1 and ω_2 and have total energies E_1 and E_2, respectively. The variations of their momenta p with positions x are shown in the figures. If $\frac{a}{b} = n^2$ and $\frac{a}{R} = n$, then the correct equations is/are

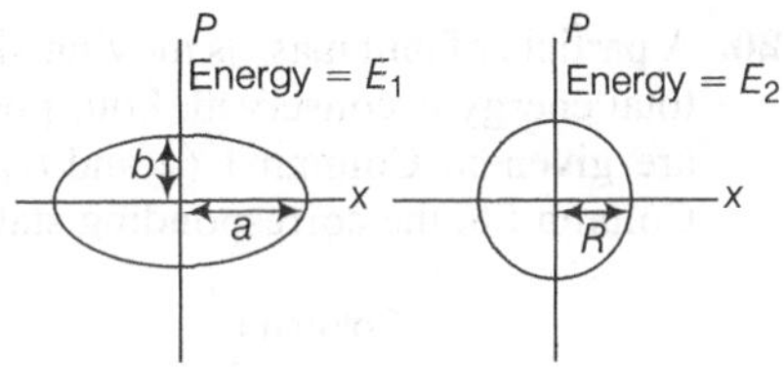

(a) $E_1\omega_1 = E_2\omega_2$ (b) $\frac{\omega_2}{\omega_1} = n^2$
(c) $\omega_1\omega_2 = n^2$ (d) $\frac{E_1}{\omega_1} = \frac{E_2}{\omega_2}$

Section 3 (Maximum Marks : 16)

- This section contains TWO questions.
- Each question contains two columns, Column I and Column II.
- Column I has four entries (A), (B), (C) and (D).
- Column II has five entries (P), (Q), (R), (S) and (T).
- Match the entries in Column I with the entries in Column II.
- One or more entries in Column I may match with one or more entries in Column II.
- The ORS contains a 4 × 5 matrix whose layout will be similar to the one shown below :

(A)	(P)	(Q)	(R)	(S)	(T)
(B)	(P)	(Q)	(R)	(S)	(T)
(C)	(P)	(Q)	(R)	(S)	(T)
(D)	(P)	(Q)	(R)	(S)	(T)

- For each entry in Column I, darken the bubbles of all the matching entries. For example, if entry (A) in Column I matches with entries (Q), (R) and (T), then darken these three bubbles in the ORS. Similarly, for entries (B), (C) and (D).
- Marking schemes:

For each entry in Column I

+ 2 If only the bubble(s) corresponding to all the correct match(es) is (are) darkened.
0 If none of the bubbles is darkened.
– 1 In all other cases.

19. **Match the nuclear processes given in Column I with the appropriate option(s) in Column II.**

	Column I		Column II
A.	Nuclear fusion	P.	absorption of thermal neutrons by ${}^{235}_{92}U$
B.	Fission in a nuclear reactor	Q.	${}^{60}_{27}Co$ nucleus
C.	β-decay	R.	Energy production in stars via hydrogen conversion to helium
D.	γ-ray emission	S.	Heavy water
		T.	Neutrino emission

20. **A particle of unit mass is moving along the *x*-axis under the influence of a force and its total energy is conserved. Four possible forms of the potential energy of the particle are given in Column I (*a* and U_0 are constants). Match the potential energies in Column I to the corresponding statements in Column II.**

	Column I		Column II
A.	$U_1(x) = \frac{U_0}{2}\left[1-\left(\frac{x}{a}\right)^2\right]^2$	P.	The force acting on the particle is zero at $x = a$
B.	$U_2(x) = \frac{U_0}{2}\left(\frac{x}{a}\right)^2$	Q.	The force acting on the particle is zero at $x = 0$
C.	$U_3(x) = \frac{U_0}{2}\left(\frac{x}{a}\right)^2 \exp\left[-\left(\frac{x}{a}\right)^2\right]$	R.	The force acting on the particle is zero at $x = -a$
D.	$U_4(x) = \frac{U_0}{2}\left[\frac{x}{a} - \frac{1}{3}\left(\frac{x}{a}\right)^3\right]$	S.	The particle experiences an attractive force towards $x = 0$ in the region $\lvert x \rvert < a$
		T.	The particle with total energy $\frac{U_0}{4}$ can oscillate about the point $x = -a$

Chemistry

Section 1 (Maximum Marks : 32)

- This section contains EIGHT questions.
- The answer to each question is a SINGLE DIGIT INTEGER ranging from 0 to 9, both inclusive.
- For each question, darken the bubble corresponding to the correct integer in ORS.
- Marking scheme :
  + 4 If the bubble corresponding to the answer is darkened.
  0 in all other cases.

1. The total number of stereoisomers that can exist for *M* is

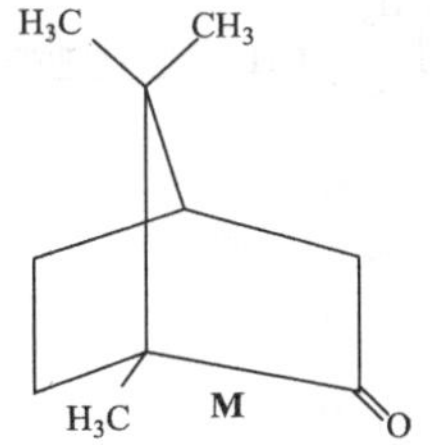

2. The number of resonance structures for *N* is

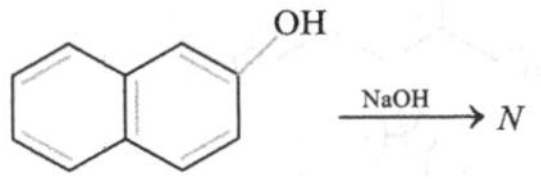

3. The total number of lone pair of electrons in N_2O_3 is

4. For the octahedral complexes of Fe^{3+} in SCN^- (thiocyanato-S) and in CN^- ligand environments, the difference between the spin only magnetic moments in Bohr magnetons (when approximated to the nearest integer) is [atomic number of Fe = 26]

5. Among the triatomic molecules/ions $BeCl_2$, N_3^-, N_2O, NO_2^+, O_3, SCl_2, ICl_2^-, I_3^- and XeF_2, the total number of linear molecules(s)/ion(s) where the hybridisation of the central atom does not have contribution from the *d*-orbital(s) is [atomic number of S = 16, Cl = 17, I = 53 and Xe = 54]

6. Not considering the electronic spin, the degeneracy of the second excited state ($n = 3$) of H-atom is 9, while the degeneracy of the second excited state of H^- is

7. All the energy released from the reaction

$X \rightarrow Y, \Delta_r G^\circ = -193\,\text{kJ mol}^{-1}$ is

used for oxidising M^+ as

$M^+ \rightarrow M^{3+} + 2e^-, E^\circ = -0.25\,\text{V}$.

Under standard conditions, the number of moles of M^+ oxidised when one mole of X is converted to Y is [F = 96500 C mol^{-}]

8. If the freezing point of a 0.01 molal aqueous solution of a cobalt (III) chloride-ammonia complex (which behaves as a strong electrolyte) is – 0.0558°C, the number of chloride(s) in the coordination sphere of the complex is

[K_f of water $= 1.86\,\text{K kg mol}^{-1}$]

Section 2 (Maximum Marks : 40)

- This section contains TEN questions.
- Each question has FOUR options (a), (b), (c) and (d), ONE OR MORE THAN ONE of these four option(s) is/are correct.
- For each question, darken the bubble(s) corresponding to the correct option(s) in the ORS.
- Marking scheme :
  + 4 If the bubble(s) corresponding to all the correct option(s) is/are darkened.
  0 If none of the bubble is darkened.

9. Compound(s) that on hydrogenation produce(s) optically inactive compound (s) is/are

(a) 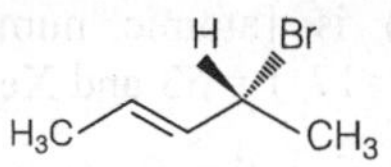

(b) 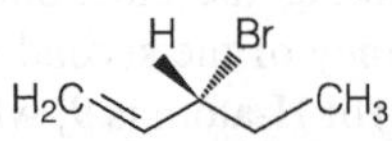

(c) 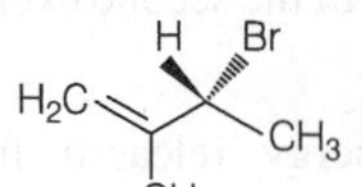

(d) 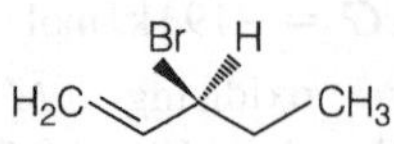

10. The major product of the following reaction is

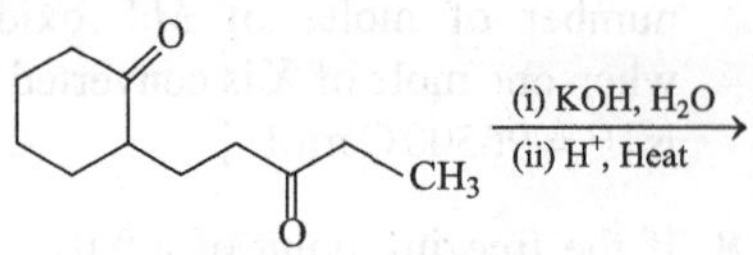

(a) 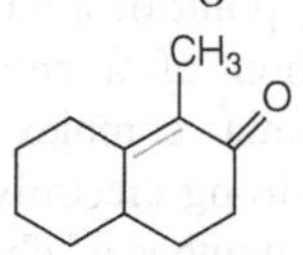

(b) 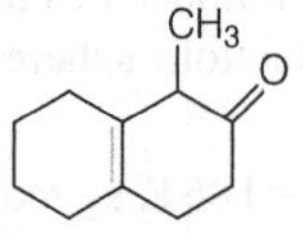

(c) 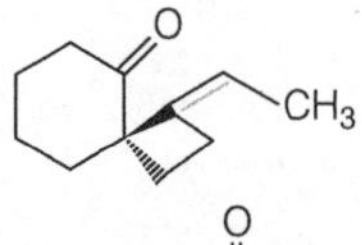

(d) 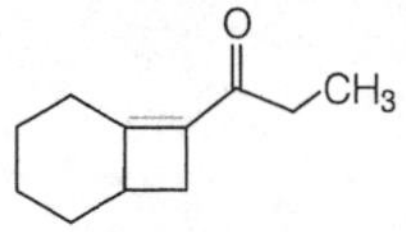

11. In the following reaction, the major product is

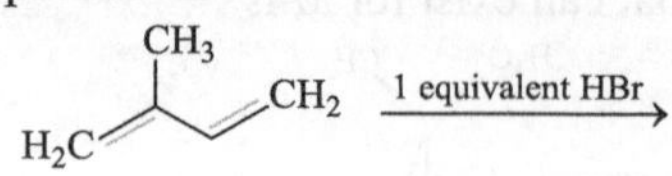

(a) 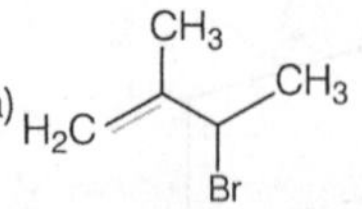

(b) 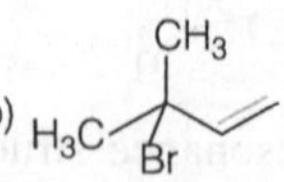

(c) 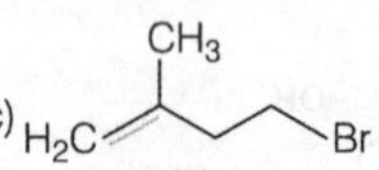

(d) 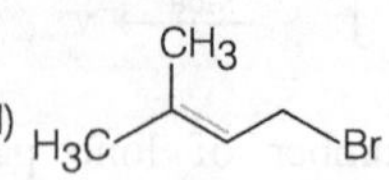

12. The structure of D-(+)-glucose is

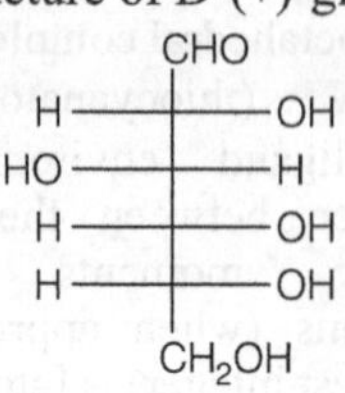

The structure of L-(−)-glucose is

(a) 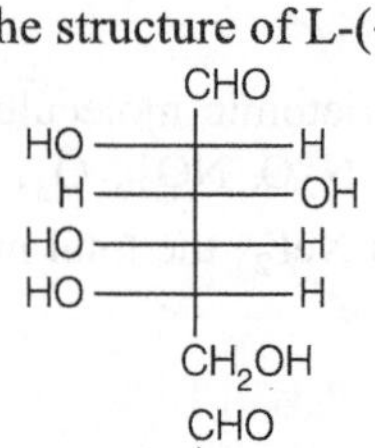

(b) 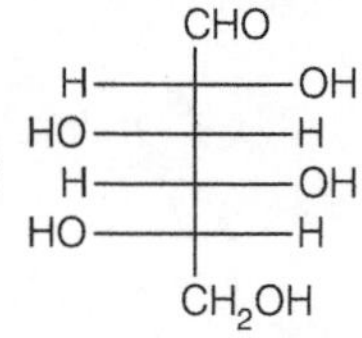

(c) 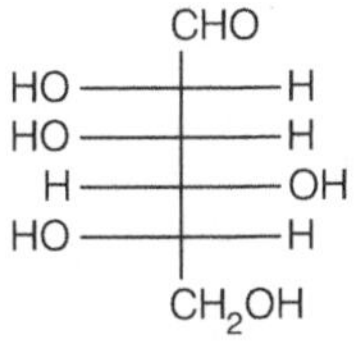

(d) 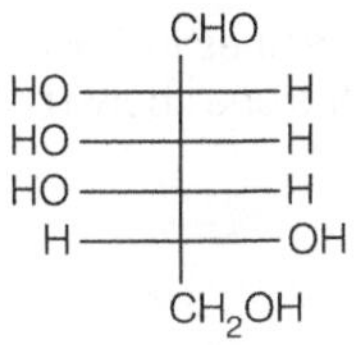

13. The major product of the reaction is

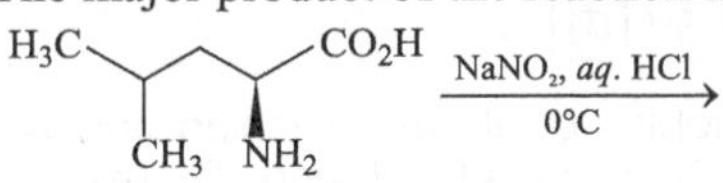

(a) 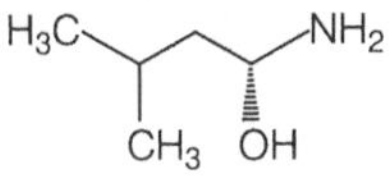

(b) 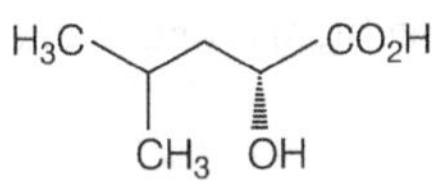

(c) 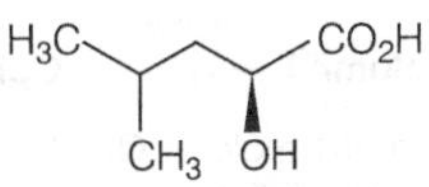

(d) 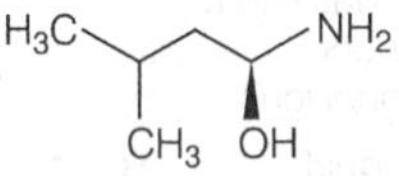

14. The correct statement(s) about Cr^{2+} and Mn^{3+} is/are [atomic number of Cr = 24 and Mn = 25]

(a) Cr^{2+} is a reducing agent
(b) Mn^{3+} is an oxidising agent
(c) both Cr^{2+} and Mn^{3+} exhibit d^4 electronic configuration
(d) when Cr^{2+} is used as a reducing agent, the chromium ion attains d^5 electronic configuration

15. Copper is purified by electrolytic refining of blister copper. The correct statement(s) about this process is/are

(a) impure Cu strip is used as cathode
(b) acidified aqueous $CuSO_4$ is used as electrolyte
(c) pure Cu deposits at cathode
(d) impurities settle as anode-mud

16. Fe^{3+} is reduced to Fe^{2+} by using

(a) H_2O_2 in presence of NaOH
(b) Na_2O_2 in water
(c) H_2O_2 in presence of H_2SO_4
(d) Na_2O_2 in presence of H_2SO_4

17. The % yield of ammonia as a function of time in the reaction,

$N_2(g) + 3H_2(g) \rightleftharpoons 2NH_3(g)$;

$\Delta H < 0$

at (p, T_1) is given below.

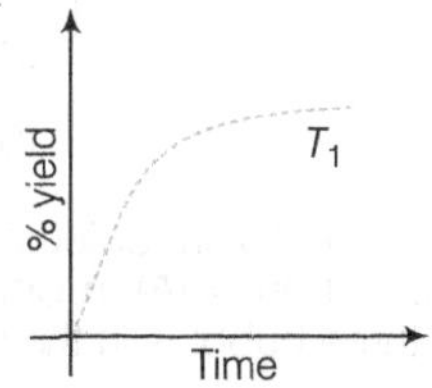

If this reaction is conducted at (p, T_1), with $T_2 > T_1$ the % yield by of ammonia as a function of time is represented by

(a) 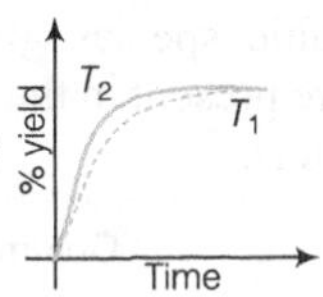

(b) 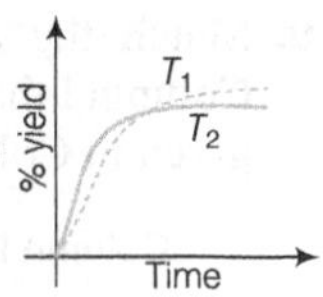

(c) 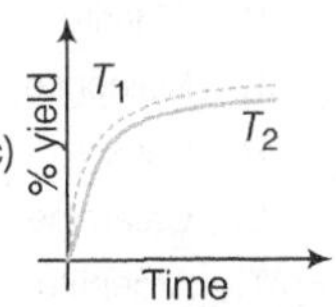

(d) 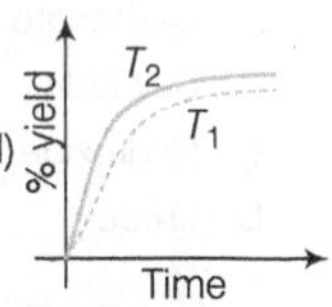

18. If the unit cell of a mineral has cubic close packed (ccp) array of oxygen atoms with m fraction of octahedral holes occupied by aluminium ions and n fraction of tetrahedral holes occupied by magnesium ions, m and n respectively, are

(a) $\frac{1}{2}, \frac{1}{8}$

(b) $1, \frac{1}{4}$

(c) $\frac{1}{2}, \frac{1}{2}$

(d) $\frac{1}{4}, \frac{1}{8}$

Section 3 (Maximum Marks : 16)

- This section contains TWO questions.
- Each question contains two columns, Column I and Column II.
- Column I has four entries (A), (B), (C) and (D).
- Column II has five entries (P), (Q), (R), (S) and (T).
- Match the entries in Column I with the entries in Column II.
- One or more entries in Column I may match with one or more entries in Column II.
- The ORS contains a 4 × 5matrix whose layout will be similar to the one shown below :
-

(A)	(P)	(Q)	(R)	(S)	(T)
(B)	(P)	(Q)	(R)	(S)	(T)
(C)	(P)	(Q)	(R)	(S)	(T)
(D)	(P)	(Q)	(R)	(S)	(T)

- For each entry in Column I, darken the bubbles of all the matching entries. For example, if entry (A) in Column I matches with entries (Q), (R) and (T), then darken these three bubbles in the ORS. Similarly, for entries (B), (C) and (D).
- Marking schemes:

For each entry in Column I

+ 2 If only the bubble(s) corresponding to all the correct match(es) is (are) darkened.

0 If none of the bubbles is darkened.

19. Match the anionic species given in Column I that are present in the ore (s) given in Column II.

Column I	Column II
A. Carbonate	P. Siderite
B. Sulphide	Q. Malachite
C. Hydroxide	R. Bauxite
D. Oxide	S. Calamine
	T. Argentite

20. Match the thermodynamic processes given under Column I with the expressions given under Column II.

Column I	Column II
A. Freezing of water at 273 K and 1 atm	P. $q = 0$
B. Expansion of 1 mole of an ideal gas into a vacuum under isolated conditions	Q. $W = 0$
C. Mixing of equal volumes of two ideal gases at constant temperature and pressure in an isolated container	R. $\Delta S_{sys} < 0$
D. Reversible heating of H_2 (g) at 1 atm from 300 K to 600 K, followed by reversible cooling to 300 K at 1 atm	S. $\Delta U = 0$
	T. $\Delta G = 0$

Mathematics

Section 1 (Maximum Marks : 32)

- This section contains EIGHT questions.
- The answer to each question is a SINGLE DIGIT INTEGER ranging from 0 to 9 both inclusive.
- For each question, darken the bubble corresponding to the correct integer in the ORS.
- Marking scheme :
  + 4 If the bubble corresponding to the answer is darkened.
  0 In all other cases.

1. Let the curve C be the mirror image of the parabola $y^2 = 4x$ with respect to the line $x + y + 4 = 0$. If A and B are the points of intersection of C with the line $y = -5$, then the distance between A and B is

2. The minimum number of times a fair coin needs to be tossed, so that the probability of getting atleast two heads is atleast 0.96, is

3. Let n be the number of ways in which 5 boys and 5 girls can stand in a queue in such a way that all the girls stand consecutively in the queue. Let m be the number of ways in which 5 boys and 5 girls can stand in a queue in such a way that exactly four girls stand consecutively in the queue. Then, the value of $\frac{m}{n}$ is

4. If the normals of the parabola $y^2 = 4x$ drawn at the end points of its latusrectum are tangents to the circle $(x-3)^2 + (y+2)^2 = r^2$, then the value of r^2 is

5. Let $f : R \to R$ be a function defined by $f(x) = \begin{cases} [x], & x \le 2 \\ 0, & x > 2 \end{cases}$, where $[x]$ denotes the greatest integer less than or equal to x. If $I = \int_{-1}^{2} \frac{xf(x^2)}{2 + f(x+1)} dx$, then the value of $(4I - 1)$ is

6. A cylindrical container is to be made from certain solid material with the following constraints : It has a fixed inner volume of V mm^3, has a 2 mm thick solid wall and is open at the top. The bottom of the container is a solid circular disc of thickness 2 mm and is of radius equal to the outer radius of the container.

If the volume of the material used to make the container is minimum, when the inner radius of the container is 10 mm, then the value of $\frac{V}{250\pi}$ is

7. Let $F(x) = \int_{x}^{x^2 + \frac{\pi}{6}} 2\cos^2 t \, dt$ for all $x \in R$ and $f : \left[0, \frac{1}{2}\right] \to [0, \infty)$ be a continuous function. For $a \in \left[0, \frac{1}{2}\right]$, if $F'(a) + 2$ is the area of the region bounded by $x = 0$, $y = 0$, $y = f(x)$ and $x = a$, then $f(0)$ is

8. The number of distinct solutions of the equation

$$\frac{5}{4}\cos^2 2x + \cos^4 x + \sin^4 x + \cos^6 x + \sin^6 x = 2$$

in the interval $[0, 2\pi]$ is

Section 2 (Maximum Marks : 40)

- This section contains TEN questions.
- Each question has FOUR options (a), (b), (c) and (d). ONE OR MORE THAN ONE of these four options are correct.
- For each question, darken the bubble(s) corresponding to all the correct options in the ORS.
- Marking scheme :
  +4 If only the bubble(s) corresponding to all the correct options are darkened.
  0 If none of the bubbles is darkened.
  –2 In all the other cases.

9. Let $y(x)$ be a solution of the differential equation
$(1+e^x)y'+ye^x=1$
If $y(0)=2$, then which of the following statement(s) is/are true?

(a) $y(-4)=0$
(b) $y(-2)=0$
(c) $y(x)$ has a critical point in the interval $(-1, 0)$
(d) $y(x)$ has no critical point in the interval $(-1, 0)$

10. Consider the family of all circles whose centres lie on the straight line $y=x$. If this family of circles is represented by the differential equation $Py''+Qy'+1=0$, where P, Q are the functions of x, y and y' (here, $y'=\frac{dy}{dx}$, $y''=\frac{d^2y}{dx^2}$), then which of the following statement(s) is/are true?

(a) $P=y+x$
(b) $P=y-x$
(c) $P+Q=1-x+y+y'+(y')^2$
(d) $P-Q=x+y-y'-(y')^2$

11. Let $g: R \to R$ be a differentiable function with $g(0)=0$, $g'(0)=0$ and $g'(1)\neq 0$.

Let $f(x)=\begin{cases}\frac{x}{|x|}g(x), & x\neq 0\\ 0, & x=0\end{cases}$

and $h(x)=e^{|x|}$ for all $x\in R$. Let $(foh)(x)$ denotes $f\{h(x)\}$ and $(hof)(x)$ denotes $h\{f(x)\}$. Then, which of the following is/are true?

(a) f is differentiable at $x=0$
(b) h is differentiable at $x=0$
(c) foh is differentiable at $x=0$
(d) hof is differentiable at $x=0$

12. Let $f(x)=\sin\left[\frac{\pi}{6}\sin\left(\frac{\pi}{2}\sin x\right)\right]$ for all $x\in R$ and $g(x)=\frac{\pi}{2}\sin x$ for all $x\in R$. Let $(fog)(x)$ denotes $f\{g(x)\}$ and $(gof)(x)$ denotes $g\{f(x)\}$. Then, which of the following is/are true?

(a) Range of f is $\left[-\frac{1}{2}, \frac{1}{2}\right]$
(b) Range of fog is $\left[-\frac{1}{2}, \frac{1}{2}\right]$
(c) $\lim_{x\to 0}\frac{f(x)}{g(x)}=\frac{\pi}{6}$
(d) There is an $x\in R$ such that $(gof)(x)=1$

13. Let ΔPQR be a triangle. Let $\mathbf{a}=\mathbf{QR}$, $\mathbf{b}=\mathbf{RP}$ and $\mathbf{c}=\mathbf{PQ}$. If $|\mathbf{a}|=12$, $|\mathbf{b}|=4\sqrt{3}$ and $\mathbf{b}\cdot\mathbf{c}=24$, then which of the following is/are true?

(a) $\frac{|\mathbf{c}|^2}{2}-|\mathbf{a}|=12$
(b) $\frac{|\mathbf{c}|^2}{2}+|\mathbf{a}|=30$
(c) $|\mathbf{a}\times\mathbf{b}+\mathbf{c}\times\mathbf{a}|=48\sqrt{3}$
(d) $\mathbf{a}\cdot\mathbf{b}=-72$

14. Let X and Y be two arbitrary, 3×3, non-zero, skew-symmetric matrices and Z be an arbitrary, 3×3, non-zero, symmetric matrix. Then, which of the following matrices is/are skew-symmetric?

(a) $Y^3Z^4-Z^4Y^3$ (b) $X^{44}+Y^{44}$
(c) $X^4Z^3-Z^3X^4$ (d) $X^{23}+Y^{23}$

15. Which of the following values of α satisfy the equation

$$\begin{vmatrix} (1+\alpha)^2 & (1+2\alpha)^2 & (1+3\alpha)^2 \\ (2+\alpha)^2 & (2+2\alpha)^2 & (2+3\alpha)^2 \\ (3+\alpha)^2 & (3+2\alpha)^2 & (3+3\alpha)^2 \end{vmatrix} = -648\alpha ?$$

(a) -4 (b) 9 (c) -9 (d) 4

16. In R^3, consider the planes $P_1 : y = 0$ and $P_2 : x + z = 1$. Let P_3 be a plane, different from P_1 and P_2, which passes through the intersection of P_1 and P_2. If the distance of the point (0, 1, 0) from P_3 is 1 and the distance of a point (α, β, γ) from P_3 is 2, then which of the following relation(s) is/are true?

(a) $2\alpha + \beta + 2\gamma + 2 = 0$
(b) $2\alpha - \beta + 2\gamma + 4 = 0$
(c) $2\alpha + \beta - 2\gamma - 10 = 0$
(d) $2\alpha - \beta + 2\gamma - 8 = 0$

17. In R^3, let L be a straight line passing through the origin. Suppose that all the points on L are at a constant distance from the two planes $P_1 : x + 2y - z + 1 = 0$ and $P_2 : 2x - y + z - 1 = 0$. Let M be the locus of the foot of the perpendiculars drawn from the points on L to the plane P_1. Which of the following point(s) lie(s) on M?

(a) $\left(0, -\frac{5}{6}, -\frac{2}{3}\right)$ (b) $\left(-\frac{1}{6}, -\frac{1}{3}, \frac{1}{6}\right)$
(c) $\left(-\frac{5}{6}, 0, \frac{1}{6}\right)$ (d) $\left(-\frac{1}{3}, 0, \frac{2}{3}\right)$

18. Let P and Q be distinct points on the parabola $y^2 = 2x$ such that a circle with PQ as diameter passes through the vertex O of the parabola. If P lies in the first quadrant and the area of ΔOPQ is $3\sqrt{2}$, then which of the following is/are the coordinates of P?

(a) $(4, 2\sqrt{2})$
(b) $(9, 3\sqrt{2})$
(c) $\left(\frac{1}{4}, \frac{1}{\sqrt{2}}\right)$
(d) $(1, \sqrt{2})$

Section 3 (Maximum Marks : 16)

- This section contains TWO questions.
- Each question contains two columns, Column I and Column II.
- Column I has four entries (A), (B), (C) and (D).
- Column II has five entries (P), (Q), (R), (S) and (T).
- Match the entries in Column I with the entries in Column II.
- One or more entries in Column I may match with one or more entries in Column II.
- The ORS contains a 4 × 5 matrix whose layout will be similar to the one shown below:

(A)	(P)	(Q)	(R)	(S)	(T)
(B)	(P)	(Q)	(R)	(S)	(T)
(C)	(P)	(Q)	(R)	(S)	(T)
(D)	(P)	(Q)	(R)	(S)	(T)

- For each entry in Column I, darken the bubbles of all the matching entries. For example, if entry (A) in Column I matches with entries (Q), (R) and (T), then darken these three bubbles in the ORS. Similarly, for entries (B), (C) and (D).
- Marking scheme :

For each entry in Column I

+2 If only the bubble(s) corresponding to all the correct matches are darkened.
0 If none of the bubbles is darkened.
−1 In all other cases.

19.

	Column I		Column II
A.	In R^2, if the magnitude of the projection vector of the vector $\alpha\hat{i} + \beta\hat{j}$ on $\sqrt{3}\,\hat{i} + \hat{j}$ is $\sqrt{3}$ and if $\alpha = 2 + \sqrt{3}\,\beta$, then possible value(s) of $\|\alpha\|$ is/are	P.	1
B.	Let a and b be real numbers such that the function $f(x) = \begin{cases} -3ax^2 - 2, & x < 1 \\ bx + a^2, & x \geq 1 \end{cases}$ is differentiable for all $x \in R$. Then, possible value(s) of a is/are	Q.	2
C.	Let $\omega\ (\neq 1)$ be a complex cube root of unity. If $(3 - 3\omega + 2\omega^2)^{4n+3} + (2 + 3\omega - 3\omega^2)^{4n+3} + (-3 + 2\omega + 3\omega^2)^{4n+3} = 0$, then the possible value(s) of n is/are	R.	3
D.	Let the harmonic mean of two positive real numbers a and b be 4. If q is a positive real number such that $a, 5, q, b$ is in arithmetic progression, then the value(s) of $\|q - 2a\|$ is/are	S.	4
		T.	5

20.

	Column I		Column II
A.	In ΔXYZ, let a, b and c be the lengths of the sides opposite to the angles X, Y and Z, respectively. If $2(a^2 - b^2) = c^2$ and $\lambda = \dfrac{\sin(X - Y)}{\sin Z}$, then possible value(s) of n for which $\cos(n\pi\lambda) = 0$, is/are	P.	1
B.	In ΔXYZ, let a, b and c be the lengths of the sides opposite to the angles X, Y and Z, respectively. If $1 + \cos 2X - 2\cos 2Y = 2 \sin X \sin Y$, then possible value(s) of $\dfrac{a}{b}$ is/are	Q.	2
C.	In R^2, let $\sqrt{3}\,\hat{i} + \hat{j}$, $\hat{i} + \sqrt{3}\,\hat{j}$ and $\beta\hat{i} + (1 - \beta)\,\hat{j}$ be the position vectors of X, Y and Z with respect to the origin O, respectively. If the distance of Z from the bisector of the acute angle of **OX** with **OY** is $\dfrac{3}{\sqrt{2}}$, then possible value(s) of $\|\beta\|$ is/are	R.	3
D.	Suppose that $F(\alpha)$ denotes the area of the region bounded by $x = 0$, $x = 2$, $y^2 = 4x$ and $y = \|\alpha x - 1\| + \|\alpha x - 2\| + \alpha x$, where $\alpha \in \{0, 1\}$. Then, the value(s) of $F(\alpha) + \dfrac{8}{3}\sqrt{2}$, when $\alpha = 0$ and $\alpha = 1$, is/are	S.	5
		T.	6

Paper 2

Physics

Section 1 (Maximum Marks : 32)

- This section contains EIGHT questions.
- The answer to each question is a SINGLE DIGIT INTEGER ranging from 0 to 9 both inclusive.
- For each question, darken the bubble corresponding to the correct integer in ORS.
- Marking scheme :

  +4 If the bubble corresponding to the answer is darkened.

  0 in all other cases.

1. Four harmonic waves of equal frequencies and equal intensities I_0 have phase angles $0, \frac{\pi}{3}, \frac{2\pi}{3}$ and π. When they are superposed, the intensity of the resulting wave is nI_0. The value of n is

2. For a radioactive material, its activity A and rate of change of its activity R are defined as $A = -\frac{dN}{dt}$ and $R = -\frac{dA}{dt}$, where $N(t)$ is the number of nuclei at time t. Two radioactive source P(mean life τ) and Q (mean life 2τ) have the same activity at $t = 0$. Their rate of change of activities at $t = 2\tau$ are R_P and R_Q, respectively. If $\frac{R_P}{R_Q} = \frac{n}{e}$, then the value of n is

3. A monochromatic beam of light is incident at 60° on one face of an equilateral prism of refractive index n and emerges from the opposite face making an angle $\theta(n)$ with the normal (see figure). For $n = \sqrt{3}$ the value of θ is 60° and $\frac{d\theta}{dn} = m$. The value of m is

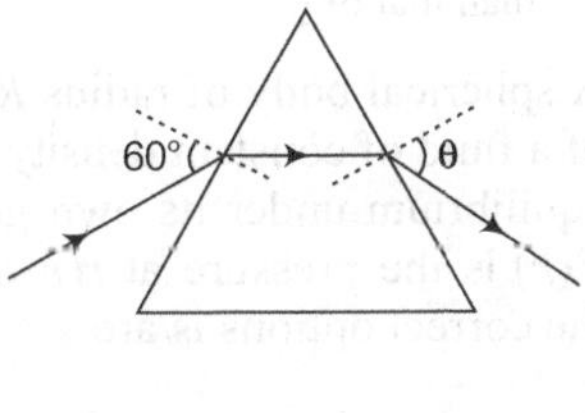

4. In the following circuit, the current through the resistor $R(= 2\Omega)$ is I amperes. The value of I is

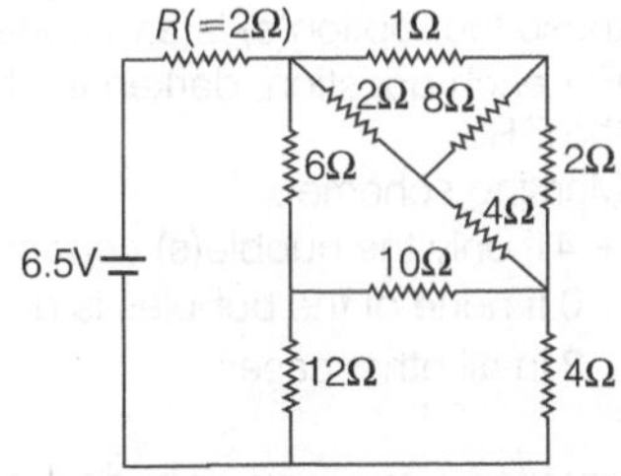

5. An electron in an excited state of Li^{2+} ion has angular momentum $\frac{3h}{2\pi}$. The de Broglie wavelength of the electron in this state is $p\pi a_0$ (where a_0 is the Bohr radius). The value of p is

6. A large spherical mass M is fixed at one position and two identical masses m are kept on a line passing through the centre of M (see figure). The point masses are connected by a rigid massless rod of length l and this assembly is free to move along the line connecting them.

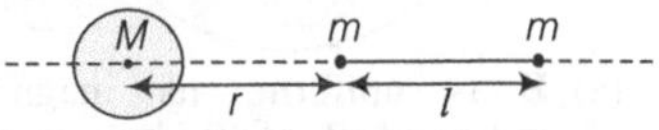

All three masses interact only through their mutual gravitational interaction. When the point mass nearer to M is at a distance $r = 3l$ from M the tension in the rod is zero for $m = k\left(\frac{M}{288}\right)$. The value of k is

7. The energy of a system as a function of time t is given as $E(t) = A^2 \exp(-\alpha t)$, where $\alpha = 0.2\,s^{-1}$. The measurement of A has an error of 1.25%. If the error in the measurement of time is 1.50%, the percentage error in the value of $E(t)$ at $t = 5\,s$ is

8. The densities of two solids spheres A and B of the same radii R vary with radial distance r as $\rho_A(r) = k\left(\frac{r}{R}\right)$ and $\rho_B(r) = k\left(\frac{r}{R}\right)^5$, respectively, where k is a constant. The moments of inertia of the individual spheres about axes passing through their centres are I_A and I_B, respectively.

If $\frac{I_B}{I_A} = \frac{n}{10}$, the value of n is

Section 2 (Maximum Marks : 32)

- This section contains EIGHT questions.
- Each question has FOUR options (a), (b), (c) and (d), ONE OR MORE THAN ONE of these four option(s) is(are) correct.
- For each question, darken the bubble(s) corresponding to all the correct option(s) in the ORS.
- Marking scheme :

  + 4 If only the bubble(s) corresponding to all the correct optin(s) is(are) darkened.

  0 If none of the bubbles is darkened.

  − 2 In all other cases.

9. Consider a uniform spherical charge distribution of radius R_1 centred at the origin O. In this distribution, a spherical cavity of radius R_2, centred at P with distance $OP = a = R_1 - R_2$ (see figure) is made. If the electric field inside the cavity at position $\mathbf{r}$ is $\mathbf{E}(\mathbf{r})$, then the correct statements is/are

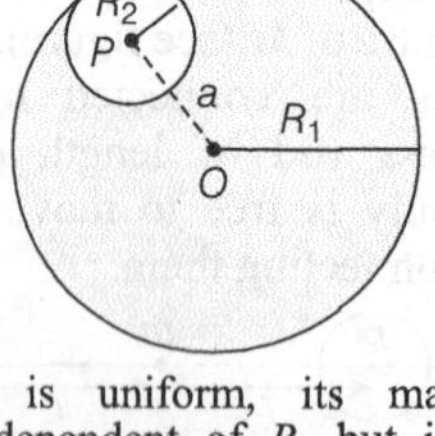

(a) $\mathbf{E}$ is uniform, its magnitude is independent of R_2 but its direction depends on $\mathbf{r}$

(b) $\mathbf{E}$ is uniform, its magnitude depends on R_2 and its direction depends on $\mathbf{r}$

(c) $\mathbf{E}$ is uniform, its magnitude is independent of 'a' but its direction depends on $\mathbf{a}$

(d) $\mathbf{E}$ is uniform and both its magnitude and direction depend on $\mathbf{a}$

10. In plotting stress *versus* strain curves for two materials P and Q, a student by mistake puts strain on the y-axis and stress on the x-axis as shown in the figure. Then, the correct statements is/are

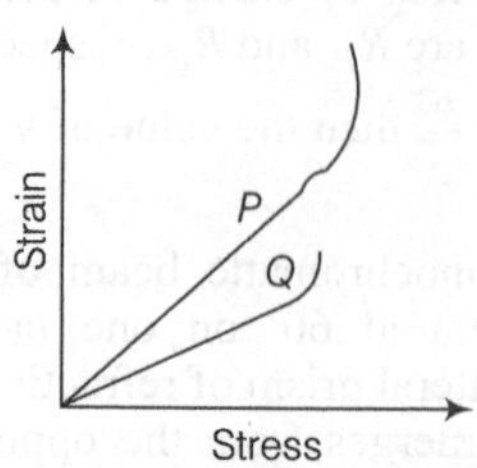

(a) P has more tensile strength than Q

(b) P is more ductile than Q

(c) P is more brittle than Q

(d) The Young's modulus of P is more than that of Q

11. A spherical body of radius R consists of a fluid of constant density and is in equilibrium under its own gravity. If $P(r)$ is the pressure at $r(r < R)$, then the correct options is/are

(a) $P(r=0)=0$

(b) $\dfrac{P\left(r=\dfrac{3R}{4}\right)}{P\left(r=\dfrac{2R}{3}\right)}=\dfrac{63}{80}$

(c) $\dfrac{P\left(r=\dfrac{3R}{5}\right)}{P\left(r=\dfrac{2R}{5}\right)}=\dfrac{16}{21}$

(d) $\dfrac{P\left(r=\dfrac{R}{2}\right)}{P\left(r=\dfrac{R}{3}\right)}=\dfrac{20}{27}$

12. A parallel plate capacitor having plates of area S and plate separation d, has capacitance C_1 in air. When two dielectrics of different relative permittivities ($\varepsilon_1 = 2$ and $\varepsilon_2 = 4$) are introduced between the two plates as shown in the figure, the capacitance becomes C_2. The ratio $\dfrac{C_2}{C_1}$ is

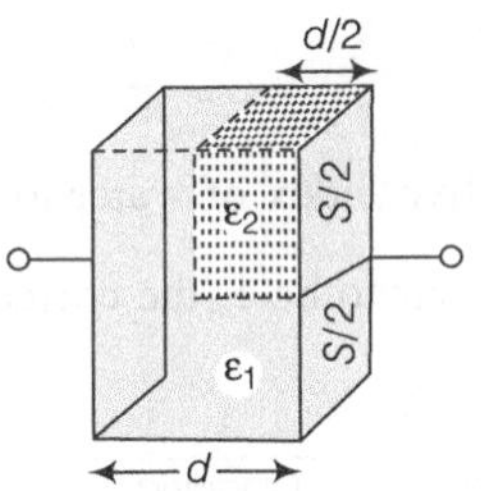

(a) $\dfrac{6}{5}$ (b) $\dfrac{5}{3}$

(c) $\dfrac{7}{5}$ (d) $\dfrac{7}{3}$

13. An ideal monoatomic gas is confined in a horizontal cylinder by a spring loaded piston (as shown in the figure). Initially the gas is at temperature T_1, pressure P_1 and volume V_1 and the spring is in its relaxed state. The gas is then heated very slowly to temperature T_2, pressure P_2 and volume V_2. During this process the piston moves out by a distance x.

Ignoring the friction between the piston and the cylinder, the correct statements is/are

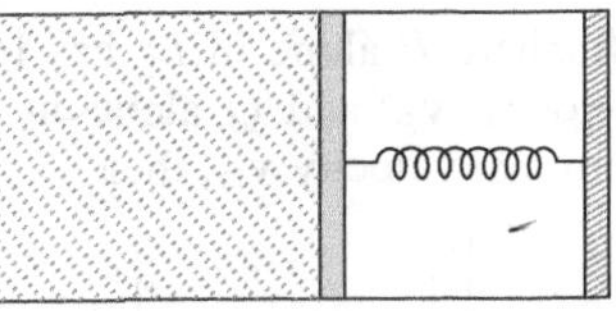

(a) If $V_2 = 2V_1$ and $T_2 = 3T_1$, then the energy stored in the spring is $\dfrac{1}{4}P_1V_1$

(b) If $V_2 = 2V_1$ and $T_2 = 3T_1$, then the change in internal energy is $3P_1V_1$

(c) If $V_2 = 3V_1$ and $T_2 = 4T_1$, then the work done by the gas is $\dfrac{7}{3}P_1V_1$

(d) If $V_2 = 3V_1$ and $T_2 = 4T_1$, then the heat supplied to the gas is $\dfrac{17}{6}P_1V_1$

14. A fission reaction is given by ${}^{236}_{92}\text{U} \rightarrow {}^{140}_{54}\text{Xe} + {}^{94}_{38}\text{Sr} + x + y$, where x and y are two particles. Considering ${}^{236}_{92}\text{U}$ to be at rest, the kinetic energies of the products are denoted by K_{Xe}, K_{Sr}, K_x(2 MeV) and K_y(2 MeV), respectively. Let the binding energies per nucleon of ${}^{236}_{92}\text{U}$, ${}^{140}_{54}\text{Xe}$ and ${}^{94}_{38}\text{Sr}$ be 7.5 MeV, 8.5 MeV and 8.5 MeV, respectively. Considering different conservation laws, the correct options is/are

(a) $x = n$, $y = n$, $K_{\text{Sr}} = 129$ MeV, $K_{\text{Xe}} = 86$ MeV

(b) $x = p$, $y = e^-$, $K_{\text{Sr}} = 129$ MeV, $K_{\text{Xe}} = 86$ MeV

(c) $x = p$, $y = n$, $K_{\text{Sr}} = 129$ MeV, $K_{\text{Xe}} = 86$ MeV

(d) $x = n$, $y = n$, $K_{\text{Sr}} = 86$ MeV, $K_{\text{Xe}} = 129$ MeV

15. Two spheres P and Q for equal radii have densities ρ_1 and ρ_2, respectively. The spheres are connected by a massless string and placed in liquids L_1 and L_2 of densities σ_1 and σ_2 and viscosities η_1 and η_2, respectively. They float in equilibrium with the sphere P in L_1 and sphere Q in L_2 and the string being taut (see figure).

If sphere P alone in L_2 has terminal velocity $\mathbf{v}_P$ and Q alone in L_1 has terminal velocity $\mathbf{v}_Q$, then

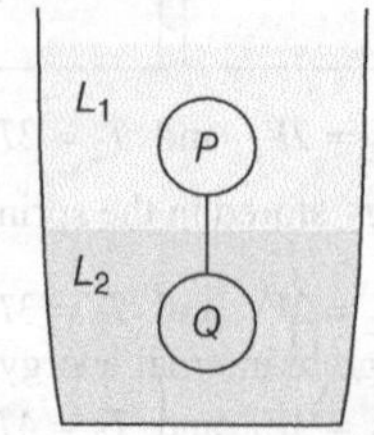

(a) $\frac{|\mathbf{v}_P|}{|\mathbf{v}_Q|} = \frac{\eta_1}{\eta_2}$ (b) $\frac{|\mathbf{v}_P|}{|\mathbf{v}_Q|} = \frac{\eta_2}{\eta_1}$

(c) $\mathbf{v}_P \cdot \mathbf{v}_Q > 0$ (d) $\mathbf{v}_P \cdot \mathbf{v}_Q < 0$

16. In terms of potential difference V, electric current I, permittivity ε_0, permeability μ_0 and speed of light c, the dimensionally correct equations is/are

(a) $\mu_0 I^2 = \varepsilon_0 V^2$ (b) $\varepsilon_0 I = \mu_0 V$

(c) $I = \varepsilon_0 c V$ (d) $\mu_0 c I = \varepsilon_0 V$

Section 3 (Maximum Marks : 16)

- This section contains TWO paragraphs.
- Based on each paragraph, there will be TWO questions.
- Each question has FOUR options (a), (b), (c) and (d). ONE OR MORE THAN ONE of these four option(s) is(are) correct.
- For each question, darken the bubble(s) corresponding to all the correct option(s) in the ORS.
- Marking scheme:
  + 4 If only the bubble(s) corresponding to all the corect option(s) is(are) darkened.
  0 If none of the bubbles is darkened.
  – 2 In all other cases.

Paragraph I

Light guidance in an optical fibre can be understood by considering a structure comprising of thin solid glass cylinder of refractive index n_1 surrounded by a medium of lower refractive index n_2. The light guidance in the structure takes place due to successive total internal reflections at the interface of the media n_1 and n_2 as shown in the figure. All rays with the angle of incidence i less than a particular value i_m are confined in the medium of refractive index n_1. The numerical aperture (NA) of the structure is defined as $\sin i_m$.

17. For two structures namely S_1 with $n_1 = \frac{\sqrt{45}}{4}$ and $n_2 = \frac{3}{2}$, and S_2 with $n_1 = \frac{8}{5}$ and $n_2 = \frac{7}{5}$ and taking the refractive index of water to be $\frac{4}{3}$ and that to air to be 1, the correct options is/are

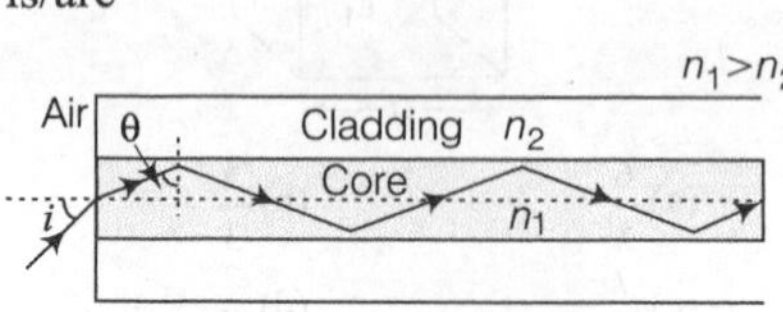

(a) NA of S_1 immersed in water is the same as that of S_2 immersed in a liquid of refractive index $\frac{16}{3\sqrt{15}}$

(b) NA of S_1 immersed in liquid of refractive index $\frac{6}{\sqrt{15}}$ is the same as that of S_2 immersed in water

(c) NA of S_1 placed in air is the same as that S_2 immersed in liquid of refractive index $\frac{4}{\sqrt{15}}$

(d) NA of S_1 placed in air is the same as that of S_2 placed in water

18. If two structures of same cross-sectional area, but different numerical apertures NA_1 and $NA_2 (NA_2 < NA_1)$ are joined longitudinally, the numerical aperture of the combined structure is

(a) $\frac{NA_1 NA_2}{NA_1 + NA_2}$ (b) $NA_1 + NA_2$

(c) NA_1 (d) NA_2

Paragraph II

In a thin rectangular metallic strip a constant current I flows along the positive x-direction, as shown in the figure. The length, width and thickness of the strip are l, w and d, respectively. A uniform magnetic field **B** is applied on the strip along the positive y-direction. Due to this, the charge carriers experience a net deflection along the z-direction. This results in accumulation of charge carriers on the surface $PQRS$ and appearance of equal and opposite charges on the face opposite to $PQRS$.

A potential difference along the z-direction is thus developed. Charge accumulation continues until the magnetic force is balanced by the electric force. The current is assumed to be uniformly distributed on the cross section of the strip and carried by electrons.

19. Consider two different metallic strips (1 and 2) of the same material. Their lengths are the same, widths are w_1 and w_2 and thicknesses are d_1 and d_2, respectively. Two points K and M are symmetrically located on the opposite faces parallel to the x-y plane (see figure). V_1 and V_2 are the potential differences between K and M in strips 1 and 2, respectively. Then, for a given current I flowing through them in a given magnetic field strength B, the correct statements is/are

(a) If $w_1 = w_2$ and $d_1 = 2d_2$, then $V_2 = 2V_1$

(b) If $w_1 = w_2$ and $d_1 = 2d_2$, then $V_2 = V_1$

(c) If $w_1 = 2w_2$ and $d_1 = d_2$, then $V_2 = 2V_1$

(d) If $w_1 = 2w_2$ and $d_1 = d_2$, then $V_2 = V_1$

20. Consider two different metallic strips (1 and 2) of same dimensions (length l, width w and thickness d) with carrier densities n_1 and n_2, respectively. Strip 1 is placed in magnetic field B_1 and strip 2 is placed in magnetic field B_2, both along positive y-directions. Then V_1 and V_2 are the potential differences developed between K and M in strips 1 and 2, respectively. Assuming that the current I is the same for both the strips, the correct options is/are

(a) If $B_1 = B_2$ and $n_1 = 2n_2$, then $V_2 = 2V_1$

(b) If $B_1 = B_2$ and $n_1 = 2n_2$, then $V_2 = V_1$

(c) If $B_1 = 2B_2$ and $n_1 = n_2$, then $V_2 = 0.5V_1$

(d) If $B_1 = 2B_2$ and $n_1 = n_2$, then $V_2 = V_1$

Chemistry

Section 1 (Maximum Marks : 32)

- This section contains EIGHT questions.
- The answer to each question is a SINGLE DIGIT INTEGER ranging from 0 to 9 both inclusive.
- For each question, darken the bubble corresponding to the correct integer in ORS.
- Marking scheme :
  +4 If the bubble corresponding to the answer is darkened.
  0 in all other cases.

1. In the complex acetylbromidodicarbonylbis (triethylphosphine) iron (II), the number of Fe—C bond (s) is

2. Among the complex ions, $[Co(NH_2-CH_2-CH_2-NH_2)_2Cl_2]^+$, $[CrCl_2(C_2O_4)_2]^{3-}$, $[Fe(H_2O)_4(OH)_2]^+$, $[Fe(NH_3)_2(CN)_4]^-$, $[Co(NH_2-CH_2-CH_2-NH_2)_2(NH_3)Cl]^{2+}$ and $[Co(NH_3)_4(H_2O)Cl]^{2+}$ the number of complex ion(s) that show(s) *cis-trans* isomerism is

3. Three moles of B_2H_6 are completely reacted with methanol. The number of moles of boron containing product formed is

4. The molar conductivity of a solution of a weak acid HX (0.01 M) is 10 times smaller than the molar conductivity of a solution of a weak acid HY (0.10 M). If $\lambda^0_{X^-} \approx \lambda^0_{Y^-}$, the difference in their pK_a values, pK_a (HX) − pK_a (HY), is (consider degree of ionisation of both acids to be <<1).

5. A closed vessel with rigid walls contains 1 mole of $^{238}_{92}$U and 1 mole of air at 298 K. Considering complete decay of $^{238}_{92}$U to $^{206}_{82}$Pb, the ratio of the final pressure to the initial pressure of the system at 298 K is

6. In dilute aqueous H_2SO_4 the complex diaquadioxalatoferrate (II) is oxidised by MnO_4^-. For this reaction, the ratio of the rate of change of $[H^+]$ to the rate of change of $[MnO_4^-]$ is

7. The number of hydroxyl group(s) in Q is

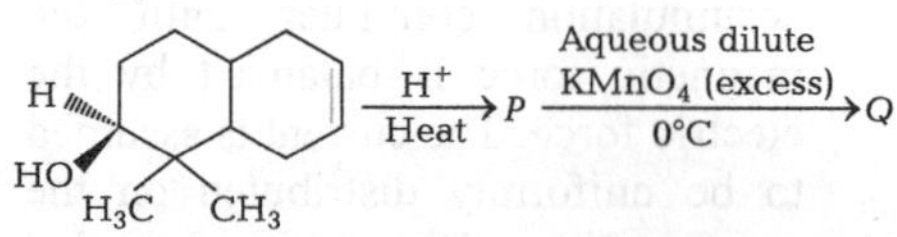

8. Among the following the number of reaction(s) that produce(s) benzaldehyde is

I. 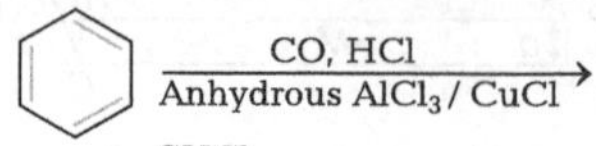

II. 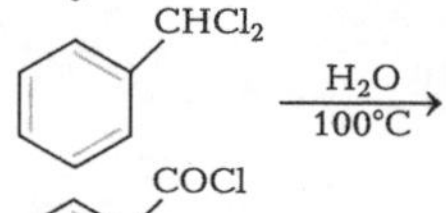

III. 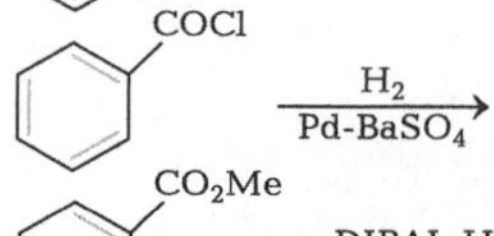

IV. 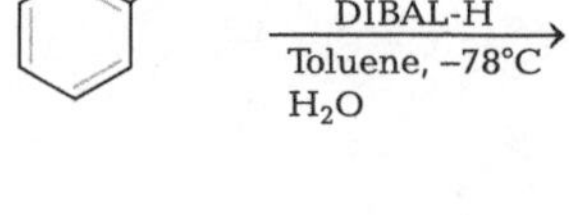

Section 2 (Maximum Marks : 32)

- This section contains EIGHT questions.
- Each question has FOUR options (a), (b), (c) and (d), ONE OR MORE THAN ONE of these four option(s) is(are) correct.
- For each question, darken the bubble(s) corresponding to all the correct option(s) in the ORS.
- Marking scheme :
  + 4 If only the bubble(s) corresponding to all the correct option(s) is(are) darkened.
  0 If none of the bubbles is darkened.
  – 2 In all other cases.

9. Under hydrolysis conditions, the compounds used for preparation of linear polymer and for chain termination, respectively are

(a) CH_3SiCl_3 and $Si(CH_3)_4$

(b) $(CH_3)_2SiCl_2$ and $(CH_3)_3SiCl$

(c) $(CH_3)SiCl_2$ and CH_3SiCl_3

(d) $SiCl_4$ and $(CH_3)_3SiCl$

10. When O_2 is adsorbed on a metallic surface, electron transfer occurs from the metal to O_2. The true statement(s) regarding this adsorption is (are)

(a) O_2 is physisorbed

(b) heat is released

(c) occupancy of $^*\pi_{2p}$ of O_2 is increased

(d) bond length of O_2 is increased

11. One mole of a monoatomic real gas satisfies the equation $p(V-b)=RT$ where, b is a constant. The relationship of interatomic potential $V(r)$ and interatomic distance r for gas is given by

(a) 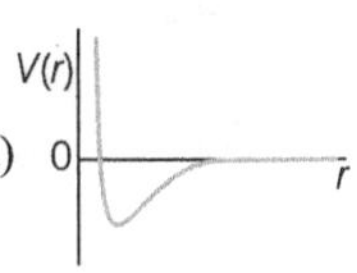

(b) 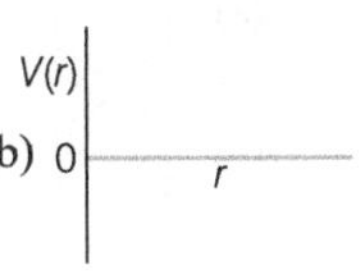

(c) 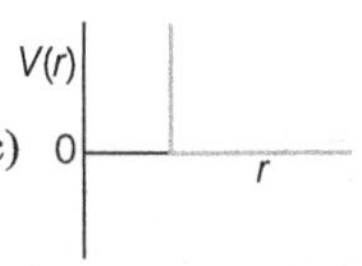

(d) 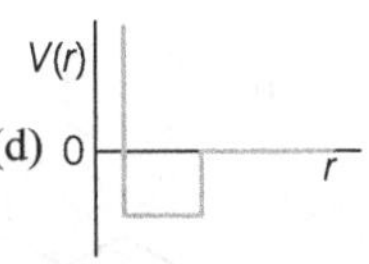

12. In the following reactions, the product S is

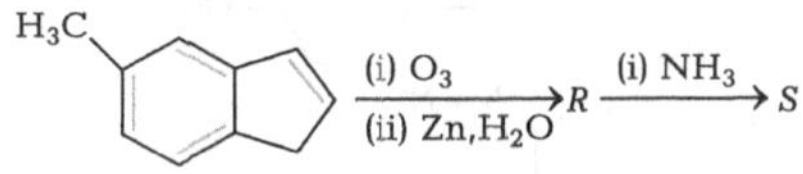

(a) 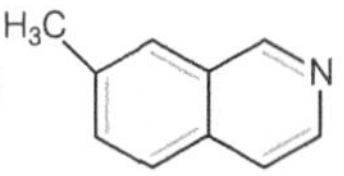

(b) 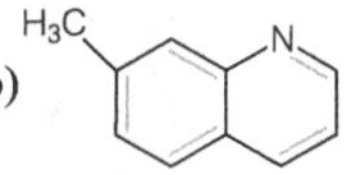

(c) 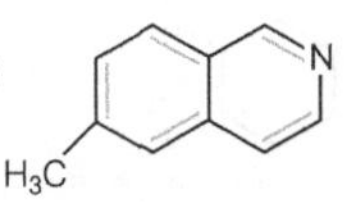

(d) 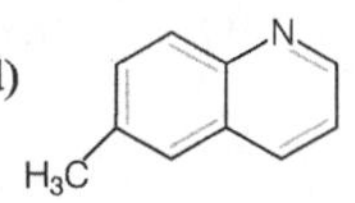

13. The major product U in the following reaction is

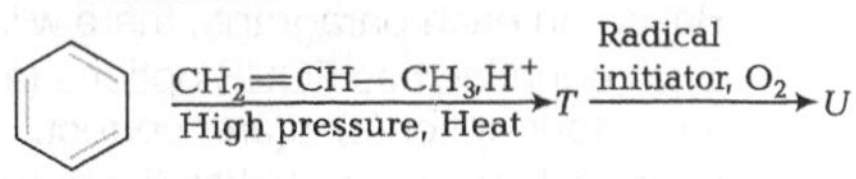

(a) 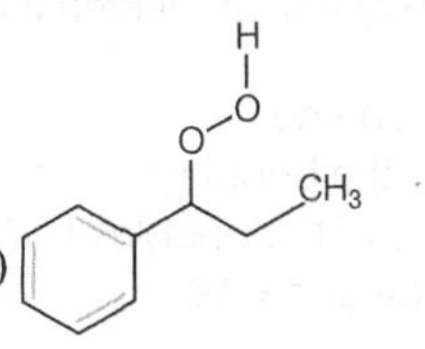

(b) 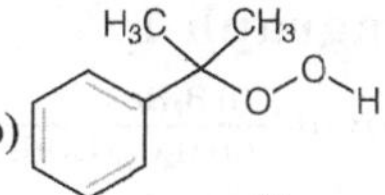

(c) 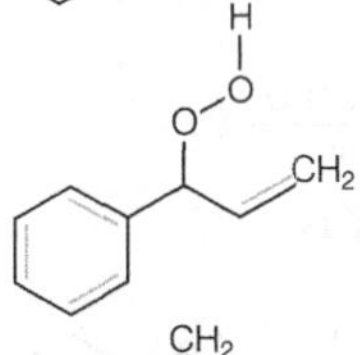

(d) 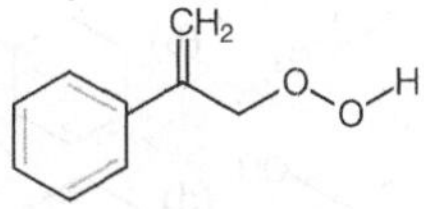

14. In the following reactions, the major product W is

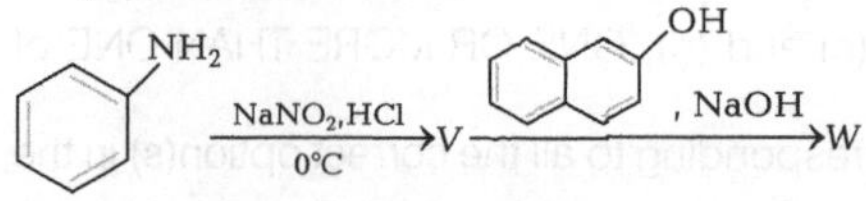

(a) 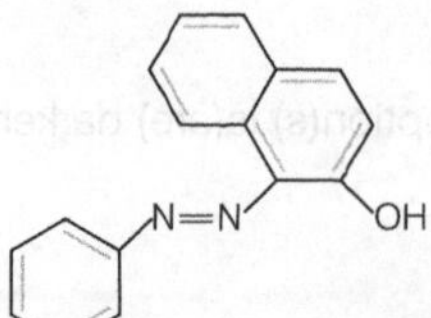

(b) 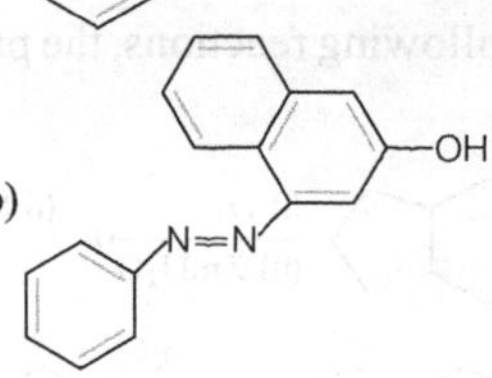

(c) 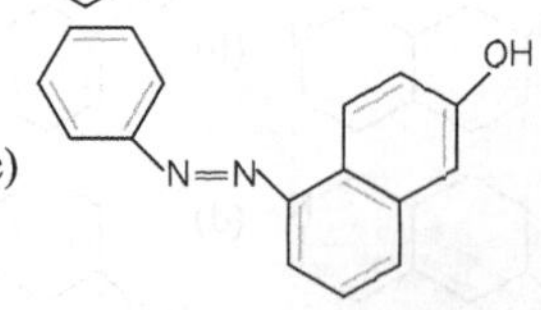

(d) 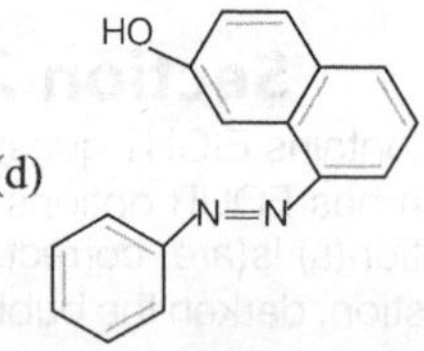

15. The correct statement(s) regarding, (i) $HClO$, (ii) $HClO_2$, (iii) $HClO_3$ and (iv) $HClO_4$ is (are)

(a) the number of $Cl = O$ bonds in (ii) and (iii) together is two

(b) the number of lone pair of electrons on Cl in (ii) and (iii) together is three

(c) the hybridisation of Cl in (iv) is sp^3

(d) amongst (i) to (iv), the strongest acid is (i)

16. The pair(s) of ions where both the ions are precipitated upon passing H_2S gas in presence of dilute HCl, is (are)

(a) Ba^{2+}, Zn^{2+} (b) Bi^{3+}, Fe^{3+}

(c) Cu^{2+}, Pb^{2+} (d) Hg^{2+}, Bi^{3+}

Section 3 (Maximum Marks : 16)

- This section contains TWO paragraphs.
- Based on each paragraph, there will be TWO questions.
- Each equation has FOUR options (a), (b), (c) and (d). ONE OR MORE THAN ONE of these four option(s) is(are) correct.
- Four each question, darken the bubble(s) corresponding to all the correct option(s) in the ORS.
- Marking scheme:
  + 4 If only the bubble(s) corresponding to all the correct option(s) is(are) darkened.
  0 If none of the bubbles is darkened.
  – 2 In all other cases.

Paragraph 1

$$C_8H_6 \xrightarrow[H_2]{Pd\text{-}BaSO_4} C_8H_8 \xrightarrow[\text{(ii) } H_2O_2, NaOH, H_2O]{\text{(i) } B_2H_6} X$$

$$C_8H_6 \xrightarrow{H_2O,\ HgSO_4, H_2SO_4} C_8H_8 \xrightarrow[\text{(ii) } H^+, \text{Heat}]{\text{(i) EtMgBr, } H_2O} Y$$

17. Compound X is

(a) 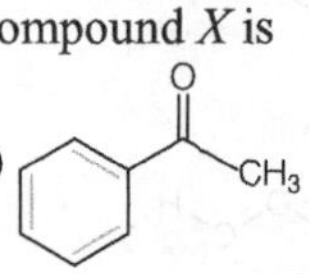
(b) 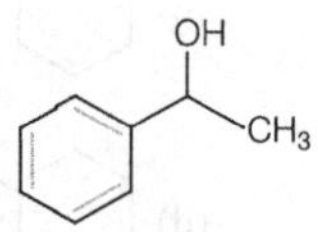

(c) 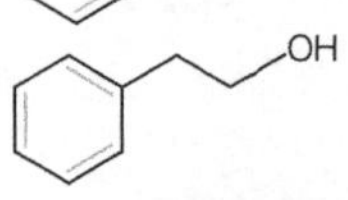
(d) 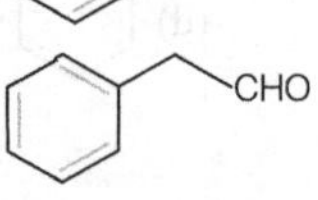

18. The major compound Y is

(a) 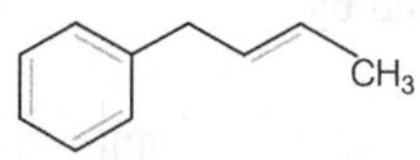

(b) 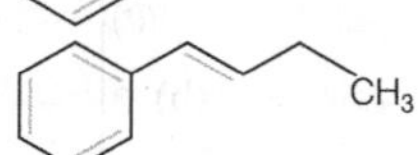

(c) 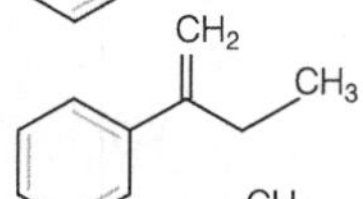

(d) 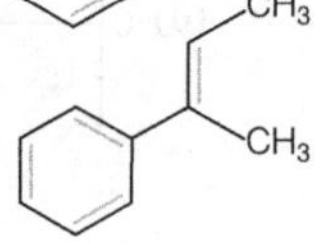

Paragraph 2

When 100 mL of 1.0 M HCl was mixed with 100 mL of 1.0 M NaOH in an insulated beaker at constant pressure, a temperature increase of 5.7°C was measured for the beaker and its contents (Expt. 1). Because the enthalpy of neutralisation of a strong acid with a strong base is a constant (-57.0 kJ mol^{-1}), this experiment could be used to measure the calorimeter constant. In a second experiment (Expt. 2), 100 mL of 2.0 M acetic acid ($K_a = 2.0 \times 10^{-5}$) was mixed with 100 mL of 1.0 M NaOH (under identical conditions to Expt. 1) where a temperature rise of 5.6°C was measured.

19. Enthalpy of dissociation (in kJ mol^{-1}) of acetic acid obtained from the Expt. 2 is

(a) 1.0
(b) 10.0
(c) 24.5
(d) 51.4

20. The pH of the solution after Expt. 2 is

(a) 2.8
(b) 4.7
(c) 5.0
(d) 7.0

Mathematics

Section 1 (Maximum Marks : 32)

- This section contains EIGHT questions.
- The answer to each question is a SINGLE DIGIT INTEGER ranging from 0 to 9 both inclusive.
- For each question, darken the bubble corresponding to the correct integer in the ORS.
- Marking scheme
  +4 If the bubble corresponding to the answer is darkened.
  0 in all other cases.

1. Suppose that all the terms of an arithmetic progression are natural numbers. If the ratio of the sum of the first seven terms to the sum of the first eleven terms is 6 : 11 and the seventh term lies in between 130 and 140, then the common difference of this AP is

2. The coefficient of x^9 in the expansion of $(1+x)(1+x^2)(1+x^3) \dots (1+x^{100})$ is

3. Suppose that the foci of the ellipse $\dfrac{x^2}{9}+\dfrac{y^2}{5}=1$ are $(f_1, 0)$ and $(f_2, 0)$, where $f_1 > 0$ and $f_2 < 0$. Let P_1 and P_2 be two parabolas with a common vertex at (0, 0) with foci at $(f_1, 0)$ and $(2f_2, 0)$, respectively. Let T_1 be a tangent to P_1 which passes through $(2f_2, 0)$ and T_2 be a tangent to P_2 which passes through $(f_1, 0)$. If m_1 is the slope of T_1 and m_2 is the slope of T_2, then the value of $\left(\dfrac{1}{m_1^2}+m_2^2\right)$ is

4. Let m and n be two positive integers greater than 1. If $\lim_{\alpha \to 0}\left(\dfrac{e^{\cos(\alpha^n)}-e}{\alpha^m}\right) = -\left(\dfrac{e}{2}\right)$, then the value of $\dfrac{m}{n}$ is

5. If $\alpha = \int_0^1 (e^{9x+3\tan^{-1}x})\left(\dfrac{12+9x^2}{1+x^2}\right)dx$, where $\tan^{-1}x$ takes only principal values, then the value of $\left(\log_e|1+\alpha| - \dfrac{3\pi}{4}\right)$ is

6. Let $f: R \to R$ be a continuous odd function, which vanishes exactly at one point and $f(1) = \frac{1}{2}$. Suppose that $F(x) = \int_{-1}^{x} f(t)dt$ for all $x \in [-1, 2]$ and $G(x) = \int_{-1}^{x} t|f\{f(t)\}|\,dt$ for all $x \in [-1, 2]$. If $\lim_{x \to 1} \frac{F(x)}{G(x)} = \frac{1}{14}$, then the value of $f\left(\frac{1}{2}\right)$ is

7. Suppose that $\mathbf{p}, \mathbf{q}$ and $\mathbf{r}$ are three non-coplanar vectors in R^3. Let the components of a vector $\mathbf{s}$ along $\mathbf{p}, \mathbf{q}$ and $\mathbf{r}$ be 4, 3 and 5, respectively. If the components of this vector $\mathbf{s}$ along $(-\mathbf{p}+\mathbf{q}+\mathbf{r})$, $(\mathbf{p}-\mathbf{q}+\mathbf{r})$ and $(-\mathbf{p}-\mathbf{q}+\mathbf{r})$ are x, y and z respectively, then the value of $2x + y + z$ is

8. For any integer k, let $\alpha_k = \cos\left(\frac{k\pi}{7}\right) + i\sin\left(\frac{k\pi}{7}\right)$, where $i = \sqrt{-1}$. The value of the expression $\frac{\sum_{k=1}^{12} |\alpha_{k+1} - \alpha_k|}{\sum_{k=1}^{3} |\alpha_{4k-1} - \alpha_{4k-2}|}$ is

Section 2 (Maximum Marks : 32)

- This section contains EIGHT questions.
- Each question has four options (a), (b), (c) and (d). ONE OR MORE THAN ONE of these four options are correct.
- For each question, darken the bubble(s) corresponding to all the correct options in the ORS.
- Marking scheme :
  + 4 If only the bubble(s) corresponding to all the correct options are darkened.
  0 If none of the bubbles is darkened.
  − 2 In all other cases.

9. Let $f, g: [-1, 2] \to R$ be continuous functions which are twice differentiable on the interval $(-1, 2)$. Let the values of f and g at the points -1, 0 and 2 be as given in the following table:

	$x = -1$	$x = 0$	$x = 2$
$f(x)$	3	6	0
$g(x)$	0	1	−1

In each of the intervals $(-1, 0)$ and $(0, 2)$, the function $(f - 3g)''$ never vanishes. Then, the correct statement(s) is/are

(a) $f'(x) - 3g'(x) = 0$ has exactly three solutions in $(-1, 0) \cup (0, 2)$

(b) $f'(x) - 3g'(x) = 0$ has exactly one solution in $(-1, 0)$

(c) $f'(x) - 3g'(x) = 0$ has exactly one solution in $(0, 2)$

(d) $f'(x) - 3g'(x) = 0$ has exactly two solutions in $(-1, 0)$ and exactly two solutions in $(0, 2)$

10. Let $f(x) = 7\tan^8 x + 7\tan^6 x - 3\tan^4 x - 3\tan^2 x$ for all $x \in \left(-\frac{\pi}{2}, \frac{\pi}{2}\right)$.

Then, the correct expression(s) is/are

(a) $\int_0^{\pi/4} x f(x)\,dx = \frac{1}{12}$

(b) $\int_0^{\pi/4} f(x)\,dx = 0$

(c) $\int_0^{\pi/4} x f(x)\,dx = \frac{1}{6}$

(d) $\int_0^{\pi/4} f(x)\,dx = 1$

11. Let $f'(x) = \dfrac{192x^3}{2+\sin^4 \pi x}$ for all $x \in R$ with $f\left(\dfrac{1}{2}\right) = 0$. If $m \le \int_{1/2}^{1} f(x)\,dx \le M$, then the possible values of m and M are

(a) $m = 13, M = 24$
(b) $m = \dfrac{1}{4}, M = \dfrac{1}{2}$
(c) $m = -11, M = 0$
(d) $m = 1, M = 12$

12. Let S be the set of all non-zero real numbers α such that the quadratic equation $\alpha x^2 - x + \alpha = 0$ has two distinct real roots x_1 and x_2 satisfying the inequality $|x_1 - x_2| < 1$. Which of the following interval(s) is/are a subset of S?

(a) $\left(-\dfrac{1}{2}, -\dfrac{1}{\sqrt{5}}\right)$ (b) $\left(-\dfrac{1}{\sqrt{5}}, 0\right)$

(c) $\left(0, \dfrac{1}{\sqrt{5}}\right)$ (d) $\left(\dfrac{1}{\sqrt{5}}, \dfrac{1}{2}\right)$

13. If $\alpha = 3\sin^{-1}\left(\dfrac{6}{11}\right)$ and $\beta = 3\cos^{-1}\left(\dfrac{4}{9}\right)$, where the inverse trigonometric functions take only the principal values, then the correct option(s) is/are

(a) $\cos\beta > 0$
(b) $\sin\beta < 0$
(c) $\cos(\alpha + \beta) > 0$
(d) $\cos\alpha < 0$

14. Let E_1 and E_2 be two ellipses whose centres are at the origin. The major axes of E_1 and E_2 lie along the X-axis and Y-axis, respectively. Let S be the circle $x^2 + (y-1)^2 = 2$. The straight line $x + y = 3$ touches the curves S, E_1 and E_2 at P, Q and R, respectively. Suppose that $PQ = PR = \dfrac{2\sqrt{2}}{3}$. If e_1 and e_2 are the eccentricities of E_1 and E_2 respectively, then the correct expression(s) is/are

(a) $e_1^2 + e_2^2 = \dfrac{43}{40}$

(b) $e_1 e_2 = \dfrac{\sqrt{7}}{2\sqrt{10}}$

(c) $|e_1^2 - e_2^2| = \dfrac{5}{8}$

(d) $e_1 e_2 = \dfrac{\sqrt{3}}{4}$

15. Consider the hyperbola H: $x^2 - y^2 = 1$ and a circle S with centre $N(x_2, 0)$. Suppose that H and S touch each other at a point $P(x_1, y_1)$ with $x_1 > 1$ and $y_1 > 0$. The common tangent to H and S at P intersects the X-axis at point M. If (l, m) is the centroid of ΔPMN, then the correct expression(s) is/are

(a) $\dfrac{dl}{dx_1} = 1 - \dfrac{1}{3x_1^2}$ for $x_1 > 1$

(b) $\dfrac{dm}{dx_1} = \dfrac{x_1}{3(\sqrt{x_1^2 - 1})}$ for $x_1 > 1$

(c) $\dfrac{dl}{dx_1} = 1 + \dfrac{1}{3x_1^2}$ for $x_1 > 1$

(d) $\dfrac{dm}{dy_1} = \dfrac{1}{3}$ for $y_1 > 0$

16. The option(s) with the values of a and L that satisfy the equation

$$\frac{\int_0^{4\pi} e^t(\sin^6 at + \cos^4 at)\,dt}{\int_0^{\pi} e^t(\sin^6 at + \cos^4 at)\,dt} = L,$$

is/are

(a) $a = 2, L = \dfrac{e^{4\pi} - 1}{e^{\pi} - 1}$

(b) $a = 2, L = \dfrac{e^{4\pi} + 1}{e^{\pi} + 1}$

(c) $a = 4, L = \dfrac{e^{4\pi} - 1}{e^{\pi} - 1}$

(d) $a = 4, L = \dfrac{e^{4\pi} + 1}{e^{\pi} + 1}$

Section 3 (Maximum Marks : 16)

- This section contains TWO paragraphs.
- Based on each paragraph, there will be TWO questions.
- Each question has FOUR options (a), (b), (c) and (d). ONE OR MORE THAN ONE of these four options are correct.
- For each question, darken the bubbles(s) corresponding to all the correct options in the ORS.

  Marking scheme :

  +4 If only the bubble(s) corresponding to all the correct options are darkened.

  0 If none of the bubbles is darkened.

  –2 In all other cases.

Paragraph 1

Let $F: R \to R$ be a thrice differentiable function. Suppose that $F(1) = 0, F(3) = -4$ and $F'(x) < 0$ for all $x \in (1, 3)$. Let $f(x) = xF(x)$ for all $x \in R$.

17. The correct statement(s) is/are

(a) $f'(1) < 0$

(b) $f(2) < 0$

(c) $f'(x) \neq 0$ for any $x \in (1, 3)$

(d) $f'(x) = 0$ for some $x \in (1, 3)$

18. If $\int_1^3 x^2 F'(x)\, dx = -12$ and $\int_1^3 x^3 F''(x)\, dx = 40$, then the correct expression(s) is/are

(a) $9f'(3) + f'(1) - 32 = 0$

(b) $\int_1^3 f(x)\, dx = 12$

(c) $9f'(3) - f'(1) + 32 = 0$

(d) $\int_1^3 f(x)\, dx = -12$

Paragraph 2

Let n_1 and n_2 be the number of red and black balls, respectively in box I. Let n_3 and n_4 be the number of red and black balls, respectively in box II.

19. One of the two boxes, box I and box II was selected at random and a ball was drawn randomly out of this box. The ball was found to be red. If the probability that this red ball was drawn from box II, is $\frac{1}{3}$, then the correct option(s) with the possible values of n_1, n_2, n_3 and n_4 is/are

(a) $n_1 = 3, n_2 = 3, n_3 = 5, n_4 = 15$

(b) $n_1 = 3, n_2 = 6, n_3 = 10, n_4 = 50$

(c) $n_1 = 8, n_2 = 6, n_3 = 5, n_4 = 20$

(d) $n_1 = 6, n_2 = 12, n_3 = 5, n_4 = 20$

20. A ball is drawn at random from box I and transferred to box II. If the probability of drawing a red ball from box I, after this transfer, is $\frac{1}{3}$, then the correct option(s) with the possible values of n_1 and n_2 is/are

(a) $n_1 = 4$ and $n_2 = 6$

(b) $n_1 = 2$ and $n_2 = 3$

(c) $n_1 = 10$ and $n_2 = 20$

(d) $n_1 = 3$ and $n_2 = 6$

Answers

Paper 1

Physics

1. (3) **2.** (7) **3.** (6) **4.** (2) **5.** (2) **6.** (7) **7.** (2)
8. (3) **9.** (d) **10.** (c) **11.** (b) **12.** (a,b,c) **13.** (a,b,d) **14.** (b)
15. (a,c) **16.** (b,c) **17.** (a,c,d) **18.** (b,d)

19. A → R or RT; B → P, S; C → Q, T; D → R

20. A → P,Q,R,T; B → Q, S; C → P,Q,R,S; D → P,R,T

Chemistry

1. (2) **2.** (9) **3.** (8) **4.** (4) **5.** (4) **6.** (3) **7.** (4)
8. (1) **9.** (b,d) **10.** (a) **11.** (d) **12.** (a) **13.** (c) **14.** (a,b,c)
15. (b,c,d) **16.** (a,b) **17.** (b) **18.** (a)

19. A →P,Q,S; B → T; C → Q; D → R

20. A →R,T; B → P,Q,S; C → P,Q,S; D → P,Q,S,T

Mathematics

1. (4) **2.** (8) **3.** (5) **4.** (2) **5.** (0) **6.** (4) **7.** (3)
8. (8) **9.** (a,c) **10.** (b,c) **11.** (a,d) **12.** (a,b,c) **13.** (a,c,d) **14.** (c,d)
15. (b,c) **16.** (b,d) **17.** (a,b) **18.** (a,d)

19. A → P,Q; B → P,Q; C → P,Q,S,T; D → Q,T

20. A → P,R,S; B → P,C; C → P,Q; D → S,T

Paper 2

Physics

1. (3) **2.** (2) **3.** (2) **4.** (1) **5.** (2) **6.** (7) **7.** (4)
8. (6) **9.** (d) **10.** (a,b) **11.** (b,c) **12.** (d) **13.** (a,b,c) **14.** (a)
15. (a,d) **16.** (a,c) **17.** (a) **18.** (d) **19.** (a,d) **20.** (a,c)

Chemistry

1. (3) **2.** (6) **3.** (6) **4.** (3) **5.** (9) **6.** (8) **7.** (4)
8. (4) **9.** (b) **10.** (b,c,d) **11.** (c) **12.** (a) **13.** (b) **14.** (a)
15. (b,c) **16.** (c,d) **17.** (c) **18.** (d) **19.** (a) **20.** (b)

Mathematics

1. (9) **2.** (8) **3.** (4) **4.** (2) **5.** (9) **6.** (7) **7.** (9)
8. (4) **9.** (b,c) **10.** (a,b) **11.** (b) **12.** (a,d) **13.** (b,c,d) **14.** (a,b)
15. (a,b,d) **16.** (a,c) **17.** (a,b,c) **18.** (a,b,c) **19.** (a,b) **20.** (c,d)

Unsolved Paper 2016

JEE Main

Joint Entrance Examination

Instructions

1. This test consists of 90 questions.
2. There are three parts in the question paper A, B, C consisting of Physics, Chemistry & Mathematics having 30 questions in each part of equal weightage. Each question is allotted 4 marks for correct response.
3. Candidates will be awarded marks as stated above in instruction no. 2 for correct response of each question. 1 mark will be deducted for indicating incorrect response of each question. No deduction from the total score will be made if no response is indicated for an item in the answer sheet.
4. There is only one correct response for each question. Filling up more than one response in any question will be treated as wrong response and marks for wrong response will be deducted according as per instructions.

A. Physics

1. A student measures the time period of 100 oscillations of a simple pendulum four times. The data set is 90s, 91s, 92s and 95s. If the minimum division in the measuring clock is 1s, then the reported mean time should be

(a) (92 ± 2s)
(b) (92 ± 5s)
(c) (92± 1.8s)
(d) (92 ± 3s)

2. A particle of mass m is moving along the side of a square of side a, with a uniform speed v in the XY-plane as shown in the figure.

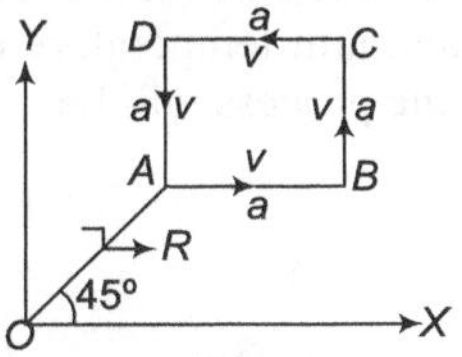

Which of the following statements is **false** for the angular momentum **L** about the origin?

(a) $\mathbf{L} = \frac{-mv}{\sqrt{2}} R\hat{\mathbf{k}}$ when the particle is moving from A to B

(b) $\mathbf{L} = mv\left(\frac{R}{\sqrt{2}} - a\right)\hat{\mathbf{k}}$ when the particle is moving from C to D

(c) $\mathbf{L} = mv\left(\frac{R}{\sqrt{2}} + a\right)\hat{\mathbf{k}}$ when the particle is moving from B to C

(d) $\mathbf{L} = \frac{mv}{\sqrt{2}} R\hat{\mathbf{k}}$ when the particle is moving from D to A

3. A point particle of mass m, moves along the uniformly rough track PQR as shown in the figure. The coefficient of friction, between the particle and the rough track equals μ. The particle is released from rest, from the point P and it comes to rest at a point R. The energies lost by the ball, over the parts PQ and QR of the track, are equal to each other, and no energy is lost when particle changes direction from PQ to QR.

The values of the coefficient of friction μ and the distance x (= QR), are respectively close to

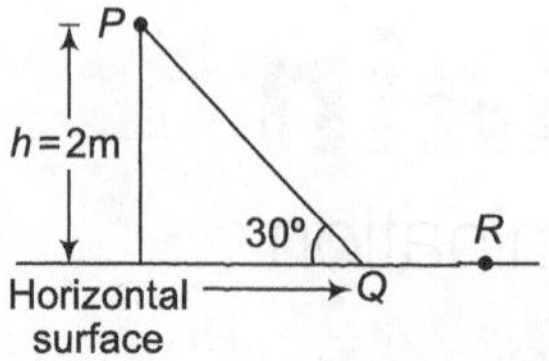

(a) 0.2 and 6.5 m (b) 0.2 and 3.5 m
(c) 0.29 and 3.5 m (d) 0.29 and 6.5 m

4. A person trying to lose weight by burning fat lifts a mass of 10 kg upto a height of 1 m 1000 times. Assume that the potential energy lost each time he lowers the mass is dissipated. How much fat will he use up considering the work done only when the weight is lifted up? Fat supplies 3.8×10^7 J of energy per kg which is converted to mechanical energy with a 20% efficiency rate. (Take, $g = 9.8 \text{ ms}^{-2}$)

(a) 2.45×10^{-3} kg (b) 6.45×10^{-3} kg
(c) 9.89×10^{-3} kg (d) 12.89×10^{-3} kg

5. A roller is made by joining together two corners at their vertices O. It is kept on two rails AB and CD which are placed asymmetrically (see the figure), with its axis perpendicular to CD and its centre O at the centre of line joining AB and CD (see the figure). It is given a light push, so that it starts rolling with its centre O moving parallel to CD in the direction shown. As it moves, the roller will tend to

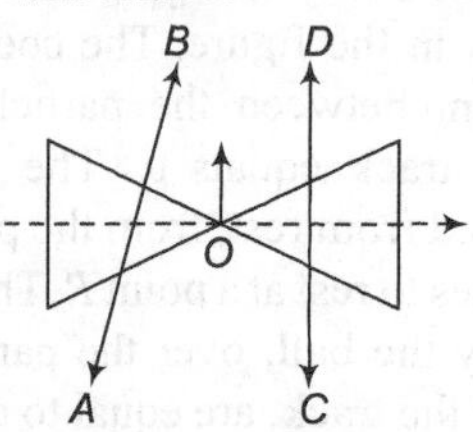

(a) turn left (b) turn right
(c) go straight
(d) turn left and right alternately

6. A satellite is revolving in a circular orbit at a height h from the Earth's surface (radius of earth $R, h << R$). The minimum increase in its orbital velocity required, so that the satellite could escape from the earth's gravitational field, is close to (neglect the effect of atmosphere)

(a) $\sqrt{2gR}$ (b) $\sqrt{gR}$
(c) $\sqrt{gR/2}$ (d) $\sqrt{gR}\,(\sqrt{2}-1)$

7. A pendulum clock loses 12 s a day if the temperature is 40°C and gains 4 s a day if the temperature is 20°C. The temperature at which the clock will show correct time, and the coefficient of linear expansion α of the metal of the pendulum shaft are, respectively

(a) 25°C, $\alpha = 1.85 \times 10^{-5}/°C$
(b) 60°C, $\alpha = 1.85 \times 10^{-4}/°C$
(c) 30°C, $\alpha = 1.85 \times 10^{-3}/°C$
(d) 55°C, $\alpha = 1.85 \times 10^{-2}/°C$

8. An ideal gas undergoes a quasi static, reversible process in which its molar heat capacity C remains constant. If during this process the relation of pressure p and volume V is given by pV^n = constant, then n is given by (here, C_p and C_V are molar specific heat at constant pressure and constant volume, respectively)

(a) $n = \dfrac{C_p}{C_V}$ (b) $n = \dfrac{C - C_p}{C - C_V}$
(c) $n = \dfrac{C_p - C}{C - C_V}$ (d) $n = \dfrac{C - C_V}{C - C_p}$

9. n moles of an ideal gas undergoes a process A and B as shown in the figure. The maximum temperature of the gas during the process will be

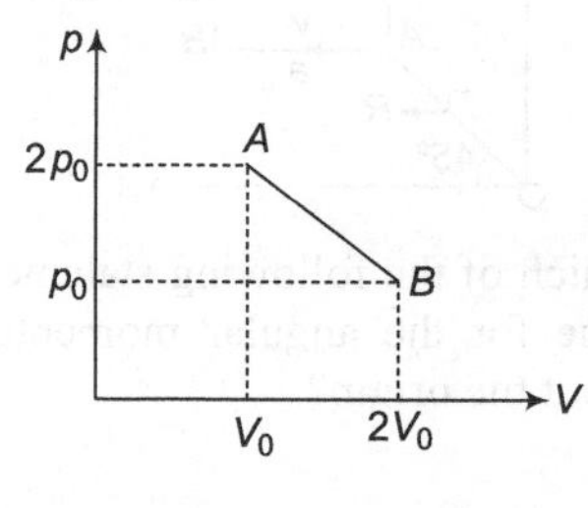

(a) $\frac{9}{4}\frac{p_0V_0}{nR}$ (b) $\frac{3}{2}\frac{p_0V_0}{nR}$

(c) $\frac{9}{2}\frac{p_0V_0}{nR}$ (d) $\frac{9p_0V_0}{nR}$

10. A particle performs simple harmonic motion with amplitude A. Its speed is trebled at the instant that it is at a distance $\frac{2}{3}A$ from equilibrium position. The new amplitude of the motion is

(a) $\frac{A}{3}\sqrt{41}$ (b) $3A$ (c) $A\sqrt{3}$ (d) $\frac{7}{3}A$

11. A uniform string of length 20 m is suspended from a rigid support. A short wave pulse is introduced at its lowest end. It starts moving up the string. The time taken to reach the support is (Take, $g = 10\,\text{ms}^{-2}$)

(a) $2\pi\sqrt{2}$ s (b) 2 s

(c) $2\sqrt{2}$ s (d) $\sqrt{2}$ s

12. The region between two concentric spheres of radii a and b, respectively (see the figure), has volume charge density $\rho = \frac{A}{r}$, where, A is a constant and r is the distance from the centre. At the centre of the spheres is a point charge Q. The value of A, such that the electric field in the region between the spheres will be constant, is

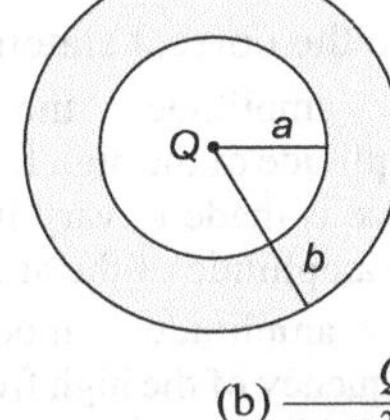

(a) $\frac{Q}{2\pi a^2}$ (b) $\frac{Q}{2\pi(b^2 - a^2)}$

(c) $\frac{2Q}{\pi(a^2 - b^2)}$ (d) $\frac{2Q}{\pi a^2}$

13. A combination of capacitors is set-up as shown in the figure. The magnitude of the electric field, due to a point charge Q (having a charge equal to the sum of the charges on the 4μF and 9μF capacitors), at a point distant 30 m from it, would equal to

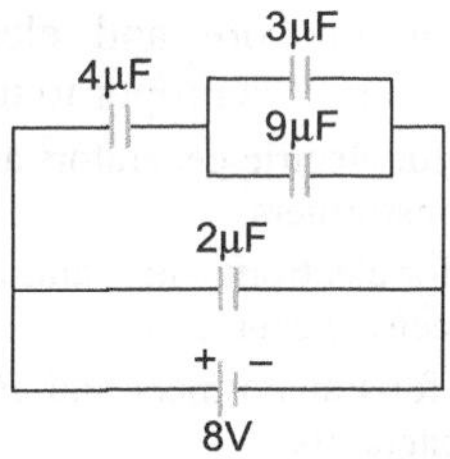

(a) 240 N/C (b) 360 N/C

(c) 420 N/C (d) 480 N/C

14. The temperature dependence of resistances of Cu and undoped Si in the temperature range 300-400 K, is best described by

(a) linear increase for Cu, linear increase for Si

(b) linear increase for Cu, exponential increase for Si

(c) linear increase for Cu, exponential decrease for Si

(d) linear decrease for Cu, linear decrease for Si

15. Two identical wires A and B, each of length l, carry the same current I. Wire A is bent into a circle of radius R and wire B is bent to form a square of side a. If B_A and B_B are the values of magnetic field at the centres of the circle and square respectively, then the ratio $\frac{B_A}{B_B}$ is

(a) $\frac{\pi^2}{8}$ (b) $\frac{\pi^2}{16\sqrt{2}}$

(c) $\frac{\pi^2}{16}$ (d) $\frac{\pi^2}{8\sqrt{2}}$

16. Hysteresis loops for two magnetic materials A and B are as given below:

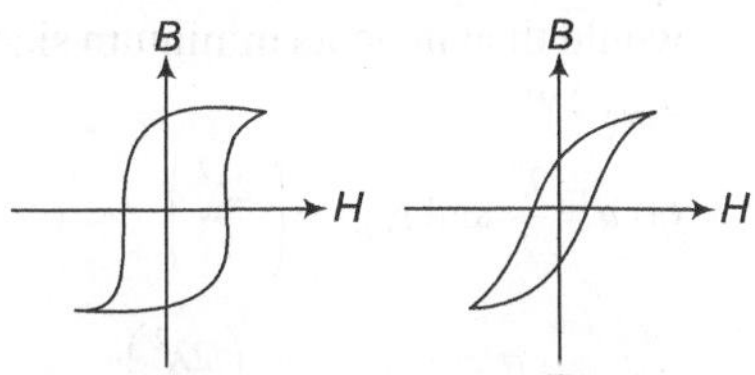

(A) (B)

These materials are used to make magnets for electric generators, transformer core and electromagnet core. Then, it is proper to use

(a) A for electric generators and transformers
(b) A for electromagnets and B for electric generators
(c) A for transformers and B for electric generators
(d) B for electromagnets and transformers

17. An arc lamp requires a direct current of 10 A at 80 V to function. If it is connected to a 220 V (rms), 50 Hz AC supply, the series inductor needed for it to work is close to

(a) 80 H (b) 0.08 H
(c) 0.044 H (d) 0.065 H

18. Arrange the following electromagnetic radiations in the order of increasing energy.

A. Blue light B. Yellow light
C. X-ray D. Radio wave

(a) D, B, A, C (b) A, B, D, C
(c) C, A, B, D (d) B, A, D, C

19. An observer looks at a distant tree of height 10 m with a telescope of magnifying power of 20. To observer the tree appears

(a) 10 times taller (b) 10 times nearer
(c) 20 times taller (d) 20 times nearer

20. The box of a pin hole camera of length L, has a hole of radius a. It is assumed that when the hole is illuminated by a parallel beam of light of wavelength λ the spread of the spot (obtained on the opposite wall of the camera) is the sum of its geometrical spread and the spread due to diffraction. The spot would then have its minimum size (say b_{min}) when

(a) $a = \frac{\lambda^2}{L}$ and $b_{min} = \left(\frac{2\lambda^2}{L}\right)$

(b) $a = \sqrt{\lambda L}$ and $b_{min} = \left(\frac{2\lambda^2}{L}\right)$

(c) $a = \sqrt{\lambda L}$ and $b_{min} = \sqrt{4\lambda L}$

(d) $a = \frac{\lambda^2}{L}$ and $b_{min} = \sqrt{4\lambda L}$

21. Radiation of wavelength λ, is incident on a photocell. The fastest emitted electron has speed v. If the wavelength is changed to $\frac{3\lambda}{4}$, the speed of the fastest emitted electron will be

(a) $> v\left(\frac{4}{3}\right)^{1/2}$ (b) $< v\left(\frac{4}{3}\right)^{1/2}$

(c) $= v\left(\frac{4}{3}\right)^{1/2}$ (d) $= v\left(\frac{3}{4}\right)^{1/2}$

22. Half-lives of two radioactive elements A and B are 20 min and 40 min, respectively. Initially, the samples have equal number of nuclei. After 80 min, the ratio of decayed numbers of A and B nuclei will be

(a) 1 : 16 (b) 4 : 1 (c) 1 : 4 (d) 5 : 4

23. If a, b, c and d are inputs to a gate and x is its output, then, as per the following time graph, the gate is

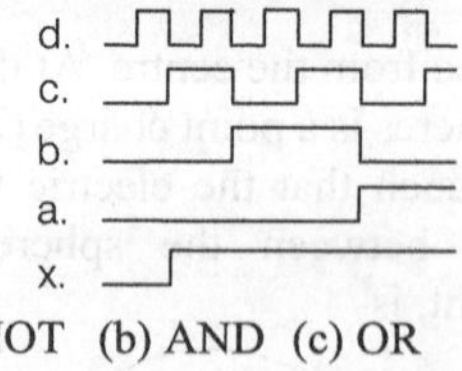

(a) NOT (b) AND (c) OR (d) NAND

24. Choose the correct statement.

(a) In amplitude modulation, the amplitude of the high frequency carrier wave is made to vary in proportion to the amplitude of the audio signal
(b) In amplitude modulation, the frequency of the high frequency carrier wave is made to vary in proportion to the amplitude of the audio signal
(c) In frequency modulation, the amplitude of the high frequency carrier wave is made to vary in proportion to the amplitude of the audio signal
(d) In frequency modulation, the amplitude of the high frequency carrier wave is made to vary in proportion to the frequency of the audio signal

25. A screw gauge with a pitch of 0.5 mm and a circular scale with 50 divisions is used to measure the thickness of a thin sheet of Aluminium. Before starting the measurement, it is found that when the two jaws of the screw gauge are brought in contact, the 45th division coincides with the main scale line and that the zero of the main scale is barely visible. What is the thickness of the sheet, if the main scale reading is 0.5 mm and the 25th division coincides with the main scale line?

(a) 0.75 mm
(b) 0.80 mm
(c) 0.70 mm
(d) 0.50 mm

26. A pipe open at both ends has a fundamental frequency f in air. The pipe is dipped vertically in water, so that half of it is in water. The fundamental frequency of the air column is now

(a) $\frac{f}{2}$ (b) $\frac{3f}{4}$ (c) $2f$ (d) f

27. A galvanometer having a coil resistance of 100 Ω gives a full scale deflection when a current of 1 mA is passed through it. The value of the resistance which can convert this galvanometer into ammeter giving a full scale deflection for a current of 10 A, is

(a) 0.01 Ω
(b) 2 Ω
(c) 0.1 Ω
(d) 3 Ω

28. In an experiment for determination of refractive index of glass of a prism by i-δ plot, it was found that a ray incident at an angle 35° suffers a deviation of 40° and that it emerges at an angle 79°. In that case, which of the following is closest to the maximum possible value of the refractive index?

(a) 1.5 (b) 1.6
(c) 1.7 (d) 1.8

29. Identify the semiconductor devices whose characteristics are as given below, in the order (a),(b),(c),(d).

(a) 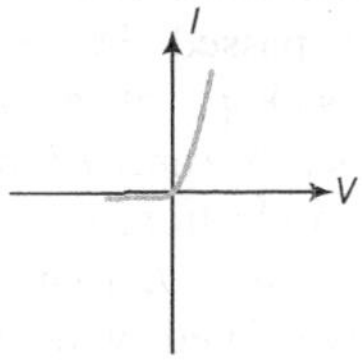

(b) 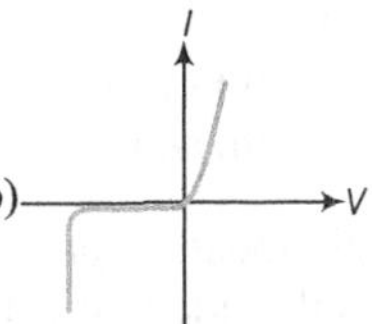

(c) 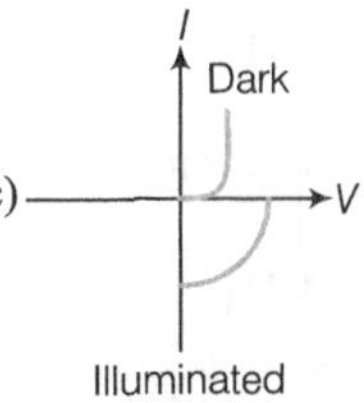

(d) 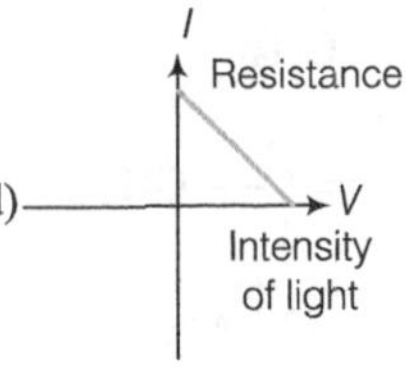

(a) Simple diode, Zener diode, Solar cell, Light dependent resistance
(b) Zener diode, Simple diode, Light dependent resistance, Solar cell
(c) Solar cell, Light dependent resistance, Zener diode, Simple diode
(d) Zener diode, Solar cell, Simple diode, Light dependent resistance

30. For a common-emitter configuration, if α and β have their usual meanings, the **incorrect** relationship between α and β is

(a) $\frac{1}{\alpha} = \frac{1}{\beta} + 1$ (b) $\alpha = \frac{\beta}{1-\beta}$

(c) $\alpha = \frac{\beta}{1+\beta}$ (d) $\alpha = \frac{\beta^2}{1+\beta^2}$

B. Chemistry

1. A stream of electrons from a heated filament was passed between two charged plates kept at a potential difference V esu. If e and m are charge and mass of an electron, respectively, then the value of h/λ (where, λ is wavelength associated with electron wave) is given by

(a) $2\,meV$ (b) $\sqrt{meV}$
(c) $\sqrt{2meV}$ (d) meV

2. 2-chloro-2-methylpentane on reaction with sodium methoxide in methanol yields

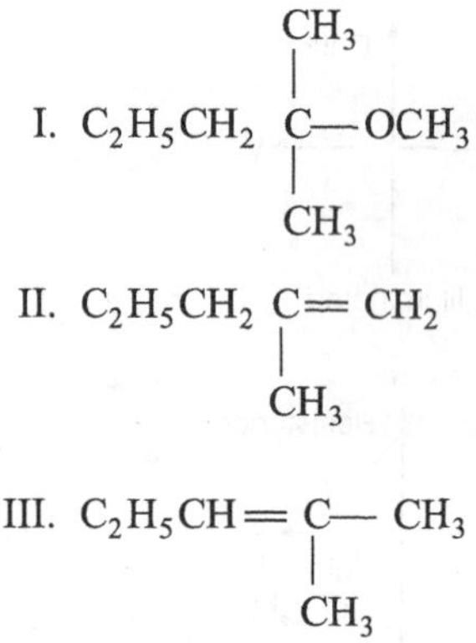

(a) Both I and III
(b) Only III
(c) Both I and II
(d) All of these

3. Which of the following compounds is metallic and ferromagnetic?

(a) CrO_2 (b) VO_2
(c) MnO_2 (d) TiO_2

4. Which of the following statements about low density polythene is false?

(a) It is a poor conductor of electricity
(b) Its synthesis required dioxygen or a peroxide initiator as a catalyst
(c) It is used in the manufacture of buckets, dustbins etc.
(d) Its synthesis requires high pressure

5. For a linear plot of $\log(x/m)$ *versus* $\log p$ in a Freundlich adsorption isotherm, which of the following statements is correct? (k and n are constants)

(a) $1/n$ appears as the intercept
(b) Only $1/n$ appears as the slope
(c) $\log\left(\frac{1}{n}\right)$ appears as the intercept
(d) Both k and $1/n$ appear in the slope term

6. The heats of combustion of carbon and carbon monoxide are -393.5 and $-283.5\ \text{kJ mol}^{-1}$, respectively. The heat of formation (in kJ) of carbon monoxide per mole is

(a) 676.5 (b) −676.5
(c) −110.5 (d) 110.5

7. The hottest region of Bunsen flame shown in the figure given below is

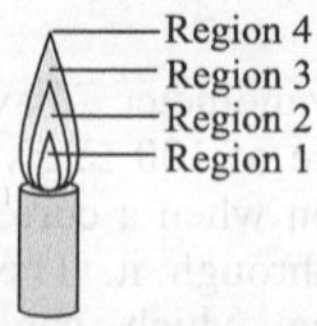

(a) region 2
(b) region 3
(c) region 4
(d) region 1

8. Which of the following is an anionic detergent?

(a) Sodium lauryl sulphate
(b) Cetyltrimethyl ammonium bromide
(c) Glyceryl oleate
(d) Sodium stearate

9. 18 g of glucose ($C_6H_{12}O_6$) is added to 178.2 g water. The vapour pressure of water (in torr) for this aqueous solution is

(a) 76.0 (b) 752.4
(c) 759.0 (d) 7.6

10. The distillation technique most suited for separating glycerol from spent lye in the soap industry is

(a) fractional distillation
(b) steam distillation
(c) distillation under reduced pressure
(d) simple distillation

11. The species in which the N-atom is in a state of *sp* hybridisation is

(a) NO_2^- (b) NO_3^- (c) NO_2 (d) NO_2^+

12. Decomposition of H_2O_2 follows a first order reaction. In 50 min, the concentration of H_2O_2 decreases from 0.5 to 0.125 M in one such decomposition. When the concentration of H_2O_2 reaches 0.05 M, the rate of formation of O_2 will be

(a) 6.93×10^{-4} mol min^{-1}
(b) 2.66 L min^{-1} at STP
(c) 1.34×10^{-2} mol min^{-1}
(d) 6.93×10^{-2} mol min^{-1}

13. The pair having the same magnetic moment is
[at. no. Cr = 24, Mn = 25, Fe = 26 and Co = 27]

(a) $[Cr(H_2O)_6]^{2+}$ and $[Fe(H_2O)_6]^{2+}$
(b) $[Mn(H_2O)_6]^{2+}$ and $[Cr(H_2O)_6]^{2+}$
(c) $[CoCl_4]^{2-}$ and $[Fe(H_2O)_6]^{2+}$
(d) $[Cr(H_2O)_6]^{2+}$ and $[CoCl_4]^{2-}$

14. The absolute configuration of

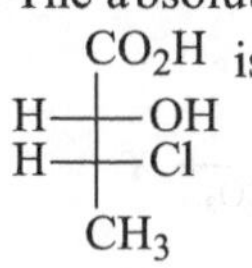

is

(a) (2S, 3R) (b) (2S, 3S)
(c) (2R, 3R) (d) (2R, 3S)

15. The equilibrium constant at 298 K for a reaction, $A + B \rightleftharpoons C + D$ is 100. If the initial concentrations of all the four species were 1 M each, then equilibrium concentration of D (in mol L^{-1}) will be

(a) 0.818 (b) 1.818
(c) 1.182 (d) 0.182

16. Which one of the following ores is best concentrated by froth floatation method?

(a) Siderite
(b) Galena
(c) Malachite
(d) Magnetite

17. At 300 K and 1 atm, 15 mL of a gaseous hydrocarbon requires 375 mL air containing 20% O_2 by volume for complete combustion. After combustion, the gases occupy 330 mL. Assuming that the water formed is in liquid form and the volumes were measured at the same temperature and pressure, the formula of the hydrocarbon is

(a) C_3H_8 (b) C_4H_8
(c) C_4H_{10} (d) C_3H_6

18. The pair in which phosphorus atoms have a formal oxidation state of +3 is

(a) pyrophosphorous and hypophosphoric acids
(b) orthophosphorous and hypophosphoric acids
(c) pyrophosphorous and pyrophosphoric acids
(d) orthophosphorous and pyrophosphorous acids

19. Which one of the following complexes shows optical isomerism?

(a) *cis* $[Co(en)_2Cl_2]Cl$
(b) *trans* $[Co(en)_2Cl_2]Cl$
(c) $[Co(NH_3)_4Cl_2]Cl$
(d) $[Co(NH_3)_3Cl_3]$

20. The reaction of zinc with dilute and concentrated nitric acid, respectively, produce

(a) NO_2 and NO
(b) NO and N_2O
(c) NO_2 and N_2O
(d) N_2O and NO_2

21. Which one of the following statements about water is false?

(a) Water can act both as an acid and as a base

(b) There is extensive intramolecular hydrogen bonding in the condensed phase
(c) Ice formed by heavy water sinks in normal water
(d) Water is oxidised to oxygen during photosynthesis

22. The concentration of fluoride, lead, nitrate and iron in a water sample from an underground lake was found to be 1000 ppb, 40 ppb, 100 ppm and 0.2 ppm, respectively. This water is unsuitable for drinking due to high concentration of
(a) lead (b) nitrate
(c) iron (d) fluoride

23. The main oxides formed on combustion of Li, Na and K in excess of air respectively are
(a) LiO_2, Na_2O_2 and K_2O
(b) Li_2O_2, Na_2O_2 and KO_2
(c) Li_2O, Na_2O_2 and KO_2
(d) Li_2O , Na_2O and KO_2

24. Thiol group is present in
(a) cystine (b) cysteine
(c) methionine (d) cytosine

25. Galvanisation is applying a coating of
(a) Cr (b) Cu
(c) Zn (d) Pb

26. Which of the following atoms has the highest first ionisation energy?
(a) Na (b) K
(c) Sc (d) Rb

27. In the Hofmann-bromamide degradation reaction, the number of moles of NaOH and Br_2 used per mole of amine produced are
(a) four moles of NaOH and two moles of Br_2
(b) two moles of NaOH and two moles of Br_2
(c) four moles of NaOH and one mole of Br_2
(d) one mole of NaOH and one mole of Br_2

28. Two closed bulbs of equal volume (V) containing an ideal gas initially at pressure p_i and temperature T_1 are connected through a narrow tube of negligible volume as shown in the figure below. The temperature of one of the bulbs is then raised to T_2. The final pressure p_f is

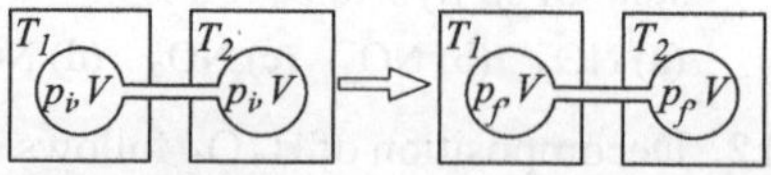

(a) $2p_i\left(\frac{T_1}{T_1+T_2}\right)$
(b) $2p_i\left(\frac{T_2}{T_1+T_2}\right)$
(c) $2p_i\left(\frac{T_1T_2}{T_1+T_2}\right)$
(d) $p_i\left(\frac{T_1T_2}{T_1+T_2}\right)$

29. The reaction of propene with HOCl ($Cl_2 + H_2O$) proceeds through the intermediate
(a) $CH_3—\overset{+}{C}H— CH_2— Cl$
(b) $CH_3— CH(OH)—\overset{+}{C}H_2$
(c) $CH_3— CHCl—\overset{+}{C}H_2$
(d) $CH_3—\overset{+}{C}H—CH_2— OH$

30. The product of the reaction given below is

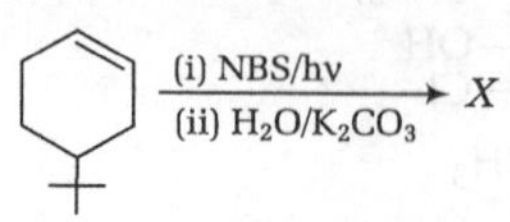

(a) 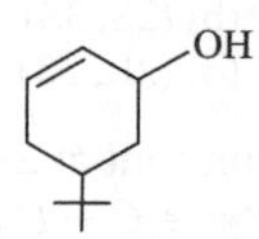

(b) 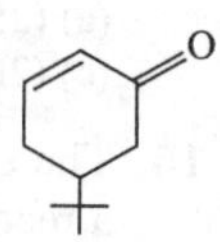

(c) 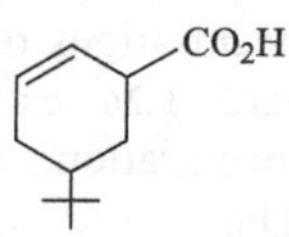

(d) 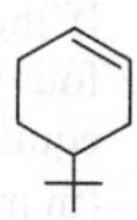

C. Mathematics

1. If $f(x)+2f\left(\frac{1}{x}\right)=3x, x\neq 0$ and $S=\{x\in R: f(x)=f(-x)\}$; then S
(a) is an empty set
(b) contains exactly one element
(c) contains exactly two elements
(d) contains more than two elements

2. A value of θ for which $\frac{2+3i\sin\theta}{1-2i\sin\theta}$ is purely imaginary, is
(a) $\frac{\pi}{3}$ (b) $\frac{\pi}{6}$
(c) $\sin^{-1}\left(\frac{\sqrt{3}}{4}\right)$ (d) $\sin^{-1}\left(\frac{1}{\sqrt{3}}\right)$

3. The sum of all real values of x satisfying the equation $(x^2-5x+5)^{x^2+4x-60}=1$ is
(a) 3 (b) – 4
(c) 6 (d) 5

4. If $A=\begin{bmatrix}5a & -b\\ 3 & 2\end{bmatrix}$ and A adj $A=AA^T$, then $5a+b$ is equal to
(a) – 1 (b) 5
(c) 4 (d) 13

5. The system of linear equations

$$x+\lambda y-z=0;$$
$$\lambda x-y-z=0;$$
$$x+y-\lambda z=0$$

has a non-trivial solution for
(a) infinitely many values of λ
(b) exactly one value of λ
(c) exactly two values of λ
(d) exactly three values of λ

6. If all the words (with or without meaning) having five letters, formed using the letters of the word SMALL and arranged as in a dictionary, then the position of the word SMALL is
(a) 46th (b) 59th
(c) 52nd (d) 58th

7. If the number of terms in the expansion of $\left(1-\frac{2}{x}+\frac{4}{x^2}\right)^n, x\neq 0$, is 28, then the sum of the coefficients of all the terms in this expansion, is
(a) 64
(b) 2187
(c) 243
(d) 729

8. If the 2nd, 5th and 9th terms of a non-constant AP are in GP, then the common ratio of this GP is
(a) $\frac{8}{5}$ (b) $\frac{4}{3}$
(c) 1 (d) $\frac{7}{4}$

9. If the sum of the first ten terms of the series

$$\left(1\frac{3}{5}\right)^2+\left(2\frac{2}{5}\right)^2+\left(3\frac{1}{5}\right)^2+4^2+\left(4\frac{4}{5}\right)^2+\ldots,$$

is $\frac{16}{5}m$, then m is equal to
(a) 102
(b) 101
(c) 100
(d) 99

10. Let $p=\lim_{x\to 0^+}(1+\tan^2\sqrt{x})^{1/2x}$, then $\log p$ is equal to
(a) 2 (b) 1
(c) $\frac{1}{2}$ (d) $\frac{1}{4}$

11. For $x\in R$, $f(x)=|\log 2-\sin x|$ and $g(x)=f(f(x))$, then
(a) g is not differentiable at $x=0$
(b) $g'(0)=\cos(\log 2)$
(c) $g'(0)=-\cos(\log 2)$
(d) g is differentiable at $x=0$ and $g'(0)=-\sin(\log 2)$

12. Consider $f(x)=\tan^{-1}\left(\sqrt{\dfrac{1+\sin x}{1-\sin x}}\right)$, $x\in\left(0,\dfrac{\pi}{2}\right)$. A normal to $y=f(x)$ at $x=\dfrac{\pi}{6}$ also passes through the point

(a) $(0, 0)$ (b) $\left(0,\dfrac{2\pi}{3}\right)$

(c) $\left(\dfrac{\pi}{6},0\right)$ (d) $\left(\dfrac{\pi}{4},0\right)$

13. A wire of length 2 units is cut into two parts which are bent respectively to form a square of side $= x$ units and a circle of radius $= r$ units. If the sum of the areas of the square and the circle so formed is minimum, then

(a) $2x = (\pi + 4)r$
(b) $(4 - \pi)x = \pi r$
(c) $x = 2r$
(d) $2x = r$

14. The integral $\displaystyle\int\frac{2x^{12}+5x^9}{(x^5+x^3+1)^3}\,dx$ is equal to

(a) $\dfrac{-x^5}{(x^5+x^3+1)^2}+C$

(b) $\dfrac{x^{10}}{2(x^5+x^3+1)^2}+C$

(c) $\dfrac{x^5}{2(x^5+x^3+1)^2}+C$

(d) $\dfrac{-x^{10}}{2(x^5+x^3+1)^2}+C$

where, C is an arbitrary constant.

15. $\displaystyle\lim_{n\to\infty}\left[\frac{(n+1)(n+2)\dots 3n}{n^{2n}}\right]^{1/n}$ is equal to

(a) $\dfrac{18}{e^4}$ (b) $\dfrac{27}{e^2}$

(c) $\dfrac{9}{e^2}$ (d) $3\log 3 - 2$

16. The area (in sq units) of the region $\{(x, y): y^2 \ge 2x$ and $x^2 + y^2 \le 4x$, $x \ge 0, y \ge 0\}$ is

(a) $\pi - \dfrac{4}{3}$ (b) $\pi - \dfrac{8}{3}$

(c) $\pi - \dfrac{4\sqrt{2}}{3}$ (d) $\dfrac{\pi}{2} - \dfrac{2\sqrt{2}}{3}$

17. If a curve $y = f(x)$ passes through the point $(1, -1)$ and satisfies the differential equation $y(1+xy)dx = x\,dy$ then $f\left(-\dfrac{1}{2}\right)$ is equal to

(a) $-\dfrac{2}{5}$ (b) $-\dfrac{4}{5}$ (c) $\dfrac{2}{5}$ (d) $\dfrac{4}{5}$

18. Two sides of a rhombus are along the lines, $x - y + 1 = 0$ and $7x - y - 5 = 0$. If its diagonals intersect at $(-1, -2)$, then which one of the following is a vertex of this rhombus?

(a) $(-3, -9)$ (b) $(-3, -8)$

(c) $\left(\dfrac{1}{3},-\dfrac{8}{3}\right)$ (d) $\left(-\dfrac{10}{3},-\dfrac{7}{3}\right)$

19. The centres of those circles which touch the circle, $x^2 + y^2 - 8x - 8y - 4 = 0$, externally and also touch the X-axis, lie on

(a) a circle
(b) an ellipse which is not a circle
(c) a hyperbola
(d) a parabola

20. If one of the diameters of the circle, given by the equation, $x^2 + y^2 - 4x + 6y - 12 = 0$, is a chord of a circle S, whose centre is at $(-3, 2)$, then the radius of S is

(a) $5\sqrt{2}$ (b) $5\sqrt{3}$
(c) 5 (d) 10

21. Let P be the point on the parabola, $y^2 = 8x$, which is at a minimum distance from the centre C of the circle, $x^2 + (y+6)^2 = 1$. Then, the equation of the circle, passing through C and having its centre at P is

(a) $x^2 + y^2 - 4x + 8y + 12 = 0$
(b) $x^2 + y^2 - x + 4y - 12 = 0$
(c) $x^2 + y^2 - \dfrac{x}{4} + 2y - 24 = 0$
(d) $x^2 + y^2 - 4x + 9y + 18 = 0$

22. The eccentricity of the hyperbola whose length of the latusrectum is equal to 8 and the length of its conjugate axis is equal to half of the distance between its foci, is

(a) $\frac{4}{3}$ (b) $\frac{4}{\sqrt{3}}$

(c) $\frac{2}{\sqrt{3}}$ (d) $\sqrt{3}$

23. The distance of the point $(1, -5, 9)$ from the plane $x - y + z = 5$ measured along the line $x = y = z$ is

(a) $3\sqrt{10}$ (b) $10\sqrt{3}$

(c) $\frac{10}{\sqrt{3}}$ (d) $\frac{20}{3}$

24. If the line, $\frac{x-3}{2} = \frac{y+2}{-1} = \frac{z+4}{3}$ lies in the plane, $lx + my - z = 9$, then $l^2 + m^2$ is equal to

(a) 26 (b) 18 (c) 5 (d) 2

25. Let $\hat{\mathbf{a}}$, $\hat{\mathbf{b}}$ and $\hat{\mathbf{c}}$ be three unit vectors such that $\hat{\mathbf{a}} \times (\hat{\mathbf{b}} \times \hat{\mathbf{c}}) = \frac{\sqrt{3}}{2}(\hat{\mathbf{b}} + \hat{\mathbf{c}})$. If $\hat{\mathbf{b}}$ is not parallel to $\hat{\mathbf{c}}$, then the angle between $\hat{\mathbf{a}}$ and $\hat{\mathbf{b}}$ is

(a) $\frac{3\pi}{4}$ (b) $\frac{\pi}{2}$

(c) $\frac{2\pi}{3}$ (d) $\frac{5\pi}{6}$

26. If the standard deviation of the numbers 2, 3, a and 11 is 3.5, then which of the following is true?

(a) $3a^2 - 26a + 55 = 0$

(b) $3a^2 - 32a + 84 = 0$

(c) $3a^2 - 34a + 91 = 0$

(d) $3a^2 - 23a + 44 = 0$

27. Let two fair six-faced dice A and B be thrown simultaneously. If E_1 is the event that die A shows up four, E_2 is the event that die B shows up two and E_3 is the event that the sum of numbers on both dice is odd, then which of the following statements is not true?

(a) E_1 and E_2 are independent

(b) E_2 and E_3 are independent

(c) E_1 and E_3 are independent

(d) E_1, E_2 and E_3 are independent

28. If $0 \le x < 2\pi$, then the number of real values of x, which satisfy the equation

$\cos x + \cos 2x + \cos 3x + \cos 4x = 0$, is

(a) 3 (b) 5

(c) 7 (d) 9

29. A man is walking towards a vertical pillar in a straight path, at a uniform speed. At a certain point A on the path, he observes that the angle of elevation of the top of the pillar is 30°. After walking for 10 min from A in the same direction, at a point B, he observes that the angle of elevation of the top of the pillar is 60°. Then, the time taken (in minutes) by him, from B to reach the pillar, is

(a) 6 (b) 10

(c) 20 (d) 5

30. The Boolean expression $(p \wedge \sim q) \vee q \vee (\sim p \wedge q)$ is equivalent to

(a) $\sim p \wedge q$

(b) $p \wedge q$

(c) $p \vee q$

(d) $p \vee \sim q$

Answers

Physics

1. (a)	**2.** (b,c)	**3.** (c)	**4.** (d)	**5.** (a)	**6.** (d)	**7.** (a)
8. (b)	**9.** (a)	**10.** (d)	**11.** (c)	**12.** (a)	**13.** (c)	**14.** (c)
15. (d)	**16.** (d)	**17.** (d)	**18.** (a)	**19.** (d)	**20.** (c)	**21.** (a)
22. (d)	**23.** (c)	**24.** (a)	**25.** (b)	**26.** (d)	**27.** (a)	**28.** (a)
29. (a)	**30.** (b,d)					

Chemistry

1. (c)	**2.** (d)	**3.** (a)	**4.** (c)	**5.** (b)	**6.** (c)	**7.** (a)
8. (a)	**9.** (b)	**10.** (c)	**11.** (d)	**12.** (a)	**13.** (a)	**14.** (a)
15. (b)	**16.** (b)	**17.** *	**18.** (d)	**19.** (a)	**20.** (d)	**21.** (b)
22. (b)	**23.** (c)	**24.** (b)	**25.** (a)	**26.** (c)	**27.** (c)	**28.** (b)
29. (a)	**30.** (a)					

Note '*" Means, none options are not correct.

Mathematics

1. (c)	**2.** (d)	**3.** (a)	**4.** (b)	**5.** (d)	**6.** (d)	**7.** (d)
8. (b)	**9.** (b)	**10.** (c)	**11.** (b)	**12.** (b)	**13.** (c)	**14.** (b)
15. (b)	**16.** (b)	**17.** (d)	**18.** (c)	**19.** (d)	**20.** (b)	**21.** (a)
22. (c)	**23.** (b)	**24.** (d)	**25.** (d)	**26.** (b)	**27.** (d)	**28.** (c)
29. (d)	**30.** (c)					

Unsolved Paper 2016

JEE Advanced

Paper 1

Physics

Section 1 (Maximum Marks : 15)

- This section contains **FIVE** questions.
- Each question has **FOUR** options (a), (b), (c) and (d). **ONLY ONE** of these four options is correct.
- Four each question, darken the bubble corresponding to the correct option in the ORS.
- For each question, marks will be awarded in one of the following categories.

Full Marks : + 3 If only the bubble corresponding to the correct option is darkened.

Zero Marks : 0 If none of the bubbles is darkened.

Negative Marks : −1 In all other cases.

1. A parallel beam of light is incident from air at an angle α on the side PQ of a right angled triangular prism of refractive index $n=\sqrt{2}$. Light undergoes total internal reflection in the prism at the face PR when α has a minimum value of 45°. The angle θ of the prism is

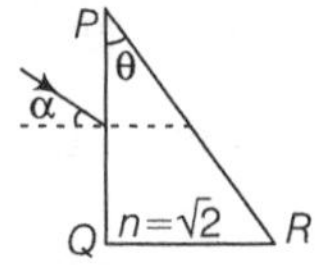

(a) 15° (b) 22.5° (c) 30° (d) 45°

2. In a historical experiment to determine Planck's constant, a metal surface was irradiated with light of different wavelengths. The emitted photoelectron energies were measured by applying a stopping potential. The relevant data for the wavelength (λ) of incident light and the corresponding stopping potential (V_0) are given below:

λ (μm)	V_0 (Volt)
0.3	2.0
0.4	1.0
0.5	0.4

Given that $c=3\times10^{8}\ \text{ms}^{-1}$ and $e=1.6\times10^{-19}$ C, Planck's constant (in units of J-s) found from such an experiment is)

(a) 6.0×10^{-34}

(b) 6.4×10^{-34}

(c) 6.6×10^{-34}

(d) 6.8×10^{-34}

3. A water cooler of storage capacity 120 litres can cool water at a constant rate of P watts. In a closed circulation system (as shown schematically in the figure), the water from the cooler is used to cool an external device that generates constantly 3 kW of heat (thermal load).

The temperature of water fed into the device cannot exceed 30°C and the entire stored 120 litres of water is initially cooled to 10°C. The entire system is thermally insulated. The minimum value of P (in watts) for which the device can be operated for 3 hours is

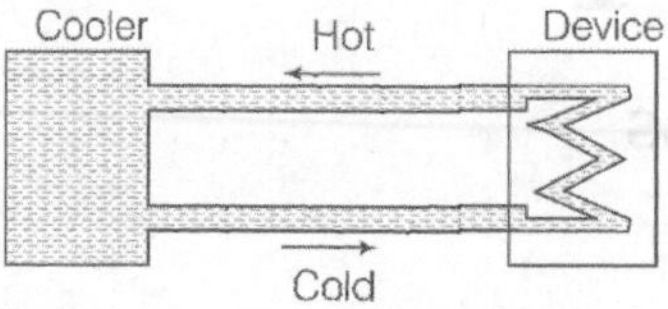

(Specific heat of water is $4.2\ \text{kJ kg}^{-1}\text{K}^{-1}$ and the density of water is $1000\ \text{kg m}^{-3}$)

(a) 1600 (b) 2067
(c) 2533 (d) 3933

4. A uniform wooden stick of mass 1.6 kg of length l rests in an inclined manner on a smooth, vertical wall of height $h\,(<l)$ such that a small portion of the stick extends beyond the wall. The reaction force of the wall on the stick is perpendicular to the stick. The stick makes an angle of 30° with the wall and the bottom of the stick is on a rough floor. The reaction of the wall on the stick is equal in magnitude to the reaction of the floor on the stick. The ratio $\frac{h}{l}$ and the frictional force f at the bottom of the stick are $(g = 10\ \text{ms}^{-2})$

(a) $\frac{h}{l}=\frac{\sqrt{3}}{16}$, $f=\frac{16\sqrt{3}}{3}$N

(b) $\frac{h}{l}=\frac{3}{16}$, $f=\frac{16\sqrt{3}}{3}$N

(c) $\frac{h}{l}=\frac{3\sqrt{3}}{16}$, $f=\frac{8\sqrt{3}}{3}$N

(d) $\frac{h}{l}=\frac{3\sqrt{3}}{16}$, $f=\frac{16\sqrt{3}}{3}$N

5. An infinite line charge of uniform electric charge density λ lies along the axis of an electrically conducting infinite cylindrical shell of radius R. At time $t = 0$, the space inside the cylinder is filled with a material of permittivity ε and electrical conductivity σ. The electrical conduction in the material follows Ohm's law. Which one of the following graphs best describes the subsequent variation of the magnitude of current density j (t) at any point in the material?

(a)
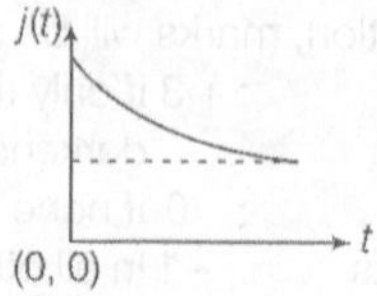

(b)
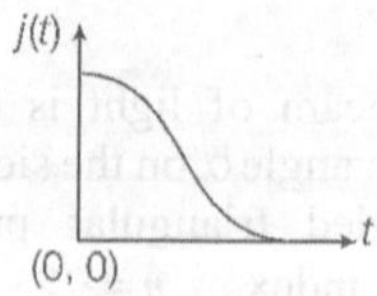

(c)
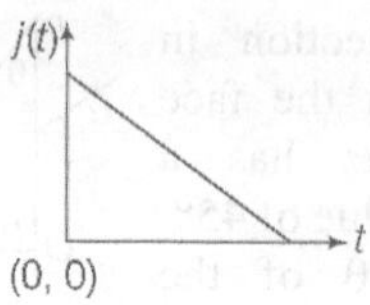

(d)
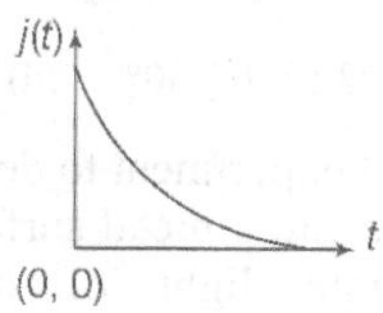

Section 2 (Maximum Marks : 32)

- This section contains **EIGHT** questions.
- Each question has **FOUR** options (a), (b) (c) and (d). **ONE OR MORE THAN ONE** of these four option(s) is (are) correct.
- For each question, darken the bubble(s) corresponding to all the correct option(s) in the ORS.
- For each questions, marks will be awarded in one of the following categories :

Full Marks	: + 4	If only the bubble(s) corresponding to all the correct option(s) is (are) darkened.
Partial Marks	: + 1	For darkening a bubble corresponding to **each correct option**, provided NO incorrect option is darkened.
Zero Marks	: 0	If none of the bubbles is darkened.
Negative Marks	: −2	In all other cases.

6. A plano-convex lens is made of material of refractive index n. When a small object is placed 30 cm away in front of the curved surface of the lens, an image of double the size of the object is produced. Due to reflection from the convex surface of the lens, another faint image is observed at a distance of 10 cm away from the lens. Which of the following statement(s) is (are) true?

(a) The refractive index of the lens is 2.5
(b) The radius of curvature of the convex surface is 45 cm
(c) The faint image is erect and real
(d) The focal length of the lens is 20 cm

7. A conducting loop in the shape of a right angled isosceles triangle of height 10 cm is kept such that the 90° vertex is very close to an infinitely long conducting wire (see the figure). The wire is electrically insulated from the loop. The hypotenuse of the triangle is parallel to the wire. The current in the triangular loop is in counterclockwise direction and increased at a constant rate of $10\,As^{-1}$. Which of the following statement(s) is (are) true?

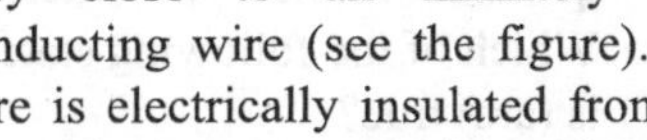

(a) There is a repulsive force between the wire and the loop
(b) If the loop is rotated at a constant angular speed about the wire, an additional emf of $\left(\frac{\mu_0}{\pi}\right)$ volt is induced in the wire
(c) The magnitude of induced emf in the wire is $\left(\frac{\mu_0}{\pi}\right)$ volt
(d) The induced current in the wire is in opposite direction to the current along the hypotenuse

8. The position vector **r** of particle of mass m is given by the following equation $\mathbf{r}(t) = \alpha t^3 \hat{\mathbf{i}} + \beta t^2 \hat{\mathbf{j}}$

where, $\alpha = \frac{10}{3}\,ms^{-3}$, $\beta = 5\,ms^{-2}$ and $m = 0.1$ kg. At $t = 1$s, which of the following statement(s) is (are) true about the particle?

(a) The velocity **v** is given by $\mathbf{v} = (10\hat{\mathbf{i}} + 10\hat{\mathbf{j}})\,ms^{-1}$
(b) The angular momentum **L** with respect to the origin is given by $\mathbf{L} = (5/3)\hat{\mathbf{k}}$ Nms
(c) The force **F** is given by $\mathbf{F} = (\hat{\mathbf{i}} + 2\hat{\mathbf{j}})$N
(d) The torque τ with respect to the origin is given by $\tau = -\frac{20}{3}\hat{\mathbf{k}}$ Nm

9. A length-scale (l) depends on the permittivity (ε) of a dielectric material, Boltzmann constant (k_B), the absolute temperature (T), the number per unit volume (n) of certain charged particles, and the charge (q) carried by each of the particles. Which of the following expression (s) for l is (are) dimensionally correct?

(a) $l=\sqrt{\left(\frac{nq^2}{\varepsilon k_B T}\right)}$ (b) $l=\sqrt{\left(\frac{\varepsilon k_B T}{nq^2}\right)}$

(c) $l=\sqrt{\left(\frac{q^2}{\varepsilon n^{2/3} k_B T}\right)}$ (d) $l=\sqrt{\left(\frac{q^2}{\varepsilon n^{1/3} k_B T}\right)}$

10. Two loudspeakers M and N are located 20m apart and emit sound at frequencies 118Hz and 121Hz, respectively. A car in initially at a point P, 1800 m away from the midpoint Q of the line MN and moves towards Q constantly at 60 km/h along the perpendicular bisector of MN. It crosses Q and eventually reaches a point R, 1800 m away from Q.

Let $\nu(t)$ represent the beat frequency measured by a person sitting in the car at time t. Let ν_P, ν_Q and ν_R be the beat frequencies measured at locations P, Q and R respectively. The speed of sound in air is $330\,\text{ms}^{-1}$. Which of the following statement(s) is (are) true regarding the sound heard by the person?

(a) The plot below represents schematically the variation of beat frequency with time

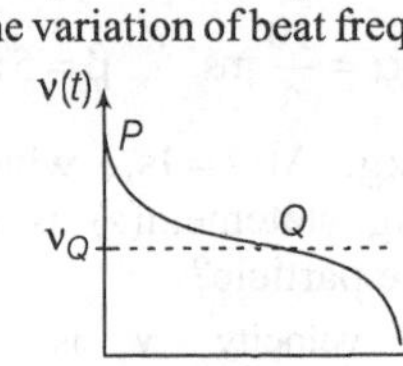

(b) The rate of change in beat frequency is maximum when the car passes through Q

(c) $\nu_P + \nu_R = 2\nu_Q$

(d) The plot below represents schematically the variations of beat frequency with time

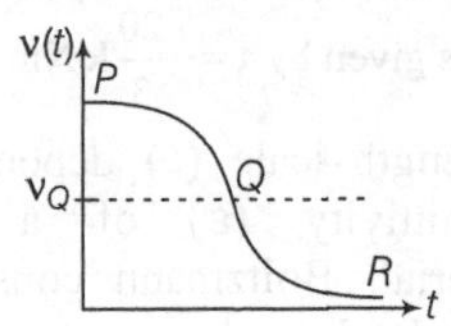

11. A transparent slab of thickness d has a refractive index $n(z)$ that increases with z. Here, z is the vertical distance inside the slab, measured from the top. The slab is placed between two media with uniform refractive indices n_1 and $n_2\,(> n_1)$, as shown in the figure. A ray of light is incident with angle θ_i from medium 1 and emerges in medium 2 with refraction angle θ_f with a lateral displacement l.

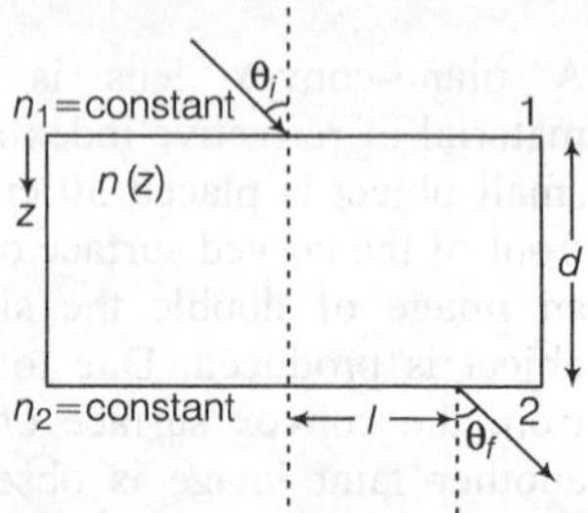

Which of the following statement(s) is (are) true?

(a) l is independent on $n(z)$

(b) $n_1 \sin\theta_i = (n_2 - n_1)\sin\theta_f$

(c) $n_1 \sin\theta_i = n_2 \sin\theta_f$

(d) l is independent of n_2

12. Highly excited states for hydrogen-like atoms (also called Rydberg states) with nuclear charge Ze are defined by their principle quantum number n, where $n >> 1$. Which of the following statement(s) is (are) true?

(a) Relative change in the radii of two consecutive orbitals does not depend on Z

(b) Relative change in the radii of two consecutive orbitals varies as $1/n$

(c) Relative change in the energy of two consecutive orbitals varies as $1/n^3$

(d) Relative change in the angular momenta of two consecutive orbitals varies as $1/n$

13. An incandescent bulb has a thin filament of tungsten that is heated to high temperature by passing an electric current. The hot filament emits black-body radiation. The filament is observed to break up at random locations after a sufficiently long time of operation due to non-uniform evaporation of tungsten from the filament. If the bulb is powered at constant voltage, which of the following statement(s) is (are) true?

(a) The temperature distribution over the filament is uniform

(b) The resistance over small sections of the filament decreases with time

(c) The filament emits more light at higher band of frequencies before it breaks up

(d) The filament consumes less electrical power towards the end of the life of the bulb

Section 3 (Maximum Marks : 15)

- This section contains **FIVE** questions.
- The answer to each question is a **SINGLE DIGIT INTEGER** ranging from 0 to 9, both inclusive.
- For each question, darken the bubble corresponding to the correct integer in the ORS.
- For each questions, marks will be awarded in one of the following categories.

Full Marks : + 3 If only the bubble corresponding to all the correct answer is darkened.
Zero Marks : 0 In all other cases.

14. A hydrogen atom in its ground state is irradiated by light of wavelength 970Å.

Taking $hc/e = 1.237 \times 10^{-6}$ eVm and the ground state energy of hydrogen atom as – 13.6 eV, the number of lines present in the emission spectrum is

15. The isotopes $^{12}_{5}\text{B}$ having a mass 12.014 u undergoes β-decay to $^{12}_{6}\text{C}$. $^{12}_{6}\text{C}$ has an excited state of the nucleus $(^{12}_{6}\text{C}^*)$ at 4.041 MeV above its ground state. If $^{12}_{5}\text{B}$ decays to $^{12}_{5}\text{C}^*$, the maximum kinetic energy of the β-particle in units of MeV is $(1\text{u} = 931.5\ \text{MeV}/c^2$, where c is the speed of light in vacuum)

16. Consider two solid spheres P and Q each of density 8 gm cm^{-3} and diameters 1 cm and 0.5 cm, respectively. Sphere P is dropped into a liquid of density 0.8 gm cm^{-3} and viscosity $\eta = 3$ poiseulles. Sphere Q is dropped into a liquid of density 1.6 gm cm^{-3} and viscosity $\eta = 2$ poiseulles. The ratio of the terminal velocities of P and Q is

17. Two inductors L_1 (inductance 1mH, internal resistance 3 Ω) and L_2 (inductance 2 mH, internal resistance 4 Ω), and a resistor R (resistance 12 Ω) are all connected in parallel across a 5V battery. The circuit is switched on at time $t = 0$. The ratio of the maximum to the minimum current (I_{max}/I_{min}) drawn from the battery is

18. A metal is heated in a furnace where a sensor is kept above the metal surface to read the power radiated (P) by the metal. The sensor has a scale that displays $\log_2(P/P_0)$, where P_0 is a constant. When the metal surface is at a temperature of 487°C, the sensor shows a value 1. Assume that the emissivity of the metallic surface remains constant. What is the value displayed by the sensor when the temperature of the metal surface is raised to 2767°C?

Chemistry

Section 1 (Maximum Marks : 15)

- This section contains **FIVE** questions.
- Each question has **FOUR** options (a), (b), (c) and (d). Only one of these four options is correct.
- For each question, darken the bubble corresponding to the correct option in the ORS.
- For each question, marks will be awarded in one of the following categories.

Full Marks	:	+ 3 If only the bubble corresponding to the correct option is darkened.
Zero Marks	:	0 If none of the bubbles is darkened.
Negative Marks	:	– 1 In all other cases.

19. P is the probability of finding the 1s electron of hydrogen atom in a spherical shell of infinitesimal thickness, dr, at a distance r from the nucleus. The volume of this shell is $4\pi r^2 dr$. The qualitative sketch of the dependence of P on r is

(a) 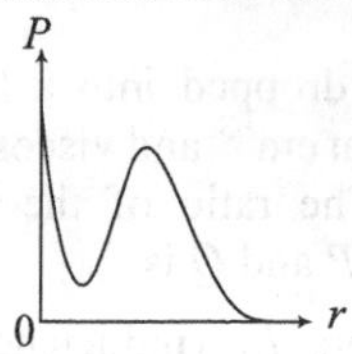

(a) 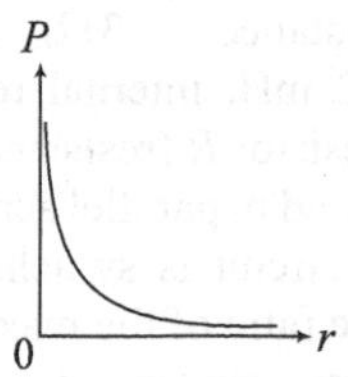

(a) 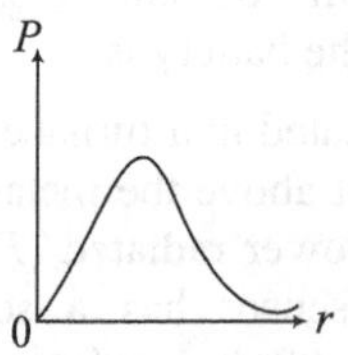

(d) 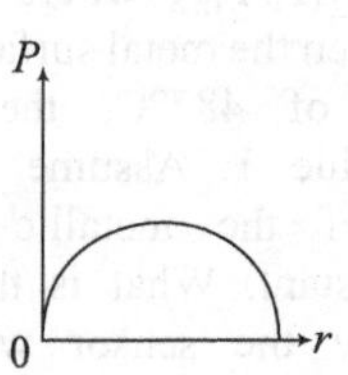

20. One mole of an ideal gas at 300 K in thermal contact with surroundings expands isothermally from 1.0 L to 2.0 L against a constant pressure of 3.0 atm. In this process, the change in entropy of surroundings (ΔS_{surr}) in J K^{-1} is (1 L atm = 101.3 J)

(a) 5.763
(b) 1.013
(c) – 1.013
(d) – 5.763

21. Among $[Ni(CO)_4]$, $[NiCl_4]^{2-}$, $[Co(NH_3)_4Cl_2]Cl$, $Na_3[CoF_6]$, Na_2O_2 and CsO_2, the total number of paramagnetic compounds is

(a) 2
(b) 3
(c) 4
(d) 5

22. The increasing order of atomic radii of the following Group 13 elements is

(a) Al < Ga < In < Tl
(b) Ga < Al < In < Tl
(c) Al < In < Ga < Tl
(d) Al < Ga < Tl < In

23. On complete hydrogenation, natural rubber produces

(a) ethylene-propylene copolymer
(b) vulcanised rubber
(c) polypropylene
(d) polybutylene

Section 2 (Maximum Marks : 32)

- This section contains **EIGHT** questions.
- Each question has **FOUR** options (a), (b), (c) and (d). **ONE OR MORE THAN ONE** of these four options is (are) correct.
- For each question, darken the bubble(s) corresponding to all the correct option(s) in the ORS.
- For each question, marks will be awarded in **one of the following categories**:

Full Marks : + 4, If only the bubble(s) corresponding to all the correct option(s) is (are) darkened.
Partial Marks : + 1, For darkening a bubble corresponding **to each correct option** provided **NO** incorrect option is darkened.
Zero Marks : 0, If none of the bubbles is darkened.
Negative Marks : − 2, In all other cases.
For example, if (a), (c) and (d) are all the correct options for a question, darkening all these three will result in + 4 marks; darkening only (a) and (d) will result in + 2 marks; and darkening (a) and (b) will result in − 2 marks, as a wrong option is also darkened.

24. The product(s) of the following reaction sequence is (are)

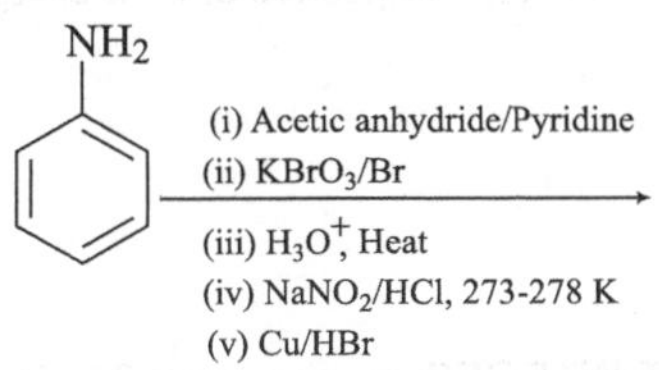

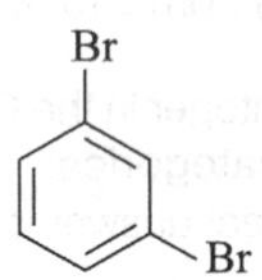

(a)

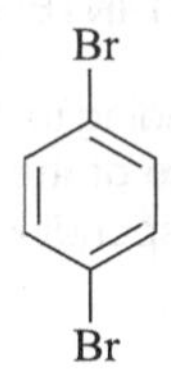

(b)

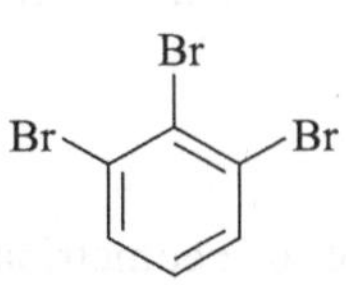

(c)

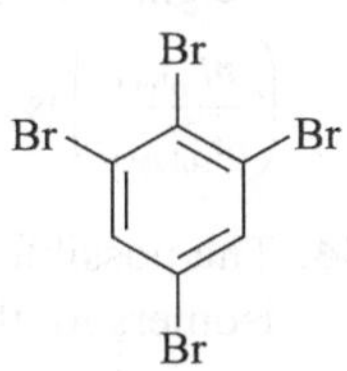

(d)

25. The correct statement(s) about the following reaction sequence is (are)

$$\text{Cumene } (C_9H_{12}) \xrightarrow[\text{(ii) } H_3O^+]{\text{(i) } O_2} P \xrightarrow{CHCl_3/NaOH} Q(\text{major}) + R(\text{minor}),\ Q \xrightarrow[PhCH_2Br]{NaOH} S$$

(a) R is steam volatile
(b) Q gives dark violet colouration with 1% aqueous $FeCl_3$ solution
(c) S gives yellow precipitate with 2, 4-dinitrophenylhydrazine
(d) S gives dark violet colouration with 1% aqueous $FeCl_3$ solution

26. The crystalline form of borax has
(a) tetranuclear $[B_4O_5(OH)_4]^{2-}$ unit
(b) all boron atoms in the same plane
(c) equal number of sp^2 and sp^3 hybridised boron atoms
(d) one terminal hydroxide per boron atom

27. The reagent(s) that can selectively precipitate S^{2-} from a mixture of S^{2-} and SO_4^{2-} in aqueous solution is (are)
(a) $CuCl_2$
(b) $BaCl_2$
(c) $Pb(OOCCH_3)_2$
(d) $Na_2[Fe(CN)_5NO]$

28. A plot of the number of neutrons (n) against the number of protons (p) of stable nuclei exhibits upward deviation from linearity for atomic number, $Z > 20$. For an unstable nucleus having n/p ratio less than 1, the possible mode(s) of decay is (are)

(a) β^-- decay (β- emission)
(b) orbital or K-electron capture
(c) neutron emission
(d) β^+-decay (positron emission)

29. Positive Tollen's test is observed for

(a) 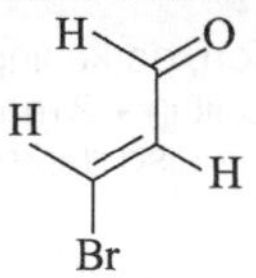 (b) 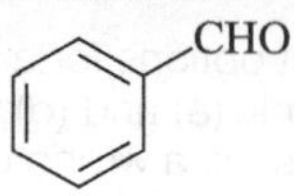

(c) 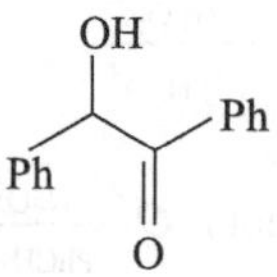 (b) 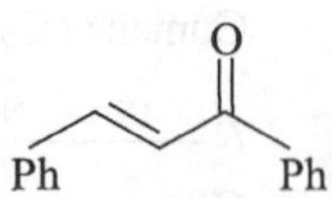

30. The compound(s) with two lone pairs of electrons on the central atom is (are)

(a) BrF_5
(b) ClF_3
(c) XeF_4
(d) SF_4

31. According to the Arrhenius equation,

(a) a high activation energy usually implies a fast reaction
(b) rate constant increases with increase in temperature. This is due to a greater number of collisions whose energy exceeds the activation energy
(c) higher the magnitude of activation energy, stronger is the temperature dependence of the rate constant
(d) the pre-exponential factor is a measure of the rate at which collisions occur, irrespective of their energy

Section 3 (Maximum Marks : 15)

- This section contains **FIVE** questions.
- The answer to each question is a **SINGLE DIGIT INTEGER** ranging from 0 to 9, both inclusive.
- For each question, darken the bubble corresponding to the correct integer in the ORS.
- For each question, marks will be awarded in **one of the following categories**:

Full Marks : +3, If only the bubble corresponding to the correct answer is darkened.

Zero marks : 0, In all other cases.

32. In the following monobromination reaction, the number of possible chiral product(s) is (are)...

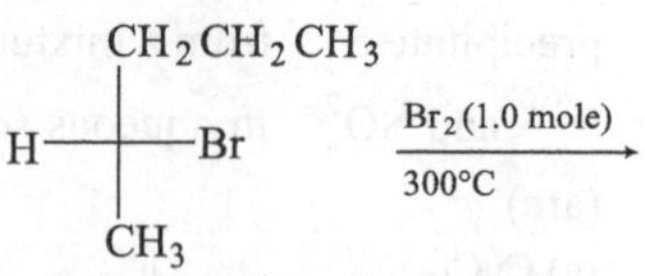

(1.0 mole) (Enantiomerically pure)

33. The mole fraction of a solute in a solution is 0.1. At 298 K, molarity of this solution is the same as its molality. Density of this solution at 298 K is 2.0 g cm^{-3}. The ratio of the molecular weights of the solute and solvent, $\left(\frac{m_{solute}}{m_{solvent}}\right)$ is ... .

34. The possible number of geometrical isomers for the complex $[CoL_2Cl_2]^-$ ($L = H_2NCH_2CH_2O^-$) is (are) ...

35. In neutral or faintly alkaline solution, 8 moles of permanganate anion quantitative oxidise thiosulphate anions to produce X moles of a sulphur containing product. The magnitude of X is ... .

36. The diffusion coefficient of an ideal gas is proportional to its mean free path and mean speed. The absolute temperature of an ideal gas is increased 4 times and its pressure is increased 2 times. As a result, the diffusion coefficient of this gas increases x times. The value of x is ...

Mathematics

Section 1 (Maximum Marks : 15)

- This section contains **FIVE** questions.
- Each question has **FOUR** options (a), (b), (c) and (d). **ONLY ONE** of these four options is correct.
- For each question, darken the bubble corresponding to the correct option in the ORS.
- For each question, marks will be awarded in one of the following categories:

Full Marks	: +3	If only the bubble corresponding to the correct option is darkened.
Zero Marks	: 0	If none of the bubbles is darkened.
Negative Marks	: −1	In all other cases.

37. A debate club consists of 6 girls and 4 boys. A team of 4 members is to be selected from this club including the selection of a captain (from among these 4 members) for the team. If the team has to include atmost one boy, the number of ways of selecting the team is

(a) 380 (b) 320
(c) 260 (d) 95

38. The least value of $\alpha \in R$ for which $4\alpha x^2 + \frac{1}{x} \geq 1$, for all $x > 0$, is

(a) $\frac{1}{64}$ (b) $\frac{1}{32}$ (c) $\frac{1}{27}$ (d) $\frac{1}{25}$

39. Let $-\frac{\pi}{6} < \theta < -\frac{\pi}{12}$. Suppose α_1 and β_1 are the roots of the equation $x^2 - 2x\sec\theta + 1 = 0$, and α_2 and β_2 are the roots of the equation $x^2 + 2x\tan\theta - 1 = 0$. If $\alpha_1 > \beta_1$ and $\alpha_2 > \beta_2$, then $\alpha_1 + \beta_2$ equals

(a) $2(\sec\theta - \tan\theta)$ (b) $2\sec\theta$
(c) $-2\tan\theta$ (d) 0

40. Let $S = \left\{x \in (-\pi, \pi) : x \neq 0, \pm\frac{\pi}{2}\right\}$. The sum of all distinct solutions of the equation $\sqrt{3}\sec x + \operatorname{cosec} x + 2(\tan x - \cot x) = 0$ in the set S is equal to

(a) $-\frac{7\pi}{9}$ (b) $-\frac{2\pi}{9}$ (c) 0 (d) $\frac{5\pi}{9}$

41. A computer producing factory has only two plants T_1 and T_2. Plant T_1 produces 20% and plant T_2 produces 80% of the total computers produced. 7% of computers produced in the factory turn out to be defective. It is knownthat P (computer turns out to be defective, given that it is produced in plant T_1) $= 10P$ (computer turns out to be defective, given that it is produced in plant T_2), where $P(E)$ denotes the probability of an event E. A computer produced in the factory is randomly selected and it does not turn out to be defective. Then, the probability that it is produced in plant T_2, is

(a) $\frac{36}{73}$ (b) $\frac{47}{79}$ (c) $\frac{78}{93}$ (d) $\frac{75}{83}$

Section 2 (Maximum Marks : 32)

- This section contains **EIGHT** questions.
- Each question has **FOUR** options (a), (b), (c) and (d). **ONE OR MORE THAN ONE** of these four option (s) is (are) correct.
- For each question, darken the bubble (s) corresponding to all the correct option(s) in the ORS.
- For each question, marks will be awarded in one of the following categories:

Full Marks : +4 If only the bubble (s) corresponding to all correct option(s) is (are) darkened.

Partial Marks : +1 For darkening a bubble corresponding **to each correct option**, provided no incorrect option is darkened.

Zero Marks : 0 If none of the dubbles is darkened

Negative Marks : –2 In all other cases.For example, if (a), (c) and (d) are all the correct options for a question, darkening all these three will result in +4 marks; darkening only (a) and (d) will result in +2 marks and darkening (a) and (b) will result

42. A solution curve of the differential equation $(x^2+xy+4x+2y+4)\frac{dy}{dx}-y^2=0$, $x>0$, passes through the point (1, 3). Then, the solution curve

(a) intersects $y=x+2$ exactly at one point

(b) intersects $y=x+2$ exactly at two points

(c) intersects $y=(x+2)^2$

(d) does not intersect $y=(x+3)^2$

43. Consider a pyramid $OPQRS$ located in the first octant $(x\geq 0, y\geq 0, z\geq 0)$ with O as origin, and OP and OR along the X-axis and the Y-axis, respectively. The base $OPQR$ of the pyramid is a square with $OP=3$. The point S is directly above the mid-point T of diagonal OQ such that $TS=3$.

Then,

(a) the acute angle between OQ and OS is $\frac{\pi}{3}$

(b) the equation of the plane containing the ΔOQS is $x-y=0$

(c) the length of the perpendicular from P to the plane containing the ΔOQS is $\frac{3}{\sqrt{2}}$

(d) the perpendicular distance from O to the straight line containing RS is $\sqrt{\frac{15}{2}}$

44. In a ΔXYZ, let x, y, z be the lengths of sides opposite to the angles X, Y, Z respectively and $2s=x+y+z$. If $\frac{s-x}{4}=\frac{s-y}{3}=\frac{s-z}{2}$ and area of incircle of the ΔXYZ is $\frac{8\pi}{3}$

then

(a) area of the ΔXYZ is $6\sqrt{6}$

(b) the radius of circumcircle of the ΔXYZ is $\frac{35}{6}\sqrt{6}$

(c) $\sin\frac{X}{2}\sin\frac{Y}{2}\sin\frac{Z}{2}=\frac{4}{35}$

(d) $\sin^2\left(\frac{X+Y}{2}\right)=\frac{3}{5}$

45. Let RS be the diameter of the circle $x^2+y^2=1$, where S is the point (1, 0). Let P be a variable point (other than R and S) on the circle and tangents to the circle at S and P meet at the point Q. The normal to the circle at P

intersects a line drawn through Q parallel to RS at point E. Then, the locus of E passes through the point(s)

(a) $\left(\frac{1}{3}, \frac{1}{\sqrt{3}}\right)$

(b) $\left(\frac{1}{4}, \frac{1}{2}\right)$

(c) $\left(\frac{1}{3}, -\frac{1}{\sqrt{3}}\right)$

(d) $\left(\frac{1}{4}, -\frac{1}{2}\right)$

46. The circle $C_1 : x^2 + y^2 = 3$ with centre at O intersects the parabola $x^2 = 2y$ at the point P in the first quadrant. Let the tangent to the circle C_1 at P touches other two circles C_2 and C_3 at R_2 and R_3, respectively. Suppose C_2 and C_3 have equal radii $2\sqrt{3}$ and centres Q_2 and Q_3, respectively. If Q_2 and Q_3 lie on the Y-axis, then

(a) $Q_2Q_3 = 12$
(b) $R_2R_3 = 4\sqrt{6}$
(c) area of the ΔOR_2R_3 is $6\sqrt{2}$
(d) area of the ΔPQ_2Q_3 is $4\sqrt{2}$

47. Let $f : R \to R, g : R \to R$ and $h : R \to R$ be differentiable functions such that $f(x) = x^3 + 3x + 2, g(f(x)) = x$ and $h(g(g(x))) = x$ for all $x \in R$. Then,

(a) $g'(2) = \frac{1}{15}$ (b) $h'(1) = 666$“
(c) $h(0) = 16$ (d) $h(g(3)) = 36$

48. Let $f : (0, \infty) \to R$ be a differentiable function such that $f'(x) = 2 - \frac{f(x)}{x}$ for all $x \in (0, \infty)$ and $f(1) \neq 1$.
Then

(a) $\lim_{x \to 0+} f'\left(\frac{1}{x}\right) = 1$

(b) $\lim_{x \to 0+} x f\left(\frac{1}{x}\right) = 2$

(c) $\lim_{x \to 0+} x^2 f'(x) = 0$

(d) $|f(x)| \leq 2$ for all $x \in (0, 2)$

49. Let $P = \begin{bmatrix} 3 & -1 & -2 \\ 2 & 0 & \alpha \\ 3 & -5 & 0 \end{bmatrix}$, where $\alpha \in R$.

Suppose $Q = [q_{ij}]$ is a matrix such that $PQ = kI$, where $k \in R, k \neq 0$ and I is the identity matrix of order 3. If $q_{23} = -\frac{k}{8}$

and $\det(Q) = \frac{k^2}{2}$

then

(a) $\alpha = 0, k = 8$
(b) $4\alpha - k + 8 = 0$
(c) $\det(P \operatorname{adj}(Q)) = 2^9$
(d) $\det(Q \operatorname{adj}(P)) = 2^{13}$

Section 3 (Maximum Marks : 12)

- This section contains **FIVE** questions.
- The answer to each question is a **SIGNLE DIGIT INTEGER** ranging from 0 to 9, both inclusive.
- For each question, darken the bubble corresponding to the correct option in the ORS.
- For each question, marks will be awarded in **one of the following categories**:

Full Marks : + 3 If only the bubble corresponding to the correct option is darkened.

Zero Marks : 0 In all other cases.

50. Let m be the smallest positive integer such that the coefficient of x^2 in the expansion of

$$(1+x)^2 + (1+x)^3 + \dots + (1+x)^{49} + (1+mx)^{50}$$

is $(3n+1)\ {}^{51}C_3$ for some positive integer n. Then, the value of n is

51. The total number of distincts $x \in R$ for which $\begin{vmatrix} x & x^2 & 1+x^3 \\ 2x & 4x^2 & 1+8x^3 \\ 3x & 9x^2 & 1+27x^3 \end{vmatrix} = 10$ is

52. Let $z = \dfrac{-1+\sqrt{3}i}{2}$, where $i = \sqrt{-1}$, and $r, s \in \{1, 2, 3\}$.

Let $P = \begin{bmatrix} (-z)^r & z^{2s} \\ z^{2s} & z^r \end{bmatrix}$ and I be the identity matrix of order 2. Then, the total number of ordered pairs (r, s) for which $P^2 = -I$ is

53. The total number of distincts $x \in [0, 1]$ for which

$$\int_0^x \frac{t^2}{1+t^4}\,dt = 2x - 1 \text{ is}$$

54. Let $\alpha, \beta \in R$ be such that $\lim\limits_{x \to 0} \dfrac{x^2 \sin(\beta x)}{\alpha x - \sin x} = 1$. Then, $6(\alpha + \beta)$ equals

Paper 2

Physics

Section 1 (Maximum Marks : 32)

- This section contains **SIX** questions.
- Each question has **FOUR** options (a), (b), (c) and (d). **ONLY ONE** of these four options is correct.
- Four each question, darken the bubble corresponding to the correct option in the ORS.
- For each question, marks will be awarded in one of the following categories.
  Full Marks : + 3 If only the bubble corresponding to the correct option is darkened.
  Zero Marks : 0 If none of the bubbles is darkened.
  Negative Marks : −1 In all other cases.

1. An accident in a nuclear laboratory resulted in deposition of a certain amount of radioactive material of half-life 18 days inside the laboratory. Tests revealed that the radiation was 64 times more than the permissible level required for safe operation of the laboratory. What is the minimum number of days after which the laboratory can be considered safe for use?

(a) 64 (b) 90 (c) 108 (d) 120

2. The electrostatic energy of Z protons uniformly distributed throughout a spherical nucleus of radius R is given by

$$E = \frac{3}{5}\frac{Z(Z-1)e^2}{4\pi\varepsilon_0 R}$$

The measured masses of the neutron, ${}^1_1\text{H}$, ${}^{15}_{7}\text{N}$ and ${}^{15}_{8}\text{O}$ are 1.008665u, 1.007825u, 15.000109u and 15.003065 u, respectively. Given that the radii of both the ${}^{15}_{7}\text{N}$ and ${}^{15}_{8}\text{O}$ nuclei are same, $1\ \text{u} = 931.5\ \text{MeV}/c^2$ (c is the speed of light) and $e^2/(4\pi\varepsilon_0) = 1.44$ MeV fm. Assuming that the difference between the binding energies of ${}^{15}_{7}\text{N}$ and ${}^{15}_{8}\text{O}$ is purely due to the electrostatic energy, the radius of either of the nuclei is ($1\ \text{fm} = 10^{-15}$ m)

(a) 2.85 fm (b) 3.03 fm
(c) 3.42 fm (d) 3.80 fm

3. A gas is enclosed in a cylinder with a movable frictionless piston. Its initial thermodynamic state at pressure $P_i = 10^5$ Pa and volume $V_1 = 10^{-3}\ \text{m}^3$ changes to a final state at $P_f = (1/32)\times 10^5$ Pa and $V_f = 8\times 10^{-3}\ \text{m}^3$ in an adiabatic quasi-static process, such that $P^3V^5 = \text{constant}$. Consider another thermodynamic process that brings the system from the same initial state to the same final state in two steps : an isobaric expansion at P_i followed by an isochoric (isovolumetric) process at volume V_f. The amount of heat supplied to the system in the two-step process is approximately

(a) 112 J (b) 294 J (c) 588 J (d) 813 J

4. There are two Vernier calipers both of which have 1 cm divided into 10 equal divisions on the main scale. The Vernier scale of one of the calipers (C_1) has 10 equal divisions that correspond to 9 main scale divisions. The Vernier scale of the other caliper (C_2) has 10 equal divisions that correspond to 11 main scale divisions.

The readings of the two calipers are shown in the figure. The measured values (in cm) by calipers C_1 and C_2 respectively, are

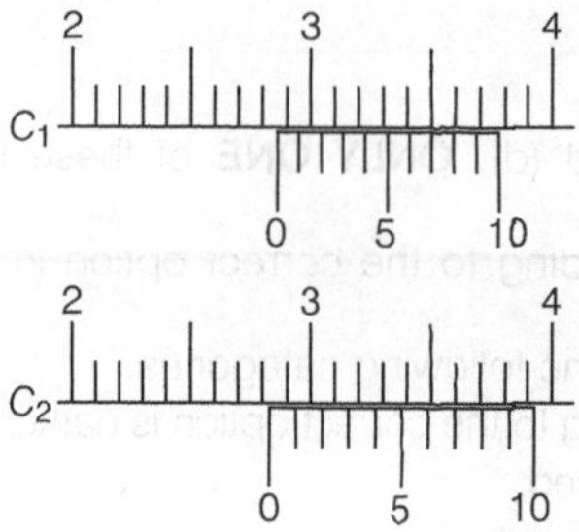

(a) 2.87 and 2.87 (b) 2.87 and 2.83
(c) 2.85 and 2.82 (d) 2.87 and 2.86

5. A small object is placed 50 cm to the left of a thin convex lens of focal length 30 cm. A convex spherical mirror of radius of curvature 100 cm is placed to the right of the lens at a distance of 50 cm. The mirror is tilted such that the axis of the mirror is at an angle $\theta = 30^\circ$ to the axis of the lens, as shown in the figure.

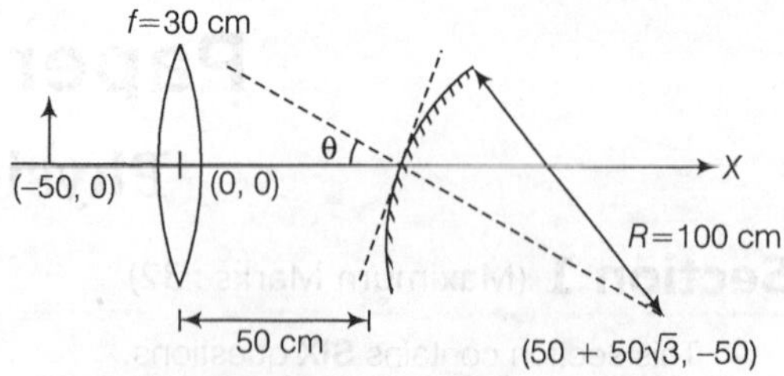

If the origin of the coordinate system is taken to be at the centre of the lens, the coordinates (in cm) of the point (x, y) at which the image is formed are

(a) $(125/3, 25/\sqrt{3})$ (b) $(50 - 25\sqrt{3}, 25)$
(c) $(0, 0)$ (d) $(25, 25\sqrt{3})$

6. The ends Q and R of two thin wires, PQ and RS, are soldered (joined) together. Initially, each of the wire has a length of 1 m 10°C. Now, the end P is maintained at 10°C, while the end S is heated and maintained at 400°C. The system is thermally insulated from its surroundings. If the thermal conductivity of wire PQ is twice that of the wire RS and the coefficient of linear thermal expansion of PQ is $1.2 \times 10^{-5}\ K^{-1}$, the change in length of the wire PQ is

(a) 0.78 mm (b) 0.90 mm
(c) 1.56 mm (d) 2.34 mm

Section 2 (Maximum Marks : 32)

- This section contains **EIGHT** questions.
- Each question has **FOUR** options (a), (b) (c) and (d). **ONE OR MORE THAN ONE** of these four option(s) is (are) correct.
- For each question, darken the bubble(s) corresponding to all the correct option(s) in the ORS.
- For each questions, marks will be awarded in one of the following categories :

  Full Marks : + 4 If only the bubble(s) corresponding to all the correct option(s) is (are) darkened.

  Partial Marks : + 1 For darkening a bubble corresponding to **each correct option**, provided NO incorrect option is darkened.

  Zero Marks : 0 If none of the bubbles is darkened.

  Negative Marks : −2 In all other cases.
- For example, if (a), (c) and (d) are all the correct options for a question, darkening all these three will result in + 4 marks ; darkening only (a) and (d) will result in + 2 marks and darkening (a) and (b) will result in −2 marks, as a wrong option is also darkened.

7. Two thin circular discs of mass m and $4m$, having radii of a and $2a$, respectively, are rigidly fixed by a massless, rigid rod of length $l = \sqrt{24}\,a$ through their centers. This assembly is laid on a firm and flat surface and set rolling without slipping on the surface so that the angular speed about the axis of the rod is ω. The angular momentum of the entire assembly about the point 'O' is **L** (see the figure). Which of the following statement(s) is (are) true?

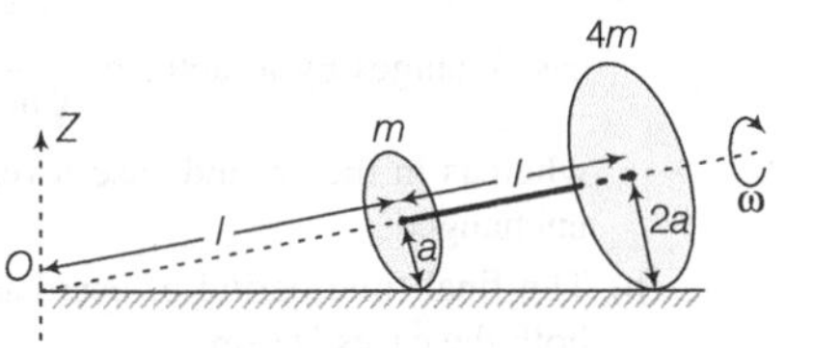

(a) The magnitude of the z-component of **L** is $55\,ma^2\,\omega$

(b) The magnitude of angular momentum of centre of mass of the assembly about the point O is $81\,ma^2\,\omega$

(c) The centre of mass of the assembly rotates about the Z-axis with an angular speed of $\frac{\omega}{5}$

(d) The magnitude of angular momentum of the assembly about its centre of mass is $17\,ma^2\frac{\omega}{2}$

8. Consider two identical galvanometers and two identical resistors with resistance R. If the internal resistance of the galvanometers $R_c < R/2$, which of the following statement(s) about anyone of the galvanometers is (are) true?

(a) The maximum voltage range is obtained when all the components are connected in series

(b) The maximum voltage range is obtained when the two resistors and one galvanometer are connected in series, and the second galvanometer is connected in parallel to the first galvanometer

(c) The maximum current range is obtained when all the components are connected in parallel

(d) The maximum current range is obtained when the two galvanometers are connected in series, and the combination is connected in parallel with both the resistors

9. A rigid wire loop of square shape having side of length L and resistance R is moving along the X-axis with a constant velocity v_0 in the plane of the paper . At $t = 0$, the right edge of the loop enters a region of length $3L$ where there is a uniform magnetic field B_0 into the plane of the paper, as shown in the figure. For sufficiently large v_0, the loop eventually crosses the region. Let x be the location of the right edge of the loop. Let $v(x)$, $I(x)$ and $F(x)$ represent the velocity of the loop, current in the loop, and force on the loop, respectively, as a function of x. Counter-clockwise current is taken as positive.

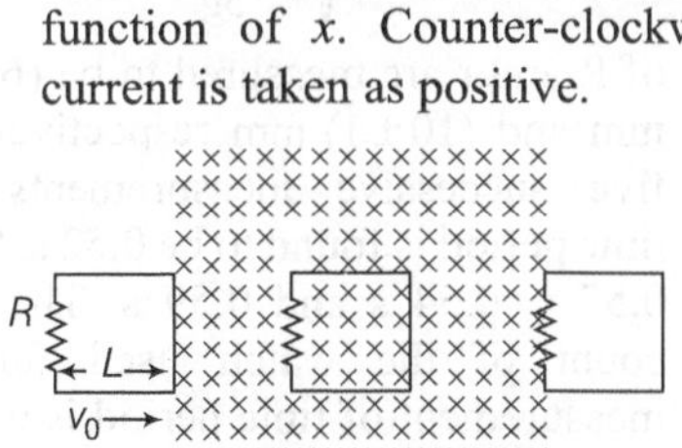

Which of the following schematic plot(s) is (are) correct? (Ignore gravity)

(a)

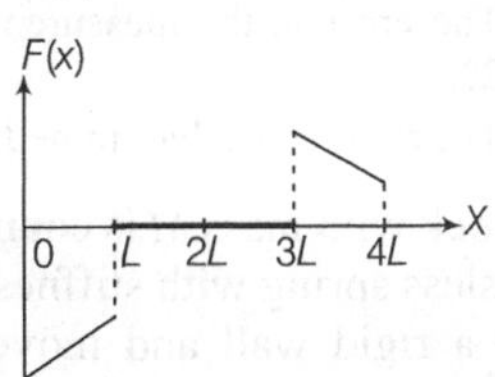

(b)
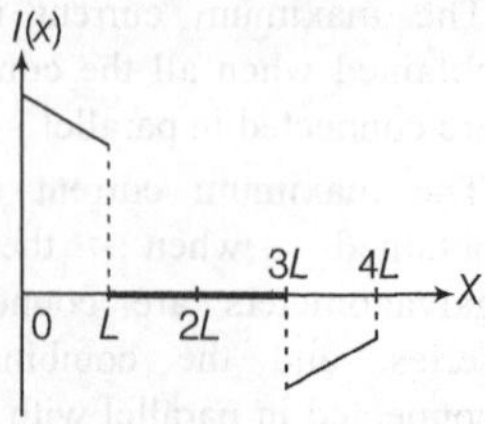

(c)
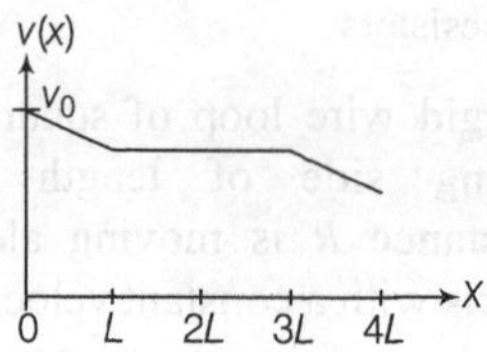

(d)
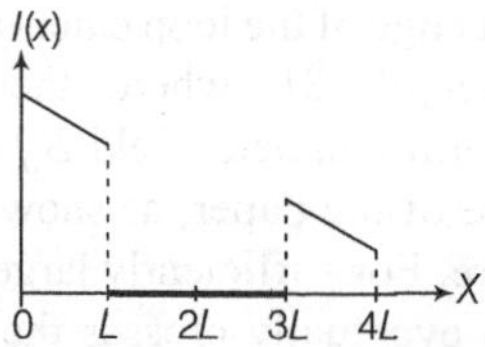

10. In an experiment to determine the acceleration due to gravity g, the formula used for the time period of a periodic motion is $T = 2\pi\sqrt{\frac{7(R-r)}{5g}}$. The values of R and r are measured to be (60 ± 1) mm and (10 ± 1) mm respectively. In five successive measurements, the time period is found to be 0.52 s, 0.56, 0.57 s, 0.54 s and 0.59 s. The least count of the watch used for the measurement of time period is 0.01 s. Which of the following statement(s) is (are) true?

(a) The error in the measurement of r is 10%

(b) The error in the measurement of T is 3.57%

(c) The error in the measurement of T is 2%

(d) The error in the determined of g is 11%

11. A block with mass M is connected by a massless spring with stiffness constant k to a rigid wall and moves without friction on a horizontal surface. The block oscillates with small amplitude A about an equilibrium position x_0. Consider two cases : (i) when the block is at x_0 and (ii) when the block is at $x = x_0 + A$. In both the cases, a particle with mass m $(< M)$ is softly placed on the block after which they stick to each other. Which of the following statement(s) is (are) true about the motion after the mass m is placed on the mass M?

(a) The amplitude of oscillation in the first case changes by a factor of $\sqrt{\frac{M}{m+M}}$, whereas in the second case it remains unchanged

(b) The final time period of oscillation in both the cases is same

(c) The total energy decreases in both the cases

(d) The instantaneous speed at x_0 of the combined masses decreases in both the cases

12. While conducting the Young's double slit experiment, a student replaced the two slits with a large opaque plate in the XY-plane containing two small holes that act as two coherent point sources (S_1, S_2) emitting light of wavelength 600 mm. The student mistakenly placed the screen parallel to the XZ-plane (for $z > 0$) at a distance $D = 3$ m from the mid-point of S_1S_2, as shown schematically in the figure. The distance between the source $d = 0.6003$ mm. The origin O is at the intersection of the screen and the line joining S_1S_2.

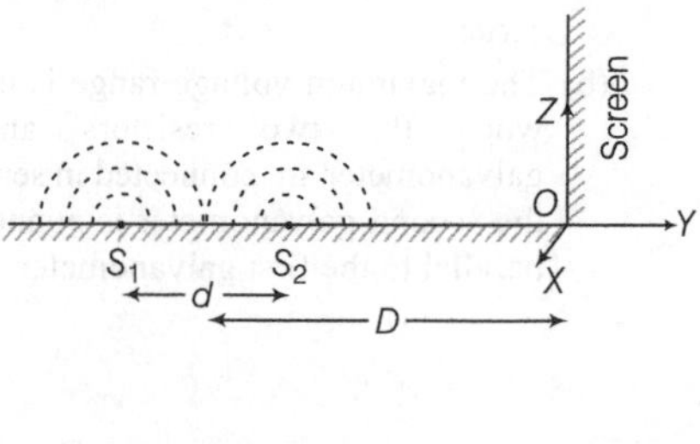

Which of the following is (are) true of the intensity pattern on the screen?

(a) Semi circular bright and dark bands centered at point O

(b) The region very close to the point O will be dark

(c) Straight bright and dark bands parallel to the X-axis

(d) Hyperbolic bright and dark bands with foci symmetrically placed about O in the x-direction

13. In the circuit shown below, the key is pressed at time $t = 0$. Which of the following statement(s) is (are) true?

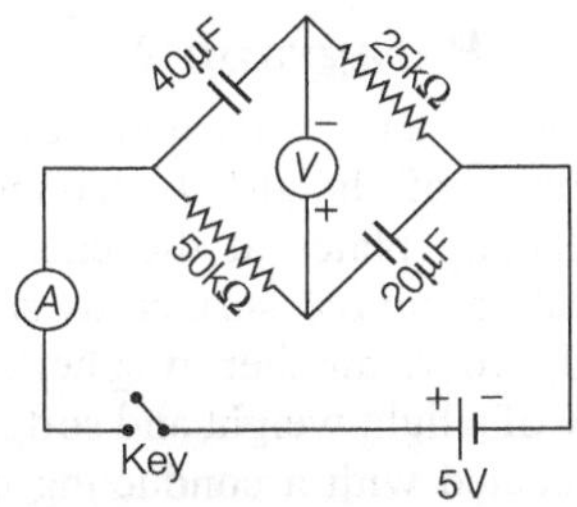

(a) The voltmeter display -5 V as soon as the key is pressed and displays $+5$ V after a long time

(b) The voltmeter will display 0 V at time $t = \ln 2$ seconds

(c) The current in the ammeter becomes $1/e$ of the initial value after 1 second

(d) The current in the ammeter becomes zero after a long time

14. Light of wavelength λ_{ph} falls on a cathode plate inside a vacuum tube as shown in the figure. The work function of the cathode surface is ϕ and the anode is a wire mesh of conducting material kept at a distance d from the cathode. A potential difference V is maintained between the electrodes. If the minimum de-Broglie wavelength of the electrons passing through the anode is λ_e, which of the following statements(s) is (are) true?

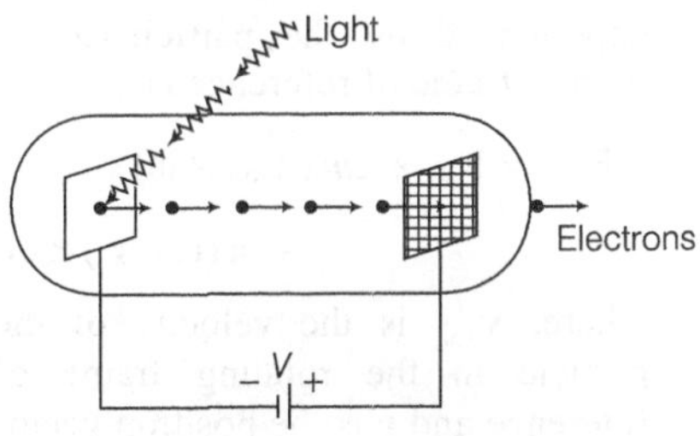

(a) λ_e increases at the same rate as λ_{ph} for $\lambda_{ph} < hc/\phi$

(b) λ_e is approximately halved, if d is doubled

(c) λ_e decreases with increase in ϕ and 0 λ_{ph}

(d) For large potential difference $(V >> \phi/e)$, λ_e is approximately halved if V is made four times

Section 3 (Maximum Marks : 16)

- This section contains **TWO** paragraphs.
- Based on each paragraph, there are **TWO** questions.
- Each question has **FOUR** options (a), (b) (c) and (d). **ONE ONLY** of these four options is correct.
- For each question, darken the bubble(s) corresponding to all the correct option(s) in the ORS.
- For each questions, marks will be awarded in one of the following categories.

Full Marks	: + 3	If only the bubble corresponding to all the correct answer is darkened.
Zero Marks	: 0	In all other cases.

Paragraph 1

A frame of the reference that is accelerated with respect to an inertial frame of reference is called a non-inertial frame of reference. A coordinate system fixed on a circular disc rotating about a fixed axis with a constant angular velocity ω is an example of a non-inertial frame of reference. The relationship between the force $\vec{F}_{rot}$ experienced by a particle of mass m moving on the rotating disc and the force $\vec{F}_{in}$ experienced by the particle in an inertial frame of reference is,

$$\vec{F}_{rot} = \vec{F}_{in} + 2m(\vec{v}_{rot} \times \vec{\omega}) + m(\vec{\omega} \times \vec{r}) \times \vec{\omega},$$

where, $\mathbf{v}_{rot}$ is the velocity of the particle in the rotating frame of reference and $\mathbf{r}$ is the position vector of the particle with respect to the centre of the disc.

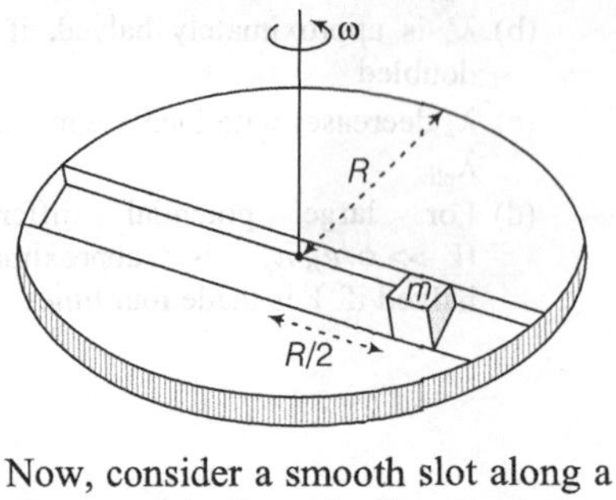

Now, consider a smooth slot along a diameter of a disc of radius R rotating counter-clockwise with a constant angular speed ω about its vertical axis through its centre. We assign a coordinate system with the origin at the centre of the disc, the X-axis along the slot, the Y-axis perpendicular to the slot and the Z-axis along the rotation axis ($\omega = \omega\hat{\mathbf{k}}$). A small block of mass m is gently placed in the slot at $\mathbf{r} = (R/2)\hat{\mathbf{i}}$ at $t = 0$ and is constrained to move only along the slot.

15. The distance r of the block at time t is

(a) $\frac{R}{2}\cos 2\omega t$ (b) $\frac{R}{2}\cos \omega t$

(c) $\frac{R}{4}(e^{\omega t} + e^{-\omega t})$ (d) $\frac{R}{4}(e^{2\omega t} + e^{-2\omega t})$

16. The net reaction of the disc on the block is

(a) $m\omega^2 R \sin \omega t\hat{\mathbf{j}} - mg\hat{\mathbf{k}}$

(b) $\frac{1}{2}m\omega^2 R(e^{\omega t} - e^{-\omega t})\hat{\mathbf{j}} + mg\hat{\mathbf{k}}$

(c) $\frac{1}{2}m\omega^2 R(e^{2\omega t} - e^{-2\omega t})\hat{\mathbf{j}} + mg\hat{\mathbf{k}}$

(d) $-m\omega^2 R \cos \omega t\hat{\mathbf{j}} - mg\hat{\mathbf{k}}$

Paragraph 2

Consider an evacuated cylindrical chamber of height h having rigid conducting plates at the ends and an insulating curved surface as shown in the figure. A number of spherical balls made of a light weight and soft material and coated with a conducting material are placed on the bottom plate. The balls have a radius $r << h$. Now, a high voltage source (HV) connected across the conducting plates such that the bottom plate is at $+V_0$ and the top plate at $-V_0$. Due to their conducting surface, the balls will get charge, will become equipotential with the plate and are repelled by it. The balls will eventually collide with the top plate, where the coefficient of restitution can be taken to be zero due to te soft nature of the material of the balls. The electric field in the chamber can be considered to be that of a parallel plate capacitor. Assume that there are no collisions between the balls and the interaction between them is negligible. (Ignore gravity)

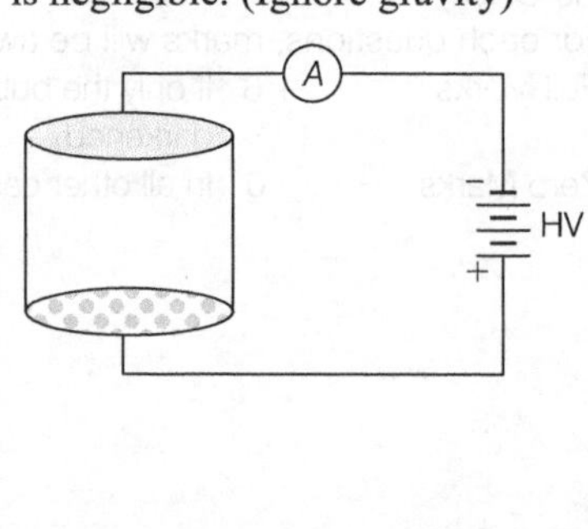

17. Which one of the following statement is correct?

(a) The balls will execute simple harmonic motion between the two plates

(b) The balls will bounce back to the bottom plate carrying the same charge they went up with

(c) The balls will stick to the top plate and remain there

(d) The balls will bounce back to the bottom plate carrying the opposite charge they went up with

18. The average current in the steady state registered by the ammeter in the circuit will be

(a) proportional to V_0^2

(b) proportional to the potential V_0

(c) zero

(d) proportions to $V_0^{1/2}$

Chemistry

Section 1 (Maximum Marks : 32)

- This section contains **SIX** questions.
- Each question has **FOUR** options (a), (b), (c) and (d). **ONLY ONE** of these four options is correct.
- Four each question, darken the bubble corresponding to the correct option in the ORS.
- For each question, marks will be awarded in one of the following categories.

Full Marks	: + 3	If only the bubble corresponding to the correct option is darkened.
Zero Marks	: 0	If none of the bubbles is darkened.
Negative Marks	: −1	In all other cases.

19. The correct order of acidity for the following compounds is

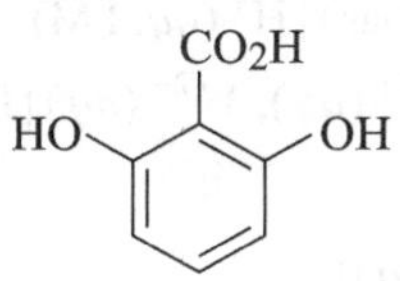

I

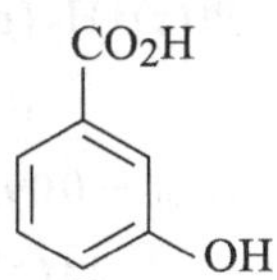

II

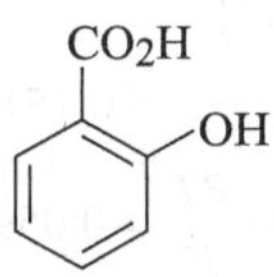

III

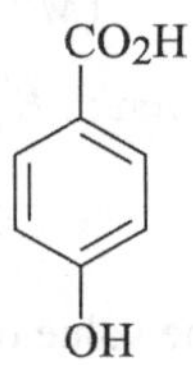

IV

(a) I > II > III > IV

(b) III > I > II > IV

(c) III > IV > II > I

(d) I > III > IV > II

20. The major product of the following reaction sequence is

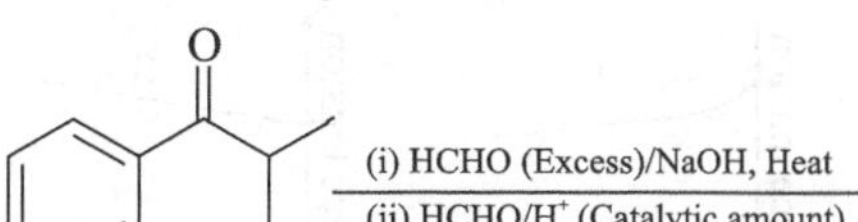

(a)

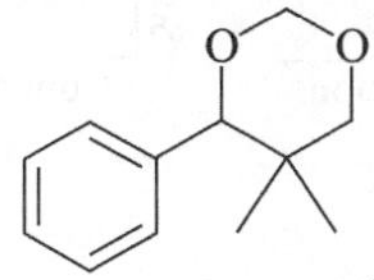

(b)

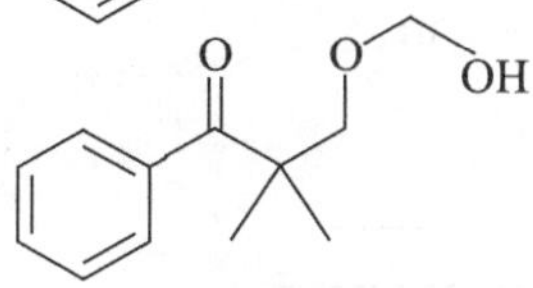

(c)

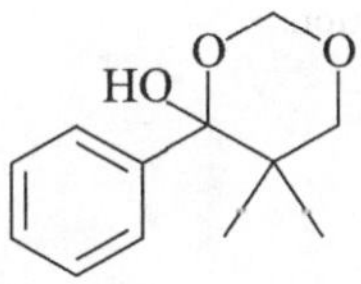

(d) 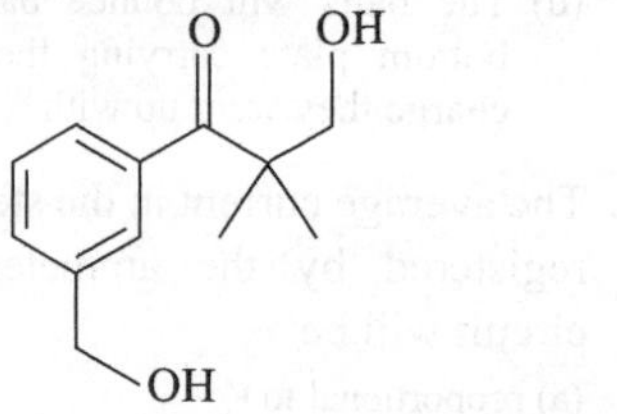

21. In the following reaction sequence in aqueous solution, the species *X*, *Y* and *Z*, respectively, are

$$S_2O_3^{2-} \xrightarrow{Ag^+} \underset{\text{(Clear solution)}}{X} \xrightarrow{Ag^+} \underset{\text{(White ppt.)}}{Y} \xrightarrow{\text{With time}} \underset{\text{(Black ppt.)}}{Z}$$

(a) $[Ag(S_2O_3)_2]^{3-}$, $Ag_2S_2O_3$, Ag_2S
(b) $[Ag(S_2O_3)_3]^{5-}$, Ag_2SO_3, Ag_2S
(c) $[Ag(SO_3)_2]^{3-}$, $Ag_2S_2O_3$, Ag
(d) $[Ag(SO_3)_3]^{3-}$, Ag_2SO_4, Ag

22. The qualitative sketches I, II and III given below show the variation of surface tension with molar concentration of three different aqueous solutions of KCl, CH_3OH and $CH_3(CH_2)_{11}OSO_3^-Na^+$ at room temperature.

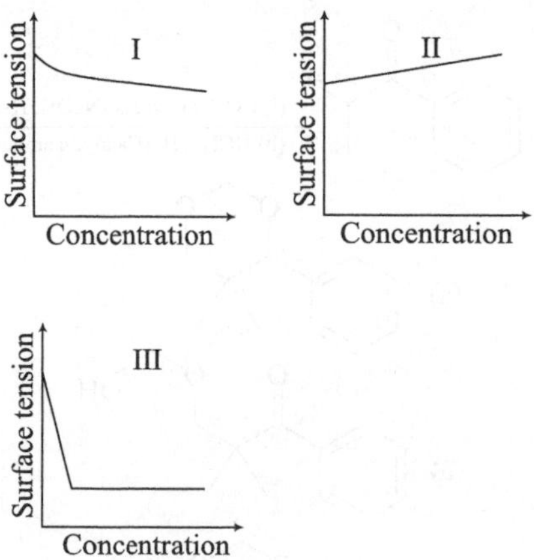

The correct assignment of the sketches is

(a) I KCl
II CH_3OH
III $CH_3(CH_2)_{11}OSO_3^-Na^+$
(b) I $CH_3(CH_2)_{11}OSO_3^-Na^+$
II CH_3OH
III KCl
(c) I KCl
II $CH_3(CH_2)_{11}OSO_3^-Na^+$
III CH_3OH
(d) I CH_3OH
II KCl
III $CH_3(CH_2)_{11}OSO_3^-Na^+$

23. The geometries of the ammonia complexes of Ni^{2+}, Pt^{2+} and Zn^{2+}, respectively, are

(a) octahedral, square planar and tetrahedral
(b) square planar, octahedral and tetrahedral
(c) tetrahedral, square planar and octahedral
(d) octahedral, tetrahedral and square planar

24. For the following electrochemical cell at 298 K,

$$Pt(s) \mid H_2(g, 1\,\text{bar}) \mid H^+(aq, 1\,M) \parallel M^{4+}(aq), M^{2+}(aq) \mid Pt(s)$$

$E_{cell} = 0.092$ V

when $\dfrac{[M^{2+}(aq)]}{[M^{4+}(aq)]} = 10^x$

Given : $E^\circ_{M^{4+}/M^{2+}} = 0.151$ V;

$2.303\dfrac{RT}{F} = 0.059$ V

The value of x is

(a) – 2 (b) – 1
(c) 1 (d) 2

Section 2 (Maximum Marks : 32)

- This section contains **EIGHT** questions.
- Each question has **FOUR** options (a), (b), (c) and (d). **ONE OR MORE THAN ONE** of these four options is (are) correct.
- For each question, darken the bubble(s) corresponding to all the correct option(s) in the ORS.
- For each question, marks will be awarded in **one of the following categories**:

Full Marks : + 4, If only the bubble corresponding to all the correct option(s) is (are) darkened.

Partial Marks : + 1, For darkening a bubble corresponding to **each correct option,** provided no incorrect option is darkened.

Zero Marks : 0, If none of the bubbles is darkened.

Negative Marks : −2, In all other cases.

For example, if (a), (c) and (d) are all the correct options for a question, darkening all these three will result in + 4 marks ; darkening only (a) and (d) will result in + 2 marks and darkening (a)and (b) will result in −2 marks, as a wrong option is also darkened.

25. According to molecular orbital theory, which of the following statements is(are) correct?

(a) C_2^{2-} is expected to be diamagnetic

(b) O_2^{2+} is expected to have a longer bond length than O_2

(c) N_2^+ and N_2^- have the same bond order

(d) He_2^+ has the same energy as two isolated He atoms

26. The correct statement(s) for cubic close packed (ccp) three dimensional structure is (are)

(a) The number of the nearest neighbours of an atom present in the topmost layer is 12

(b) The packing efficiency of atom is 74%

(c) The number of octahedral and tetrahedral voids per atom are 1 and 2, respectively

(d) The unit cell edge length is $2\sqrt{2}$ times the radius of the atom

27. Reagent(s) which can be used to bring about the following transformation is (are)

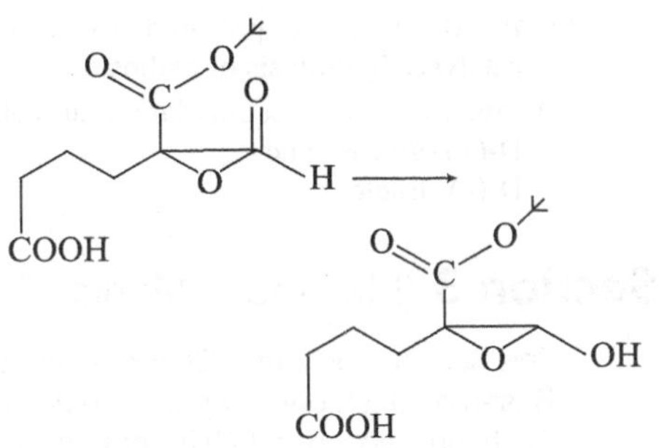

(a) $LiAlH_4$ in $(C_2H_5)_2O$

(b) BH_3 in THF

(c) $NaBH_4$ in C_2H_5OH

(d) Raney Ni/H_2 in THF

28. Extraction of copper from copper pyrite ($CuFeS_2$) involves

(a) crushing followed by concentration of the ore by froth-floatation

(b) removal of iron as slag

(c) self reduction step to produce 'blister copper' following evolution of SO_2

(d) refining of 'blister copper' by carbon reduction

29. The nitrogen containing compound produced in the reaction of HNO_3 with P_4O_{10}

(a) can also be prepared by reaction of P_4 and HNO_3
(b) is diamagnetic
(c) contains one N—N bond
(d) reacts with Na metal producing a brown gas

30. Mixture(s) showing positive deviation from Raoult's law at 35°C is (are)

(a) carbon tetrachloride + methanol
(b) carbon disulphide + acetone
(c) benzene + toluene
(d) phenol + aniline

31. For 'invert sugar', the correct statement(s) is (are)

(Given: specific rotations of (+) - sucrose, (+) - maltose, L-(−)-glucose and L-(+)-fructose in aqueous solution are $+66°$, $+140°$, $-52°$ and $92°$, respectively)

(a) Invert sugar is prepared by acid catalysed hydrolysis of maltose
(b) Invert sugar is an equimolar mixture of D-(+) -glucose and D-(−)- fructose
(c) Specific rotation of invert sugar is $-20°$
(d) On reaction with Br_2 water, invert sugar forms saccharic acid as one of the products

32. Among the following reactions(s), which gives(give) *tert*-butyl benzene as the major product?

(a) 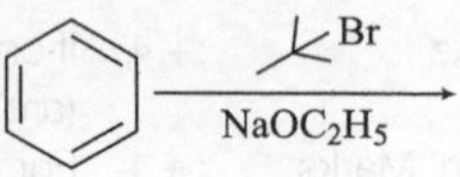

(b) 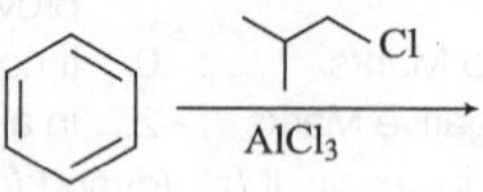

(c) 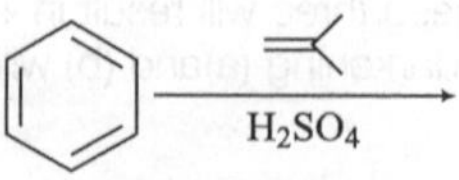

(d) 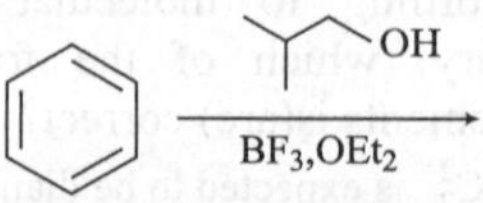

Section 3 (Maximum Marks : 12)

- This section contains **FOUR** questions.
- Based on each paragraph, there are TWO questions.
- Each question has FOUR options (a), (b), (c) and (d). ONLY ONE of these four options is correct.
- Four each question, darken the bubble corresponding to the correct option in the ORS.
- For each question, marks will be awarded in **one of the following categories**:

Full Marks : +3, If only the bubble corresponding to the correct answer is darkened.

Paragraph for Questions 33-34

- Thermal decomposition of gaseous X_2 to gaseous X at 298 K takes place according to the following equation:

$$X_2(g) \rightleftharpoons 2X(g)$$

- The standard reaction Gibbs energy, $\Delta_r G°$, of this reaction is positive. At the start of the reaction, there is one mole of X_2 and no X. As the reaction proceeds, the number of moles of X formed is given by β. Thus, $\beta_{\text{equilibrium}}$ is the number of moles of X formed at equilibrium. The reaction is carried out at a constant total pressure of 2 bar. Consider the gases to behave ideally. (Given, $R = 0.083$ L bar K^{-1} mol^{-1})

33. The equilibrium constant K_p for this reaction at 298 K, in terms of $\beta_{\text{equilibrium}}$ is

(a) $\dfrac{8\beta^2_{\text{equilibrium}}}{2-\beta_{\text{equilibrium}}}$ (b) $\dfrac{8\beta^2_{\text{equilibrium}}}{4-\beta^2_{\text{equilibrium}}}$

(c) $\dfrac{4\beta^2_{\text{equilibrium}}}{2-\beta_{\text{equilibrium}}}$ (d) $\dfrac{4\beta^2_{\text{equilibrium}}}{4-\beta^2_{\text{equilibrium}}}$

34. The incorrect statement among the following for this reaction, is

(a) Decrease in the total pressure will result in the formation of more moles of gaseous X

(b) At the start of the reaction, dissociation of gaseous X_2 takes place spontaneously

(c) $\beta_{\text{equilibrium}} = 0.7$

(d) $K_C < 1$

Paragraph for Questions 35-36

- Treatment of compound O with $KMnO_4/H^+$ gave P, which on heating with ammonia gave Q. The compound Q on treatment with $Br_2/NaOH$ produced R. On strong heating, Q gave S, which on further treatment with ethyl 2-bromopropanoate in the presence of KOH followed by acidification, gave a compound T.

(O)

35. The compound R is

(a) 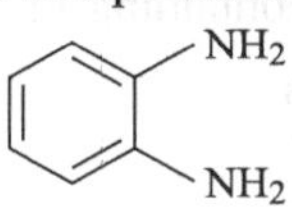

(b) 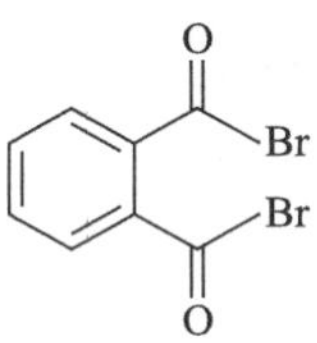

(c) 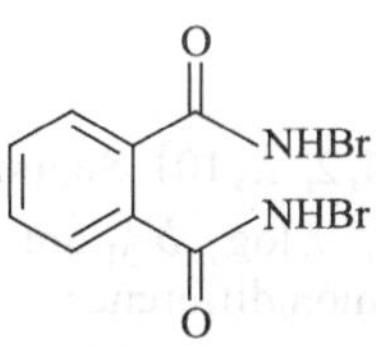

(d) 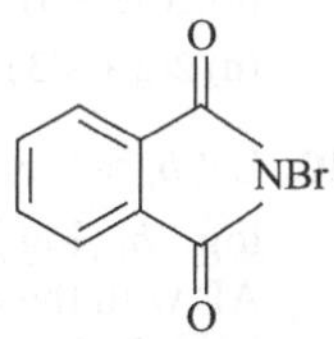

36. The compound T is

(a) glycine

(b) alanine

(c) valine

(d) serine

Mathematics

Section 1 (Maximum Marks : 15)

- This section contains **SIX** questions.
- Each question has **FOUR** options (a), (b), (c) and (d). **ONLY ONE** of these four options is correct.
- For each question, darken the bubble corresponding to the correct option in the ORS.

For each question, marks will be awarded in one of the following categories :

Full Marks : +3 If only the bubble corresponding to the correct option is darkened.

Zero Marks : 0 If none of the bubbles is darkened.

Negative Marks : −1 In all other cases.

37. Let $P = \begin{bmatrix} 1 & 0 & 0 \\ 4 & 1 & 0 \\ 16 & 4 & 1 \end{bmatrix}$ and I be the identity matrix of order 3. If $Q = [q_{ij}]$ is a matrix, such that $P^{50} - Q = I$, then $\dfrac{q_{31} + q_{32}}{q_{21}}$ equals

(a) 52 (b) 103 (c) 201 (d) 205

38. Area of the region $\{(x, y)\} \in R^2 : y \geq \sqrt{|x+3|}$, $5y \leq (x+9) \leq 15\}$ is equal to

(a) $\frac{1}{6}$ (b) $\frac{4}{3}$

(c) $\frac{3}{2}$ (d) $\frac{5}{3}$

39. The value of $\sum_{k=1}^{13} \frac{1}{\sin\left(\frac{\pi}{4} + \frac{(k-1)\pi}{6}\right)\sin\left(\frac{\pi}{4} + \frac{k\pi}{6}\right)}$ is equal to

(a) $3 - \sqrt{3}$

(b) $2(3 - \sqrt{3})$

(c) $2(\sqrt{3} - 1)$

(d) $2(2 + \sqrt{3})$

40. Let $b_i > 1$ for $i = 1, 2, \ldots, 101$. Suppose $\log_e b_1, \log_e b_2, \ldots, \log_e b_{101}$ are in AP with the common difference $\log_e 2$. Suppose $a_1, a_2, \ldots, a_{101}$ are in AP, such that $a_1 = b_1$ and $a_{51} = b_{51}$. If $t = b_1 + b_2 + \ldots + b_{51}$ and $s = a_1 + a_2 + \ldots + a_{51}$, then

(a) $s > t$ and $a_{101} > b_{101}$

(b) $s > t$ and $a_{101} < b_{101}$

(c) $s < t$ and $a_{101} > b_{101}$

(d) $s < t$ and $a_{101} < b_{101}$

41. The value of $\int_{-\pi/2}^{\pi/2} \frac{x^2 \cos x}{1+e^x} dx$ is equal to

(a) $\frac{\pi^2}{4} - 2$ (b) $\frac{\pi^2}{4} + 2$

(c) $\pi^2 - e^{-\pi/2}$ (d) $\pi^2 + e^{\pi/2}$

42. Let P be the image of the point (3, 1, 7) with respect to the plane $x - y + z = 3$. Then, the equation of the plane passing through P and containing the straight line $\frac{x}{1} = \frac{y}{2} = \frac{z}{1}$ is

(a) $x + y - 3z = 0$

(b) $3x + z = 0$

(c) $x - 4y + 7z = 0$

(d) $2x - y = 0$

Section 2 (Maximum Marks : 32)

- This section contains **EIGHT** questions.
- Each question has **FOUR** options (a), (b), (c) and (d). **ONE OR MORE THAN ONE** of these four option(s) is(are) correct.
- For each question, darken the bubble(s) corresponding to all the correct option(s) in the ORS.
- For each question, marks will be awarded in one of the following categories :

Full Marks	: +4	If only the bubble(s) corresponding to all correct option(s) is (are) darkened.
Partial Marks	: +1	For darkening a bubble corresponding **to each correct option**, provided no incorrect option is darkened.
Zero Marks	: 0	If none of the dubbles is darkened.
Negative Marks	: −2	In all other cases.

43. Let $a, b \in R$ and $f : R \to R$ be defined by $f(x) = a\cos(|x^3 - x|) + b|x|\sin(|x^3 + x|)$. Then, f is

(a) differentiable at $x = 0$, if $a = 0$ and $b = 1$

(b) differentiable at $x = 1$, if $a = 1$ and $b = 0$

(c) not differentiable at $x = 0$, if $a = 1$ and $b = 0$

(d) not differentiable at $x = 1$, if $a = 1$ and $b = 1$

44. Let $f(x)$

$$= \lim_{n\to\infty}\left[\frac{n^n(x+n)\left(x+\frac{n}{2}\right)\dots\left(x+\frac{n}{n}\right)}{n!(x^2+n^2)\left(x^2+\frac{n^2}{4}\right)\dots\left(x^2+\frac{n^2}{n^2}\right)}\right]^{\frac{x}{n}},$$

for all $x = 0$. Then

(a) $f\left(\frac{1}{2}\right)\geq f(1)$ (b) $f\left(\frac{1}{3}\right)\leq f\left(\frac{2}{3}\right)$

(c) $f'(2)\leq 0$ (d) $\frac{f'(3)}{f(3)}\geq\frac{f'(2)}{f(2)}$

45. Let $f: R\rightarrow(0,\infty)$ and $g: R\rightarrow R$ be twice differentiable functions such that f'' and g'' are continuous functions on R.

Suppose $f'(2)=g(2)=0$, $f''(2)\neq 0$ and $g'(2)\neq 0$. If $\lim_{x\to 2}\frac{f(x)g(x)}{f'(x)g'(x)}=1$, then

(a) f has a local minimum at $x = 2$
(b) f has a local maximum at $x = 2$
(c) $f''(2) > f(2)$
(d) $f(x)-f''(x)=0$, for atleast one $x\in R$

46. Let $\hat{u}=u_1\hat{i}+u_2\hat{j}+u_3\hat{k}$ be a unit vector in R^3 and $\hat{w}=\frac{1}{\sqrt{6}}(\hat{i}+\hat{j}+2\hat{k})$. Given that there exists a vector $\vec{v}$ in R^3, such that $|\hat{u}+\vec{v}|=1$ and $\hat{w}\cdot(\hat{u}+\vec{v})=1$.

Which of the following statement(s) is/are correct?

(a) There is exactly one choice for such $\vec{v}$
(b) There are infinitely many choices for such $\vec{v}$
(c) If $\hat{u}$ lies in the XY-plane, then $|u_1|=|u_2|$
(d) If $\hat{u}$ lies in the XY-plane, then $2|u_1|=|u_3|$

47. Let P be the point on the parabola $y^2=4x$, which is at the shortest distance from the centre S of the circle $x^2+y^2-4x-16y+64=0$. Let Q be the point on the circle dividing the line segment SP internally. Then,

(a) $SP=2\sqrt{5}$
(b) $SQ:QP=(\sqrt{5}+1):2$
(c) the x-intercept of the normal to the parabola at P is 6
(d) the slope of the tangent to the circle at Q is $\frac{1}{2}$

48. Let $a, b\in R$ and $a^2+b^2\neq 0$. Suppose $S=\left\{z\in C: z=\frac{1}{a+ibt}, t\in R, t\neq 0\right\}$, where $i=\sqrt{-1}$. If $z=x+iy$ and $z\in S$, then (x, y) lies on

(a) the circle with radius $\frac{1}{2a}$ and centre $\left(\frac{1}{2a}, 0\right)$ for $a>0, b\neq 0$
(b) the circle with radius $-\frac{1}{2a}$ and centre $\left(-\frac{1}{2a}, 0\right)$ for $a<0$, $b\neq 0$
(c) the X-axis for $a\neq 0, b=0$
(d) the Y-axis for $a=0, b\neq 0$

49. Let $a, \lambda, \mu\in R$. Consider the system of linear equations $ax+2y=\lambda$ and $3x-2y=\mu$. Which of the following statement(s) is/are correct?

(a) If $a=-3$, then the system has infinitely many solutions for all values of λ and μ
(b) If $a\neq -3$, then the system has a unique solution for all values of λ and μ
(c) If $\lambda+\mu=0$, then the system has infinitely many solutions for $a=-3$
(d) If $\lambda+\mu\neq 0$, then the system has no solution for $a=-3$

50. Let $f:\left[-\frac{1}{2}, 2\right] \rightarrow R$ and $g:\left[-\frac{1}{2}, 2\right] \rightarrow R$ be functions defined by $f(x)=[x^2-3]$ and $g(x)=|x|\, f(x)+|4x-7|\, f(x)$, where $[y]$ denotes the greatest integer less than or equal to y for $y \in R$. Then,

(a) f is discontinuous exactly at tree points in $\left[-\frac{1}{2}, 2\right]$

(b) f is discontinuous exactly at four points in $\left[-\frac{1}{2}, 2\right]$

(c) g is not differentiable exactly at four points in $\left(-\frac{1}{2}, 2\right)$

(d) g is not differentiable exactly at five points in $\left(-\frac{1}{2}, 2\right)$

Section 3 (Maximum Marks : 12)

- This section contains **TWO** paragraphs.
- Based on each paragraph, there are **TWO** questions.
- Each question has **FOUR** options (a), (b), (c) and (d). **ONLY ONE** of these four options is correct.
- For each question, darken the bubble corresponding to the correct option in the ORS.
- For each question, marks will be awarded in **one of the following categories**:

  Full Marks : + 3 If only the bubble corresponding to the correct option is darkened.

  Zero Marks : 0 In all other cases.

Paragraph 1

Football teams T_1 and T_2 have to play two games against each other. It is assumed that the outcomes of the two games are independent. The probabilities of T_1 winning, drawing and losing a game against T_2 are $\frac{1}{2}, \frac{1}{6}$ and $\frac{1}{3}$, respectively. Each team gets 3 points for a win, 1 point for a draw and 0 point for a loss in a game. Let X and Y denote the total points scored by teams T_1 and T_2, respectively, after two games.

51. $P(X > Y)$ is

(a) $\frac{1}{4}$ (b) $\frac{5}{12}$ (c) $\frac{1}{2}$ (d) $\frac{7}{12}$

52. $P(X = Y)$ is

(a) $\frac{11}{36}$ (b) $\frac{1}{3}$ (c) $\frac{13}{36}$ (d) $\frac{1}{2}$

Paragraph 2

Let $F_1(x_1, 0)$ and $F_2(x_2, 0)$, for $x_1 < 0$ and $x_2 > 0$, be the foci of the ellipse $\frac{x^2}{9}+\frac{y^2}{8}=1$. Suppose a parabola having vertex at the origin and focus at F_2 intersects the ellipse at point M in the first quadrant and at point N in the fourth quadrant.

53. The orthocentre of ΔF_1MN is

(a) $\left(-\frac{9}{10}, 0\right)$ (b) $\left(\frac{2}{3}, 0\right)$

(c) $\left(\frac{9}{10}, 0\right)$ (d) $\left(\frac{2}{3}, \sqrt{6}\right)$

54. If the tangents to the ellipse at M and N meet at R and the normal to the parabola at M meets the X-axis at Q, then the ratio of area of ΔMQR to area of the quadrilateral MF_1NF_2 is

(a) 3 : 4 (b) 4 : 5 (c) 5 : 8 (d) 2 : 3

Answers

Paper 1

Physics

1. (a) 2. (b) 3. (b) 4. (d) 5. (d) 6. (a, d) 7. (a,c)
8. (a,b,d) 9. (b,d) 10. (b,c,d) 11. (a,c,d) 12. (a,b,d) 13. (c, d) 14. (6)
15. (9) 16. (3) 17. (8) 18. (9)

Chemistry

19. (c) 20. (c) 21. (b) 22. (b) 23. (a) 24. (b) 25. (b,c)
26. (a,c,d) 27. (a) 28. (b,d) 29. (a,b,c) 30. (b,c) 31. (b,c,d) 32. (5)
33. (9) 34. (5) 35. (6) 36. (4)

Mathematics

37. (a) 38. (c) 39. (c) 40. (c) 41. (c) 42. (a, d) 43. (b,c,d)
44. (a,c,d) 45. (a,c) 46. (a,b,c) 47. (b, c) 48. (a) 49. (b, c) 50. (5)
51. (2) 52. (1) 53. (1)

Paper 2

Physics

1. (c) 2. (c) 3. (c) 4. (b) 5. (d) 6. (a) 7. (c,d)
8. (a,c) 9. (b,c) 10. (a,b,d) 11. (a,b,d) 12. (a,b) 13. (a,b,c,d) 14. (d)
15. (c) 16. (b) 17. (d) 18. (a)

Chemistry

19. (a) 20. (a) 21. (a) 22. (d) 23. (a) 24. (d) 25. (a,c)
26. (b,c,d) 27. (c) 28. (a,b,c) 29. (b,d) 30. (a,b) 31. (b,c) 32. (b,c,d)
33. (b) 34. (c) 35. (a) 36. (b)

Mathematics

37. (b) 38. (c) 39. (c) 40. (b) 41. (a) 42. (c) 43. (a,b)
44. (b,c) 45. (a,d) 46. (b,c) 47. (a,c,d) 48. (a,c,d) 49. (b,c,d) 50. (b,c)
51. (b) 52. (c) 53. (a) 54. (c)

UNSOLVED PAPER 2017

JEE Main

Instructions

1. This test consists of 90 questions.
2. There are three parts in the question paper consisting of Physics, Chemistry & Mathematics having 30 questions in each part of equal weightage. Each question is allotted 4 marks for correct response.
3. Candidates will be awarded marks as stated above in instruction no. 2 for correct response of each question. 1 mark will be deducted for indicating incorrect response of each question. No deduction from the total score will be made if no response is indicated for an item in the answer sheet.
4. There is only one correct response for each question. Filling up more than one response in any question will be treated as wrong response and marks for wrong response will be deducted according as per instructions.

Physics

1. An observer is moving with half the speed of light towards a stationary microwave source emitting waves at frequency 10 GHz. What is the frequency of the microwave measured by the observer? (speed of light $= 3 \times 10^8\ ms^{-1}$)

(a) 12.1 GHz
(b) 17.3 GHz
(c) 15.3 GHz
(d) 10.1 GHz

2. The following observations were taken for determining surface tension T of water by capillary method. Diameter of capillary, $d = 1.25 \times 10^{-2}$ m rise of water, $h = 1.45 \times 10^{-2}$m. Using $g = 9.80$ m/s^2 and the simplified relation $T = \frac{rhg}{2} \times 10^3$N/m, the possible error in surface tension is closest to

(a) 1.5% (b) 2.4%
(c) 10% (d) 0.15%

3. Some energy levels of a molecule are shown in the figure. The ratio of the wavelengths $r = \lambda_1 / \lambda_2$ is given by

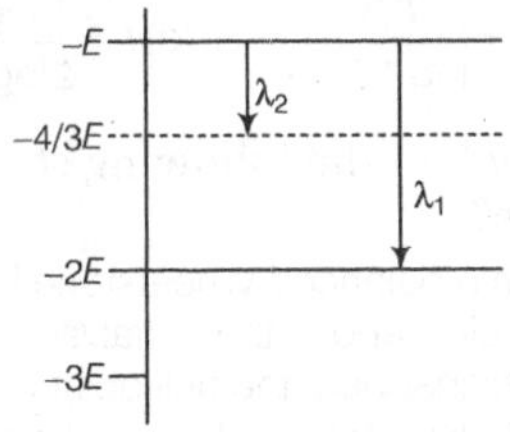

(a) $r = \frac{2}{3}$ (b) $r = \frac{3}{4}$
(c) $r = \frac{1}{3}$ (d) $r = \frac{4}{3}$

4. A body of mass $m = 10^{-2}$kg is moving in a medium and experiences a frictional force $F = -kv^2$. Its initial speed is $v_0 = 10$ ms^{-1}. If, after 10 s, its energy is $\frac{1}{8}mv_0^2$, the value of k will be

(a) 10^{-3}kgs^{-1} (b) 10^{-4}kgm^{-1}
(c) 10^{-1}kgm^{-1}s^{-1} (d) 10^{-3}kgm^{-1}

5. C_p and C_V are specific heats at constant pressure and constant volume, respectively. It is observed that $C_p - C_V = a$ for hydrogen gas $C_p - C_V = b$ for nitrogen gas. The correct relation between a and b is

(a) $a = b$ (b) $a = 14b$

(c) $a = 28b$ (d) $a = \frac{1}{14} b$

6. The moment of inertia of a uniform cylinder of length l and radius R about its perpendicular bisector is I. What is the ratio l/R such that the moment of inertia is minimum?

(a) $\frac{\sqrt{3}}{2}$ (b) 1 (c) $\frac{3}{\sqrt{2}}$ (d) $\sqrt{\frac{3}{2}}$

7. A radioactive nucleus A with a half-life T, decays into a nucleus B. At $t = 0$, there is no nucleus B. After sometime t, the ratio of the number of B to that of A is 0.3. Then, t is given by

(a) $t = T \frac{\log 1.3}{\log_e 2}$ (b) $t = T \log 1.3$

(c) $t = \frac{T}{\log 1.3}$ (d) $t = \frac{T \log_e 2}{2 \log 1.3}$

8. Which of the following statements is false?

(a) In a balanced Wheatstone bridge, if the cell and the galvanometer are exchanged, the null point is disturbed
(b) A rheostat can be used as a potential divider
(c) Kirchhoff's second law represents energy conservation
(d) Wheatstone bridge is the most sensitive when all the four resistances are of the same order of magnitude

9. A capacitance of 2 μF is required in an electrical circuit across a potential difference of 1kV. A large number of 1 μF capacitors are available which can withstand a potential difference of not more than 300 V. The minimum number of capacitors required to achieve this is

(a) 16 (b) 24 (c) 32 (d) 2

10. In the given circuit diagram, when the current reaches steady state in the circuit, the charge on the capacitor of capacitance C will be

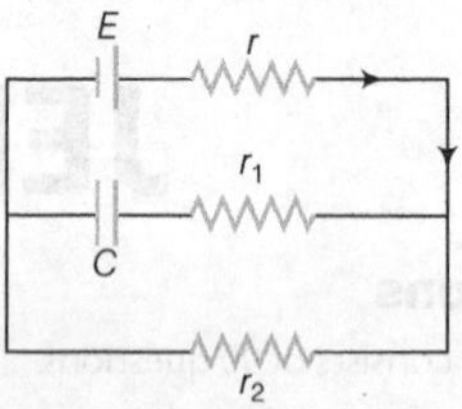

(a) $CE \frac{r_1}{(r_2 + r)}$

(b) $CE \frac{r_2}{(r + r_2)}$

(c) $CE \frac{r_1}{(r_1 + r)}$

(d) CE

11. In the below circuit, the current in each resistance is

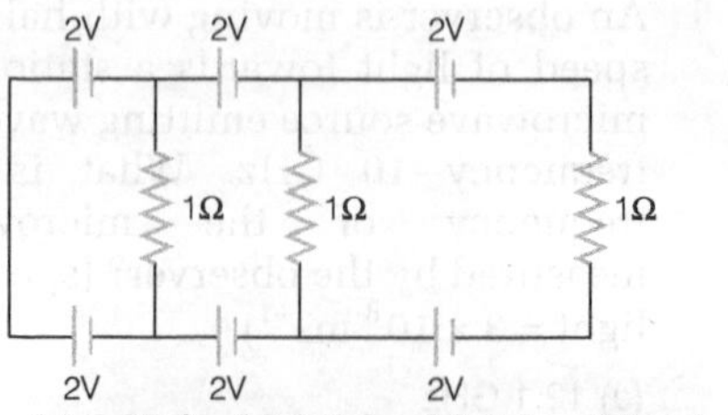

(a) 0.25 A (b) 0.5 A (c) 0 A (d) 1 A

12. In amplitude modulation, sinusoidal carrier frequency used is denoted by ω_c and the signal frequency is denoted by ω_m. The bandwidth $(\Delta\omega_m)$ of the signal is such that $\Delta\omega_m << \omega_c$. Which of the following frequencies is not contained in the modulated wave?

(a) ω_c (b) $\omega_m + \omega_c$

(c) $\omega_c - \omega_m$ (d) ω_m

13. In a common emitter amplifier circuit using an n-p-n transistor, the phase difference between the input and the output voltages will be

(a) 90° (b) 135°

(c) 180° (d) 45°

14. A copper ball of mass 100 g is at a temperature T. It is dropped in a copper calorimeter of mass 100 g, filled with 170 g of water at room temperature. Subsequently, the temperature of the system is found to be 75°C. T is (Given, room temperature = 30°C, specific heat of copper = 0.1 cal/g°C)

(a) 885°C (b) 1250°C (c) 825°C (d) 800°C

15. In a Young's double slit experiment, slits are separated by 0.5 mm and the screen is placed 150 cm away. A beam of light consisting of two wavelengths, 650 nm and 520 nm, is used to obtain interference fringes on the screen. The least distance from the common central maximum to the point where the bright fringes due to both the wavelengths coincide, is

(a) 7.8 mm (b) 9.75 mm

(c) 15.6 mm (d) 1.56 mm

16. An electric dipole has a fixed dipole moment $\mathbf{p}$, which makes angle θ with respect to X-axis. When subjected to an electric field $\mathbf{E}_1 = E\hat{\mathbf{i}}$, it experiences a torque $\mathbf{T}_1 = \tau\hat{\mathbf{k}}$. When subjected to another electric field $\mathbf{E}_2 = \sqrt{3}E_1\hat{\mathbf{j}}$, it experiences a torque $\mathbf{T}_2 = -\mathbf{T}_1$. The angle θ is

(a) 45° (b) 60° (c) 90° (d) 30°

17. A slender uniform rod of mass M and length l is pivoted at one end so that it can rotate in a vertical plane (see the figure). There is negligible friction at the pivot. The free end is held vertically above the pivot and then released. The angular acceleration of the rod when it makes an angle θ with the vertical, is

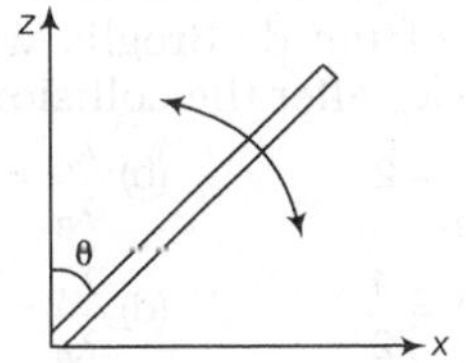

(a) $\frac{2g}{3l}\sin\theta$ (b) $\frac{3g}{2l}\cos\theta$

(c) $\frac{2g}{3l}\cos\theta$ (d) $\frac{3g}{2l}\sin\theta$

18. An external pressure P is applied on a cube at 0°C so that it is equally compressed from all sides. K is the bulk modulus of the material of the cube and α is its coefficient of linear expansion. Suppose we want to bring the cube to its original size by heating. The temperature should be raised by

(a) $\frac{P}{\alpha K}$ (b) $\frac{3\alpha}{PK}$ (c) $3PK\alpha$ (d) $\frac{P}{3\alpha K}$

19. A diverging lens with magnitude of focal length 25 cm is placed at a distance of 15 cm from a converging lens of magnitude of focal length 20 cm. A beam of parallel light falls on the diverging lens. The final image formed is

(a) virtual and at a distance of 40 cm from convergent lens

(b) real and at a distance of 40 cm from the divergent lens

(c) real and at a distance of 6 cm from the convergent lens

(d) real and at a distance of 40 cm from convergent lens

20. An electron beam is accelerated by a potential difference V to hit a metallic target to produce X-rays. It produces continuous as well as characteristic X-rays. If λ_{min} is the smallest possible wavelength of X-rays in the spectrum, the variation of $\log \lambda_{min}$ with $\log V$ is correctly represented in

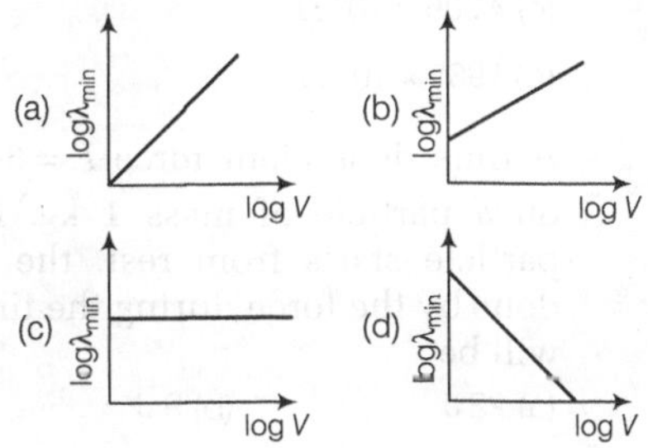

21. The temperature of an open room of volume 30 m^3 increases from 17°C to 27°C due to the sunshine. The atmospheric pressure in the room remains 1×10^5 Pa. If n_i and n_f are the number of molecules in the room before and after heating, then $n_f - n_i$ will be

(a) 1.38×10^{23}

(b) 2.5×10^{25}

(c) -2.5×10^{25}

(d) -1.61×10^{23}

22. In a coil of resistance 100 Ω, a current is induced by changing the magnetic flux through it as shown in the figure. The magnitude of change in flux through the coil is

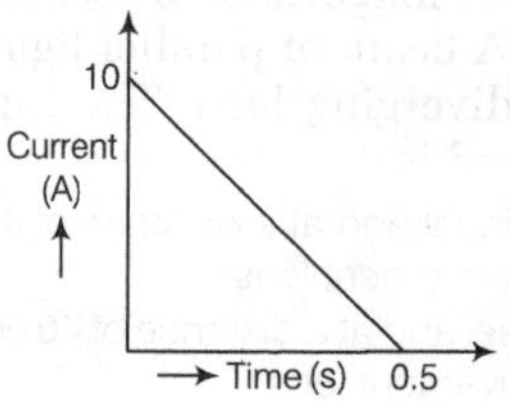

(a) 225 Wb (b) 250 Wb

(c) 275 Wb (d) 200 Wb

23. When a current of 5 mA is passed through a galvanometer having a coil of resistance 15 Ω, it shows full scale deflection. The value of the resistance to be put in series with the galvanometer to convert it into a voltmeter of range 0-10 V is

(a) 2.045×10^3 Ω

(b) 2.535×10^3 Ω

(c) 4.005×10^3 Ω

(d) 1.985×10^3 Ω

24. A time dependent force $F = 6t$ acts on a particle of mass 1 kg. If the particle starts from rest, the work done by the force during the first 1 s will be

(a) 22 J (b) 9 J

(c) 18 J (d) 4.5 J

25. A magnetic needle of magnetic moment 6.7×10^{-2} Am^2 and moment of inertia 7.5×10^{-6} kg m^2 is performing simple harmonic oscillations in a magnetic field of 0.01 T. Time taken for 10 complete oscillations is

(a) 8.89 s (b) 6.98 s

(c) 8.76 s (d) 6.65 s

26. The variation of acceleration due to gravity g with distance d from centre of the Earth is best represented by (R = Earth's radius)

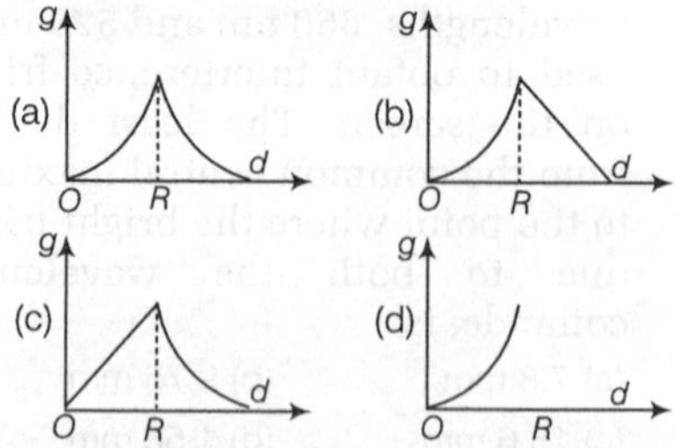

27. A body is thrown vertically upwards. Which one of the following graphs correctly represent the velocity *vs* time?

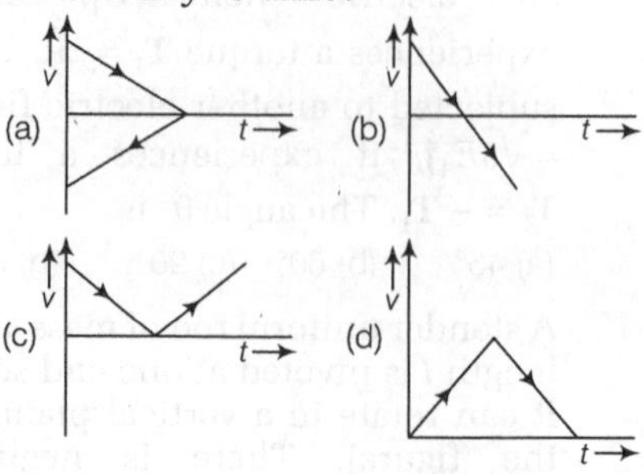

28. A particle A of mass m and initial velocity v collides with a particle B of mass $\frac{m}{2}$ which is at rest. The collision is head on, and elastic. The ratio of the de-Broglie wavelengths λ_A to λ_B after the collision is

(a) $\frac{\lambda_A}{\lambda_B} = 2$ (b) $\frac{\lambda_A}{\lambda_B} = \frac{2}{3}$

(c) $\frac{\lambda_A}{\lambda_B} = \frac{1}{2}$ (d) $\frac{\lambda_A}{\lambda_B} = \frac{1}{3}$

29. A particle is executing simple harmonic motion with a time period T. At time $t = 0$, it is at its position of equilibrium. The kinetic energy-time graph of the particle will look, like

(a) 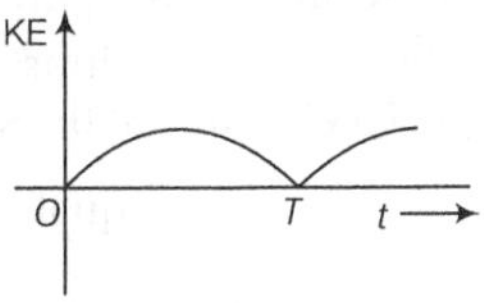

(b) 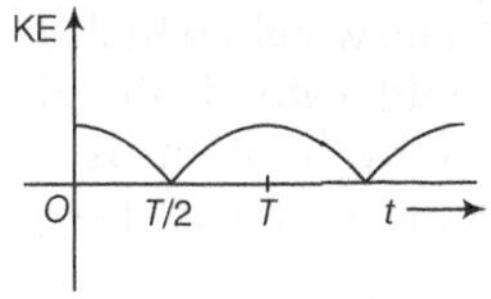

(c) 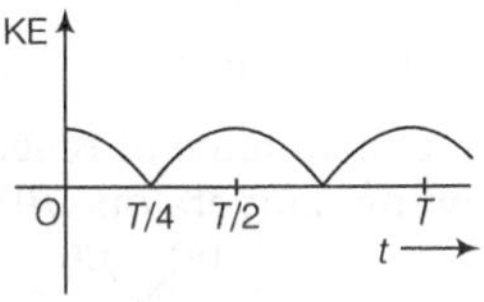

(d) 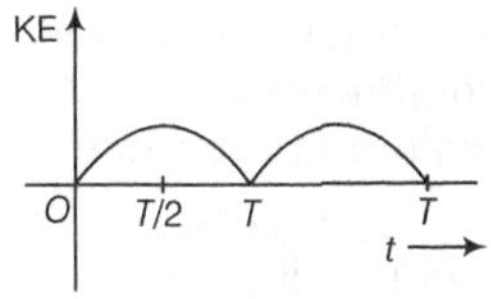

30. A man grows into a giant such that his linear dimensions increase by a factor of 9. Assuming that his density remains same, the stress in the leg will change by a factor of

(a) $\frac{1}{9}$ (b) 81 (c) $\frac{1}{81}$ (d) 9

Chemistry

31. Which of the following compounds will give significant amount of *meta*-product during mononitration reaction?

(a) 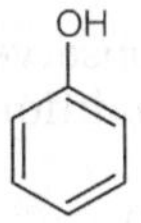

(b) 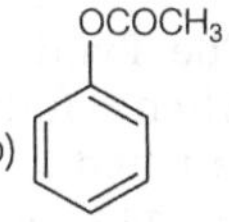

(c) 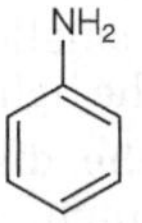

(d) 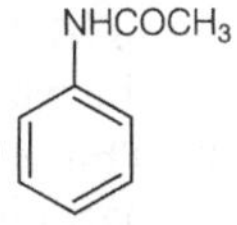

32. ΔU is equal to

(a) isochoric work (b) isobaric work

(c) adiabatic work (d) isothermal work

33. The increasing order of reactivity of the following halides for the S_N1 reaction is

I. $CH_3CH(Cl)CH_2CH_3$

II. $CH_3CH_2CH_2Cl$

III. p $H_3CO—C_6H_4—CH_2Cl$

(a) (III) < (II) < (I) (b) (II) < (I) < (III)

(c) (I) < (III) < (II) (d) (II) < (III) < (I)

34. The radius of the second Bohr orbit for hydrogen atom is (Planck's constant $(h) = 6.6262 \times 10^{-34}$ Js; mass of electron $= 9.1091 \times 10^{-31}$ kg ; charge of electron $(e) = 1.60210 \times 10^{-19}$ C; permitivity of vacuum $(\epsilon_0) = 8.854185 \times 10^{-12}$ $kg^{-1}m^{-3}A^2$)

(a) 1.65 Å

(b) 4.76 Å

(c) 0.529 Å

(d) 2.12 Å

35. pK_a of a weak acid (HA) and pK_b of a weak base (BOH) are 3.2 and 3.4, respectively. The pH of their salt (AB) solution is

(a) 7.2 (b) 6.9 (c) 7.0 (d) 1.0

36. The formation of which of the following polymers involves hydrolysis reaction?

(a) Nylon-6 (b) Bakelite

(c) Nylon-6, 6 (d) Terylene

37. The most abundant elements by mass in the body of a healthy human adult are Oxygen (61.4%); Carbon (22.9%), Hydrogen (10.0 %); and Nitrogen (2.6%). The weight which a 75 kg person would gain if all ^1H atoms are replaced by ^2H atoms is

(a) 15 kg (b) 37.5 kg (c) 7.5 kg (d) 10 kg

38. Which of the following, upon treatment with *tert*-BuONa followed by addition of bromine water, fails to decolourise the colour of bromine?

(a) 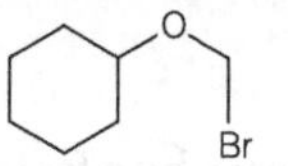

(b) 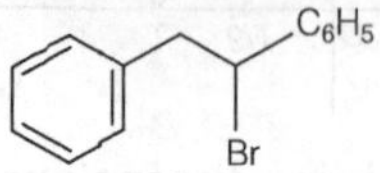

(c) 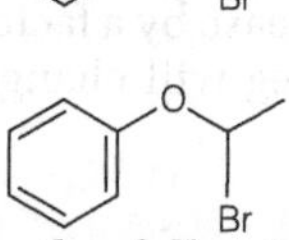

(d) 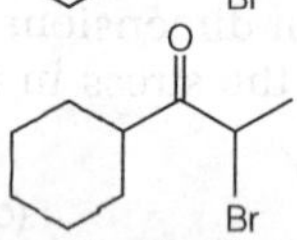

39. In the following reactions, ZnO is respectively acting as a/an

(i) $ZnO + Na_2O \longrightarrow Na_2ZnO_2$

(ii) $ZnO + CO_2 \longrightarrow ZnCO_3$

(a) base and acid (b) base and base
(c) acid and acid (d) acid and base

40. Both lithium and magnesium display several similar properties due to the diagonal relationship; however, the one which is incorrect is

(a) Both form basic carbonates
(b) Both form soluble bicarbonates
(c) Both form nitrides
(d) nitrates of both Li and Mg yield NO_2 and O_2 on heating

41. 3-methyl-pent-2-ene on reaction with HBr in presence of peroxide forms an addition product. The number of possible stereoisomers for the product is

(a) six (b) zero (c) two (d) four

42. A metal crystallises in a face centred cubic structure. If the edge length of its unit cell is '*a*', the closest approach between two atoms in metallic crystal will be

(a) $2a$ (b) $2\sqrt{2}\,a$ (c) $\sqrt{2}\,a$ (d) $\frac{a}{\sqrt{2}}$

43. Two reactions R_1 and R_2 have identical pre- exponential factors. Activation energy of R_1 exceeds that of R_2 by 10 kJ mol^{-1}. If k_1 and k_2 are rate constants for reactions R_1 and R_2, respectively at 300 K, then $\ln\left(\frac{k_2}{k_1}\right)$ is equal to

($R = 8.314$ J mol^{-1}K^{-1})

(a) 8 (b) 12 (c) 6 (d) 4

44. The correct sequence of reagents for the following conversion will be

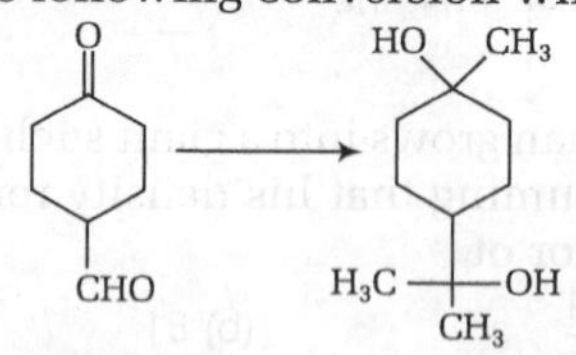

(a) $[Ag(NH_3)_2]^+ OH^-$, H^+/CH_3OH, CH_3MgBr

(b) CH_3MgBr, H^+/CH_3OH, $[Ag(NH_3)_2]^+ OH^-$

(c) CH_3MgBr, $[Ag(NH_3)_2]^+ OH^-$, H^+/CH_3OH

(d) $[Ag(NH_3)_2]^+ OH^-$, CH_3MgBr, H^+/CH_3OH

45. The Tyndall effect is observed only when following conditions are satisfied

1. The diameter of the dispersed particles is much smaller than the wavelength of the light used.
2. The diameter of the dispersed particle is not much smaller than the wavelength of the light used.
3. The refractive indices of the dispersed phase and dispersion medium are almost similar in magnitude.
4. The refractive indices of the dispersed phase and dispersion medium differ greatly in magnitude.

(a) 1 and 4 (b) 2 and 4
(c) 1 and 3 (d) 2 and 3

46. Which of the following compounds will behave as a reducing sugar in an aqueous KOH solution?

(a)
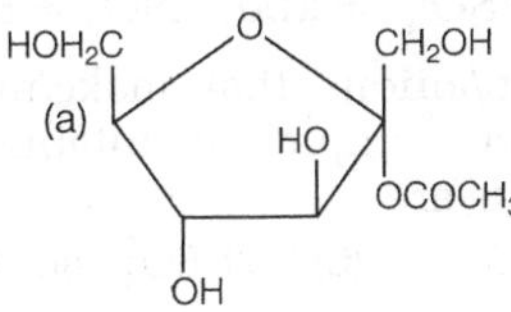

(b)
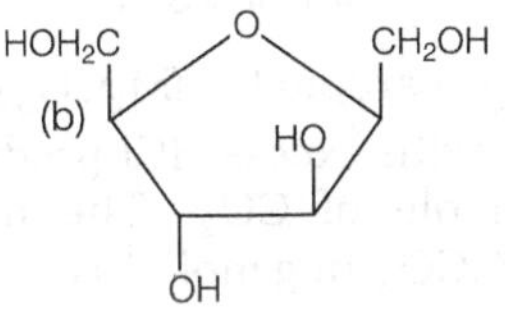

(c)
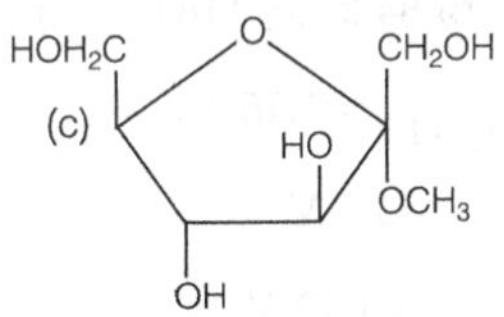

(d)
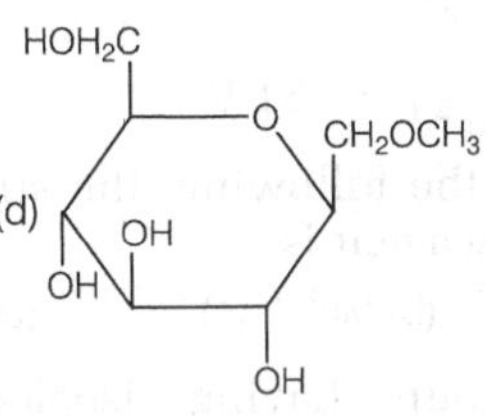

47. Given,

$C_{(graphite)} + O_2(g) \longrightarrow CO_2(g);$

$\Delta_r H° = -393.5 \text{ kJ mol}^{-1}$

$H_2(g) + \frac{1}{2}O_2(g) \longrightarrow H_2O(l);$

$\Delta_r H° = -285.8 \text{ kJ mol}^{-1}$

$CO_2(g) + 2H_2O(l) \longrightarrow CH_4(g) + 2O_2(g);$

$\Delta_r H° = +890.3 \text{ kJ mol}^{-1}$

Based on the above thermochemical equations, the value of $\Delta_r H°$ at 298 K for the reaction,

$C_{(graphite)} + 2H_2(g) \longrightarrow CH_4(g)$ will be

(a) $+78.8 \text{ kJ mol}^{-1}$ (b) $+144.0 \text{ kJ mol}^{-1}$

(c) $-74.8 \text{ kJ mol}^{-1}$ (d) $-144.0 \text{ kJ mol}^{-1}$

48. Which of the following reactions is an example of a redox reaction?

(a) $XeF_4 + O_2F_2 \longrightarrow XeF_6 + O_2$

(b) $XeF_2 + PF_5 \longrightarrow [XeF]^+PF_6^-$

(c) $XeF_6 + H_2O \longrightarrow XeOF_4 + 2HF$

(d) $XeF_6 + 2H_2O \longrightarrow XeO_2F_2 + 4HF$

49. The products obtained when chlorine gas reacts with cold and dilute aqueous NaOH are

(a) ClO^- and ClO_3^- (b) ClO_2^- and ClO_3^-

(c) Cl^- and ClO^- (d) Cl^- and ClO_2^-

50. The major product obtained in the following reaction is

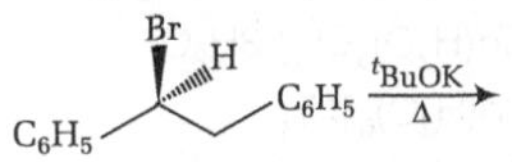

(a) $(\pm)\ C_6H_5CH(O^tBu)CH_2C_6H_5$

(b) $C_6H_5CH{=}CHC_6H_5$

(c) $(+)\ C_6H_5CH(O^tBu)CH_2C_6H_5$

(d) $(-)C_6H_5CH(O^tBu)CH_2C_6H_5$

51. Sodium salt of an organic acid '*X*' produces effervescence with conc. H_2SO_4. '*X*' reacts with the acidified aqueous $CaCl_2$ solution to give a white precipitate which decolourises acidic solution of $KMnO_4$. '*X*' is

(a) C_6H_5COONa (b) $HCOONa$

(c) CH_3COONa (d) $Na_2C_2O_4$

52. Which of the following species is not paramagnetic?

(a) NO (b) CO

(c) O_2 (d) B_2

53. The freezing point of benzene decreases by 0.45°C when 0.2 g of acetic acid is added to 20 g of benzene. If acetic acid associates to form a dimer in benzene, percentage association of acetic acid in benzene will be (K_f for benzene = 5.12 K kg mol^{-1})

(a) 64.6 % (b) 80.4 %

(c) 74.6 % (d) 94.6 %

54. Which of the following molecules is least resonance stabilised?

(a) 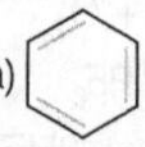
(b) 
(c) 
(d) 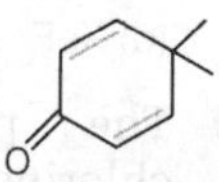

55. On treatment of 100 mL of 0.1 M solution of $CoCl_3.6H_2O$ with excess of $AgNO_3$; 1.2×10^{22} ions are precipitated. The complex is

(a) $[Co(H_2O)_4Cl_2]\,Cl.2H_2O$
(b) $[Co(H_2O)_3Cl_3].3H_2O$
(c) $[Co(H_2O)_6]Cl_3$
(d) $[Co(H_2O)_5Cl]\,Cl_2.H_2O$

56. The major product obtained in the following reaction is

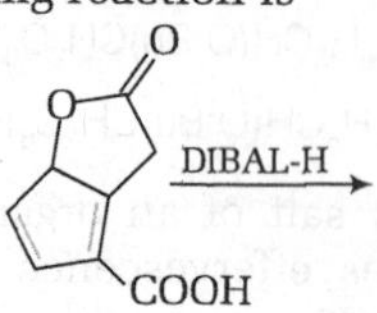

(a) 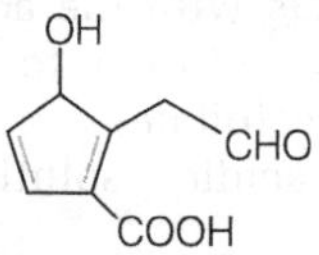
(b) 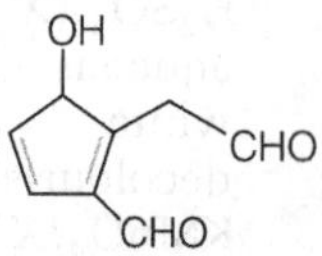
(c) 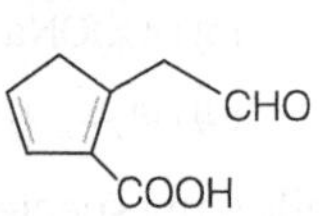
(d) 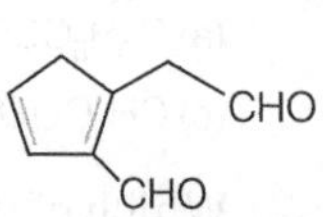

57. A water sample has ppm level concentration of following anions

$F^- = 10$; $SO_4^{2-} = 100$; $NO_3^- = 50$

the anion/anions that make/makes the water sample unsuitable for drinking is/are

(a) Only NO_3^-
(b) Both SO_4^{2-} and NO_3^-
(c) Only F^-
(d) Only SO_4^{2-}

58. 1 g of a carbonate (M_2CO_3) on treatment with excess HCl produces 0.01186 mole of CO_2. The molar mass of M_2CO_3 in g mol^{-1} is

(a) 1186 (b) 84.3 (c) 118.6 (d) 11.86

59. Given, $E^\circ_{Cl_2/Cl^-} = 1.36$ V,

$E^\circ_{Cr^{3+}/Cr} = -0.74$ V

$E^\circ_{Cr_2O_7^{2-}/Cr^{3+}} = 1.33$ V,

$E^\circ_{MnO_4^-/Mn^{2+}} = 1.51$ V

Among the following, the strongest reducing agent is

(a) Cr (b) Mn^{2+} (c) Cr^{3+} (d) Cl^-

60. The group having isoelectronic species is

(a) O^{2-}, F^-, Na^+, Mg^{2+}
(b) O^-, F^-, Na, Mg^+
(c) O^{2-}, F^-, Na, Mg^{2+}
(d) O^-, F^-, Na^+, Mg^{2+}

Mathematics

61. If S is the set of distinct values of b for which the following system of linear equations

$$x + y + z = 1, \Rightarrow x + ay + z = 1$$
and $\quad ax + by + z = 0$

has no solution, then S is

(a) an infinite set
(b) a finite set containing two or more elements
(c) singleton set
(d) a empty set

62. The statement $(p \to q) \to [(\sim p \to q) \to q]$ is

(a) a tautology
(b) equivalent to $\sim p \to q$
(c) equivalent to $p \to \sim q$
(d) a fallacy

63. If $5(\tan^2 x - \cos^2 x) = 2\cos 2x + 9$, then the value of $\cos 4x$ is

(a) $-\frac{3}{5}$ (b) $\frac{1}{3}$ (c) $\frac{2}{9}$ (d) $-\frac{7}{9}$

64. For three events A, B and C, if P(exactly one of A or B occurs) $= P$(exactly one of B or C occurs) $= P$ (exactly one of C or A occurs) $= \frac{1}{4}$ and P (all the three events occur simultaneously) $= \frac{1}{16}$, then the probability that atleast one of the events occurs, is

(a) $\frac{7}{32}$ (b) $\frac{7}{16}$ (c) $\frac{7}{64}$ (d) $\frac{3}{16}$

65. Let ω be a complex number such that $2\omega + 1 = z$, where $z = \sqrt{-3}$. If

$$\begin{vmatrix} 1 & 1 & 1 \\ 1 & -\omega^2 - 1 & \omega^2 \\ 1 & \omega^2 & \omega^7 \end{vmatrix} = 3k, \text{ then } k \text{ is}$$

equal to

(a) $-z$ (b) z (c) -1 (d) 1

66. Let k be an integer such that the triangle with vertices $(k, -3k)$, $(5, k)$ and $(-k, 2)$ has area 28 sq units. Then, the orthocentre of this triangle is at the point

(a) $\left(2, -\frac{1}{2}\right)$ (b) $\left(1, \frac{3}{4}\right)$

(c) $\left(1, -\frac{3}{4}\right)$ (d) $\left(2, \frac{1}{2}\right)$

67. If 20 m of wire is available for fencing off a flower-bed in the form of a circular sector, then the maximum area (in sq m) of the flower-bed is

(a) 12.5 (b) 10 (c) 25 (d) 30

68. The area (in sq units) of the region $\{(x, y) : x \geq 0, x + y \leq 3, x^2 \leq 4y$ and $y \leq 1 + \sqrt{x}\}$ is

(a) $\frac{59}{12}$ (b) $\frac{3}{2}$ (c) $\frac{7}{3}$ (d) $\frac{5}{2}$

69. If the image of the point $P(1, -2, 3)$ in the plane $2x + 3y - 4z + 22 = 0$ measured parallel to the line $\frac{x}{1} = \frac{y}{4} = \frac{z}{5}$ is Q, then PQ is equal to

(a) $3\sqrt{5}$ (b) $2\sqrt{42}$ (c) $\sqrt{42}$ (d) $6\sqrt{5}$

70. For $x \in \left(0, \frac{1}{4}\right)$, if the derivative of $\tan^{-1}\left(\frac{6x\sqrt{x}}{1 - 9x^3}\right)$ is $\sqrt{x} \cdot g(x)$, then $g(x)$ equals

(a) $\frac{9}{1 + 9x^3}$ (b) $\frac{3x\sqrt{x}}{1 - 9x^3}$

(c) $\frac{3x}{1 - 9x^3}$ (d) $\frac{3}{1 + 9x^3}$

71. If $(2 + \sin x)\frac{dy}{dx} + (y + 1)\cos x = 0$ and $y(0) = 1$, then $y\left(\frac{\pi}{2}\right)$ is equal to

(a) $\frac{1}{3}$ (b) $-\frac{2}{3}$ (c) $-\frac{1}{3}$ (d) $\frac{4}{3}$

72. Let a vertical tower AB have its end A on the level ground. Let C be the mid-point of AB and P be a point on the ground such that $AP = 2AB$. If $\angle BPC = \beta$, then $\tan \beta$ is equal to

(a) $\frac{6}{7}$ (b) $\frac{1}{4}$

(c) $\frac{2}{9}$ (d) $\frac{4}{9}$

73. If $A = \begin{bmatrix} 2 & -3 \\ -4 & 1 \end{bmatrix}$, then adj $(3A^2 + 12A)$ is equal to

(a) $\begin{bmatrix} 72 & -84 \\ -63 & 51 \end{bmatrix}$ (b) $\begin{bmatrix} 51 & 63 \\ 84 & 72 \end{bmatrix}$

(c) $\begin{bmatrix} 51 & 84 \\ 63 & 72 \end{bmatrix}$ (d) $\begin{bmatrix} 72 & -63 \\ -84 & 51 \end{bmatrix}$

74. For any three positive real numbers a, b and c, if $9(25a^2 + b^2) + 25(c^2 - 3ac) = 15b(3a + c)$, then

(a) b, c and a are in GP

(b) b, c and a are in AP

(c) a, b and c are in AP

(d) a, b and c are in GP

75. The distance of the point $(1, 3, -7)$ from tho plane passing through the point $(1, -1, -1)$ having normal perpendicular to both the lines

$\frac{x-1}{1}=\frac{y+2}{-2}=\frac{z-4}{3}$ and $\frac{x-2}{2}=\frac{y+1}{-1}=\frac{z+7}{-1}$, is

(a) $\frac{20}{\sqrt{74}}$ units (b) $\frac{10}{\sqrt{83}}$ units
(c) $\frac{5}{\sqrt{83}}$ units (d) $\frac{10}{\sqrt{74}}$ units

76. Let $I_n=\int \tan^n x\,dx\,(n>1)$. If $I_4+I_6=a\tan^5 x+bx^5+C$, where C is a constant of integration, then the ordered pair (a,b) is equal to

(a) $\left(-\frac{1}{5},1\right)$ (b) $\left(\frac{1}{5},0\right)$
(c) $\left(\frac{1}{5},-1\right)$ (d) $\left(-\frac{1}{5},0\right)$

77. The eccentricity of an ellipse whose centre is at the origin is 1/2. If one of its directrices is $x=-4$, then the equation of the normal to it at $\left(1,\frac{3}{2}\right)$ is

(a) $2y-x=2$ (b) $4x-2y=1$
(c) $4x+2y=7$ (d) $x+2y=4$

78. If a hyperbola passes through the point $P(\sqrt{2},\sqrt{3})$ and has foci at $(\pm 2,0)$, then the tangent to this hyperbola at P also passes through the point

(a) $(3\sqrt{2},2\sqrt{3})$ (b) $(2\sqrt{2},3\sqrt{3})$
(c) $(\sqrt{3},\sqrt{2})$ (d) $(-\sqrt{2},-\sqrt{3})$

79. The function $f:R\to\left[-\frac{1}{2},\frac{1}{2}\right]$ defined as $f(x)=\frac{x}{1+x^2}$ is

(a) invertible
(b) injective but not surjective
(c) surjective but not injective
(d) neither injective nor surjective

80. $\lim_{x\to\pi/2}\frac{\cot x-\cos x}{(\pi-2x)^3}$ equals

(a) $\frac{1}{24}$ (b) $\frac{1}{16}$ (c) $\frac{1}{8}$ (d) $\frac{1}{4}$

81. Let $\mathbf{a}=2\hat{\mathbf{i}}+\hat{\mathbf{j}}-2\hat{\mathbf{k}}$, $\mathbf{b}=\hat{\mathbf{i}}+\hat{\mathbf{j}}$ and $\mathbf{c}$ be a vector such that $|\mathbf{c}-\mathbf{a}|=3$, $|(\mathbf{a}\times\mathbf{b})\times\mathbf{c}|=3$ and the angle between $\mathbf{c}$ and $\mathbf{a}\times\mathbf{b}$ is 30°. Then, $\mathbf{a}\cdot\mathbf{c}$ is equal to

(a) $\frac{25}{8}$ (b) 2 (c) 5 (d) $\frac{1}{8}$

82. The normal to the curve $y(x-2)(x-3)=x+6$ at the point, where the curve intersects the Y-axis passes through the point

(a) $\left(-\frac{1}{2},-\frac{1}{2}\right)$ (b) $\left(\frac{1}{2},\frac{1}{2}\right)$
(c) $\left(\frac{1}{2},-\frac{1}{3}\right)$ (d) $\left(\frac{1}{2},\frac{1}{3}\right)$

83. If two different numbers are taken from the set {0, 1, 2, 3, ..., 10}, then the probability that their sum as well as absolute difference are both multiple of 4, is

(a) $\frac{6}{55}$ (b) $\frac{12}{55}$ (c) $\frac{14}{45}$ (d) $\frac{7}{55}$

84. A man X has 7 friends, 4 of them are ladies and 3 are men. His wife Y also has 7 friends, 3 of them are ladies and 4 are men. Assume X and Y have no common friends. Then, the total number of ways in which X and Y together can throw a party inviting 3 ladies and 3 men, so that 3 friends of each of X and Y are in this party, is

(a) 485 (b) 468 (c) 469 (d) 484

85. The value of $({}^{21}C_1-{}^{10}C_1)+({}^{21}C_2-{}^{10}C_2)+({}^{21}C_3-{}^{10}C_3)+({}^{21}C_4-{}^{10}C_4)+\ldots+({}^{21}C_{10}-{}^{10}C_{10})$ is

(a) $2^{21}-2^{11}$ (b) $2^{21}-2^{10}$
(c) $2^{20}-2^{9}$ (d) $2^{20}-2^{10}$

86. A box contains 15 green and 10 yellow balls. If 10 balls are randomly drawn one-by-one with replacement, then the variance of the number of green balls drawn is

(a) $\frac{12}{5}$ (b) 6 (c) 4 (d) $\frac{6}{25}$

87. Let $a, b, c \in R$. If $f(x) = ax^2 + bx + c$ be such that $a + b + c = 3$ and $f(x + y) = f(x) + f(y) + xy$, $\forall\ x, y \in R$, then $\sum_{n=1}^{10} f(n)$ is equal to

(a) 330 (b) 165 (c) 190 (d) 255

88. The radius of a circle having minimum area, which touches the curve $y = 4 - x^2$ and the lines $y = |x|$, is

(a) $2(\sqrt{2} + 1)$ (b) $2(\sqrt{2} - 1)$
(c) $4(\sqrt{2} - 1)$ (d) $4(\sqrt{2} + 1)$

89. For a positive integer n, if the quadratic equation, $x(x + 1) + (x + 1)(x + 2) + \ldots + (\overline{x + n - 1})(x + n) = 10n$ has two consecutive integral solutions, then n is equal to

(a) 12 (b) 9
(c) 10 (d) 11

90. $\int_{\pi/4}^{3\pi/4} \frac{dx}{1 + \cos x}$ is equal to

(a) -2 (b) 2
(c) 4 (d) -1

Answers

Physics

1. (b)	**2.** (a)	**3.** (c)	**4.** (b)	**5.** (b)	**6.** (d)	**7.** (a)	**8.** (a)	**9.** (c)	**10.** (b)
11. (c)	**12.** (d)	**13.** (c)	**14.** (a)	**15.** (a)	**16.** (b)	**17.** (d)	**18.** (d)	**19.** (d)	**20.** (d)
21. (c)	**22.** (b)	**23.** (d)	**24.** (d)	**25.** (d)	**26.** (c)	**27.** (b)	**28.** (a)	**29.** (c)	**30.** (d)

Chemistry

31. (c)	**32.** (c)	**33.** (b)	**34.** (d)	**35.** (b)	**36.** (a)	**37.** (c)	**38.** (a)	**39.** (d)	**40.** (a)
41. (d)	**42.** (d)	**43.** (d)	**44.** (a)	**45.** (b)	**46.** (a)	**47.** (c)	**48.** (a)	**49.** (c)	**50.** (b)
51. (d)	**52.** (b)	**53.** (d)	**54.** (d)	**55.** (d)	**56.** (a)	**57.** (c)	**58.** (b)	**59.** (a)	**60.** (a)

Mathematics

61. (d)	**62.** (a)	**63.** (d)	**64.** (b)	**65.** (a)	**66.** (d)	**67.** (c)	**68.** (d)	**69.** (b)	**70.** (a)
71. (a)	**72.** (c)	**73.** (b)	**74.** (c)	**75.** (b)	**76.** (b)	**77.** (b)	**78.** (b)	**79.** (c)	**80.** (b)
81. (b)	**82.** (b)	**83.** (a)	**84.** (a)	**85.** (d)	**86.** (a)	**87.** (a)	**88.** (c)	**89.** (d)	**90.** (b)

UNSOLVED PAPER 2017

JEE Advanced

Paper 1

Physics

Section 1 (Maximum Marks : 28)

- This section contains **SEVEN** questions.
- Each question has **FOUR** options (a), (b), (c) and (d). **ONE OR MORE THAN ONE** of these four options is (are) correct.
- For each question, darken the bubble(s) corresponding to all the correct option(s) in the ORS.
- For each question, marks will be awarded in one of the following categories:

Full Marks : + 4 If only the bubble(s) corresponding to all the correct option(s) is (are) darkened.

Partial Marks : + 1 For darkening a bubble corresponding **to each correct option**, provided NO incorrect option is darkened

Zero Marks : 0 If none of the bubbles is darkened.

Negative Marks : −2 In all other cases.

1. In the circuit shown, $L = 1\ \mu H$, $C = 1\ \mu F$ and $R = 1\ k\Omega$. They are connected in series with an AC source $V = V_0 \sin \omega t$ as shown. Which of the following options is/are correct?

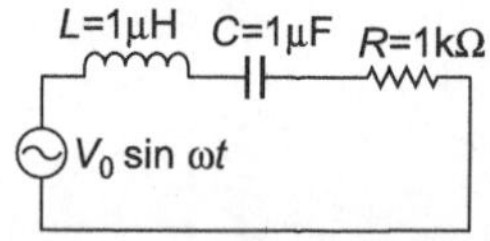

(a) At $\omega \sim 0$ the current flowing through the circuit becomes nearly zero

(b) The frequency at which the current will be in phase with the voltage is independent of R

(c) The current will be in phase with the voltage if $\omega = 10^4$ rads^{-1}

(d) At $\omega >> 10^6$ rads^{-1}, the circuit behaves like a capacitor

2. For an isosceles prism of angle A and refractive index μ, it is found that the angle of minimum deviation $\delta_m = A$. Which of the following options is/are correct?

(a) For the angle of incidence $i_1 = A$, the ray inside the prism is parallel to the base of the prism

(b) At minimum deviation, the incident angle i_1 and the refracting angle r_1 at the first refracting surface are related by $r_1 = \left(\frac{i_i}{2}\right)$

(c) For this prism, the emergent ray at the second surface will be tangential to the surface when the angle of incidence at the first surface is $i_1 = \sin^{-1}\left[\sin A \sqrt{4\cos^2 \frac{A}{2} - 1} - \cos A\right]$

(d) For this prism, the refractive index μ and the angle prism A are related as $A = \frac{1}{2}\cos^{-1}\left(\frac{\mu}{2}\right)$

3. A circular insulated copper wire loop is twisted to form two loops of area A and $2A$ as shown in the figure. At the point of crossing the wires remain electrically insulated from each other. The entire loop lies in the plane (of the paper). A uniform magnetic field **B** points into the plane of the paper. At $t = 0$, the loop starts rotating about the common diameter as axis with a constant angular velocity ω in the magnetic field. Which of the following options is/are correct?

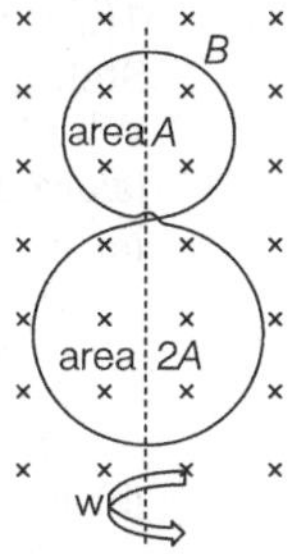

(a) the emf induced in the loop is proportional to the sum of the areas of the two loops
(b) The rate of change of the flux is maximum when the plane of the loops is perpendicular to plane of the paper
(c) The net emf induced due to both the loops is proportional to $\cos \omega t$
(d) The amplitude of the maximum net emf induced due to both the loops is equal to the amplitude of maximum emf induced in the smaller loop alone

4. A flat plane is moving normal to its plane through a gas under the action of a constant force F. The gas is kept at a very low pressure. The speed of the plate v is much less than the average speed u of the gas molecules. Which of the following options is/are true?

(a) At a later time the external force F balances the resistive force
(b) The plate will continue to move with constant non-zero acceleration, at all times
(c) The resistive force experienced by the plate is proportional to v
(d) The pressure difference between the leading and trailing faces of the plate is proportional to uv

5. A block of mass M has a circular cut with a frictionless surface as shown. The block rests on the horizontal frictionless surfaced of a fixed table. Initially the right edge of the block is at $x = 0$, in a coordinate system fixed to the table. A point mass m is released from rest at the topmost point of the path as shown and it slides down. When the mass loses contact with the block, its position is x and the velocity is v. At that instant, which of the following option is/are correct?

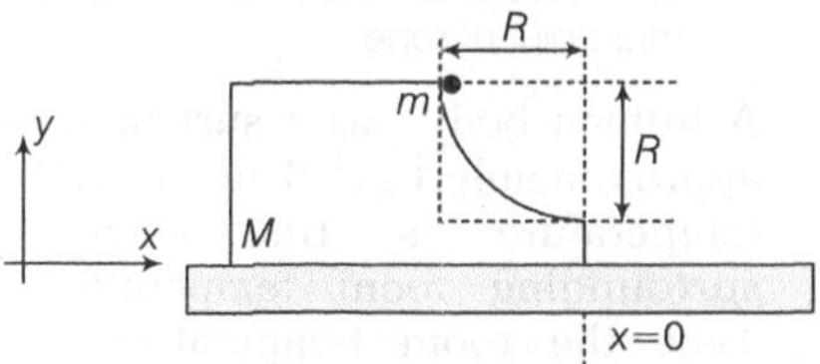

(a) The velocity of the point mass m is $v = \sqrt{\dfrac{2gR}{1 + \dfrac{m}{M}}}$
(b) The x component of displacement of the centre of mass of the block M is $-\dfrac{mR}{M + m}$
(c) The position of the point mass is $x = -\sqrt{2}\dfrac{mR}{M + m}$
(d) The velocity of the block M is $V = -\dfrac{m}{M}\sqrt{2gR}$

6. A block M hangs vertically at the bottom end of a uniform rope of constant mass per unit length. The top end of the rope is attached to a fixed rigid support at O. A transverse wave pulse (Pulse 1) of wavelength λ_0 is produced at point O on the rope. The pulse takes time T_{OA} to reach point A. If the wave pulse of wavelength λ_0 is produced at point A (Pulse 2) without disturbing the position of M it takes time T_{AO} to reach point O. Which of the following options is/are correct?

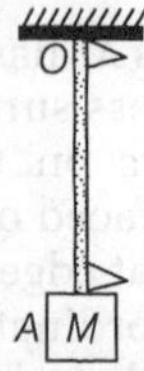

(a) The time $T_{AO} = T_{OA}$
(b) The wavelength of Pulse 1 becomes longer when it reaches point A
(c) The velocity of any pulse along the rope is independent of its frequency and wavelength
(d) The velocities of the two pulses (Pulse 1 and Pulse 2) are the same at the mid-point of rope

7. A human body has a surface area of approximately 1 m^2. The normal body temperature is 10K above the surrounding room temperature T_0. Take the room temperature to be $T_0 = 300$ K. For $T_0 = 300$ K, the value of $\sigma T_0^4 = 460$ Wm^{-2} (where σ is the Stefan Boltzmann constant). Which of the following options is/are correct?

(a) If the body temperature rises significantly, then the peak in the spectrum of electromagnetic radiation emitted by the body would shift to longer wavelengths
(b) If the surrounding temperature reduces by a small amount $\Delta T_0 << T_0$, then to maintain the same body temperature the same (living) human being needs to radiate $\Delta W = 4\sigma T_0^3 \Delta T_0$ more energy per unit time
(c) The amount of energy radiated by the body in 1s is close to 60 J
(d) Reducing the exposed surface area of the body (e.g. by curling up) allows humans to maintain the same body temperature while reducing the energy lost by radiation

Section 2 (Maximum Marks : 15)

- This section contains **FIVE** questions
- The answer to each question is a **SINGLE DIGIT INTEGER** ranging from 0 to 9, both inclusive.
- For each question, darken the bubble corresponding to the correct integer in the ORS.
- For each question, marks will be awarded in one of the following categories:

Full Marks : + 3 If only the bubble corresponding to all the correct answer is darkened.

Zero Marks : 0 In all other cases.

8. An electron in a hydrogen atom undergoes a transition from an orbit with quantum number n_i to another with quantum number n_f. V_i and V_f are respectively the initial and final potential energies of the electron. If $\frac{V_i}{V_f} = 6.25$, then the smallest possible n_f is

9. A drop of liquid of radius $R = 10^{-2}$ m having surface tension $S = \frac{0.1}{4\pi}$ Nm^{-1} divides itself into K identical drops. In this process the total change in the surface energy $\Delta U = 10^{-3}$ J. If $K = 10^{\alpha}$, then the value of α is

10. A stationary source emits sound of frequency $f_0 = 492$ Hz. The sound is reflected by a large car approaching the source with a speed of 2 ms^{-1}. The reflected signal is received by the source and superposed with the original. What will be the beat frequency of the resulting signal in Hz? (Given that the speed of sound in air is 330 ms^{-1} and the car reflects the sound at the frequency it has received).

11. ^{131}I is an isotope of Iodine that β decays to an isotope of Xenon with a half-life of 8 days. A small amount of a serum labelled with ^{131}I is injected into the blood of a person. The activity of the amount of ^{131}I injected was 2.4×10^5 Becquerel (Bq). It is known that the injected serum will get distributed uniformly in the blood stream in less than half an hour. After 11.5 h, 2.5 ml of blood is drawn from the person's body, and gives an activity of 115 Bq. The total volume of blood in the person's body, in litres is approximately (you may use $e^2 \approx 1 + x$ for $|x| << 1$ and $\ln 2 \approx 0.7$).

12. A monochromatic light is travelling in a medium of refractive index $n = 1.6$. It enters a stack of glass layers from the bottom side at an angle $\theta = 30°$. The interfaces of the glass layers are parallel to each other. The refractive indices of different glass layers are monotonically decreasing as $n_m = n - m\,\Delta n$, where n_m is the refractive index of the mth slab and $\Delta n = 0.1$ (see the figure). The ray is refracted out parallel to the interface between the $(m-1)$th and mth slabs from the right side of the stack. What is the value of m ?

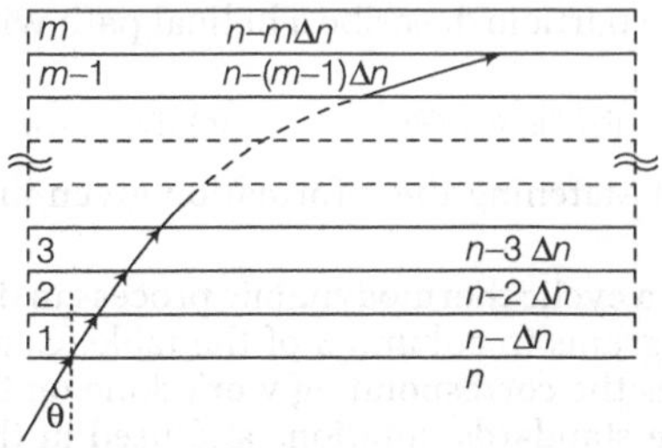

Section 3 (Maximum Marks : 18)

- This section contains **SIX** questions of matching type.
- The section contains **TWO** tables (each having 3 columns and 4 rows)
- Based on each table, there are **THREE** questions.
- Each question has **FOUR** options [a], [b], [c] and [d]. **ONLY ONE** of these four options is correct.
- For each question, darken the bubble corresponding to the correct option in the ORS.
- For each question, Marks will be awarded in one of the following categories:

Full Marks : + 3 If only the bubble corresponding to the correct option is darkened

Zero Marks : 0 If none of the bubbles is darkened

Negative Marks : −1 In all other cases

Directions (Q.Nos. 13-15) Matching the information given in the three columns of the following table.

A charged particle (electron or proton) is introduced at the origin ($x = 0, y = 0, z = 0$) with a given initial velocity **v**. A uniform electric field **E** and a uniform magnetic field **B** exist everywhere. The velocity **v**, electric field **E** and magnetic field **B** are given in columns 1, 2 and 3, respectively. The quantities E_0, B_0 are positive in magnitude.

	Column 1		Column 2		Column 3
(I)	Electron with $\mathbf{v} = 2\frac{E_0}{B_0}\hat{x}$	(i)	$\mathbf{E} = E_0\hat{z}$	(P)	$\mathbf{B} = -B_0\hat{x}$
(II)	Election with $\mathbf{v} = \frac{E_0}{B_0}\hat{y}$	(ii)	$\mathbf{E} = -E_0\hat{y}$	(Q)	$\mathbf{B} = B_0\hat{x}$
(III)	Proton with $\mathbf{v} = 0$	(iii)	$\mathbf{E} = -E_0\hat{x}$	(R)	$\mathbf{B} = B_0\hat{y}$
(IV)	Proton with $\mathbf{v} = 2\frac{E_0}{B_0}\hat{x}$	(iv)	$\mathbf{E} = E_0\hat{x}$	(S)	$\mathbf{B} = B_0\hat{z}$

13. In which case would the particle move in a straight line along the negative direction of Y-axis (i.e. move along $-\hat{y}$)?

(a) (IV) (ii) (S) (b) (II) (iii) (Q) (c) (III), (ii) (R) (d) (III) (ii) (P)

14. In which case will the particle move in a straight line with constant velocity?

(a) (II) (iii) (S) (b) (III) (iii) (P) (c) (IV) (i) (S) (d) (III) (ii) (R)

15. In which case will the particle describe a helical path with axis along the positive z-direction?

(a) (II) (ii) (R) (b) (III) (iii) (P) (c) (IV) (i) (S) (d) (IV) (ii) (R)

Directions (Q.Nos. 16-18) Matching the information given in the three columns of the following table.

An ideal gas is undergoing a cyclic thermodynamic process in different ways as shown in the corresponding p-V diagrams in column 3 of the table. Consider only the path from state 1 to state 2. W denotes the corresponding work done on the system. The equations and plots in the table have standards notations and used in thermodynamic processes. Here γ is the ratio of heat capacities at constant pressure and constant volume. The number of moles in the gas is n.

	Column 1		Column 2		Column 3
(I)	$W_{1\to 2} = \frac{1}{\gamma - 1}(p_2V_2 - p_1V_1)$	(i)	Isothermal	(P)	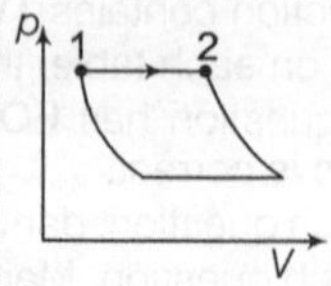
(II)	$W_{1\to 2} = -pV_2 + pV_1$	(ii)	Isochoric	(Q)	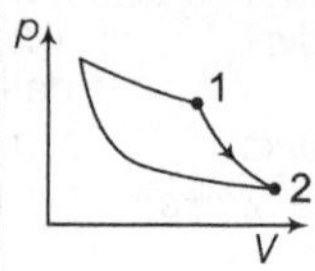
(III)	$W_{1\to 2} = 0$	(iii)	Isobaric	(R)	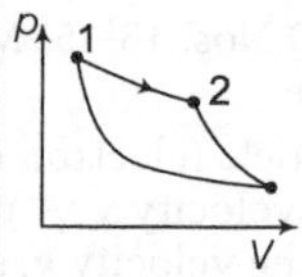
(IV)	$W_{1\to 2} = -nRT\ln\left(\frac{V_2}{V_1}\right)$	(iv)	Adiabatic	(S)	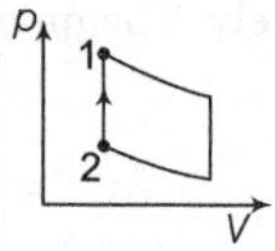

16. Which one of the following options correctly represents a thermodynamic process that is used as a correction in the determination of the speed of sound in an ideal gas?

(a) (IV) (ii) (R)
(b) (I) (ii) (Q)
(c) (I), (iv) (Q)
(d) (III) (iv) (R)

17. Which of the following options is the only correct representation of a process in which $\Delta U = \Delta Q - p\Delta V$?

(a) (II) (iii) (S)
(b) (II) (iii) (P)
(c) (III) (iii) (P)
(d) (II) (iv) (R)

18. Which one of the following options is the correct combination?

(a) (II) (iv) (P)
(b) (III) (ii) (S)
(c) (II) (iv) (R)
(d) (IV) (ii) (S)

Chemistry

Section 1 (Maximum Marks : 28)

- This section contains **SEVEN** questions.
- Each question has **FOUR** options (a), (b), (c) and (d). **ONE OR MORE THAN ONE** of these four option(s) is(are) correct.
- For each question, darken the bubble(s) corresponding to all the correct option(s) in the ORS.
- For each question, marks will be awarded in **one of the following categories** :
  Full Marks : + 4 If only the bubble(s) corresponding to all the correct option(s) is(are) darkened.
  Partial Marks : +1 For darkening a bubble corresponding to each correct option, provided NO incorrect option is darkened.
  Zero Marks : 0 If none of the bubbles is darkened.
  Negative Marks : −2 In all other cases.
- For example, if (a), (c) and (d) are all the correct options for a question, darkening all these three will get + 4 marks; darkening only (a) and (d) will get +2 marks and darkening (a) and (b) will get − 2 marks, as a wrong option is also darkened.

1. An ideal gas is expanded form (p_1, V_1, T_1) to (p_2, V_2, T_2) under different conditions. The correct statement(s) among the following is (are)

(a) The work done by the gas is less when it is expanded reversibly from V_1 to V_2 under adiabatic conditions as compared to that when expanded reversibly form V_1 to V_2 under isothermal conditions.

(b) The change in internal energy of the gas is (i) zero, if it is expanded reversibly with $T_1 = T_2$, and (ii) positive, if it is expanded reversibly under adiabatic conditions with $T_1 \neq T_2$

(c) If the expansion is carried out freely, it is simultaneously both isothermal as well as adiabatic

(d) The work done on the gas is maximum when it is compressed irrversibly from (p_2, V_2) to (p_1, V_1) against constant pressure p_1

2. The IUPAC name(s) of the following compound is (are)

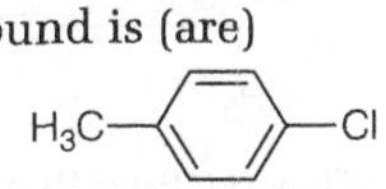

(a) 4-methylchlorobenzene
(b) 4-chlorotoluene
(c) 1-chloro-4-methylbenzene
(d) 1-methyl-4-chlorobenzene

3. For a solution formed by mixing liquids L and M, the vapour pressure of L plotted against the mole fraction of M in solution is shown in the following figure. Here x_L and x_M represent mole fractions of L and M,

respectively, in the solution. The correct statement(s) applicable to this system is (are)

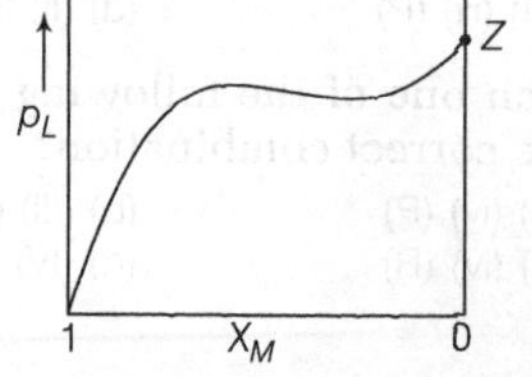

(a) The point Z represents vapour pressure of pure liquid M and Raoult's law is obeyed from $x_L = 0$ to $x_L = 1$

(b) Attractive intermolecular interactions between L - L in pure liquid L and M - M in pure liquid M are stronger than those between L - M when mixed in solution

(c) The point Z represents vapour pressure of pure liquid M and Raoult's law is obeyed when $x_L \to 0$

(d) The point Z represents vapour pressure of pure liquid L and Raoult's law is obeyed when $x_L \to 1$

4. The correct statement(s) for the following addition reactions is (are)

(i)
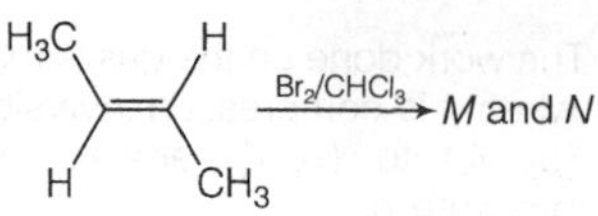

(i)
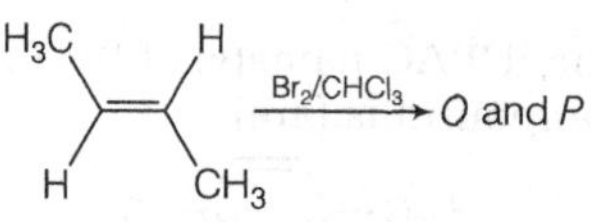

(a) (M and O) and (N and P) are two pairs of enantiomers

(b) Bromination proceeds through *trans*-addition in both the reactions

(c) O and P are identical molecules

(d) (M and O) and (N and P) two pairs of diastereomers

5. Addition of excess aqueous ammonia to a pink coloured aqueous solution of $MCl_2 \cdot 6H_2O(X)$ and NH_4Cl gives an octahedral complex Y in the presence of air. In aqueous solution, complex Y behaves as 1 : 3 electrolyte. The reaction of X with excess HCl at room temperature results in the formation of a blue colured complex Z. The calculated spin only magnetic moment of X and Z is 3.87 B.M., whersas it is zero for complex Y.

Among the following options, which statement(s) is (are) correct?

(a) The hybridisation of the central metal ion in Y is d^2sp^3

(b) Addition of silver nitrate to Y given only two equivalents of silver chloride

(c) When X and Y are in equilibrium at 0°C, the colour of the solution is pink

(d) Z is a tetrahedral complex

6. The correct statement(s) about the oxoacids, $HClO_4$ and $HClO$, is (are)

(a) The central atom in both $HClO_4$ and $HClO$ is sp^3-hybridised

(b) $HClO_4$ is formed in the reaction between Cl_2 and H_2O

(c) The conjugate base of $HClO_4$ is weaker base than H_2O

(d) $HClO_4$ is more acidic than $HClO$ because of the resonance stabilisation of its anion

7. The colour of the X_2 molecules of group 17 elements changes gradually from yellow to violet down the group. This is due to

(a) decrease in $\pi^* - \sigma^*$ gap down the group

(b) decrease in ionisation energy down the group

(c) the physical state of X_2 at room temperature changes from gas to solid down the group

(d) decreases in HOMO-LUMO gap down the group

Section 2 (Maximum Marks : 15)

- This section contains **FIVE** questions.
- The answer to each question is a **SINGLE DIGIT INTEGER** ranging from 0 to 9, both inclusive.
- For each question, darken the bubble corresponding to the correct integer in the ORS.
- For each question, marks will be awarded in one of th following categories :

Full Marks : + 3 If only the bubble corresponding to all the correct answer is darkened.

Zero Marks : 0 In all other cases.

8. The sum of the number of lone pairs of electrons on each central atom in the following species is

$[TeBr_6]^{2-}$, $[BrF_2]^+$, SNF_3 and $[XeF_3]^-$

(Atomic numbers : N = 7, F = 9, S = 16, Br = 35, Te = 52, Xe = 54)

9. Among the following, the number of aromatic compound(s) is

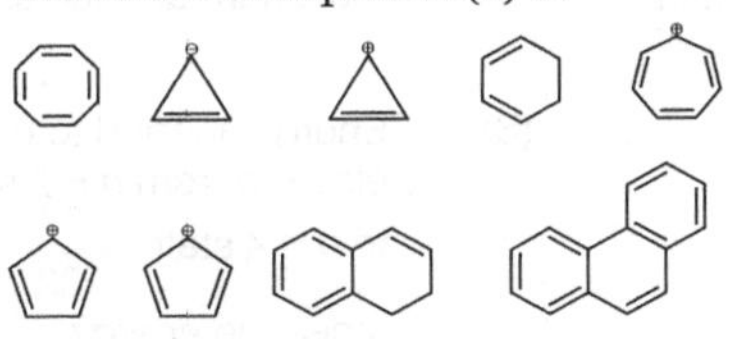

10. AmongH_2, He_2^+, Li_2, Be_2, B_2, C_2, N_2, and $F_2O_2^-$, the number of diamagnetic species is

(Atomic numbers : H = 1, He = 2, Li = 3, Be = 4, B = 5, C = 6, N = 7, O = 8, F = 9)

11. A crystalline solid of a pure substance has a face-centred cubic structure with a cell edge of 400 pm. If the density of the substance in the crystal is 8 g cm^{-3}, then the number of atoms present in 256 g of the crystal is $N \times 10^{24}$. The value of N is

12. The conductance of a 0.0015 M aqueous solution of a weak monobasic acid was determined by using a conductivity cell consisting of platinised Pt electrodes. The distance between the electrodes is 120 cm with an area of cross section of 1 cm^2. The conductance of this solution was found to be 5×10^{-7} S. The pH of the solution is 4. The value of limiting molar conductivity (Λ_m°) of this weak monobasic acid in aqueous solution is $Z \times 10^2$ S cm^{-1} mol^{-1}. The value of Z is

Section 3 (Maximum Marks : 18)

- This section contains **SIX** question of matching type.
- This section contains **TWO** tables (each having 3 columns and 4 rows)
- Based on each table, there are **THREE** questions.
- Each question has **FOUR** options (a), (b), (c) and (d). **ONLY ONE** of these options is correct.
- For each question, darken the bubble corresponding to the correct option is the ORS.
- For each question, marks will be awarded in **one of the following categories**:

Full Marks : + 3 If only the bubble corresponding to the correct option is darkened

Zero Marks : 0 If none of the bubbles is darkened

Negative Marks : –1 In all othe cases

Answer Q. 13, Q. 14 and Q. 15 by appropriately matching the information given in the three columns of the following table.

The wave function, ψ_n, l, m_l is a mathematical function whose value depends upon spherical polar coordinates (r, θ, ϕ) of the electron and characterised by the quantum number n, l and m_l. Here r is distance from nucleus, θ is colatitude and ϕ is azimuth. In the mathematical functions given in the Table, Z is atomic number and a_0 is Bohr radius.

Column 1		Column 2		Column 3	
(I)	1s-orbital	(i)	$\psi_{n,l,m_l} \propto \left(\frac{Z}{a_0}\right)^{\frac{3}{2}} e^{-\left(\frac{Zr}{a_0}\right)}$	(P)	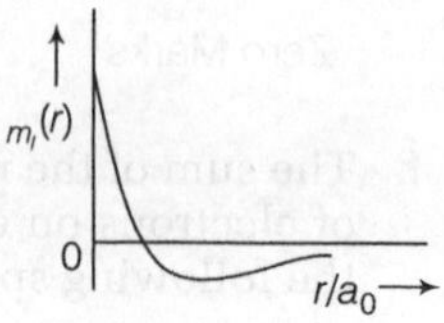
(II)	2s-orbital	(ii)	One radial node	(Q)	Probability density at nucleus $\propto \frac{1}{a_0^3}$
(III)	$2p_z$-orbital	(iii)	$\psi_{n,l,m_l} \propto \left(\frac{Z}{a_0}\right)^{\frac{5}{2}} re^{-\left(\frac{Zr}{a_0}\right)} \cos\theta$	(R)	Probability density is maximum at nucleus
(IV)	$3d_z^2$-orbital	(iv)	*xy*-plane is a nodal plane	(S)	Energy needed to excite electron from $n = 2$ state to $n = 4$ state is $\frac{27}{32}$ times the energy needed to excite electron from $n = 2$ state to $n = 6$ state

13. For He^+ ion, the only **INCORRECT** combination is

(a) (I) (i) (S) (b) (II) (ii) (Q) (c) (I) (iii) (R) (d) (I) (i) (R)

14. For the given orbital in Column 1, the Only **CORRECT** combination for any hydrogen-like species is

(a) (II) (ii) (P) (b) (I) (ii) (S) (c) (IV) (iv) (R) (d) (III) (iii) (P)

15. For hydrogen atom, the only **CORRECT** combination is

(a) (I) (i) (P) (b) (I) (iv) (R) (c) (II) (i) (Q) (d) (I) (i) (S)

Answer Q. 16, Q. 17 and Q. 18 by appropriately matching the information given in the three columns of the following table.

Column 1, 2 and 3 contain starting materials, reaction conditions, and type of reactions, respectively.

Column 1	Column 2	Column 3
(I) Toluene	(i) $NaOH/Br_2$	(P) Condensation
(II) Acetophenone	(ii) $Br_2/h\nu$	(Q) Carboxylation
(III) Benzaldehyde	(iii) $(CH_3CO)_2O/CH_3COOK$	(R) Substitution
(IV) Phenol	(iv) $NaOH/CO_2$	(S) Haloform

16. The only **CORRECT** combination in which the reaction proceeds through radical mechanism is

(a) (IV) (i) (Q) (b) (III) (ii) (P)
(c) (II) (iii) (R) (d) (I) (ii) (R)

17. For the synthesis of benzoic acid, the only CORRECT combination is

(a) (II) (i) (S) (b) (I) (iv) (Q)
(c) (IV) (ii) (P) (d) (III) (iv) (R)

18. The only **CORRECT** combination that gives two different carboxylic acids is

(a) (IV) (iii) (Q) (b) (II) (iv) (R)
(c) (I) (i) (S) (d) (III) (iii) (P)

Mathematics

Section 1 (Maximum Marks : 28)

- This section contains **SEVEN** questions.
- Each question has **FOUR** options (a), (b), (c) and (d). **ONE OR MORE THAN ONE** of these four option(s) is(are) correct.
- For each question, darken the bubble(s) corresponding to all the correct option(s) in the ORS.
- For each question, marks will be awarded in one of the following categories:

Full Marks : + 4 If only the bubble(s) corresponding to all the correct option(s) is(are) darkened.
Partial Marks : +1 For darkening a bubble corresponding to each correct option, provided NO incorrect option is darkened.
Zero Marks : 0 If none of the bubbles is darkened.
Negative Marks : −2 In all other cases.

- For example, if (a), (c) and (d) are all the correct options for a question, darkening all these three will get + 4 marks; darkening only (a) and (d) will get +2 marks and darkening (a) and (b) will get − 2 marks, as a wrong option is also darkened.

1. Let X and Y be two events such that $P(X) = \frac{1}{3}, P(X|Y) = \frac{1}{2}$ and $P(Y|X) = \frac{2}{5}$. Then

(a) $P(Y) = \frac{4}{15}$ (b) $P(X'|Y) = \frac{1}{2}$
(c) $P(X \cup Y) = \frac{2}{5}$ (d) $P(X \cap Y) = \frac{1}{5}$

2. Let $f : R \to (0, 1)$ be a continuous function. Then, which of the following function(s) has (have) the value zero at some point in the interval $(0, 1)$?

(a) $e^x - \int_0^x f(t) \sin t \, dt$
(b) $f(x) + \int_0^{\frac{\pi}{2}} f(t) \sin t \, dt$
(c) $x - \int_0^{\frac{\pi}{2} - x} f(t) \cos t \, dt$
(d) $x^9 - f(x)$

3. Let a, b, x and y be real numbers such that $a - b = 1$ and $y \neq 0$. If the complex number $z = x + iy$ satisfies $\text{Im}\left(\frac{az + b}{z + 1}\right) = y$, then which of the following is(are) possible value(s) of x?

(a) $1 - \sqrt{1 + y^2}$
(b) $-1 - \sqrt{1 - y^2}$
(c) $1 + \sqrt{1 + y^2}$
(d) $-1 + \sqrt{1 - y^2}$

4. If $2x - y + 1 = 0$ is a tangent to the hyperbola $\frac{x^2}{a^2} - \frac{y^2}{16} = 1$ then which of the following CANNOT be sides of a right angled triangle?

(a) $a, 4, 1$ (b) $2a, 4, 1$
(c) $a, 4, 2$ (d) $2a, 8, 1$

5. Let $[x]$ be the greatest integer less than or equals to x. Then, at which of the following point(s) the function $f(x) = x\cos(\pi(x+[x]))$ is discontinuous ?

(a) $x = -1$ (b) $x = 1$
(c) $x = 0$ (d) $x = 2$

6. Which of the following is(are) NOT the square of a 3×3 matrix with real entries?

(a) $\begin{bmatrix} 1 & 0 & 0 \\ 0 & 1 & 0 \\ 0 & 0 & -1 \end{bmatrix}$ (b) $\begin{bmatrix} 1 & 0 & 0 \\ 0 & -1 & 0 \\ 0 & 0 & -1 \end{bmatrix}$

(c) $\begin{bmatrix} -1 & 0 & 0 \\ 0 & -1 & 0 \\ 0 & 0 & -1 \end{bmatrix}$ (d) $\begin{bmatrix} 1 & 0 & 0 \\ 0 & 1 & 0 \\ 0 & 0 & 1 \end{bmatrix}$

7. If a chord, which is not a tangent, of the parabola $y^2 = 16x$ has the equation $2x + y = p$, and mid-point (h, k), then which of the following is(are) possible value(s) of p, h and k?

(a) $p = -1, h = 1, k = -3$
(b) $p = 2, h = 3, k = -4$
(c) $p = -2, h = 2, k = -4$
(d) $p = 5, h = 4, k = -3$

Section 2 (Maximum Marks : 15)

- This section contains **FIVE** questions.
- The answer to each question is a **SINGLE DIGIT INTEGER** ranging from 0 to 9, both inclusive.
- For each question, darken the bubble corresponding to the correct integer in the ORS.
- For each question, marks will be awarded in one of the following categories:
  Full Marks : +3 If only the bubble corresponding to the correct answer is darkened.
  Zero Marks : 0 In all other cases.

8. For a real number α, if the system

$$\begin{bmatrix} 1 & \alpha & \alpha^2 \\ \alpha & 1 & \alpha \\ \alpha^2 & \alpha & 1 \end{bmatrix} \begin{bmatrix} x \\ y \\ z \end{bmatrix} = \begin{bmatrix} 1 \\ -1 \\ 1 \end{bmatrix}$$

of linear equations, has infinitely many solutions, then $1 + \alpha + \alpha^2 =$

9. The sides of a right angled triangle are in arithmetic progression. If the triangle has area 24, then what is the length of its smallest side?

10. Let $f : R \to R$ be a differentiable function such that $f(0) = 0, f\left(\frac{\pi}{2}\right) = 3$ and $f'(0) = 1$.

If $g(x) = \int_x^{\frac{\pi}{2}} [f'(t) \operatorname{cosec} t - \cot t \operatorname{cosec} t \, f(t)] \, dt$ for $x \in \left(0, \frac{\pi}{2}\right]$, then $\lim_{x \to 0} g(x) =$

11. For how many values of p, the circle $x^2 + y^2 + 2x + 4y - p = 0$ and the coordinate axes have exactly three common points?

12. Words of length 10 are formed using the letters A, B, C, D, E, F, G, H, I, J. Let x be the number of such words where no letter is repeated; and let y be the number of such words where exactly one letter is repeated twice and no other letter is repeated. Then, $\frac{y}{9x} =$

Section 3 (Maximum Marks : 18)

- This section contains **SIX** questions of matching type.
- This section contains **TWO** tables (each having 3 columns and 4 rows).
- Based on each table, there are **THREE** questions.
- Each question has **FOUR** options (a), (b), (c) and (d). **ONLY ONE** of these four options is correct.
- For each question, darken the bubble corresponding to the correct option in the ORS.
- For each question, marks will be awarded in one of the following categories:

Full Marks : + 3 If only the bubble corresponding to the correct option is darkened.
Zero Marks : 0 If none of the bubbles is darkened.
Negative Marks : – 1 In all other cases.

Directions (Q.Nos. 13-15) by appropriately matching the information given in the three columns of the following table.

Columns 1, 2 and 3 contain conics, equations of tangents to the conics and points of contact, respectively.

	Column-1		Column-2		Column-3
(I)	$x^2 + y^2 = a^2$	(i)	$my = m^2x + a$	(P)	$\left(\frac{a}{m^2}, \frac{2a}{m}\right)$
(II)	$x^2 + a^2y^2 = a^2]$	(ii)	$y = mx + a\sqrt{m^2 + 1}$	(Q)	$\left(\frac{-ma}{\sqrt{m^2+1}}, \frac{a}{\sqrt{m^2+1}}\right)$
(III)	$y^2 = 4ax$	(iii)	$y = mx + \sqrt{a^2m^2 - 1}$	(R)	$\left(\frac{-a^2m}{\sqrt{a^2m^2+1}}, \frac{1}{\sqrt{a^2m^2+1}}\right)$
(IV)	$x^2 - a^2y^2 = a^2$	(iv)	$y = mx + \sqrt{a^2m^2 + 1}$	(S)	$\left(\frac{-a^2m}{\sqrt{a^2m^2-1}}, \frac{-1}{\sqrt{a^2m^2-1}}\right)$

13. For $a = \sqrt{2}$, if a tangent is drawn to a suitable conic (Column 1) at the point of contact $(-1, 1)$, then which of the following options is the only CORRECT combination for obtaining its equation?

(a) (I) (ii) (Q) (b) (I) (i) (P)
(c) (III) (i) (P) (d) (II) (ii) (Q)

14. The tangent to a suitable conic (Column 1) at $\left(\sqrt{3}, \frac{1}{2}\right)$ is found to be $\sqrt{3}x + 2y = 4$, then which of the following options is the only CORRECT combination?

(a) (IV) (iv) (S) (b) (II) (iv) (R)
(c) (IV) (iii) (S) (d) (II) (iii) (R)

15. If a tangent to a suitable conic (Column 1) is found to be $y = x + 8$ and its point of contact is (8, 16), then which of the following options is the only CORRECT combination?

(a) (III) (i) (P) (b) (I) (ii) (Q)
(c) (II) (iv) (R) (d) (III) (ii) (Q)

Directions (Q.Nos. 16-18) by appropriately matching the information given in the three columns of the following table.

Let $f(x) = x + \log_e x - x \log_e x, x \in (0, \infty)$

Column 1 contains information about zeros of $f(x)$, $f'(x)$ and $f''(x)$.

Column 2 contains information about the limiting behaviour of $f(x)$, $f'(x)$ and $f''(x)$ at infinity.

Column 3 contains information about increasing/decreasing nature of $f(x)$ and $f'(x)$.

	Column-1		Column-2		Column-3
(I)	$f(x) = 0$ for some $x \in (1, e^2)$	(i)	$\lim_{x \to \infty} f(x) = 0$	(P)	f is increasing in (0, 1)
(II)	$f'(x) = 0$ for some $x \in (1, e)$	(ii)	$\lim_{x \to \infty} f(x) = -\infty$	(Q)	f is decreasing in (e, e^2)
(III)	$f'(x) = 0$ for some $x \in (0, 1)$	(iii)	$\lim_{x \to \infty} f'(x) = -\infty$	(R)	f' is increasing in (0, 1)
(IV)	$f''(x) = 0$ for some $x \in (1, e)$	(iv)	$\lim_{x \to \infty} f''(x) = 0$	(S)	f' is decreasing in (e, e^2)

16. Which of the following options is the only INCORRECT combination?

(a) (I) (iii) (P)
(b) (II) (iv) (Q)
(c) (II) (iii) (P)
(d) (III) (i) (R)

17. Which of the following options is the only CORRECT combination?

(a) (I) (ii) (R)
(b) (III) (iv) (P)
(c) (II) (iii) (S)
(d) (IV) (i) (S)

18. Which of the following options is the only CORRECT combination?

(a) (III) (iii) (R)
(b) (IV) (iv) (S)
(c) (II) (ii) (Q)
(d) (I) (i) (P)

Paper 2

Physics

Section 1 (Maximum Marks : 21)

- This section contains **SEVEN** questions.
- Each question has **FOUR** options (a), (b), (c) and (d). **ONLY ONE** of these four options is correct.
- For each question, darken the bubble corresponding to the correct option in the ORS.
- For each question, marks will be awarded in one of the following categories.

Full Marks : + 3 If only the bubble corresponding to the correct option is darkened.

Zero Marks : 0 If none of the bubbles is darkened.

Negative Marks : −1 In all other cases

1. Consider regular polygons with number of sides $n = 3, 4, 5$ ...... as shown in the figure. The centre of mass of all the polygons is at height h from the ground. They roll on a horizontal surface about the leading vertex without slipping and sliding as depicted. The maximum increase in height of the locus of the centre of mass for each each polygon is Δ. Then, Δ depends on n and h as

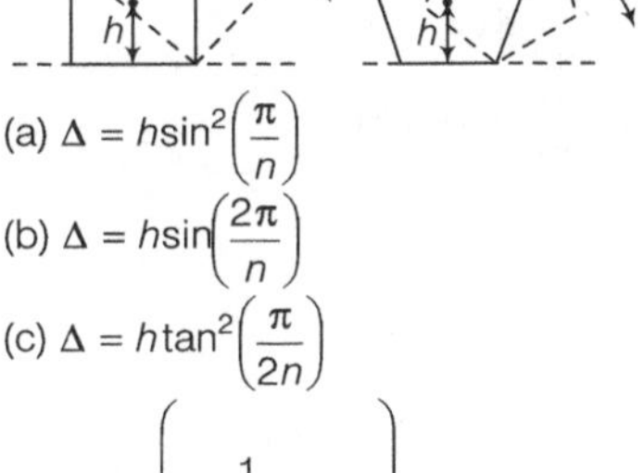

(a) $\Delta = h\sin^2\left(\frac{\pi}{n}\right)$

(b) $\Delta = h\sin\left(\frac{2\pi}{n}\right)$

(c) $\Delta = h\tan^2\left(\frac{\pi}{2n}\right)$

(d) $\Delta = h\left(\frac{1}{\cos\left(\frac{\pi}{n}\right)} - 1\right)$

2. Consider an expanding sphere of instantaneous radius R whose total mass remains constant. The expansion is such that the instantaneous density ρ remains uniform throughout the volume. The rate of fractional change in density $\left(\frac{1}{\rho}\frac{d\rho}{dt}\right)$ is constant. The velocity v of any point of the surface of the expanding sphere is proportional to

(a) R (b) $\frac{1}{R}$

(c) R^3 (d) $R^{\frac{2}{3}}$

3. A photoelectric material having work-function ϕ_0 is illuminated with light of wavelength $\lambda \left(\lambda < \frac{hc}{\phi_0}\right)$. The fastest photoelectron has a de-Broglie wavelength λ_d. A change in wavelength of the incident light by $\Delta\lambda$ results in a change $\Delta\lambda_d$ in λ_d. Then, the ratio $\frac{\Delta\lambda_d}{\Delta\lambda}$ is proportional to

(a) $\frac{\lambda_d^2}{\lambda^2}$ (b) $\frac{\lambda_d}{\lambda}$

(c) $\frac{\lambda_d^3}{\lambda}$ (d) $\frac{\lambda_d^3}{\lambda^2}$

4. Three vectors **P**, **Q** and **R** are shown in the figure. Let S be any point on the vector **R**. The distance between the points P and S is $b[\mathbf{R}]$. The general relation among vectors **P**, **Q** and **S** is

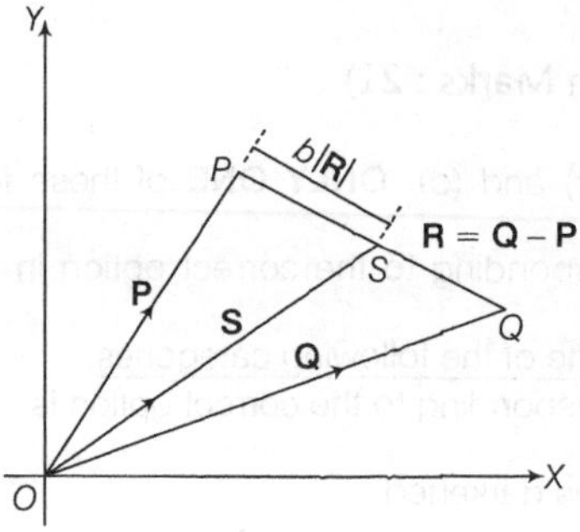

(a) $\mathbf{S} = (1 - b^2)\mathbf{P} + b\mathbf{Q}$
(b) $\mathbf{S} = (b - 1)\mathbf{P} + b\mathbf{Q}$
(c) $\mathbf{S} = (1 - b)\mathbf{P} + b\mathbf{Q}$
(d) $\mathbf{S} = (1 - b)\mathbf{P} + b^2\mathbf{Q}$

5. A person measures the depth of a well by measuring the time interval between dropping a stone and receiving the sound of impact with the bottom of the well. The error in his measurement of time is $\delta T = 0.01$s and he measures the depth of the well to be $L = 20$ m. Take the acceleration due to gravity $g = 10$ ms^{-2} and the velocity of sound is 300 ms^{-1}. Then the fractional error in the measurement, $\frac{\delta L}{L}$, is closest to

(a) 1% (b) 5%
(c) 3% (d) 0.2%

6. A rocket is launched normal to the surface of the Earth, away from the Sun, along the line joining the Sun and the Earth. The Sun is 3×10^5 times heavier than the Earth and is at a distance 2.5×10^4 times larger than the radius of Earth. The escape velocity from Earth's gravitational field is $v_e = 11.2$ km s^{-1}. The minimum initial velocity (v_s) required for the rocket to be able to leave the Sun-Earth system is closest to

(Ignore the rotation and revolution of the Earth and the presence of any other planet)

(a) $v_s = 72$ km s^{-1}
(b) $v_s = 22$ km s^{-1}
(c) $v_s = 42$ km s^{-1}
(d) $v_s = 62$ km s^{-1}

7. A symmetric star shaped conducting wire loop is carrying a steady state current I as shown in the figure. The distance between the diametrically opposite vertices of the star is $4a$. The magnitude of the magnetic field at the center of the loop is

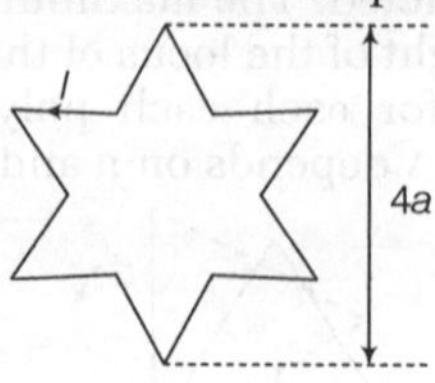

(a) $\frac{\mu_0 I}{4\pi a} 6[\sqrt{3} - 1]$
(b) $\frac{\mu_0 I}{4\pi a} 6[\sqrt{3} + 1]$
(c) $\frac{\mu_0 I}{4\pi a} 3[\sqrt{3} - 1]$
(d) $\frac{\mu_0 I}{4\pi a} 3[2 - \sqrt{3}]$

Section 2 (Maximum Marks : 28)

- This section contains **SEVEN** questions.
- Each question has **FOUR** options (a), (b) (c) and (d). **ONE OR MORE THAN ONE** of these four option(s) is (are) correct.
- For each question, darken the bubble(s) corresponding to all the correct option(s) in the ORS.
- For each questions, marks will be awarded in one of the following categories :

Full Marks : + 4 If only the bubble(s) corresponding to all the correct option(s) is (are) darkened.

Partial Marks : + 1 For darkening a bubble corresponding **to each correct option,** provided NO incorrect option is darkened.

Zero Marks : 0 If none of the bubbles is darkened.

Negative Marks : −2 In all other cases.

- For example, if (a), (c) and (d) are all the correct options for a question, darkening all these three will get in + 4 marks ; darkening only (a) and (d) will get + 2 marks and darkening (a) and (b) will result in −2 marks, as a wrong option is also darkened.

8. A wheel of radius R and mass M is placed at the bottom of a fixed step of height R as shown in the figure. A constant force is continuously applied on the surface of the wheel so that it just climbs the step without slipping. Consider the torque τ about an axis normal to the plane of the paper passing through the point Q. Which of the following options is/are correct?

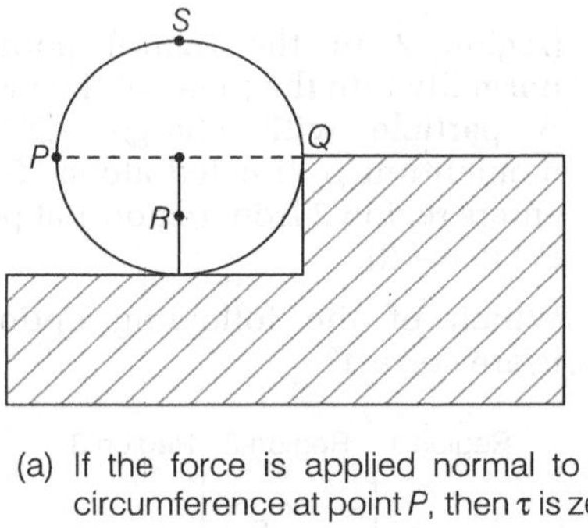

(a) If the force is applied normal to the circumference at point P, then τ is zero

(b) If the force is applied tangentially at point S, then $\tau \neq 0$ but the wheel never climbs the step

(c) If the force is applied at point P tangentially, then τ decreases continuously as the wheel climbs

(d) If the force is applied normal to the circumference at point X, then τ is constant

9. Two coherent monochromatic point sources S_1 and S_2 of wavelength $\lambda = 600$ nm are placed symmetrically on either side of the centre of the circle as shown. The sources are separated by a distance $d = 1.8$ mm. This arrangement produces interference fringes visible as alternate bright and dark spots on the circumference of the circle. The angular separation between two consecutive bright spots is $\Delta\theta$. Which of the following options is/are correct?

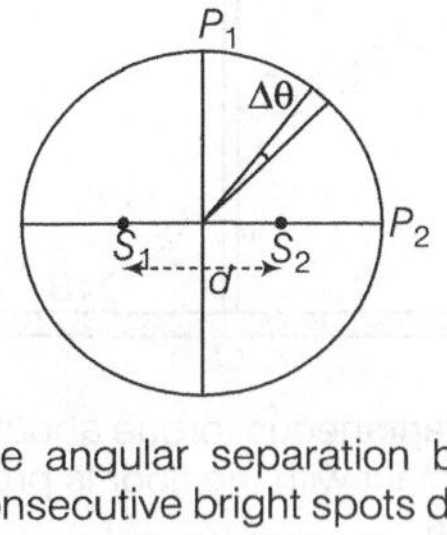

(a) The angular separation between two consecutive bright spots decreases as we move from P_1 to P_2 along the first quadrant

(b) A dark spot will be formed at the point P_2

(c) The total number of fringes produced between P_1 and P_2 in the first quadrant is close to 3000

(d) At P_2 the order of the fringe will be maximum

10. A point charge $+Q$ is placed just outside an imaginary hemispherical surface of radius R as shown in the figure. Which of the following statements is/are correct?

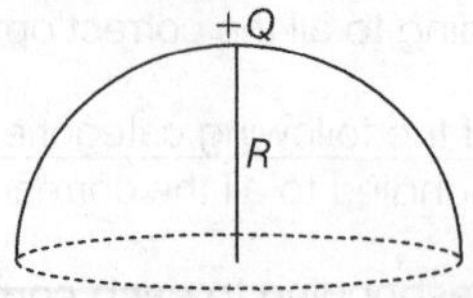

(a) The electric flux passing through the curved surface of the hemisphere is $-\frac{Q}{2\varepsilon_0}\left(1-\frac{1}{\sqrt{2}}\right)$.

(b) The component of the electric field normal to the flat surface is constant over the surface

(c) Total flux through the curved and the flat surfaces is $\frac{Q}{\varepsilon_0}$

(d) The circumference of the flat surface is an equipotential

11. A rigid uniform bar AB of length L is slipping from its vertical position on a frictionless floor (as shown in the figure). At some instant of time, the angle made by the bar with the vertical is θ. Which of the following statements about its motion is/are correct?

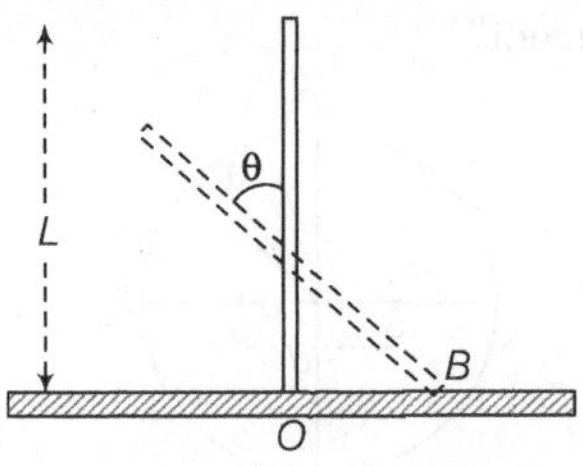

(a) Instantaneous torque about the point in contact with the floor is proportional to $\sin\theta$

(b) The trajectory of the point A is parabola

(c) The mid-point of the bar will fall vertically downward

(d) When the bar makes an angle θ with the vertical, the displacement of its mid-point from the initial position is proportional to $(1-\cos\theta)$

12. A source of constant voltage V is connected to a resistance R and two ideal inductors L_1 and L_2 through a switch S as shown. There is no mutual inductance between the two inductors. The switch S is initially open. At $t = 0$, the switch is closed and current begins to flow. Which of the following options is/are correct?

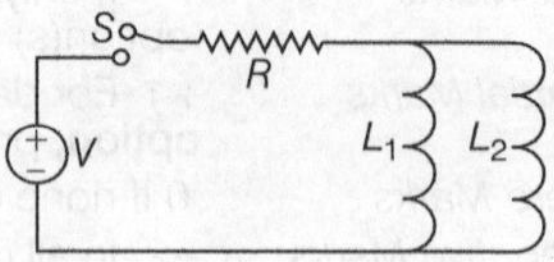

(a) After a long time, the current through L_1 will be $\frac{V}{R}\frac{L_2}{L_1+L_2}$

(b) After a long time, the current through L_2 will be $\frac{V}{R}\frac{L_1}{L_1+L_2}$

(c) The ratio of the currents through L_1 and L_2 is fixed at all times $(t > 0)$

(d) At $t = 0$, the current through the resistance R is $\frac{V}{R}$

13. A uniform magnetic field B exists in the region between $x = 0$ and $x = \frac{3R}{2}$ (region 2 in the figure) pointing normally into the plane of the paper. A particle with charge $+Q$ and momentum p directed along X-axis enters region 2 from region 1 at point P_1 $(y = -R)$.

Which of the following option(s) is/are correct?

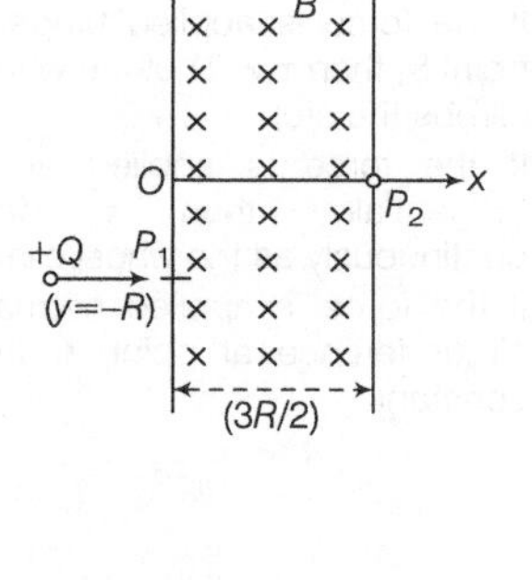

(a) When the particle re-enters region 1 through the longest possible path in region 2, the magnitude of the change in its linear momentum between point P_1 and the farthest point from Y-axis is $\frac{p}{\sqrt{2}}$

(b) For $B = \frac{8}{13}\frac{p}{QR}$, the particle will enter region 3 through the point P_2 on X-axis

(c) For $B > \frac{2}{3}\frac{p}{QR}$, the particle will re-enter region 1

(d) For a fixed B, particles of same charge Q and same velocity v, the distance between the point P_1 and the point of re-entry into region 1 is inversely proportional to the mass of the particle

14. The instantaneous voltages at three terminals marked X, Y and Z are given by $V_X = V_0 \sin \omega t$,

$$V_Y = V_0 \sin\left(\omega t + \frac{2\pi}{3}\right) \text{ and}$$

$$V_Z = V_0 \sin\left(\omega t + \frac{4\pi}{3}\right).$$

An ideal voltmeter is configured to read rms value of the potential difference between its terminals. It is connected between points X and Y and then between Y and Z. The reading(s) of the voltmeter will be

(a) $V_{YZ}^{rms} = V_0\sqrt{\frac{1}{2}}$

(b) $V_{XY}^{rms} = V_0\sqrt{\frac{3}{2}}$

(c) independent of the choice of the two terminals

(d) $V_{XY}^{rms} = V_0$

Section 3 (Maximum Marks : 12)

- This section contains **TWO** paragraphs.
- Based on each paragraph, there are **TWO** questions.
- Each question has **FOUR** options (a), (b) (c) and (d). **ONLY ONE** of these four options is correct.
- For each question, darken the bubble corresponding to all the correct option in the ORS.
- For each questions, marks will be awarded in one of the following categories.

  Full Marks : + 3 If only the bubble corresponding to all the correct answer is darkened.

  Zero Marks : 0 In all other cases.

Paragraph 1

Consider a simple RC circuit as shown in Figure 1.

Process 1 In the circuit the switch S is closed at $t = 0$ and the capacitor is fully charged to voltage V_0 (i.e. charging continues for time $T >> RC$). In the process some dissipation (E_D) occurs across the resistance R. The amount of energy finally stored in the fully charged capacitor is E_c.

Process 2 In a different process the voltage is first set to $\frac{V_0}{3}$ and maintained for a charging time $T >> RC$. Then, the voltage is raised to $\frac{2V_0}{3}$ without discharging the capacitor and again maintained for a time $T >> RC$. The process is repeated one more time by raising the voltage to V_0 and the capacitor is charged to the same final voltage V_0 as in Process 1.

These two processes are depicted in Figure 2.

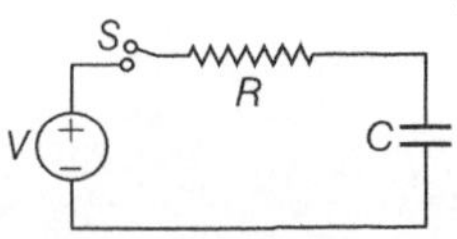

Figure 1

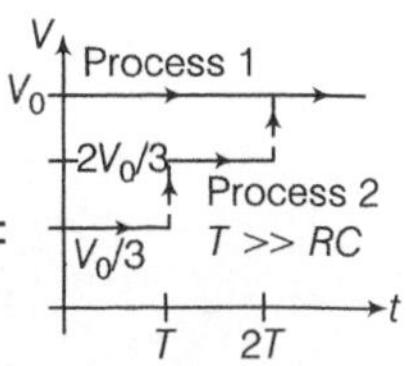

Figure 2

15. In Process 1, the energy stored in the capacitor E_C and heat dissipated across resistance E_D are related by

(a) $E_C = E_D \ln 2$ (b) $E_C = E_D$

(c) $E_C = 2E_D$ (d) $E_C = \frac{1}{2} E_D$

16. In Process 2, total energy dissipated across the resistance E_D is

(a) $E_D = \frac{1}{3}\left(\frac{1}{2} CV_0^2\right)$

(b) $E_D = 3\left(\frac{1}{2} CV_0^2\right)$

(c) $E_D = 3CV_0^2$

(d) $E_D = \frac{1}{2} CV_0^2$

Paragraph 2

One twirls a circular ring (of mass M and radius R) near the tip of one's finger as shown in Figure 1. In the process the finger never loses contact with the inner rim of the ring. The finger traces out the surface of a cone, shown by the dotted line. The radius of the path traced out by the point where the ring and the finger is in contact is r. The finger rotates with an angular velocity ω_0. The rotating ring rolls without slipping on the outside of a smaller circle described by the point where the ring and the finger is in contact (Figure 2). The coefficient of friction between the ring and the finger is μ and the acceleration due to gravity is g.

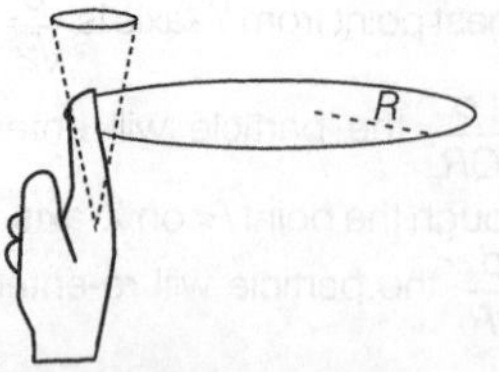

Figure 1

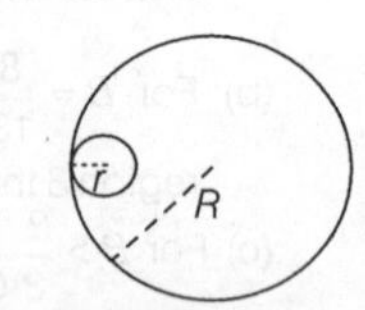

Figure 2

17. The total kinetic energy of the ring is

(a) $M\omega_0^2(R-r)^2$

(b) $\frac{1}{2} M\omega_0^2(R-r)^2$

(c) $M\omega_0^2 R^2$

(d) $\frac{3}{2} M\omega_0^2(R-r)^2$

18. The minimum value of ω_0 below which the ring will drop down is

(a) $\sqrt{\frac{g}{2\mu(R-r)}}$

(b) $\sqrt{\frac{3g}{2\mu(R-r)}}$

(c) $\sqrt{\frac{g}{\mu(R-r)}}$

(d) $\sqrt{\frac{2g}{\mu(R-r)}}$

Chemistry

Section 1 (Maximum Marks : 21)

- This section contain **SEVEN** questions
- Each question has FOUR options (a), (b), (c) and (d). **ONLY ONE** of these four options is correct.
- For each question, darken the bubble corresponding to the correct option in the ORS.
- For each question, marks will be awarded in one of the following categories :

Full Marks : + 3 If only the bubble corresponding to the correct option is darkened.

Partial Marks : +1 For darkening a bubble corresponding to each correct option, provided NO incorrect option is darkened.

Zero Marks : 0 If none of the bubbles is darkened.

Negative Marks : −2 In all other cases.

1. The order of basicity among the following compounds is

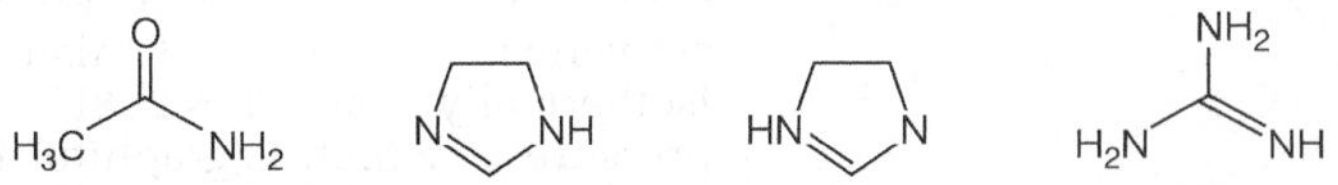

(I) (II) (III) (IV)

(a) II>I>IV>III (b) I>IV>III>I (c) IV>I>II>I (d) IV>I>II>III

2. Pure water freezes at 273 K and 1 bar. The addition of 34.5 g of ethanol to 500 g of water changes the freezing point of the solution. Use the freezing point depression constant of water as 2 K kg mol^{-1}. The figures shown below represent plots of vapour pressure (V.P.) *versus* temperature (T). [Molecular weight of ethanol is 46 g mol^{-1}]

Among the following, the option representing change in the freezing point is

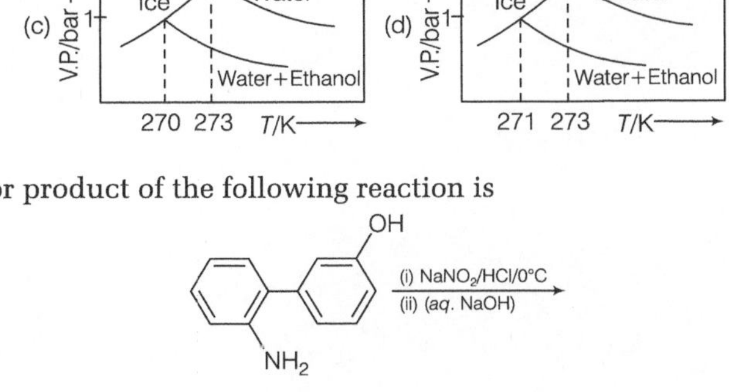

3. The major product of the following reaction is

(a) 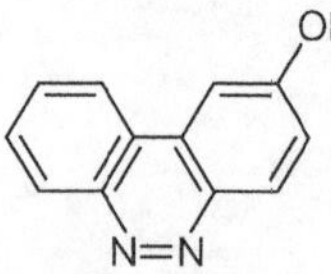

(c) 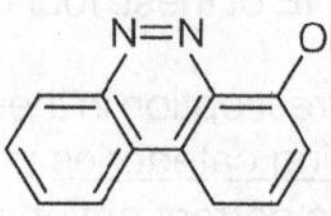

(b) 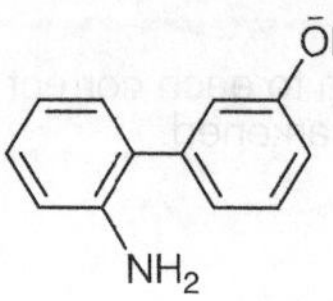

(d) 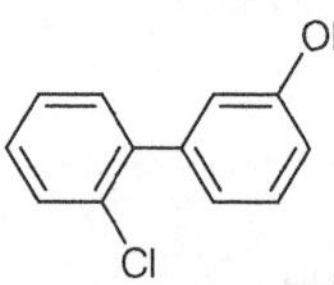

4. For the following cell,

$Zn(s) \mid ZnSO_4(aq) \mid\mid CuSO_4(aq) \mid Cu(s)$

when the concentration of Zn^{2+} is 10 times the concentration of Cu^{2+}, the expression for ΔG (in J mol^{-1}) is

[F is Faraday constant; R is gas constant; T is temperature; $E°$ (cell) = 1.1 V]

(a) 2.303 RT +1.1 F (b) 1.1 F

(c) 2.303 RT − 2.2 F (d) −2.2 F

5. The order of the oxidation state of the phosphorus atom in H_3PO_2, H_3PO_4, H_3PO_3 and $H_4P_2O_6$ is

(a) $H_3PO_4 > H_3PO_2 > H_3PO_3 > H_4P_2O_6$

(b) $H_3PO_4 > H_4P_2O_6 > H_3PO_3 > H_3PO_2$

(c) $H_3PO_2 > H_3PO_3 > H_4P_2O_6 > H_3PO_4$

(d) $H_3PO_3 > H_3PO_2 > H_3PO_4 > H_4P_2O_6$

6. The standard state Gibbs free energies of formation of C(graphite) and C(diamond) at T = 298 K are

$\Delta_f G°$ [C(graphite)] = 0 kJ mol^{-1}

$\Delta_f G°$ [C(diamond)] = 2.9 kJ mol^{-1}

The standard state means that the pressure should be 1 bar, and substance should be pure at a given temperature. The conversion of graphite [C(graphite)] to diamond [C(diamond)] reduces its volume by 2×10^{-6} m^3 mol^{-1}. If C(graphite) is converted to C(diamond) isothermally at T = 298K, the pressure at which C(graphite) is in equilibrium with C(diamond), is

[Useful information : 1 J = 1 kg $m^2 s^{-2}$,
1 Pa = 1 kg m^{-1} s^{-2}; 1 bar = 10^5 Pa]

(a) 58001 bar

(b) 1450 bar

(c) 14501 bar

(d) 29001 bar

7. Which of the following combination will produce H_2 gas?

(a) Fe metal and conc. HNO_3

(b) Cu metal and conc. HNO_3

(c) Au metal and NaCN (aq) in the presence of air

(d) Zn metal and NaOH (aq)

Section 2 (Maximum Marks : 28)

- This section contains **SEVEN** questions.
- Each question has FOUR option (a), (b), (c) and (d). **ONE OR MORE THAN ONE** of these four option(s) is(are) correct.
- For each question, darken the bubble(s) corresponding to all correct option(s) in the ORS.
- For each question, marks will be awarded in one of the following categories :

Full Marks : + 4 If only the bubble(s) corresponding to all the correct option(s) is(are)

Partial Marks : +1 For darkening a bubble corresponding to each correct option, provided NO incorrect option is darkened.

Zero Marks : 0 If none of the bubbles is darkened.

Negative Marks : −2 In all other cases

- For example, if (a), (c) and (d) are all the correct option for a question, darkening all these three will get +4 marks; darkening only (a) and (d) will get +2 marks and darkening (a) and (b) will get −2 marks, as a wrong option is also darkened.

8. For a reaction taking place in a container in equilibrium with its surroundings, the effect of temperature on its equilibrium constant K in terms of change in entropy is described by

(a) With increase in temperature, the value of K for endothermic reaction increases because unfavourable change in entropy of the surroundings decreases

(b) With increase in temperature, the value of K for exothermic reaction decreases because favourable change in entropy of the surrounding decreases

(c) With increase in temperature, the value of K for endothermic reaction increases because the entropy change of the system is negative

(d) With increase in temperature, the value of K for exothermic reaction decreases because the entropy change of the system is positive

9. In a bimolecular reaction, the steric factor P was experimentally determined to be 4.5. the correct option(s) among the following is(are)

(a) The activation energy of the reaction is unaffected by the value of the steric factor

(b) Experimentally determined value of frequency factor is higher than that predicted by Arrhenius equation

(c) The value of frequency factor predicted by Arrhenius equation is higher than that determined experimentally

(d) Since $P = 4.5$, the reaction will not proceed unless an effective catalyst is used

10. For the following compounds, the correct statement(s) with respect to nucleophilic substitution reaction is(are)

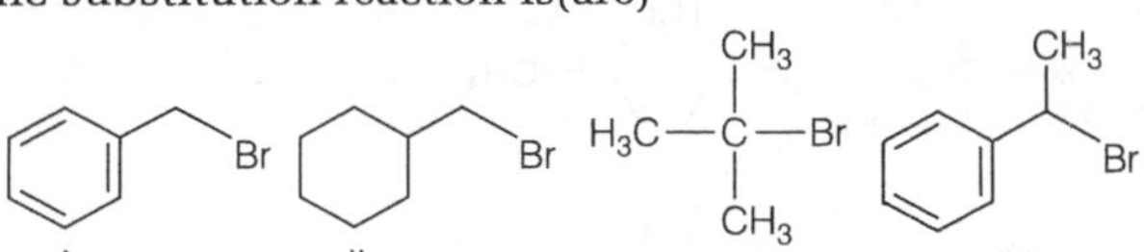

I II III IV

(a) Compound IV undergoes inversion of configuration

(b) The order of reactivity for I, III and IV is : IV > I > III

(c) I and III follow S_N1 mechanism

(d) I and II follow S_N1 mechanism

11. Among the following, the correct statement(s) is(are)

(a) $Al(CH_3)_3$ has the three-centre two-electron bonds in its dimeric structure
(b) The Lewis acidity of BCl_3 is greater than that of $AlCl_3$
(c) $AlCl_3$ has the three-centre two-electron bonds in its dimeric structure
(d) BH_3 has the three-centre two-electron bonds in its dimeric structure

12. The option(s) with only amphoteric oxides is(are)

(a) NO, B_2O_3, PbO, SnO_2
(b) Cr_2O_3, CrO, SnO, PbO
(c) Cr_2O_3, BeO, SnO, SnO_2
(d) ZnO, Al_2O_3, PbO, PbO_2

13. The correct statement(s) about surface properties is(are)

(a) The critical temperatures of ethane and nitrogen are 563 K and 126 K, respectively. The adsorption of ethane will be more than that of nitrogen of same amount of activated charcoal at a given temperature
(b) Cloud is an emulsion type of colloid in which liquid is dispersed phase and gas is dispersion medium
(c) Adsorption is accompanied by decrease in enthalpy and decrease in entropy of the system
(d) Brownian motion of colloidal particles does not depend on the size of the particles but depends on viscosity of the solution

14. Compound P and R upon ozonolysis produce Q and S, respectively. The molecular formula of Q and S is C_8H_8O. Q undergoes Cannizzaro reaction but not haloform reaction, whereas S undergoes haloform reaction but not Cannizzaro reaction.

(i) $P \xrightarrow[\text{(ii) Zn/H}_2\text{O}]{\text{(i) O}_3/\text{CH}_2\text{Cl}_2} \underset{(C_8H_8O)}{Q}$ (ii) $R \xrightarrow[\text{(ii) Zn/H}_2\text{O}]{\text{(i) O}_3/\text{CH}_2\text{Cl}_2} \underset{(C_8H_8O)}{S}$

The option(s) with suitable combination of P and R, respectively, is(are)

(a) 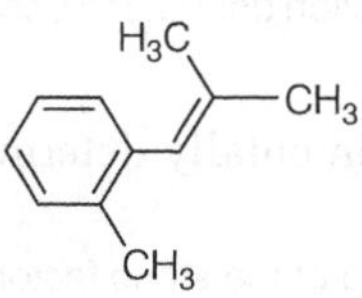 and 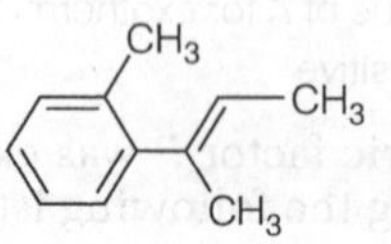

(b) 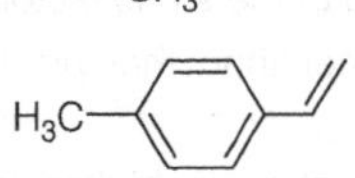 and 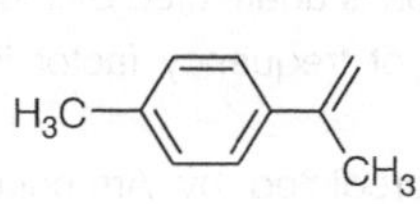

(c)  and 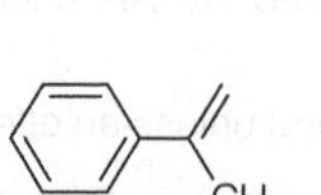

(d) 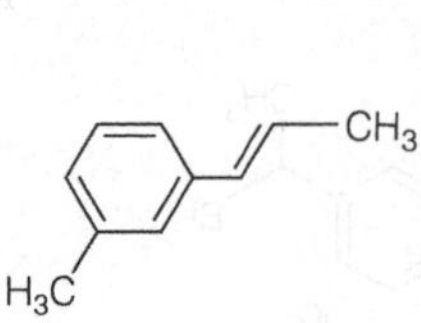 and 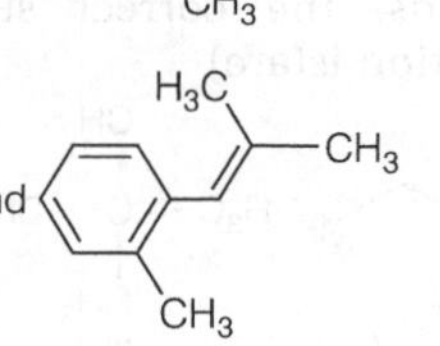

Section 3 (Maximum Marks : 12)

- This section contains **TWO** paragraphs.
- Based on each paragraph, there are **TWO** questions.
- Each question has **FOUR** options (a), (b), (c) and (d). **ONLY ONE** of these four options is correct.
- For each question, darken the bubble corresponding to the correct integer in the ORS.
- For each question, marks will be awarded in one of the following categories :

Full Marks : + 3 If only the bubble corresponding to the correct option is darkened

Zero Marks : 0 If none of the bubbles is darkened

Paragraph 1

Upon heating $KClO_3$ in presence of catalytic amount of MnO_2, a gas *W* is formed. Excess amount of *W* reacts with white phosphorus to give *X*. The reaction of *X* with pure HNO_3 gives *Y* and *Z*.

15. *Y* **and** *Z* **are, respectively**

(a) N_2O_4 and HPO_3
(b) N_2O_4 and H_3PO_3
(c) N_2O_3 and H_3PO_4
(d) N_2O_5 and HPO_3

16. *W* **and** *X* **are, respectively**

(a) O_2 and P_4O_{10}
(b) O_2 and P_4O_6
(c) O_3 and P_4O_6
(d) O_3 and P_4O_{10}

Paragraph 2

The reaction of compound *P* with CH_3MgBr (excess) in $(C_2H_5)_2O$ followed by addition of H_2O gives *Q*. The compound *Q* on treatment with H_2SO_4 at 0° C gives *R*. The reaction of *R* with CH_3COCl in the presence of anhydrous $AlCl_3$ in CH_2Cl_2 followed by treatment with H_2O produces compound *S*. [Et in compound *P* is ethyl group]

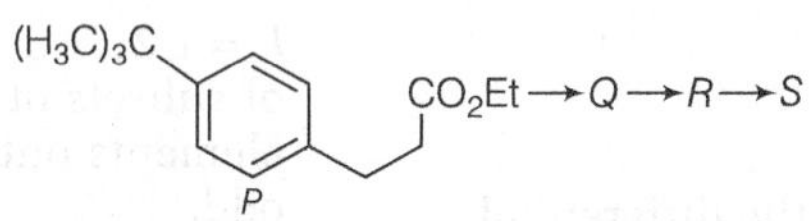

P

17. The product *S* **is**

(a) 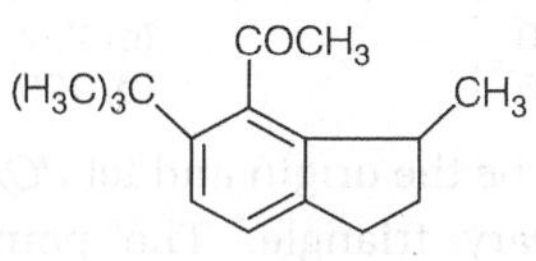

(b) 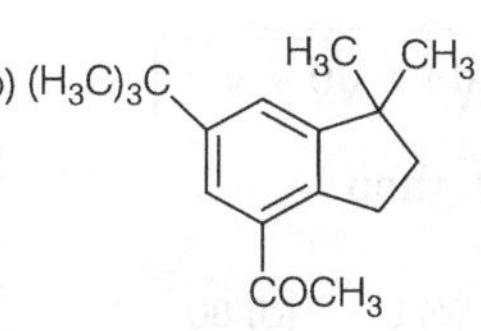

(c) 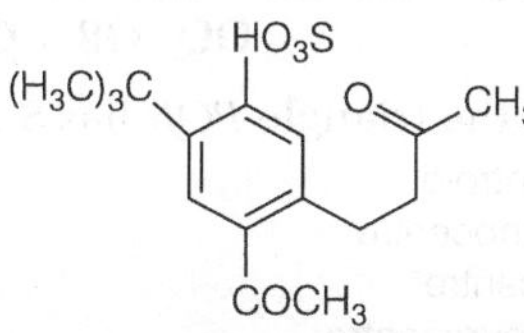

(d) 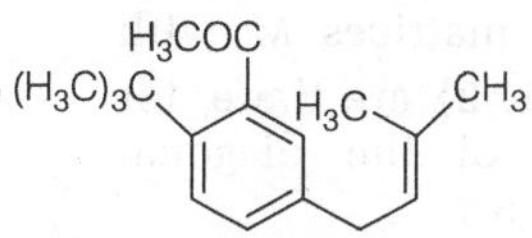

18. The reactions, Q to R and R to S, are
(a) Aromatic sulfonation and Friedel-Crafts acylation
(b) Friedel-Crafts alkylation and Friedel-Crafts acylation
(c) Friedel-Crafts alkylation, dehydration and Friedel-Crafts acylation
(d) Dehydration and Friedel-Crafts acylation

Mathematics

Section 1 (Maximum Marks : 21)

- This section contains **SEVEN** questions.
- Each question has **FOUR** options (a), (b), (c) and (d). **ONLY ONE** of these four options is correct.
- For each question, darken the bubble corresponding to the correct option in the ORS.
- For each question, marks will be awarded in one of the following categories :
  Full Marks : +3 If only the bubble corresponding to the correct option is darkened.
  Zero Marks : 0 If none of the bubbles is darkened.
  Negative Marks : −1 In all other cases.

1. If $f : R \to R$ is a twice differentiable function such that $f''(x) > 0$ for all $x \in R$, and $f\left(\frac{1}{2}\right) = \frac{1}{2}$, $f(1) = 1$, then
(a) $f'(1) \le 0$
(b) $f'(1) > 1$
(c) $0 < f'(1) \le \frac{1}{2}$
(d) $\frac{1}{2} < f'(1) \le 1$

2. If $y = y(x)$ satisfies the differential equation
$$8\sqrt{x}\left(\sqrt{9+\sqrt{x}}\right)dy = \left(\sqrt{4+\sqrt{9+\sqrt{x}}}\right)^{-1}$$
$dx, x > 0$ and $y(0) = \sqrt{7}$, then $y(256) =$
(a) 16 (b) 3 (c) 9 (d) 80

3. How many 3×3 matrices M with entries from {0, 1, 2} are there, for which the sum of the diagonal entries of $M^T M$ is 5 ?
(a) 198 (b) 162
(c) 126 (d) 135

4. Three randomly chosen nonnegative integers x, y and z are found to satisfy the equation $x + y + z = 10$. Then the probability that z is even, is
(a) $\frac{1}{2}$ (b) $\frac{36}{55}$ (c) $\frac{6}{11}$ (d) $\frac{5}{11}$

5. Let $S = \{1, 2, 3, \ldots, 9\}$. For $k = 1, 2, \ldots 5$, let N_k be the number of subsets of S, each containing five elements out of which exactly k are odd. Then $N_1 + N_2 + N_3 + N_4 + N_5 =$
(a) 210 (b) 252
(c) 126 (d) 125

6. Let O be the origin and let PQR be an arbitrary triangle. The point S is such that
$$\mathbf{OP} \cdot \mathbf{OQ} + \mathbf{OR} \cdot \mathbf{OS} = \mathbf{OR} \cdot \mathbf{OP} + \mathbf{OQ} \cdot \mathbf{OS} = \mathbf{OQ} \cdot \mathbf{OR} + \mathbf{OP} \cdot \mathbf{OS}$$
Then the triangle PQR has S as its
(a) centroid
(b) orthocentre
(c) incentre
(d) circumcentre

7. The equation of the plane passing through the point (1, 1, 1) and perpendicular to the planes $2x + y - 2z = 5$ and $3x - 6y - 2z = 7$ is

(a) $14x + 2y - 15z = 1$
(b) $-14x + 2y + 15z = 3$
(c) $14x - 2y + 15z = 27$
(d) $14x + 2y + 15z = 31$

Section 2 (Maximum Marks : 28)

- This section contains **SEVEN** questions.
- Each question has **FOUR** options (a), (b), (c) and (d). **ONE OR MORE THAN ONE** of these four option(s) is(are) correct.
- For each question, darken the bubble(s) corresponding to all the correct option(s) in the ORS.
- For each question, marks will be awarded in one of the following categories :

Full Marks : +4 If only the bubble(s) corresponding to all the correct option(s) is(are) darkened.
Partial Marks : +1 For darkening a bubble corresponding to each correct option, provided NO incorrect option is darkened.
Zero Marks : 0 If none of the bubbles is darkened.
Negative Marks : −2 In all other cases.

- For example, if (a), (c) and (d) are all the correct options for a question, darkening all these three will get + 4 marks; darkening only (a) and (d) will get +2 marks and darkening (a) and (b) will get − 2 marks, as a wrong option is also darkened.

8. If $f : R \to R$ is a differentiable function such that $f'(x) > 2f(x)$ for all $x \in R$, and $f(0) = 1$ then

(a) $f(x) > e^{2x}$ in $(0, \infty)$
(b) $f'(x) < e^{2x}$ in $(0, \infty)$
(c) $f(x)$ is increasing in $(0, \infty)$
(d) $f(x)$ is decreasing in (0∞)

9. If $I = \sum_{k=1}^{98} \int_k^{k+1} \frac{k+1}{x(x+1)} dx$, then

(a) $I > \log_e 99$
(b) $I < \log_e 99$
(c) $I < \frac{49}{50}$
(d) $I > \frac{49}{50}$

10. If the line $x = \alpha$ divides the area of region $R = \{(x, y) \in R^2 : x^3 \le y \le x, 0 \le x \le 1\}$ into two equal parts, then

(a) $2\alpha^4 - 4\alpha^2 + 1 = 0$
(b) $\alpha^4 + 4\alpha^2 - 1 = 0$
(c) $\frac{1}{2} < \alpha < 1$
(d) $0 < \alpha \le \frac{1}{2}$

11. Let α and β be non zero real numbers such that $2(\cos\beta - \cos\alpha) + \cos\alpha \cos\beta = 1$. Then which of the following is/are true?

(a) $\sqrt{3} \tan\left(\frac{\alpha}{2}\right) - \tan\left(\frac{\beta}{2}\right) = 0$
(b) $\tan\left(\frac{\alpha}{2}\right) - \sqrt{3} \tan\left(\frac{\beta}{2}\right) = 0$
(c) $\tan\left(\frac{\alpha}{2}\right) + \sqrt{3} \tan\left(\frac{\beta}{2}\right) = 0$
(d) $\sqrt{3} \tan\left(\frac{\alpha}{2}\right) + \tan\left(\frac{\beta}{2}\right) = 0$

12. Let

$$f(x) = \frac{1 - x(1 + |1 - x|)}{|1 - x|} \cos\left(\frac{1}{1 - x}\right)$$

for $x \ne 1$. Then

(a) $\lim_{x \to 1^+} f(x) = 0$
(b) $\lim_{x \to 1^-} f(x)$ does not exist
(c) $\lim_{x \to 1^-} f(x) = 0$
(d) $\lim_{x \to 1^+} f(x)$ does not exist

13. If $g(x) = \int_{\sin x}^{\sin(2x)} \sin^{-1}(t)\, dt$, then

(a) $g'\left(-\frac{\pi}{2}\right) = 2\pi$

(b) $g'\left(-\frac{\pi}{2}\right) = -2\pi$

(c) $g'\left(\frac{\pi}{2}\right) = 2\pi$

(d) $g'\left(\frac{\pi}{2}\right) = -2\pi$

14. If $f(x) = \begin{vmatrix} \cos(2x) & \cos(2x) & \sin(2x) \\ -\cos x & \cos x & -\sin x \\ \sin x & \sin x & \cos x \end{vmatrix}$, then

(a) $f(x)$ attains its minimum at $x = 0$

(b) $f(x)$ attains its maximum at $x = 0$

(c) $f'(x) = 0$ at more than three points in $(-\pi, \pi)$

(d) $f'(x) = 0$ at exactly three points in $(-\pi, \pi)$

Section 3 (Maximum Marks : 12)

- This section contains **TWO** paragraphs.
- Based on each paragraph, there are **TWO** questions.
- Each question has **FOUR** options (a), (b), (c) and (d). **ONLY ONE** of these four options is correct.
- For each question, darken the bubble corresponding to the correct option in the ORS.
- For each question, marks will be awarded in one of the following categories:

Full Marks : + 3 If only the bubble corresponding to the correct option is darkened.

Zero Marks : 0 In all other cases.

Paragraph 1

Let O be the origin and $\mathbf{OX}, \mathbf{OY}, \mathbf{OZ}$ be three unit vectors in the directions of the sides $\mathbf{QR}, \mathbf{RP}, \mathbf{PQ}$ respectively, of a triangle PQR.

15. If the triangle PQR varies, then the minimum value of $\cos(P+Q) + \cos(Q+R) + \cos(R+P)$ is

(a) $-\frac{3}{2}$ (b) $\frac{3}{2}$

(c) $\frac{5}{3}$ (d) $-\frac{5}{3}$

16. $|\mathbf{OX} \times \mathbf{OY}| =$

(a) $\sin(P+Q)$ (b) $\sin(P+R)$

(c) $\sin(Q+R)$ (d) $\sin 2R$

Paragraph 2

Let p, q be integers and let α, β be the roots of the equation, $x^2 - x - 1 = 0$ where $\alpha \neq \beta$. For $n = 0, 1, 2, \ldots\ldots$, let $a_n = p\alpha^n + q\beta^n$.

FACT : If a and b are rational numbers and $a + b\sqrt{5} = 0$, then $a = 0 = b$.

17. $a_{12} =$

(a) $a_{11} + 2a_{10}$

(b) $2a_{11} + a_{10}$

(c) $a_{11} - a_{10}$

(d) $a_{11} + a_{10}$

18. If $a_4 = 28$, then $p + 2q =$

(a) 14 (b) 7

(c) 21 (d) 12

Answers

Paper 1

Physics

1. (a,b)	**2.** (a,b,c)	**3.** (b,d)	**4.** (a,c,d)	**5.** (a,b)
6. (a,c,d)	**7.** (b,c,d)	**8.** (5)	**9.** (6)	**10.** (6)
11. (5)	**12.** (8)	**13.** (c)	**14.** (a)	**15.** (c)
16. (c)	**17.** (b)	**18.** (d)		

Chemistry

1. (a,c,d)	**2.** (b,c)	**3.** (b,d)	**4.** (b,d)	**5.** (a,c,d)
6. (a,c,d)	**7.** (a,d)	**8.** (6)	**9.** (5)	**10.** (6)
11. (2)	**12.** (6)	**13.** (c)	**14.** (a)	**15.** (d)
16. (d)	**17.** (a)	**18.** (d)		

Mathematics

1. (a,b)	**2.** (c,d)	**3.** (b,d)	**4.** (a,b,c)	**5.** (a,b,d)
6. (a,c)	**7.** (b)	**8.** (1)	**9.** (6)	**10.** (2)
11. (2)	**12.** (5)	**13.** (a)	**14.** (b)	**15.** (a)
16. (d)	**17.** (c)	**18.** (c)		

Paper 2

Physics

1. (d)	**2.** (a)	**3.** (d)	**4.** (c)	**5.** (a)
6. (c)	**7.** (a)	**8.** (a,c)	**9.** (c,d)	**10.** (a,d)
11. (a,c,d)	**12.** (a,b,c)	**13.** (b,c)	**14.** (b,c)	**15.** (b)
16. (a)	**17.** (*)	**18.** (c)		

Chemistry

1. (d)	**2.** (a)	**3.** (a)	**4.** (c)	**5.** (b)
6. (c)	**7.** (d)	**8.** (a,b)	**9.** (a,b)	**10.** (c,d)
11. (a,b,d)	**12.** (c,d)	**13.** (a,c)	**14.** (c,d)	**15.** (d)
16. (a)	**17.** (b)	**18.** (b)		

Mathematics

1. (b)	**2.** (b)	**3.** (a)	**4.** (c)	**5.** (c)
6. (b)	**7.** (d)	**8.** (a,c)	**9.** (b,d)	**10.** (a,c)
11. (b,c)	**12.** (c,d)	**13.** (*)	**14.** (b,c)	**15.** (a)
16. (a)	**17.** (d)	**18.** (d)		

JEE Main

Solved Paper 2018

INSTRUCTIONS

1. This test consists of 90 questions.
2. Each question is allotted 4 marks for correct response.
3. Candidates will be awarded marks as stated above in instruction no. 2 for correct response of each question. 1 mark will be deducted for indicating incorrect response of each question. No deduction from the total score will be made if no response is indicated for an item in the answer sheet.
4. There is only one correct response for each question. Filling up more than one response in any question will be treated as wrong response and marks for wrong response will be deducted according as per instructions.

Physics

1. The density of a material in the shape of a cube is determined by measuring three sides of the cube and its mass. If the relative errors in measuring the mass and length are respectively 1.5% and 1%, the maximum error in determining the density is

(a) 2.5% (b) 3.5%
(c) 4.5% (d) 6%

2. All the graphs below are intended to represent the same motion. One of them does it incorrectly. Pick it up.

(a)
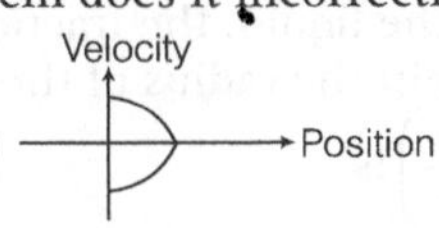

(b)
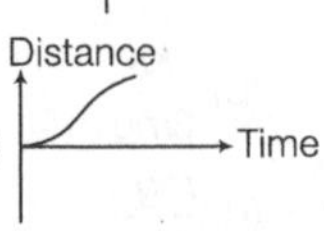

(c)
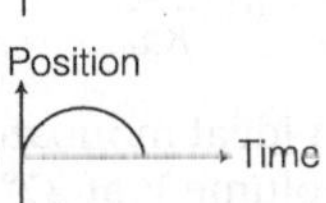

(d)
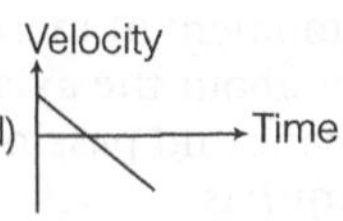

3. Two masses $m_1 = 5$ kg and $m_2 = 10$ kg connected by an inextensible string over a frictionless pulley, are moving as shown in the figure. The coefficient of friction of horizontal surface is 0.15. The minimum weight m that should be put on top of m_2 to stop the motion is

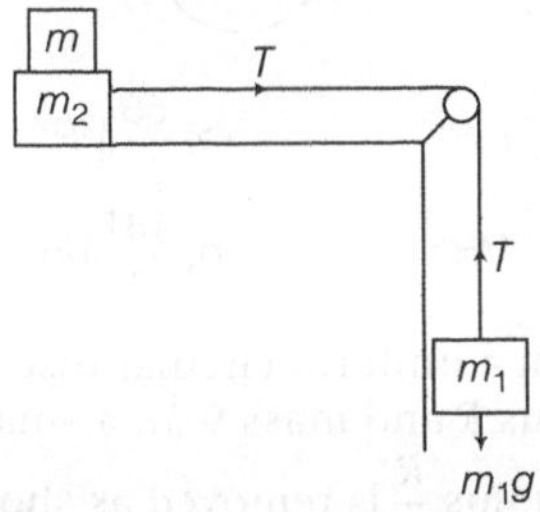

(a) 18.3 kg (b) 27.3 kg
(c) 43.3 kg (d) 10.3 kg

4. A particle is moving in a circular path of radius a under the action of an attractive potential $U = -\frac{k}{2r^2}$. Its total energy is

(a) $-\frac{k}{4a^2}$ (b) $\frac{k}{2a^2}$

(c) zero (d) $-\frac{3}{2}\frac{k}{a^2}$

5. In a collinear collision, a particle with an initial speed v_0 strikes a stationary particle of the same mass. If the final total kinetic energy is 50% greater than the original kinetic energy, the magnitude of the relative velocity between the two particles after collision, is

(a) $\frac{v_0}{4}$ (b) $\sqrt{2}\, v_0$

(c) $\frac{v_0}{2}$ (d) $\frac{v_0}{\sqrt{2}}$

6. Seven identical circular planar discs, each of mass M and radius R are welded symmetrically as shown in the figure. The moment of inertia of the arrangement about the axis normal to the plane and passing through the point P is

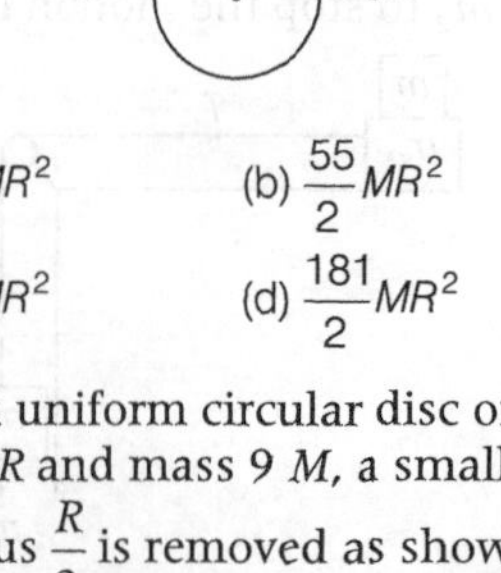

(a) $\frac{19}{2}MR^2$ (b) $\frac{55}{2}MR^2$

(c) $\frac{73}{2}MR^2$ (d) $\frac{181}{2}MR^2$

7. From a uniform circular disc of radius R and mass $9\,M$, a small disc of radius $\frac{R}{3}$ is removed as shown in the figure. The moment of inertia of the remaining disc about an axis perpendicular to the plane of the disc and passing through centre of disc is

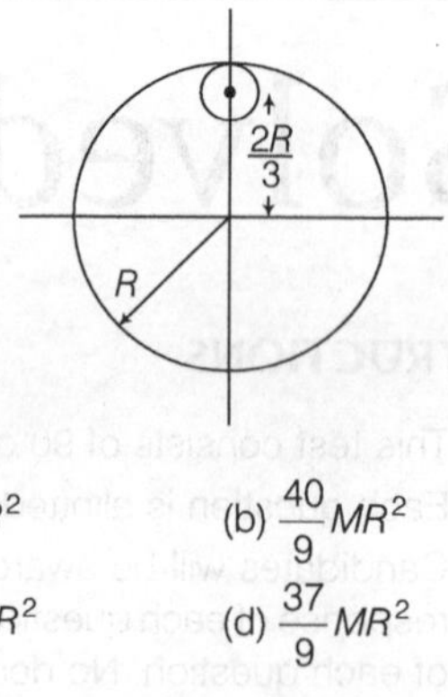

(a) $4MR^2$ (b) $\frac{40}{9}MR^2$

(c) $10MR^2$ (d) $\frac{37}{9}MR^2$

8. A particle is moving with a uniform speed in a circular orbit of radius R in a central force inversely proportional to the nth power of R. If the period of rotation of the particle is T, then :

(a) $T \propto R^{3/2}$ for any n

(b) $T \propto R^{\frac{n}{2}+1}$

(c) $T \propto R^{(n+1)/2}$

(d) $T \propto R^{n/2}$

9. A solid sphere of radius r made of a soft material of bulk modulus K is surrounded by a liquid in a cylindrical container. A massless piston of area a floats on the surface of the liquid, covering entire cross-section of cylindrical container. When a mass m is placed on the surface of the piston to compress the liquid, the fractional decrement in the radius of the sphere, $\left(\frac{dr}{r}\right)$ is

(a) $\frac{Ka}{mg}$ (b) $\frac{Ka}{3mg}$

(c) $\frac{mg}{3Ka}$ (d) $\frac{mg}{Ka}$

10. Two moles of an ideal monoatomic gas occupies a volume V at 27°C. The gas expands adiabatically to a volume 2 V. Calculate (i) the final temperature of the gas and (ii) change in its internal energy.

(a) (i) 189 K (ii) 2.7 kJ

(b) (i) 195 K (ii) −2.7 kJ

(c) (i) 189 K (ii) −2.7 kJ

(d) (i) 195 K (ii) 2.7 kJ

11. The mass of a hydrogen molecule is 3.32×10^{-27} kg. If 10^{23} hydrogen molecules strike per second, a fixed wall of area 2 cm^2 at an angle of 45° to the normal and rebound elastically with a speed of 10^3 m/s, then the pressure on the wall is nearly

(a) 2.35×10^3 N/m^2 (b) 4.70×10^3 N/m^2

(c) 2.35×10^2 N/m^2 (d) 4.70×10^2 N/m^2

12. A silver atom in a solid oscillates in simple harmonic motion in some direction with a frequency of 10^{12} per second. What is the force constant of the bonds connecting one atom with the other? (Take, molecular weight of silver $= 108$ and Avogadro number $= 6.02 \times 10^{23}$ g mol^{-1})

(a) 6.4 N/m (b) 7.1 N/m

(c) 2.2 N/m (d) 5.5 N/m

13. A granite rod of 60 cm length is clamped at its middle point and is set into longitudinal vibrations. The density of granite is 2.7×10^3 kg/m^3 and its Young's modulus is 9.27×10^{10} Pa. What will be the fundamental frequency of the longitudinal vibrations?

(a) 5 kHz (b) 2.5 kHz

(c) 10 kHz (d) 7.5 kHz

14. Three concentric metal shells A, B and C of respective radii a, b and c $(a < b < c)$ have surface charge densities $+\sigma$, $-\sigma$ and $+\sigma$, respectively. The potential of shell B is

(a) $\frac{\sigma}{\varepsilon_0}\left[\frac{a^2-b^2}{a}+c\right]$ (b) $\frac{\sigma}{\varepsilon_0}\left[\frac{a^2-b^2}{b}+c\right]$

(c) $\frac{\sigma}{\varepsilon_0}\left[\frac{b^2-c^2}{b}+a\right]$ (d) $\frac{\sigma}{\varepsilon_0}\left[\frac{b^2-c^2}{c}+a\right]$

15. A parallel plate capacitor of capacitance 90 pF is connected to a battery of emf 20 V. If a dielectric material of dielectric constant $K = \frac{5}{3}$ is inserted between the plates, the magnitude of the induced charge will be

(a) 1.2 nC (b) 0.3 nC

(c) 2.4 nC (d) 0.9 nC

16. In an AC circuit, the instantaneous emf and current are given by

$$e = 100 \sin 30\,t,\ i = 20 \sin\left(30\,t - \frac{\pi}{4}\right)$$

In one cycle of AC, the average power consumed by the circuit and the wattless current are, respectively

(a) 50 , 10 (b) $\frac{1000}{\sqrt{2}}$, 10

(c) $\frac{50}{\sqrt{2}}$, 0 (d) 50 , 0

17. Two batteries with emf 12 V and 13 V are connected in parallel across a load resistor of 10 Ω. The internal resistances of the two batteries are 1 Ω and 2 Ω, respectively. The voltage across the load lies between

(a) 11.6 V and 11.7 V

(b) 11.5 V and 11.6 V

(c) 11.4 V and 11.5 V

(d) 11.7 V and 11.8 V

18. An electron, a proton and an alpha particle having the same kinetic energy are moving in circular orbits of radii r_e, r_p, r_α respectively, in a uniform magnetic field B. The relation between r_e, r_p, r_α is

(a) $r_e > r_p = r_\alpha$ (b) $r_e < r_p = r_\alpha$

(c) $r_e < r_p < r_\alpha$ (d) $r_e < r_\alpha < r_p$

19. The dipole moment of a circular loop carrying a current I, is m and the magnetic field at the centre of the loop is B_1. When the dipole moment is doubled by keeping the current constant, the magnetic field at the centre of the loop is B_2. The ratio $\frac{B_1}{B_2}$ is

(a) 2 (b) $\sqrt{3}$

(c) $\sqrt{2}$ (d) $\frac{1}{\sqrt{2}}$

20. For an R-L-C circuit driven with voltage of amplitude v_m and frequency $\omega_0 = \frac{1}{\sqrt{LC}}$, the current exhibits resonance. The quality factor, Q is given by

(a) $\frac{\omega_0 L}{R}$ (b) $\frac{\omega_0 R}{L}$

(c) $\frac{R}{\omega_0 C}$ (d) $\frac{CR}{\omega_0}$

21. An EM wave from air enters a medium. The electric fields are $\mathbf{E}_1 = E_{01}\hat{\mathbf{x}}\cos\left[2\pi\nu\left(\frac{z}{c} - t\right)\right]$ in air and $\mathbf{E}_2 = E_{02}\hat{\mathbf{x}}\cos[k(2z - ct)]$ in medium, where the wave number k and frequency ν refer to their values in air. The medium is non-magnetic.

If ε_{r_1} and ε_{r_2} refer to relative permittivities of air and medium respectively, which of the following options is correct?

(a) $\frac{\varepsilon_{r_1}}{\varepsilon_{r_2}} = 4$ (b) $\frac{\varepsilon_{r_1}}{\varepsilon_{r_2}} = 2$

(c) $\frac{\varepsilon_{r_1}}{\varepsilon_{r_2}} = \frac{1}{4}$ (d) $\frac{\varepsilon_{r_1}}{\varepsilon_{r_2}} = \frac{1}{2}$

22. Unpolarised light of intensity I passes through an ideal polariser A. Another identical polariser B is placed behind A. The intensity of light beyond B is found to be $\frac{I}{2}$. Now, another identical polariser C is placed between A and B. The intensity beyond B is now found to be $\frac{1}{8}$. The angle between polariser A and C is

(a) 0° (b) 30°

(c) 45° (d) 60°

23. The angular width of the central maximum in a single slit diffraction pattern is 60°. The width of the slit is 1 μm. The slit is illuminated by monochromatic plane waves. If another slit of same width is made near it, Young's fringes can be observed on a screen placed at a distance 50 cm from the slits. If the observed fringe width is 1 cm, what is slit separation distance? (i.e. distance between the centres of each slit.)

(a) 25 μm (b) 50 μm

(c) 75 μm (d) 100 μm

24. An electron from various excited states of hydrogen atom emit radiation to come to the ground state. Let λ_n, λ_g be the de-Broglie wavelength of the electron in the nth state and the ground state, respectively. Let Λ_n be the wavelength of the emitted photon in the transition from the nth state to the ground state. For large n, (A, B are constants)

(a) $\Lambda_n \approx A + \frac{B}{\lambda_n^2}$ (b) $\Lambda_n \approx A + B\lambda_n^2$

(c) $\Lambda_n^2 \approx A + B\lambda_n^2$ (d) $\Lambda_n^2 \approx \lambda$

25. If the series limit frequency of the Lyman series is ν_L, then the series limit frequency of the Pfund series is

(a) 25 ν_L (b) 16 ν_L

(c) $\frac{\nu_L}{16}$ (d) $\frac{\nu_L}{25}$

26. It is found that, if a neutron suffers an elastic collinear collision with deuterium at rest, fractional loss of its energy is P_d; while for its similar collision with carbon nucleus at rest, fractional loss of energy is P_c. The values of P_d and P_c are respectively

(a) (.89, .28) (b) (.28, .89)
(c) (0, 0) (d) (0, 1)

27. The reading of the ammeter for a silicon diode in the given circuit is

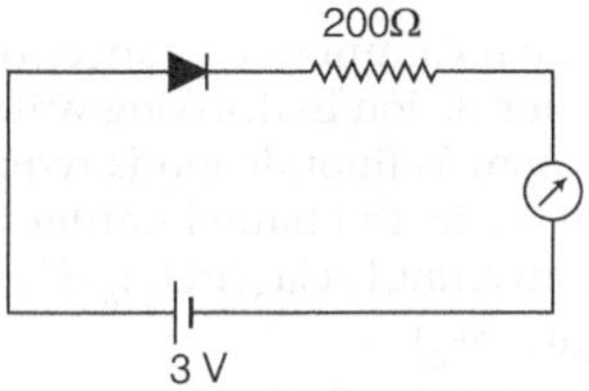

(a) 0 (b) 15 mA
(c) 11.5 mA (d) 13.5 mA

28. A telephonic communication service is working at carrier frequency of 10 GHz. Only 10% of it is utilised for transmission. How many telephonic channels can be transmitted simultaneously, if each channel requires a bandwidth of 5 kHz?

(a) 2×10^3 (b) 2×10^4
(c) 2×10^5 (d) 2×10^6

29. In a potentiometer experiment, it is found that no current passes through the galvanometer when the terminals of the cell are connected across 52 cm of the potentiometer wire. If the cell is shunted by a resistance of 5 Ω, a balance is found when the cell is connected across 40 cm of the wire. Find the internal resistance of the cell.

(a) 1 Ω (b) 1.5 Ω
(c) 2 Ω (d) 2.5 Ω

30. On interchanging the resistances, the balance point of a meter bridge shifts to the left by 10 cm. The resistance of their series combination is 1 kΩ. How much was the resistance on the left slot before interchanging the resistances?

(a) 990 Ω (b) 505 Ω
(c) 550 Ω (d) 910 Ω

Chemistry

31. The ratio of mass per cent of C and H of an organic compound ($C_xH_yO_z$) is 6 : 1. If one molecule of the above compound ($C_xH_yO_z$) contains half as much oxygen as required to burn one molecule of compound C_xH_y completely to CO_2 and H_2O. The empirical formula of compound $C_xH_yO_z$ is

(a) $C_3H_6O_3$ (b) C_2H_4O
(c) $C_3H_4O_2$ (d) $C_2H_4O_3$

32. Which type of 'defect' has the presence of cations in the interstitial sites?

(a) Schottky defect
(b) Vacancy defect
(c) Frenkel defect
(d) Metal deficiency defect

33. According to molecular orbital theory, which of the following will not be a viable molecule?

(a) He_2^{2+} (b) He_2^+
(c) H_2^- (d) H_2^{2-}

34. Which of the following lines correctly show the temperature dependence of equilibrium constant, K, for an exothermic reaction?

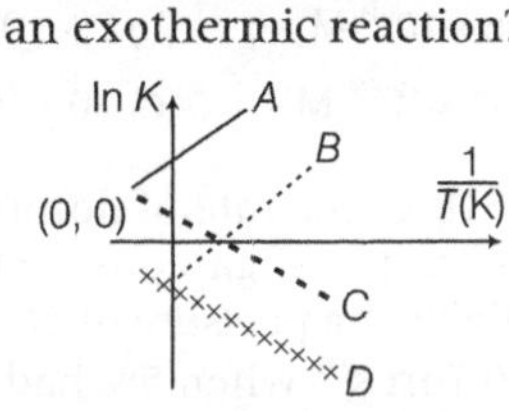

(a) A and B (b) B and C
(c) C and D (d) A and D

35. The combustion of benzene (l) gives $CO_2(g)$ and $H_2O(l)$. Given that heat of combustion of benzene at constant volume is -3263.9 kJ mol^{-1} at 25° C; heat of combustion (in kJ mol^{-1}) of benzene at constant pressure will be ($R = 8.314$ $JK^{-1}mol^{-1}$)

(a) 4152.6 (b) −452.46
(c) 3260 (d) −3267.6

36. For 1 molal aqueous solution of the following compounds, which one will show the highest freezing point?

(a) $[Co(H_2O)_6]Cl_3$
(b) $[Co(H_2O)_5Cl]Cl_2 \cdot H_2O$
(c) $[Co(H_2O)_4Cl_2]Cl \cdot 2H_2O$
(d) $[Co(H_2O)_3Cl_3] \cdot 3H_2O$

37. An aqueous solution contains 0.10 M H_2S and 0.20 M HCl. If the equilibrium constants for the formation of HS^- from H_2S is 1.0×10^{-7} and that of S^{2-} from HS^- ions is 1.2×10^{-13} then the concentration of S^{2-} ions in aqueous solution is

(a) 5×10^{-8} (b) 3×10^{-20}
(c) 6×10^{-21} (d) 5×10^{-19}

38. An aqueous solution contains an unknown concentration of Ba^{2+}. When 50 mL of a 1 M solution of Na_2SO_4 is added, $BaSO_4$ just begins to precipitate. The final volume is 500 mL. The solubility product of $BaSO_4$ is 1×10^{-10}. What is the original concentration of Ba^{2+}?

(a) 5×10^{-9} M (b) 2×10^{-9} M
(c) 1.1×10^{-9} M (d) 1.0×10^{-10} M

39. At 518°C, the rate of decomposition of a sample of gaseous acetaldehyde, initially at a pressure of 363 Torr, was 1.00 Torr s^{-1} when 5% had reacted and 0.5 Torr s^{-1} when 33% had reacted. The order of the reaction is

(a) 2 (b) 3 (c) 1 (d) 0

40. How long (approximate) should water be electrolysed by passing through 100 amperes current so that the oxygen released can completely burn 27.66 g of diborane? (Atomic weight of $B = 10.8\ \mu$)

(a) 6.4 hours (b) 0.8 hours
(c) 3.2 hours (d) 1.6 hours

41. The recommended concentration of fluoride ion in drinking water is up to 1 ppm as fluoride ion is required to make teeth enamel harder by converting $[3Ca_3(PO_4)_2 \cdot Ca(OH)_2]$ to :

(a) $[CaF_2]$
(b) $[3(CaF_2) \cdot Ca(OH)_2]$
(c) $[3Ca_3(PO_4)_2 \cdot CaF_2]$
(d) $[3\{Ca_3(PO_4)_2\} \cdot CaF_2]$

42. Which of the following compounds contain(s) no covalent bond(s)?

$KCl, PH_3, O_2, B_2H_6, H_2SO_4$

(a) KCl, B_2H_6, PH_3 (b) KCl, H_2SO_4
(c) KCl (d) KCl, B_2H_6

43. Which of the following are Lewis acids?

(a) PH_3 and BCl_3 (b) $AlCl_3$ and $SiCl_4$
(c) PH_3 and $SiCl_4$ (d) BCl_3 and $AlCl_3$

44. Total number of lone pair of electron in I_3^- ion is

(a) 3 (b) 6 (c) 9 (d) 12

45. Which of the following salts is the most basic in aqueous solution?

(a) $Al(CN)_3$ (b) CH_3COOK
(c) $FeCl_3$ (d) $Pb(CH_3COO)_2$

46. Hydrogen peroxide oxidises $[Fe(CN)_6]^{4-}$ to $[Fe(CN)_6]^{3-}$ in acidic medium but reduces $[Fe(CN)_6]^{3-}$ to $[Fe(CN)_6]^{4-}$ in alkaline medium. The other products formed are, respectively.

(a) $(H_2O + O_2)$ and H_2O
(b) $(H_2O + O_2)$ and $(H_2O + OH^-)$
(c) H_2O and $(H_2O + O_2)$
(d) H_2O and $(H_2O + OH^-)$

47. The oxidation states of Cr, in $[Cr(H_2O)_6]Cl_3$, $[Cr(C_6H_6)_2]$, and $K_2[Cr(CN)_2(O)_2(O_2)(NH_3)]$ respectively are

(a) +3, +4 and +6 (b) +3, +2 and +4
(c) +3, 0 and +6 (d) +3, 0 and +4

48. The compound that does not produce nitrogen gas by the thermal decomposition is

(a) $Ba(N_3)_2$ (b) $(NH_4)_2Cr_2O_7$
(c) NH_4NO_2 (d) $(NH_4)_2SO_4$

49. When metal '*M*' is treated with NaOH, a white gelatinous precipitate '*X*' is obtained, which is soluble in excess of NaOH. Compound '*X*' when heated strongly gives an oxide which is used in chromatography as an adsorbent. The metal '*M*' is

(a) Zn (b) Ca
(c) Al (d) Fe

50. Consider the following reaction and statements :

$$[Co(NH_3)_4Br_2]^+ + Br^- \longrightarrow [Co(NH_3)_3Br_3] + NH_3$$

I. Two isomers are produces if the reactant complex ion is a *cis*-isomer.
II. Two isomers are produced if the reactant complex ion is a *trans*-isomer.
III. Only one isomer is produced if the reactant complex ion is a *trans*-isomer.
IV. Only one isomer is produced if the reactant complex ion is a *cis*-isomer.

The correct statements are

(a) (I) and (II) (b) (I) and (III)
(c) (III) and (IV) (d) (II) and (IV)

51. Glucose on prolonged heating with HI gives

(a) *n*-hexane (b) 1-hexene
(c) Hexanoic acid (d) 6-iodohexanal

52. The *trans*-alkenes are formed by the reduction of alkynes with

(a) H_2-Pd/C, $BaSO_4$ (b) $NaBH_4$
(c) Na/liq. NH_3 (d) Sn-HCl

53. Which of the following compounds will be suitable for Kjeldahl's method for nitrogen estimation?

(a) 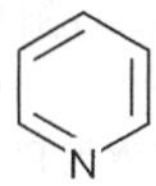

(b) 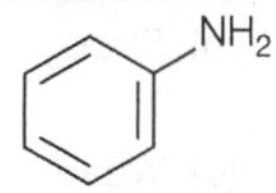

(c) 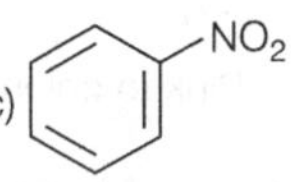

(d) 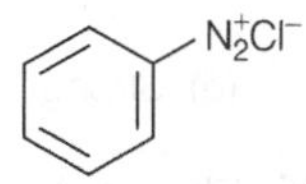

54. Phenol on treatment with CO_2 in the presence of NaOH followed by acidification produces compound *X* as the major product. *X* on treatment with $(CH_3CO)_2O$ in the presence of catalytic amount of H_2SO_4 produces:

(a) 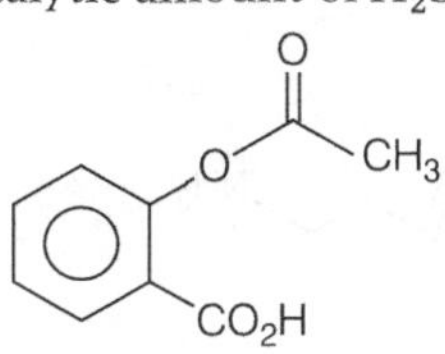

(b) 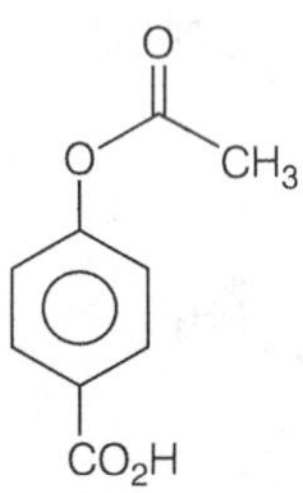

(c) 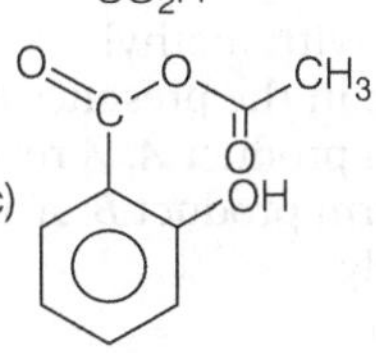

(d) 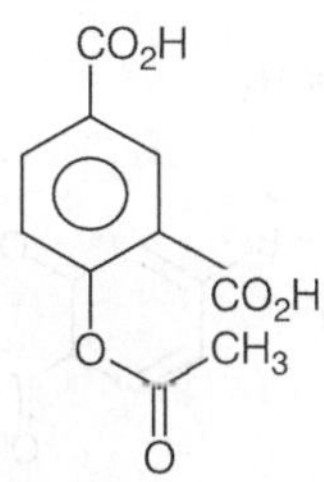

55. An alkali is titrated against an acid with methyl orange as indicator, which of the following is a correct combination?

	Base	Acid	End point
(a)	Weak	Strong	Colourless to pink
(b)	Strong	Strong	Pinkish red to yellow
(c)	Weak	Strong	Yellow to pinkish red
(d)	Strong	Strong	Pink to colourless

56. The predominant form of histamine present in human blood is (pK_a, Histidine = 6.0)

(a) 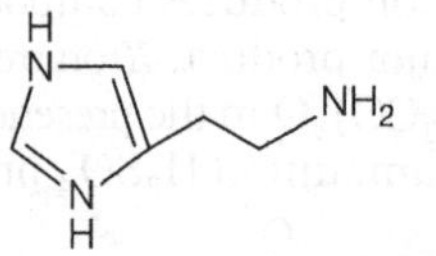

(b) 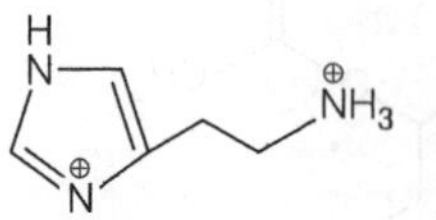

(c) 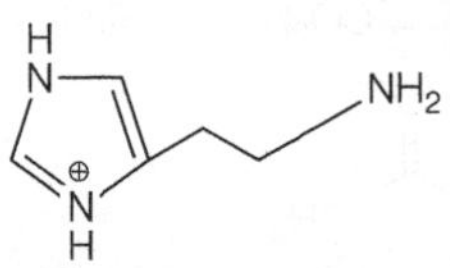

(d) 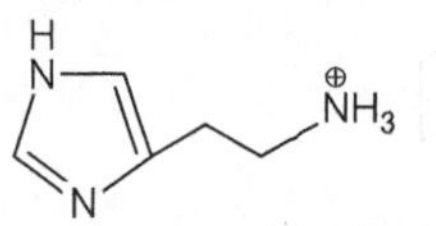

57. Phenol reacts with methyl chloroformate in the presence of NaOH to form product *A*. *A* reacts with Br_2 to form product *B*. *A* and *B* are respectively

(a) 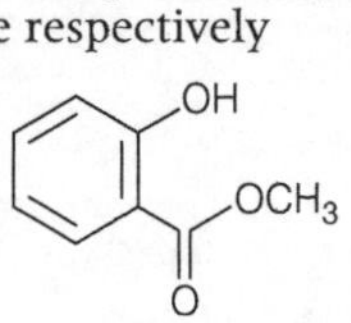

and 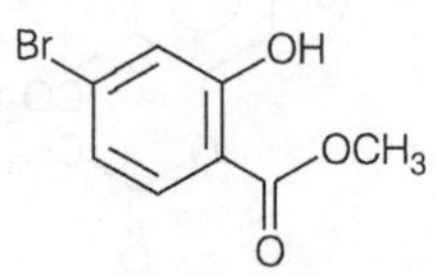

(b) 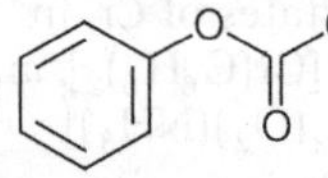

and 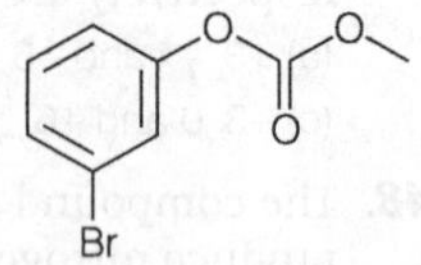

(c) 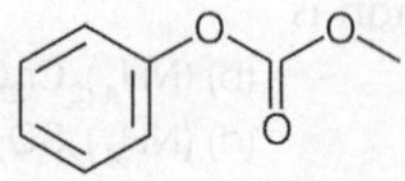

and 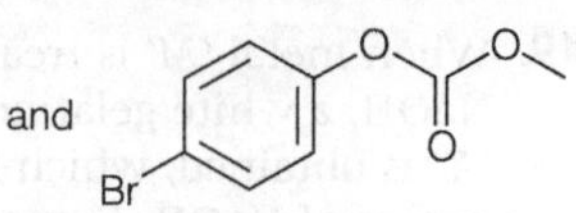

(d) 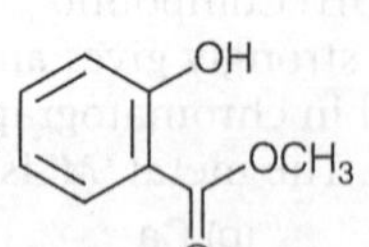

and 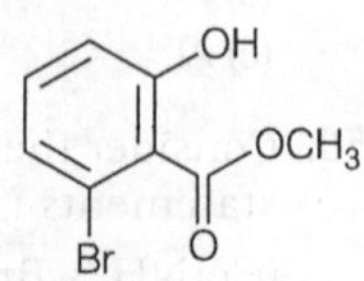

58. The increasing order of basicity of the following compounds is

I. 

II. 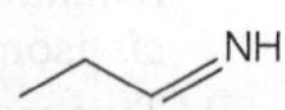

III. 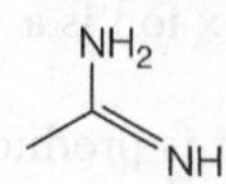

IV. 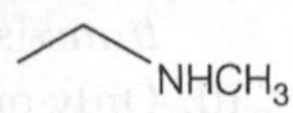

(a) (I) < (II) < (III) < (IV)
(b) (II) < (I) < (III) < (IV)
(c) (II) < (I) < (IV) < (III)
(d) (IV) < (II) < (I) < (III)

59. The major product formed in the following reaction is

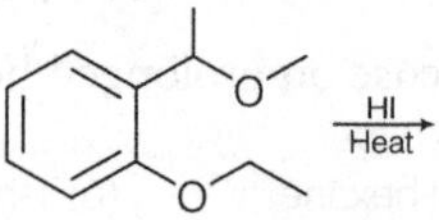

(a) 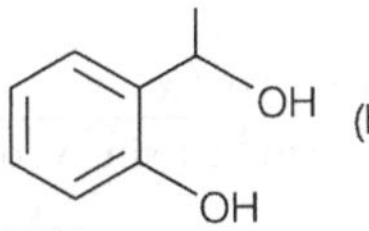

(b) 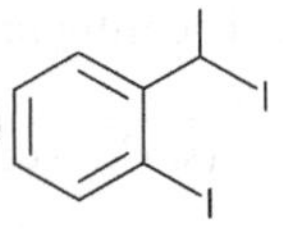

(c) 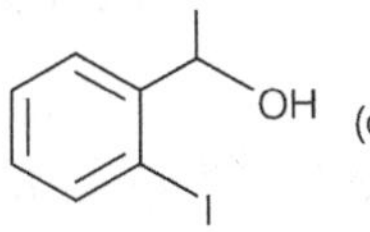

(d) 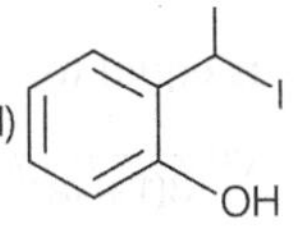

60. The major product of the following reaction is

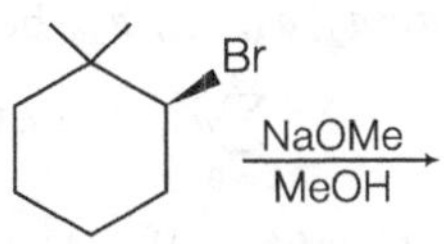

(a) 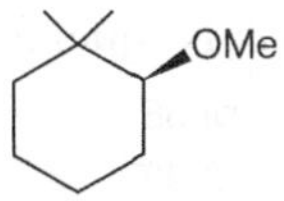

(b) 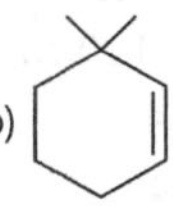

(c) 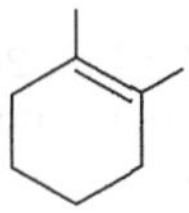
(d) 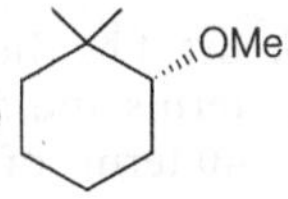

Mathematics

61. Two sets A and B are as under $A = \{(a, b) \in \mathbf{R} \times \mathbf{R} : |a - 5| < 1$ and $|b - 5| < 1\}$; $B = \{(a, b) \in \mathbf{R} \times \mathbf{R} :$ $4(a-6)^2 + 9(b-5)^2 \leq 36\}$. Then,

(a) $B \subset A$
(b) $A \subset B$
(c) $A \cap B = \phi$ (an empty set)
(d) neither $A \subset B$ nor $B \subset A$

62. Let $S = \{x \in \mathbf{R} : x \geq 0$ and $2|\sqrt{x} - 3| + \sqrt{x}(\sqrt{x} - 6) + 6 = 0$. Then, S

(a) is an empty set
(b) contains exactly one element
(c) contains exactly two elements
(d) contains exactly four elements

63. If $\alpha, \beta \in \mathbf{C}$ are the distinct roots of the equation $x^2 - x + 1 = 0$, then $\alpha^{101} + \beta^{107}$ is equal to

(a) –1 (b) 0
(c) 1 (d) 2

64. If $\begin{vmatrix} x-4 & 2x & 2x \\ 2x & x-4 & 2x \\ 2x & 2x & x-4 \end{vmatrix}$ $= (A + Bx)(x - A)^2$, then the ordered pair (A, B) is equal to

(a) (–4, – 5) (b) (–4, 3)
(c) (4, 5) (d) (4, 5)

65. If the system of linear equations

$$x + ky + 3z = 0$$
$$3x + ky - 2z = 0$$
$$2x + 4y - 3z = 0$$

has a non-zero solution (x, y, z), then $\frac{xz}{y^2}$ is equal to

(a) –10 (b) 10
(c) –30 (d) 30

66. From 6 different novels and 3 different dictionaries, 4 novels and 1 dictionary are to be selected and arranged in a row on a shelf, so that the dictionary is always in the middle. The number of such arrangements is

(a) atleast 1000
(b) less than 500
(c) atleast 500 but less than 750
(d) atleast 750 but less than 1000

67. The sum of the coefficients of all odd degree terms in the expansion of $\left(x + \sqrt{x^3 - 1}\right)^5 + \left(x - \sqrt{x^3 - 1}\right)^5$, $(x > 1)$ is

(a) –1 (b) 0
(c) 1 (d) 2

68. Let $a_1, a_2, a_3, \ldots, a_{49}$ be in AP such that $\sum_{k=0}^{12} a_{4k+1} = 416$ and $a_9 + a_{43} = 66$. If $a_1^2 + a_2^2 + \ldots + a_{17}^2 = 140\text{ m}$, then m is equal to

(a) 66 (b) 68
(c) 34 (d) 33

69. Let A be the sum of the first 20 terms and B be the sum of the first 40 terms of the series

$1^2 + 2\cdot 2^2 + 3^2 + 2\cdot 4^2 + 5^2 + 2\cdot 6^2 + \ldots$

If $B - 2A = 100\lambda$, then λ is equal to

(a) 232 (b) 248
(c) 464 (d) 496

70. For each $t \in \mathbf{R}$, let $[t]$ be the greatest integer less than or equal to t. Then,

$$\lim_{x \to 0^+} x\left(\left[\frac{1}{x}\right] + \left[\frac{2}{x}\right] + \ldots + \left[\frac{15}{x}\right]\right)$$

(a) is equal to 0
(b) is equal to 15
(c) is equal to 120
(d) does not exist (in **R**)

71. Let $S = (t \in \mathbf{R} : f(x) = |x - \pi| \cdot (e^{|x|} - 1)\sin|x|$ is not differentiable at $t\}$. Then, the set S is equal to

(a) ϕ (an empty set) (b) $\{0\}$
(c) $\{\pi\}$ (d) $\{0, \pi\}$

72. If the curves $y^2 = 6x, 9x^2 + by^2 = 16$ intersect each other at right angles, then the value of b is

(a) 6 (b) $\frac{7}{2}$
(c) 4 (d) $\frac{9}{2}$

73. Let $f(x) = x^2 + \frac{1}{x^2}$ and $g(x) = x - \frac{1}{x}$, $x \in \mathbf{R} - \{-1, 0, 1\}$. If $h(x) = \frac{f(x)}{g(x)}$, then the local minimum value of $h(x)$ is

(a) 3 (b) -3
(c) $-2\sqrt{2}$ (d) $2\sqrt{2}$

74. The integral

$$\int \frac{\sin^2 x \cos^2 x}{(\sin^5 x + \cos^3 x \sin^2 x + \sin^3 x \cos^2 x + \cos^5 x)^2} dx$$

is equal to

(a) $\frac{1}{3(1 + \tan^3 x)} + C$ (b) $\frac{-1}{3(1 + \tan^3 x)} + C$

(c) $\frac{1}{1 + \cot^3 x} + C$ (d) $\frac{-1}{1 + \cot^3 x} + C$

(where C is a constant of integration)

75. The value of $\int_{-\pi/2}^{\pi/2} \frac{\sin^2 x}{1 + 2^x} dx$ is

(a) $\frac{\pi}{8}$ (b) $\frac{\pi}{2}$
(c) 4π (d) $\frac{\pi}{4}$

76. Let $g(x) = \cos x^2, f(x) = \sqrt{x}$ and α, β $(\alpha < \beta)$ be the roots of the quadratic equation $18x^2 - 9\pi x + \pi^2 = 0$. Then, the area (in sq units) bounded by the curve $y = (gof)(x)$ and the lines $x = \alpha$, $x = \beta$ and $y = 0$, is

(a) $\frac{1}{2}(\sqrt{3} - 1)$ (b) $\frac{1}{2}(\sqrt{3} + 1)$
(c) $\frac{1}{2}(\sqrt{3} - \sqrt{2})$ (d) $\frac{1}{2}(\sqrt{2} - 1)$

77. Let $y = y(x)$ be the solution of the differential equation

$\sin x \frac{dy}{dx} + y\cos x = 4x, x \in (0, \pi)$.

If $y\left(\frac{\pi}{2}\right) = 0$, then $y\left(\frac{\pi}{6}\right)$ is equal to

(a) $\frac{4}{9\sqrt{3}}\pi^2$ (b) $\frac{-8}{9\sqrt{3}}\pi^2$
(c) $-\frac{8}{9}\pi^2$ (d) $-\frac{4}{9}\pi^2$

78. A straight line through a fixed point $(2, 3)$ intersects the coordinate axes at distinct points P and Q. If O is the origin and the rectangle $OPRQ$ is completed, then the locus of R is

(a) $3x + 2y = 6$ (b) $2x + 3y = xy$
(c) $3x + 2y = xy$ (d) $3x + 2y = 6xy$

79. Let the orthocentre and centroid of a triangle be $A(-3, 5)$ and $B(3, 3)$, respectively. If C is the circumcentre of this triangle, then the radius of the circle having line segment AC as diameter, is

(a) $\sqrt{10}$ (b) $2\sqrt{10}$

(c) $3\sqrt{\frac{5}{2}}$ (d) $\frac{3\sqrt{5}}{2}$

80. If the tangent at $(1, 7)$ to the curve $x^2 = y - 6$ touches the circle $x^2 + y^2 + 16x + 12y + c = 0$, then the value of c is

(a) 195 (b) 185

(c) 85 (d) 95

81. Tangent and normal are drawn at $P(16, 16)$ on the parabola $y^2 = 16x$, which intersect the axis of the parabola at A and B, respectively. If C is the centre of the circle through the points P, A and B and $\angle CPB = \theta$, then a value of $\tan\theta$ is

(a) $\frac{1}{2}$ (b) 2

(c) 3 (d) $\frac{4}{3}$

82. Tangents are drawn to the hyperbola $4x^2 - y^2 = 36$ at the points P and Q. If these tangents intersect at the point $T(0, 3)$, then the area (in sq units) of ΔPTQ is

(a) $45\sqrt{5}$ (b) $54\sqrt{3}$

(c) $60\sqrt{3}$ (d) $36\sqrt{5}$

83. If L_1 is the line of intersection of the planes $2x - 2y + 3z - 2 = 0$, $x - y + z + 1 = 0$ and L_2 is the line of intersection of the planes $x + 2y - z - 3 = 0$, $3x - y + 2z - 1 = 0$, then the distance of the origin from the plane, containing the lines L_1 and L_2 is

(a) $\frac{1}{4\sqrt{2}}$ (b) $\frac{1}{3\sqrt{2}}$

(c) $\frac{1}{2\sqrt{2}}$ (d) $\frac{1}{\sqrt{2}}$

84. The length of the projection of the line segment joining the points $(5, -1, 4)$ and $(4, -1, 3)$ on the plane, $x + y + z = 7$ is

(a) $\frac{2}{\sqrt{3}}$ (b) $\frac{2}{3}$

(c) $\frac{1}{3}$ (d) $\sqrt{\frac{2}{3}}$

85. Let $\mathbf{u}$ be a vector coplanar with the vectors $\mathbf{a} = 2\hat{\mathbf{i}} + 3\hat{\mathbf{j}} - \hat{\mathbf{k}}$ and $\mathbf{b} = \hat{\mathbf{j}} + \hat{\mathbf{k}}$. If $\mathbf{u}$ is perpendicular to $\mathbf{a}$ and $\mathbf{u} \cdot \mathbf{b} = 24$, then $|\mathbf{u}|^2$ is equal to

(a) 336 (b) 315

(c) 256 (d) 84

86. A bag contains 4 red and 6 black balls. A ball is drawn at random from the bag, its colour is observed and this ball along with two additional balls of the same colour are returned to the bag. If now a ball is drawn at random from the bag, then the probability that this drawn ball is red, is

(a) $\frac{3}{10}$ (b) $\frac{2}{5}$

(c) $\frac{1}{5}$ (d) $\frac{3}{4}$

87. If $\sum_{i=1}^{9}(x_i - 5) = 9$ and $\sum_{i=1}^{9}(x_i - 5)^2 = 45$, then the standard deviation of the 9 items $x_1, x_2, \ldots, x_9$ is

(a) 9 (b) 4

(c) 2 (d) 3

88. If sum of all the solutions of the equation

$$8\cos x \cdot \left(\cos\left(\frac{\pi}{6} + x\right) \cdot \cos\left(\frac{\pi}{6} - x\right) - \frac{1}{2}\right) = 1$$

in $[0, \pi]$ is $k\pi$, then k is equal to

(a) $\frac{2}{3}$ (b) $\frac{13}{9}$

(c) $\frac{8}{9}$ (d) $\frac{20}{9}$

89. PQR is a triangular park with $PQ = PR = 200$ m. A TV tower stands at the mid-point of QR. If the angles of elevation of the top of the tower at P, Q and R are respectively 45°, 30° and 30°, then the height of the tower (in m) is

(a) 100 (b) 50

(c) $100\sqrt{3}$ (d) $50\sqrt{2}$

90. The boolean expression $\sim(p \vee q) \vee (\sim p \wedge q)$ is equivalent to

(a) $\sim p$ (b) p

(c) q (d) $\sim q$

Answers

Physics

1.	(c)	2.	(b)	3.	(b)	4.	(c)	5.	(b)	6.	(d)	7.	(a)	8.	(c)	9.	(c)	10.	(c)
11.	(a)	12.	(b)	13.	(a)	14.	(b)	15.	(a)	16.	(b)	17.	(b)	18.	(b)	19.	(c)	20.	(a)
21.	(c)	22.	(c)	23.	(a)	24.	(a)	25.	(d)	26.	(a)	27.	(c)	28.	(c)	29.	(b)	30.	(c)

Chemistry

31.	(d)	32.	(c)	33.	(d)	34.	(a)	35.	(d)	36.	(d)	37.	(b)	38.	(c)	39.	(a)	40.	(c)
41.	(c)	42.	(c)	43.	(d)	44.	(c)	45.	(b)	46.	(c)	47.	(c)	48.	(d)	49.	(c)	50.	(b)
51.	(a)	52.	(c)	53.	(b)	54.	(a)	55.	(c)	56.	(d)	57.	(c)	58.	(c)	59.	(d)	60.	(b)

Mathematics

61.	(b)	62.	(c)	63.	(c)	64.	(c)	65.	(b)	66.	(a)	67.	(d)	68.	(c)	69.	(b)	70.	(c)
71.	(a)	72.	(d)	73.	(d)	74.	(b)	75.	(d)	76.	(a)	77.	(c)	78.	(c)	79.	(c)	80.	(d)
81.	(b)	82.	(a)	83.	(b)	84.	(d)	85.	(a)	86.	(b)	87.	(c)	88.	(b)	89.	(a)	90.	(a)

JEE ADVANCED

UNSOLVED PAPER 2018

Paper 1

PHYSICS

Section 1 (Maximum Marks : 24)

- This section contains **SIX (06)** questions.
- Each question has **FOUR** options for correct answer(s). **ONE OR MORE THAN ONE** of these four option(s) is (are) correct options(s).
- For each question, choose the correct options(s) to answer the question.
- Answer to each question will be evaluated according to the following marking scheme:

Full Marks : **+4** If only (all) the correct option(s) is (are) chosen.
Partial Marks : **+3** If all the four options are correct but ONLY three options are chosen.
Partial Marks : **+2** If three or more options are correct but ONLY two options are chosen, both of which are correct options.
Partial Marks : **+1** If two or more options are correct but ONLY one option is chosen and it is a correct option.
Zero Marks : **0** If none of the options is chosen (i.e. the question is unanswered).
Negative Marks : **−2** In all other cases.

- **For example:** If first, third and fourth are the ONLY three correct options for a question with second option being an incorrect option; selecting only all the three correct options will result in + 4 marks. Selecting only two of the three correct options (e.g. the first and fourth options), without selecting any incorrect option (second option in this case), will result in +2 marks. Selecting only one of the three correct options (either first or third or fourth option), without selecting any incorrect option (second option in this case), will result in +1 marks. Selecting any incorrect option(s) (second option in this case), with or without selection of any correct option(s) will result in –2 marks.

1. The potential energy of mass m at a distance r from a fixed point O is given by $V(r) = kr^2 / 2$, where k is a positive constant of appropriate dimensions. This particle is moving in a circular orbit of radius R about the point O. If v is the speed of the particle and L is the magnitude of its angular momentum about O, which of the following statements is (are) true ?

(a) $v = \sqrt{\frac{k}{2m}}R$ (b) $v = \sqrt{\frac{k}{m}}R$

(c) $L = \sqrt{mk}R^2$ (d) $L = \sqrt{\frac{mk}{2}}R^2$

2. Consider a body of mass 1.0 kg at rest at the origin at time $t = 0$. A force

$\mathbf{F} = (\alpha t\hat{\mathbf{i}} + \beta\hat{\mathbf{j}})$ is applied on the body, where $\alpha = 1.0\,\text{Ns}^{-1}$ and $\beta = 1.0\,\text{N}$. The torque acting on the body about the origin at time $t = 1.0\,\text{s}$ is τ. Which of the following statements is (are) true ?

(a) $|\tau| = \frac{1}{3}\,\text{N-m}$

(b) The torque τ is in the direction of the unit vector $+\hat{\mathbf{k}}$

(c) The velocity of the body at $t = 1$ s is $\mathbf{v} = \frac{1}{2}(\hat{\mathbf{i}} + 2\hat{\mathbf{j}})\,\text{ms}^{-1}$

(d) The magnitude of displacement of the body at $t = 1$ s is $\frac{1}{6}$ m

3. A uniform capillary tube of inner radius r is dipped vertically into a beaker filled with water. The water rises to a height h in the capillary tube above the water surface in the beaker. The surface tension of water is σ. The angle of contact between water and the wall of the capillary tube is θ. Ignore the mass of water in the meniscus. Which of the following statements is (are) true ?

(a) For a given material of the capillary tube, h decreases with increase in r

(b) For a given material of the capillary tube, h is independent of σ

(c) If this experiment is performed in a lift going up with a constant acceleration, then h decreases

(d) h is proportional to contact angle θ

4. In the figure below, the switches S_1 and S_2 are closed simultaneously at $t = 0$ and a current starts to flow in the circuit. Both the batteries have the same magnitude of the electromotive force (emf) and the polarities are as indicated in the figure. Ignore mutual inductance between the inductors. The current I in the middle wire reaches its maximum magnitude I_{max} at time $t = \tau$. Which of the following statements is (are) true ?

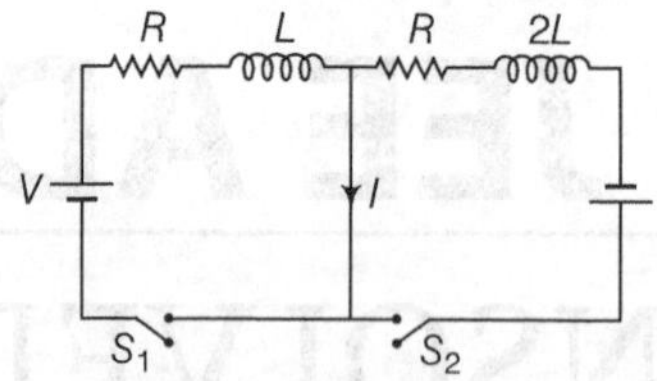

(a) $I_{max} = \frac{V}{2R}$ (b) $I_{max} = \frac{V}{4R}$

(c) $\tau = \frac{L}{R}\ln 2$ (d) $\tau = \frac{2L}{R}\ln 2$

5. Two infinitely long straight wires lie in the xy-plane along the lines $x = \pm R$. The wire located at $x = +R$ carries a constant current I_1 and the wire located at $x = -R$ carries a constant current I_2. A circular loop of radius R is suspended with its centre at $(0, 0, \sqrt{3}R)$ and in a plane parallel to the xy-plane. This loop carries a constant current I in the clockwise direction as seen from above the loop. The current in the wire is taken to be positive, if it is in the $+\hat{\mathbf{j}}$-direction. Which of the following statements regarding the magnetic field **B** is (are) true?

(a) If $I_1 = I_2$, then **B** cannot be equal to zero at the origin (0,0,0)

(b) If $I_1 > 0$ and $I_2 < 0$, then **B** can be equal to zero at the origin (0,0,0)

(c) If $I_1 < 0$ and $I_2 > 0$, then **B** can be equal to zero at the origin (0,0,0)

(d) If $I_1 = I_2$, then the z-component of the magnetic field at the centre of the loop is $\left(-\frac{\mu_0 I}{2R}\right)$

6. One mole of a monoatomic ideal gas undergoes a cyclic process as shown in the figure (where, V is the volume and T is the temperature). Which of the statements below is (are) true ?

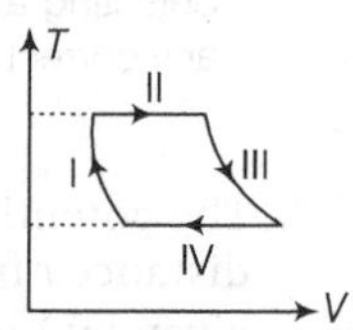

(a) Process I is an isochoric process

(b) In process II, gas absorbs heat

(c) In process IV, gas releases heat

(d) Processes I and III are not isobaric

Section 2 (Maximum Marks : 24)

- This section contains **EIGHT (08)** questions. The answer to each question is a **NUMERICAL VALUE.**
- Four each question, enter the correct numerical value (in decimal notation, truncated/rounded-off to the **second decimal place;** e.g. 6.25, 7.00.– 0.33, –.30, 30.27, –127.30) using the mouse and the on-screen virtual numeric keypad in the place designated to enter the answer.
- Answer to each question will be evaluated according to the following marking scheme:
  Full Marks : **+3** If ONLY the correct numerical value is entered as answer.
  Zero Marks : **0** In all other cases.

7. Two vectors **A** and **B** are defined as $\mathbf{A} = a\hat{\mathbf{i}}$ and $\mathbf{B} = a(\cos\omega t\hat{\mathbf{i}} + \sin\omega t\hat{\mathbf{j}})$, where a is a constant and $\omega = \pi/6$ rad s^{-1}. If $|\mathbf{A}+\mathbf{B}| = \sqrt{3}|\mathbf{A}-\mathbf{B}|$ at time $t = \tau$ for the first time, the value of τ, in seconds, is ....

8. Two men are walking along a horizontal straight line in the same direction. The main in front walks at a speed 1.0 ms^{-1} and the man behind walks at a speed 2.0 ms^{-1}. A third man is standing at a height 12 m above the same horizontal line such that all three men are in a vertical plane. The two walking men are blowing identical whistles which emit a sound of frequency 1430 Hz. The speed of sound in air 330 ms^{-1}. At the instant, when the moving men are 10 m apart, the stationary man is equidistant from them. The frequency of beats in Hz, heard by the stationary man at this instant, is ....

9. A ring and a disc are initially at rest, side by side, at the top of an inclined plane which makes an angle 60° with the horizontal. They start to roll without slipping at the same instant of time along the shortest path. If the time difference between their reaching the ground is $(2-\sqrt{3})/\sqrt{10}$ s, then the height of the top of the inclined plane, in metres, is .... (Take, $g = 10\,ms^{-2}$)

10. A spring block system is resting on a frictionless floor as shown in the figure. The spring constant is 2.0 Nm^{-1} and the mass of the block is 2.0 kg . Ignore the mass of the spring. Initially, the spring is in an unstretched condition. Another block of mass 1.0 kg moving with a speed of 2.0 ms^{-1} collides elastically with the first block. The collision is such that the 2.0 kg block does not hit the wall. The distance, in metres, between the two blocks when the spring returns to its unstretched position for the first time after the collision is ....

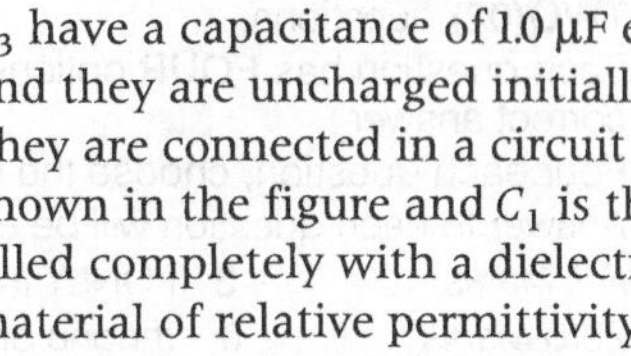

11. Three identical capacitors C_1, C_2 and C_3 have a capacitance of 1.0 μF each and they are uncharged initially. They are connected in a circuit as shown in the figure and C_1 is then filled completely with a dielectric material of relative permittivity ε_r. The cell electromotive force (emf) $V_0 = 8$ V. First the switch S_1 is closed while the switch S_2 is kept open. When the capacitor C_3 is fully charged, S_1 is opened and S_2 is closed simultaneously. When all the capacitors reach equilibrium, the charge on C_3 is found to be $5\mu C$. The value of ε_r =............

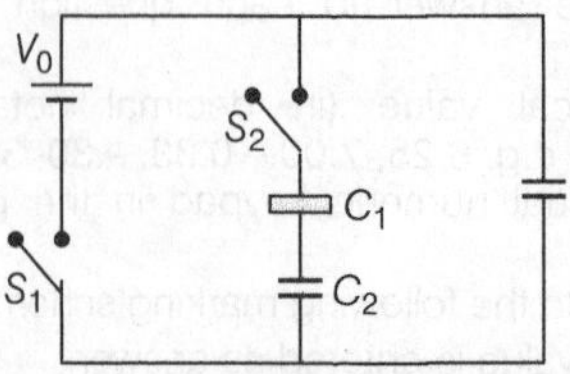

12. In the *xy*-plane, the region $y > 0$ has a uniform magnetic field $B_1\hat{\mathbf{k}}$ and the region $y < 0$ has another uniform magnetic field $B_2\hat{\mathbf{k}}$. A positively charged particle is projected from the origin along the positive *Y*-axis with speed $v_0 = \pi\ \text{ms}^{-1}$ at $t = 0$, as shown in figure. Neglect gravity in this problem. Let $t = T$ be the time when the particle crosses the *X*-axis from below for the first time. If $B_2 = 4B_1$, the average speed of the particle, in ms^{-1}, along the *X*-axis in the time interval T is........... .

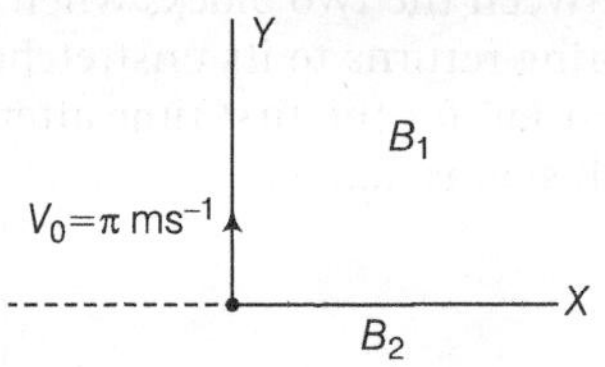

13. Sunlight of intensity $1.3\ \text{kWm}^{-2}$ is incident normally on a thin convex lens of focal length 20 cm. Ignore the energy loss of light due to the lens and assume that the lens aperture size is much smaller than its focal length. The average intensity of light, in kW m^{-2}, at a distance 22 cm from the lens on the other side is ....

14. Two conducting cylinders of equal length but different radii are connected in series between two heat baths kept at temperatures $T_1 = 300\ \text{K}$ and $T_2 = 100\ \text{K}$, as shown in the figure. The radius of the bigger cylinder is twice that of the smaller one and the thermal conductivities of the materials of the smaller and the larger cylinders are K_1 and K_2, respectively. If the temperature at the junction of the cylinders in the steady state is 200 K, then K_1/K_2 =......... .

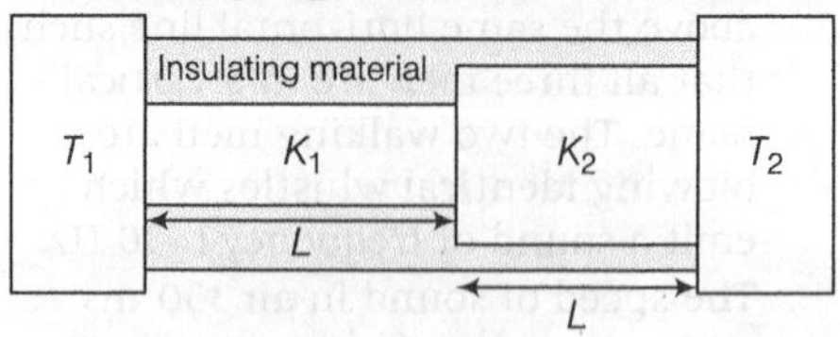

Section 3 (Maximum Marks : 12)

- This section contains **TWO (02)** paragraphs. Based on each paragraph, there are **TWO(02)** questions.
- Each question has **FOUR** options. **ONLY ONE** of these four options corresponds to the correct answer.
- Four each question, choose the option corresponding to the correct answer.
- Answer to each question will be evaluated according to the following marking scheme:

Full Marks : **+3** If ONLY the correct option is chosen.
Zero Marks : **0** If none of the options is chosen (i.e. the question is unanswered).
Negative Marks : **−1** In all other cases.

Paragraph "X"

In electromagnetic theory, the electric and magnetic phenomena are related to each other. Therefore, the dimensions of electric and magnetic quantities must also be related to each other. In the questions below, [E] and [B] stand for dimensions of electric and magnetic fields respectively, while $[\varepsilon_0]$ and $[\mu_0]$ stand for dimensions of the permittivity and permeability of free space, respectively. [L] and [T] are dimensions of length and time, respectively. All the quantities are given in SI units.

(There are two questions based on PARAGRAPH "X", the question given below is one of them)

15. The relation between $[E]$ and $[B]$ is

(a) $[E] = [B]\ [L]\ [T]$

(b) $[E] = [B][L]^{-1}[T]$

(c) $[E] = [B][L][T]^{-1}$

(d) $[E] = [B][L]^{-1}[T]^{-1}$

16. The relation between $[\varepsilon_0]$ and $[\mu_0]$ is

(a) $[\mu_0] = [\varepsilon_0][L]^2[T]^{-2}$

(b) $[\mu_0] = [\varepsilon_0][L]^{-2}[T]^2$

(c) $[\mu_0] = [\varepsilon_0]^{-1}[L]^2[T]^{-2}$

(d) $[\mu_0] = [\varepsilon_0]^{-1}[L]^{-2}[T]^2$

Paragraph "A"

If the measurement errors in all the independent quantities are known, then it is possible to determine the error in any dependent quantity. This is done by the use of series expansion and truncating the expansion at the first power of the error. For example, consider the relation $z = x/y$. If the errors in x, y and z are Δx, Δy and Δz respectively, then

$$z \pm \Delta z = \frac{x \pm \Delta x}{y \pm \Delta y}$$

$$= \frac{x}{y}\left(1 \pm \frac{\Delta x}{x}\right)\left(1 \pm \frac{\Delta y}{y}\right)^{-1}$$

The series expansion for $\left(1 \pm \frac{\Delta y}{y}\right)^{-1}$, to first power in $\Delta y/y$, is $1 \mp (\Delta y/y)$. The relative errors in independent variables are always added. So, the error in z will be

$$\Delta z = z\left(\frac{\Delta x}{x} + \frac{\Delta y}{y}\right)$$

The above derivation makes the assumption that $\Delta x/x \ll 1$, $\Delta y/y \ll 1$. Therefore, the higher powers of these quantities are neglected.

(There are two questions based on PARAGRAPH "A", the question given below is one of them)

17. Consider the ratio $r = \frac{(1-a)}{(1+a)}$ to be determined by measuring a dimensionless quantity a. If the error in the measurement of a is Δa ($\Delta a/a \ll 1$), then what is the error Δr in determining r?

(a) $\frac{\Delta a}{(1+a)^2}$

(b) $\frac{2\Delta a}{(1+a)^2}$

(c) $\frac{2\Delta a}{(1-a)^2}$

(d) $\frac{2a\Delta a}{(1-a^2)}$

18. In an experiment, the initial number of radioactive nuclei is 3000. It is found that 1000 ± 40 nuclei decayed in the first 1.0s. For $|x| \ll 1$, $\ln(1+x) = x$ up to first power in x. The error $\Delta\lambda$, in the determination of the decay constant λ in s^{-1}, is

(a) 0.04

(b) 0.03

(c) 0.02

(d) 0.01

CHEMISTRY

Section 1 (Maximum Marks : 24)

- This section contains **SIX (06)** questions.
- Each question has **FOUR** options for correct answer(s). **ONE OR MORE THAN ONE** of these four option(s) is (are) correct options(s).
- For each question, choose the correct options(s) to answer the question.
- Answer to each question will be evaluated according to the following marking scheme:

Full Marks : **+4** If only (all) the correct option(s) is (are) chosen.
Partial Marks : **+3** If all the four options are correct but ONLY three options are chosen.
Partial Marks : **+2** If three or more options are correct but ONLY two options are chosen, both of which are correct options.
Partial Marks : **+1** If two or more options are correct but ONLY one option is chosen and it is a correct option.
Zero Marks : **0** If none of the options is chosen (i.e. the question is unanswered).
Negative Marks : **−2** In all other cases.

- **For example:** If first, third and fourth are the ONLY three correct options for a question with second option being an incorrect option; selecting only all the three correct options will result in + 4 marks. Selecting only two of the three correct options (e.g. the first and fourth options), without selecting any incorrect option (second option in this case), will result in +2 marks. Selecting only one of the three correct options (either first or third or fourth option), without selecting any incorrect option (second option in this case), will result in +1 marks. Selecting any incorrect option(s) (second option in this case), with or without selection of any correct option(s) will result in −2 marks.

1. The compound(s) which generate(s) N_2 gas upon thermal decomposition below 300°C is (are)

(a) NH_4NO_3 (b) $(NH_4)_2Cr_2O_7$
(c) $Ba(N_3)_2$ (d) Mg_3N_2

2. The correct statement (s) regarding the binary transition metal carbonyl compounds is (are) (Atomic numbers : Fe = 26, Ni = 28)

(a) Total number of valence shell electrons at metal centre in $Fe(CO)_5$ or $Ni(CO)_4$ is 16
(b) These are predominantly low spin in nature
(c) Metal-carbon bond strengthens when the oxidation state of the metal is lowered
(d) The carbonyl C—O bond weakens when the oxidation state of the metal is increased

3. Based on the compounds of group 15 elements, the correct statement(s) is (are)

(a) Bi_2O_5 is more basic than N_2O_5
(b) NF_3 is more covalent than BiF_3
(c) PH_3 boils at lower temperature than NH_3
(d) The N—N single bond is stronger than the P—P single bond

4. In the following reaction sequence, the correct structure (s) of X is (are)

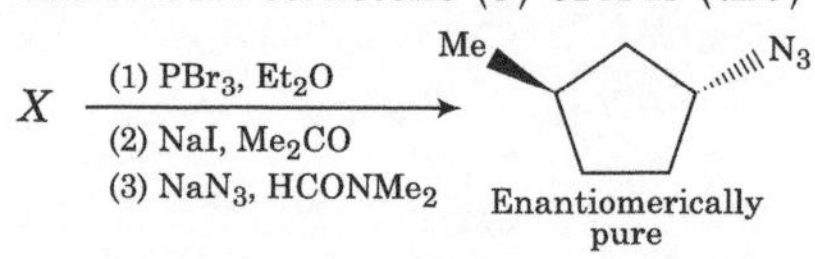

(a)
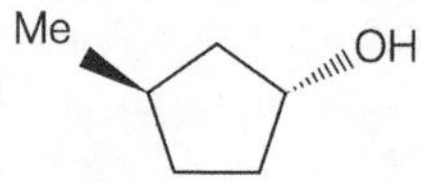

(b)
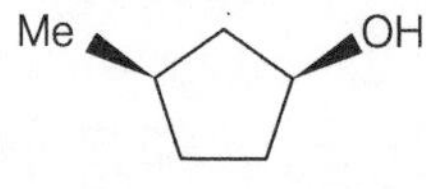

(c) 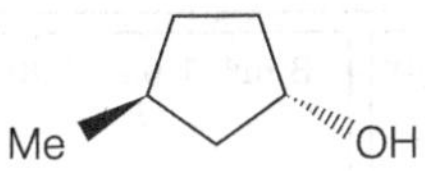

(d) 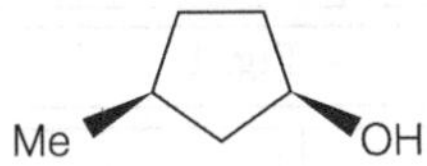

5. The reaction(s) leading to the formation of 1,3,5-trimethylbenzene is (are)

(a) 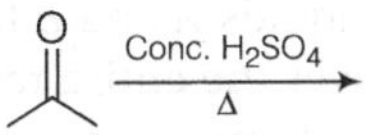

(b) 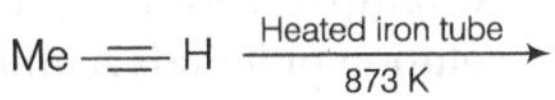

(c) 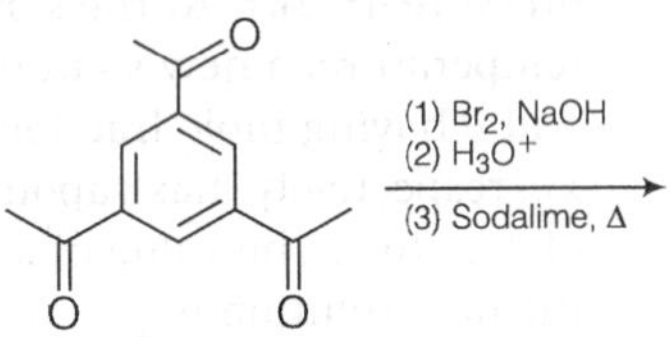

(d) 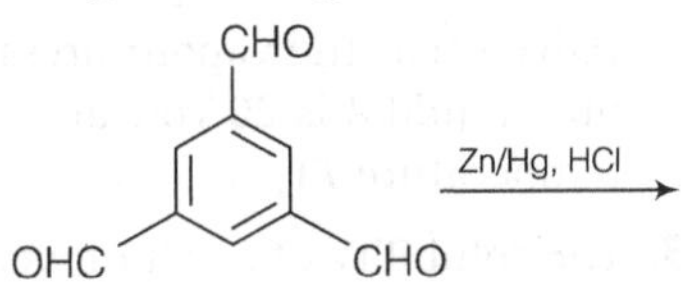

6. A reversible cyclic process for an ideal gas is shown below. Here, p, V and T are pressure, volume and temperature, respectively. The thermodynamic parameters q, w, H and U are heat, work, enthalpy and internal energy, respectively.

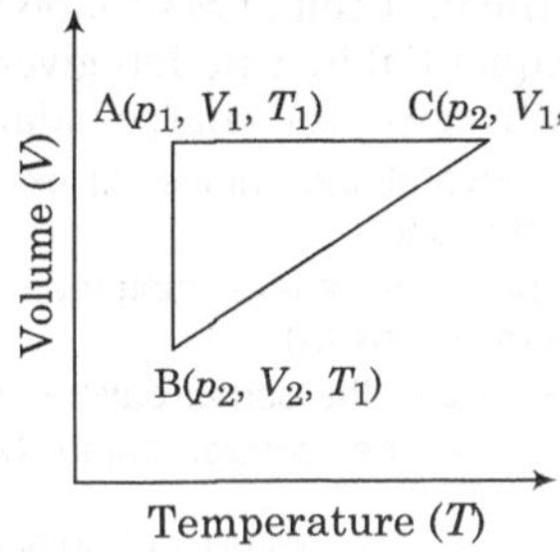

The correct options is (are)

(a) $q_{AC} = \Delta U_{BC}$ and $w_{AB} = p_2(V_2 - V_1)$

(b) $w_{BC} = p_2(V_2 - V_1)$ and $q_{BC} = \Delta H_{AC}$

(c) $\Delta H_{CA} < \Delta U_{CA}$ and $q_{AC} = \Delta U_{BC}$

(d) $q_{BC} = \Delta H_{AC}$ and $\Delta H_{CA} > \Delta U_{CA}$

Section 2 (Maximum Marks : 24)

- This section contains **EIGHT (08)** questions. The answer to each questions is a **NUMERICAL VALUE.**
- Four each question, enter the correct numerical value (in decimal notation, truncated/rounded-off to the **second decimal place;** e.g. 6.25, 7.00.– 0.33, –.30, 30.27, –127.30) using the mouse and the on screen virtual numeric keypad in the place designated to enter the answer.
- Answer to each question will be evaluated according to the following marking scheme

Full Marks : + 3 If ONLY the correct numerical value is entered as answer.

Zero Marks : 0 In all other cases.

7. Among the species given below, the total number of diamagnetic species is____

H atom, NO_2 monomer, O_2^- (superoxide), dimeric sulphur in vapour phase,

Mn_3O_4, $(NH_4)_2[FeCl_4]$, $(NH_4)_2[NiCl_4]$, K_2MnO_4, K_2CrO_4

8. The ammonia prepared by treating ammonium sulphate with calcium hydroxide is completely used by $NiCl_2 \cdot 6H_2O$ to form a stable coordination compound. Assume that both the reactions are 100% complete. If 1584 g of ammonium sulphate and 952 g of $NiCl_2 \cdot 6H_2O$ are used in the preparation, the combined weight (in grams) of gypsum and the nickel-ammonia coordination compound thus produced is____

(Atomic weights in g mol^{-1} :
H = 1, N = 14, O = 16, S = 32,
Cl = 35.5, Ca = 40, Ni = 59)

9. Consider an ionic solid MX with NaCl structure. Construct a new structure (Z) whose unit cell is constructed from the unit cell of MX following the sequential instruction given below. Neglect the charge balance.

(a) Remove all the anions (X) except the central one

(b) Replace all the face centered cations (M) by anions (X)

(c) Remove all the corner cations (M)

(d) Replace the central anion (X) with cation (M)

The value of $\left(\frac{\text{Number of anions}}{\text{Number of cations}}\right)$ in Z is ___

10. For the electrochemical cell,

$Mg(s)|Mg^{2+}(aq, 1\text{ M})||Cu^{2+}(aq, 1\text{ M})|Cu(s)$

The standard emf of the cell is 2.70 V at 300 K. When the concentration of Mg^{2+} is changed to x M, the cell potential changes to 2.67 V at 300 K. The value of x is _____.

(Given, $\frac{F}{R} = 11500\text{ K V}^{-1}$, where F is the Faraday constant and R is the gas contant, In (10) = 2.30)

11. A closed tank has two compartments A and B, both filled with oxygen (assumed to be ideal gas). The partition separating the two compartments is fixed and is a perfect heat insulator (Fig. 1). If the old partition is replaced by a new partition which can slide and conduct heat but does not allow the gas to leak across (Fig. 2), the volume (in m^3) of the compartment A after the system attains equilibrium is ____.

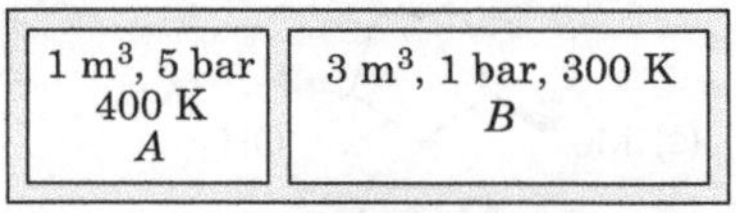

Fig. 1

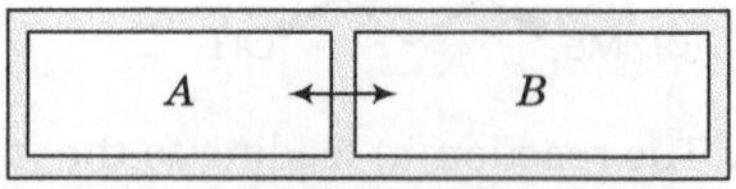

Fig. 2

12. Liquids A and B form ideal solution over the entire range of composition. At temperature T, equimolar binary solution of liquids A and B has vapour pressure 45 torr. At the same temperature, a new solution of A and B having mole fractions x_A and x_B, respectively, has vapour pressure of 22.5 torr. The value of x_A / x_B in the new solution is _____.

(Given that the vapour pressure of pure liquid A is 20 Torr at temperature T)

13. The solubility of a salt of weak acid (AB) at pH 3 is $Y \times 10^{-3}$ mol L^{-1}. The value of Y is _____

(Given that the value of solubility product of AB (K_{sp}) $= 2 \times 10^{-10}$ and the value of ionisation constant of HB (K_a) $= 1 \times 10^{-8}$)

14. The plot given below shows $p - T$ curves (where p is the pressure and T is the temperature) for two solvents X and Y and isomolal solution of NaCl in these solvents. NaCl completely dissociates in both the solvents.

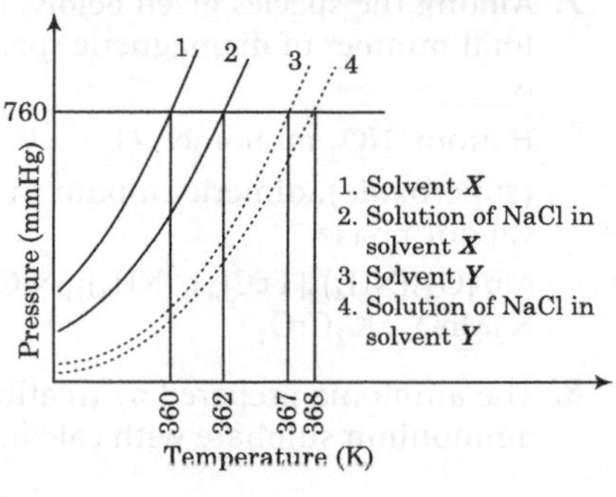

On addition of equal number of moles of a non-volatile solute S in equal amount (in kg) of these solvents, the elevation of boiling point of solvent X is three times that of solvent Y. Solute S is known to undergo dimerisation in these solvents. If the degree of dimerisation is 0.7 in solvent Y, the degree of dimerisation in solvent X is ____.

Section 3 (Maximum Marks : 12)

- This section contains **TWO (02)** paragraphs. Based on each paragraph, there are **TWO(02)** questions.
- Each question has **FOUR** options. **ONLY ONE** of these four options corresponds to the correct answer.
- Four each question, choose the option corresponding to the correct answer.
- Answer to each question will be evaluated according to the following marking scheme:

Full Marks : +3 If ONLY the correct option is chosen.
Zero Marks : 0 If none of the options is chosen (i.e. the question is unanswered).
Negative Marks : −1 In all other cases.

Paragraph "X"

Treatment of benzene with CO/ HCl in the presence of anhydrous $AlCl_3$ / CuCl followed by reaction with Ac_2O/ NaOAc gives compound X as the major product. Compound X upon reaction with Br_2 / Na_2CO_3 followed by heating at 473 K with moist KOH furnishes Y as the major product. Reaction of X with H_2 / Pd-C, followed by H_3PO_4 treatment gives Z as the major product.

15. The compound Y is

(a) 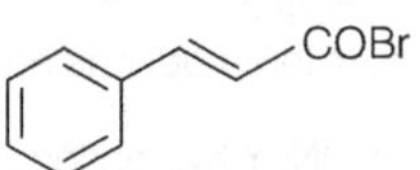

(b) 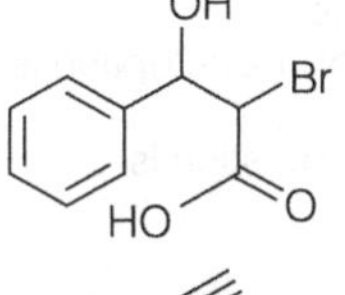

(c) 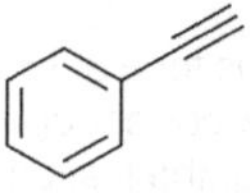

(d) 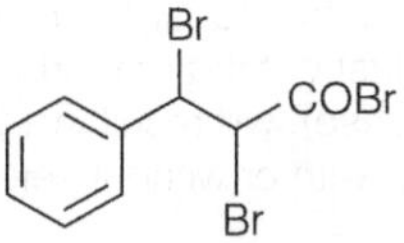

16. The compound Z is

(a) 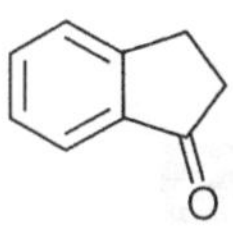

(b) 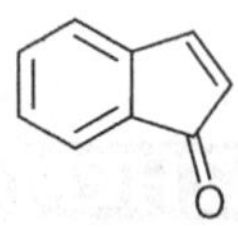

(c) 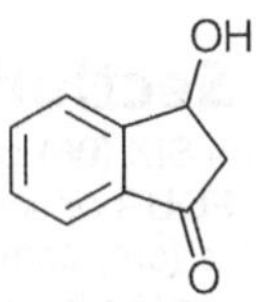

(d) 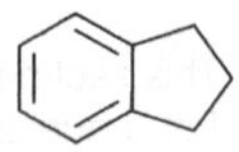

Paragraph "A"

An organic acid $P(C_{11}H_{12}O_2)$ can easily be oxidised to a dibasic acid which reacts with ethylene glycol to produce a polymer dacron. Upon ozonolysis, P gives an aliphatic ketone as one of the products. P undergoes the following reaction sequences to furnish R via Q. The compound P also undergoes another set of reactions to produce S.

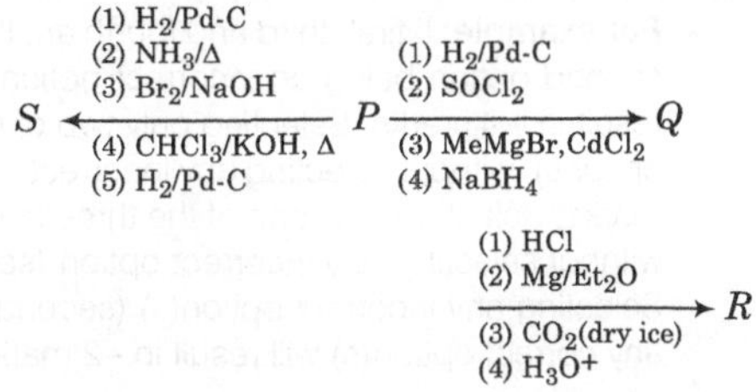

17. The Compound R is

(a) 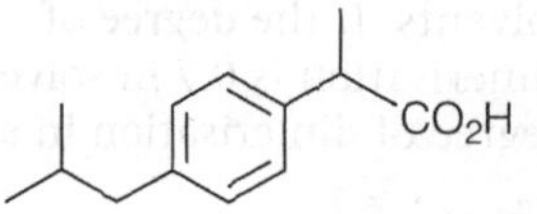

(b) 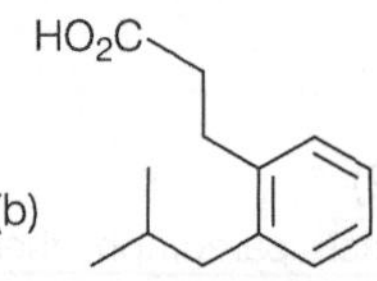

(c) 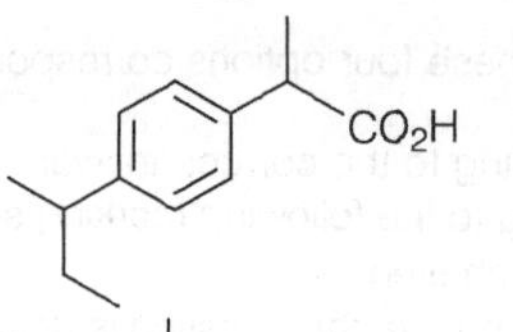

(d) 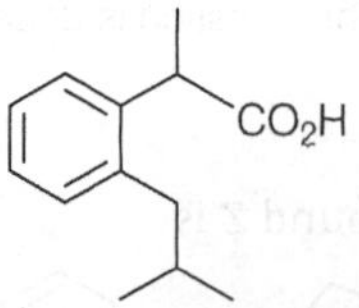

18. The compound S is

(a) 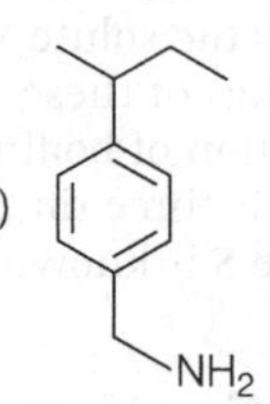

(b) 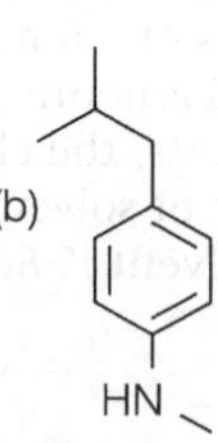

(c) 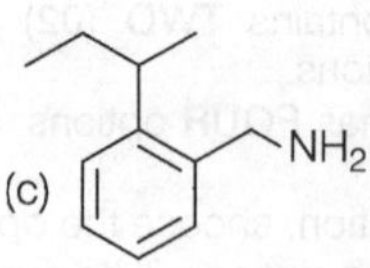

(d) 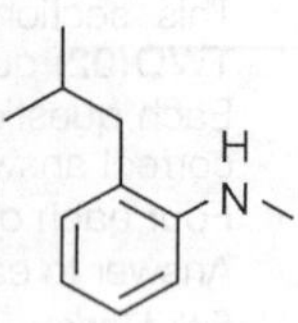

MATHEMATICS

Section 1 (Maximum Marks : 24)

- This section contains **SIX (06)** questions.
- Each question has **FOUR** options for correct answer(s). **ONE OR MORE THAN ONE** of these four option(s) is (are) correct options(s).
- For each question, choose the correct options(s) to answer the question.
- Answer to each question will be evaluated according to the following marking scheme:

Full Marks : **+4** If only (all) the correct option(s) is (are) chosen.
Partial Marks : **+3** If all the four options are correct but ONLY three options are chosen.
Partial Marks : **+2** If three or more options are correct but ONLY two options are chosen, both of which are correct options.
Partial Marks : **+1** If two or more options are correct but ONLY one option is chosen and it is a correct option.
Zero Marks : **0** If none of the options is chosen (i.e. the question is unanswered).
Negative Marks : **–2** In all other cases.

- **For example:** If first, third and fourth are the ONLY three correct options for a question with second option being an incorrect option; selecting only all the three correct options will result in + 4 marks. Selecting only two of the three correct options (e.g. the first and fourth options), without selecting any incorrect option (second option in this case), will result in +2 marks. Selecting only one of the three correct options (either first or third or fourth option), without selecting any incorrect option (second option in this case), will result in +1 marks. Selecting any incorrect option(s) (second option in this case), with or without selection of any correct option(s) will result in –2 marks.

1. For a non-zero complex number z, let $\arg(z)$ denote the principal argument with $-\pi < \arg(z) \le \pi$. Then, which of the following statement(s) is (are) FALSE ?

(a) $\arg(-1-i) = \frac{\pi}{4}$, where $i = \sqrt{-1}$

(b) The function $f : R \to (-\pi, \pi]$, defined by $f(t) = \arg(-1 + it)$ for all $t \in R$, is continuous at all points of R, where $i = \sqrt{-1}$.

(c) For any two non-zero complex numbers z_1 and z_2, $\arg\left(\frac{z_1}{z_2}\right) - \arg(z_1) + \arg(z_2)$ is an integer multiple of 2π.

(d) For any three given distinct complex numbers z_1, z_2 and z_3, the locus of the point z satisfying the condition $\arg\left(\frac{(z-z_1)(z_2-z_3)}{(z-z_3)(z_2-z_1)}\right) = \pi$, lies on a straight line.

2. In a ΔPQR, let $\angle PQR = 30°$ and the sides PQ and QR have lengths $10\sqrt{3}$ and 10, respectively. Then, which of the following statement(s) is (are) TRUE?

(a) $\angle QPR = 45°$

(b) The area of the ΔPQR is $25\sqrt{3}$ and $\angle QRP = 120°$

(c) The radius of the incircle of the ΔPQR is $10\sqrt{3} - 15$

(d) The area of the circumcircle of the ΔPQR is 100π

3. Let $P_1 : 2x + y - z = 3$ and $P_2 : x + 2y + z = 2$ be two planes. Then, which of the following statement(s) is (are) TRUE?

(a) The line of intersection of P_1 and P_2 has direction ratios $1, 2, -1$

(b) The line $\frac{3x-4}{9} = \frac{1-3y}{9} = \frac{z}{3}$ is perpendicular to the line of intersection of P_1 and P_2

(c) The acute angle between P_1 and P_2 is $60°$

(d) If P_3 is the plane passing through the point $(4, 2, -2)$ and perpendicular to the line of intersection of P_1 and P_2, then the distance of the point $(2, 1, 1)$ from the plane P_3 is $\frac{2}{\sqrt{3}}$

4. For every twice differentiable function $f : R \to [-2, 2]$ with $(f(0))^2 + (f'(0))^2 = 85$, which of the following statement(s) is (are) TRUE ?

(a) There exist $r, s \in R$, where $r < s$, such that f is one-one on the open interval (r, s)

(b) There exists $x_0 \in (-4, 0)$ such that $|f'(x_0)| \le 1$

(c) $\lim_{x \to \infty} f(x) = 1$

(d) There exists $\alpha \in (-4, 4)$ such that $f(\alpha) + f''(\alpha) = 0$ and $f'(\alpha) \ne 0$

5. Let $f : R \to R$ and $g : R \to R$ be two non-constant differentiable functions. If $f'(x) = (e^{(f(x) - g(x))})\, g'(x)$ for all $x \in R$ and $f(1) = g(2) = 1$, then which of the following statement(s) is (are) TRUE?

(a) $f(2) < 1 - \log_e 2$

(b) $f(2) > 1 - \log_e 2$

(c) $g(1) > 1 - \log_e 2$

(d) $g(1) < 1 - \log_e 2$

6. Let $f : [0, \infty) \to R$ be a continuous function such that $f(x) = 1 - 2x + \int_0^x e^{x-t} f(t)\, dt$ for all $x \in [0, \infty)$. Then, which of the following statement(s) is (are) TRUE?

(a) The curve $y = f(x)$ passes through the point $(1, 2)$

(b) The curve $y = f(x)$ passes through the point $(2, -1)$

(c) The area of the region $\{(x, y) \in [0, 1] \times R : f(x) \le y \le \sqrt{1 - x^2}\}$ is $\frac{\pi - 2}{4}$

(d) The area of the region $\{(x, y) \in [0, 1] \times R : f(x) \le y \le \sqrt{1 - x^2}\}$ is $\frac{\pi - 1}{4}$

Section 2 (Maximum Marks : 24)

- This section contains **EIGHT (08)** questions. The answer to each question is a **NUMERICAL VALUE.**
- Four each question, enter the correct numerical value (in decimal notation, truncated/rounded-off to the **second decimal place;** e.g. 6.25, 7.00.– 0.33, –.30, 30.27, –127.30) using the mouse and the on-screen virtual numeric keypad in the place designated to enter the answer.
- Answer to each question will be evaluated according to the following marking scheme:
  Full Marks : **+3** If ONLY the correct numerical value is entered as answer.
  Zero Marks : **0** In all other cases.

7. The value of

$$((\log_2 9)^2)^{\frac{1}{\log_2 (\log_2 9)}} \times (\sqrt{7})^{\frac{1}{\log_4 7}} \text{ is}$$

....

8. The number of 5 digit numbers which are divisible by 4, with digits from the set $\{1, 2, 3, 4, 5\}$ and the repetition of digits is allowed, is ....

9. Let X be the set consisting of the first 2018 terms of the arithmetic progression 1, 6, 11, ..., and Y be the set consisting of the first 2018 terms of the arithmetic progression 9, 16, 23, ... Then, the number of elements in the set $X \cup Y$ is ....

10. The number of real solutions of the equation

$$\sin^{-1}\left(\sum_{i=1}^{\infty} x^{i+1} - x\sum_{i=1}^{\infty}\left(\frac{x}{2}\right)^i\right) = \frac{\pi}{2} - \cos^{-1}\left(\sum_{i=1}^{\infty}\left(-\frac{x}{2}\right)^i - \sum_{i=1}^{\infty}(-x)^i\right)$$

lying in the interval $\left(-\frac{1}{2}, \frac{1}{2}\right)$ is ....

(Here, the inverse trigonometric functions $\sin^{-1} x$ and $\cos^{-1} x$ assume values in $\left[-\frac{\pi}{2}, \frac{\pi}{2}\right]$ and $[0, \pi]$, respectively.)

11. For each positive integer n, let

$$y_n = \frac{1}{n}((n+1)(n+2)\dots(n+n))^{\frac{1}{n}}.$$

For $x \in R$, let $[x]$ be the greatest integer less than or equal to x. If $\lim_{n \to \infty} y_n = L$, then the value of $[L]$ is ....

12. Let **a** and **b** be two unit vectors such that $\mathbf{a} \cdot \mathbf{b} = 0$. For some $x, y \in R$, let $\mathbf{c} = x\mathbf{a} + y\mathbf{b} + (\mathbf{a} \times \mathbf{b})$. If $|\mathbf{c}| = 2$ and the vector **c** is inclined at the same angle α to both **a** and **b**, then the value of $8\cos^2 \alpha$ is ....

13. Let a, b, c be three non-zero real numbers such that the equation

$$\sqrt{3}\, a \cos x + 2b \sin x = c, x \in \left[-\frac{\pi}{2}, \frac{\pi}{2}\right]$$

, has two distinct real roots α and β with $\alpha + \beta = \frac{\pi}{3}$. Then, the value of $\frac{b}{a}$ is ....

14. A farmer F_1 has a land in the shape of a triangle with vertices at $P(0, 0)$, $Q(1, 1)$ and $R(2, 0)$. From this land, a neighbouring farmer F_2 takes away the region which lies between the sides PQ and a curve of the form $y = x^n$ $(n > 1)$. If the area of the region taken away by the farmer F_2 is exactly 30% of the area of ΔPQR, then the value of n is ....

Section 3 (Maximum Marks : 12)

- This section contains **TWO (02)** paragraphs. Based on each paragraph, there are **TWO(02)** questions.
- Each question has **FOUR** options. **ONLY ONE** of these four options corresponds to the correct answer.
- Four each question, choose the option corresponding to the correct answer.
- Answer to each question will be evaluated according to the following marking scheme:

Full Marks : **+3** If ONLY the correct option is chosen.
Zero Marks : **0** If none of the options is chosen (i.e. the question is unanswered).
Negative Marks : **−1** In all other cases.

Paragraph X

Let S be the circle in the XY-plane defined by the equation $x^2 + y^2 = 4$.

(There are two questions based on Paragraph "X", the question given below is one of them)

15. Let E_1E_2 and F_1F_2 be the chords of S passing through the point P_0 $(1, 1)$ and parallel to the X-axis and the Y-axis, respectively. Let G_1G_2 be the chord of S passing through P_0 and having slope -1. Let the tangents to S at E_1 and E_2 meet at E_3, then tangents to S at F_1 and F_2 meet at F_3, and the tangents to S at G_1 and G_2 meet at G_3. Then, the points E_3, F_3 and G_3 lie on the curve

(a) $x + y = 4$
(b) $(x-4)^2 + (y-4)^2 = 16$
(c) $(x-4)(y-4) = 4$
(d) $xy = 4$

16. Let P be a point on the circle S with both coordinates being positive. Let the tangent to S at P intersect the coordinate axes at the points M and N. Then, the mid-point of the line segment MN must lie on the curve

(a) $(x + y)^2 = 3xy$
(b) $x^{2/3} + y^{2/3} = 2^{4/3}$
(c) $x^2 + y^2 = 2xy$
(d) $x^2 + y^2 = x^2y^2$

Paragraph A

There are five students S_1, S_2, S_3, S_4 and S_5 in a music class and for them there are five seats R_1, R_2, R_3, R_4 and R_5 arranged in a row, where initially the seat R_i is allotted to the student S_i, $i = 1, 2, 3, 4, 5$. But, on the examination day, the five students are randomly allotted the five seats.

(There are two questions based on Paragraph "A", the question given below is one of them)

17. The probability that, on the examination day, the student S_1 gets the previously allotted seat R_1, and NONE of the remaining students gets the seat previously allotted to him/her is

(a) $\frac{3}{40}$ (b) $\frac{1}{8}$
(c) $\frac{7}{40}$ (d) $\frac{1}{5}$

18. For $i = 1, 2, 3, 4$, let T_i denote the event that the students S_i and S_{i+1} do NOT sit adjacent to each other on the day of the examination. Then, the probability of the event $T_1 \cap T_2 \cap T_3 \cap T_4$ is

(a) $\frac{1}{15}$ (b) $\frac{1}{10}$
(c) $\frac{7}{60}$ (d) $\frac{1}{5}$

Paper 2

PHYSICS

Section 1 (Maximum Marks : 24)

- This section contains **SIX (06)** questions.
- Each question has **FOUR** options for correct answer(s). **ONE OR MORE THAN ONE** of these four option(s) is (are) correct options(s).
- For each question, choose the correct options(s) to answer the question.
- Answer to each question will be evaluated according to the following marking scheme:

Full Marks	: +4	If only (all) the correct option(s) is (are) chosen.
Partial Marks	: +3	If all the four options are correct but ONLY three options are chosen.
Partial Marks	: +2	If three or more options are correct but ONLY two options are chosen, both of which are correct options.
Partial Marks	: +1	If two or more options are correct but ONLY one option is chosen and it is a correct option.
Zero Marks	: 0	If none of the options is chosen (i.e. the question is unanswered).
Negative Marks	: –2	In all other cases.

- **For example:** If first, third and fourth are the ONLY three correct options for a question with second option being an incorrect option; selecting only all the three correct options will result in + 4 marks. Selecting only two of the three correct options (e.g. the first and fourth options), without selecting any incorrect option (second option in this case), will result in +2 marks. Selecting only one of the three correct options (either first or third or fourth option), without selecting any incorrect option (second option in this case), will result in +1 marks. Selecting any incorrect option(s) (second option in this case), with or without selection of any correct option(s) will result in –2 marks.

1. A particle of mass m is initially at rest at the origin. It is subjected to a force and starts moving along the X-axis. Its kinetic energy K changes with time as $dK / dt = \gamma t$, where γ is a positive constant of appropriate dimensions. Which of the following statements is (are) true?

(a) The force applied on the particle is constant

(b) The speed of the particle is proportional to time

(c) The distance of the particle from the origin increases linearly with time

(d) The force is conservative

2. Consider a thin square plate floating on a viscous liquid in a large tank. The height h of the liquid in the tank is much less than the width of the tank. The floating plate is pulled horizontally with a constant velocity u_0. Which of the following statements is (are) true?

(a) The resistive force of liquid on the plate is inversely proportional to h

(b) The resistive force of liquid on the plate is independent of the area of the plate

(c) The tangential (shear) stress on the floor of the tank increases with u_0

(d) The tangential (shear) stress on the plate varies linearly with the viscosity η of the liquid

3. An infinitely long thin non-conducting wire is parallel to the Z-axis and carries a uniform line charge density λ. It pierces a thin non-conducting spherical shell of radius R in such a way that the arc PQ subtends an angle 120° at the centre O of the spherical shell, as shown in the figure. The permittivity of free space is ε_0. Which of the following statements is (are) true?

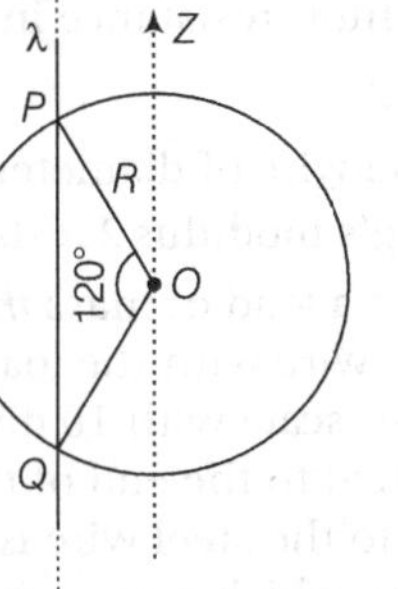

(a) The electric flux through the shell is $\sqrt{3}\, R\lambda / \varepsilon_0$

(b) The z-component of the electric field is zero at all the points on the surface of the shell

(c) The electric flux through the shell is $\sqrt{2}\, R\lambda / \varepsilon_0$

(d) The electric field is normal to the surface of the shell at all points

4. A wire is bent in the shape of a right angled triangle and is placed in front of a concave mirror of focal length f as shown in the figure. Which of the figures shown in the four options qualitatively represent(s) the shape of the image of the bent wire? (These figures are not to scale.)

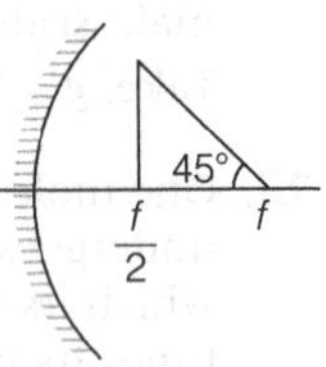

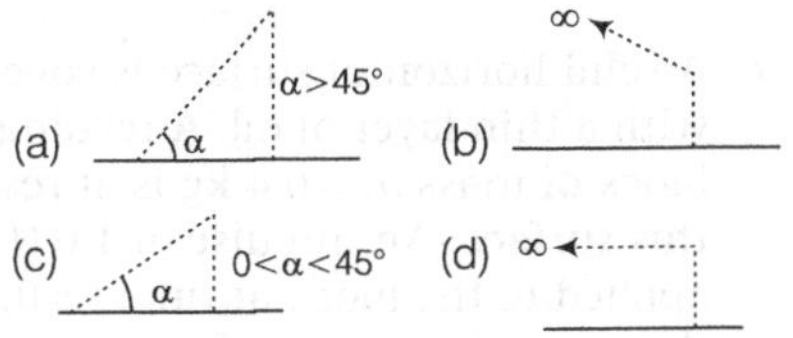

5. In a radioactive decay chain, ${}^{232}_{90}\text{Th}$ nucleus decays to ${}^{212}_{82}\text{Pb}$ nucleus. Let N_α and N_β be the number of α and β- particles respectively, emitted in this decay process. Which of the following statements is (are) true?

(a) $N_\alpha = 5$ (b) $N_\alpha = 6$
(c) $N_\beta = 2$ (d) $N_\beta = 4$

6. In an experiment to measure the speed of sound by a resonating air column, a tuning fork of frequency 500 Hz is used. The length of the air column is varied by changing the level of water in the resonance tube. Two successive resonances are heard at air columns of length 50.7 cm and 83.9 cm. Which of the following statements is (are) true?

(a) The speed of sound determined from this experiment is 332 m s^{-1}

(b) The end correction in this experiment is 0.9 cm

(c) The wavelength of the sound wave is 66.4 cm

(d) The resonance at 50.7 cm corresponds to the fundamental harmonic

Section 2 (Maximum Marks : 24)

- This section contains **EIGHT (08)** questions. The answer to each question is a **NUMERICAL VALUE.**
- Four each question, enter the correct numerical value (in decimal notation, truncated/rounded-off to the **second decimal place;** e.g. 6.25, 7.00.– 0.33, –.30, 30.27, –127.30) using the mouse and the on-screen virtual numeric keypad in the place designated to enter the answer.
- Answer to each question will be evaluated according to the following marking scheme:

Full Marks : **+3** If ONLY the correct numerical value is entered as answer.

Zero Marks : **0** In all other cases.

7. A solid horizontal surface is covered with a thin layer of oil. A rectangular block of mass $m = 0.4$ kg is at rest on this surface. An impulse of 1.0 N s is applied to the block at time $t = 0$, so that it starts moving along the X-axis with a velocity $v(t) = v_0 e^{-t/\tau}$, where v_0 is a constant and $\tau = 4$ s. The displacement of the block, in metres, at $t = \tau$ is .... (Take, $e^{-1} = 0.37$).

8. A ball is projected from the ground at an angle of 45° with the horizontal surface. It reaches a maximum height of 120 m and returns to the ground. Upon hitting the ground for the first time, it loses half of its kinetic energy. Immediately after the bounce, the velocity of the ball makes an angle of 30° with the horizontal surface. The maximum height it reaches after the bounce, in metres, is ....

9. A particle of mass 10^{-3}kg and charge 1.0 C is initially at rest. At time $t = 0$, the particle comes under the influence of an electric field $\mathbf{E}(t) = E_0 \sin \omega t\, \hat{\mathbf{i}}$, where $E_0 = 1.0\ \text{NC}^{-1}$ and $\omega = 10^3$ rad s^{-1}. Consider the effect of only the electrical force on the particle. Then, the maximum speed in m s^{-1}, attained by the particle at subsequent times is ....

10. A moving coil galvanometer has 50 turns and each turn has an area $2 \times 10^{-4}\text{m}^2$. The magnetic field produced by the magnet inside the galvanometer is 0.02 T. The torsional constant of the suspension wire is 10^{-4}N - m rad^{-1}. When a current flows through the galvanometer, a full scale deflection occurs, if the coil rotates by 0.2 rad. The resistance of the coil of the galvanometer is 50 Ω. This galvanometer is to be converted into an ammeter capable of measuring current in the range 0 – 1.0 A. For this purpose, a shunt resistance is to be added in parallel to the galvanometer. The value of this shunt resistance in ohms, is ....

11. A steel wire of diameter 0.5 mm and Young's modulus 2×10^{11} N m^{-2} carries a load of mass m. The length of the wire with the load is 1.0 m. A vernier scale with 10 divisions is attached to the end of this wire. Next to the steel wire is a reference wire to which a main scale, of least count 1.0 mm, is attached. The 10 divisions of the vernier scale correspond to 9 divisions of the main scale. Initially, the zero of vernier scale coincides with the zero of main scale. If the load on the steel wire is increased by 1.2 kg, the vernier scale division which coincides with a main scale division is .... Take, $g = 10$ ms^{-2} and $\pi = 3.2$).

12. One mole of a monoatomic ideal gas undergoes an adiabatic expansion in which its volume becomes eight times its initial value. If the initial temperature of the gas is 100 K and the universal gas constant $R = 8.0$ j mol^{-1} K^{-1}, the decrease in its internal energy in joule, is ....

13. In a photoelectric experiment, a parallel beam of monochromatic light with power of 200 W is incident on a perfectly absorbing cathode of work function 6.25 eV. The frequency of light is just above the threshold frequency, so that the photoelectrons are emitted with negligible kinetic energy. Assume that the photoelectron emission efficiency is 100%. A potential difference of 500 V is applied between the cathode and the anode. All the emitted electrons are incident normally on the anode and are absorbed. The anode experiences a force $F = n \times 10^{-4}$N due to the impact of the electrons. The value of n is .... (Take mass of the electron, $m_e = 9 \times 10^{-31}$kg and eV $= 1.6 \times 10^{-19}$ J)

14. Consider a hydrogen-like ionised atom with atomic number Z with a single electron. In the emission spectrum of this atom, the photon emitted in the $n = 2$ to $n = 1$ transition has energy 74.8 eV higher than the photon emitted in the $n = 3$ to $n = 2$ transition. The ionisation energy of the hydrogen atom is 13.6 eV. The value of Z is ....

Section 3 (Maximum Marks : 12)

- This section contains **TWO (02)** paragraphs. Based on each paragraph, there are **TWO(02)** questions.
- Each question has **FOUR** options. **ONLY ONE** of these four options corresponds to the correct answer.
- Four each question, choose the option corresponding to the correct answer.
- Answer to each question will be evaluated according to the following marking scheme:

Full Marks : **+3** If ONLY the correct option is chosen.
Zero Marks : **0** If none of the options is chosen (i.e. the question is unanswered).
Negative Marks : **−1** In all other cases.

15. The electric field E is measured at a point $P(0, 0, d)$ generated due to various charge distributions and the dependence of E on d is found to be different for different charge distributions. List-I contains different relations between E and d. List-II describes different electric charge distributions, along with their locations. Match the functions in List-I with the related charge distributions in List-II.

List-I	List-II
P. E is independent of d	1. A point charge Q at the origin
Q. $E \propto \frac{1}{d}$	2. A small dipole with point charges Q at $(0, 0, l)$ and $-Q$ at $(0, 0, -1)$. (Take, $2l << d$)
R. $E \propto \frac{1}{d^2}$	3. An infinite line charge coincident with the X-axis, with uniform lincar charge density λ.
S. $E \propto \frac{1}{d^3}$	4. Two infinite wires carrying a uniform linear charge density parallel to the X- axis. The one along $(y = 0, z = l)$ has a charge density $+\lambda$ and the one along $(y = 0, z = -l)$ has a charge density $-\lambda$. (Take, $2l << d$).
	5. Infinite plane charge coincident with the xy-plane with uniform surface charge density.

(a) P → 5; Q → 3,4; R → 1; S → 2
(b) P → 5; Q → 3; R → 1, 4; S → 2
(c) P → 5; Q → 3; R → 1, 2; S → 4
(d) P → 4; Q → 2, 3; R → 1; S → 5

16. A planet of mass M, has two natural satellites with masses m_1 and m_2. The radii of their circular orbits are R_1 and R_2, respectively. Ignore the gravitational force between the satellites. Define v_1, L_1, K_1 and T_1 to be respectively, the orbital speed, angular momentum, kinetic energy and time period of revolution of satellite 1; and v_2, L_2, K_2 and T_2 to be the corresponding quantities of satellite 2. Given, $m_1 / m_2 = 2$ and $R_1 / R_2 = 1/4$, match the ratios in List-I to the numbers in List-II.

List-I	List-II
P. $\frac{v_1}{v_2}$	1. $\frac{1}{8}$
Q. $\frac{L_1}{L_2}$	2. 1
R. $\frac{K_1}{K_2}$	3. 2
S. $\frac{T_1}{T_2}$	4. 8

(a) P $\rightarrow$ 4; Q $\rightarrow$ 2; R $\rightarrow$ 1; S $\rightarrow$ 3
(b) P $\rightarrow$ 3; Q $\rightarrow$ 2; R $\rightarrow$ 4; S $\rightarrow$ 1
(c) P $\rightarrow$ 2; Q $\rightarrow$ 3; R $\rightarrow$ 1; S $\rightarrow$ 4
(d) P $\rightarrow$ 2; Q $\rightarrow$ 3; R $\rightarrow$ 4; S $\rightarrow$ 1

17. One mole of a monoatomic ideal gas undergoes four thermodynamic processes as shown schematically in the pV-diagram below. Among these four processes, one is isobaric, one is isochoric, one is isothermal and one is adiabatic. Match the processes mentioned in List-I with the corresponding statements in List-II.

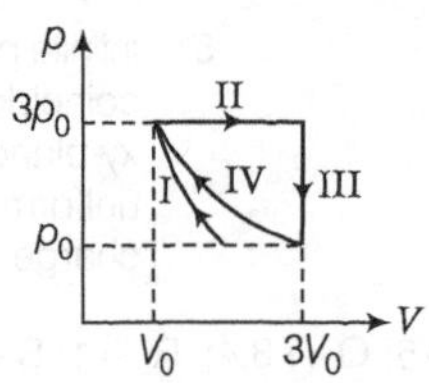

List-I	List-II
P. In process I	1. Work done by the gas is zero
Q. In process II	2. Temperature of the gas remains unchanged
R. In process III	3. No heat is exchanged between the gas and its surroundings
S. In process IV	4. Work done by the gas is $6p_0V_0$

(a) P $\rightarrow$ 4; Q $\rightarrow$ 3; R $\rightarrow$ 1; S $\rightarrow$ 2
(b) P $\rightarrow$ 1; Q $\rightarrow$ 3; R $\rightarrow$ 2; S $\rightarrow$ 4
(c) P $\rightarrow$ 3; Q $\rightarrow$ 4; R $\rightarrow$ 1; S $\rightarrow$ 2
(d) P $\rightarrow$ 3; Q $\rightarrow$ 4; R $\rightarrow$ 2; S $\rightarrow$ 1

18. In the List-I below, four different paths of a particle are given as functions of time. In these functions, α and β are positive constants of appropriate dimensions and $\alpha \neq \beta$. In each case, the force acting on the particle is either zero or conservative. In List-II, five physical quantities of the particle are mentioned: **p** is the linear momentum, **L** is the angular momentum about the origin, K is the kinetic energy, U is the potential energy and E is the total energy. Match each path in List-I with those quantities in List-II, which are conserved for that path.

List-I	List-II
P. $\mathbf{r}(t) = \alpha t\,\mathbf{i} + \beta t\,\mathbf{j}$	1. **p**
Q. $\mathbf{r}(t) = \alpha \cos \omega t\,\mathbf{i} + \beta \sin \omega t\,\mathbf{j}$	2. **L**
R. $\mathbf{r}(t) = \alpha (\cos \omega t\,\mathbf{i} + \sin \omega t\,\mathbf{j})$	3. K
S. $\mathbf{r}(t) = \alpha t\,\mathbf{i} + \frac{\beta}{2} t^2\,\mathbf{j}$	4. U
	5. E

(a) P $\rightarrow$ 1, 2, 3, 4, 5; Q $\rightarrow$ 2, 5; R $\rightarrow$ 2, 3, 4, 5; S $\rightarrow$ 5
(b) P $\rightarrow$ 1, 2, 3, 4, 5; Q $\rightarrow$ 3, 5; R $\rightarrow$ 2, 3, 4, 5; S $\rightarrow$ 2, 5
(c) P $\rightarrow$ 2, 3, 4; Q $\rightarrow$ 5; R $\rightarrow$ 1, 2, 4; S $\rightarrow$ 2, 5
(d) P $\rightarrow$ 1, 2, 3, 5; Q $\rightarrow$ 2, 5; R $\rightarrow$ 2, 3, 4, 5; S $\rightarrow$ 2, 5

CHEMISTRY

Section 1 (Maximum Marks : 24)

- This section contains **SIX (06)** questions.
- Each question has **FOUR** options for correct answers. **ONE OR MORE THAN ONE** of these four options is are correct options(s).
- For each question, choose the correct options to answer the question.
- Answer to each question will be evaluated according to the following marking scheme:

Full Marks : +4 If only all the correct options is are chosen.

Partial Marks : +3 If all the four options are correct but ONLY three options are chosen.

Partial Marks : +2 If three or more options are correct but ONLY two options are chosen, both of which are correct options

Partial Marks : +1 If two or more options are correct but ONLY one option is chosen and it is a correct option.

Zero Marks : 0 If none of the options is chosen (i.e. the question is unanswered).

Negative Marks : −2 In all other cases.

- **For example:** If first, third and fourth are the ONLY three correct options for a question with second option being an incorrect option; selecting only all the three correct options will result in +4 marks. Selecting only two of the three correct options (e.g. the first and fourth options), without selecting any incorrect options (second option in this case), will result in +2 marks. Selecting only one of the three correct options (either first or third or fourth option), without selecting any incorrect optin (second option in this case), will result in +1 marks. Selecting any incorrect options (second option in this case), with or without selection of any correct option(s) will result in −2 marks.

1. The correct option(s) regarding the complex $[Co(en)(NH_3)_3(H_2O)]^{3+}$ $(en = H_2NCH_2CH_2NH_2)$ is (are)

(a) It has two geometrical isomers

(b) It will have three geometrical isomers, if bidentate 'en' is replaced by two cyanide ligands

(c) It is paramagnetic

(d) It absorbs light at longer wavelength as compared to $[Co(en)(NH_3)_4]^{3+}$

2. The correct option(s) to distinguish nitrate salts to Mn^{2+} and Cu^{2+} taken separately is (are)

(a) Mn^{2+} shows the characteristic green colour in the flame test

(b) Only Cu^{2+} shows the formation of precipitate by passing H_2S in acidic medium

(c) Only Mn^{2+} shows the formation of precipitate by passing H_2S in faintly basic medium

(d) Cu^{2+}/Cu has higher reduction potential than Mn^{2+}/Mn (measured under similar conditions)

3. Aniline reacts with mixed acid (conc. HNO_3 and conc. H_2SO_4) at 288 K to give P (51%), Q (47%) and R (2%). The major product(s) of the following sequence is (are)

$$R \xrightarrow[\substack{(3)\ H_3O^+ \\ (4)\ NaNO_2,\ HCl/273\text{-}278\ K \\ (5)\ EtOH,\ \Delta}]{\substack{(1)\ Ac_2O,\ pyridine \\ (2)\ Br_2,\ CH_3CO_2H}} S \xrightarrow[\substack{(3)\ NaNO_2,\ HCl/273\text{-}278\ K \\ (4)\ H_3PO_2}]{\substack{(1)\ Sn/HCl \\ (2)\ Br_2/H_2O\ (excess)}}$$

Major product(s)

(a) 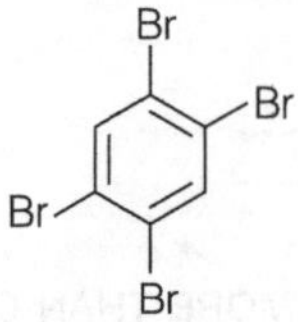 (b) 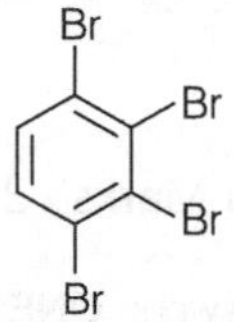

(c) 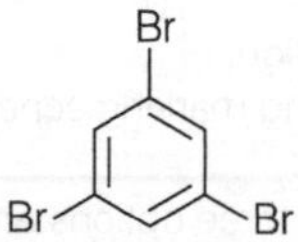 (d) 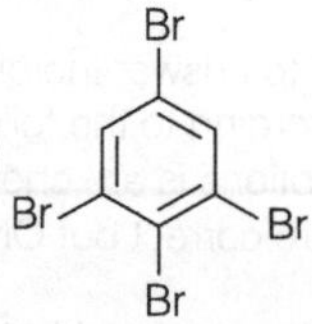

4. The Fischer presentation of D-glucose is given below.

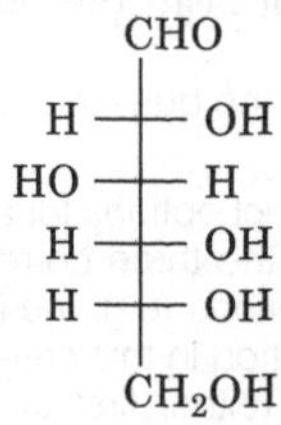

D-glucose

The correct structure(s) of β-L-glucopyranose is (are)

(a) 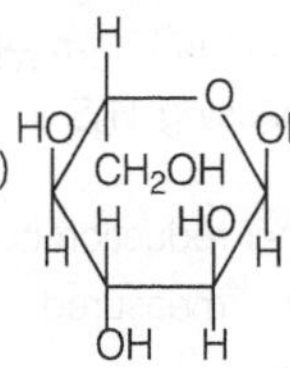 (b) 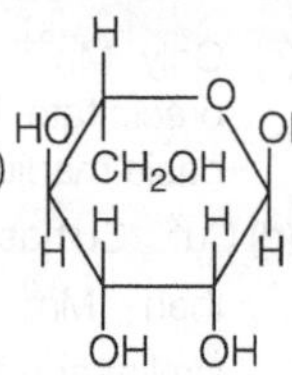

(c) 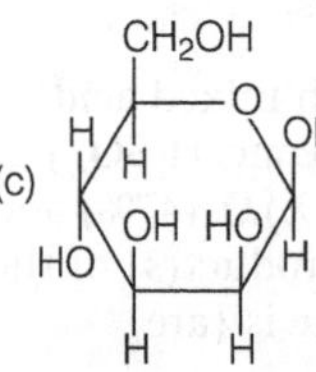 (d) 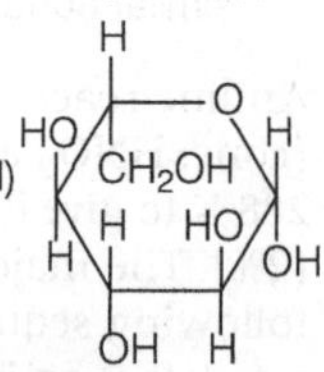

5. For a first order reaction $A(g) \longrightarrow 2B(g) + C(g)$ at constant volume and 300 K, the total pressure at the beginning ($t = 0$) and at time t are p_0 and p_t, respectively.

Initially, only A is present with concentration $[A]_0$, and $t_{1/3}$ is the time required for the partial pressure of A to reach $1/3^{rd}$ of its initial value. The correct option(s) is (are)

(Assume that all these gases behave as ideal gases)

(a) 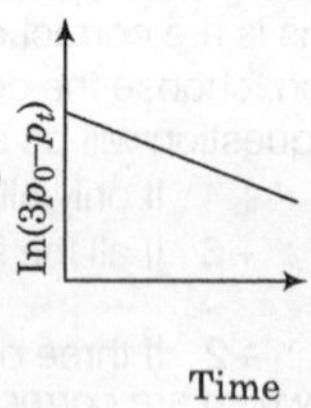 (b) 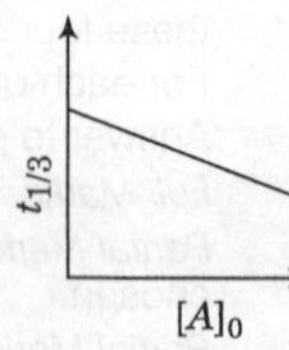

(c) 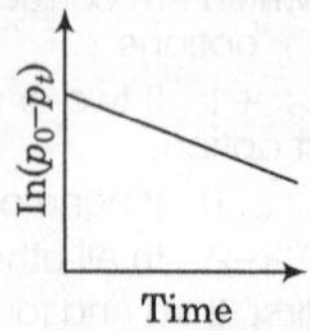 (d) 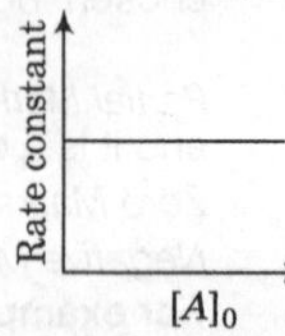

6. For a reaction, $A \rightleftharpoons P$, the plots of $[A]$ and $[P]$ with time at temperatures T_1 and T_2 are given below.

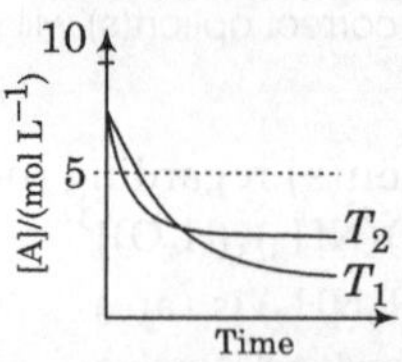 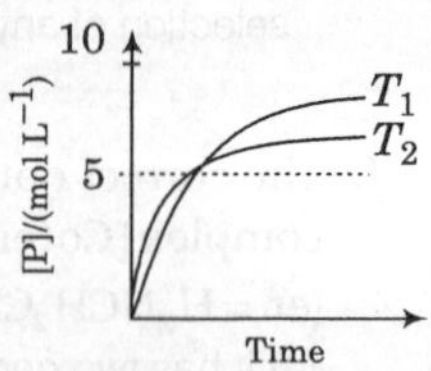

If $T_2 > T_1$, the correct statement(s) is are

(Assume $\Delta H^\ominus$ and $\Delta S^\ominus$ are independent of temperature and ratio of $\ln K$ at T_1 to $\ln K$ at T_2 is greater than T_2/T_1. Here H, S, G and K are enthalpy, entropy, Gibbs energy and equilibrium constant, respectively.)

(a) $\Delta H^\ominus < 0, \Delta S^\ominus < 0$

(b) $\Delta G^\ominus < 0, \Delta H^\ominus > 0$

(c) $\Delta G^\ominus < 0, \Delta S^\ominus < 0$

(d) $\Delta G^\ominus < 0, \Delta S^\ominus > 0$

Section 2 (Maximum Marks : 24)

- This section contains **EIGHT (08)** questions. The answer to each questions is a **NUMERICAL VALUE.**
- Four each question, enter the correct numerical value (in decimal notation, truncated/rounded-off to the **second decimal place;** e.g. 6.25, 7.00.– 0.33, –.30, 30.27, –127.30) using the mouse and the on screen virtual numeric keypad in the place designated to enter the answer.
- Answer to each question will be evaluated according to the following marking scheme:
  Full Marks : + 3 If ONLY the correct numerical value is entered as answer.
  Zero Marks : 0 In all other cases.

7. The total number of compounds having at least one bridging oxo group among the molecules given below is ....

N_2O_3, N_2O_5, P_4O_6, P_4O_7, $H_4P_2O_5$, $H_5P_3O_{10}$, $H_2S_2O_3$, $H_2S_2O_5$

8. Galena (an ore) is partially oxidised by passing air through it at high temperature. After some time, the passage of air is stopped, but the heating is continued in a closed furnace such that the content undergo self-reduction. The weight (in kg) of Pb produced per kg of O_2 consumed is ....

(Atomic weights in g mol^{-1} : $O = 16, S = 32, Pb = 207$)

9. To measure the quantity of $MnCl_2$ dissolved in an aqueous solution, it was completely converted to $KMnO_4$ using the reaction,

$$MnCl_2 + K_2S_2O_8 + H_2O \longrightarrow KMnO_4 + H_2SO_4 + HCl$$

(equation not balanced).

Few drops of concentrated HCl were added to this solution and gently warmed. Further, oxalic acid (225 mg) was added in portions till the colour of the permanganate ion disappeared. The quantity of $MnCl_2$ (in mg) present in the initial solution is ....

(Atomic weights in g mol^{-1}: $Mn = 55, Cl = 35.5$)

10. For the given compound *X*, the total number of optically active stereoisomers is ....

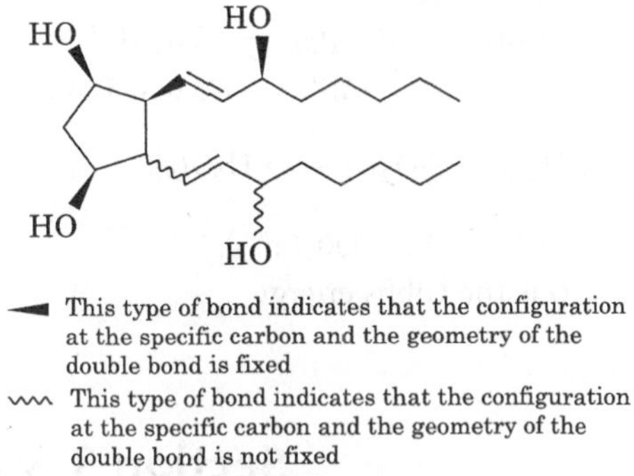

11. In the following reaction sequence, the amount of *D* (in gram) formed from 10 moles of acetophenone is ....

(Atomic weights in g mol^{-1} : $H = 1, C = 12, N = 14, O = 16, Br = 80$. The yield (%) corresponding to the product in each step is given in the parenthesis)

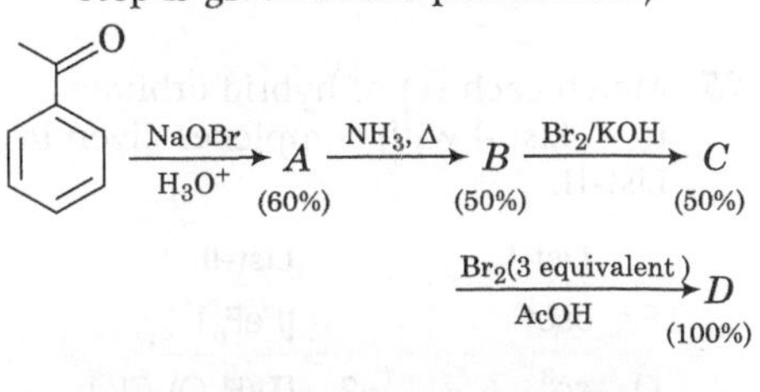

12. The surface of copper gets tarnished by the formation of copper oxide. N_2 gas was passed to prevent the oxide formation during heating of copper at 1250 K. However, the N_2 gas contains 1 mole % of water vapour as impurity. The water vapour oxidises copper as per the reaction given below

$$2Cu(g) + H_2O(g) \longrightarrow Cu_2O(s) + H_2(g)$$

p_{H_2} is the minimum partial pressure of H_2 (in bar) needed to prevent the oxidation at 1250 K. The value of $\ln(p_{H_2})$ is ....

(Given : total pressure = 1 bar, R (universal gas constant) = 8 J K^{-1} mol^{-1}, ln(10) = 2.30 Cu(s) and $Cu_2O(s)$ are mutually immiscible.)

At 1250 K :

$$2Cu(s) + 1/2\, O_2(g) \longrightarrow Cu_2O(s); \quad \Delta G^\ominus = -78{,}000 \text{ J mol}^{-1}$$

$$H_2(g) + \frac{1}{2}O_2(g) \longrightarrow H_2O(g); \quad \Delta G^\ominus = -1{,}78{,}000 \text{ J mol}^{-1};$$

G is the Gibbs energy

13. Consider the following reversible reaction,

$$A(g) + B(g) \rightleftharpoons AB(g)$$

The activation energy of the backward reaction exceeds that of the forward reaction by $2RT$ (in J mol^{-1}). If the pre-exponential factor of the forward reaction is 4 times that of the reverse reaction, the absolute value of $\Delta G^\ominus$ (in J mol^{-1}) for the reaction at 300 K is ....

(Given ; ln(2) = 0.7 RT = 2500 J mol^{-1} at 300 K and G is the Gibbs energy)

14. Consider an electrochemical cell : $A(s) \mid A^{n+}(aq, 2M) \parallel B^{2n+}(aq, 1M) \mid B(s)$. The value of $\Delta H^\ominus$ for the cell reaction is twice of $\Delta G^\ominus$ at 300 K. If the emf of the cell is zero, the $\Delta S^\ominus$ (in J K^{-1} mol^{-1}) of the cell reaction per mole of B formed at 300 K is ... .

(Given : ln(2) = 0.7, R (universal gas constant) = 8.3 J K^{-1} mol^{-1}. H, S and G are enthalpy, entropy and Gibbs energy, respectively.)

Section 3 (Maximum Marks : 12)

- This section contains **FOUR (04)** questions.
- Each question has **TWO (02)** matching lists : **LIST-I** and **LIST-II**.
- **FOUR** options are given representing matching of elements from **LIST-I** and **LIST-II. ONLY ONE** of these four options corresponds to a correct matching.
- For each question, choose the option corresponding to the correct matching.
- For each question, marks will be awarded according to the following marking scheme:
  Full Marks : +3 If ONLY the option corresponding to the correct matching is chosen.
  Zero Marks : 0 If none of the options is chosen (i.e. the question is unanswered).
  Negative Marks : −1 In all other cases.

15. Match each set of hybrid orbitals from List–I with complexes given in List-II.

List–I	List–II
P. dsp^2	1. $[FeF_6]^{4-}$
Q. sp^3	2. $[Ti(H_2O)_3Cl_3]$
R. sp^3d^2	3. $[Cr(NH_3)_6]^{3+}$
S. d^2sp^3	4. $[FeCl_4]^{2-}$
	5. $[Ni(CO)_4]$
	6. $[Ni(CN)_4]^{2-}$

The correct option is

(a) $P \rightarrow 5$; $Q \rightarrow 4, 6$; $R \rightarrow 2, 3$; $S \rightarrow 1$
(b) $P \rightarrow 5,6$; $Q \rightarrow 4$; $R \rightarrow 3$; $S \rightarrow 1,2$
(c) $P \rightarrow 6$; $Q \rightarrow 4, 5$; $R \rightarrow 1$; $S \rightarrow 2, 3$
(d) $P \rightarrow 4,6$, $Q \rightarrow 5, 6$; $R \rightarrow 1,2$; $S \rightarrow 3$

16. The desired product X can be prepared by reacting the major product of the reactions in List-I with one or more appropriate reagents in List-II.

(given, order of migratory aptitude : aryl > alkyl > hydrogen)

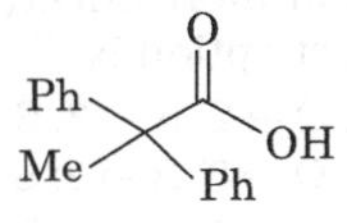

X

	List–I		List–II
P.	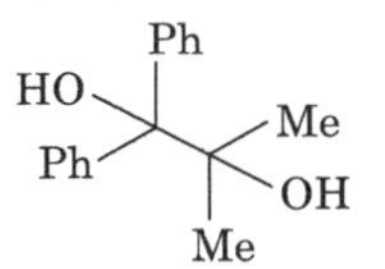 + H_2SO_4	1.	I_2, NaOH
Q.	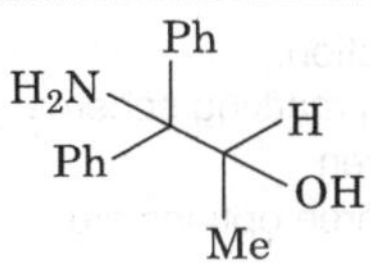 + HNO_2	2.	$[Ag(NH_3)_2]OH$
R.	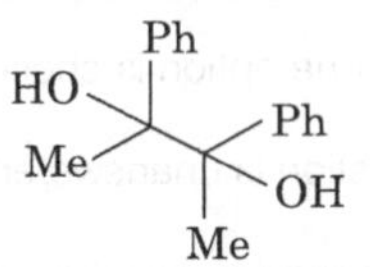 + H_2SO_4	3.	Fehling solution
S.	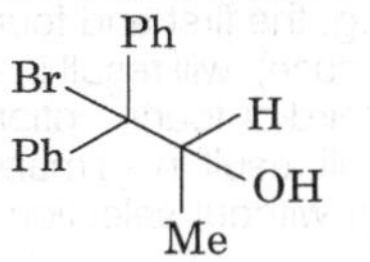 + $AgNO_3$	4.	HCHO, NaOH
		5.	NaOBr

The correct option is

(a) P → 1; Q → 2, 3; R → 1, 4; S → 2,4
(b) P → 1, 5; Q → 3,4; R → 4, 5; S → 3
(c) P → 1, 5; Q → 3,4; R → 5; S → 2,4
(d) P → 1, 5; Q → 2, 3; R → 1,5; S → 2,3

17. List-I contains reactions and List-II contains major products.

	List-I		List-II
P.	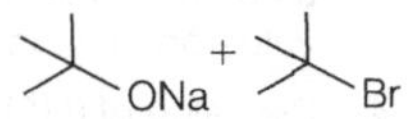 →	1.	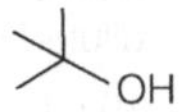
Q.	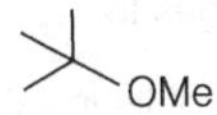 + HBr →	2.	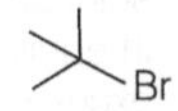
R.	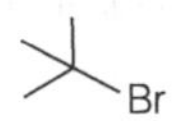 + NaOMe →	3.	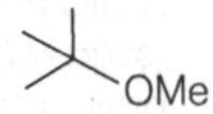
S.	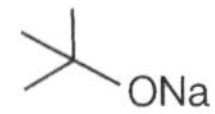 + MeBr →	4.	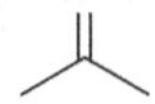
		5.	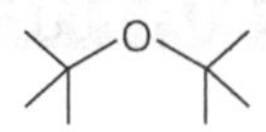

Match each reaction in List-I with one or more products in List-II and choose the correct option.

(a) P → 1, 5; Q → 2; R → 3; S → 4
(b) P → 1, 4; Q → 2; R → 4; S → 3
(c) P → 1, 4; Q → 1,2; R → 3,4; S → 4
(d) P → 4, 5; Q → 4; R → 4; S → 3,4

18. Dilution processes of different aqueous solutions, with water, are given in List-I. The effects of dilution of the solution on $[H^+]$ are given in List-II.

(**Note** Degree of dissociation (α) of weak acid and weak base is << 1; degree of hydrolysis of salt << 1; $[H^+]$ represents the concentration of H^+ ions)

	List-I		List-II
P.	(10 mL of 0.1 M NaOH + 20 mL of 0.1 M acetic acid) diluted to 60 mL	1.	the value of $[H^+]$ does not change on dilution
Q.	(20 mL of 0.1 M NaOH + 20 mL of 0.1 M acetic acid) diluted to 80 mL	2.	the value of $[H^+]$ changes to half of its initial value on dilution

	List-I		List-II
R.	(20 mL of 0.1M HCl + 20 mL of 0.1 M ammonia solution) diluted to 80 mL	3.	the value of $[H^+]$ changes to two times of its initial value on dilution.
S.	10 mL saturated solution of $Ni(OH)_2$ in equilibrium with exces solid $Ni(OH)_2$ is diluted to 20 mL (solid $Ni(OH)_2$ is still present after dilution).	4.	the value of $[H^+]$ changes to $\frac{1}{\sqrt{2}}$ times of its initial value on dilution
		5.	the value of $[H^+]$ changes to $\sqrt{2}$ times of its initial value on dilution

Match each process given in List-I with one or more effect(s) in List-II. The correct option is

(a) P→4; Q→2; R→3; S→1
(b) P→4; Q→3; R→2; S→3
(c) P→1; Q→4; R→5; S→3
(d) P→1; Q→5; R→4; S→1

MATHEMATICS

Section 1 (Maximum Marks : 24)

- This section contains **SIX (06)** questions.
- Each question has **FOUR** options for correct answer(s). **ONE OR MORE THAN ONE** of these four option(s) is (are) correct options(s).
- For each question, choose the correct options(s) to answer the question.
- Answer to each question will be evaluated according to the following marking scheme:

Full Marks	: **+4**	If only (all) the correct option(s) is (are) chosen.
Partial Marks	: **+3**	If all the four options are correct but ONLY three options are chosen.
Partial Marks	: **+2**	If three or more options are correct but ONLY two options are chosen, both of which are correct options.
Partial Marks	: **+1**	If two or more options are correct but ONLY one option is chosen and it is a correct option.
Zero Marks	: **0**	If none of the options is chosen (i.e. the question is unanswered).
Negative Marks	: **−2**	In all other cases.

- **For example:** If first, third and fourth are the ONLY three correct options for a question with second option being an incorrect option; selecting only all the three correct options will result in + 4 marks. Selecting only two of the three correct options (e.g. the first and fourth options), without selecting any incorrect option (second option in this case), will result in +2 marks. Selecting only one of the three correct options (either first or third or fourth option), without selecting any incorrect option (second option in this case), will result in +1 marks. Selecting any incorrect option(s) (second option in this case), with or without selection of any correct option(s) will result in –2 marks.

1. For any positive integer n, define $f_n:(0,\infty)\rightarrow R$ as

$$f_n(x) = \sum_{j=1}^{n} \tan^{-1}\left(\frac{1}{1+(x+j)(x+j-1)}\right)$$

for all $x\in(0,\infty)$. (Here, the inverse trigonometric function $\tan^{-1}x$ assumes values in $\left(-\frac{\pi}{2},\frac{\pi}{2}\right)$). Then, which of the following statement(s) is (arc) TRUE?

(a) $\sum_{j=1}^{5}\tan^2(f_j(0))=55$

(b) $\sum_{j=1}^{10}(1+f'_j(0))\sec^2(f_j(0))=10$

(c) For any fixed positive integer n, $\lim\limits_{x\to\infty}\tan(f_n(x))=\frac{1}{n}$

(d) For any fixed positive integer n, $\lim\limits_{x\to\infty}\sec^2(f_n(x))=1$

2. Let T be the line passing through the points $P(-2,7)$ and $Q(2,-5)$. Let F_1 be the set of all pairs of circles (S_1, S_2) such that T is tangent to S_1 at P and tangent to S_2 at Q, and also such that S_1 and S_2 touch each other at a point, say M. Let E_1 be the set representing the locus of M as the pair (S_1, S_2) varies in F_1. Let the set of all straight line segments joining a pair of distinct points of E_1 and passing through the point $R(1,1)$ be F_2. Let E_2 be the set of the mid-points of the line segments in the set F_2. Then, which of the following statement(s) is (are) TRUE?

(a) The point $(-2, 7)$ lies in E_1

(b) The point $\left(\frac{4}{5}, \frac{7}{5}\right)$ does NOT lie in E_2

(c) The point $\left(\frac{1}{2}, 1\right)$ lies in E_2

(d) The point $\left(0, \frac{3}{2}\right)$ does NOT lie in E_1

3. Let S be the set of all column matrices $\begin{bmatrix} b_1 \\ b_2 \\ b_3 \end{bmatrix}$ such that $b_1, b_2, b_3 \in R$ and the system of equations (in real variables)

$$-x+2y+5z = b_1$$
$$2x-4y+3z=b_2$$
$$x-2y+2z= b_3$$

has at least one solution. Then, which of the following system(s) (in real variables) has (have) at least one solution for each $\begin{bmatrix} b_1 \\ b_2 \\ b_3 \end{bmatrix} \in S$?

(a) $x+2y+3z=b_1$, $4y+5z=b_2$ and $x+2y+6z=b_3$

(b) $x+y+3z=b_1$, $5x+2y+6z=b_2$ and $-2x-y-3z=b_3$

(c) $-x+2y-5z=b_1$, $2x-4y+10z=b_2$ and $x-2y+5z=b_3$

(d) $x+2y+5z=b_1$, $2x+3z=b_2$ and $x+4y-5z=b_3$

4. Consider two straight lines, each of which is tangent to both the circle $x^2+y^2 = (1/2)$ and the parabola $y^2=4x$. Let these lines intersect at the point Q. Consider the ellipse whose centre is at the origin $O(0,0)$ and whose semi-major axis is OQ. If the length of the minor axis of this ellipse is $\sqrt{2}$, then which of the following statement(s) is (are) TRUE?

(a) For the ellipse, the eccentricity is $1/\sqrt{2}$ and the length of the latus rectum is 1

(b) For the ellipse, the eccentricity is 1/2 and the length of the latus rectum is 1/2

(c) The area of the region bounded by the ellipse between the lines $x=\frac{1}{\sqrt{2}}$ and $x=1$ is $\frac{1}{4\sqrt{2}}(\pi-2)$

(d) The area of the region bounded by the ellipse between the lines $x=\frac{1}{\sqrt{2}}$ and $x=1$ is $\frac{1}{16}(\pi-2)$

5. Let s, t, r be non-zero complex numbers and L be the set of solutions $z=x+iy$ $(x, y \in R, i=\sqrt{-1})$ of the equation $sz+t\bar{z}+r=0$, where $\bar{z}=x-iy$. Then, which of the following statement(s) is (are) TRUE?

(a) If L has exactly one element, then $|s| \neq |t|$

(b) If $|s|=|t|$, then L has infinitely many elements

(c) The number of elements in $L \cap \{z: |z-1+i|=5\}$ is at most 2

(d) If L has more than one element, then L has infinitely many elements

6. Let $f:(0,\pi)\rightarrow R$ be a twice differentiable function such that $\lim\limits_{t\rightarrow x}\dfrac{f(x)\sin t-f(t)\sin x}{t-x}=\sin^2 x$ for all $x\in(0,\pi)$.

If $f\left(\dfrac{\pi}{6}\right)=-\dfrac{\pi}{12}$, then which of the following statement(s) is (are) TRUE?

(a) $f\left(\dfrac{\pi}{4}\right)=\dfrac{\pi}{4\sqrt{2}}$

(b) $f(x)<\dfrac{x^4}{6}-x^2$ for all $x\in(0,\pi)$

(c) There exists $\alpha\in(0,\pi)$ such that $f'(\alpha)=0$

(d) $f''\left(\dfrac{\pi}{2}\right)+f\left(\dfrac{\pi}{2}\right)=0$

Section 2 (Maximum Marks : 24)

- This section contains **EIGHT (08)** questions. The answer to each question is a **NUMERICAL VALUE.**
- Four each question, enter the correct numerical value (in decimal notation, truncated/rounded-off to the **second decimal place;** e.g. 6.25, 7.00.–0.33, –.30, 30.27, –127.30) using the mouse and the on-screen virtual numeric keypad in the place designated to enter the answer.
- Answer to each question will be evaluated according to the following marking scheme:

Full Marks : **+3** If ONLY the correct numerical value is entered as answer.

Zero Marks : **0** In all other cases.

7. The value of the integral $\displaystyle\int_0^{1/2}\frac{1+\sqrt{3}}{((x+1)^2(1-x)^6)^{1/4}}\,dx$ is ....

8. Let P be a matrix of order 3×3 such that all the entries in P are from the set $\{-1,0,1\}$. Then, the maximum possible value of the determinant of P is ....

9. Let X be a set with exactly 5 elements and Y be a set with exactly 7 elements. If α is the number of one-one functions from X to Y and β is the number of onto functions from Y to X, then the value of $\dfrac{1}{5!}(\beta-\alpha)$ is ....

10. Let $f:R\rightarrow R$ be a differentiable function with $f(0)=0$. If $y=f(x)$ satisfies the differential equation $\dfrac{dy}{dx}=(2+5y)(5y-2)$, then the value of $\lim\limits_{x\rightarrow-\infty}f(x)$ is ....

11. Let $f:R\rightarrow R$ be a differentiable function with $f(0)=1$ and satisfying the equation $f(x+y)=f(x)f'(y)+f'(x)f(y)$ for all $x,y\in R$.

Then, the value of $\log_e(f(4))$ is ....

12. Let P be a point in the first octant, whose image Q in the plane $x+y=3$ (that is, the line segment PQ is perpendicular to the plane $x+y=3$ and the mid-point of PQ lies in the plane $x+y=3$) lies on the Z-axis. Let the distance of P from the X-axis be 5. If R is the image of P in the XY-plane, then the length of PR is ....

13. Consider the cube in the first octant with sides OP, OQ and OR of length 1, along the X-axis, Y-axis and Z-axis, respectively, where $O(0,0,0)$ is the origin. Let $S\left(\dfrac{1}{2},\dfrac{1}{2},\dfrac{1}{2}\right)$ be the centre of the cube and T be the vertex of the cube opposite to the origin O such that S lies on the diagonal OT. If $\mathbf{p}=\mathbf{SP}$, $\mathbf{q}=\mathbf{SQ}$, $\mathbf{r}=\mathbf{SR}$ and $\mathbf{t}=\mathbf{ST}$, then the value of $|(\mathbf{p}\times\mathbf{q})\times(\mathbf{r}\times\mathbf{t})|$ is ..... .

14. Let $X = ({}^{10}C_1)^2 + 2({}^{10}C_2)^2 + 3({}^{10}C_3)^2 + \ldots + 10({}^{10}C_{10})^2$, where ${}^{10}C_r$, $r \in \{1, 2, \ldots, 10\}$ denote binomial coefficients. Then, the value of $\frac{1}{1430} X$ is ....

Section 3 (Maximum Marks : 12)

- This section contains **FOUR (04)** questions.
- Each question has **TWO (02)** matching lists : **LIST-I** and **LIST-II.**
- **FOUR** options are given representing matching of elements from **LIST-I** and **LIST-II. ONLY ONE** of these four options corresponds to a correct matching.
- For each question, choose the option corresponding to the correct matching.
- For each question, marks will be awarded according to the following marking scheme.
  Full Marks : **+3** If ONLY the option corresponding to the correct matching is chosen.
  Zero Marks : **0** If none of the options is chosen (i.e. the question is unanswered).
  Negative Marks : **−1** In all other cases.

15. Let $E_1 = \{x \in R : x \neq 1 \text{ and } \frac{x}{x-1} > 0\}$ and

$$E_2 = \left\{ x \in E_1 : \sin^{-1}\left(\log_e\left(\frac{x}{x-1}\right)\right) \text{ is a real number} \right\}$$

(Here, the inverse trigonometric function $\sin^{-1} x$ assumes values in $\left[-\frac{\pi}{2}, \frac{\pi}{2}\right]$.).Let $f : E_1 \to R$ be the function defined by $f(x) = \log_e\left(\frac{x}{x-1}\right)$ and $g : E_2 \to R$ be the function defined by $g(x) = \sin^{-1}\left(\log_e\left(\frac{x}{x-1}\right)\right)$.

	List-I		List-II
P.	The range of f is	1.	$\left(-\infty, \frac{1}{1-e}\right] \cup \left[\frac{e}{e-1}, \infty\right)$
Q.	The range of g contains	2.	$(0,1)$
R.	The domain of f contains	3.	$\left[-\frac{1}{2}, \frac{1}{2}\right]$
S.	The domain of g is	4.	$(-\infty, 0) \cup (0, \infty)$
		5.	$\left(-\infty, \frac{e}{e-1}\right]$
		6.	$(-\infty, 0) \cup \left(\frac{1}{2}, \frac{e}{e-1}\right]$

The correct option is

(a) P → 4; Q → 2; R → 1; S → 1
(b) P → 3; Q → 3; R → 6; S → 5
(c) P → 4; Q → 2; R → 1; S → 6
(d) P → 4; Q → 3; R → 6; S → 5

16. In a high school, a committee has to be formed from a group of 6 boys $M_1, M_2, M_3, M_4, M_5, M_6$ and 5 girls G_1, G_2, G_3, G_4, G_5.

(i) Let α_1 be the total number of ways in which the committee can be formed such that the committee has 5 members, having exactly 3 boys and 2 girls.

(ii) Let α_2 be the total number of ways in which the committee can be formed such that the committee has at least 2 members, and having an equal number of boys and girls.

(iii) Let α_3 be the total number of ways in which the committee can be formed such that the committee has 5 members, at least 2 of them being girls.

(iv) Let α_4 be the total number of ways in which the committee can be formed such that the committee has 4 members, having at least 2 girls such that both M_1 and G_1 are NOT in the committee together.

List-I		List-II
P. The value of α_1 is	1.	136
Q. The value of α_2 is	2.	189
R. The value of α_3 is	3.	192
S. The value of α_4 is	4.	200
	5.	381
	6.	461

The correct option is

(a) P $\rightarrow$ 4; Q $\rightarrow$ 6; R $\rightarrow$ 2; S $\rightarrow$ 1
(b) P $\rightarrow$ 1; Q $\rightarrow$ 4; R $\rightarrow$ 2; S $\rightarrow$ 3
(c) P $\rightarrow$ 4; Q $\rightarrow$ 6; R $\rightarrow$ 5; S $\rightarrow$ 2
(d) P $\rightarrow$ 4; Q $\rightarrow$ 2; R $\rightarrow$ 3; S $\rightarrow$ 1

17. Let $H: \frac{x^2}{a^2} - \frac{y^2}{b^2} = 1$, where $a > b > 0$, be a hyperbola in the XY-plane whose conjugate axis LM subtends an angle of 60° at one of its vertices N. Let the area of the ΔLMN be $4\sqrt{3}$.

	List-I		List-II
P.	The length of the conjugate axis of H is	1.	8
Q.	The eccentricity of H is	2.	$\frac{4}{\sqrt{3}}$
R.	The distance between the foci of H is	3.	$\frac{2}{\sqrt{3}}$
S.	The length of the latus rectum of H is	4.	4

The correct option is

(a) P $\rightarrow$ 4; Q $\rightarrow$ 2; R $\rightarrow$ 1; S $\rightarrow$ 3
(b) P $\rightarrow$ 4; Q $\rightarrow$ 3; R $\rightarrow$ 1; S $\rightarrow$ 2
(c) P $\rightarrow$ 4; Q $\rightarrow$ 1; R $\rightarrow$ 3; S $\rightarrow$ 2
(d) P $\rightarrow$ 3; Q $\rightarrow$ 4; R $\rightarrow$ 2; S $\rightarrow$ 1

18. Let $f_1 : R \rightarrow R$, $f_2 : \left(-\frac{\pi}{2}, \frac{\pi}{2}\right) \rightarrow R$, $f_3 : (-1, e^{\pi/2} - 2) \rightarrow R$ and $f_4 : R \rightarrow R$ be functions defined by

(i) $f_1(x) = \sin(\sqrt{1 - e^{-x^2}})$,

(ii) $f_2(x) = \begin{cases} \frac{|\sin x|}{\tan^{-1} x} & \text{if } x \neq 0 \\ 1 & \text{if } x = 0 \end{cases}$, where the inverse trigonometric function $\tan^{-1} x$ assumes values in $\left(-\frac{\pi}{2}, \frac{\pi}{2}\right)$,

(iii) $f_3(x) = [\sin(\log_e(x+2))]$, where for $t \in R$, $[t]$ denotes the greatest integer less than or equal to t,

(iv) $f_4(x) = \begin{cases} x^2 \sin\left(\frac{1}{x}\right) & \text{if } x \neq 0 \\ 0 & \text{if } x = 0 \end{cases}$

List-I		List-II
P. The function f_1 is	1.	NOT continuous at $x = 0$
Q. The function f_2 is	2.	continuous at $x = 0$ and NOT differentiable at $x = 0$
R. The function f_3 is	3.	differentiable at $x = 0$ and its derivative is NOT continuous at $x = 0$
S. The function f_4 is	4.	differentiable at $x = 0$ and its derivative is continuous at $x = 0$

The correct option is

(a) P $\rightarrow$ 2; Q $\rightarrow$ 3; R $\rightarrow$ 1; S $\rightarrow$ 4
(b) P $\rightarrow$ 4; Q $\rightarrow$ 1; R $\rightarrow$ 2; S $\rightarrow$ 3
(c) P $\rightarrow$ 4; Q $\rightarrow$ 2; R $\rightarrow$ 1; S $\rightarrow$ 3
(d) P $\rightarrow$ 2; Q $\rightarrow$ 1; R $\rightarrow$ 4; S $\rightarrow$ 3

Answers

Paper 1

Physics

1. (b, c) 2. (a, c) 3. (a, c) 4. (b, d) 5. (a, b, d) 6. (b, c, d) 7. 2.00 s
8. 5.00 Hz 9. 0.75 m 10. 2.09 m 11. 1.50 12. 2.00 m/s 13. 130 kW/m^2
14. 4.00 15. (c) 16. (d) 17. (b) 18. (c)

Chemistry

1. (b, c) 2. (b, c) 3. (a, b, c) 4. (b) 5. (a, b, d) 6. (b, c) 7. (1)
8. (2992) 9. (3) 10. (10) 11. (2.22) 12. (19) 13. (4.47) 14. (0.05) 15. (c)
16. (a) 17. (a) 18. (b)

Mathematics

1. (a, b, d) 2. (b, c, d) 3. (c, d) 4. (a, b, d) 5. (b,c)
6. (b, c) 7. (8) 8. (625) 9. (3748) 10. (2) 11. (1) 12. (3)
13. (0.5) 14. (4) 15. (a) 16. (d) 17. (a) 18. (c)

Paper 2

Physics

1. (a, b) 2. (a, c, d) 3. (a, b) 4. (d) 5. (a, c) 6. (a, c) 7. 6.30 m
8. 30.00 m 9. 2.00 ms^{-1} 10. 5.55 Ω 11. 3.00 12. 900.00 J 13. 24.00
14. 3.00 15. (b) 16. (b) 17. (c) 18. (a)

Chemistry

1. (a, b, d) 2. (b, d) 3. (d) 4. (d) 5. (a, d) 6. (a, c) 7. (6)
8. (6.47 kg) 9. (126 mg) 10. (7) 11. (495 g) 12. (– 14.6)
13. (+ 8500 J/mol) 14. (– 11.62 JK^{-1} mol^{-1}) 15. (c) 16. (d) 17. (b) 18. (d)

Mathematics

1. (d) 2. (a, d) 3. (a, d) 4. (a, c) 5. (a, c, d) 6. (b, c, d) 7. (2) 8. (4)
9. (119) 10. (0.4) 11. (2) 12. (8) 13. (0.5) 14. (646) 15. (a)
16. (c) 17. (b) 18. (d)

JEE ADVANCED
UNSOLVED PAPER 2019

Duration : 3 Hours | **Max. Marks** : 360

Paper 1

PHYSICS

Section 1 (Maximum Marks : 12)

- This section contains **FOUR (04)** questions.
- Each question has **FOUR** options. **ONLY ONE** of these four options is the correct answer.
- For each question, choose the option corresponding to the correct answer.
- Answer to each question will be evaluated according to the following marking scheme:

Full Marks : **+3** If ONLY the correct options is chosen.
Zero Marks : **0** If none of the options is chosen. (i.e., the question is unanswered)
Negative Marks : **−1** In all other cases.

1. In a radioactive sample, ${}^{40}_{19}K$ nuclei either decay into stable ${}^{40}_{20}Ca$ nuclei with decay constant 4.5×10^{-10} per year or into stable ${}^{40}_{18}Ar$ nuclei with decay constant 0.5×10^{-10} per year. Given that in this sample all the stable ${}^{40}_{20}Ca$ and ${}^{40}_{18}Ar$ nuclei are produced by the ${}^{40}_{19}K$ nuclei only. In time $t \times 10^9$ years, if the ratio of the sum of stable ${}^{40}_{20}Ca$ and ${}^{40}_{18}Ar$ nuclei to the radioactive ${}^{40}_{19}K$ nuclei is 99, the value of t will be
[Given: In $10 = 2.3$]

(a) 9.2 (b) 1.15 (c) 4.6 (d) 2.3

2. A current carrying wire heats a metal rod. The wire provides a constant power (P) to the rod. The metal rod is enclosed in an insulated container. It is observed that the temperature (T) in the metal rod changes with time (t) as

$T(t) = T_0\left(1 + \beta t^{\frac{1}{4}}\right)$, where, β is a constant with appropriate dimension while T_0 is a constant with dimension of temperature. The heat capacity of the metal is

(a) $\dfrac{4P(T(t) - T_0)^4}{\beta^4 T_0^5}$ (b) $\dfrac{4P(T(t) - T_0)^3}{\beta^4 T_0^4}$

(c) $\dfrac{4P(T(t) - T_0)}{\beta^4 T_0^2}$ (d) $\dfrac{4P(T(t) - T_0)^2}{\beta^4 T_0^3}$

3. A thin spherical insulating shell of radius R carries a uniformly distributed charge such that the potential at its surface is V_0. A hole with a small area $\alpha 4\pi R^2$ ($\alpha << 1$) is made on the shell without affecting the rest of the shell. Which one of the following statements is correct?

(a) The ratio of the potential at the center of the shell to that of the point at $\frac{1}{2}R$ from center towards the hole will be $\frac{1-\alpha}{1-2\alpha}$.

(b) The potential at the center of the shell is reduced by $2\alpha V_0$.

(c) The magnitude of electric field at the center of the shell is reduced by $\frac{\alpha V_0}{2R}$.

(d) The magnitude of electric field at a point, located on a line passing through the hole and shell's centre, on a distance $2R$ from the center of the spherical shell will be reduced by $\frac{\alpha V_0}{2R}$.

4. Consider a spherical gaseous cloud of mass density $\rho(r)$ in free space where r is the radial distance from its center. The gaseous cloud is made of particles of equal mass m moving in circular orbits about the common center with the same kinetic energy K. The force acting on the particles is their mutual gravitational force. If $\rho(r)$ is constant in time, the particle number density $n(r) = \rho(r) / m$ is [G is universal gravitational constant]

(a) $\frac{K}{6\pi r^2 m^2 G}$ (b) $\frac{K}{\pi r^2 m^2 G}$

(c) $\frac{3K}{\pi r^2 m^2 G}$ (d) $\frac{K}{2\pi r^2 m^2 G}$

Section 2 (Maximum Marks : 32)

- This section contains **EIGHT (08)** questions.
- Each question has **FOUR** options. **ONE OR MORE THAN ONE** of these four option(s) is (are) correct options(s).
- For each question, choose the options(s) corresponding to (all) the correct answer(s).
- Answer to each question will be evaluated according to the following marking scheme:

Full Marks : **+4** If only (all) the correct option(s) is (are) chosen.

Partial Marks : **+3** If all the four options are correct but ONLY three options are chosen.

Partial Marks : **+2** If three or more options are correct but ONLY two options are chosen, both of which are correct.

Partial Marks : **+1** If two or more options are correct but ONLY one option is chosen and it is a correct option;

Zero Marks : 0 If none of the option is chosen (i.e., the question is unanswered);

Negative Marks : **−1** In all other cases.

- For example, in a question, if (A), (B) and (D) are the ONLY three options corresponding to correct answers, then
  choosing ONLY (A), (B) and (D) will get + 4 marks
  choosing ONLY (A) and (B) will get + 2 marks
  choosing ONLY (A) and (D) will get + 2 marks
  choosing ONLY (B) and (D) will get + 2 marks
  choosing ONLY (A) will get + 1 marks
  choosing ONLY (B) will get + 1 marks
  choosing ONLY (D) will get + 1 marks
  choosing no option (i.e., the question is unanswered) will get 0 marks; and
  choosing any other combination of options will get – 1 mark.

5. A cylindrical capillary tube of 0.2 mm radius is made by joining two capillaries T_1 and T_2 of different materials having water contact angles of 0° and 60°, respectively. The capillary tube is dipped vertically in water in two different configurations, case I and II as shown in figure. Which of the following option(s) is(are) correct? [Surface tension of water $= 0.075\ N/m$, density of water $= 1000\ kg/m^3$, take $g = 10\ m/s^2$]

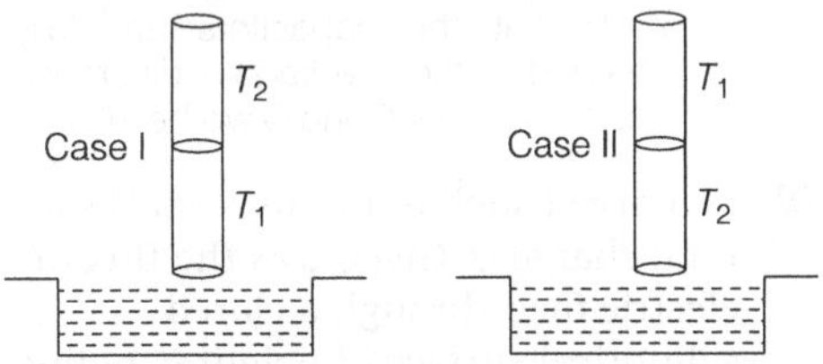

(a) For case I, if the joint is kept at 8 cm above the water surface, the height of water column in the tube will be 7.5 cm. (Neglect the weight of the water in the meniscus).

(b) For case I, if the capillary joint is 5 cm above the water surface, the height of water column raised in the tube will be more than 8.75 cm. (Neglect the weight of the water in the meniscus).

(c) The correction in the height of water column raised in the tube, due to weight of water contained in the meniscus, will be different for both cases.

(d) For case II, if the capillary joint is 5 cm above the water surface, the height of water column raised in the tube will be 3.75 cm. (Neglect the weight of the water in the meniscus).

6. A conducting wire of parabolic shape, initially $y = x^2$, is moving with velocity $\mathbf{v} = v_0\hat{\mathbf{i}}$ in a non-uniform magnetic field $\mathbf{B} = B_0\left(1 + \left(\frac{y}{L}\right)^{\beta}\right)\hat{\mathbf{k}}$, as shown in figure. If V_0, B_0, L and β are positive constants and $\Delta\phi$ is the potential difference developed between the ends of the wire, then the correct statement(s) is/are

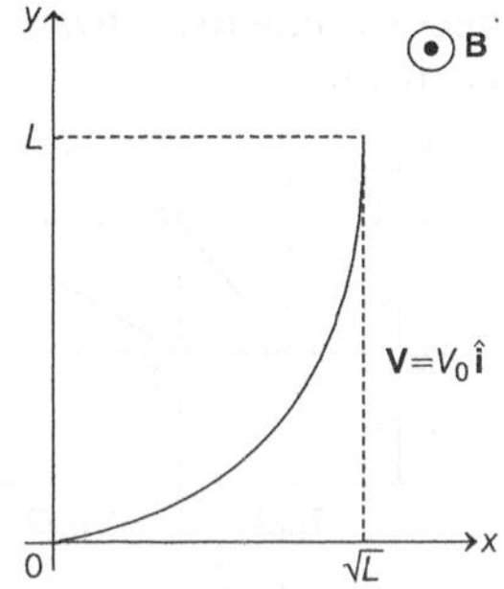

(a) $|\Delta\phi| = \frac{4}{3}B_0V_0L$ for $\beta = 2$

(b) $|\Delta\phi|$ remains the same if the parabolic wire is replaced by a straight wire, $y = x$ initially, of length $\sqrt{2}L$

(c) $|\Delta\phi| = \frac{1}{2}B_0V_0L$ for $\beta = 0$

(d) $|\Delta\phi|$ is proportional to the length of the wire projected on the y-axis.

7. A thin convex lens is made of two materials with refractive indices n_1 and n_2, as shown in the figure. The radius of curvature of the left and right spherical surfaces are equal. f is the focal length of the lens when $n_1 = n_2 = n$. The focal length is $f + \Delta f$ when $n_1 = n$ and $n_2 = n + \Delta n$. Assuming $\Delta n << (n - 1)$ and $1 < n < 2$, the correct statement(s) is/are

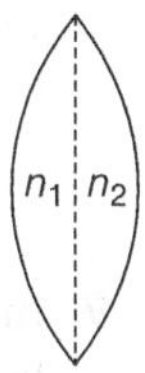

(a) If $\frac{\Delta n}{n} < 0$ then $\frac{\Delta f}{f} > 0$

(b) For $n = 1.5$, $\Delta n = 10^{-3}$ and $f = 20$ cm, the value of $|\Delta f|$ will be 0.02 cm (round off to 2nd decimal place).

(c) $\left|\frac{\Delta f}{f}\right| < \left|\frac{\Delta n}{n}\right|$

(d) The relation between $\frac{\Delta f}{f}$ and $\frac{\Delta n}{n}$ remains unchanged if both the convex surfaces are replaced by concave surfaces of the same radius of curvature.

8. One mole of a monatomic ideal gas goes through a thermodynamic cycle, as shown in the volume versus temperature (V-T) diagram. The correct statement(s) is/are [R is the gas constant]

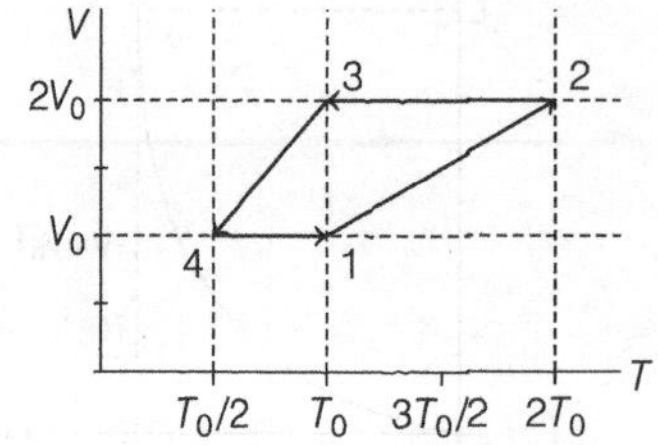

(a) Work done in this thermodynamic cycle $(1 \to 2 \to 3 \to 4 \to 1)$ is $|W| = \frac{1}{2} RT_0$.

(b) The ratio of heat transfer during processes $1 \to 2$ and $2 \to 3$ is $\left|\frac{Q_{1 \to 2}}{Q_{2 \to 3}}\right| = \frac{5}{3}$.

(c) The above thermodynamic cycle exhibits only isochoric and adiabatic processes.

(d) The ratio of heat transfer during processes $1 \to 2$ and $3 \to 4$ is $\left|\frac{Q_{1 \to 2}}{Q_{3 \to 4}}\right| = \frac{1}{2}$.

9. In the circuit shown, initially there is no charge on the capacitors and keys S_1 and S_2 are open. The values of the capacitors are $C_1 = 10\ \mu F$, $C_2 = 30\ \mu F$ and $C_3 = C_4 = 80\ \mu F$.

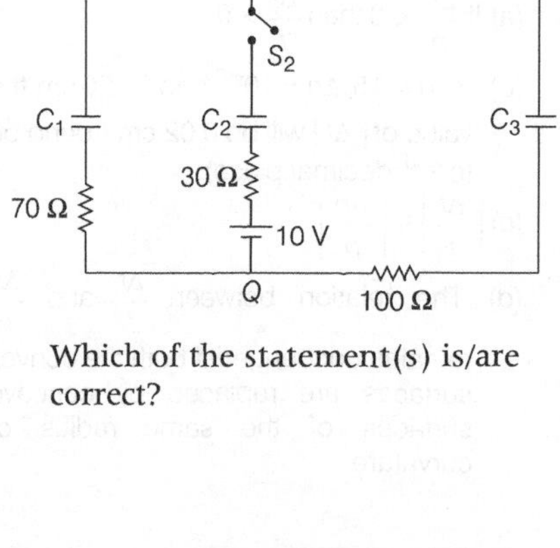

Which of the statement(s) is/are correct?

(a) The key S_1 is kept closed for long time such that capacitors are fully charged. Now, key S_2 is closed, at this time, the instantaneous current across 30 Ω resistor (between points P and Q) will be 0.2 A (round off to 1st decimal place).

(b) If key S_1 is kept closed for long time such that capacitors are fully charged, the voltage across the capacitor C_1 will be 4V.

(c) At time $t = 0$, the key S_1 is closed, the instantaneous current in the closed circuit will be 25 mA.

(d) If key S_1 is kept closed for long time such that the capacitors are fully charged, the voltage difference between points P and Q will be 10 V.

10. A charged shell of radius R carries a total charge Q. Given ϕ as the flux of electric field through a closed cylindrical surface of height h, radius r and with its center same as that of the shell. Here, center of the cylinder is a point on the axis of the cylinder which is equidistant from its top and bottom surfaces. Which of the following option(s) is/are correct? [ϵ_0 is the permittivity of free space]

(a) If $h > 2R$ and $r = 4R/5$ then $\phi = Q/5\epsilon_0$

(b) If $h > 2R$ and $r = 3R/5$ then $\phi = Q/5\epsilon_0$

(c) If $h < 8R/5$ and $r = 3R/5$ then $\phi = 0$

(d) If $h > 2R$ and $r > R$ then $\phi = Q/\epsilon_0$

11. Two identical moving coil galvanometers have 10 Ω resistance and full scale deflection at 2μA current. One of them is converted into a voltmeter of 100 mV full scale reading and the other into an ammeter of 1 mA full scale current using appropriate resistors. These are then used to measure the voltage and current in the Ohm's law experiment with $R = 1000\ \Omega$ resistor by using an ideal cell. Which of the following statement(s) is/are correct?

(a) The resistance of the voltmeter will be $100\ k\Omega$.

(b) The resistance of the ammeter will be $0.02\ \Omega$.
(round off to 2nd decimal place).

(c) If the ideal cell is replaced by a cell having internal resistance of 5Ω then the measured value of R will be more than $1000\ \Omega$.

(d) The measured value of R will be $978\ \Omega < R < 982\ \Omega$.

12. Let us consider a system of units in which mass and angular momentum are dimensionless. If length has dimension of L, which of the following statement(s) is/are correct?

(a) The dimension of force is $[L]^{-3}$.

(b) The dimension of power is $[L]^{-5}$.

(c) The dimension of energy is $[L]^{-2}$.

(d) The dimension of linear momentum is $[L]^{-1}$.

Section 3 (Maximum Marks : 18)

- This section contains **SIX (06)** questions. The answer to each question is a **NUMERICAL VALUE.**
- Four each question, enter the correct numerical value of the answer using the mouse and the on-screen virtual numeric keypad in the place designated to enter the answer. If the numerical value has more than two decimal places, truncate/round-off the value to TWO decimal places.
- Answer to each question will be evaluated according to the following marking scheme:

13. A parallel plate capacitor of capacitance C has spacing d between two plates having area A. The region between the plates is filled with N dielectric layers, parallel to its plates, each with thickness, $\delta = \frac{d}{N}$. The dielectric constant of the mth layer is $K_m = K\left(1 + \frac{m}{N}\right)$. For a very large $N (> 10^3)$, the capacitance C is $\alpha\left(\frac{K\varepsilon_0 A}{d \ln 2}\right)$.
The value of α will be ....

[ϵ_0 is the permittivity of free space.]

14. A planar structure of length L and width W is made of two different optical media of refractive indices $n_1 = 1.5$ and $n_2 = 1.44$ as shown in figure. If $L >> W$, a ray entering from end AB will emerge from end CD only if the total internal reflection condition is met inside the structure. For $L = 9.6$ m, if the incident angle θ is varied, the maximum time taken by a ray to exit the plane CD is $t \times 10^{-9}$ s, where, t is ....

[Speed of light, $c = 3 \times 10^8$ m/s]

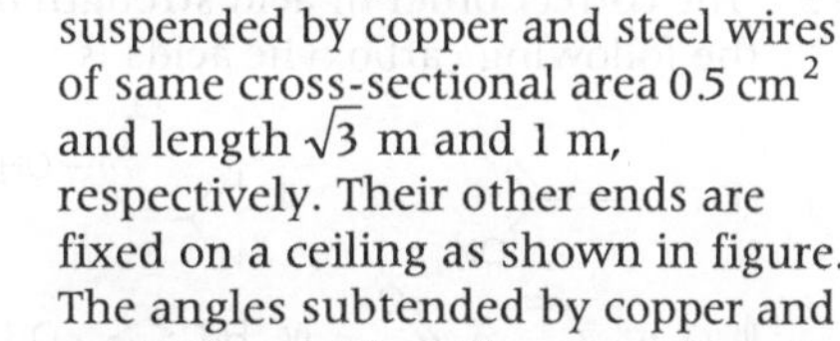

15. A block of weight 100 N is suspended by copper and steel wires of same cross-sectional area $0.5\ cm^2$ and length $\sqrt{3}$ m and 1 m, respectively. Their other ends are fixed on a ceiling as shown in figure. The angles subtended by copper and steel wires with ceiling are 30° and 60°, respectively.

If elongation in copper wire is (Δl_C) and elongation in steel wire is (Δl_S), then the ratio $\frac{\Delta l_C}{\Delta l_S}$ is ....

[Young's modulus for copper and steel are 1×10^{11} N / m^2 and 2×10^{11} N/m^2 respectively.]

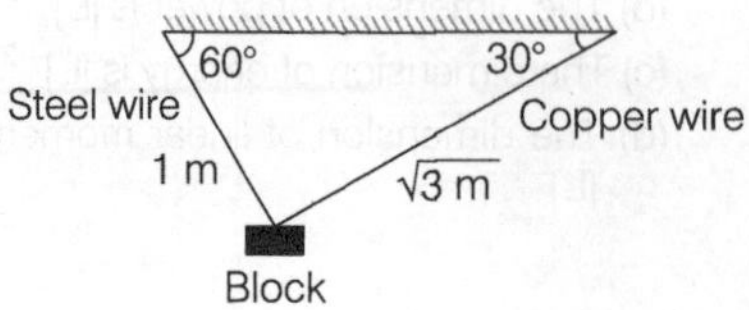

16. A liquid at 30°C is poured very slowly into a Calorimeter that is at temperature of 110°C. The boiling temperature of the liquid is 80°C. It is found that the first 5 gm of the liquid completely evaporates. After pouring another 80 gm of the liquid the equilibrium temperature is found to be 50°C. The ratio of the latent heat of the liquid to its specific heat will be .... °C.

[Neglect the heat exchange with surrounding]

17. A particle is moved along a path AB-BC-CD-DE-EF-FA, as shown in figure, in presence of a force $\mathbf{F} = (\alpha y\hat{\mathbf{i}} + 2\alpha x\hat{\mathbf{j}})$ N, where x and y are in meter and $\alpha = -1\,\text{Nm}^{-1}$. The work done on the particle by this force **F** will be .... Joule.

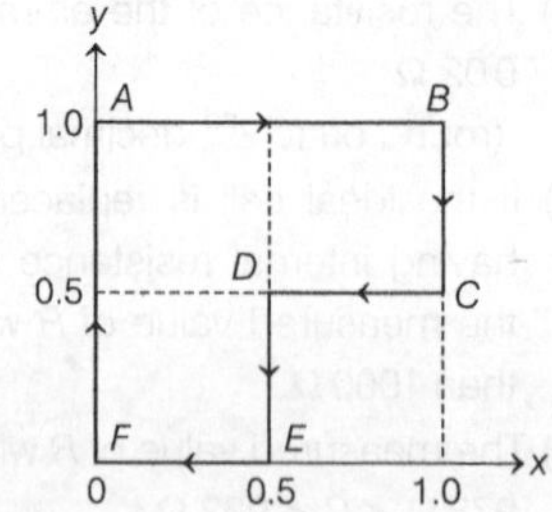

18. A train S1, moving with a uniform velocity of 108 km/h, approaches another train S2 standing on a platform. An observer O moves with a uniform velocity of 36 km/h towards S2, as shown in figure.

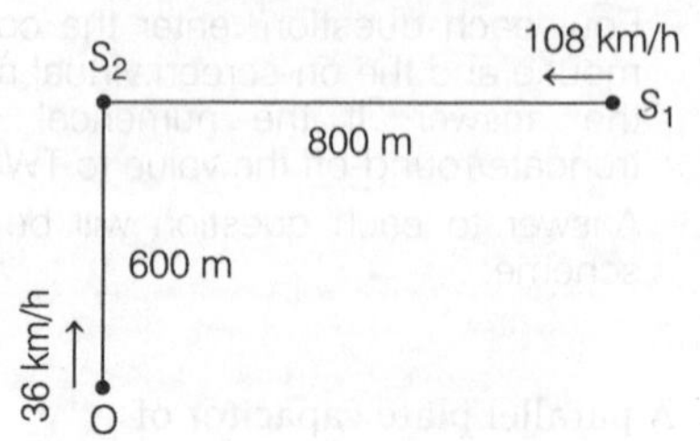

Both the trains are blowing whistles of same frequency 120 Hz. When O is 600 m away from S2 and distance between S1 and S2 is 800 m, the number of beats heard by O is .... [Speed of the sound = 330 m / s ....]

CHEMISTRY

Section 1 (Maximum Marks : 12)

Instructions: Same as given in Physics.

19. The correct order of acid strength of the following carboxylic acids is

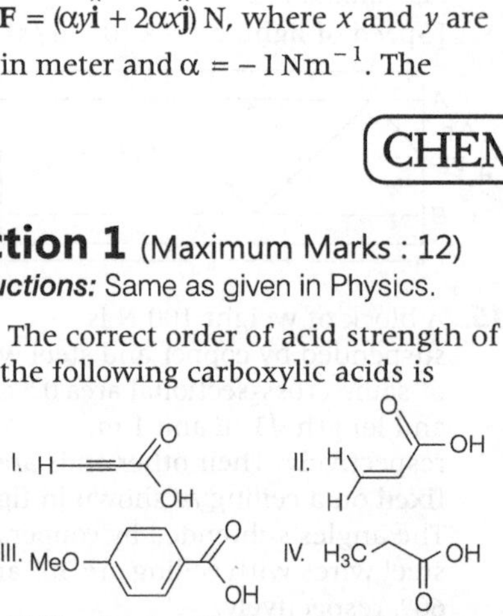

(a) III > II > I > IV (b) I > II > III > IV
(c) II > I > IV > III (d) I > III > II > IV

20. The green colour produced in the borax bead test of a chromium (III) salt is due to

(a) Cr_2O_3 (b) CrB
(c) $Cr(BO_2)_3$ (d) $Cr_2(B_4O_7)_3$

21. Calamine, malachite, magnetite and cryolite, respectively, are

(a) $ZnCO_3$, $CuCO_3$, Fe_2O_3, Na_3AlF_6
(b) $ZnSO_4$, $CuCO_3$, Fe_2O_3, AlF_3
(c) $ZnSO_4$, $Cu(OH)_2$, Fe_3O_4, Na_3AlF_6
(d) $ZnCO_3$, $CuCO_3 \cdot Cu(OH)_2$, Fe_3O_4, Na_3AlF_6

22. Molar conductivity (Λ_m) of aqueous solution of sodium stearate, which behaves as a strong electrolyte, is recorded at varying concentrations (C) of sodium stearate. Which one of the following plots provides the correct representation of micelle formation in the solution?

(critical micelle concentration (CMC) is marked with an arrow in the figures)

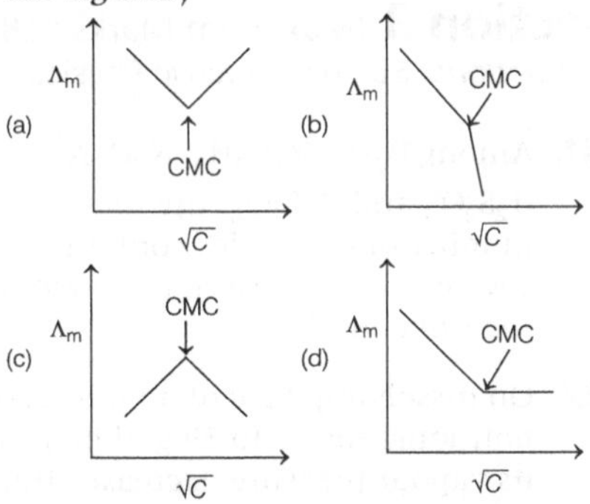

Section 2 (Maximum Marks : 32)

Instructions: Same as given in Physics.

23. Choose the reaction(s) from the following options, for which the standard enthalpy of reaction is equal to the standard enthalpy of formation.

(a) $2C(g) + 3H_2(g) \longrightarrow C_2H_6(g)$

(b) $2H_2(g) + O_2(g) \longrightarrow 2H_2O(l)$

(c) $\frac{3}{2}O_2(g) \longrightarrow O_3(g)$

(d) $\frac{1}{8}S_8(s) + O_2(g) \longrightarrow SO_2(g)$

24. A tin chloride Q undergoes the following reactions (not balanced)

$$Q + Cl^- \longrightarrow X$$
$$Q + Me_3N \longrightarrow Y$$
$$Q + CuCl_2 \longrightarrow Z + CuCl$$

X is a monoanion having pyramidal geometry. Both Y and Z are neutral compounds.

Choose the correct option(s).

(a) There is a coordinate bond in Y

(b) The central atom in Z has one lone pair of electrons

(c) The oxidation state of the central atom in Z is + 2

(d) The central atom in X is sp^3 hybridised

25. In the decay sequence.

$${}^{238}_{92}U \xrightarrow{-x_1} {}^{234}_{90}Th \xrightarrow{-x_2}$$
$${}^{234}_{91}Pa \xrightarrow{-x_3} {}^{234}Z \xrightarrow{-x_4} {}^{230}_{90}Th$$

x_1, x_2, x_3 and x_4 are particles/ radiation emitted by the respective isotopes. The correct option(s) is(are)

(a) Z is an isotope of uranium

(b) x_2 is β^-

(c) x_1 will deflect towards negatively charged plate

(d) x_3 is γ-ray

26. Which of the following statement(s) is(are) correct regarding the root mean square speed (U_{rms}) and average translational kinetic energy (E_{av}) of a molecule in a gas at equilibrium?

(a) U_{rms} is inversely proportional to the square root of its molecular mass

(b) U_{rms} is doubled when its temperature is increased four times

(c) E_{av} is doubled when its temperature is increased four times

(d) E_{av} at a given temperature does not depend on its molecular mass

27. Choose the correct option(s) for the following set of reactions.

$$C_6H_{10}O \xrightarrow[\text{(ii) } H_2O]{\text{(i) MeMgBr}} Q \xrightarrow{\text{Conc. HCl}} S \text{ (major)}$$

$$Q \xrightarrow{20\%\ H_3PO_4,\ 360\ K} R \text{ (major)}$$

$$T \text{ (major)} \xleftarrow[\text{(ii) } Br_2,\ h\nu]{\text{(i) } H_2,\ Ni} R \text{ (major)} \xrightarrow[\Delta]{\text{HBr, benzoyl peroxide}} U \text{ (major)}$$

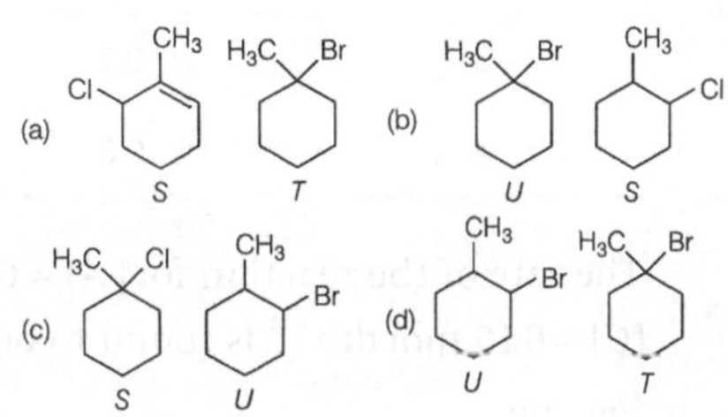

28. Which of the following statement(s) is(are) true?

(a) The two six-membered cyclic hemiacetal forms of D-(+)-glucose are called anomers

(b) Oxidation of glucose with bromine water gives glutamic acid

(c) Monosaccharides cannot be hydrolysed to given polyhydroxy aldehydes and ketones

(d) Hydrolysis of sucrose gives dextrorotatory glucose and laevorotatory fructose

29. Fusion of MnO_2 with KOH in presence of O_2 produces a salt W. Alkaline solution of W upon electrolytic oxidation yields another salt X. The manganese containing ions present in W and X, respectively, are Y and Z. Correct statement(s) is (are)

(a) Both Y and Z are coloured and have tetrahedral shape

(b) Y is diamagnetic in nature while Z is paramagnetic

(c) In both Y and Z, π-bonding occurs between p-orbitals of oxygen and d-orbitals of manganese

(d) In aqueous acidic solution, Y undergoes disproportionation reaction to give Z and MnO_2

30. Each of the following options contains a set of four molecules. Identify the option(s) where all four molecules posses permanent dipole moment at room temperature.

(a) SO_2, C_6H_5Cl, H_2Se, BrF_5 (b) $BeCl_2$, CO_2, BCl_3, $CHCl_3$

(c) NO_2, NH_3, $POCl_3$, CH_3Cl (d) BF_3, O_3, SF_6, XeF_6

Section 3 (Maximum Marks : 18)

Instructions: Same as given in Physics.

31. Among B_2H_6, $B_3N_3H_6$, N_2O, N_2O_4, $H_2S_2O_3$ and $H_2S_2O_8$, the total number of molecules containing covalent bond between two atoms of the same kind is ....

32. On dissolving 0.5 g of a non-volatile non-ionic solute to 39 g of benzene, its vapour pressure decreases from 650 mmHg to 640 mmHg. The depression of freezing point of benzene (in K) upon addition of the solute is ....

(Given data : Molar mass and the molal freezing point depression constant of benzene are 78 g mol^{-1} and 5.12 K kg mol^{-1}, respectively).

33. Consider the kinetic data given in the following table for the reaction $A + B + C \longrightarrow$ Product

Experiment No.	$[A]$ (mol dm^{-3})	$[B]$ (mol dm^{-3})	$[C]$ (mol dm^{-3})	Rate of reaction (mol $dm^{-3}s^{-1}$)
1	0.2	0.1	0.1	6.0×10^{-5}
2	0.2	0.2	0.1	6.0×10^{-5}
3	0.2	0.1	0.2	1.2×10^{-4}
4	0.3	0.1	0.1	9.0×10^{-5}

The rate of the reaction for $[A] = 0.15$ mol dm^{-3}, $[B] = 0.25$ mol dm^{-3} and $[C] = 0.15$ mol dm^{-3} is found to be $Y \times 10^{-5}$ mol $dm^{-3}s^{-1}$. The value of Y is ....

34. For the following reaction, the equilibrium constant K_c at 298 K is 1.6×10^{17}.

$$Fe^{2+}(aq) + S^{2-}(aq) \rightleftharpoons FeS(s)$$

When equal volumes of 0.06 M $Fe^{2+}(aq)$ and 0.2 M $S^{2-}(aq)$ solutions are mixed, the equilibrium concentration of $Fe^{2+}(aq)$ is found by $Y \times 10^{-17}$ M. The value of Y is ....

35. At 143 K, the reaction of XeF_4 with O_2F_2 produces a xenon compound Y. The total number of lone pair(s) of electrons present on the whole molecule of Y is ....

36. Schemes 1 and 2 describe the conversion of P to Q and R to S, respectively. Scheme 3 describes the synthesis of T from Q and S. The total number of Br atoms in a molecule of T is ....

Scheme 1

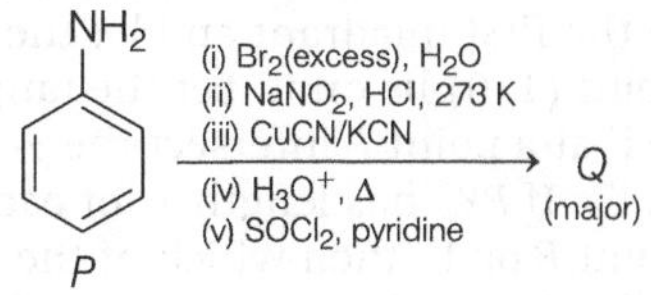

Scheme 2

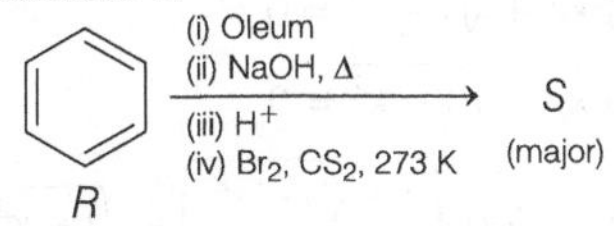

Scheme 3

$$S \xrightarrow[\text{(ii) } Q]{\text{(i) NaOH, } \Delta} \underset{\text{(major)}}{T}$$

MATHEMATICS

Section 1 (Maximum Marks : 12)

Instructions: Same as given in Physics.

37. Let S be the set of all complex numbers z satisfying $|z - 2 + i| \geq \sqrt{5}$. If the complex number z_0 is such that $\dfrac{1}{|z_0 - 1|}$ is the maximum of the set $\left\{\dfrac{1}{|z-1|} : z \in S\right\}$, then the principal argument of $\dfrac{4 - z_0 - \bar{z}_0}{z_0 - \bar{z}_0 + 2i}$ is

(a) $\frac{\pi}{4}$ (b) $\frac{3\pi}{4}$ (c) $-\frac{\pi}{2}$ (d) $\frac{\pi}{2}$

38. Let $M = \begin{bmatrix} \sin^4\theta & -1-\sin^2\theta \\ 1+\cos^2\theta & \cos^4\theta \end{bmatrix} = \alpha I + \beta M^{-1}$,

where $\alpha = \alpha(\theta)$ and $\beta = \beta(\theta)$ are real numbers, and I is the 2×2 identity matrix. If α^* is the minimum of the set $\{\alpha(\theta) : \theta \in [0, 2\pi)\}$ and β^* is the minimum of the set $\{\beta(\theta) : \theta \in [0, 2\pi)\}$, then the value of $\alpha^* + \beta^*$ is

(a) $-\frac{17}{16}$ (b) $-\frac{31}{16}$ (c) $-\frac{37}{16}$ (d) $-\frac{29}{16}$

39. A line $y = mx + 1$ intersects the circle $(x-3)^2 + (y+2)^2 = 25$ at the points P and Q. If the midpoint of the line segment PQ has x-coordinate $-\frac{3}{5}$, then which one of the following options is correct?

(a) $6 \leq m < 8$ (b) $-3 \leq m < -1$
(c) $4 \leq m < 6$ (d) $2 \leq m < 4$

40. The area of the region $\{(x, y) : xy \leq 8, 1 \leq y \leq x^2\}$ is

(a) $8 \log_e 2 - \frac{14}{3}$ (b) $8 \log_e 2 - \frac{7}{3}$
(c) $16 \log_e 2 - \frac{14}{3}$ (d) $16 \log_e 2 - 6$

Section 2 (Maximum Marks : 32)

Instructions: Same as given in Physics.

41. Let Γ denote a curve $y = y(x)$ which is in the first quadrant and let the point $(1, 0)$ lie on it. Let the tangent to Γ at a point P intersect the y-axis at Y_P. If PY_P has length 1 for each point P on Γ, then which of the following options is/are correct?

(a) $xy' + \sqrt{1-x^2} = 0$

(b) $xy' - \sqrt{1-x^2} = 0$

(c) $y = \log_e\left(\dfrac{1+\sqrt{1-x^2}}{x}\right) - \sqrt{1-x^2}$

(d) $y = -\log_e\left(\dfrac{1+\sqrt{1-x^2}}{x}\right) + \sqrt{1-x^2}$

42. Define the collections $\{E_1, E_2, E_3, \dots\}$ of ellipses and $\{R_1, R_2, R_3, \dots\}$ of rectangles as follows :

$$E_1 : \frac{x^2}{9} + \frac{y^2}{4} = 1$$

R_1 : rectangle of largest area, with sides parallel to the axes, inscribed in E_1;

E_n : ellipse $\dfrac{x^2}{a_n^2} + \dfrac{y^2}{b_n^2} = 1$ of the largest area inscribed in $R_{n-1}, n > 1$;

R_n : rectangle of largest area, with sides parallel to the axes, inscribed in $E_n, n > 1$.

Then which of the following options is/are correct?

(a) The eccentricities of E_{18} and E_{19} are NOT equal.

(b) The distance of a focus from the centre in E_9 is $\dfrac{\sqrt{5}}{32}$.

(c) $\sum_{n=1}^{N}$(area of R_n) < 24, for each positive integer N.

(d) The length of latusrectum of E_9 is $\dfrac{1}{6}$

43. In a non-right-angled triangle ΔPQR, let p, q, r denote the lengths of the sides opposite to the angles at P, Q, R respectively. The median from R meets the side PQ at S, the perpendicular from P meets the side QR at E, and RS and PE intersect at O. If $p = \sqrt{3}, q = 1$, and the radius of the circumcircle of the ΔPQR equals 1, then which of the following options is/are correct?

(a) Length of $OE = \dfrac{1}{6}$

(b) Length of $RS = \dfrac{\sqrt{7}}{2}$

(c) Area of $\Delta SOE = \dfrac{\sqrt{3}}{12}$

(d) Radius of incircle of $\Delta PQR = \dfrac{\sqrt{3}}{2}(2-\sqrt{3})$

44. Let α and β be the roots of $x^2 - x - 1 = 0$, with $\alpha > \beta$. For all positive integers n, define

$$a_n = \frac{\alpha^n - \beta^n}{\alpha - \beta}, n \geq 1,$$

$$b_1 = 1 \text{ and } b_n = a_{n-1} + a_{n+1}, n \geq 2$$

Then which of the following options is/are correct?

(a) $\sum_{n=1}^{\infty} \dfrac{b_n}{10^n} = \dfrac{8}{89}$

(b) $b_n = \alpha^n + \beta^n$ for all $n \geq 1$

(c) $a_1 + a_2 + a_3 + \dots + a_n = a_{n+2} - 1$ for all $n \geq 1$

(d) $\sum_{n=1}^{\infty} \dfrac{a_n}{10^n} = \dfrac{10}{89}$

45. Let L_1 and L_2 denote the lines

$$\mathbf{r} = \hat{\mathbf{i}} + \lambda(-\hat{\mathbf{i}} + 2\hat{\mathbf{j}} + 2\hat{\mathbf{k}}), \lambda \in R$$

and $\mathbf{r} = \mu(2\hat{\mathbf{i}} - \hat{\mathbf{j}} + 2\hat{\mathbf{k}}), \mu \in R$

respectively. If L_3 is a line which is perpendicular to both L_1 and L_2 and cuts both of them, then which of the following options describe(s) L_3?

(a) $\mathbf{r} = \frac{2}{9}(2\hat{\mathbf{i}} - \hat{\mathbf{j}} + 2\hat{\mathbf{k}}) + t(2\hat{\mathbf{i}} + 2\hat{\mathbf{j}} - \hat{\mathbf{k}}),$ $t \in R$

(b) $\mathbf{r} = \frac{1}{3}(2\hat{\mathbf{i}} + \hat{\mathbf{k}}) + t(2\hat{\mathbf{i}} + 2\hat{\mathbf{j}} - \hat{\mathbf{k}}), t \in R$

(c) $\mathbf{r} = \frac{2}{9}(4\hat{\mathbf{i}} + \hat{\mathbf{j}} + \hat{\mathbf{k}}) + t(2\hat{\mathbf{i}} + 2\hat{\mathbf{j}} - \hat{\mathbf{k}}), t \in R$

(d) $\mathbf{r} = t(2\hat{\mathbf{i}} + 2\hat{\mathbf{j}} - \hat{\mathbf{k}}), t \in R$

46. There are three bags B_1, B_2 and B_3. The bag B_1 contains 5 red and 5 green balls, B_2 contains 3 red and 5 green balls, and B_3 contains 5 red and 3 green balls. Bags B_1, B_2 and B_3 have probabilities $\frac{3}{10}, \frac{3}{10}$ and $\frac{4}{10}$ respectively of being chosen. A bag is selected at random and a ball is chosen at random from the bag. Then which of the following options is/are correct?

(a) Probability that the chosen ball is green, given that the selected bag is B_3, equals $\frac{3}{8}$.

(b) Probability that the selected bag is B_3, given that the chosen ball is green, equals $\frac{5}{13}$.

(c) Probability that the chosen ball is green equals $\frac{39}{80}$.

(d) Probability that the selected bag is B_3 and the chosen ball is green equals $\frac{3}{10}$.

47. Let $M = \begin{bmatrix} 0 & 1 & a \\ 1 & 2 & 3 \\ 3 & b & 1 \end{bmatrix}$ and

$$\text{adj } M = \begin{bmatrix} -1 & 1 & -1 \\ 8 & -6 & 2 \\ -5 & 3 & -1 \end{bmatrix}$$

where a and b are real numbers. Which of the following options is/are correct?

(a) $\det(\text{adj } M^2) = 81$

(b) If $M \begin{bmatrix} \alpha \\ \beta \\ \gamma \end{bmatrix} = \begin{bmatrix} 1 \\ 2 \\ 3 \end{bmatrix}$, then $\alpha - \beta + \gamma = 3$

(c) $(\text{adj } M)^{-1} + \text{adj } M^{-1} = -M$

(d) $a + b = 3$

48. Let $f : R \to R$ be given by

$$f(x) = \begin{cases} x^5 + 5x^4 + 10x^3 + 10x^2 + 3x + 1, & x < 0; \\ x^2 - x + 1, & 0 \le x < 1; \\ \frac{2}{3}x^3 - 4x^2 + 7x - \frac{8}{3}, & 1 \le x < 3; \\ (x-2)\log_e(x-2) - x + \frac{10}{3}, & x \ge 3; \end{cases}$$

Then which of the following options is/are correct?

(a) f is increasing on $(-\infty, 0)$

(b) f' is NOT differentiable at $x = 1$

(c) f is onto

(d) f' has a local maximum at $x = 1$

Section 3 (Maximum Marks : 18)

Instructions: Same as given in Physics.

49. Let S be the sample space of all 3×3 matrices with entries from the set $\{0, 1\}$. Let the events E_1 and E_2 be given by

$E_1 = \{A \in S : \det A = 0\}$ and

$E_2 = \{A \in S :$ sum of entries of A is $7\}$.

If a matrix is chosen at random from S, then the conditional probability $P(E_1 \mid E_2)$ equals ....

50. Let the point B be the reflection of the point $A(2, 3)$ with respect to the line $8x - 6y - 23 = 0$. Let Γ_A and Γ_B be circles of radii 2 and 1 with centres A and B respectively. Let T be a common tangent to the circles Γ_A and Γ_B such that both the circles are on the same side of T. If C is the point of intersection of T and the line passing through A and B, then the length of the line segment AC is ....

51. Let $\omega \neq 1$ be a cube root of unity. Then the minimum of the set $\{|a + b\omega + c\omega^2|^2 : a, b, c \text{ distinct non-zero integers}\}$ equals ....

52. If $I = \frac{2}{\pi} \int_{-\pi/4}^{\pi/4} \frac{dx}{(1 + e^{\sin x})(2 - \cos 2x)}$

then $27I^2$ equals ....

53. Three lines are given by $\mathbf{r} = \lambda\hat{\mathbf{i}}, \lambda \in R$, $\mathbf{r} = \mu(\hat{\mathbf{i}} + \hat{\mathbf{j}}), \mu \in R$ and $\mathbf{r} = \nu(\hat{\mathbf{i}} + \hat{\mathbf{j}} + \hat{\mathbf{k}}), \nu \in R$

Let the lines cut the plane $x + y + z = 1$ at the points A, B and C respectively. If the area of the triangle ABC is Δ then the value of $(6\Delta)^2$ equals ....

54. Let $AP(a ; d)$ denote the set of all the terms of an infinite arithmetic progression with first term a and common difference $d > 0$. If $AP(1 ; 3) \cap AP(2 ; 5) \cap AP(3 ; 7) = AP(a ; d)$, then $a + d$ equals ....

Paper 2

PHYSICS

Section 1 (Maximum Marks : 32)

- This section contains **EIGHT (08)** questions.
- Each question has **FOUR** options for correct answer(s). **ONE OR MORE THAN ONE** of these four option(s) is (are) correct options(s).
- For each question, choose the correct options(s) to answer the question.
- Answer to each question will be evaluated according to the following marking scheme:

  Full Marks : **+4** If only (all) the correct option(s) is (are) chosen.

  Partial Marks : **+3** If all the four options are correct but ONLY three options are chosen.

  Partial Marks : **+2** If three or more options are correct but ONLY two options are chosen, both of which are correct options.

  Partial Marks : **+1** If two or more options are correct but ONLY one option is chosen and it is a correct option.

  Zero Marks : **0** If none of the options is chosen (i.e. the question is unanswered).

  Negative Marks : **−1** In all other cases.
- **For example:** in a question, if (A), (B) and (D) are the ONLY three options corresponding to correct answer, then
  choosing ONLY (A), (B) and (D) will get +4 marks;
  choosing ONLY (A) and (B) will get +2 marks;
  choosing ONLY (A) and (D) will get +2 marks;
  choosing ONLY (B) and (D) will get +2 marks;
  choosing ONLY (A) will get +1 mark;
  choosing ONLY (B) will get +1 mark;
  choosing ONLY (D) will get +1 mark;
  choosing no option (i.e. the question is unanswered) will get 0 marks; and
  choosing any other combination of options will −1 mark.

1. An electric dipole with dipole moment $\frac{p_0}{\sqrt{2}}(\hat{\mathbf{i}}+\hat{\mathbf{j}})$ is held fixed at the origin O in the presence of a uniform electric field of magnitude E_0.

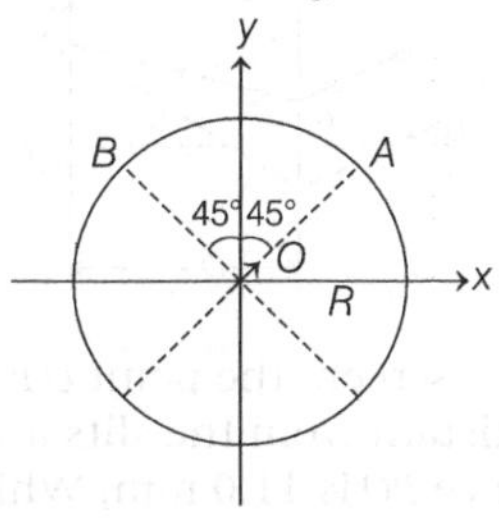

If the potential is constant on a circle of radius R centered at the origin as shown in figure, then the correct statement(s) is/are,
(ϵ_0 is the permittivity of the free space, $R >>$ dipole size)

(a) The magnitude of total electric field on any two points of the circle will be same.

(b) Total electric field at point B is $\vec{\mathbf{E}}_B = 0$

(c) $R = \left(\frac{p_0}{4\pi\epsilon_0 E_0}\right)^{1/3}$

(d) Total electric field at point A is $\vec{\mathbf{E}}_A = \sqrt{2}E_0(\hat{\mathbf{i}}+\hat{\mathbf{j}})$

2. A small particle of mass m moving inside a heavy, hollow and straight tube along the tube axis undergoes elastic collision at two ends. The tube has no friction and it is closed at one end by a flat surface while the other end is fitted with a heavy movable flat piston as shown in figure.

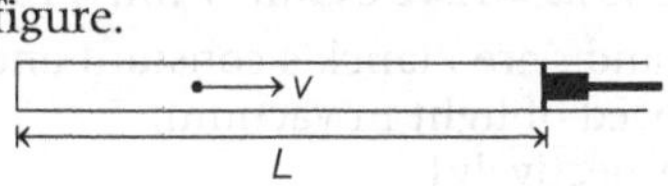

When the distance of the piston from closed end is $L = L_0$, the particle speed is $v = v_0$. The piston is moved inward at a very low speed V such that $V << \frac{dL}{L}v_0$, where dL is the infinitesimal displacement of the piston. Which of the following statement(s) is/are correct?

(a) After each collision with the piston, the particle speed increases by $2V$.

(b) If the piston moves inward by dL, the particle speed increases by $2v\frac{dL}{L}$.

(c) The particle's kinetic energy increases by a factor of 4 when the piston is moved inward from L_0 to $1/2\ L_0$.

(d) The rate at which the particle strikes the piston is v/L.

3. Three glass cylinders of equal height $H = 30$ cm and same refractive index $n = 1.5$ are placed on a horizontal surface as shown in figure.

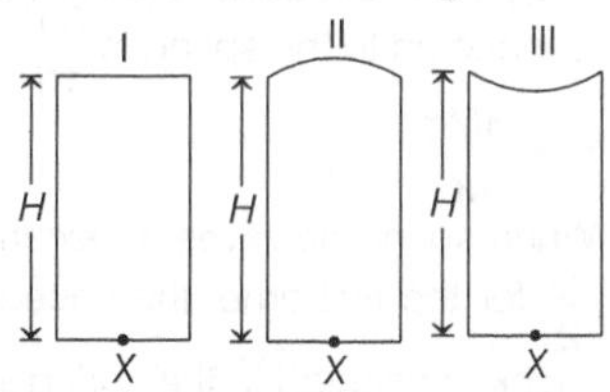

Cylinder I has a flat top, cylinder II has s convex top and cylinder III has a concave top. The radii of curvature of the two curved tops are same ($R = 3$ m). If H_1, H_2, and H_3 are the apparent depths of a point X on the bottom of the three cylinders, respectively, the correct statement(s) is/are

(a) $H_2 > H_1$

(b) $H_3 > H_1$

(c) $0.85\text{ cm} < (H_2 - H_1) < 0.9\text{ cm}$

(d) $H_2 > H_3$

4. A block of mass $2M$ is attached to a massless spring with spring-constant k.

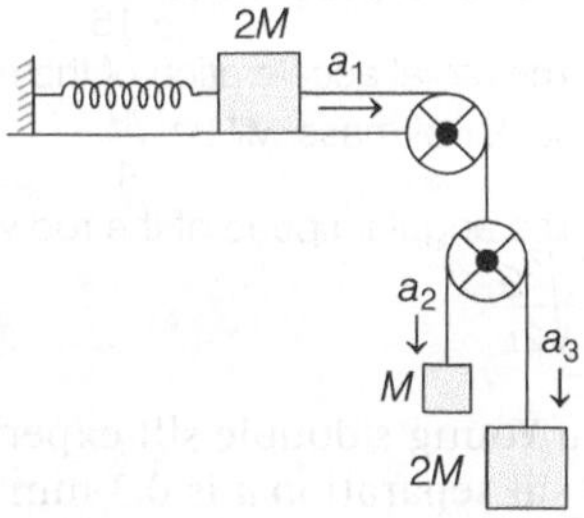

This block is connected to two other blocks of masses M and $2M$ using two massless pulleys and strings. The accelerations of the blocks are a_1, a_2 and a_3 as shown in the figure. The system is released from rest with the spring in its unstretched state. The maximum extension of the spring is x_0. Which of the following option(s) is/are correct?
[g is the acceleration due to gravity. Neglect friction]

(a) $a_2 - a_1 = a_1 - a_3$

(b) At an extension of $\frac{x_0}{4}$ of the spring, the magnitude of acceleration of the block connected to the spring is $\frac{3g}{10}$.

(c) $x_0 = \frac{4Mg}{k}$

(d) When spring achieves an extension of $\frac{x_0}{2}$ for the first time, the speed of the block connected to the spring is $3g\sqrt{\frac{M}{5k}}$.

5. A thin and uniform rod of mass M and length L is held vertical on a floor with large friction. The rod is released from rest so that it falls by rotating about its contact-point with the floor without slipping. Which of the following statement(s) is/are correct, when the rod makes an angle 60° with vertical?
[g is the acceleration due to gravity]

(a) The angular acceleration of the rod will be $\frac{2g}{L}$.

(b) The normal reaction force from the floor on the rod will be $\frac{Mg}{16}$.

(c) The radial acceleration of the rod's center of mass will be $\frac{3g}{4}$.

(d) The angular speed of the rod will be $\sqrt{\frac{3g}{2L}}$.

6. In a Young's double slit experiment, the slit separation d is 0.3 mm and the screen distance D is 1 m. A parallel beam of light of wavelength 600 nm is incident on the slits at angle α as shown in figure.

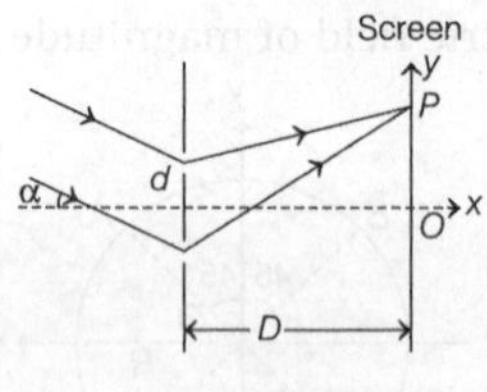

On the screen, the point O is equidistant from the slits and distance PO is 11.0 mm. Which of the following statement(s) is/are correct?

(a) For $\alpha = 0$, there will be constructive interference at point P.

(b) For $\alpha = \frac{0.36}{\pi}$ degree, there will be destructive interference at point P.

(c) For $\alpha = \frac{0.36}{\pi}$ degree, there will be destructive interference at point O.

(d) Fringe spacing depends on α.

7. A free hydrogen atom after absorbing a photon of wavelength λ_a gets excited from the state $n = 1$ to the state $n = 4$. Immediately after that the electron jumps to $n = m$ state by emitting a photon of wavelength λ_e. Let the change in momentum of atom due to the absorption and the emission be Δp_a and Δp_e, respectively. If $\frac{\lambda_a}{\lambda_e} = \frac{1}{5}$, which of the option(s) is/are correct?
[Use $hc = 1242$ eVnm; 1 nm $= 10^{-9}$ m, h and c are Planck's constant and speed of light in vacuum, respectively]

(a) The ratio of kinetic energy of the electron in the state $n = m$ to the state, $n = 1$ is $\frac{1}{4}$

(b) $m = 2$

(c) $\frac{\Delta p_a}{\Delta p_e} = \frac{1}{2}$

(d) $\lambda_e = 418$ nm

8. A mixture of ideal gas containing 5 moles of monatomic gas and 1 mole of rigid diatomic gas is initially at pressure P_0, volume V_0, and temperature T_0. If the gas mixture is adiabatically compressed to a volume $\frac{V_0}{4}$, then the correct statement(s) is/are
(Given, $2^{1.2} = 2.3$; $2^{3.2} = 9.2$; R is a gas constant)

(a) The final pressure of the gas mixture after compression is in between $9P_0$ and $10P_0$.
(b) The average kinetic energy of the gas mixture after compression is in between $18\,RT_0$ and $19\,RT_0$.
(c) Adiabatic constant of the gas mixture is 1.6.
(d) The work $|W|$ done during the process is $13RT_0$.

Section 2 (Maximum Marks : 18)

- This section contains **SIX (06)** questions. The answer to each question is a **NUMERICAL VALUE.**
- For each question, enter the correct numerical value of the answer using the mouse and the on-screen virtual numeric keypad in the place designated to enter the answer. If the numerical value has more than two decimal places, **truncate/round-off** the value to **TWO** decimal places.
- Answer to each question will be evaluated according to the following marking scheme:
  Full Marks : **+3** If ONLY the correct numerical value is entered as answer.
  Zero Marks : **0** In all other cases.

9. Suppose a $^{226}_{88}$Ra nucleus at rest and in ground state undergoes α-decay to a $^{222}_{86}$Rn nucleus in its excited state. The kinetic energy of the emitted α particle is found to be 4.44 MeV. $^{222}_{86}$Rn nucleus then goes to its ground state by γ-decay. The energy of the emitted γ photon is .... keV.

[Given : atomic mass of $^{226}_{88}$Ra $= 226.005$ u, atomic mass of $^{222}_{86}$Rn $= 222.000$ u, atomic mass of α particle $= 4.000$ u, $1\text{ u} = 931\text{ MeV}/c^2$, c is speed of the light]

10. A monochromatic light is incident from air on a refracting surface of a prism of angle 75° and refractive index $n_0 = \sqrt{3}$. The other refracting surface of the prism is coated by a thin film of material of refractive index n as shown in figure. The light suffers total internal reflection at the coated prism surface for an incidence angle of $\theta \le 60°$. The value of n^2 is ....

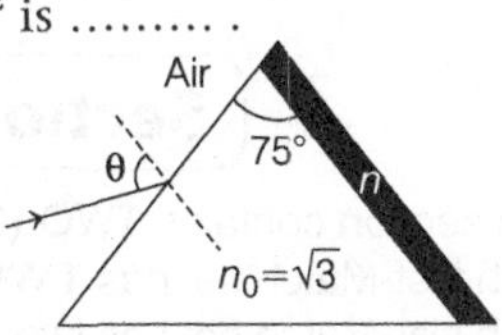

11. A perfectly reflecting mirror of mass M mounted on a spring constitutes a spring-mass system of angular frequency Ω such that $\frac{4\pi M\Omega}{h} = 10^{24}$ m^{-2} with h as Planck's constant. N photons of wavelength $\lambda = 8\pi \times 10^{-6}$ m strike the mirror simultaneously at normal incidence such that the mirror gets displaced by 1 μm.

If the value of N is $x \times 10^{12}$, then the value of x is ....
[Consider the spring as massless]

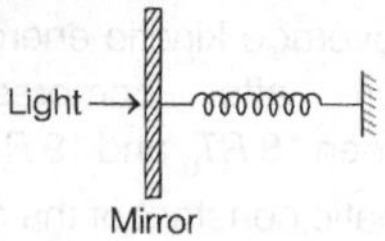

12. An optical bench has 1.5 m long scale having four equal divisions in each cm. While measuring the focal length of a convex lens, the lens is kept at 75 cm mark of the scale and the object pin is kept at 45 cm mark. The image of the object pin on the other side of the lens overlaps with image pin that is kept at 135 cm mark. In this experiment, the percentage error in the measurement of the focal length of the lens is ....

13. A ball is thrown from ground at an angle θ with horizontal and with an initial speed u_0. For the resulting projectile motion, the magnitude of average velocity of the ball up to the point when it hits the ground for the first time is V_1. After hitting the ground, the ball rebounds at the same angle θ but with a reduced speed of $\frac{u_0}{\alpha}$. Its motion continues for a long time as shown in figure. If the magnitude of average velocity of the ball for entire duration of motion is $0.8\,V_1$, the value of α is ....

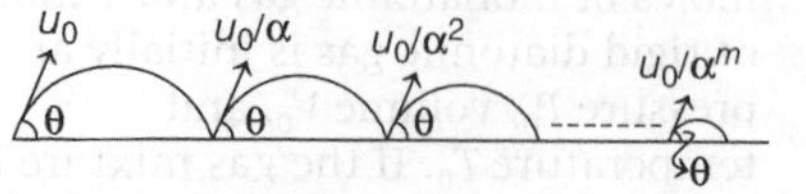

14. A 10 cm long perfectly conducting wire PQ is moving with a velocity 1 cm/s on a pair of horizontal rails of zero resistance. One side of the rails is connected to an inductor $L = 1$ mH and a resistance $R = 1\ \Omega$ as shown in figure. The horizontal rails, L and R lie in the same plane with a uniform magnetic field $B = 1$ T perpendicular to the plane. If the key S is closed at certain instant, the current in the circuit after 1 millisecond is $x \times 10^{-3}$ A, where the value of x is ....
[Assume the velocity of wire PQ remains constant (1 cm/s) after key S is closed. Given, $e^{-1} = 0.37$, where e is base of the natural logarithm]

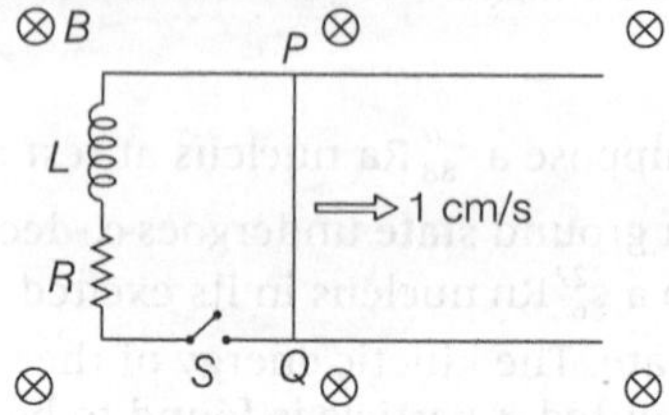

Section 3 (Maximum Marks : 12)

- This section contains **TWO (02)** List-Match sets.
- Each List-Match set has **TWO (02)** Multiple Choice Questions.
- Each List-Match set has two lists : List-I and List-II
- List-I has **Four** entries (I), (II), (III) and (IV) and **List-II** has **Six** entries (P), (Q), (R), (S), (T) and (U).
- **FOUR** options are given in each Multiple Choice Question based on **List-I** and **List-II** and **ONLY ONE** of these four options satisfies the condition asked in the Multiple Choice Questions.
- Answer to each question will be evaluated according to the following marking scheme :
Full Marks : **+3** If ONLY the option corresponding to the correct combination is chosen:
Zero Marks : **0** If none of the options is chosen (i.e. the question is unanswered).
Negative Marks : **–1** In all other cases.

Answer the following by appropriately matching the lists based on the information given in the paragraph

A musical instrument is made using four different metal strings, 1, 2, 3 and 4 with mass per unit length μ, 2μ, 3μ and 4μ respectively. The instrument is played by vibrating the strings by varying the free length in between the range L_0 and $2L_0$. It is found that in string-1 (μ) at free length L_0 and tension T_0 the fundamental mode frequency is f_0.

List-I gives the above four strings while list-II lists the magnitude of some quantity.

List-I	List-II
(I) String-1 (μ)	(P) 1
(II) String-2 (2μ)	(Q) 1/2
(III) String-3 (3μ)	(R) $1/\sqrt{2}$
(IV) String-4 (4μ)	(S) $1/\sqrt{3}$
	(T) 3/16
	(U) 1/16

15. If the tension in each string is T_0, the correct match for the highest fundamental frequency in f_0 units will be

(a) I $\rightarrow$ P, II $\rightarrow$ Q, III $\rightarrow$ T, IV $\rightarrow$ S
(b) I $\rightarrow$ P, II $\rightarrow$ R, III $\rightarrow$ S, IV $\rightarrow$ Q
(c) I $\rightarrow$ Q, II $\rightarrow$ S, III $\rightarrow$ R, IV $\rightarrow$ P
(d) I $\rightarrow$ Q, II $\rightarrow$ P, III $\rightarrow$ R, IV $\rightarrow$ T

16. The length of the strings 1, 2, 3 and 4 are kept fixed at $L_0, \frac{3L_0}{2}, \frac{5L_0}{4}$ and $\frac{7L_0}{4}$, respectively. Strings 1, 2, 3 and 4 are vibrated at their 1st, 3rd, 5th and 14th harmonics, respectively such that all the strings have same frequency.

The correct match for the tension in the four strings in the units of T_0 will be

(a) I $\rightarrow$ P, II $\rightarrow$ R, III $\rightarrow$ T, IV $\rightarrow$ U
(b) I $\rightarrow$ P, II $\rightarrow$ Q, III $\rightarrow$ R, IV $\rightarrow$ T
(c) I $\rightarrow$ P, II $\rightarrow$ Q, III $\rightarrow$ T, IV $\rightarrow$ U
(d) I $\rightarrow$ T, II $\rightarrow$ Q, III $\rightarrow$ R, IV $\rightarrow$ U

Answer the following by appropriately matching the lists based on the information given in the paragraph

In a thermodynamic process on an ideal monatomic gas, the infinitesimal heat absorbed by the gas is given by $T\Delta X$. where, T is temperature of the system and ΔX is the infinitesimal change in a thermodynamic quantity X of the system. For a mole of monatomic ideal gas,

$$X = \frac{3}{2}R\ln\left(\frac{T}{T_A}\right) + R\ln\left(\frac{V}{V_A}\right)$$

Here, R is gas constant, V is volume of gas, T_A and V_A are constants.

The List-I below gives some quantities involved in a process and List-II gives some possible values of these quantities.

List-I	List-II
(I) Work done by the system in process $1 \rightarrow 2 \rightarrow 3$	(P) $\frac{1}{3}RT_0\ln 2$
(II) Change in internal energy in process $1 \rightarrow 2 \rightarrow 3$	(Q) $\frac{1}{3}RT_0$
(III) Heat absorbed by the system in process $1 \rightarrow 2 \rightarrow 3$	(R) RT_0
(IV) Heat absorbed by the system in process $1 \rightarrow 2$	(S) $\frac{4}{3}RT_0$
	(T) $\frac{1}{3}RT_0(3 + \ln 2)$
	(U) $\frac{5}{6}RT_0$

17. If the process on one mole of monatomic ideal gas is as shown in the TV-diagram with $P_0V_0 = \frac{1}{3}RT_0$, the correct match is,

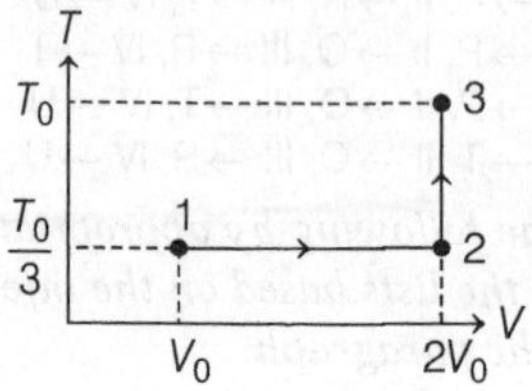

(a) I → P, II → R, III → T, IV → S
(b) I → P, II → T, III → Q, IV → T
(c) I → S, II → T, III → Q, IV → U
(d) I → P, II → R, III → T, IV → P

18. If the process carried out on one mole of monoatomic ideal gas is as shown in the PV-diagram with $p_0V_0 = \frac{1}{3}RT_0$, the correct match is,

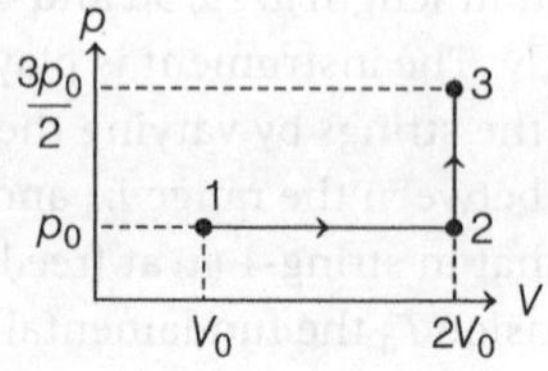

(a) I → S, II → R, III → Q, IV → T
(b) I → Q, II → R, III → P, IV → U
(c) I → Q, II → S, III → R, IV → U
(d) I → Q, II → R, III → S, IV → U

CHEMISTRY

Section 1 (Maximum Marks : 32)

Instructions: Same as given in Physics.

19. Choose the correct option(s) for the following reaction sequence

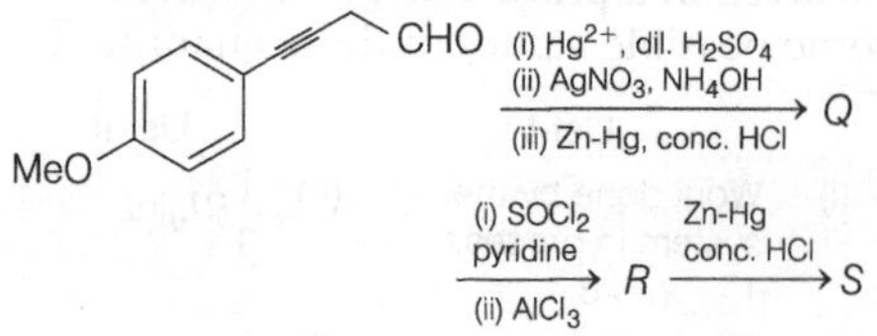

Consider Q, R and S as major products.

(a)

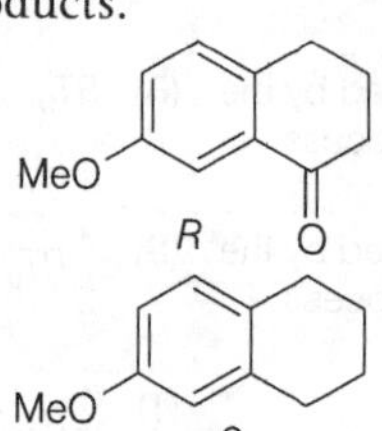

R, S

(b)

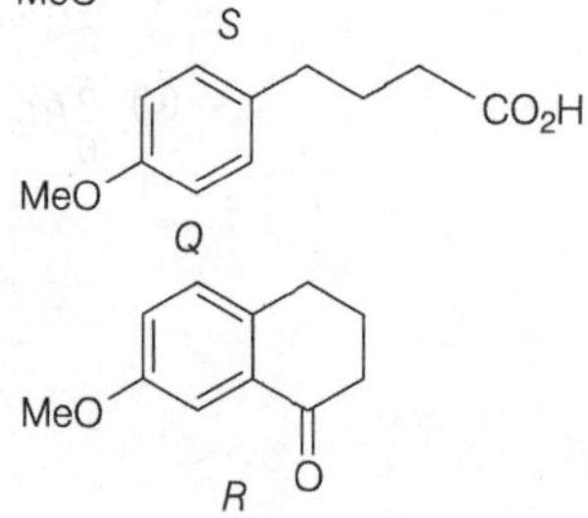

Q, R

(c)

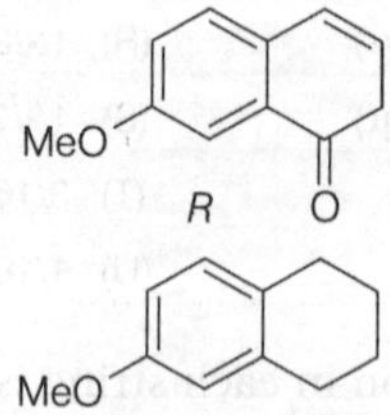

R, S

(d)

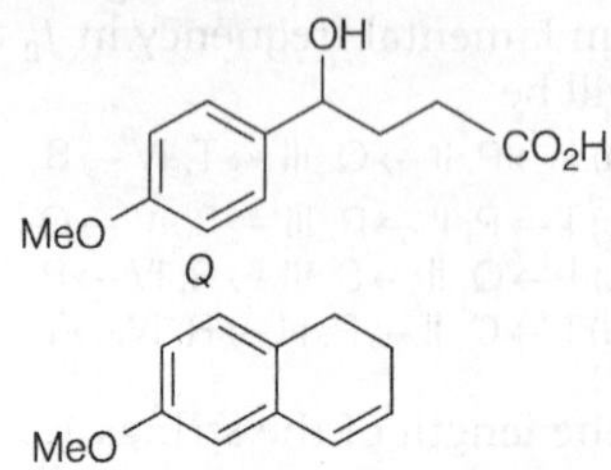

Q, S

20. Choose the correct option(s) that give(s) an aromatic compound as the major product.

(a)

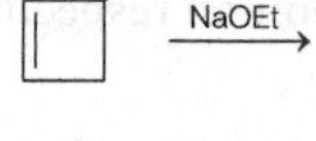

(b)

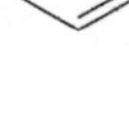

+ Cl_2 (excess) $\xrightarrow{\text{UV, 500 K}}$

(c) 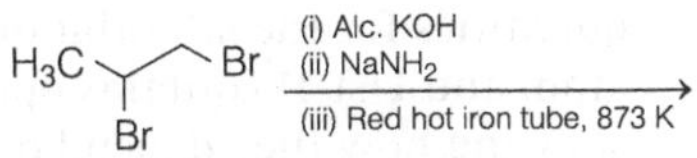

(d) 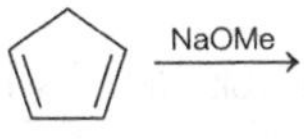

21. The ground state energy of hydrogen atom is −13.6 eV. Consider an electronic state Ψ of He^+ whose energy, azimuthal quantum number and magnetic quantum number are −3.4 eV, 2 and 0, respectively.

Which of the following statement(s) is(are) true for the state Ψ?

(a) It is a 4*d* state
(b) The nuclear charge experienced by the electron in this state is less than 2*e*, where *e* is the magnitude of the electronic charge
(c) It has 2 angular nodes
(d) It has 3 radial nodes

22. Consider the following reactions (unbalanced).

$$Zn + \text{Hot conc. } H_2SO_4 \longrightarrow G + R + X$$
$$Zn + \text{conc. NaOH} \longrightarrow T + Q$$
$$G + H_2S + NH_4OH \longrightarrow Z \text{ (a precipitate)} + X + Y$$

Choose the correct option(s).

(a) The oxidation state of Zn in *T* is +1
(b) *R* is a V-shaped molecule
(c) Bond order of *Q* is 1 in its ground state
(d) *Z* is dirty white in colour

23. With reference to *aqua-regia*, choose the correct option(s).

(a) *Aqua-regia* is prepared by mixing conc. HCl and conc. HNO_3 in 3 : 1 (*v* / *v*) ratio
(b) The yellow colour of *aqua-regia* is due to the presence of NOCl and Cl_2
(c) Reaction of gold with *aqua-regia* produces an anion having Au in +3 oxidation state
(d) Reaction of gold with *aqua regia* produces NO_2 in the absence of air

24. Choose the correct option(s) from the following.

(a) Teflon is prepared by heating tetrafluoroethene in presence of a persulphate catalyst at high pressure
(b) Natural rubber is polyisoprene containing *trans* alkene units
(c) Cellulose has only α-D-glucose units that are joined by glycosidic linkages
(d) Nylon-6 has amide linkages

25. The cyanide process of gold extraction involves leaching out gold from its ore with CN^- in the presence of *Q* in water to form *R*. Subsequently, *R* is treated with *T* to obtain Au and *Z*. Choose the correct option(s).

(a) *Q* is O_2 (b) *Z* is $[Zn(CN)_4]^{2-}$
(c) *T* is Zn (d) *R* is $[Au(CN)_4]^-$

26. Which of the following reactions produce(s) propane as a major product?

(a) 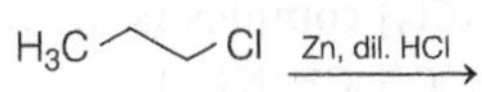

(b) 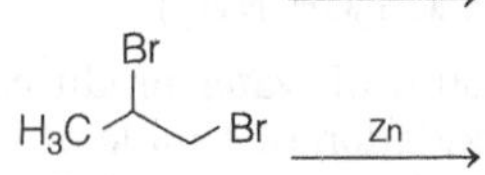

(c) 

(d) 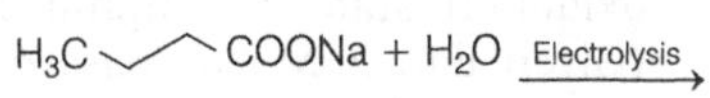

Section 2 (Maximum Marks : 18)

Instructions: Same as given in Physics.

27. The decomposition reaction

$$2N_2O_5(g) \xrightarrow{\Delta} 2N_2O_4(g) + O_2(g)$$ is started in a closed cylinder under isothermal isochoric condition at an initial pressure of 1 atm. After $Y \times 10^3$ s, the pressure inside the cylinder is found to be 1.45 atm. If the rate constant of the reaction is $5 \times 10^{-4}\ s^{-1}$, assuming ideal gas behaviour, the value of *Y* is ....

28. The mole fraction of urea in an aqueous urea solution containing 900 g of water is 0.05. If the density of the solution is 1.2 g cm^{-3}, then molarity of urea solution is ......
(Given data : Molar masses of urea and water are 60 g mol^{-1} and 18 g mol^{-1}, respectively)

29. Total number of hydroxyl groups present in a molecule of the major product P is ....

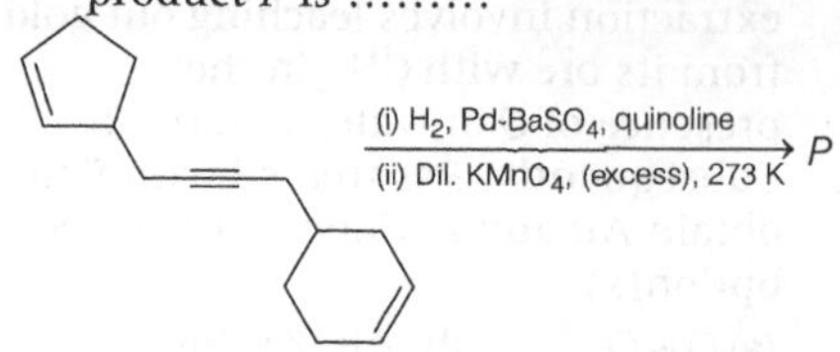

30. Total number of *cis* N—Mn—Cl bond angles (that is Mn—N and Mn—Cl bonds in *cis* positions) present in a molecule of *cis* [$Mn(en)_2Cl_2$] complex is ....
(en = $NH_2CH_2CH_2NH_2$)

31. The amount of water produced (in g) in the oxidation of 1 mole of rhombic sulphur by conc. HNO_3 to a compound with the highest oxidation state of sulphur is ......
(Given data : Molar mass of water = 18 g mol^{-1})

32. Total number of isomers considering both structural and stereoisomers of cyclic ethers with the molecular formula C_4H_8O is ....

Section 3 (Maximum Marks : 12)

Instructions: Same as given in Physics.

Answer the following by appropriately matching the lists based on the information given in the paragraph.

33. Consider the Bohr's model of a one-electron atom where the electron moves around the nucleus. In the following List-I contains some quantities for the nth orbit of the atom and List-II contains options showing how they depend on n.

	List-I		List-II
(I)	Radius of the n th orbit	(P)	$\propto n^{-2}$
(II)	Angular momentum of the electron in the nth orbit	(Q)	$\propto n^{-1}$
(III)	Kinetic energy of the electron in the nth orbit	(R)	$\propto n^{0}$
(IV)	Potential energy of the electron in the nth orbit	(S)	$\propto n^{1}$
		(T)	$\propto n^{2}$
		(U)	$\propto n^{1/2}$

Which of the following options has the correct combination considering List-I and List-II?

(a) (III), (P) (b) (III), (S)
(c) (IV), (U) (d) (IV), (Q)

34. Consider the Bohr's model of a one-electron atom where the electron moves around the nucleus. In the following List-I contains some quantities for the nth orbit of the atom and List-II contains options showing how they depend on n.

	List-I		List-II
(I)	Radius of the nth orbit	(P)	$\propto n^{-2}$
(II)	Angular momentum of the electron in the nth orbit	(Q)	$\propto n^{-1}$
(III)	Kinetic energy of the electron in the nth orbit	(R)	$\propto n^{0}$
(IV)	Potential energy of the electron in the nth orbit	(S)	$\propto n^{1}$
		(T)	$\propto n^{2}$
		(U)	$\propto n^{1/2}$

Which of the following options has the correct combination considering List-I and List-II?

(a) (II), (R) (b) (I), (P)
(c) (I), (T) (d) (II), (Q)

35. List-I includes starting materials and reagents of selected chemical reactions. List-II gives structures of compounds that may be formed as intermediate products and/or final products from the reactions of List-I.

	List-I	Reagents		List-II
(I)	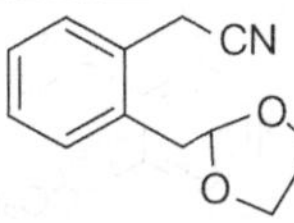	(i) DIBAL-H (ii) Dil. HCl → (iii) $NaBH_4$ (iv) Conc. H_2SO_4	(P)	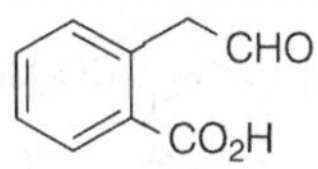
(II)	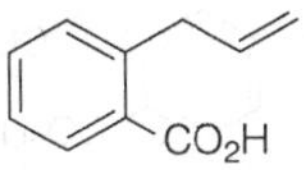	(i) O_3 (ii) Zn, H_2O → (iii) $NaBH_4$ (iv) Conc. H_2SO_4	(Q)	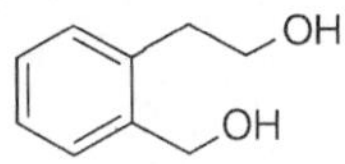
(III)	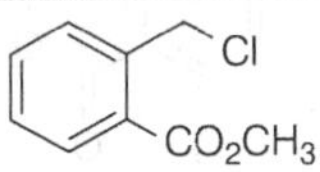	(i) KCN (ii) H_3O^+, Δ → (iii) $LiAlH_4$ (iv) Conc. H_2SO_4	(R)	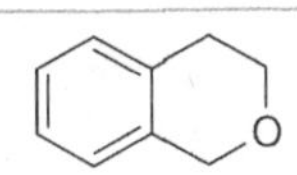
(IV)	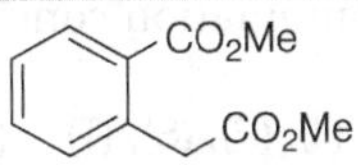	(iii) $LiAlH_4$ → (iv) Conc. H_2SO_4	(S)	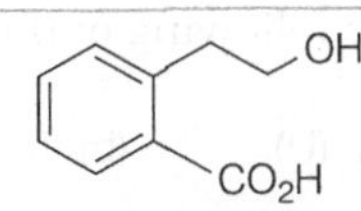
			(T)	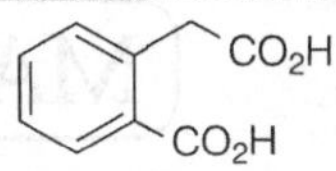
			(U)	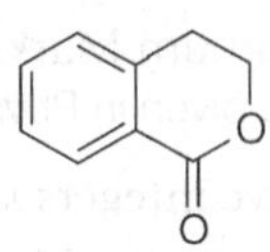

Which of the following options has correct combination considering List-I and List-II?

(a) (III), (S), (R) (b) (IV), (Q), (R) (c) (III), (T), (U) (d) (IV), (Q), (U)

36. List-I includes starting materials and reagents of selected chemical reactions. List-II gives structures of compounds that may be formed as intermediate products and/or final products from the reactions of List-I.

	List-I	Reagents		List-II
(I)	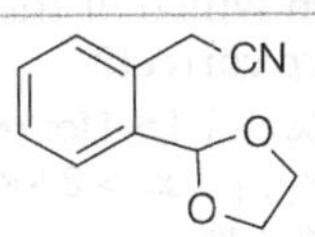	(i) DIBAL-H (ii) Dil. HCl → (iii) $NaBH_4$ (iv) Conc. H_2SO_4	(P)	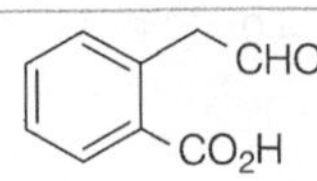
(II)	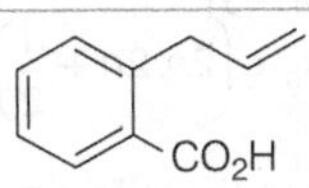	(i) O_3 (ii) Zn, H_2O → (iii) $NaBH_4$ (iv) Conc. H_2SO_4	(Q)	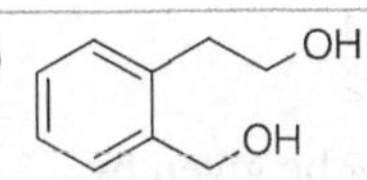

List-I		List-II	
(III)	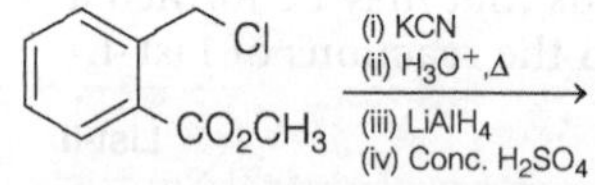	(R)	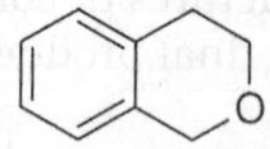
(IV)	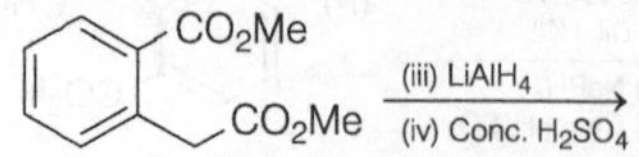	(S)	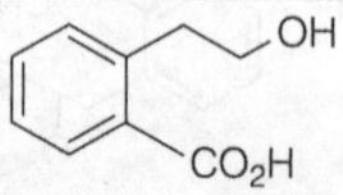
		(T)	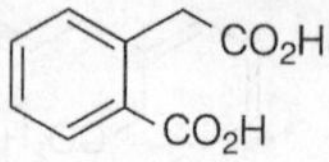
		(U)	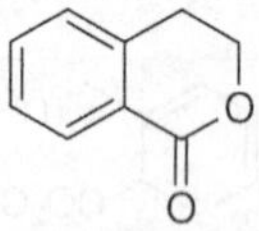

Which of the following options has correct combination considering List-I and List-II?

(a) (II), (P), (S), (U) (b) (I), (Q), (T), (U) (c) (II), (P), (S), (T) (d) (I), (S), (Q), (R)

MATHEMATICS

Section 1 (Maximum Marks : 32)

Instructions: Same as given in Physics.

37. For non-negative integers n, let

$$f(n)=\frac{\sum_{k=0}^{n}\sin\left(\frac{k+1}{n+2}\pi\right)\sin\left(\frac{k+2}{n+2}\pi\right)}{\sum_{k=0}^{n}\sin^2\left(\frac{k+1}{n+2}\pi\right)}$$

Assuming $\cos^{-1}x$ takes values in $[0,\pi]$, which of the following options is/are correct?

(a) If $\alpha=\tan(\cos^{-1}f(6))$, then $\alpha^2+2\alpha-1=0$

(b) $f(4)=\frac{\sqrt{3}}{2}$

(c) $\sin(7\cos^{-1}f(5))=0$

(d) $\lim_{n\to\infty}f(n)=\frac{1}{2}$

38. Let $f:R\to R$ be given by $f(x)=(x-1)(x-2)(x-5)$. Define

$$F(x)=\int_0^x f(t)dt,\ x>0$$

Then which of the following options is/are correct?

(a) $F(x)\neq 0$ for all $x\in(0,5)$

(b) F has a local maximum at $x=2$

(c) F has two local maxima and one local minimum in $(0,\infty)$

(d) F has a local minimum at $x=1$

39. Let, $f(x)=\frac{\sin\pi x}{x^2}$, $x>0$

Let $x_1<x_2<x_3<\ldots<x_n<\ldots$ be all the points of local maximum of f and $y_1<y_2<y_3<\ldots<y_n<\ldots$ be all the points of local minimum of f.

Then which of the following options is/are correct?

(a) $|x_n-y_n|>1$ for every n

(b) $x_{n+1}-x_n>2$ for every n

(c) $x_1<y_1$

(d) $x_n\in\left(2n,2n+\frac{1}{2}\right)$ for every n

40. Three lines $L_1:\mathbf{r}=\lambda\hat{\mathbf{i}},\ \lambda\in R$,

$L_2:\mathbf{r}=\hat{\mathbf{k}}+\mu\hat{\mathbf{j}},\ \mu\in R$ and

$L_3:\mathbf{r}=\hat{\mathbf{i}}+\hat{\mathbf{j}}+v\hat{\mathbf{k}},\ v\in R$ are given.

For which point(s) Q on L_2 can we find a point P on L_1 and a point R on L_3 so that P, Q and R are collinear?

(a) $\hat{\mathbf{k}}$ (b) $\hat{\mathbf{k}}+\hat{\mathbf{j}}$ (c) $\hat{\mathbf{k}}+\frac{1}{2}\hat{\mathbf{j}}$ (d) $\hat{\mathbf{k}}-\frac{1}{2}\hat{\mathbf{j}}$

41. For $a \in R$, $|a| > 1$, let $\lim\limits_{n\to\infty}\left(\dfrac{1+\sqrt[3]{2}+\ldots+\sqrt[3]{n}}{n^{7/3}\left(\dfrac{1}{(an+1)^2}+\dfrac{1}{(an+2)^2}+\ldots+\dfrac{1}{(an+n)^2}\right)}\right)=54$

Then the possible value(s) of a is/are

(a) -6 (b) 7 (c) 8 (d) -9

42. Let $f:R$ be a function. We say that f has PROPERTY 1 if $\lim\limits_{h\to 0}\dfrac{f(h)-f(0)}{\sqrt{|h|}}$ exists and is finite, and PROPERTY 2 if $\lim\limits_{h\to 0}\dfrac{f(h)-f(0)}{h^2}$ exists and is finite.

Then which of the following options is/are correct?

(a) $f(x)=\sin x$ has PROPERTY 2 (b) $f(x)=x^{2/3}$ has PROPERTY 1

(c) $f(x)=|x|$ has PROPERTY 1 (d) $f(x)=x|x|$ has PROPERTY 2

43. Let $x \in R$ and let $P=\begin{bmatrix}1&1&1\\0&2&2\\0&0&3\end{bmatrix}$, $Q=\begin{bmatrix}2&x&x\\0&4&0\\x&x&6\end{bmatrix}$ and $R=PQP^{-1}$, the which of the following options is/are correct?

(a) There exists a real, number x such that $PQ=QP$

(b) For $x=0$, if $R\begin{bmatrix}1\\a\\b\end{bmatrix}=6\begin{bmatrix}1\\a\\b\end{bmatrix}$, then $a+b=5$

(c) For $x=1$, there exists a unit vector $\alpha\hat{\mathbf{i}}+\beta\hat{\mathbf{j}}+\gamma\hat{\mathbf{k}}$ for which $R\begin{bmatrix}\alpha\\\beta\\\gamma\end{bmatrix}=\begin{bmatrix}0\\0\\0\end{bmatrix}$

(d) $\det R=\det\begin{bmatrix}2&x&x\\0&4&0\\x&x&5\end{bmatrix}+8$, for all $x\in R$

44. Let $P_1=I=\begin{bmatrix}1&0&0\\0&1&0\\0&0&1\end{bmatrix}$, $P_2=\begin{bmatrix}1&0&0\\0&0&1\\0&1&0\end{bmatrix}$, $P_3=\begin{bmatrix}0&1&0\\1&0&0\\0&0&1\end{bmatrix}$, $P_4=\begin{bmatrix}0&1&0\\0&0&1\\1&0&0\end{bmatrix}$,

$P_5=\begin{bmatrix}0&0&1\\1&0&0\\0&1&0\end{bmatrix}$, $P_6=\begin{bmatrix}0&0&1\\0&1&0\\1&0&0\end{bmatrix}$ and $X=\sum\limits_{k=1}^{6}P_k\begin{bmatrix}2&1&3\\1&0&2\\3&2&1\end{bmatrix}P_k^T$

where, P_k^T denotes the transpose of the matrix P_k. Then which of the following option is/are correct?

(a) X is a symmetric matrix

(b) The sum of diagonal entries of X is 18

(c) $X - 30I$ is an invertible matrix

(d) If $X\begin{bmatrix}1\\1\\1\end{bmatrix} = \alpha\begin{bmatrix}1\\1\\1\end{bmatrix}$, then $\alpha = 30$

Section 2 (Maximum Marks : 18)

Instructions: Same as given in Physics.

45. Let $\mathbf{a} = 2\hat{\mathbf{i}} + \hat{\mathbf{j}} - \hat{\mathbf{k}}$ and $\mathbf{b} = \hat{\mathbf{i}} + 2\hat{\mathbf{j}} + \hat{\mathbf{k}}$ be two vectors. Consider a vector $\mathbf{c} = \alpha\mathbf{a} + \beta\mathbf{b}$, $\alpha, \beta \in R$. If the projection of $\mathbf{c}$ on the vector $(\mathbf{a} + \mathbf{b})$ is $3\sqrt{2}$, then the minimum value of $(\mathbf{c} - (\mathbf{a} \times \mathbf{b}))\cdot\mathbf{c}$ equals ....

46. Let $|X|$ denote the number of elements in a set X. Let $S = \{1, 2, 3, 4, 5, 6\}$ be a sample space, where each element is equally likely to occur. If A and B are independent events associated with S, then the number of ordered pairs (A, B) such that $1 \le |B| < |A|$, equals ....

47. Suppose

$$\det\begin{bmatrix}\sum_{k=0}^{n} k & \sum_{k=0}^{n} {}^nC_k k^2 \\ \sum_{k=0}^{n} {}^nC_k k & \sum_{k=0}^{n} {}^nC_k 3^k\end{bmatrix} = 0 \text{ holds}$$

for some positive integer n. Then $\sum_{k=0}^{n} \frac{{}^nC_k}{k+1}$ equals ....

48. Five persons A, B, C, D and E are seated in a circular arrangement. If each of them is given a hat of one of the three colours red, blue and green, then the number of ways of distributing the hats such that the persons seated in adjacent seats get different coloured hats is ....

49. The value of the integral

$$\int_0^{\pi/2} \frac{3\sqrt{\cos\theta}}{(\sqrt{\cos\theta} + \sqrt{\sin\theta})^5}\, d\theta \text{ equals}$$

....

50. The value of

$$\sec^{-1}\left(\frac{1}{4}\sum_{k=0}^{10} \sec\left(\frac{7\pi}{12} + \frac{k\pi}{2}\right) \sec\left(\frac{7\pi}{12} + \frac{(k+1)\pi}{2}\right)\right)$$

in the interval $\left[-\frac{\pi}{4}, \frac{3\pi}{4}\right]$ equals......

Section 3 (Maximum Marks : 12)

Instructions: Same as given in Physics.

Answer the following by appropriately matching the lists based on the information given in the paragraph.

51. Let $f(x) = \sin(\pi\cos x)$ and $g(x) = \cos(2\pi\sin x)$ be two functions defined for $x > 0$. Define the following sets whose elements are written in the increasing order :

$X = \{x : f(x) = 0\}$, $Y = \{x : f'(x) = 0\}$

$Z = \{x : g(x) = 0\}$, $W = \{x : g'(x) = 0\}$

List-I contains the sets X, Y, Z and W. List-II contains some information regarding these sets.

List-I		List-II	
(I)	X	(P)	$\supseteq\left\{\frac{\pi}{2}, \frac{3\pi}{2}, 4\pi, 7\pi\right\}$
(II)	Y	(Q)	an arithmetic progression
(III)	Z	(R)	NOT an arithmetic progression
(IV)	W	(S)	$\supseteq\left\{\frac{\pi}{6}, \frac{7\pi}{6}, \frac{13\pi}{6}\right\}$
		(T)	$\supseteq\left\{\frac{\pi}{3}, \frac{2\pi}{3}, \pi\right\}$
		(U)	$\supseteq\left\{\frac{\pi}{6}, \frac{3\pi}{4}\right\}$

Which of the following is the only CORRECT combination?

(a) (IV), (P), (R), (S)
(b) (III), (P), (Q), (U)
(c) (III), (R), (U)
(d) (IV), (Q), (T)

52. Let $f(x) = \sin(\pi \cos x)$ and $g(x) = \cos(2\pi \sin x)$ be two functions defined for $x > 0$. Define the following sets whose elements are written in the increasing order :

$X = \{x : f(x) = 0\}$, $Y = \{x : f'(x) = 0\}$

$Z = \{x : g(x) = 0\}$, $W = \{x : g'(x) = 0\}$

List-I contains the sets X, Y, Z and W. List-II contains some information regarding these sets.

List-I		List-II	
(I)	X	(P)	$\supseteq \left\{\frac{\pi}{2}, \frac{3\pi}{2}, 4\pi, 7\pi\right\}$
(II)	Y	(Q)	an arithmetic progression
(III)	Z	(R)	NOT an arithmetic progression
(IV)	W	(S)	$\supseteq \left\{\frac{\pi}{6}, \frac{7\pi}{6}, \frac{13\pi}{6}\right\}$
		(T)	$\supseteq \left\{\frac{\pi}{3}, \frac{2\pi}{3}, \pi\right\}$
		(U)	$\supseteq \left\{\frac{\pi}{6}, \frac{3\pi}{4}\right\}$

Which of the following is the only CORRECT combination?

(a) (II), (Q), (T) (b) (II), (R), (S)
(c) (I), (P), (R) (d) (I), (Q), (U)

53. Let the circles $C_1 : x^2 + y^2 = 9$ and $C_2 : (x-3)^2 + (y-4)^2 = 16$, intersect at the points X and Y. Suppose that another circle $C_3 : (x-h)^2 + (y-k)^2 = r^2$ satisfies the following conditions :

(i) Centre of C_3 is collinear with the centres of C_1 and C_2.
(ii) C_1 and C_2 both lie inside C_3 and
(iii) C_3 touches C_1 at M and C_2 at N.

Let the line through X and Y intersect C_3 at Z and W, and let a common tangent of C_1 and C_3 be a tangent to the parabola $x^2 = 8\alpha y$.

There are some expression given in the List-I whose values are given in List-II below.

List-I		List-II	
(I)	$2h + k$	(P)	6
(II)	$\dfrac{\text{Length of } ZW}{\text{Length of } XY}$	(Q)	$\sqrt{6}$
(III)	$\dfrac{\text{Area of triangle } MZN}{\text{Area of triangle } ZMW}$	(R)	$\frac{5}{4}$
(IV)	α	(S)	$\frac{21}{5}$
		(T)	$2\sqrt{6}$
		(U)	$\frac{10}{3}$

Which of the following is the only INCORRECT combination?

(a) (III), (R) (b) (IV), (S)
(c) (I), (P) (d) (IV), (U)

54. Let the circle $C_1 : x^2 + y^2 = 9$ and $C_2 : (x-3)^2 + (y-4)^2 = 16$, intersect at the points X and Y. Suppose that another circle $C_3 : (x-h)^2 + (y-k)^2 = r^2$ satisfies the following conditions :

(i) centre of C_3 is collinear with the centers of C_1 and C_2.
(ii) C_1 and C_2 both lie inside C_3, and
(iii) C_3 touches C_1 at M and C_2 at N.

Let the line through X and Y intersect C_3 at Z and W, and let a common tangent of C_1 and C_3 be a tangent to the parabola $x^2 = 8\alpha y$.

There are some expression given in the List-I whose values are given in List-II below.

	List-I		List-II
(I)	$2h + k$	(P)	6
(II)	$\dfrac{\text{Length of } ZW}{\text{Length of } XY}$	(Q)	$\sqrt{6}$
(III)	$\dfrac{\text{Area of triangle } MZN}{\text{Area of triangle } ZMW}$	(R)	$\dfrac{5}{4}$
(IV)	α	(S)	$\dfrac{21}{5}$
		(T)	$2\sqrt{6}$
		(U)	$\dfrac{10}{3}$

Which of the following is the only CORRECT combination?

(a) (II), (T) (b) (I), (S) (c) (II), (Q) (d) (I), (U)

Answers

Paper 1

1.	a	2.	b	3.	a	4.	d	5.	a,c,d
6.	a,b,d	7.	a,b,d	8.	a,b	9.	b,c	10.	b,c,d
11.	b,d	12.	a,c,d	13.	1.00	14.	50.00	15.	2.00
16.	270.00	17.	0.75	18.	8.13	19.	b	20.	c
21.	d	22.	b	23.	c,d	24.	a,d	25.	a,b,c
26.	a,b,d	27.	c,d	28.	a,c,d	29.	a,c,d	30.	a,c
31.	4.00	32.	1.02	33.	6.75	34.	8.9	35.	19
36.	4	37.	c	38.	d	39.	d	40.	c
41.	a,c	42.	c,d	43.	a,b,d	44.	b,c,d	45.	a,b,c
46.	a,c	47.	b,c,d	48.	b,c,d	49.	0.50	50.	10
51.	3.00	52.	4.0	53.	0.75	54.	157.00		

Paper 2

1.	c,d	2.	a,c	3.	a,d	4.	a	5.	b,c,d
6.	b	7.	a,b	8.	a,c,d	9.	135.00	10.	1.50
11.	1.00	12.	1.38 & 1.39	13.	4.00	14.	0.63	15.	b
16.	c	17.	d	18.	d	19.	a,b	20.	c,d
21.	a,c	22.	b,c,d	23.	a,b,c	24.	a,d	25.	a,b,c
26.	a,c	27.	2.3	28.	2.98 mole	29.	6	30.	6
31.	288	32.	10.0	33.	a	34.	c	35.	b
36.	a	37.	a,b,c	38.	a,b,d	39.	a,b,d	40.	c,d
41.	c,d	42.	b,c	43.	b,d	44.	a,b,d	45.	18
46.	1523	47.	620	48.	30	49.	0.5	50.	0
51.	a	52.	a	53.	b	54.	c		

JEE ADVANCED

Solved Paper 2020

Paper 1

PHYSICS

Section 1 (Maximum Marks : 18)

- This section contains **SIX (06)** questions.
- Each question has **FOUR** options. **ONLY ONE** of these four options is the correct answer.
- For each question, choose the option corresponding to the correct answer.
- Answer to each question will be evaluated according to the following marking scheme:

Full Marks : **+3** If ONLY the correct options is chosen.
Zero Marks : **0** If none of the options is chosen. (i.e., the question is unanswered)
Negative Marks : **−1** In all other cases.

1. A football of radius R is kept on a hole of radius r $(r < R)$ made on a plank kept horizontally. One end of the plank is now lifted so that it gets tilted making an angle θ from the horizontal as shown in the figure below. The maximum value of θ so that the football does not start rolling down the plank satisfies (figure is schematic and not drawn to scale)

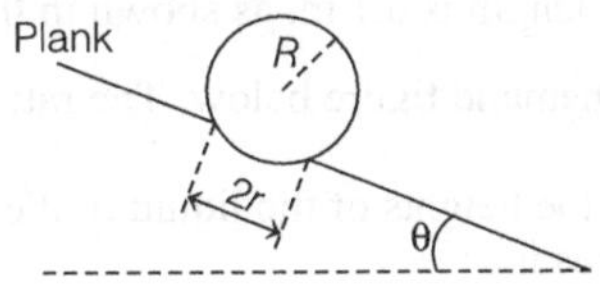

(a) $\sin\theta = \frac{r}{R}$ (b) $\tan\theta = \frac{r}{R}$

(c) $\sin\theta = \frac{r}{2R}$ (d) $\cos\theta = \frac{r}{2R}$

2. A light disc made of aluminium (a nonmag- netic material) is kept horizontally and is free to rotate about its axis as shown in the figure. A strong magnet is held vertically at a point above the disc away from its axis. On revolving the magnet about the axis of the disc, the disc will (figure is schematic and not drawn to scale)

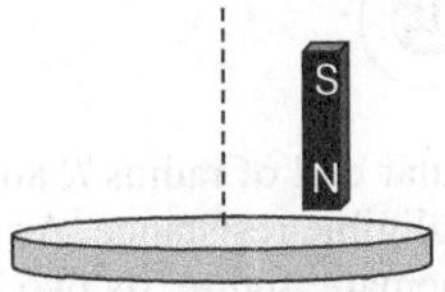

(a) rotate in the direction opposite to the direction of magnet's motion
(b) rotate in the same direction as the direction of magnet's motion
(c) not rotate and its temperature will remain unchanged
(d) not rotate but its temperature will slowly rise.

3. A small roller of diameter 20 cm has an axle of diameter 10 cm (see figure below on the left). It is on a horizontal floor and a meter scale is positioned horizontally on its axle with one edge of the scale on top of the axle (see figure on the right). The scale is now pushed slowly on the axle so that it moves without slipping on the axle, and the roller starts rolling without slipping. After the roller has moved 50 cm, the position of the scale will look like (figures are schematic and not drawn to scale)

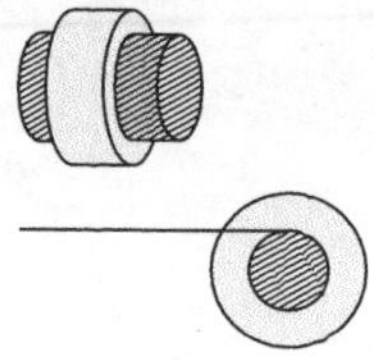

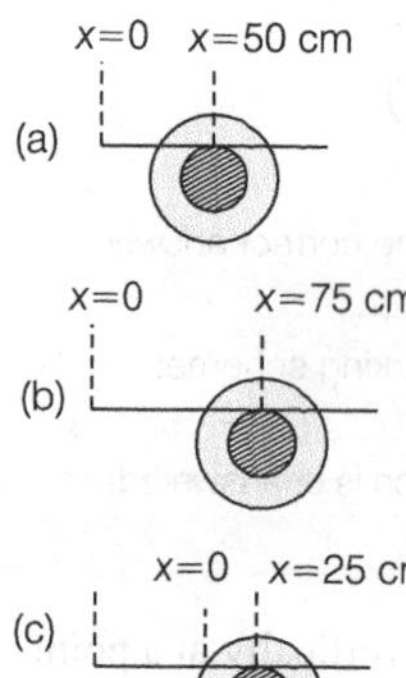

4. A circular coil of radius R and N turns has negligible resistance. As shown in the schematic figure, its two ends are connected to two wires and it is hanging by those wires with its plane being vertical. The wires are connected to a capacitor with charge Q through a switch. The coil is in a horizontal uniform magnetic field B_0 parallel to the plane of the coil. When the switch is closed, the capacitor gets discharged through the coil in a very short time. By the time the capacitor is discharged fully, magnitude of the angular momentum gained by the coil will be (assume that the discharge time is so short that the coil has hardly rotated during this time)

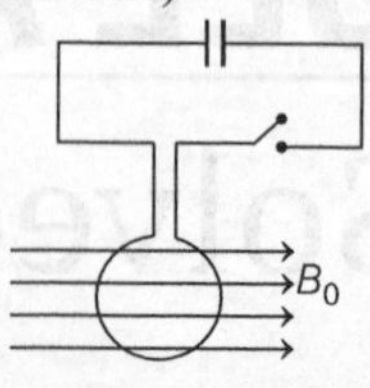

(a) $\frac{\pi}{2} NQB_0R^2$ (b) πNQB_0R^2

(c) $2\pi NQB_0R^2$ (d) $4\pi NQB_0R^2$

5. A parallel beam of light strikes a piece of transparent glass having cross section as shown in the figure below. Correct shape of the emergent wavefront will be (figures are schematic and not drawn to scale)

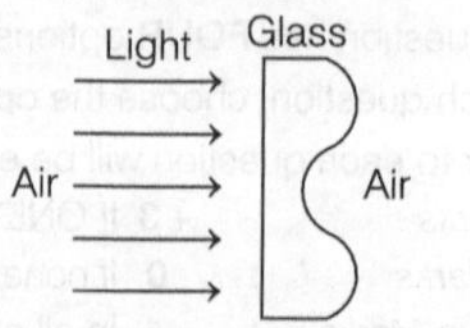

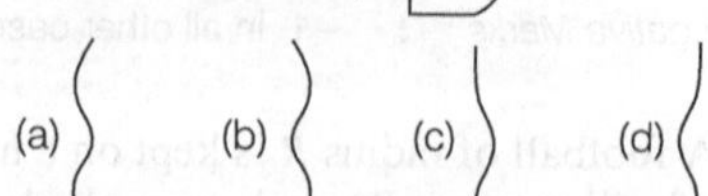

6. An open-ended U-tube of uniform cross-sectional area contains water (density 10^3 kg m^{-3}). Initially the water level stands at 0.29 m from the bottom in each arm. Kerosene oil (a water-immiscible liquid) of density 800 kg m^{-3} is added to the left arm until its length is 0.1 m, as shown in the schematic figure below. The ratio $\left(\frac{h_1}{h_2}\right)$ of the heights of the liquid in the two arms is

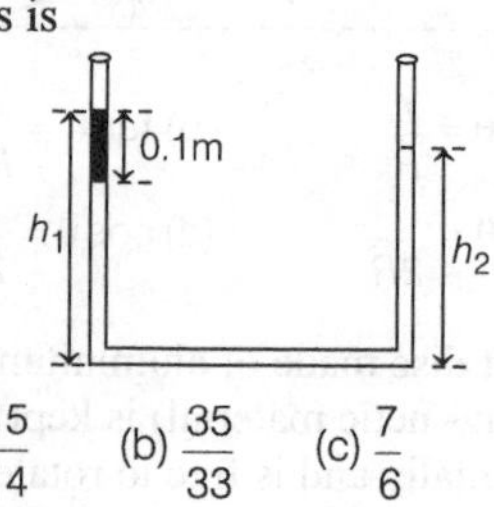

(a) $\frac{15}{14}$ (b) $\frac{35}{33}$ (c) $\frac{7}{6}$ (d) $\frac{5}{4}$

Section 2 (Maximum Marks : 24)

- This section contains **SIX (06)** questions.
- Each question has **FOUR** options. **ONE OR MORE THAN ONE** of these four option(s) is (are) correct options(s).
- For each question, choose the options(s) corresponding to (all) the correct answer(s).
- Answer to each question will be evaluated according to the following marking scheme:

Full Marks : **+4** If only (all) the correct option(s) is (are) chosen.
Partial Marks : **+3** If all the four options are correct but ONLY three options are chosen.
Partial Marks : **+2** If three or more options are correct but ONLY two options are chosen, both of which are correct.
Partial Marks : **+1** If two or more options are correct but ONLY one option is chosen and it is a correct option;
Zero Marks : **0** If none of the option is chosen (i.e., the question is unanswered);
Negative Marks : **−1** In all other cases.

7. A particle of mass m moves in circular orbits with potential energy $V(r) = Fr$, where F is a positive constant and r is its distance from the origin. Its energies are calculated using the Bohr model. If the radius of the particle's orbit is denoted by R and its speed and energy are denoted by v and E, respectively, then for the nth orbit (here h is the Planck's constant)

(a) $R \propto n^{1/3}$ and $v \propto n^{2/3}$

(b) $R \propto n^{2/3}$ and $v \propto n^{1/3}$

(c) $E = \frac{3}{2}\left(\frac{n^2h^2F^2}{4\pi^2 m}\right)^{1/3}$

(d) $E = 2\left(\frac{n^2h^2F^2}{4\pi^2 m}\right)^{1/3}$

8. The filament of a light bulb has surface area 64 mm^2. The filament can be considered as a black body at temperature 2500 K emitting radiation like a point source when viewed from far. At night the light bulb is observed from a distance of 100 m. Assume the pupil of the eyes of the observer to be circular with radius 3 mm. Then (Take Stefan-Boltzmann constant $= 5.67 \times 10^{-8}$ Wm^{-2}K^{-4}, Wien's displacement constant $= 2.90 \times 10^{-3}$ m - K, Planck's constant $= 6.63 \times 10^{-34}$ Js, speed of light in vacuum $= 3.00 \times 10^8$ ms^{-1})

(a) power radiated by the filament is in the range 642 W to 645 W

(b) radiated power entering into one eye of the observer is in the range 3.15×10^{-8} W to 3.25×10^{-8} W

(c) the wavelength corresponding to the maximum intensity of light is 1160 nm

(d) taking the average wavelength of emitted radiation to be 1740 nm, the total number of photons entering per second into one eye of the observer is in the range 2.75×10^{11} to 2.85×10^{11}

9. Sometimes it is convenient to construct a system of units so that all quantities can be expressed in terms of only one physical quantity. In one such system, dimensions of different quantities are given in terms of a quantity X as follows: [position] = $[X^\alpha]$; [speed] = $[X^\beta]$; [acceleration] = $[X^p]$; [linear momentum] = $[X^q]$; [force] = $[X^r]$. Then

(a) $\alpha + p = 2\beta$ (b) $p + q - r = \beta$

(c) $p - q + r = \alpha$ (d) $p + q + r = \beta$

10. A uniform electric field, $\mathbf{E} = -400\sqrt{3}\hat{y}$ NC^{-1} is applied in a region. A charged particle of mass m carrying positive charge q is projected in this region with an initial speed of $2\sqrt{10} \times 10^6$ ms^{-1}. This particle is aimed to hit a target T, which is 5 m away from its entry point into the field as shown schematically in the figure.

Take $\frac{q}{m} = 10^{10}\ \text{Ckg}^{-1}$. Then

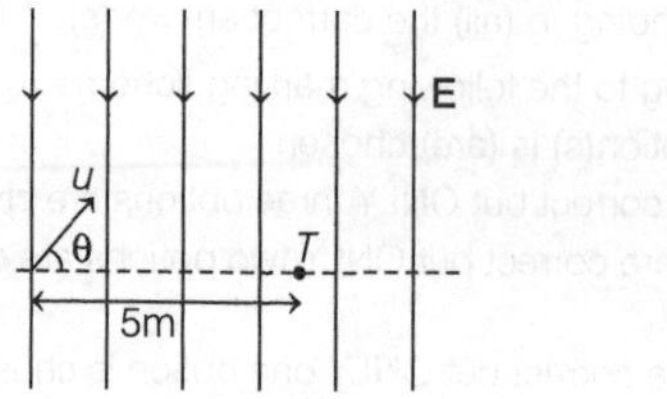

(a) the particle will hit T if projected at an angle 45° from the horizontal

(b) the particle will hit T if projected either at an angle 30° or 60° from the horizontal

(c) time taken by the particle to hit T could be $\sqrt{\frac{5}{6}}$ μs as well as $\sqrt{\frac{5}{2}}$ μs

(d) time taken by the particle to hit T is $\sqrt{\frac{5}{3}}$ μs

11. Shown in the figure is a semicircular metallic strip that has thickness t and resistivity ρ. Its inner radius is R_1 and outer radius is R_2. If a voltage V_0 is applied between its two ends, a current I flows in it. In addition, it is observed that a transverse voltage ΔV develops between its inner and outer surfaces due to purely kinetic effects of moving electrons (ignore any role of the magnetic field due to the current). Then (figure is schematic and not drawn to scale)

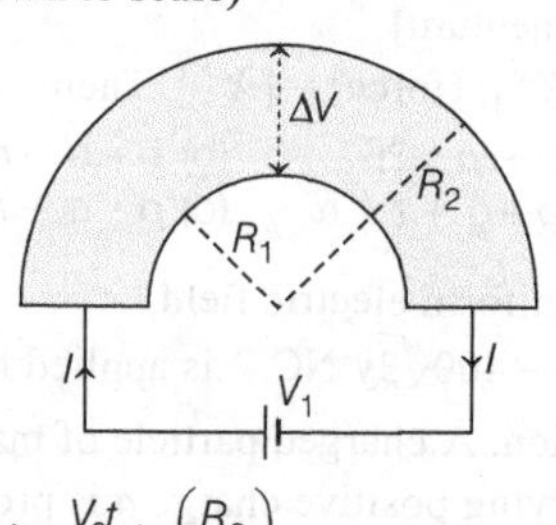

(a) $I = \frac{V_0 t}{\pi\rho} \ln\left(\frac{R_2}{R_1}\right)$

(b) the outer surface is at a higher voltage than the inner surface

(c) the outer surface is at a lower voltage than the inner surface

(d) $\Delta V \propto I^2$

12. As shown schematically in the figure, two vessels contain water solutions (at temperature T) of potassium permanganate ($KMnO_4$) of different concentrations n_1 and n_2 ($n_1 > n_2$) molecules per unit volume with $\Delta n = (n_1 - n_2) << n_1$. When they are connected by a tube of small length l and cross-sectional area S, $KMnO_4$ starts to diffuse from the left to the right vessel through the tube. Consider the collection of molecules to behave as dilute ideal gases and the difference in their partial pressure in the two vessels causing the diffusion. The speed v of the molecules is limited by the viscous force $-\beta v$ on each molecule, where β is a constant. Neglecting all terms of the order $(\Delta n)^2$, which of the following is/are correct? (k_B is the Boltzmann constant)

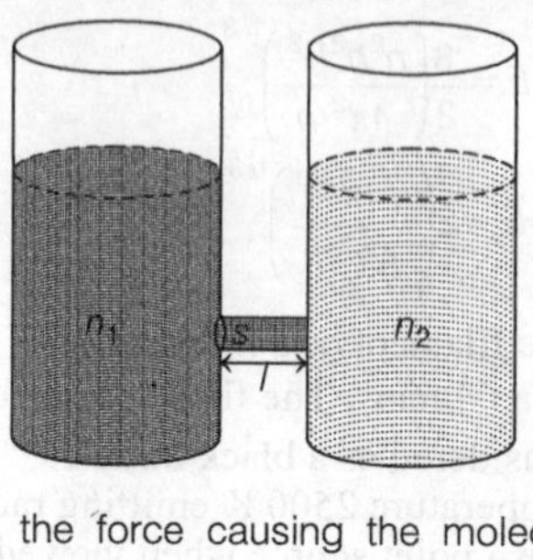

(a) the force causing the molecules to move across the tube is $\Delta n k_B T S$

(b) force balance implies $n_1 \beta v l = \Delta n k_B T$

(c) total number of molecules going across the tube per sec is $\left(\frac{\Delta n}{l}\right)\left(\frac{k_B T}{\beta}\right) S$

(d) rate of molecules getting transferred through the tube does not change with time

Section 3 (Maximum Marks : 24)

- This section contains **SIX (06)** questions. The answer to each question is a **NUMERICAL VALUE.**
- Four each question, enter the correct numerical value of the answer using the mouse and the on-screen virtual numeric keypad in the place designated to enter the answer. If the numerical value has more than two decimal places, truncate/round-off the value to TWO decimal places.
- Answer to each question will be evaluated according to the following marking scheme:

Full Marks : **+4** If ONLY the correct numerical value is entered.

Zero Marks : **0** In all other cases.

13. Put a uniform meter scale horizontally on your extended index fingers with the left one at 0.00 cm and the right one at 90.00 cm. When you attempt to move both the fingers slowly towards the center, initially only the left finger slips with respect to the scale and the right finger does not. After some distance, the left finger stops and the right one starts slipping. Then the right finger stops at a distance x_R from the center (50.00 cm) of the scale and the left one starts slipping again. This happens because of the difference in the frictional forces on the two fingers. If the coefficients of static and dynamic friction between the fingers and the scale are 0.40 and 0.32, respectively, the value of x_R (in cm) is ....

14. When water is filled carefully in a glass, one can fill it to a height h above the rim of the glass due to the surface tension of water. To calculate h just before water starts flowing, model the shape of the water above the rim as a disc of thickness h having semicircular edges, as shown schematically in the figure. When the pressure of water at the bottom of this disc exceeds what can be withstood due to the surface tension, the water surface breaks near the rim and water starts flowing from there. If the density of water, its surface tension and the acceleration due to gravity are 10^3 kg m^{-3}, 0.07 Nm^{-1} and 10 ms^{-2}, respectively, the value of h (in mm) is ....

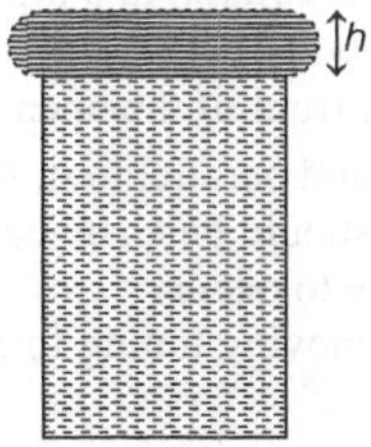

15. One end of a spring of negligible unstretched length and spring constant k is fixed at the origin (0,0). A point particle of mass m carrying a positive charge q is attached at its other end. The entire system is kept on a smooth horizontal surface. When a point dipole **p** pointing towards the charge q is fixed at the origin, the spring gets stretched to a length l and attains a new equilibrium position (see figure below). If the point mass is now displaced slightly by $\Delta l \ll l$ from its equilibrium position and released, it is found to oscillate at frequency $\frac{1}{\delta}\sqrt{\frac{k}{m}}$.

The value of δ is ______.

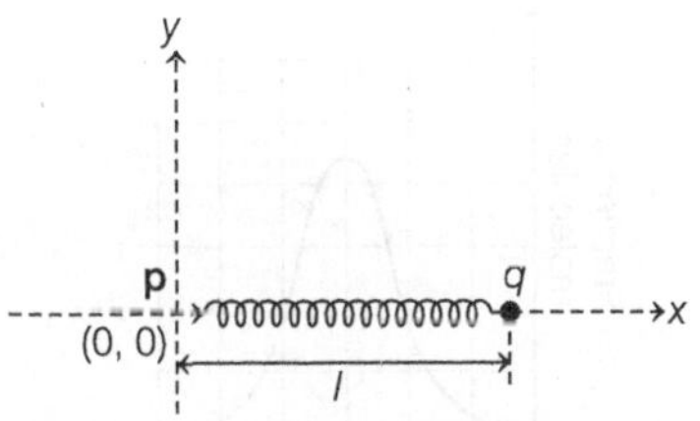

16. Consider one mole of helium gas enclosed in a container at initial pressure P_1 and volume V_1. It expands isothermally to volume $4V_1$. After this, the gas expands adiabatically and its volume becomes $32V_1$. The work done by the gas during isothermal and adiabatic expansion processes are W_{iso} and W_{adia}, respectively. If the ratio $\frac{W_{iso}}{W_{adia}} = f \ln 2$, then f is________.

17. A stationary tuning fork is in resonance with an air column in a pipe. If the tuning fork is moved with a speed of $2\ ms^{-1}$ in front of the open end of the pipe and parallel to it, the length of the pipe should be changed for the resonance to occur with the moving tuning fork. If the speed of sound in air is $320\ ms^{-1}$, the smallest value of the percentage change required in the length of the pipe is ________.

18. A circular disc of radius R carries surface charge density $\sigma(r) = \sigma_0\left(1 - \frac{r}{R}\right)$, where σ_0 is a constant and r is the distance from the center of the disc. Electric flux through a large spherical surface that encloses the charged disc completely is ϕ_0. Electric flux through another spherical surface of radius $\frac{R}{4}$ and concentric with the disc is ϕ. Then the ratio $\frac{\phi_0}{\phi}$ is________.

CHEMISTRY

Section 1 (Maximum Marks : 18)

- This section contains **SIX (06)** questions.
- Each question has **FOUR** options. **ONLY ONE** of these four options is the correct answer.
- For each question, choose the option corresponding to the correct answer.
- Answer to each question will be evaluated according to the following marking scheme:

Full Marks : **+3** If only the correct option is chosen.
Zero Marks : **0** If none of the options is chosen. (i.e. the question is unanswered);
Negative Marks : **−1** In all other cases.

19. If the distribution of molecular speeds of a gas is as per the figure shown below, then the ratio of the most probable, the average, and the root mean square speeds, respectively, is

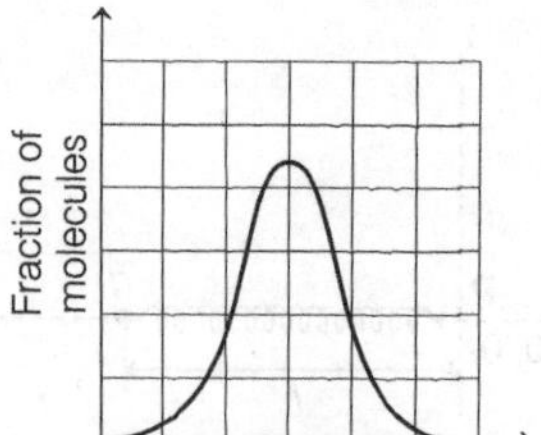

(a) 1 : 1 : 1 (b) 1 : 1 : 1.224
(c) 1 : 1.128 : 1.224 (d) 1 : 1.128 : 1

20. Which of the following liberates O_2 upon hydrolysis?

(a) Pb_3O_4 (b) KO_2
(c) Na_2O_2 (d) Li_2O_2

21. A colourless aqueous solution contains nitrates of two metals, X and Y. When it was added to an aqueous solution of NaCl, a white precipitate was formed. This precipitate was found to be partly soluble in hot water to give a residue P and a solution Q. The residue P was soluble in aqueous NH_3 and also in excess sodium thiosulphate. The hot solution Q gave a yellow precipitate with KI. The metals X and Y, respectively, are

(a) Ag and Pb
(b) Ag and Cd
(c) Cd and Pb
(d) Cd and Zn

22. Newman projections P, Q, R and S are shown below :

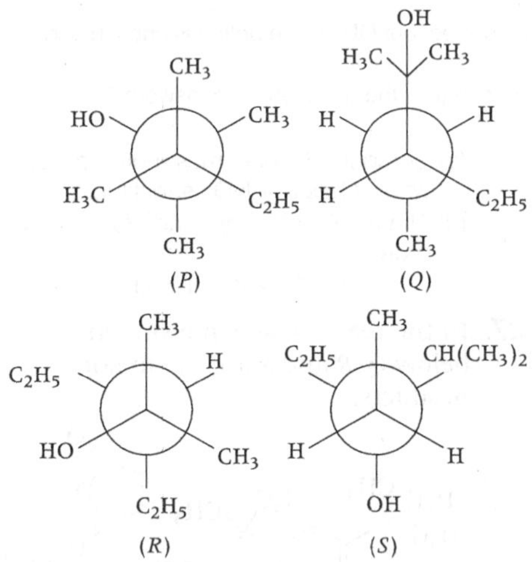

Which one of the following options represents identical molecules?

(a) P and Q
(b) Q and S
(c) Q and R
(d) R and S

23. Which one of the following structures has the IUPAC name 3-ethynyl-2-hydroxy-4-methylhex-3-en-5-ynoic acid?

(a) 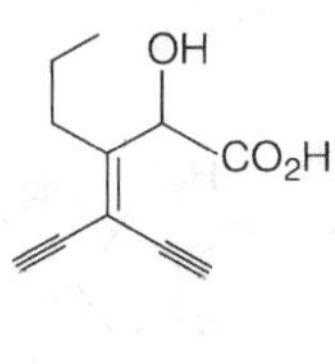 (b) 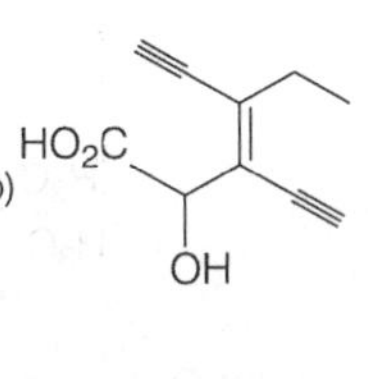

(c) 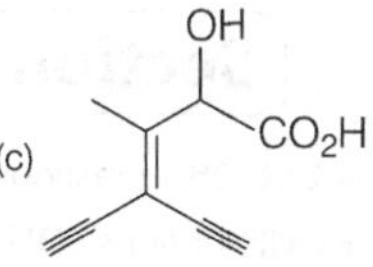 (d) 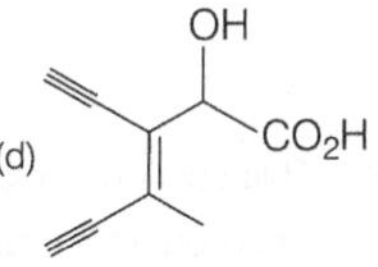

24. The Fischer projection of D-erythrose is shown below :

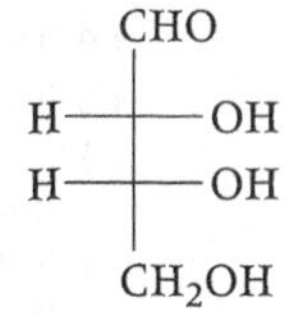

D-erythrose

D-erythrose and its isomers are listed as P, Q, R, and S in Column-I. Choose the correct relationship of P, Q, R, and S with D-erythrose from Column II.

	Column I		Column II
P.	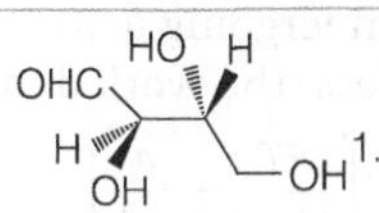	1.	Diastereomer
Q.	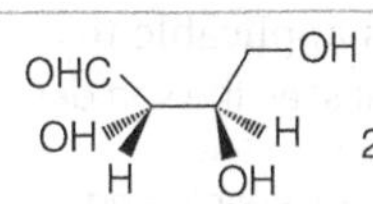	2.	Identical
R.	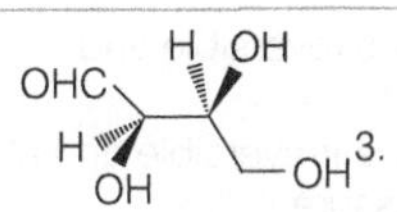	3.	Enantiomer
S.	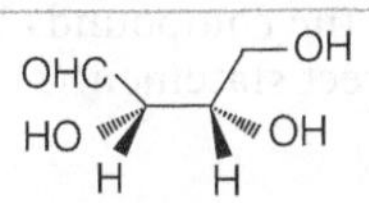		

Codes

	P	Q	R	S
(a)	2	3	2	2
(b)	3	1	1	2
(c)	2	1	1	3
(d)	2	3	3	1

Section 2 (Maximum Marks : 24)

- This section contains **SIX (06)** questions.
- Each question has **FOUR** options. **ONE OR MORE THAN ONE** of these four option(s) is (are) correct answer(s).
- For each question, choose the option(s) corresponding to (all) the correct answer(s).
- Answer to each question will be evaluated according to the following marking scheme:

Full Marks	:	**+4** If only (all) the correct option(s) is (are) chosen.
Partial Marks	:	**+3** If all the four options are correct but ONLY three options are chosen.
Partial Marks	:	**+2** If three or more options are correct but ONLY two options are chosen, both of which are correct.
Partial Marks	:	**+1** If two or more options are correct but ONLY one option is chosen and it is a correct option;
Zero Marks	:	**0** If none of the option is chosen (i.e., the question is unanswered);

25. In thermodynamics, the p-V work done is given by

$$w = -\int dV\, p_{\text{ext}}$$

For a system undergoing a particular process, the work done is

$$w = -\int dV\left(\frac{RT}{V-b} - \frac{a}{V^2}\right)$$

This equation is applicable to a

(a) system that satisfies the van der Waals' equation of state

(b) process that is reversible and isothermal

(c) process that is reversible and adiabatic

(d) process that is irreversible and at constant pressure

26. With respect to the compounds I-V. Choose the correct statement(s).

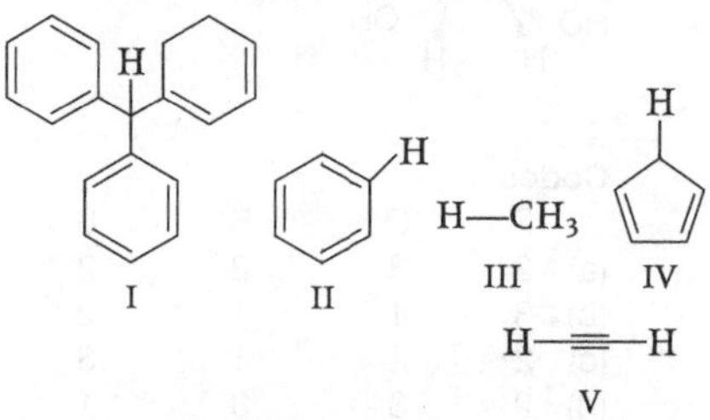

(a) The acidity of compound I is due to delocalisation in the conjugate base.

(b) The conjugate base of compound IV is aromatic.

(c) Compound II becomes more acidic, when it has a —NO_2 substituent.

(d) The acidity of compounds follows the order

I > IV > V > II > III.

27. In the reaction scheme shown below Q, R and S are the major products.

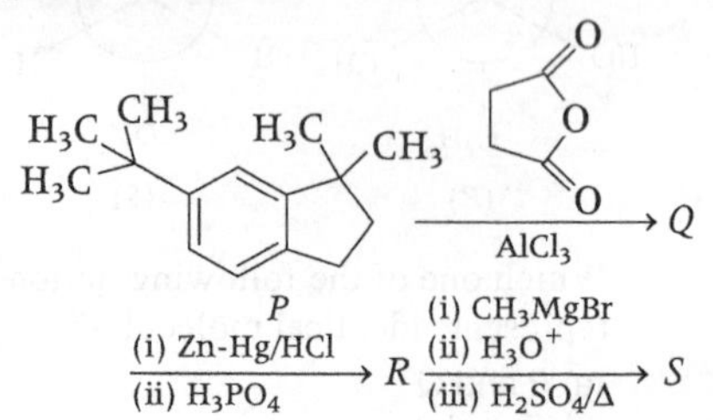

The correct structure of

(a) S is

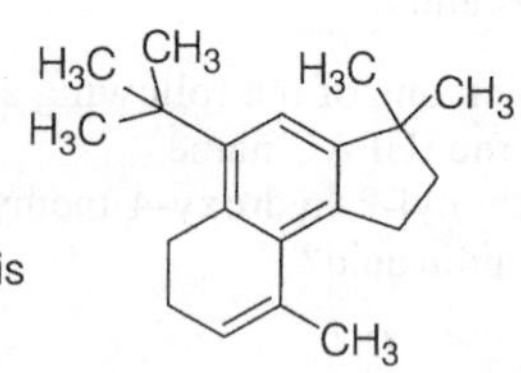

(b) Q is

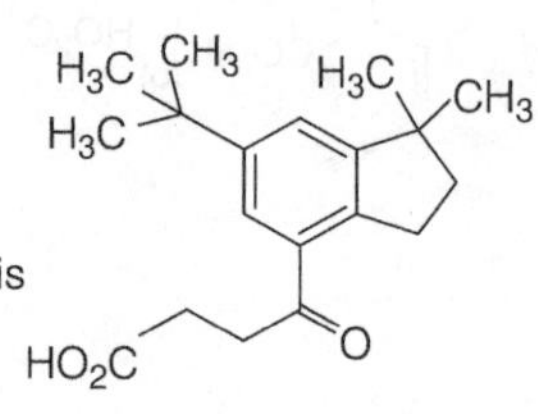

(c) R is 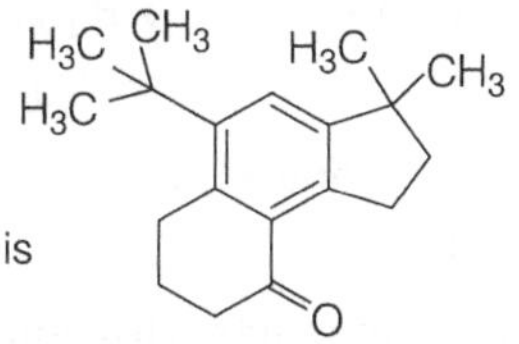

(d) S is 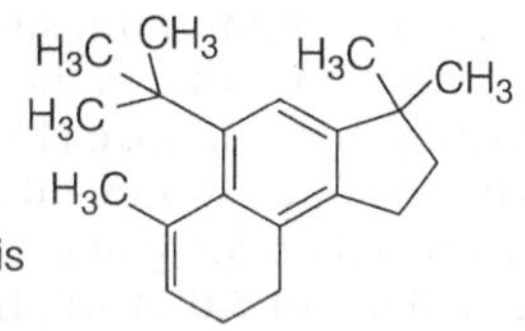

28. Choose the correct statement(s) among the following:

(a) $[FeCl_4]^-$ has tetrahedral geometry.
(b) $[Co(en)(NH_3)_2Cl_2]^+$ has 2 geometrical isomers.
(c) $[FeCl_4]^-$ has higher spin-only magnetic moment than $[Co(en)(NH_3)_2Cl_2]^+$.
(d) The cobalt ion in $[Co(en)(NH_3)_2Cl_2]^+$ has sp^3d^2 hybridisation.

29. With respect to hypochlorite, chlorate and perchlorate ions, choose the correct statement(s).

(a) The hypochlorite ion is the strongest conjugate base.
(b) The molecular shape of only chlorate ion is influenced by the lone pair of electrons of Cl.
(c) The hypochlorite and chlorate ions disproportionate to give rise to identical set of ions.
(d) The hypochlorite ion oxidises the sulphite ion.

30. The cubic unit cell structure of a compound containing cation M and anion X is shown below. When compared to the anion, the cation has smaller ionic radius. Choose the correct statement(s).

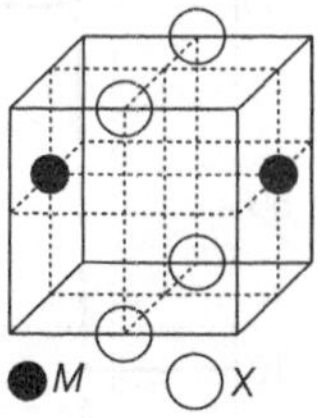

(a) The empirical formula of the compound is MX.
(b) The cation M and anion X have different coordination geometries.
(c) The ratio of M — X bond length of the cubic unit cell edge length is 0.866.
(d) The ratio of the ionic radii of cation M to anion X is 0.414.

Section 3 (Maximum Marks : 24)

- This section contains **SIX (06)** questions. The answer to each question is a **NUMERICAL VALUE.**
- Four each question, enter the correct numerical value of the answer using the mouse and the on-screen virtual numeric keypad in the place designated to enter the answer. If the numerical value has more than two decimal places, truncate/round-off the value to TWO decimal places.
- Answer to each question will be evaluated according to the following marking scheme:
  Full Marks : **+4** If ONLY the correct numerical value is entered.
  Zero Marks : **0** In all other cases.

31. 5.00 mL of 0.10 M oxalic acid solution taken in a conical flask is titrated against NaOH from a burette using phenolphthalein indicator. The volume of NaOH required for the appearance of permanent faint pink color is tabulated below for five experiments. What is the concentration, in molarity, of the NaOH solution?

Exp. No.	Vol. of NaOH (mL)
1	12.5
2	10.5
3	9.0
4	9.0
5	9.0

32. Consider the reaction, $A \rightleftharpoons B$ at 1000 K. At time t', the temperature of the system was increased to 2000 K and the system was allowed to reach equilibrium. Throughout this experiment the partial pressure of A was maintained at 1 bar. Given, below is the plot of the partial pressure of B with time. What is the ratio of the standard Gibbs energy of the reaction at 1000 K to that at 2000 K?

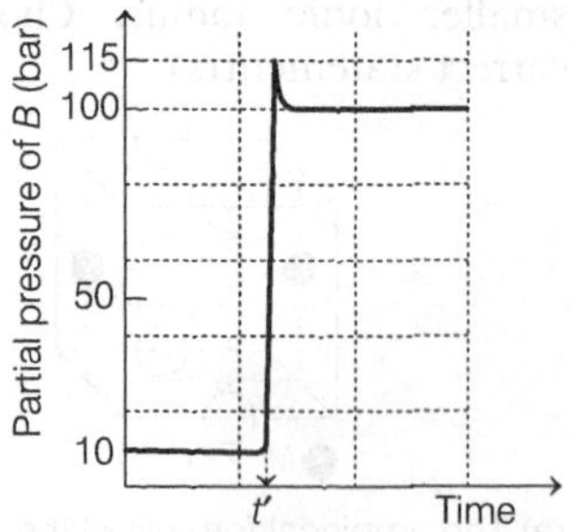

33. Consider a 70% efficient hydrogen-oxygen fuel cell working under standard conditions at 1 bar and 298 K. Its cell reaction is

$$H_2(g) + \frac{1}{2}O_2(g) \longrightarrow H_2O(l)$$

The work derived from the cell on the consumption of 1.0×10^{-3} mole of $H_2(g)$ is used to compress 1.00 mole of a monoatomic ideal gas in a thermally insulated container. What is the change in the temperature (in K) of the ideal gas?

The standard reduction potentials for the two half-cells are given below:

$O_2(g) + 4H^+(aq) + 4e^- \longrightarrow 2H_2O(l)$, $E° = 1.23$ V,

$2H^+(aq) + 2e^- \longrightarrow H_2(g)$, $E° = 0.00$ V

Use, $F = 96500$ C mol^{-1}, $R = 8.314$ J mol^{-1} K^{-1}.

34. Aluminium reacts with sulphuric acid to form aluminium sulphate and hydrogen. What is the volume of hydrogen gas in litre (L) produced at 300 K and 1.0 atm pressure, when 5.4 g of aluminium and 50.0 mL of 5.0 M sulphuric acid are combined for the reaction?
(Use molar mass of aluminium as 27.0 g mol^{-1}, $R = 0.082$ atm L mol^{-1} K^{-1})

35. ${}^{238}_{92}U$ is known to undergo radioactive decay to form ${}^{206}_{82}Pb$ by emitting alpha and beta particles. A rock initially contained 68×10^{-6} g of ${}^{238}_{92}U$. If the number of alpha particles that it would emit during its radioactive decay of ${}^{238}_{92}U$ to ${}^{206}_{82}Pb$ in three half-lives is $Z \times 10^{18}$, then what is the value of Z?

36. In the following reaction, compound Q is obtained from compound P *via* an ionic intermediate.

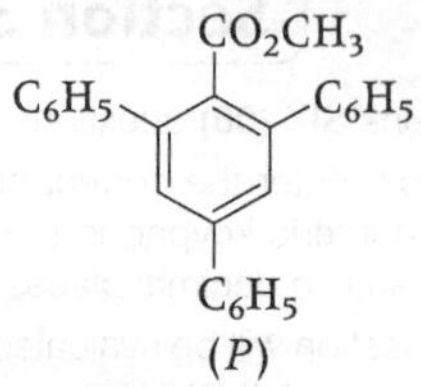

(P)

$\xrightarrow[\text{}]{\text{Conc. } H_2SO_4}$ Q (A coloured compound)

What is the degree of unsaturation of Q?

MATHEMATICS

Section 1 (Maximum Marks : 18)

- This section contains **SIX (06)** questions.
- Each question has **FOUR** options. **ONLY ONE** of these four options is the correct answer.
- For each question, choose the option corresponding to the correct answer.
- Answer to each question will be evaluated according to the following marking scheme:

Full Marks : **+3** If ONLY the correct options is chosen.
Zero Marks : **0** If none of the options is chosen. (i.e. the question is unanswered)
Negative Marks : **−1** In all other cases.

37. Suppose a,b denote the distinct real roots of the quadratic polynomial $x^2+20x-2020$ and suppose c,d denote the distinct complex roots of the quadratic polynomial $x^2-20x+2020$. Then the value of $ac(a-c)+ad(a-d)+bc(b-c)+bd(b-d)$ is

(a) 0 (b) 8000
(c) 8080 (d) 16000

38. If the function $f:\mathbf{R}\to\mathbf{R}$ is defined by $f(x)=|x|(x-\sin x)$, then which of the following statements is TRUE?

(a) f is one-one, but NOT onto
(b) f is onto, but NOT one-one
(c) f is BOTH one-one and onto
(d) f is NEITHER one-one NOR onto

39. Let the functions $f:\mathbf{R}\to\mathbf{R}$ and $g:\mathbf{R}\to\mathbf{R}$ be defined by

$$f(x)=e^{x-1}-e^{-|x-1|}$$

and $g(x)=\frac{1}{2}(e^{x-1}+e^{1-x})$.

Then the area of the region in the first quadrant bounded by the curves $y=f(x)$, $y=g(x)$ and $x=0$ is

(a) $(2-\sqrt{3})+\frac{1}{2}(e-e^{-1})$

(b) $(2+\sqrt{3})+\frac{1}{2}(e-e^{-1})$

(c) $(2-\sqrt{3})+\frac{1}{2}(e+e^{-1})$

(d) $(2+\sqrt{3})+\frac{1}{2}(e+e^{-1})$

40. Let a,b and λ be positive real numbers. Suppose P is an end point of the latus rectum of the parabola $y^2=4\lambda x$, and suppose the ellipse $\frac{x^2}{a^2}+\frac{y^2}{b^2}=1$ passes through the point P. If the tangents to the parabola and the ellipse at the point P are perpendicular to each other, then the eccentricity of the ellipse is

(a) $\frac{1}{\sqrt{2}}$ (b) $\frac{1}{2}$
(c) $\frac{1}{3}$ (d) $\frac{2}{5}$

41. Let C_1 and C_2 be two biased coins such that the probabilities of getting head in a single toss are $\frac{2}{3}$ and $\frac{1}{3}$, respectively. Suppose α is the number of heads that appear when C_1 is tossed twice, independently, and suppose β is the number of heads that appear when C_2 is tossed twice, independently. Then the probability that the roots of the quadratic polynomial $x^2-\alpha x+\beta$ are real and equal, is

(a) $\frac{40}{81}$ (b) $\frac{20}{81}$
(c) $\frac{1}{2}$ (d) $\frac{1}{4}$

42. Consider the rectangles lying in the region

$$\left\{(x, y) \in \mathrm{R} \times \mathrm{R} : 0 \le x \le \frac{\pi}{2} \text{ and } 0 \le y \le 2\sin(2x)\right\}$$

and having one side on the X-axis. The area of the rectangle which has the maximum perimeter among all such rectangles, is

(a) $\frac{3\pi}{2}$ (b) π (c) $\frac{\pi}{2\sqrt{3}}$ (d) $\frac{\pi\sqrt{3}}{2}$

Section 2 (Maximum Marks : 24)

- This section contains **SIX (06)** questions.
- Each question has **FOUR** options. **ONE OR MORE THAN ONE** of these four option(s) is (are) correct answer(s).
- For each question, choose the options(s) corresponding to (all) the correct answer(s).
- Answer to each question will be evaluated according to the following marking scheme:

Full Marks	:	**+4** If only (all) the correct option(s) is (are) chosen.
Partial Marks	:	**+3** If all the four options are correct but ONLY three options are chosen.
Partial Marks	:	**+2** If three or more options are correct but ONLY two options are chosen, both of which are correct.
Partial Marks	:	**+1** If two or more options are correct but ONLY one option is chosen and it is a correct option;
Zero Marks	:	**0** If none of the options is chosen (i.e., the question is unanswered);
Negative Marks	:	**−2** In all other cases.

43. Let the function $f : \mathbf{R} \to \mathbf{R}$ be defined by $f(x) = x^3 - x^2 + (x-1)\sin x$ and let $g : \mathbf{R} \to \mathbf{R}$ be an arbitrary function. Let $fg : \mathbf{R} \to \mathbf{R}$ be the product function defined by $(fg)(x) = f(x)g(x)$. Then which of the following statements is/are TRUE?

(a) If g is continuous at $x = 1$, then fg is differentiable at $x = 1$

(b) If fg is differentiable at $x = 1$, then g is continuous at $x = 1$

(c) If g is differentiable at $x = 1$, then fg is differentiable at $x = 1$

(d) If fg is differentiable at $x = 1$, then g is differentiable at $x = 1$

44. Let M be a 3×3 invertible matrix with real entries and let I denote the 3×3 identity matrix. If $M^{-1} = \mathrm{adj}(\mathrm{adj}\, M)$, then which of the following statements is/are ALWAYS TRUE?

(a) $M = I$ (b) $\det M = 1$

(c) $M^2 = I$ (d) $(\mathrm{adj}\, M)^2 = I$

45. Let S be the set of all complex numbers z satisfying $|z^2 + z + 1| = 1$. Then which of the following statements is/are TRUE?

(a) $\left|z + \frac{1}{2}\right| \le \frac{1}{2}$ for all $z \in S$

(b) $|z| \le 2$ for all $z \in S$

(c) $\left|z + \frac{1}{2}\right| \ge \frac{1}{2}$ for all $z \in S$

(d) The set S has exactly four elements

46. Let x, y and z be positive real numbers. Suppose x, y and z are the lengths of the sides of a triangle opposite to its angles X, Y and Z, respectively. If

$$\tan\frac{X}{2} + \tan\frac{Z}{2} = \frac{2y}{x + y + z},$$

then which of the following statements is/are TRUE?

(a) $2Y = X + Z$ (b) $Y = X + Z$

(c) $\tan\frac{X}{2} = \frac{x}{y + z}$ (d) $x^2 + z^2 - y^2 = xz$

47. Let L_1 and L_2 be the following straight lines.

$$L_1 : \frac{x-1}{1} = \frac{y}{-1} = \frac{z-1}{3} \text{ and}$$

$$L_2 : \frac{x-1}{-3} = \frac{y}{-1} = \frac{z-1}{1}.$$

Suppose the straight line

$$L : \frac{x-\alpha}{l} = \frac{y-1}{m} = \frac{z-\gamma}{-2}$$

lies in the plane containing L_1 and L_2 and passes through the point of intersection of L_1 and L_2. If the line L bisects the acute angle between the lines L_1 and L_2, then which of the following statements is/are TRUE?

(a) $\alpha - \gamma = 3$ (b) $l + m = 2$
(c) $\alpha - \gamma = 1$ (d) $l + m = 0$

48. Which of the following inequalities is/are TRUE?

(a) $\int_0^1 x\cos x dx \geq \frac{3}{8}$

(b) $\int_0^1 x\sin x dx \geq \frac{3}{10}$

(c) $\int_0^1 x^2 \cos x dx \geq \frac{1}{2}$

(d) $\int_0^1 x^2 \sin x dx \geq \frac{2}{9}$

Section 3 (Maximum Marks : 24)

- This section contains **SIX (06)** questions. The answer to each question is a **NUMERICAL VALUE.**
- For each question, enter the correct numerical value of the answer using the mouse and the on-screen virtual numeric keypad in the place designated to enter the answer. If the numerical value has more than two decimal places, **truncate/round-off** the value to **TWO** decimal places.
- Answer to each question will be evaluated according to the following marking scheme:

Full Marks : **+4** If ONLY the correct numerical value is entered.
Zero Marks : **0** In all other cases.

49. Let m be the minimum possible value of $\log_3(3^{y_1} + 3^{y_2} + 3^{y_3})$, where y_1, y_2, y_3 are real numbers for which $y_1 + y_2 + y_3 = 9$. Let M be the maximum possible value of $(\log_3 x_1 + \log_3 x_2 + \log_3 x_3)$, where x_1, x_2, x_3 are positive real numbers for which $x_1 + x_2 + x_3 = 9$. Then the value of $\log_2(m^3) + \log_3(M^2)$ is...... .

50. Let $a_1, a_2, a_3, ...$ be a sequence of positive integers in arithmetic progression with common difference 2. Also, let $b_1, b_2, b_3, ...$ be a sequence of positive integers in geometric progression with common ratio 2. If $a_1 = b_1 = c$, then the number of all possible values of c, for which the equality $2(a_1 + a_2 + ... + a_n) = b_1 + b_2 + ... + b_n$ holds for some positive integer n, is ......

51. Let $f:[0,2] \to \mathbf{R}$ be the function defined by

$$f(x) = (3 - \sin(2\pi x))\sin\left(\pi x - \frac{\pi}{4}\right) - \sin\left(3\pi x + \frac{\pi}{4}\right).$$

If $\alpha, \beta \in [0,2]$ are such that $\{x \in [0,2] : f(x) \geq 0\} = [\alpha, \beta]$, then the value of $\beta - \alpha$ is ......

52. In a triangle PQR, let $\mathbf{a} = \mathbf{QR}$, $\mathbf{b} = \mathbf{RP}$ and $\mathbf{c} = \mathbf{PQ}$. If $|\mathbf{a}| = 3$, $|\mathbf{b}| = 4$ and $\frac{\mathbf{a}\cdot(\mathbf{c}-\mathbf{b})}{\mathbf{c}\cdot(\mathbf{a}-\mathbf{b})} = \frac{|\mathbf{a}|}{|\mathbf{a}|+|\mathbf{b}|}$, then the value of $|\mathbf{a} \times \mathbf{b}|^2$ is ....

53. For a polynomial $g(x)$ with real coefficients, let m_g denote the number of distinct real roots of $g(x)$.

Suppose S is the set of polynomials with real coefficients defined by

$$S = \{(x^2 - 1)^2 (a_0 + a_1 x + a_2 x^2 + a_3 x^3) : a_0, a_1, a_2, a_3 \in \mathbf{R}\};$$

For a polynomial f, let f' and f'' denote its first and second order derivatives, respectively. Then the minimum possible value of $(m_{f'} + m_{f''})$, where $f \in S$, is ....

54. Let e denote the base of the natural logarithm. The value of the real number a for which the right hand limit

$$\lim_{x \to 0^+} \frac{(1-x)^{\frac{1}{x}} - e^{-1}}{x^a}$$

is equal to a non zero real number, is ....

Paper 2

PHYSICS

Section 1 (Maximum Marks : 18)

- This section contains **SIX (06)** questions.
- The answer to each question is a **SINGLE DIGIT INTEGER** ranging from 0 to 9, **BOTH INCLUSIVE**.
- For each question, enter the correct integer corresponding to the answer using the mouse and the on-screen virtual numeric keypad in the place designated to enter the answer.
- Answer to each question will be evaluated according to the following marking scheme:

Full Marks : **+3** If only (all) the correct option(s) is (are) chosen.
Zero Marks : **0** If the question is unanswered).
Negative Marks : **−1** In all other cases.

1. A large square container with thin transparent vertical walls and filled with water (refractive index $\frac{4}{3}$) is kept on a horizontal table. A student holds a thin straight wire vertically inside the water 12 cm from one of its corners, as shown schematically in the figure. Looking at the wire from this corner, another student sees two images of the wire, located symmetrically on each side of the line of sight as shown. The separation (in cm) between these images is ________.

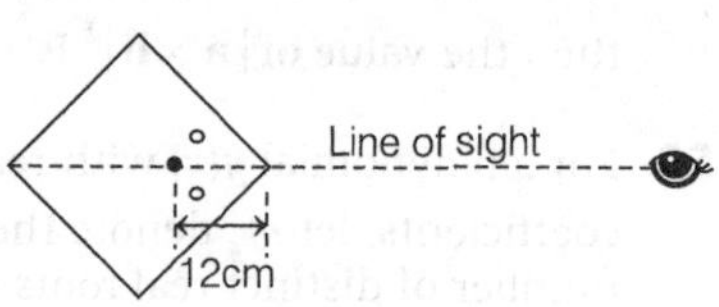

2. A train with cross-sectional area S_t is moving with speed v_0 inside a long tunnel of cross-sectional area $S_0 (S_0 = 4S_t)$. Assume that almost all the air (density ρ) in front of the train flows back between its sides and the walls of the tunnel. Also, the air flow with respect to the train is steady and laminar. Take the ambient pressure and that inside the train to be p_0. If the pressure in the region between the sides of the train and the tunnel walls is p then $p_0 - p = \frac{7}{2N} \rho v_t^2$. The value of N is________.

3. Two large circular discs separated by a distance of 0.01 m are connected to a battery via a switch as shown in the figure. Charged oil drops of density 900 kg m^{-3} are released through a tiny hole at the center of the top disc. Once some oil drops achieve terminal velocity, the switch is closed to apply a voltage of 200 V across the discs. As a result, an oil drop of radius 8×10^{-7} m stops moving vertically and floats between the discs. The number of electrons present in this oil drop is ________. (neglect the buoyancy force, take acceleration due to gravity $= 10\ ms^{-2}$ and charge on an electron (e) $= 1.6 \times 10^{-19}$ C)

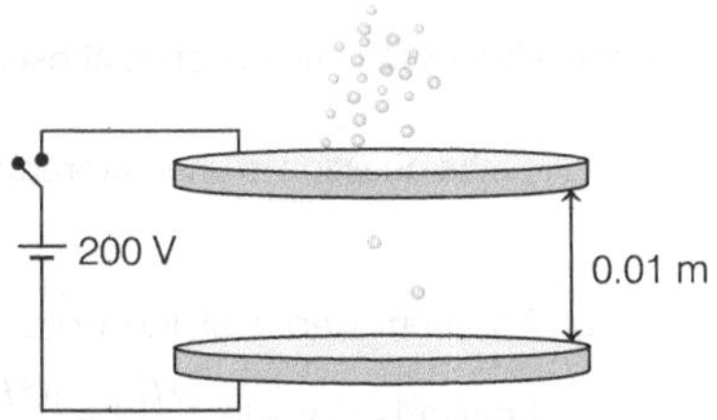

4. A hot air balloon is carrying some passengers, and a few sandbags of mass 1 kg each so that its total mass is 480 kg. Its effective volume giving the balloon its buoyancy is V. The balloon is floating at an equilibrium height of 100 m. When N number of sandbags are thrown out, the balloon rises to a new equilibrium height close to 150 m with its volume V remaining unchanged. If the variation of the density of air with height h from the ground is $pho(h) = \rho_0 e^{-\frac{h}{h_0}}$, where $\rho_0 = 1.25$ kg m^{-3} and $h_0 = 6000$ m, the value of N is ________.

5. A point charge q of mass m is suspended vertically by a string of length l. A point dipole of dipole moment **p** is now brought towards q from infinity so that the charge moves away. The final equilibrium position of the system including the direction of the dipole, the angles and distances is shown in the figure below. If the work done in bringing the dipole to this position is $N \times (mgh)$, where g is the acceleration due to gravity, then the value of N is ________. (Note that for three coplanar forces keeping a point mass in equilibrium, $\frac{F}{\sin\theta}$ is the same for all forces, where F is any one of the forces and θ is the angle between the other two forces)

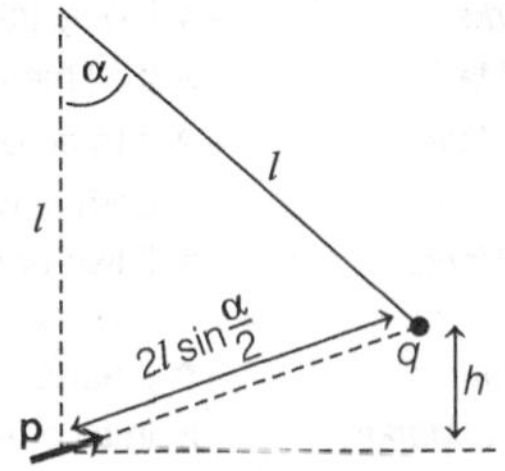

6. A thermally isolated cylindrical closed vessel of height 8 m is kept vertically. It is divided into two equal parts by a diathermic (perfect thermal conductor) frictionless partition of mass 8.3 kg. Thus the partition is held initially at a distance of 4 m from the top, as shown in the schematic figure below. Each of the two parts of the vessel contains 0.1 mole of an ideal gas at temperature 300 K. The partition is now released and moves without any gas leaking from one part of the vessel to the other. When equilibrium is reached, the distance of the partition from the top (in m) will be ________ (take the acceleration due to gravity $= 10\ ms^{-2}$ and the universal gas constant= 9.3 J mol^{-1} K^{-1}).

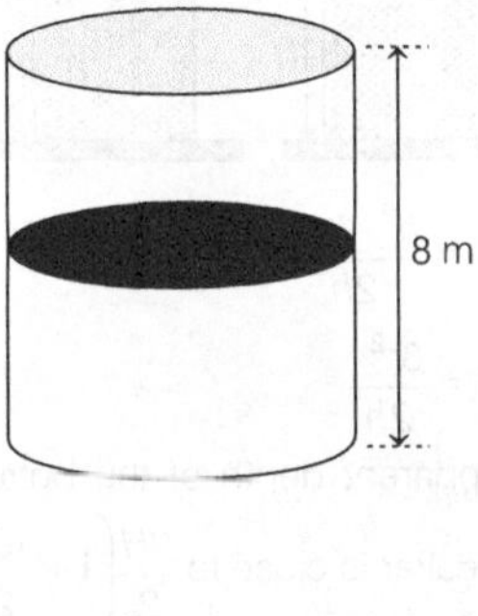

Section 2 (Maximum Marks : 24)

- This section contains **SIX (06)** questions.
- Each question has **FOUR** options. **ONE OR MORE THAN ONE** of these four option(s) is (are) correct options(s).
- For each question, choose the options(s) corresponding to (all) the correct answer(s).
- Answer to each question will be evaluated according to the following marking scheme:

Full Marks	:	**+4** If only (all) the correct option(s) is (are) chosen.
Partial Marks	:	**+3** If all the four options are correct but ONLY three options are chosen.
Partial Marks	:	**+2** If three or more options are correct but ONLY two options are chosen, both of which are correct.
Partial Marks	:	**+1** If two or more options are correct but ONLY one option is chosen and it is a correct option;
Zero Marks	:	**0** If none of the option is chosen (i.e., the question is unanswered);
Negative Marks	:	**−2** In all other cases.

7. A beaker of radius r is filled with water $\left(\text{refractive index} \frac{4}{3}\right)$ up to a height H as shown in the figure on the left. The beaker is kept on a horizontal table rotating with angular speed ω. This makes the water surface curved so that the difference in the height of water level at the center and at the circumference of the beaker is $h\,(h << H, h << r)$, as shown in the figure on the right. Take this surface to be approximately spherical with a radius of curvature R. Which of the following is/are correct? (g is the acceleration due to gravity)

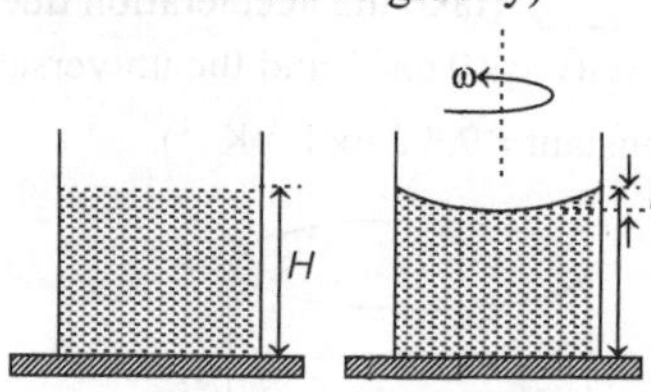

(a) $R = \dfrac{h^2 + r^2}{2h}$

(b) $R = \dfrac{3r^2}{2h}$

(c) Apparent depth of the bottom of the beaker is close to $\dfrac{3H}{2}\left(1 + \dfrac{\omega^2 H}{2g}\right)^{-1}$

(d) Apparent depth of the bottom of the beaker is close to $\dfrac{3H}{4}\left(1 + \dfrac{\omega^2 H}{4g}\right)^{-1}$

8. A student skates up a ramp that makes an angle 30° with the horizontal. He/she starts (as shown in the figure) at the bottom of the ramp with speed v_0 and wants to turn around over a semicircular path xyz of radius R during which he/she reaches a maximum height h (at point y) from the ground as shown in the figure. Assume that the energy loss is negligible and the force required for this turn at the highest point is provided by his/her weight only. Then (g is the acceleration due to gravity)

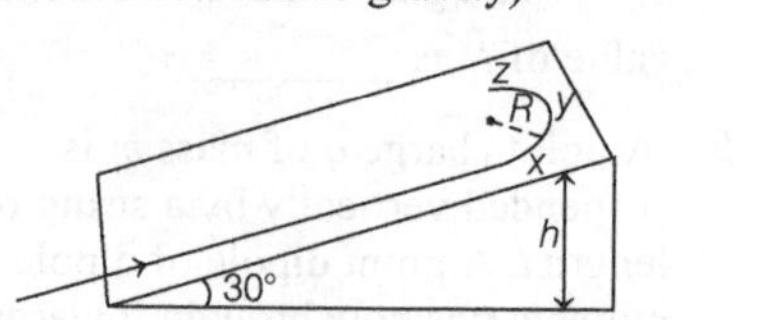

(a) $v_0^2 - 2gh = \dfrac{1}{2}gR$

(b) $v_0^2 - 2gh = \dfrac{\sqrt{3}}{2}gR$

(c) the centripetal force required at points x and z is zero

(d) the centripetal force required is maximum at points x and z

9. A rod of mass m and length L, pivoted at one of its ends, is hanging vertically. A bullet of the same mass moving at speed v strikes the rod horizontally at a distance x from its pivoted end and gets embedded in it. The combined system now rotates with angular speed ω about the pivot. The maximum angular speed ω_M is achieved for $x = x_M$. Then

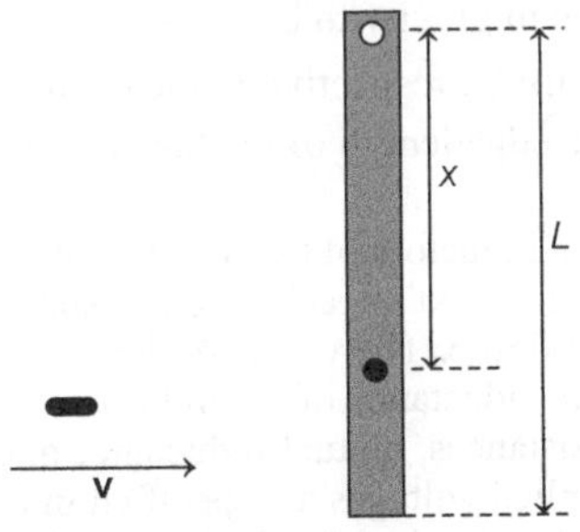

(a) $\omega = \dfrac{3vx}{L^2 + 3x^2}$ (b) $\omega = \dfrac{12vx}{L^2 + 12x^2}$

(c) $x_M = \dfrac{L}{\sqrt{3}}$ (d) $\omega_M = \dfrac{v}{2L}\sqrt{3}$

10. In an X-ray tube, electrons emitted from a filament (cathode) carrying current I hit a target (anode) at a distance d from the cathode. The target is kept at a potential V higher than the cathode resulting in emission of continuous and characteristic X-rays. If the filament current I is decreased to $\dfrac{I}{2}$, the potential difference V is increased to $2V$, and the separation distance d is reduced to $\dfrac{d}{2}$, then

(a) the cut-off wavelength will reduce to half, and the wavelengths of the characteristic X-rays will remain the same

(b) the cut-off wavelength as well as the wavelengths of the characteristic X-rays will remain the same

(c) the cut-off wavelength will reduce to half, and the intensities of all the X-rays will decrease

(d) the cut-off wavelength will become two times larger, and the intensity of all the X-rays will decrease

11. Two identical non-conducting solid spheres of same mass and charge are suspended in air from a common point by two non-conducting, massless strings of same length. At equilibrium, the angle between the strings is α. The spheres are now immersed in a dielectric liquid of density 800 kg m^{-3} and dielectric constant 21. If the angle between the strings remains the same after the immersion, then

(a) electric force between the spheres remains unchanged

(b) electric force between the spheres reduces

(c) mass density of the spheres is 840 kg m^{-3}

(d) the tension in the strings holding the spheres remains unchanged

12. Starting at time $t = 0$ from the origin with speed 1 ms^{-1}, a particle follows a two-dimensional trajectory in the x-y plane so that its coordinates are related by the equation $y = \dfrac{x^2}{2}$. The x and y components of its accel- eration are denoted by a_x and a_y, respectively. Then

(a) $a_x = 1\text{ ms}^{-2}$ implies that when the particle is at the origin, $a_y = 1\text{ ms}^{-2}$

(b) $a_x = 0$ implies $a_y = 1\text{ ms}^{-2}$ at all times

(c) at $t = 0$, the particle's velocity points in the x-direction

(d) $a_x = 0$ implies that at $t = 1$ s, the angle between the particle's velocity and the x axis is 45°

Section 3 (Maximum Marks : 24)

- This section contains **SIX (06)** questions. The answer to each question is a **NUMERICAL VALUE.**
- Four each question, enter the correct numerical value of the answer using the mouse and the on-screen virtual numeric keypad in the place designated to enter the answer. If the numerical value has more than two decimal places, truncate/round-off the value to TWO decimal places.
- Answer to each question will be evaluated according to the following marking scheme:

Full Marks : **+4** If ONLY the correct numerical value is entered.
Zero Marks : **0** In all other cases.

13. A spherical bubble inside water has radius R. Take the pressure inside the bubble and the water pressure to be p_0. The bubble now gets compressed radially in an adiabatic manner so that its radius becomes $(R-a)$. For $a << R$ the magnitude of the work done in the process is given by $(4\pi p_0 R a^2)X$, where X is a constant and $\gamma = C_p/C_V = 41/30$. The value of X is ______.

14. In the balanced condition, the values of the resistances of the four arms of a Wheatstone bridge are shown in the figure below. The resistance R_3 has temperature coefficient $0.0004°C^{-1}$. If the temperature of R_3 is increased by $100°C$, the voltage developed between S and T will be ______ volt.

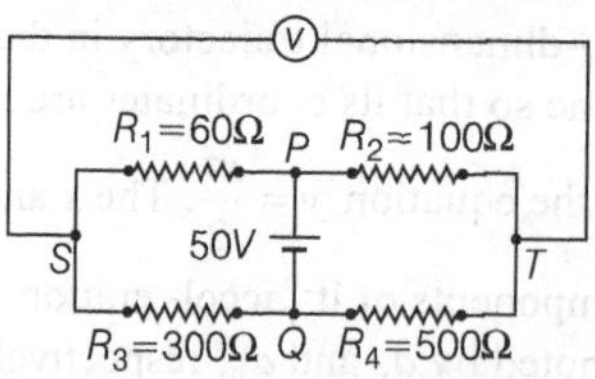

15. Two capacitors with capacitance values $C_1 = 2000 \pm 10$ pF and $C_2 = 3000 \pm 15$ pF are connected in series. The voltage applied across this combination is $C = 5.00 \pm 0.02$ V. The percentage error in the calculation of the energy stored in this combination of capacitors is ______.

16. A cubical solid aluminium (bulk modulus $= -V\dfrac{dP}{dV} = 70$ GPa) block has an edge length of 1 m on the surface of the earth. It is kept on the floor of a 5 km deep ocean. Taking the average density of water and the acceleration due to gravity to be 10^3 kg m^{-3} and 10 ms^{-2}, respectively, the change in the edge length of the block in mm is ______.

17. The inductors of two LR circuits are placed next to each other, as shown in the figure. The values of the self-inductance of the inductors, resistances, mutual-inductance and applied voltages are specified in the given circuit. After both the switches are closed simultaneously, the total work done by the batteries against the induced EMF in the inductors by the time the currents reach their steady state values is ______ mJ.

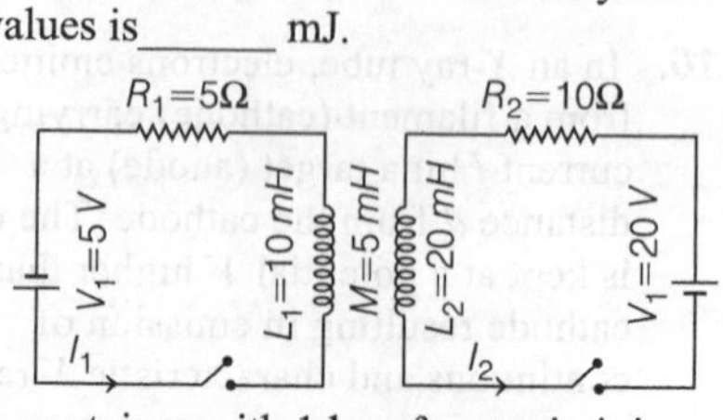

18. A container with 1 kg of water in it is kept in sunlight, which causes the water to get warmer than the surroundings. The average energy per unit time per unit area received due to the sunlight is 700 Wm^{-2} and it is absorbed by the water over an effective area of 0.05 m^2. Assuming that the heat loss from the water to the surroundings is governed by Newton's law of cooling, the difference (in °C) in the temperature of water and the surroundings after a long time will be ______. (Ignore effect of the container and take constant for Newton's law of cooling = 0.001 s^{-1}, Heat capacity of water = 4200 J kg^{-1}K^{-1})

CHEMISTRY

Section 1 (Maximum Marks : 18)

- This section contains **SIX (06)** questions.
- The answer to each question is a **SINGLE DIGIT INTEGER** ranging from 0 TO 9. **BOTH INCLUSIVE**.
- For each question, enter the correct integer corresponding to the answer using the mouse and the on-screen virtual numeric keypad in the place designated to enter the answer.
- Answer to each question will be evaluated according to the following marking scheme:

Full Marks : **+3** If ONLY the correct integer is entered.
Zero Marks : **0** If the question is unanswered.
Negative Marks : **−1** In all other cases.

19. The 1st, 2nd and 3rd ionisation enthalpies, I_1, I_2, and I_3, of four atoms with atomic numbers n, $n+1$, $n+2$, and $n+3$, where $n < 10$, are tabulated below. What is the value of n?

Atomic number	Ionisation enthalpy (kJ/mol)		
	I_1	I_2	I_3
n	1681	3374	6050
$n+1$	2081	3952	6122
$n+2$	496	4562	6910
$n+3$	738	1451	7733

20. Consider the following compounds in the liquid form :
O_2, HF, H_2O, NH_3, H_2O_2, CCl_4, $CHCl_3$, C_6H_6, C_6H_5Cl

When a charged comb is brought near their flowing stream, how many of them show deflection as per the following figure?

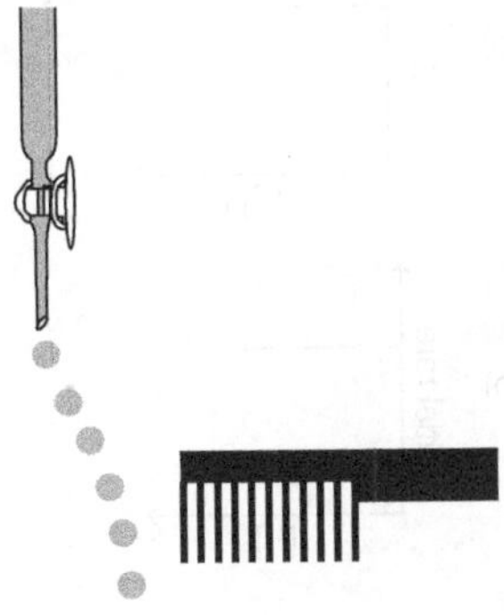

21. In the chemical reaction between stoichiometric quantities of $KMnO_4$ and KI in weakly basic solution, what is the number of moles of I_2 released for 4 moles of $KMnO_4$ consumed?

22. An acidified solution of potassium chromate was layered with an equal volume of amyl alcohol. When it was shaken after the addition of 1 mL of 3% H_2O_2, a blue alcohol layer was obtained. The blue colour is due to the formation of a chromium (VI) compound 'X'. What is the number of oxygen atoms bonded to chromium through only single bond in a molecule of X?

23. The structure of a peptide is given below.

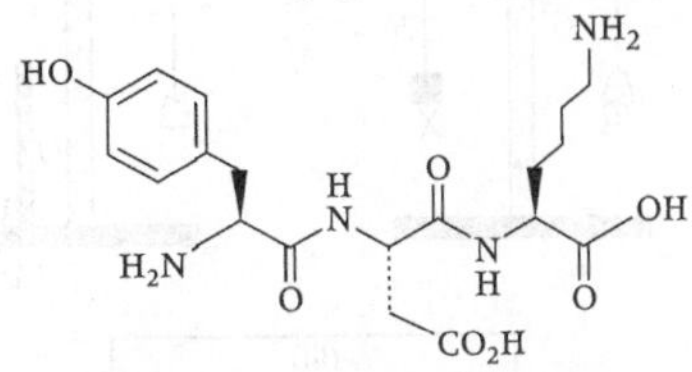

If the absolute values of the net charge of the peptide at pH = 2, pH = 6, and pH = 11 are $|Z_1|$, $|Z_2|$ and $|Z_3|$, respectively, then what is $|Z_1| + |Z_2| + |Z_3|$?

24. An organic compound ($C_8H_{10}O_2$) rotates plane-polarised light. It produces pink color with neutral $FeCl_3$ solution. What is the total number of all the possible isomers for this compound?

Section 2 (Maximum Marks : 24)

- This section contains **SIX (06)** questions.
- Each question has **FOUR** options. **ONE OR MORE THAN ONE** of these four option(s) is (are) correct asnswer(s).
- For each question, choose the option(s) corresponding to (all) the correct answer(s).
- Answer to each question will be evaluated according to the following marking scheme:

Full Marks	:	+4 If only (all) the correct option(s) is (are) chosen.
Partial Marks	:	+3 If all the four options are correct but ONLY three options are chosen.
Partial Marks	:	+2 If three or more options are correct but ONLY two options are chosen, both of which are correct.
Partial Marks	:	+1 If two or more options are correct but ONLY one option is chosen and it is a correct option.
Zero Marks	:	0 If none of the options is chosen (i.e. the question is unanswered).
Negative Marks	:	−2 In all other cases.

25. **In an experiment, *m* grams of a compound *X* (gas/liquid/solid) taken in a container is loaded in a balance as shown in figure I below.**

In the presence of a magnetic field, the pan with *X* is either deflected upwards (figure II), or deflected downwards (figure III), depending on the compound *X*. Identify the correct statement(s).

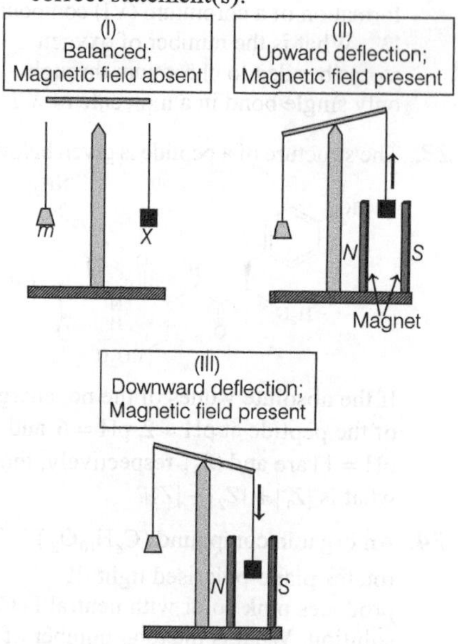

(a) If *X* is $H_2O(l)$, deflection of the pan is upwards.

(b) If *X* is $K_4[Fe(CN)_6](s)$, deflection of the pan is upwards.

(c) If *X* is $O_2(g)$, deflection of the pan is downwards.

(d) If *X* is $C_6H_6(l)$, deflection of the pan is downwards.

26. **Which of the following plots is(are) correct for the given reaction?**

($[P]_0$ is the initial concentration of *P*)

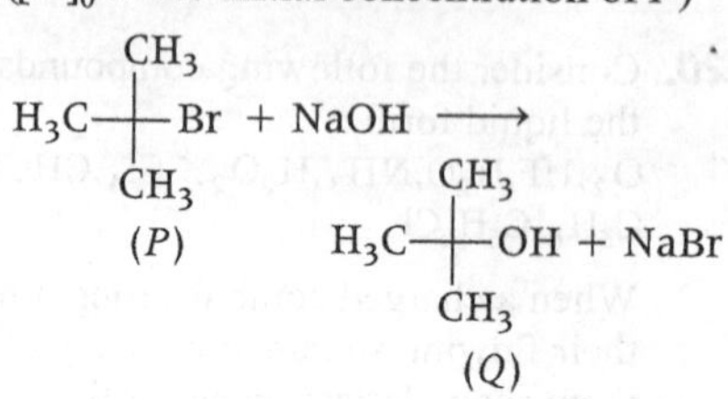

(a)

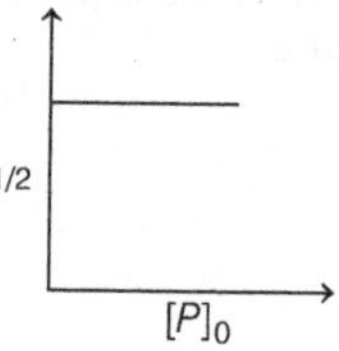

(b)

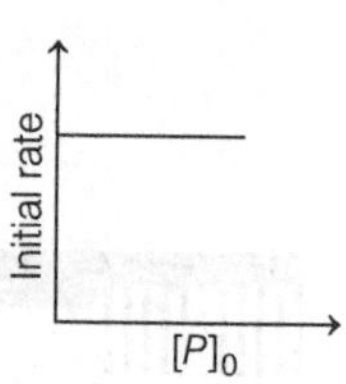

(c) 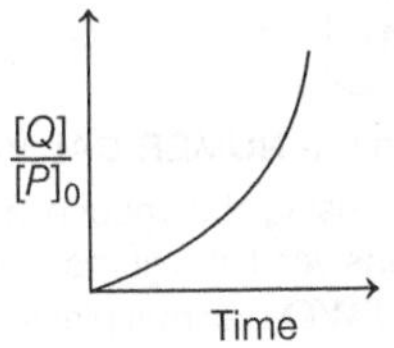

(d) 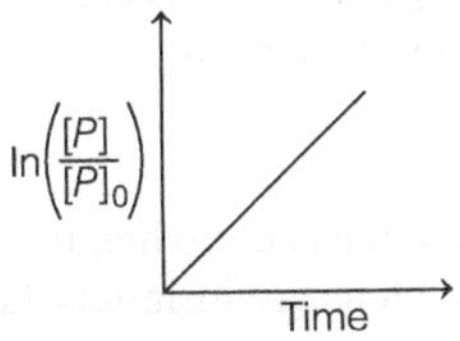

27. Which among the following statement(s) is(are) true for the extraction of aluminium from bauxite?

(a) Hydrated Al_2O_3 precipitates, when CO_2 is bubbled through a solution of sodium aluminate.

(b) Addition of Na_3AlF_6 lowers the melting point of alumina.

(c) CO_2 is evolved at the anode during electrolysis.

(d) The cathode is a steel vessel with a lining of carbon.

28. Choose the correct statement(s) among the following:

(a) $SnCl_2 \cdot 2H_2O$ is a reducing agent.

(b) SnO_2 reacts with KOH to form $K_2[Sn(OH)_6]$.

(c) A solution of $PbCl_2$ in HCl contains Pb^{2+} and Cl^- ions.

(d) The reaction of Pb_3O_4 with hot dilute nitric acid to give PbO_2 is a redox reaction.

29. Consider the following four compounds, I, II, III, and IV.

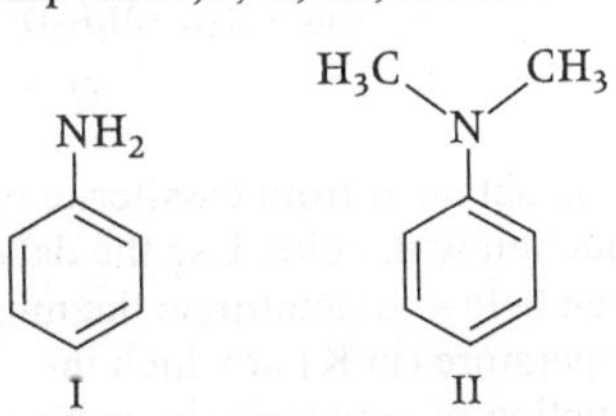

I II

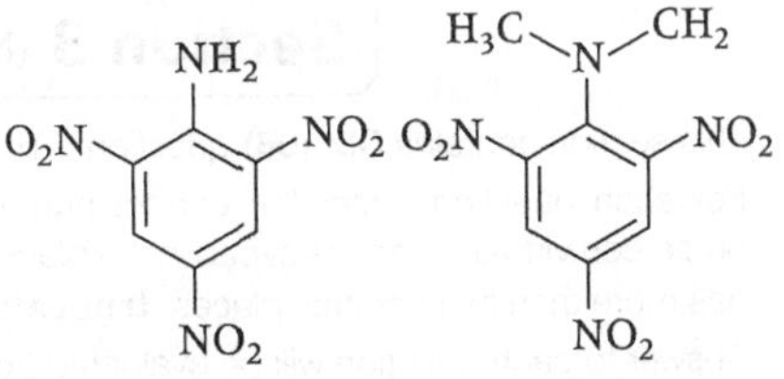

III IV

Choose the correct statement(s).

(a) The order of basicity is II > I > III > IV.

(b) The magnitude of pK_b difference between I and II is more than that between III and IV.

(c) Resonance effect is more in III than in IV.

(d) Steric effect makes compound IV more basic than III.

30. Consider the following transformations of a compound *P*.

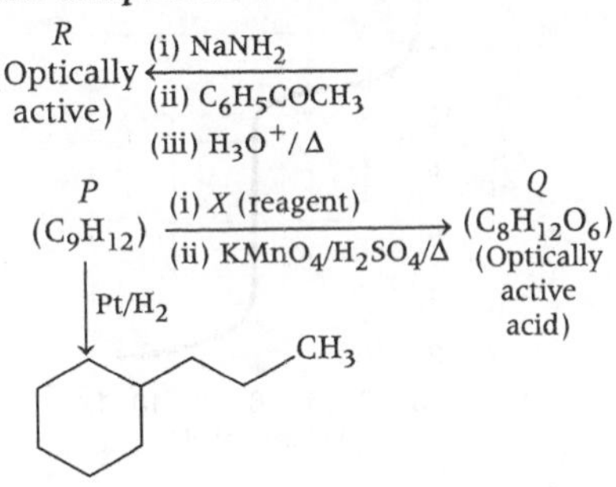

Choose the correct option(s).

(a) *P* is 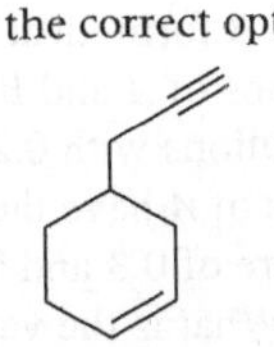

(b) *X* is Pd-C/quinoline/H_2

(c) *P* is 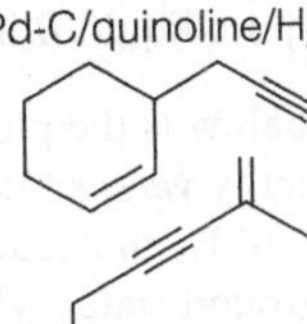

(d) *R* is 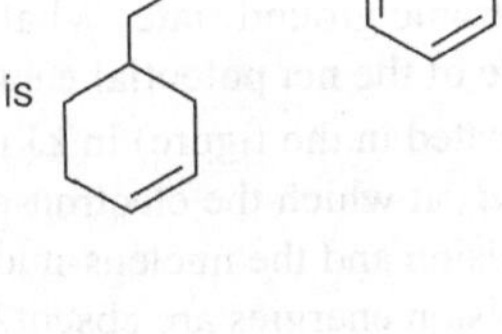

Section 3 (Maximum Marks : 24)

- This section contains **SIX (06)** questions. The answer to each question is a **NUMERICAL VALUE.**
- For each question, enter the correct numerical value of the answer using the mouse and the on-screen virtual numeric keypad in the place designated to enter the answer. If the numerical value has more than two decimal places, **truncate/round-off** the value to **TWO** decimal places.
- Answer to each question will be evaluated according to the following marking scheme:

Full Marks : **+4** If only the correct numerical value is entered as answer.

Zero Marks : **0** In all other cases.

31. A solution of 0.1 M weak base (B) is titrated with 0.1 M of a strong acid (HA). The variation of pH of the solution with the volume of HA added is shown in the figure below. What is the pK_b of the base? The neutralisation reaction is given by

$$B + HA \longrightarrow BH^+ + A^-.$$

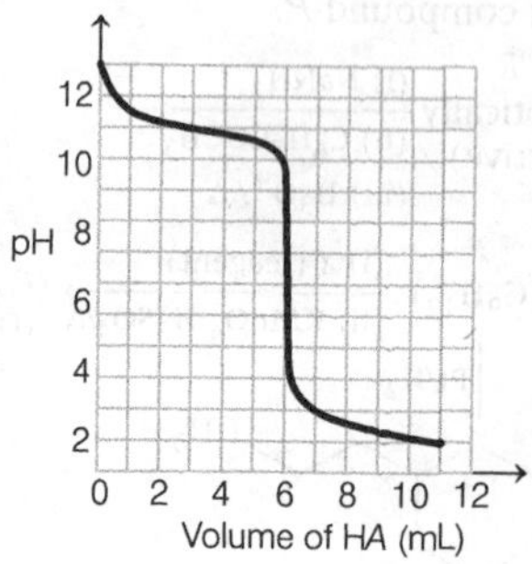

32. Liquids A and B form ideal solution for all compositions of A and B at 25°C. Two such solutions with 0.25 and 0.50 mole fractions of A have the total vapour pressure of 0.3 and 0.4 bar, respectively. What is the vapour pressure of pure liquid B in bar?

33. The figure below is the plot of potential energy *versus* internuclear distance (d) of H_2 molecule in the electronic ground state. What is the value of the net potential energy E_0 (as indicated in the figure) in kJ mol^{-1}, for $d = d_0$ at which the electron-electron repulsion and the nucleus-nucleus repulsion energies are absent? As reference, the potential energy of H atom is taken as zero when its electron and the nucleus are infinitely far apart. Use Avogardo constant as 6.023×10^{23} mol^{-1}.

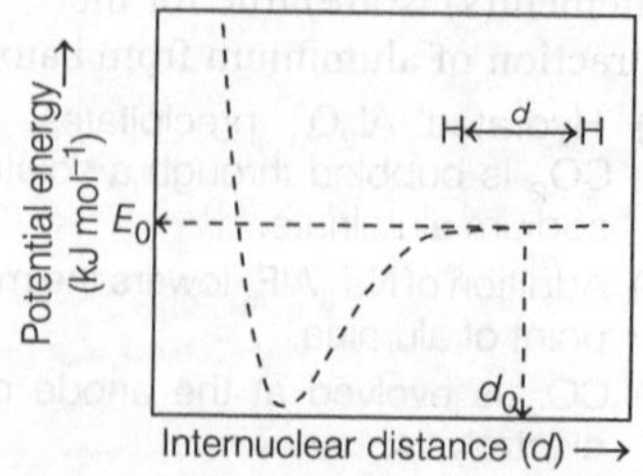

34. Consider the reaction sequence from P to Q shown below. The overall yield of the major product Q from P is 75%. What is the amount in grams of Q obtained from 9.3 mL of P? (Use density of P = 1.00 g mL^{-1}; Molar mass of C = 12.0, H = 1.0, O = 16.0 and N = 14.0 g mol^{-1})

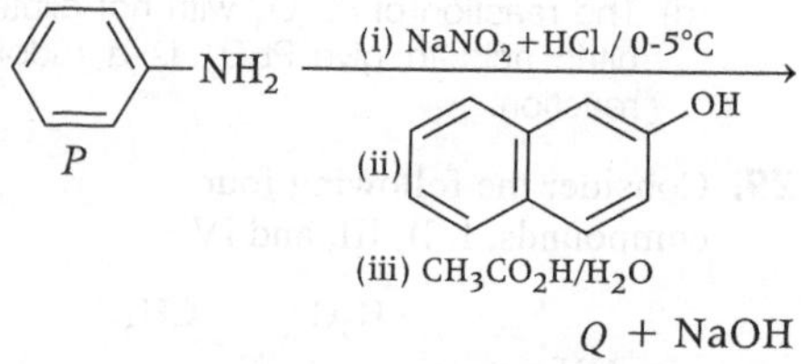

35. Tin is obtained from cassiterite by reduction with coke. Use the data given below to determine the minimum temperature (in K) at which the reduction of cassiterite by coke would take place.

At 298 K : $\Delta_f H°[SnO_2(s)]$

$= -581.0$ kJ mol^{-1},

$$\Delta_f H°[(CO_2)(g)] = -394.0 \text{ kJ mol}^{-1}$$
$$S°[SnO_2(s)] = 56.0 \text{J K}^{-1}\text{mol}^{-1}$$
$$S°[Sn(s)] = 52.0 \text{JK}^{-1} \text{mol}^{-1}$$
$$S°[C(s)] = 6.0 \text{J K}^{-1} \text{mol}^{-1}$$
$$S°[CO_2(g)] = 210.0 \text{JK}^{-1} \text{mol}^{-1}$$

Assume that, the enthalpies and the entropies are temperature independent.

36. An acidified solution of 0.05 M Zn^{2+} is saturated with 0.1 M H_2S. What is the minimum molar concentration (M) of H^+ required to prevent the precipitation of ZnS?
Use $K_{sp}(ZnS) = 1.25 \times 10^{-22}$ and overall dissociation constant of H_2S, $K_{net} = K_1K_2 = 1 \times 10^{-21}$.

MATHEMATICS

Section 1 (Maximum Marks : 18)

- This section contains **SIX (06)** questions.
- The answer to each question is a **SINGLE DIGIT INTEGER** ranging from 0 TO 9. **BOTH INCLUSIVE.**
- For each question, enter the correct integer corresponding to the answer using the mouse and the on-screen virtual numeric keypad in the place designated to enter the answer.
- Answer to each question will be evaluated according to the following marking scheme:

Full Marks : **+3** If ONLY the correct integer is entered.
Zero Marks : **0** If the question is unanswered.
Negative Marks : **−1** In all other cases.

37. For a complex number z, let Re(z) denote the real part of z. Let S be the set of all complex numbers z satisfying $z^4 - |z|^4 = 4iz^2$, where $i = \sqrt{-1}$. Then the minimum possible value of $|z_1 - z_2|^2$, where $z_1, z_2 \in S$ with $\text{Re}(z_1) > 0$ and $\text{Re}(z_2) < 0$ is …… .

38. The probability that a missile hits a target successfully is 0.75. In order to destroy the target completely, at least three successful hits are required. Then the minimum number of missiles that have to be fired so that the probability of completely destroying the target is **NOT** less than 0.95, is ……… .

39. Let O be the centre of the circle $x^2 + y^2 = r^2$, where $r > \frac{\sqrt{5}}{2}$. Suppose PQ is a chord of this circle and the equation of the line passing through P and Q is $2x + 4y = 5$. If the centre of the circumcircle of the triangle OPQ lies on the line $x + 2y = 4$, then the value of r is ……… .

40. The trace of a square matrix is defined to be the sum of its diagonal entries. If A is a 2×2 matrix such that the trace of A is 3 and the trace of A^3 is − 18, then the value of the determinant of A is ……… .

41. Let the functions $f : (-1, 1) \to \mathbf{R}$ and $g : (-1, 1) \to (-1, 1)$ be defined by
$$f(x) = |2x - 1| + |2x + 1| \text{ and}$$
$$g(x) = x - [x],$$
where $[x]$ denotes the greatest integer less than or equal to x. Let $f \circ g$: $(-1, 1) \to$ R be the composite function

defined by $(f \circ g)(x) = f(g(x))$. Suppose c is the number of points in the interval $(-1, 1)$ at which $f \circ g$ is **NOT** continuous, and suppose d is the number of points in the interval $(-1, 1)$ at which $f \circ g$ is NOT differentiable. Then the value of $c + d$ is ....

42. The value of the limit

$$\lim_{x \to \frac{\pi}{2}} \frac{4\sqrt{2}(\sin 3x + \sin x)}{\left(2\sin 2x \sin\frac{3x}{2} + \cos\frac{5x}{2}\right) - \left(\sqrt{2} + \sqrt{2}\cos 2x + \cos\frac{3x}{2}\right)}$$

is ....

Section 2 (Maximum Marks : 24)

- This section contains **SIX (06)** questions.
- Each question has **FOUR** options. **ONE OR MORE THAN ONE** of these four option(s) is (are) correct asnswer(s).
- For each question, choose the option(s) corresponding to (all) the correct answer(s).
- Answer to each question will be evaluated according to the following marking scheme:

Full Marks	:	**+4** If only (all) the correct option(s) is (are) chosen.
Partial Marks	:	**+3** If all the four options are correct but ONLY three options are chosen.
Partial Marks	:	**+2** If three or more options are correct but ONLY two options are chosen, both of which are correct.
Partial Marks	:	**+1** If two or more options are correct but ONLY one option is chosen and it is a correct option.
Zero Marks	:	**0** If none of the options is chosen (i.e. the question is unanswered).
Negative Marks	:	**−2** In all other cases.

43. Let b be a nonzero real number. Suppose $f: \mathbf{R} \to \mathbf{R}$ is a differentiable function such that $f(0) = 1$. If the derivative f' of f satisfies the equation

$$f'(x) = \frac{f(x)}{b^2 + x^2}$$

for all $x \in \mathbf{R}$, then which of the following statements is/are TRUE?

(a) If $b > 0$, then f is an increasing function
(b) If $b < 0$, then f is a decreasing function
(c) $f(x)f(-x) = 1$ for all $x \in \mathbf{R}$
(d) $f(x) - f(-x) = 0$ for all $x \in \mathbf{R}$

44. Let a and b be positive real numbers such that $a > 1$ and $b < a$. Let P be a point in the first quadrant that lies on the hyperbola $\frac{x^2}{a^2} - \frac{y^2}{b^2} = 1$. Suppose the tangent to the hyperbola at P passes through the point $(1, 0)$, and suppose the normal to the hyperbola at P cuts off equal intercepts on the coordinate axes. Let Δ denote the area of the triangle formed by the tangent at P, the normal at P and the X-axis. If e denotes the eccentricity of the hyperbola, then which of the following statements is/are TRUE?

(a) $1 < e < \sqrt{2}$ (b) $\sqrt{2} < e < 2$
(c) $\Delta = a^4$ (d) $\Delta = b^4$

45. Let $f: \mathbf{R} \to \mathbf{R}$ and $g: \mathbf{R} \to \mathbf{R}$ be functions satisfying

$$f(x + y) = f(x) + f(y) + f(x)f(y)$$

and $f(x) = xg(x)$

for all $x, y \in \mathbf{R}$. If $\lim_{x \to 0} g(x) = 1$, then which of the following statements is/are TRUE?

(a) f is differentiable at every $x \in \mathbf{R}$
(b) If $g(0) = 1$, then g is differentiable at every $x \in \mathbf{R}$
(c) The derivative $f'(1)$ is equal to 1
(d) The derivative $f'(0)$ is equal to 1

46. Let $\alpha, \beta, \gamma, \delta$ be real numbers such that $\alpha^2+\beta^2+\gamma^2 \neq 0$ and $\alpha+\gamma=1$. Suppose the point $(3, 2, -1)$ is the mirror image of the point $(1, 0, -1)$ with respect to the plane $\alpha x+\beta y+\gamma z=\delta$. Then which of the following statements is/are TRUE?

(a) $\alpha+\beta=2$ (b) $\delta-\gamma=3$
(c) $\delta+\beta=4$ (d) $\alpha+\beta+\gamma=\delta$

47. Let a and b be positive real numbers. Suppose $\mathbf{PQ}=a\hat{\mathbf{i}}+b\hat{\mathbf{j}}$ and $\mathbf{PS}=a\hat{\mathbf{i}}-b\hat{\mathbf{j}}$ are adjacent sides of a parallelogram $PQRS$. Let $\mathbf{u}$ and $\mathbf{v}$ be the projection vectors of $\mathbf{w}=\hat{\mathbf{i}}+\hat{\mathbf{j}}$ along $\mathbf{PQ}$ and $\mathbf{PS}$, respectively. If $|\mathbf{u}|+|\mathbf{v}|=|\mathbf{w}|$ and if the area of the parallelogram $PQRS$ is 8, then which of the following statements is/are TRUE?

(a) $a+b=4$
(b) $a-b=2$
(c) The length of the diagonal PR of the parallelogram $PQRS$ is 4
(d) $\mathbf{w}$ is an angle bisector of the vectors $\mathbf{PQ}$ and $\mathbf{PS}$

48. For non-negative integers s and r, let

$$\binom{s}{r}=\begin{cases}\dfrac{s!}{r!(s-r)!} & \text{if } r \leq s,\\ 0 & \text{if } r>s\end{cases}$$

For positive integers m and n, let

$$g(m, n)=\sum_{p=0}^{m+n} \frac{f(m, n, p)}{\binom{n+p}{p}}$$

where for any non-negative integer p,

$$f(m, n, p)=\sum_{i=0}^{p}\binom{m}{i}\binom{n+i}{p}\binom{p+n}{p-i}.$$

Then which of the following statements is/are TRUE?

(a) $g(m, n)=g(n, m)$ for all positive integers m, n
(b) $g(m, n+1)=g(m+1, n)$ for all positive integers m, n
(c) $g(2m, 2n)=2g(m, n)$ for all positive integers m, n
(d) $g(2m, 2n)=(g(m, n))^2$ for all positive integers m, n

Section 3 (Maximum Marks : 24)

- This section contains **SIX (06)** questions. The answer to each question is a **NUMERICAL VALUE.**
- For each question, enter the correct numerical value of the answer using the mouse and the on-screen virtual numeric keypad in the place designated to enter the answer. If the numerical value has more than two decimal places, **truncate/round-off** the value to **TWO** decimal places.
- Answer to each question will be evaluated according to the following marking scheme:

Full Marks : **+4** If ONLY the correct numerical value is entered.
Zero Marks : **0** In all other cases.

49. An engineer is required to visit a factory for exactly four days during the first 15 days of every month and it is mandatory that no two visits take place on consecutive days. Then the number of all possible ways in which such visits to the factory can be made by the engineer during 1-15 June 2021 is ....

50. In a hotel, four rooms are available. Six persons are to be accommodated in these four rooms in such a way that each of these rooms contains at least one person and at most two persons. Then the number of all possible ways in which this can be done is ....

51. Two fair dice, each with faces numbered 1, 2, 3, 4, 5 and 6, are rolled together and the sum of the numbers on the faces is observed. This process is repeated till the sum is either a prime number or a perfect square. Suppose the sum turns out to be a perfect square before it turns out to be a prime number. If p is the probability that this perfect square is an odd number, then the value of $14p$ is ....

52. Let the function $f:[0,1] \to \mathrm{R}$ be defined by

$$f(x) = \frac{4^x}{4^x + 2}$$

Then the value of

$$f\left(\frac{1}{40}\right) + f\left(\frac{2}{40}\right) + f\left(\frac{3}{40}\right) + \dots + f\left(\frac{39}{40}\right) - f\left(\frac{1}{2}\right)$$

is ....

53. Let $f: \mathbf{R} \to \mathbf{R}$ be a differentiable function such that its derivative f' is continuous and $f(\pi) = -6$.
If $F:[0, \pi] \to \mathbf{R}$ is defined by $F(x) = \int_0^x f(t)dt$, and if

$$\int_0^{\pi} (f'(x) + F(x))\cos x \, dx = 2$$

then the value of $f(0)$ is ....

54. Let the function $f:(0, \pi) \to \mathbf{R}$ be defined by

$$f(\theta) = (\sin\theta + \cos\theta)^2 + (\sin\theta - \cos\theta)^4$$

Suppose the function f has a local minimum at θ precisely when $\theta \in \{\lambda_1 \pi, \dots, \lambda_r \pi\}$, where $0 < \lambda_1 < \dots \lambda_r < 1$. Then the value of $\lambda_1 + \dots + \lambda_r$ is ....

Answers

Paper 1

1.	a	2.	b	3.	b	4.	b	5.	a	6.	b	7.	b,c	8.	b,c,d	9.	a,b	10.	b,c
11.	a,c,d	12.	a,b,c	13.	25.60	14.	3.74	15.	3.14	16.	1.77 to 1.78	17.	0.62 to 0.63	18.	6.40	19.	b	20.	b
21.	a	22.	c	23.	d	24.	c	25.	a,b,c	26.	a,b,c	27.	b,d	28.	a,c	29.	a,b,d	30.	a,c
31.	0.11	32.	0.25	33.	13.32	34.	6.15	35.	1.2	36.	18	37.	d	38.	c	39.	a	40.	a
41.	b	42.	c	43.	a,c	44.	b,c,d	45.	b,c	46.	b,c	47.	a,b	48.	a,b,d	49.	8.00	50.	3.00
51.	1.00	52.	108.00	53.	5.00	54.	1.00												

Paper 2

1.	*	2.	9	3.	6	4.	4	5.	2	6.	*	7.	a,d	8.	a,d	9.	a,c,d	10.	a,c
11.	b,c	12.	a,b,c,d	13.	2.05	14.	0.27	15.	1.3	16.	0.24	17.	55.00	18.	8.33	19.	9	20.	6
21.	6	22.	4	23.	5	24.	6	25.	a,b,c	26.	a	27.	a,b,c,d	28.	a,b	29.	c,d	30.	b,c
31.	3.00	32.	0.2	33.	−5242.41	34.	18.60	35.	935	36.	0.2	37.	8	38.	6	39.	2	40.	5
41.	4	42.	8	43.	a,c	44.	a,d	45.	a,b,d	46.	a,b,c	47.	a,c	48.	a,b,d	49.	495.00	50.	1080.00
51.	8.00	52.	19.00	53.	4.00	54.	0.50												

Note (*) None of the option is correct.